Biological Indicators
of Aquatic Ecosystem Stress

Biological Indicators
of Aquatic Ecosystem Stress

Edited by

S. Marshall Adams

Environmental Sciences Division
Oak Ridge National Laboratory
Oak Ridge, Tennessee

American Fisheries Society
Bethesda, Maryland
2002

Suggested citation formats

Entire book

Adams, S. M., editor. 2002. Biological indicators of aquatic ecosystem stress. American Fisheries Society, Bethesda, Maryland.

Chapter within the book

Beyers, D. W., and J. A. Rice. 2002. Evaluating stress in fish using bioenergetics-based stressor-response models. Pages 289–320 *in* S. M. Adams, editor. Biological indicators of aquatic ecosystem stress. American Fisheries Society, Bethesda, Maryland.

Cover design by Tracy Early. Background image: St. Joe River, Idaho, photo by Ted Bjornn, Idaho Cooperative Fish and Wildlife Research Unit, University of Idaho, Moscow. Inset photos, from top: Beef cattle grazing in Kern County, California, USDA photo by Ron Nichols. Bonneville Dam, photo courtesy of USGS. Weyerhauser paper mills and Reynolds metal plant are both located in Longview, Washington on the Columbia River, intense industrial concentration causes visible pollution, photo courtesy of National Archives. No till planting of corn near Plymouth, Iowa, USDA photo by Gene Alexander.

Printed in the United States of America on acid-free paper.

Library of Congress Control Number 2002101549
ISBN 1-888569-28-X (hardcover)
ISBN 1-888569-43-3 (softcover)

American Fisheries Society
5410 Grosvenor Lane, Suite 110
Bethesda, Maryland 20814-2199
USA

Distributed exclusively outside North and Central America by
CABI Publishing, CAB International, Wallingford, Oxon OX10 8DE, U.K.

ISBN 0 85199 629 9 (hardcover)
ISBN 0 85199 630 2 (paperback)

Table of Contents

Contributors

S. Marshall Adams (Chapter 1): Environmental Sciences Division, Oak Ridge National Laboratory, Oak Ridge, Tennessee 37830, USA

Mary R. Arkoosh (Chapter 6): Environmental Conservation Division, Northwest Fisheries Science Center, National Marine Fisheries Service, National Oceanic and Atmospheric Administration, 2030 SE Marine Science Drive, Newport, Oregon 97365, USA

Martin J. Attrill (Chapter 12): Benthic Ecology Research Group, Plymouth Environmental Research Centre (Department of Biological Sciences), University of Plymouth, Drake Circus, Plymouth, Devon, PL4 8AA, U.K; E-mail: mattrill@plymouth.ac.uk

Bruce A. Barton (Chapter 4): Department of Biology and Missouri River Institute, University of South Dakota, Vermillion, South Dakota 57069, USA

Daniel W. Beyers (Chapter 8): Larval Fish Laboratory, Department of Fishery and Wildlife Biology, Colorado State University, Fort Collins, Colorado 80523, USA; E-mail: danb@lamar.colostate.edu

Tracy K. Collier (Chapter 14): Ecotoxicology and Environmental Fish Health Program, Environmental Conservation Division, Northwest Fisheries Science Center, National Marine Fisheries Service, National Oceanic and Atmospheric Administration, 2030 SE Marine Science Drive, Newport, Oregon 97365, USA

John L. Curnutt (Chapter 13): Florida Caribbean Science Center, U.S. Geological Survey, Biological Resources Division, Florida International University, Miami, Florida 33199, USA

Donald L. DeAngelis (Chapter 13): Florida Caribbean Science Center, U.S. Geological Survey, Biological Resources Division, Department of Biology, University of Miami, Coral Gables, Florida 33124, USA

Richard T. Di Giulio (Chapter 2): Nicholas School of the Environment, Duke University, Durham, North Carolina 27708-0328, USA

John W. Fournie (Chapter 7): U.S. Environmental Protection Agency, National Health and Environmental Effects Research Laboratory, Gulf Ecology Division, 1 Sabine Island Drive, Gulf Breeze, Florida 32561, USA; E-mail: fournie.john@epa.gov

Mark S. Greeley, Jr. (Chapter 9): Environmental Sciences Division, Oak Ridge National Laboratory, Oak Ridge, Tennessee 37831-6038, USA

Peter V. Hodson (Chapter 16): School of Environmental Studies, Queen's University, Kingston, Ontario K7L 3N6, Canada; E-mail: hodsonp@biology.queens.ca

Lyndal L. Johnson (Chapter 14): Ecotoxicology and Environmental Fish Health Program, Environmental Conservation Division, Northwest Fisheries Science Center, National Marine Fisheries Service, National Oceanic and Atmospheric Administration, 2030 SE Marine Science Drive, Newport, Oregon 97365, USA

Edward E. Little (Chapter 11): U.S. Geological Survey, Columbia Environmental Research Center, 4200 New Haven Road, Columbia, Missouri 65201, USA; E-mail: edward_little@usgs.gov

Mark S. Myers (Chapter 7): Environmental Conservation Division, Northwest Fisheries Science Center, 2725 Montlake Boulevard E., Seattle, Washington 98112, USA; E-mail: mark.s.myers@noaa.gov

John D. Morgan (Chapter 4): Faculty of Science and Technology, Malaspina University-College, Nanaimo, British Columbia V9R 5S5, Canada

Michael Power (Chapter 10): Department of Biology, University of Waterloo, Waterloo, Ontario N2L 3G1, Canada

Charles D. Rice, Ph.D. (Chapter 6): CIET/ENTOX, Clemson University, P. O. Box 709, 509 Westinghouse Road, Pendleton, South Carolina 29670, USA; E-mail: cdrice@clemson.edu

James A. Rice (Chapter 8): Department of Zoology, North Carolina State University, Raleigh, North Carolina 27695-7617, USA; E-mail: jim_rice@ncsu.edu

Daniel Schlenk (Chapter 2): Aquatic Ecotoxicology, Department of Environmental Sciences, University of California, Riverside, California 92521, USA

Eric P. Smith (Chapter 15): Department of Statistics, Virginia Polytechnic Institute and State University, Blacksburg, Virginia 24061, USA

Christopher W. Theodorakis (Chapter 3 and Chapter 5): The Institute of Environmental and Human Health, Department of Environmental Toxicology, Texas Tech University, Lubbock, Texas 79416, USA

Mathilakath M. Vijayan (Chapter 4): Department of Biology, University of Waterloo, Waterloo, Ontario N2L 3G1, Canada

Isaac I. Wirgin (Chapter 3 and Chapter 5): Nelson Institute of Environmental Medicine, New York University Medical Center, Tuxedo, New York 10987, USA

Symbols and Abbreviations

The following symbols and abbreviations may be found in this book without definition. Also undefined are standard mathematical and statistical symbols given in most dictionaries.

A	ampere	ft^3/s	cubic feet per second (0.0283 m^3/s)
AC	alternating current		
Bq	becquerel	g	gram
C	coulomb	G	giga (10^9, as a prefix)
°C	degrees Celsius	gal	gallon (3.79 L)
cal	calorie	Gy	gray
cd	candela	h	hour
cm	centimeter	ha	hectare (2.47 acres)
Co.	Company	hp	horsepower (746 W)
Corp.	Corporation	Hz	hertz
cov	covariance	in	inch (2.54 cm)
DC	direct current; District of Columbia	Inc.	Incorporated
		i.e.	(id est) that is
D	dextro (as a prefix)	IU	international unit
d	day	J	joule
d	dextrorotatory	K	Kelvin (degrees above absolute zero)
df	degrees of freedom		
dL	deciliter	k	kilo (10^3, as a prefix)
E	east	kg	kilogram
E	expected value	km	kilometer
e	base of natural logarithm (2.71828...)	*l*	levorotatory
		L	levo (as a prefix)
e.g.	(exempli gratia) for example	L	liter (0.264 gal, 1.06 qt)
eq	equivalent	lb	pound (0.454 kg, 454g)
et al.	(et alii) and others	lm	lumen
etc.	et cetera	log	logarithm
eV	electron volt	Ltd.	Limited
F	filial generation; Farad	M	mega (10^6, as a prefix); molar (as a suffix or by itself)
°F	degrees Fahrenheit		
fc	footcandle (0.0929 lx)	m	meter (as a suffix or by itself); milli (10^{23}, as a prefix)
ft	foot (30.5 cm)		

mi	mile (1.61 km)	s	second
min	minute	T	tesla
mol	mole	tris	tris(hydroxymethyl)-
N	normal (for chemistry); north		aminomethane (a buffer)
	(for geography); newton	U.K	United Kingdom
N	sample size	U.S.	United States (adjective)
NS	not significant	USA	United States of America
n	ploidy; nanno (10^{29}, as		(noun)
	a prefix)	V	volt
o	ortho (as a chemical prefix)	V, Var	variance (population)
oz	ounce (28.4 g)	var	variance (sample)
P	probability	W	watt (for power); west (for
p	para (as a chemical prefix)		geography)
p	pico (10^{212}, as a prefix)	Wb	weber
Pa	pascal	yd	yard (0.914 m, 91.4 cm)
pH	negative log of hydrogen ion	α	probability of type I error
	activity		(false rejection of null
ppm	parts per million		hypothesis)
qt	quart (0.946 L)	β	probability of type II error
R	multiple correlation or		(false acceptance of null
	regression coefficient		hypothesis)
r	simple correlation or	Ω	ohm
	regression coefficient	μ	micro (10^{26}, as a prefix)
rad	radian	$'$	minute (angular)
S	siemens (for electrical	$''$	second (angular)
	conductance);	o	degree (temperature as a
	south (for geography)		prefix, angular as a suffix)
SD	standard deviation	%	per cent (per hundred)
SE	standard error	‰	per mille (per thousand)

Biological Indicators of Aquatic Ecosystem Stress: Introduction and Overview

S. MARSHALL ADAMS

Introduction

The interest in and use of bioindicators (including biomarkers and biocriteria) for use in environmental assessment has increased steadily during the last decade. Many state agencies in the United States, which function as custodians of water quality management programs under the Federal Clean Water Act (CWA) for example, have incorporated various biological measures into their bioassessment programs to evaluate the quality of surface water resources. Chemical water quality criteria developed through laboratory toxicity tests on standard test organisms have traditionally been used as surrogates for determining attainment of the biologically based goals of the CWA. Chemical criteria were originally developed to set discharge effluent and water quality standards and also to avoid some of the early-recognized problems with measurement of biological parameters in the field, such as those associated with high variability. With a greater variety of biological assessment tools now available, an improved understanding of ecosystem structure and function, and an increased ability to interpret biological data, biocriteria have become more attractive and useful for assessing the effects of environmental stressors on biological systems.

Relying on chemical criteria alone for assessing the status of surface water integrity can, in many instances, inaccurately portray the biological and ecological condition of aquatic systems. For example, impairment, as revealed by use of biological indicators, was evident in 50% of the 645 stream and river segments analyzed in Ohio, whereas no impairments were observed based on chemical indicators (Yoder and Rankin 1998). Use of chemical criteria alone to assess the effects of water quality on ecological systems can lead to an incomplete foundation for legislation related to resource policy because it does not include broader ecological measures (Yoder and Rankin 1998; Barbour et al. 1996). Laboratory-based chemical criteria usually consider only one influential factor (e.g., toxicant) at a time, do not include multiple chemical exposures, and are often restricted to parameters that are convenient to measure. More importantly, however, chemical criteria alone fail to reflect all the other factors in the environment that can impair aquatic ecosystems, such as sedimentation, alterations in habitat and natural flow regimes, varying temperature and oxygen regimes, and changes in ecological factors such as food availability and predator–prey interactions. Biological criteria, on the other hand, possess several attributes that are desirable for assessing the quality of surface water resources. Some types of biocriteria are not only reflective of chemical exposure but also have the capacity to integrate many of the physical, chemical, and biological stressors that operate in aquatic ecosystems. In addition, many biocriteria are capable of integrating the effects of stressors on organisms, both spatially and temporally, and are thus more suited for measuring and interpreting the possible effects of multiple stressors on aquatic ecosystems. Bioindicators, therefore, can reflect environmental problems that might otherwise be missed, or underestimated, by approaches that rely on chemical criteria alone, simply because they provide the opportunity to recognize and account for natural ecological conditions and variability.

Biocriteria can also be used to assess damage or injury to natural resources from environmental stressors. In marine systems, pollution has been defined as "the environmental damage caused by wastes discharged into the sea" (Clark et al. 2001). This definition inherently implies that environmental damage has to be demonstrated in order to prove that a site is polluted. Within this context, chemical criteria and biomarkers of chemical exposure alone cannot be used to assess environmental damage. For example, measuring the levels of a chemical in the environment is basically documenting the level of contamination, while biomarker responses, even though they may provide some indication of damage at the cellular and subcellular level, do not provide assessments of environmental damage at higher levels of biological organization. Biologically relevant endpoints at these higher levels of organization, which are included as a component of the ecological risk assessment process, are typically used as the basis of regulatory and management decisions. In the United States, the Natural Resources Damage Assessment (NRDA) process, which is included under three contemporary environmental statutes (the CWA, the Comprehensive Environmental Response, Compensation and Liability Act, and the Oil Pollution Act), imposes liability for damages to natural resources from release of hazardous substances into the environment. Assessing biological damage or injury requires the use of methods that demonstrate measurable biological responses. Some of the endpoints approved for use within the NRDA process, however, can be characterized as biomarkers (see definition to follow) and, therefore, may not be entirely appropriate for assessing damage at ecological significant levels. Biomarkers at lower levels of biological organization are potentially very useful for assessing stress effects, but they must be correlated and calibrated against higher-level bioindicator responses (Chapter 12). Biomarkers that have been calibrated and correlated with higher-level effects, such as population- and community-level attributes, can indeed serve as valid bioindicators (McCarty and Munkittrick 1996; Adams et al. 2000).

Recognizing the positive attributes of biocriteria and some of the limitations of traditional chemical criteria, the U.S. Environmental Protection Agency (EPA) recently issued technical and programmatic implementation guidance for development of biocriteria in environmental monitoring and assessment programs (U.S. EPA 1996, 1999). The legal authority for developing and providing guidance for biocriteria comes from Section 303(C)(2)(B) of the Clean Water Act (CWA), which requires individual states to adopt these criteria based on bioassessments. In addition, Section 304(a)(8) of the CWA also directs the EPA to develop and publish guidance in the area of biocriteria. Certain biocriteria, such as the Biological Monitoring Working Party (BMWP), are also being applied in the United Kingdom, Australia, and several countries within the European community. This integrative index of benthic invertebrate community integrity (the BMWP) is routinely used along with chemical measures to assess water quality (Hawkes 1998). Therefore, as applied within the environmental management and regulatory framework of these countries, biocriteria can be generally defined as narrative or numeric expressions that describe the reference biological integrity (structure and function) of aquatic communities inhabiting waters of a given designated aquatic life use. Within this definition, then, biocriteria can be generally regarded as regulatory-based measurements and have as their main purpose the documentation of the numbers and kinds of organisms present in an aquatic system.

Because biocriteria, including biomarkers and bioindicators, have become increasingly popular bioassessment tools, it is important to have a comprehensive document that provides guidance relative to the design, measurement, and application of various biocriteria in aquatic ecosystems. The main purpose of this book, therefore, is to provide a comprehensive reference and guide relative to the various biological endpoints that can be measured and used to assess the effects of environmental stressors on aquatic organisms, populations, and communities. The topics addressed by the various chapters in this book are not limited to the strict definition of biocriteria as defined above for regulatory purposes. This book, however, addresses all major levels of biological organization from the biomolecular to the community and landscape levels. Guidance provided by this book can be used in biological monitoring and assessment studies for evaluating the effects of environmental stressors on the integrity of aquatic ecosystems.

Biocriteria, Biomarkers, and Bioindicators

To understand how the various types of biological endpoints addressed in this book can best be applied in field situations, the distinctions between biocriteria, biomarkers, and bioindicators should be clarified. Biocriteria, as defined within the context of regulatory applications, are regarded as the numbers and kinds of organisms present in the aquatic system of interest. This definition or use is generally restricted to measurements and studies at the population and community levels of biological organization and usually includes integrative indices of community health such as the index of biotic integrity (IBI), the stream condition index (SCI), the invertebrate community index (ICI), and the biological monitoring working party score (BMWP).

As applied in this document, biomarkers are considered as functional measures of exposure to environmental stressors, which are usually expressed at the suborganismal level of biological organization (Benson and DiGiulio 1992; Huggett et al. 1992; NRC 1987). Biomarkers, such as molecular, biochemical, and even physiological endpoints, are used primarily to indicate that an organism has been exposed to a stressor such as a xenobiotic chemical. Evidence of biological exposure to a stressor has been more broadly defined by the U.S. EPA (1991) as those endpoints that measure the apparent effects of stressors, including chemical water quality criteria, whole effluent toxicity tests, tissue residues, and biomarkers. Thus, in addition to the more traditional measures of exposure (e.g., chemical tissue residues, acute and chronic toxicity tests), biomarkers are also regarded here as measures of exposure. Bioindicators, on the other hand, are defined less precisely than biomarkers and can be viewed as either structural entities, such as sentinel species (Van Gestel and Van Brummelen 1996), or they can be considered functionally as biological effects endpoints at higher levels of organization (Adams 1990a; Engle and Vaughan 1996). Within this context, then, some bioindicators are included within the definition of biocriteria because bioindicators also include population- and community-level attributes in addition to organism-level and ecosystem- and landscape-level responses. As used by the U.S. EPA (1991) in the Environmental Monitoring and Assessment Program, response indicators are considered surrogates for bioindicators and are operationally defined as composite measures of the cumulative effects of stress and exposure and also include the more direct measures of community and population responses. The main features of biomarkers and bioindicators may be reflected in a single definition that considers a bioindicator as "an anthropogenically induced variation in biochemical, physiological, or ecological components or processes, structures, or functions that has been either statistically correlated or causally linked, in at least a semiquantitative manner, to biological effects at one or more of the organism, population, community, or ecosystem levels of biological organization" (McCarty and Munkittrick 1996). Thus, a biomarker may be operationally considered a bioindicator or even a biocriteria if it can be causally related or linked to a biologically significant endpoint at the organism level or above (Adams et al. 2001).

Biomarkers and bioindicators have their own unique set of advantages and limitations relative to their value and use for assessing the effects of stress on aquatic ecosystems. Table 1 summarizes the major features of biomarkers and bioindicators relative to their advantages and limitations for use in field bioassessment studies. In general, biomarkers are used to indicate exposure of an organism to a stressor, and bioindicators are used primarily as indicators of stress effects at higher levels of organization mainly because of their composite or integrative nature. The main attributes of biomarkers and bioindicators that are important for consideration in the design of bioassessment studies are sensitivity and specificity to stressors, relationship to cause, response variability, temporal scales of response, and ecological or biological significance (Table 1). In general, biomarkers are stressor sensitive and rapidly responding endpoints that help to identify the mechanistic basis of causal relationships between a stressor and its effect. The primary limitations of biomarkers, however, are that they are generally characterized by a relative high response variability (i.e., coefficient of variation is relatively high because response parameters of individuals are typically more variable compared with the more integrative attributes of communities such as diversity, for example), rarely integrate effects of stressors over long periods of time and, most importantly, generally have low ecological relevance. On the other hand, bioindicators, including traditional biocriteria, provide little useful information for helping to understand the underlying causal mechanisms between stressors and effects because their sensitivity and specificity to stressors is low and they tend to integrate the effects of multiple stressors over large spatial and temporal scales (Adams 1990a;

Table 1. Major features of biomarkers and bioindicators relative to their advantages and limitations for use in field bioassessment studies.

Major features	Biomarkers	Bioindicators
Types of response	Subcellular, cellular	Individual through community
Primary indicators of	Exposure	Effects
Sensitivity to stressors	High	Low
Relationship to cause	High	Low
Response variability	High	Low–moderate
Specificity to stressors	Moderate–high	Low
Time scale of response	Short	Long
Ecological relevance	Low	High

Depledge and Fossi 1994). Although bioindicators (and biocriteria) have a relatively low degree of response variability and high ecological relevance or significance, they have little value in helping to identify the underlying cause of observed changes in ecosystems. Thus, when designing and conducting field bioassessment programs, a variety of endpoints should be used that represent a range of spatial and temporal response scales and also include a large range of spatial and temporal sensitivities and specificities to different stressors. The complexity of natural systems, their inherent high variability, and the influence of multiple environmental factors (or stressors) on ecosystems suggest that no single measure (or perhaps even a few measures) is adequate for assessing the effects of multiple stressors on the status or integrity of aquatic ecosystems. An appropriate suite of endpoints is required for determining the biological significance of stress and understanding the underlying cause or mechanistic basis of observed effects (Hodson 1990; Attrill and Depledge 1997). In many instances, simply documenting that a change has occurred in a system or measuring such a change with a few biological parameters may not be adequate. It is also necessary to understand the mechanistic basis of an effect or change if more informed decisions are to be made regarding effective management and mitigation practices in disturbed ecosystems. Overreliance on any one or a few indicators can result in environmental regulation that is less accurate and either under- or overprotective of water resources (Yoder and Rankin 1998). A credible and genuinely cost-effective approach to water quality management should, therefore, include an appropriate mix of chemical, physical, and biological indicators, with each being used in their respective roles as environmental stressor (i.e., contaminant, eutrophication), exposure response (i.e., biomarkers), and effects response (i.e., bioindicators).

The concept of applying a suite of endpoints in bioassessment studies is illustrated in Figure 1, which shows that a combination of both rapidly responding and sensitive biomarkers and the more ecological relevant bioindicators (including biocriteria) should be incorporated in field bioassessment designs. This figure also illustrates that given limited resources (finances and personnel, etc.) for conducting bioassessment studies, the number and types of measurements that can be taken is limited and those responses that are measured should perhaps focus at the organismal level. In addition to an emphasis at the organismal level, study designs should also include a few measures at both the lower levels (i.e., biomarkers) and higher levels (i.e., bioindicators, biocriteria) of biological organization. With such a design, organism-level responses can serve as an intermediate or pivotal response point by which the mechanistic basis of effects at lower levels (biomarkers) can be causally linked to ecologically relevant measures at the population and community levels (bioindicators/biocriteria). This concept of causal relationships between levels of organization is also shown in Figure 2, where increasing levels of biological organization result in decreasing mecha-

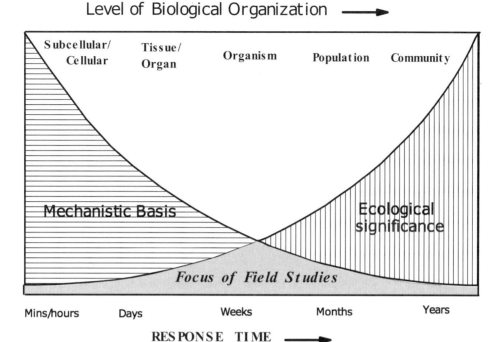

Figure 1. Field bioassessment studies should include a combination of both rapidly responding sensitive biomarkers and the more ecologically relevant bioindicators with a focus at the individual organism level. Organism-level responses provide a pivotal point through which mechanistic understanding and the ecological consequences of stressors can be linked.

nistic understanding but increasing levels of ecological significance. Thus, a selected suite of measures along this continuum of levels of organization is recommended in the design of aquatic ecosystem bioassessment studies.

Scope of Book

Given the above background, the primary purpose of this book is to provide practical information and guidance for improving our ability to assess and predict the effects of environmental stressors on the integrity of aquatic ecosystems. Even though the title of this book focuses on bioindicators, within the strict definition of terms, both biomarkers and bioindicators are addressed by their respective topics within the various chapters of this book. For example, Chapter 2 addresses molecular endpoints that, within their stricter definition, are biomarkers, but molecular responses may also function as bioindicators within the context of using an integrated suite of responses over several levels of biological organization to establish possible causality. For the purpose of this discussion and within the context of this book, there-

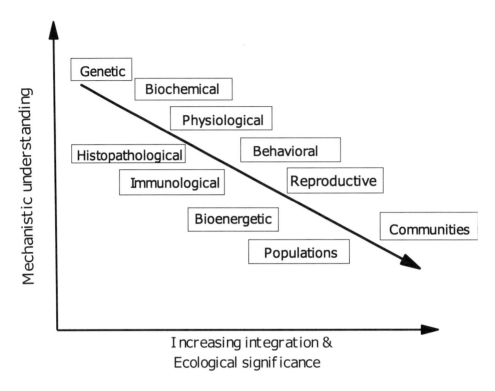

Figure 2. Increasing levels of biological organization result in decreasing mechanistic understanding but increasing levels of ecological significance. A selected suite of measures along this continuum of levels of organization is recommended in the design of bioassessment studies in aquatic ecosystems.

fore, the term bioindicators is used in a general sense to capture all the biological measures that are used or could be applied in field bioassessment studies, including biomarkers, bioindicators, and biocriteria. Each of the chapters in this book provides practical guidance for the experimental design and application of specific types or groups of bioindicators in field bioassessment studies.

When the original book was published on the use of bioindicators of stress in fish (Adams 1990b), the concept and application of biomarkers and bioindicators in field studies was just starting to gain increased popularity and use. The emergence of bioindicators and biocriteria in field bioassessment studies during this period was also evidenced by other books and papers published on this topic, including McCarthy and Shugart (1990), Niimi (1990), Huggett et al. (1992), Peakall (1992), and Fossi and Leonzio (1993), and also by the appearance of some environmental journals devoted primarily to this topic such as the journals *Biomarkers* and *Ecotoxicology*. Over the last decade, bioindicators have been increasingly applied in field studies, new biomarkers have been developed, and others have been validated in a variety of field applications (Adams et al. 2001). This book represents an up-dated and more comprehensive version of the original works published during the early 1990s on the use and application of bioindicators in bioassessment studies.

The sequence of chapters in this book is arranged in order of increasing biological organization from the molecular and biochemical levels (chapters 2 and 3) to the community and landscape levels (chapters 12 and 13). Several chapter topics in this document are similar to those in the original book, *Bioindicators of Stress in Fish* (Adams 1990b), but the topics of these chapters have been updated to represent more recent methods, approaches, and applications. This updated version also includes new contributions in several areas including biomolecular biomarkers and population genetics; behavioral indicators; integration of population, community, and landscape indicators; the assessment of stress effects across levels of organization; statistical design and application; and finally a synthesis that attempts to integrate all the major emergent points across all chapters and topics.

Chapters 2–13, which deal with different groups of bioindicators arranged into major sections including (1) identification and description of the particular endpoints measured for each indicator type and the relevance of each indicator type for use in the measurement and assessment of stress effects in aquatic systems, (2) discussion of the advantages and limitations of each group of indicators for use in assessing the effects of stress on the biotic integrity of aquatic systems, (3) presentation of a few representative examples describing how these specific indicators have been applied in the laboratory and in the field, and (4) summary and synthesis tables at the end of each chapter that present each of the major endpoints and their distinguishing attributes. Even though most chapters have focused on fish as examples for applying bioindicators in field studies, other examples of aquatic vertebrates and invertebrates are also provided throughout. Many of the principles, approaches, and examples given for fish in these chapters can be equally applied to other taxonomic groups of aquatic organisms. In addition, in most cases, material presented in these chapters is equally applicable to marine, estuarine, and freshwater systems. Excluding some of the specific biomarkers of exposure, many of the methods presented in these chapters should be relevant to a variety of environmental stressors including chemical pollutants, temperature, dissolved oxygen, siltation, and alterations in food and habitat availability. Many of the biocriteria and case history examples provided throughout these chapters, however, focus on chemical pollutants and their effects because relatively more is known about this class of stressors, and biological effects of chemicals have been the subject of rather intense research investigations over the past three decades. It is our intent that this book will provide practical guidance relative to the design and application of field bioassessment studies for improving the predictive capability of stress effects on aquatic ecosystems. Such predictions could ultimately serve as the basis of more informed management and regulatory decisions for improving the integrity and status of aquatic ecosystems.

References

Adams, S. M. 1990a. Status and use of bioindicators for evaluating effects of chronic stress on fish. Pages 1–9 in S. M. Adams, editor. Biological indicators of stress in fish. American Fisheries Society, Symposium 8, Bethesda, Maryland.

Adams, S. M., editor. 1990b. Biological indicators of stress in fish. American Fisheries Society, Symposium 8, Bethesda Maryland.

Adams, S. M., J. P. Giesy, L. A. Tremblay, and C. T. Easton. 2001. The use of biomarkers in ecological risk assessment: recommendations from the Christchurch conference on biomarkers in ecotoxicology. Biomarkers 6:1–6.

Adams, S. M., M. S. Greeley, and M. G. Ryon. 2000. Evaluating effects of contaminants on fish health at multiple levels of biological organization: extrapolating from lower to higher levels. Human and Ecological Risk Assessment 6:15–27.

Attrill, M. J., and M. H. Depledge. 1997. Community and population indicators of ecosystem health: targeting links between levels of biological organization. Aquatic Toxicology 38:183–197.

Barbour, M. T., J. Gerritsen, G. E. Griffity, R. Frydenborg, E. McCarron, J. S. White, and M. L. Bastian. 1996. A framework for biological criteria for Florida streams using benthic macroinvertebrates. Journal of the North American Benthological Society 15:185–211.

Benson, W. H., and R. T. DiGiulio. 1992. Biomarkers in hazard assessment of contaminated sediments. Pages 241–265 in G. A. Burton, editor. Sediment toxicity assessment. Lewis Publishers, Boca Raton, Florida.

Clark, R. B., C. Frid, and M. J. Attrill. 2001. Marine pollution. 5th edition. Oxford University Press, Oxford.

Depledge, M. H., and M. C. Fossi. 1994. The role of biomarkers in environmental assessment (2). Invertebrates. Ecotoxicology 3:161–172.

Engle, D. W., and D. S. Vaughan. 1996. Biomarkers, natural variability, and risk assessment: can they coexist? Human and Ecological Risk Assessment 2:257–262.

Fossi, C., and C. Leonzio. 1993. Nondestructive biomarkers in vertebrates. Lewis Publishers, Boca Raton, Florida.

Hawkes, H. A. 1998. Origin and development of the biological monitoring working party score systems. Water Research 32:964–968.

Hodson, P. V. 1990. Indicators of ecosystem health at the species level and the example of selenium effects on fish. Environmental Monitoring and Assessment 15:241–254.

Huggett, R. J., R. A. Kimerle, P. M. Mehrle, and H. L. Bergman, editors. 1992. Biomarkers: biochemical, physiological, and histological markers of anthropogenic stress. Lewis Publishers, Boca Raton, Florida.

McCarthy, J. F., and L. R. Shugart, editors. 1990. Biomarkers of environmental contamination. Lewis Publishers, Baco Raton, Florida.

McCarty, L. S., and K. R. Munkittrick. 1996. Environmental biomarkers in aquatic toxicology: fiction, fantasy, or functional? Human and Ecological Risk Assessment 2: 268–274.

Niimi, A. J. 1990. Review of biochemical methods and other indicators to assess fish health in aquatic ecosystems containing toxic chemicals. Journal of Great Lakes Research 16:529–541.

NRC (National Research Council). 1987. Committee on biological markers. Environmental Health Perspectives 74:3–9.

Peakall, D. B., editor. 1992. Animal biomarkers as pollution indicators. Ecotoxicological series no. 1. Chapman and Hall, London.

U.S. Environmental Protection Agency (EPA). 1991. Environmental monitoring and assessment program. EMAP- surface waters monitoring and research strategy-fiscal year 1991, EPA/600/3-91/022. Office of Research and Development, Environmental Protection Laboratory, Corvallis, Oregon.

U.S. Environmental Protection Agency (EPA). 1996. Biological criteria: technical guidance for streams and small rivers, revised edition. U.S. Environmental Protection Agency, Office of Science and Technology, Document No. EPA/822/B-96/001.

U.S. Environmental Protection Agency (EPA). 1999. Rapid bioassessment protocols for use in streams and wadeable rivers: periphyton, benthic macroinvertebrates, and fish. U.S. Environmental Protection Agency, Office of Science and Technology, Document No. EPA.-B-99-002.

Van Gestel, C. A. M., and T. C. Van Brummelen. 1996. Incorporation of the biomarker concept in ecotoxicology calls for a redefinition of terms. Ecotoxicology 5:217–225.

Yoder, C. O., and E. T. Rankin. 1998. The role of biological indicators in a state water quality management process. Environmental Monitoring and Assessment 51:61–88.

2

Biochemical Responses as Indicators of Aquatic Ecosystem Health

Daniel Schlenk and Richard T. Di Giulio

Introduction

One of the basic premises of toxicology is that toxicity results from the interaction of excessive concentrations of the toxicant with a receptor or molecular target. This receptor may be a specific protein that undergoes a conformational change and translocates to the nucleus (e.g., aryl hydrocarbon receptor [AhR]). The target may also be one or more molecules that signal a cascade of reactions eventually altering gene expression in an attempt to rid the cell of the toxin or toxicant or defend against its adverse effects. Consequently, biochemical responses are typically the first line of defense in the cell following exposure to xenobiotics. Biochemical responses following interactions between the toxicant and the target usually alter expression of one or more proteins (often the receptor itself) that either sequester the chemical, make the chemical more excretable through biotransformation, or induce molecular signals that can eventually initiate more of a systemic physiological response in an organism. Since biochemical responses occur at one of the lowest levels of biological organization, the responses are extremely sensitive to toxicant exposure and also are extremely variable and plastic among individuals in a given population. Because of this response variability, it is often difficult to ascertain whether biochemical responses are indicative of stress or hormetic adaptation.

Since most biochemical responses to toxicant insult are primary defense mechanisms, biochemical endpoints can be very informative in helping to identify causal mechanisms potentially responsible for any biological effects ultimately realized at higher levels of organization. In addition, since a biochemical response is dependent upon interaction of the toxicant with a molecular target, biochemical effects can also be used to ascertain bioavailability of absorbed toxicants that greatly overshadow the significance of basic residue analyses. Lastly, since many biochemical responses are receptor driven and receptors are highly selective for specific ligands, analyses of exposure to specific classes of stressors can also be performed.

Described below are several biochemical responses that have been employed successfully to address chemical exposures and their effects in aquatic systems. The science underlying these responses, their strengths and weaknesses, examples of field applications, and brief descriptions of available methodologies are discussed. Additionally, case studies are presented that highlight the use of biochemical markers in aquatic research and include many of the responses described herein.

Biochemical Markers of Exposure

Cytochrome P450 1A

Cytochrome P450 monooxygenases (CYP) are a large family of enzymes that are found in virtually all organisms. Having more than 900 genes identified throughout phylogeny, the nomenclature of CYP, and, hence, identity of genera or species-specific isoforms has been difficult to obtain. Isoforms of this enzyme family have diverse functions and often overlapping substrate specificities. One particular subfamily of CYP that appears to have a selectivity for planar aromatic hydrocarbons is cytochrome P450 1A (CYP1A). Cytochrome P450 1A catalyzes the monooxygenation of an assortment of chemicals including polycyclic aromatic hydrocarbons (PAH), polychlorinated biphenyls (PCB), furans, and dioxins (2,3,7,8 tetrachloro-dibenzodioxin [TCDD]). Cytochrome P450 1A isoforms occur in all vertebrates, and, structurally-homologous proteins have been identified in a number of invertebrate species that share some catalytic similarities with the vertebrate isoforms (Livingstone et al. 1997; Peters et al. 1999; Synder 2000).

In vertebrates, CYP1A is primarily regulated by the AhR (a cytosolic protein that binds CYP1A substrates), is translocated to the nucleus, and initiates transcription of several genes including CYP1A isoforms. Regulation of CYP1A-like proteins in invertebrates is still unclear, as, at the time of this writing, no homologous invertebrate AhR has been identified (Hahn 1998). Since CYP1A is transcriptionally regulated, measurements of expression can include CYP1A mRNA, as well as the protein and its catalytic activity. Two substrates that have been predominantly used to measure catalytic activity include ethoxyresorufin and benzo(a)pyrene (BaP). In vertebrates, ethoxyresorufin is preferentially dealkylated (ethoxyresorufin-0-deethylase [EROD]) to the fluorescent 7-hydroxy-resorufin by CYP1A isoforms. A rapid 96-well assay has been developed and is relatively easy to perform given access to an appropriate spectrofluorimeter or spectrophotometer (Hahn et al. 1996). In addition, hydroxylation of BaP, also known as aryl hydrocarbon hydroxylase (AHH), has also been used to measure CYP1A catalytic activity (Williams and Buhler 1984; Collier et al. 1995a). However, AHH typically involves the use of a radioactive substrate and the ability to measure radioactive isotopes, although recent advances in liquid chromatography mass spectroscopy methods may allow nonradiometric determination.

Catalytic activities have historically been the simplest methods to measure CYP1A. Unfortunately, several phenomena can limit exclusive use of catalytic activity for CYP1A measurement. The first of these is the sample preparation. Catalytic activities of enzymes, especially CYP isoforms, are extremely sensitive to heat denaturation. Consequently, the tissue sample must be dissected immediately upon the death of the organism and frozen and stored at temperatures that maintain activity (at least –70°C). If the tissue is thawed prior to preparative homogenization or stored in temperatures above –70°C, activity can be destroyed. A second phenomenon that often reduces catalytic activities of CYP are coexposure to various inorganic or organometal agents (Fent et al. 1998). In most contaminated environments, a mixture of metals and organics is present. A third factor that should also be considered when evaluating catalytic activities of CYP1A is the persistent nature of the environmental substrates present in the animal and the potential for competitive inhibition of the catalytic assay. It has been demonstrated that animals having excessive PCB body burdens may also have lower EROD or AHH activities (Stegeman and Hahn 1994). Lastly, another factor that should be considered when measuring CYP1A catalytic activity, especially in invertebrates, is the difference in substrate specificities between vertebrate and invertebrate CYP1A. Although AHH activities have been shown to correlate with CYP1A-like protein expressions in some molluscan species, EROD activity, if detectable, has not been shown to consistently correlate with isoform expression (D. Livingstone, personal communication). A common mistake is to assume that all substrate specificities are identical between all species. This is clearly not the case.

Given that catalytic activities have the potential for diminishment under a variety of environmental and test situations, it is suggested that measurements of CYP1A protein concentrations be included with activity in an environmental assessment. Enzyme-linked immunosorbent assay (ELISA) kits with CYP1A specific antibodies are now commercially available to measure CYP1A in most species of fish (Goksoyr and Husoy 1992). When catalytic activity is measured in conjunction with protein expression, mechanistic-based assessments can be used to identify causative contaminants. For example, if induction of CYP1A protein is observed in a migratory fish species from a potentially contaminated site compared with a reference site without a concomitant increase in catalytic activity, it is possible that either competitive substrates or inhibiting metals are bioavailable in the sampled organism. These relationships may allow assessors to identify specific classes of chemicals that may be present without performing expensive nontargeted chemical residue evaluations or toxicity identification evaluations (TIE).

Although there are numerous studies validating CYP1A expression as a measure of bioavailable AhR-ligands (Stegeman 1993; Stegeman and Hahn 1994; Bucheli and Fent 1995; Sarasquete and Segner 2000), few studies have shown that CYP1A expression is an indicator of stress. The processes and outcomes of biotransformation are complex. Clearly, the cell risks continual bioactivation of some compounds (i.e., PAHs) with the aim of metabolite elimination. Cytochrome P450 1A clearly bioactivates a smaller percentage of the PAHs to mutagenic metabolites compared with the majority of non-toxic metabolites (Williams and Buhler 1984). Many of these mutagenic metabolites are conjugated by phase II biotransformation pathways (see below, and Figure 1) to nontoxic derivatives. It is also interesting to note that neoplastic cells tend to downregulate phase I biotransformation pathways (i.e., CYP) and upregulate phase II pathways (i.e., glutathione S-transferase [GST]).

Clear associations have been observed in the field between PAH exposure, CYP1A expression, DNA adducts, and liver lesions in several fish species (Collier et al. 1996; Reichert et al. 1998). Other disease associations with CYP1A include immune suppression (Collier et al. 1998) and reproductive disorders in fish primarily by Ah-ligands (Andersson et al. 1988; Johnson et al. 1992; Collier et al. 1995a, 1998). Other evidence relating CYP1A as an indicator of stress includes reduction of activity and expression in nonmigratory organisms residing in severely contaminated waterways, such as the Elizabeth River (Van Veld et al. 1991, 1992; Cooper et al. 1998) or New Bedford

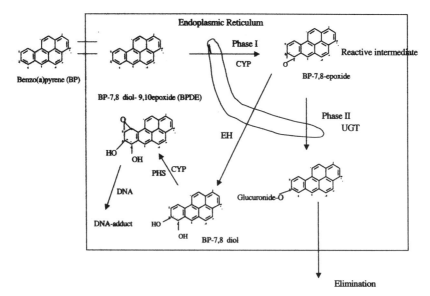

Figure 1. Phase I and Phase II metabolism of benzo(a) pyrene (BaP).

Harbor (Elskus et al. 1999). If animals of the same species from an unimpacted site are placed in these systems, CYP1A induction readily occurs (Van Veld et al. 1992; Elskus et al. 1999). It is unclear whether downregulation of CYP1A in surviving organisms in these environments provides a selective advantage from xenobiotic activation or is merely an effect of cells becoming more neoplastic in character (Myers et al. 1998). Alternatively, cells may be less able to mount an induction response because the metabolic processes responsible for protein synthesis have been damaged. Even though these relationships exist between contaminant loading and activity, it is very difficult to repeat these studies in the laboratory. Thus, determination of threshold levels of CYP1A expression for ecological risk assessments is fraught with uncertainties. For example, CYP1A is highly influenced by hormonal signals that are extremely variable within feral fish and invertebrates (Stegeman and Hahn 1994). There are also tremendous differences in basal expression of CYP1A between species and even populations within the same species (e.g., Elizabeth River and New Beford Harbor *Fundulus* spp.). In addition, when sampling the animals from the field, it is unclear how long CYP1A has been induced. It is likely that sustained induction will have different effects than that from episodic induction of any indicator protein. For these reasons, it is unlikely that CYP1A will ever be used as a threshold indicator for a hazard assessment evaluation. However, its use as an indicator of exposure and bioavailability of planar aromatic hydrocarbons can be a cost-effective mechanism of screening potentially contaminated areas for subsequent residue analysis and TIE analyses. Perhaps the best use of CYP1A, in conjunction with biliary fluorescent aromatic compounds (FAC), is to provide an assessment of exposure to PAHs in fish. Unlike halogenated aromatic hydrocarbons, which tend to accumulate in biota, PAHs are readily metabolized in fish and eliminated through the bile. Thus, body burden residues cannot be used to accurately assess exposure to PAHs. In summary, CYP1A is a relatively sensitive and, therefore, rapidly responding biochemical endpoint of exposure and, in combination with other indicators (see below), may provide some indication of adverse effects in aquatic organisms. However, observation of induction alone does not equate with adverse effects, thus, limiting its use as a threshold or indicator for hazard assessments.

Phase II Enzymes

Phase II enzymes catalyze the conjugation of various endogenous (e.g., sex steroids) and exogenous (i.e., many pollutants) substrates with several highly hydrophilic compounds that occur at high levels in cells. The purpose of these reactions is to increase the water solubility of the substrate and, thereby, facilitate its biliary or urinary excretion. Phase II reactions often occur in concert with phase I reactions, such as those catalyzed by CYP1A described above. For example, CYP1A can catalyze the epoxidation or hydroxylation of an unreactive and lipophilic compound such as BaP; this oxidized BaP metabolite then becomes a suitable substrate for phase II conjugations (see Figure 1).

Three major groups of phase II enzymes occur in animals. These are the glutathione S-transferases (GST), the uridinediphosphate glucuronosyltransferases (UDPGT), and the sulfotransferases (ST). These enzymes are named for the endogenous water soluble conjugant employed by each enzyme group—glucuronic acid (UDPGT), glutathione (GSH), and sulfate (ST). Among aquatic organisms, these enzymes have been studied most in fishes, and an excellent review of piscine phase II enzymes is provided by George (1994). In the context of aquatic toxicology, phase II pathways are considered of great importance because the resulting product (the conjugated xenobiotic), such as the glucuronidated metabolite of BaP (Figure 1), is, in the great majority of cases, nontoxic and readily excreted. As explained above, this is in contrast to phase I metabolism, which frequently results in the activation of the contaminant to a more reactive and more toxic product. For example, CYP1A catalyzes the oxidation of BaP to products that, while serving as suitable phase II substrates, can also bind to DNA and lead to mutations (see Chapter 3). Thus, the balance between phase I activation reactions and phase II conjugation pathways can underlie the toxicity of many organic xenobiotics such as the PAHs (Figure 1).

The use of phase II enzymes as biomarkers in fishes (and likely other aquatic animals), however, is less clear than is the case for CYP1A. Phase II enzyme activities (mainly GST and UDPGT) have been measured in laboratory exposures and in field studies involving pollution gradients. Laboratory examples include Lemaire et al. (1996), Petrivalsky et al. (1997), and Novi et al. (1998). In these studies, results with potential inducers of GST and UDPGT were variable; inductions, when observed, were typically less than twofold. Given the variability exhibited in most measurements taken in the course of field studies, such ranges in induction are of limited value, for example, in associating biomarkers with a pollution gradient. However, significant differences in enzyme activities between polluted and reference sites have been reported for some field studies, including Lenartova et al. (2000), Armknecht et al. (1998), Vigano et al. (1998), and Tuvikene et al. (1999)

Measures of phase II enzymes may be useful in the context of the balance between phase I activations and phase II detoxifications described earlier. For example, progeny of a population of killifish *Fundulus heteroclitus* that inhabits a highly PAH-polluted site in a Virginia estuary is resistant to the acute toxicity of sediments from that site versus progeny of killifish from a nearby clean site (Williams 1994). As described earlier, tissues of fish from the polluted site exhibit reduced CYP1A inducibility as compared with fish from the reference site (Van Veld et al. 1991). These fish, however, do exhibit greater GST activity than do reference fish (Armknecht et al. 1998). These differences in activity are reminiscent of biochemical changes observed in cancer cells, which appear to confer drug resistance to these cells (Farber 1990). Thus, these changes in killifish may serve as adaptations to living in an environment contaminated with chemicals such as PAHs for which the toxicity is enhanced by CYP1A but diminished by GST. In this light, concomitant measures of phase II enzymes and CYP1A may provide a marker for relative sensitivity to such types of chemicals.

Analysis of GST activity is usually performed spectrophotometrically according to the assay developed by Habig and Jakoby (1981). This method is convenient in that it is dependent on the direct change in absorption of the substrate when it is conjugated with GSH. It is essential to have a good quality spectrophotometer due to the high absorbance of some of the solutions. Habig and Jakoby (1981) list conditions for nine different spectrophotometric assays according to the substrate used. Modifications of the Habig and Jakoby (1981) method have been employed using different substrates by numerous researchers including James et al. (1979) and Ramsdell and Eaton (1990). Additionally, Gallagher et al. (2000) describe a GST assay for use with a 96-well microplate reader that is modified from Habig and Jakoby (1981).

Glutathione S-transferase can also be identified using an antibody approach. Gallagher et al. (2000) describe a method using GST antiserum raised against affinity-purified striped bass *Morone saxatilis* GST. In short, cystolic GST from striped bass was purified using GST affinity chromatography, and GST activity was verified by the presence of activity toward chlorodinitrobenzene. Glutathione S-transferase antibody-containing serum was obtained from rabbits that were immunized against the GSH affinity-purified fraction of striped bass cytosol. Cytosolic proteins from fish were separated on sodium dodecylsulfate–polyacrylanide gels. The membrane was incubated overnight to prevent nonspecific binding. The primary antibody–antiserum complex was visualized using a commercial chemiluminescense kit and detected by autoradiography.

Uridinediphosphate glucuronosyltransferase activity is typically measured in microsomes from fish liver using p-nitrophenol, 1-napthol, or 4-hydroxybiphenyl as substrates (van der Oost et al. 1996; Novi et al. 1998; Vigano et al. 1998). Triton X-100 treated microsomal protein is incubated under the appropriate assay conditions (usually, buffer, magnesium chloride, substrate, and UDP glucuronic acid). After incubation, the reaction is stopped, the supernatant is centrifuged, and the amount of substrate remaining is measured spectroscopically.

Methods of assaying ST activity include a fluorometric method, a radiochemical extraction method, and a radiochemical thin layer chromatography method. Each of these methods is outlined thoroughly in Tong and James (2000).

Fluorescent Aromatic Compounds

Polycyclic aromatic hydrocarbons are typically biotransformed and excreted through the bile as oxidized conjugates of the original parent compound in most vertebrates (Thakker et al. 1985). Since PAHs are fluorescent, bile may be collected from individual animals and FACs can be measured to determine exposure to PAHs (Krahn et al. 1984). This endpoint is particularly valuable in fish as biotransformation of PAHs prevents the use of body burden residues for the estimation of exposure (See Cytochrome P450 1A section). Fluorescent aromatic compounds have been used in several fish species as biomarkers of exposure to PAHs (Collier et al. 1995a, 1996, 1998). Direct correlations between FACs and PAH concentrations in sediment, PAH-DNA adducts, and CYP1A were observed in English sole *Pleuronectes vetulus* from Eagle Harbor in Puget Sound (French et al. 1996) and in oyster toadfish *Opsanus tau* from segments of the Elizabeth River (Collier et al. 1995b). In addition, inverse relationships were also observed between ovarian development and FACs in female English sole from Puget Sound, Washington (Johnson et al. 1992). A direct correlation was observed between FACs and liver damage in Atlantic flounder *Platichthys flesus* caged in contaminated waterways in Norway (Beyer et al. 1996). Although analysis of these metabolites is possibly one of the most specific indicators for exposure to aromatic hydrocarbons, size of the animal and instrument access (flurometer) can be limiting factors of the assay.

Oxidative Stress

Oxidative stress refers broadly to the deleterious effects of reactive oxygen species (ROS) that occur when the production of ROS overwhelms endogenous antioxidant defense systems of cells. For an excellent treatise on this subject, the reader is referred to Halliwell and Gutteridge (1999). Reactive oxygen species include the one-, two-, and three-electron reduction products of molecular oxygen (O_2); these are the superoxide anion radical (O_2^{-}), hydrogen peroxide (H_2O_2), and the hydroxyl radical ($^{\bullet}OH$), respectively. A radical is a chemical species with an unshared electron occupying an orbital, a feature that makes many radicals inherently reactive. While not a radical itself, H_2O_2 can readily accept an electron from a reduced transition metal, thereby generating the extremely reactive $^{\bullet}OH$; this is referred to as the Fenton reaction when the metal is ferrous iron (Fe^{3+}). The four-electron reduction product of O_2 is water, a product of aerobic respiration arising from the use of O_2 as the terminal electron acceptor by mitochondria. Other important ROS in biology include singlet oxygen ($^1O_2$) and nitric oxide (NO). Cellular electron transport chains such as those in mitochondria, chloroplasts, and microsomes that are involved in oxygen metabolism comprise important endogenous sources of ROS; that is, these systems are not 100%

efficient, and some "leakiness" of partially reduced oxygen species (i.e., ROS) occurs normally. Additionally, the respiratory burst produced by phagocytes such as macrophages and neutrophils produces ROS that underlie the microbiocidal activity of these cells—an example of a beneficial aspect of ROS production (for the host organisms, anyway).

In addition to the endogenous production of ROS, environmental contaminants can elicit oxidative stress by several mechanisms (Halliwell and Gutteridge 1999). A variety of chemicals can participate in redox cycling, including quinones and diols, nitroaromatics, aromatic azo dyes, bipyridyls (e.g., paraquat), and transition metals (such as iron, copper, and manganese). In the redox cycling of a quinone, for example, the parent quinone accepts an electron from a cellular reductant, such as reduced nicotinicadenine diphosphate (NADPH), giving rise to the semiquinone radical. While potentially damaging itself, this radical can readily donate its unpaired electron to O_2, which regenerates the parent quinone and completes the redox cycle (see Figure 2). This "futile" cycling can continue, with two negative consequences at each turn of the cycle: the oxidation of a high energy intracellular reductant and the production of $O_2^{\cdot-}$ and other ROS. These species, particularly $^{\cdot}OH$, are very reactive and generally nonspecific with respect to molecular targets. Targets include membrane lipids resulting in lipid peroxidation of the cell, DNA resulting in oxidized nucleotides and resultant instabilities in nucleotide sequences, and proteins resulting in alterations in function of enzymes, for example.

In addition to redox cycling, environmental contaminants can exert oxidative stress through other mechanisms. For example, some compounds such as coplanar PCBs and dioxins are potent AhR agonists (and hence

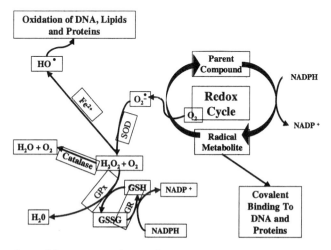

Figure 2. Adapted from Di Giulio et al. 1995. Overview of redox cycling, illustrating free radical production, antioxidant defenses, and toxicological consequences.

CYP1A inducers) but poor substrates for CYP1A-mediated oxidations. Such compounds can enhance ROS production via increased microsomal electron transport activity and by interference with catalytic oxidation of the recalcitrant substrate. Chemical-mediated phototoxicity comprises another important mechanism. For example, some PAHs such as BaP, fluoranthene, and anthracene are activated to a higher energy state by sunlight in the ultraviolet light B range (Arfsten et al. 1996). This energy can be readily transferred to O_2 yielding $^1O_2$. In aquatic organisms, cells that accumulate such compounds and are exposed to sunlight can undergo severe oxidative stress. For example, photoactivated PAHs are believed to be acutely toxic to fish via membrane oxidations in gill cells that receive sunlight exposure (Weinstein and Oris 1999). Interference with antioxidant defense systems, described below, is another important mechanism of oxidative stress. For example, metals such as cadmium (Cd), which are strong thiol ligands can bind to GSH, thereby eliminating its antioxidant function.

Antioxidant defense systems occur in all aerobic organisms in order to defend cells from the deleterious effects of ROS that arise naturally. In that light, the development of antioxidant defenses can be viewed as the "cost" incurred by the use of the highly efficient terminal electron acceptor, O_2. In addition to providing protection from ROS associated with aerobic metabolism, antioxidant defenses can serve to protect cells from ROS resulting from exposures to environmental contaminants through mechanisms described earlier. Importantly, conditions that enhance the flux of ROS, including exposures to redox-active chemicals, for example, can increase antioxidant defense system components through inductions of antioxidant enzymes and associated proteins involved in antioxidant defense.

Antioxidants can be divided into two broad categories: enzymatic and nonenzymatic. Key antioxidant enzymes include the following (for detailed descriptions, see Figure 2 and Halliwell and Gutteridge 1999):

Superoxide dismuatases catalyze the reaction between two molecules of $O_2^{\cdot -}$ to yield H_2O_2 and O_2:

$$2O_2^{\cdot -} + 2H^+ \circ H_2O_2 + O_2$$

Catalases catalyze the reaction between two molecules of H_2O_2 to yield two molecules of H_2O and O_2 (i.e., also a dismutation):

$$2\,H_2O_2 \circ 2H_2O + O_2$$

Glutathione peroxidases (GPx) catalyze the reductions of peroxides (both H_2O_2 and lipid peroxides, ROOH) to yield the corresponding alcohol (H_2O in the case of H_2O_2, ROH in the case of ROOH), and H_2O. This enzyme employs GSH to provide reducing equivalents; 2 GSH are oxidized to glutathione disulfide (GSSG):

$$ROOH + 2GSH \circ ROH + GSSG$$

Glutathione reductase (GR) reduces GSSG to GSH at the expense of NADPH provided from the pentose phosphate pathway:

$$GSSG + NADPH \circ 2GSH + NADP^+$$

Antioxidant defenses also include a number of nonenzymatic components that scavenge ROS. These include compounds synthesized by animals in vivo such as glutathione, uric acid, and melanins, and compounds derived from the diet such as α-tocopherol (vitamin E), ascorbic acid (vitamin C), and β-carotene (vitamin A).

When the flux of ROS exceeds the capacity of antioxidant defenses to effectively detoxify them, oxidative damage to cells can occur. Reactive oxygen species are generally nonspecific with respect to intracellular targets; a highly reactive ROS such as ˙OH can very rapidly oxidize biological molecules such as DNA, lipids, and proteins in close proximity to its point of origin. Oxidations of DNA result in oxidized DNA bases (such as 8-hydroxydeoxyguanosine), apurinic and apyrmidinic sites (i.e., the damaged base creates an instability that leads to loss of the base from its nucleotide sequence), and strand breaks. Oxidations of unsaturated lipids proceed via the process of lipid peroxidation, a chain reaction in which one molecule of ˙OH can produce many molecules of lipid peroxide via lipid peroxyl radical intermediates. Many proteins can also be oxidized by ROS, leading to various structural and functional changes. An example that has been exploited as a biomarker is the oxidation of the ferrous iron ($Fe2^+$) center of hemoglobin to the ferric ($Fe3^+$) state. This oxidized product is referred to as methemoglobin and is readily measured in blood samples.

There has been considerable interest in the use of responses associated with oxidative stress as biomarkers for environmental contaminants. This is in part due to its role in the fundamental biology of aerobic organisms and because it comprises a mechanism of action for many chemicals representing diverse structures. However, in contrast to many biomarkers for other endpoints (such as CYP1A, metallothioneins [MT] and acetylcholinesterase [AchE] also described herein), there is no single measure that by itself confirms the occurrence of oxidative stress or prestress adaptations. From the foregoing discussion, it is apparent that many potential measures of adaptive responses (i.e., increases in antioxidant defense system components) and deleterious responses (i.e., oxidations of DNA, lipids, and proteins) exist. This complicates the practical implementation of markers for oxidative stress in biomonitoring programs. Moreover, many measures associated with oxidative stress have not been consistently useful in the biomarker context, despite support from theory and laboratory studies. This is more problematic for aquatic organisms for which relevant research has been far more limited than is the case for mammalian models.

Nonetheless, oxidative stress comprises a very important fundamental mode of chemical toxicity, and some aquatic field studies have employed various measures with success. For example, Stephensen et al. (2000) observed higher activities of hepatic catalase and GR in shorthorn sculpin *Myoxocephalus scorpius* from some polluted harbors versus a reference site in Iceland. However, these differences were only observed for males and were inconsistent with respect to pollution within the different harbors examined. Such inconsistencies in response appear typical for antioxidant enzyme activity measurements in fish in both laboratory and field studies. At present, measures of glutathione show more promise than antioxidant enzyme activities as measures of prooxidant exposure in fish. For example, increased concentrations of total liver glutathione (GSH + GSSG), relative to reference sites, have been observed in Eurasian perch *Perca fluviatilis* inhabiting lakes receiving bleached pulp mill effluents in Sweden (Lindstrom-Seppa and Oikari 1990), in English sole collected from a PCB- and PAH-polluted portion of the Puget Sound in Washington (Nishimoto et al. 1995), and in brown bullhead *Ameiurus nebulosus* from a PAH-polluted section of the Buffalo River in New York (Eufemia et al. 1997).

The most widely employed spectrophotometric methods for glutathione measurements are based on Griffith (1980). In this assay, GSH reacts with 5,5'-dithiobis(2-nitrobenzoic acid) (DTNB) to produce the thionitrobenzoate (TNB) anion, a chromophore that absorbs strongly at 405 nm. The assay is specific for GSH (versus other thiols that also react with DTNB) by the inclusion of GR and NADPH (cofactor for GR). The GSH–TNB product reacts with GSH to produce GSSG and the chromphore TNB⁻; the concentration of GSH + GSSG proportional to the rate of TNB⁻ formation driven by GR. In order to measure GSSG specifically, the procedure is the same except that the sample is first incubated with 2-vinylpyridine (2VP). Glutathione is derivatized by 2VP, preventing its reaction with DTNB and, thereby, making the assay specific for GSSG. The fraction of GSH in a sample can then be estimated by subtracting GSSG measured in an aliquot with 2VP from the total glutahione measured in an aliquot without 2VP. For large sample sizes, typical in monitoring programs, the microplate method of Baker et al. (1990), based upon the Griffith (1980) assay, is recommended. The reader is also referred to Anderson (1996) for a clear explanation of this and other glutathione measurements.

A more accurate but more labor intensive analysis of glutathione is afforded by high-performance liquid chromatography (HPLC). This approach is recommended for laboratories proficient in HPLC, particularly where accurate measures of GSSG are desired. In healthy liver cells, for example, GSSG typically comprises less than 10% of total glutathione (Halliwell and Gutteridge 1999). While apparently unexploited in aquatic field studies, increases in the GSSG:GSH ratio have been employed as a measure of oxidative stress in mammals. Senft et al. (2000) describe a HPLC method for glutathione analysis employing fluorescence detection; this technique requires derivatization of GSH by 0-phthaladehyde. The use of electrochemical detection, as described by Lakritz et al. (1997), provides sensitive analysis without derivatization

Careful handling of samples is essential for accurate measures of glutathione; the peptide is readily degraded and GSH is readily oxidized to GSSG. Following sacrifice, tissue samples are preferably assayed immediately or immediately flash-frozen (e.g., in field studies, placed in vials and immersed in liquid nitrogen). Assays should be performed as rapidly as possible and various reaction mixtures kept on ice except as required.

Metallothioneins

Metallothioneins (MT) are relatively small cytosolic proteins that help regulate essential metal homeostasis, as well as the cellular redox potential. Having a significantly high percentage of thiol groups, MTs can covalently interact with most transition metals and scavenge free radical species generated from oxidative stress (Kagi and Schaffer 1988; Lazo et al. 1998; Klaassen et al. 1999). Genetic knockout studies in mammals have demonstrated the importance of these compounds in protection against oxidative damage and toxicities resulting from excessive metal exposures (e.g., Cd and arsenic; for a review in mammals see Klaassen et al. 1999). Since the proteins are small and fairly easy to purify, they have been identified in numerous species of plants and animals with conserved cystiene residues in specific locations of proteins isolated from *Neurospora* to *Homo sapiens* (Kagi and Schaffer 1988). Since expression of MTs is transcriptionally controlled by intracellular metal concentrations (primarily zinc), several groups have used expression of MT as an indicator of metal exposure in numerous vertebrate and invertebrate aquatic organisms (for reviews see Roesijadi 1992; Olsson 1996; Viarengo et al. 1999). These proteins can be easily measured using size-exclusion chromatography or polarography, or through metal saturation and spectrophotometric assays following partial chromatographic analyses (Schlenk and Brouwer 1991; Olsson 1996; Schlenk et al. 1997b). At the time of this writing, antibodies for fish or invertebrate isoforms of MT are not currently commercially available; consequently, the majority of studies measuring MT protein used the procedures noted above. Unfortunately, many of these analyses have probably overestimated levels of MT expression as metal saturation, and spectrophotometric assays fail to differentiate other low molecular weight, heat stable, cellular thiols, such as glutathione, which are typically present is significantly greater concentrations than MT. Because of this feature, it is imperative to chromatographically separate the low molecular weight fractions of cells prior to thiol analyses. With the exception of antibody-based

ELISA methods and polarography, size-exclusion chromatography followed by ion-exchange chromatography with subsequent metal measurement in collected fractions is the most accurate but also the most time-consuming method for protein quantification.

Because of the excessive time required for protein measurement, several groups have resorted to measuring MT mRNA through Northern blot or quantitative reverse transcriptase polymerase chain reaction (qRT-PCR; Kaplan et al. 1995b; Schlenk et al. 1997a). Although the metal content is not measured in these procedures, MT expression can be measured quickly without the use of atomic absorption spectrophotometers or similar instruments. Although molecular probes for analyses are not commercially available, cDNA sequences encoding many isoforms from fish and invertebrates are published. Primers can be constructed from these published sequences and used to amplify species-specific cDNAs, which can be applied in Northern slot blot analyses for multiple samples. The major disadvantage of measuring transcription of MT is that many toxicants inhibit transcription or translation in cells, thus, a lack of MT, or any other protein induction, may be the result of the animal being overwhelmed by toxicity.

Similar to CYP1A, MT expression has the clear advantage over metal residual analyses of providing an evaluation of metal bioavailability with sampled organisms. However, like CYP1A, MT expression is extremely variable in feral organisms, especially adults. In addition, because of their role in maintaining cellular redox potentials, induction of MT may occur in the absence of metal exposure, especially during periods of acute oxidative stress (Schlenk and Rice 1998). Examples in mammals include extracellular inflammation, starvation, and oxidative stress resulting from disease or chemical exposures (Sato and Bremner 1993). Prolonged exposure to the oxidant potassium permanganate actually reduced MT mRNA expression in channel catfish *Ictalurus punctatus* (Schlenk et al. 2000). Thus, MTs should never be used alone without another indicator of cellular damage, primarily oxidative damage.

Several field studies have demonstrated relationships between exposure to metals and MT expression (Hogstrand et al. 1991; Farag et al. 1995; Kaplan et al. 1995a; Schlenk et al. 1995; Tom et al. 1999; Wong et al. 2000) although few have observed relationships with stress. In fact, MT induction has been observed following arsenical exposure in channel catfish but without increases in lipid peroxidation or gluathione depletion (Schlenk, et al. 1997b). In contrast, direct relationships between lipid peroxidation and MT expression were observed in trout from the Clark Fork River (Farag, et al. 1995). Metallothionein is one of a multitude of acute phase stress proteins. However, its excessive presence in various species of metal-resistant fish indicates that that expression may not consistently correlate with anthropogenic stress (Kille et al. 1991; Kille and Olsson 1994). In addition, since MT is under transcriptional control, overall reductions in RNA synthesis or protein synthesis by any number of toxicants may prevent correlations with stress or metal exposure.

In summary, MT can be a useful measurement of metal bioavailability in the field when used in well-studied species (i.e., trout, channel catfish, turbot, mussel, oyster, crab) and in conjunction with other endpoints, specifically measurements of oxidative damage. When sampling fish, sexually mature animals should be avoided, especially females since estradiol downregulates MT expression (Olsson et al. 1995). Caging studies with fish should also be avoided, as handling stress tends to reduce signal-to-noise ratios (Baer and Thomas 1990). Overall, extreme caution should be used when attempting to apply MT as a bioindicator of stress. Although several laboratory- and aquacultural-based studies indicate relationships with stress, feral variability and the documented incidence of adaptive expression of MT in animals surviving long-term exposure to stressors limits its usefulness as a consistent stress endpoint.

Multidrug Resistance

The multidrug (or xenobiotic) resistance (MDR) protein family, also known as the P-glycoprotein family, are transport proteins spanning the plasma membrane in numerous cells including blood cells (hemocytes) and tissues of the intestine and liver of several aquatic organisms (Kurelec 1992, 1996; Minier et al. 1999). Induced by a plethora of xenobiotics, MDR proteins essentially pump the chemicals out of the cell and are partly responsible for the resistance often developed in neoplastic cells during chemotherapy. Activity can be easily measured using fluorescent dyes followed by inhibition by verapamil, which is a selective inhibitor of the transporter (Kurelec 1992). Multidrug resistance appears to be conserved throughout phylogeny such that antibodies to human MDRs often recognize homologous proteins in fish and invertebrates (Kurelec 1992). Advantages of using such a measure include the relative ease of measurement (either activity or expression), its rapid induction following exposure to environmentally relevant concentrations of xenobiotics, and its relationship to neoplastic transformation of cells (Cooper et al. 1998). Thus, MDR expression appears to only be present when cells are exposed to "unwanted" chemicals and, as such, may be considered a biochemical indicator of stress in cells.

The major disadvantage of MDRs is the lack of selectivity with regard to toxicants. Similar to the CYP monooxygenase system (specifically CYP3A), which has a widespread substrate specificity, metals and organics can each be substrates for the transporter and induce expression (Eufemia and Epel 2000). Unfortunately, when MDR is expressed in a feral organism, it can be uncertain what agent is responsible for the effect. Therefore, causal relationships to specific chemical classes are limited. In addition, since the transporter is driven by ATP, any inhibitor of electron transport or compound that impairs ATP synthesis will reduce activity. As with the CYP1A system, measuring both activity and protein expression is warranted and may provide more data with regard to toxicant exposure.

When used with other neoplastic markers, such as induced phase II enzyme activity (i.e., GST), or proliferating cellular nuclear antigens, MDR can be useful in a diagnostic capability for neoplasia or even as a biomarker of susceptibility. Determination of susceptibility is necessary in reducing uncertainty in risk assessment, as it is imperative that the most sensitive sentinel species be used in bioassays. Therefore, MDR may be useful in this regard as further research will elucidate its role in cellular toxicity.

Porphyrins

Porphyrins are intermediate products of the heme biosynthetic pathway. Heme prosthetic groups are used by numerous proteins (e.g., hemoglobin, CYP) within biological systems. Several lines of evidence indicate oxidative stress may be caused by incomplete monooxygenation by CYP of chlorinated aromatic hydrocarbons. The close proximity of heme-containing proteins (i.e., CYP) to this oxidative damage may lead to heme destruction and subsequent porphyrin formation (Hahn and Chandran 1996). The primary advantage of using porphyrins is the occurrence of the metabolites in blood, which can be analyzed by HPLC under specific wavelengths of fluorescence or absorbance. Consequently, porphyrin measurements can be made in animals without the death of the organism (nondestructive sampling), which has allowed numerous studies to be carried out in birds (Melancon et al. 1992). However, porphyrin profiles have also been proposed as biochemical indicators of chlorinated aromatic hydrocarbon exposure and effects in lake whitefish *Coregonus clupeaformis* exposed to bleached kraft pulp mill effluent (Xu et al. 1994). Cytochrome P450 1A and porphyrins have been measured in fish hepatoma cell lines and have been used as bioassays to measure the bioavailable occurrence of PAHs and PCBs in sediment, oil shale, and water extracts in PAH- and PCB-contaminated sites (Huuskonen et al. 1998, 2000). Other confirmatory proteins and biometabolites that could be used, in addition to porphyrins, to confirm exposure and effects of planar halogenated aromatic hydrocarbons are heat stress protein 30 (HSP30; heme oxygenase), CYP1A (heme protein), and bilirubin and biliverdin (endogenous Ah agonists). Use of porphyrins, largely has been overlooked in fish, but deserves more attention because of the potential for nonlethal measurement in feral organisms. However, little is known regarding the baseline values and what effects species, gender, development, or season may have on production.

Retinoids

A third group of biometabolites that have shown promise as indicators of stress and exposure are retinoids. Retinoids are derived from vitamin A, which has been shown to be depleted following exposure to planar halogenated aromatic compounds (Peakall 1992; Spear et al. 1992). The mechanism of depletion is unclear, but one study demonstrated dose-dependent stimulation of hepatic retinoic acid hydroxylation and glucuronidation in brook trout *Salvelinus fontinalis* following exposure to a coplanar PCB (3,3',4,4' tetrachlorobiphenyl) (Boyer et al. 2000). Enhancing the metabolism and elimination of retinoids may be one mechanism of depletion. Retinoids were also depleted from brown bullheads in PAH-contaminated sites in the Great Lakes (Arcand-Hoy and Metcalfe 1999). In the white sucker *Catostomus commersoni,* a direct relationship between maternal retinoid loss, CYP1A activity in the liver, and prevalence of embryonic malformations in animals collected from the same contaminated location were observed (Branchaud et al. 1995). After a 3-year exposure to PAH-contaminated sediment, retinol concentrations in plasma and liver were significantly reduced in Atlantic flounder (Besselink et al. 1998). The inverse relationship between retinoids and CYP1A expression and the importance of vitamin A in animal health indicate retinoids may provide significant information regarding the adverse effects of planar aromatic hydrocarbons. However, similar to the situation with porphyrins, little is known regarding thresholds of vitamin A levels necessary for animal health or about the baseline variance of vitamin A levels in multiple species of fish. Consequently, their use as absolute bioindicators is still limited.

Cholinesterases

Acetylcholinesterase (AChE) catalyzes the hydrolysis (and, hence, inactivation) of the neurotransmitter acetylcholine (ACh) in cholinergic nerves (Ecobichon 1996; see Figure 3). Inhibition of AChE resulting in overaccumulation of ACh and prolonged electrical activity at nerve endings comprises a key mechanism of toxicity for organophosphorous (OP) and carbamate insecticides that are currently major-use insecticides. Acetylcholinesterase inhibition by these compounds occurs through the formation of a covalent bond between the active site of the enzyme and the insecticide. The consequences of AChE inhibition depend upon the nervous tissue affected and include cardiovascular, respiratory, and gastrointestinal dysfunction (parasympathetic autonomic tissue); hyperactivity, lethargy, and unconsciousness (central nervous system); and muscle weakness, respiratory collapse, and paralysis (somatic motor nerve fibers). Acetylcholinesterase and other cholinesterase activities (such as butyrylcholinesterase, which preferentially hydrolyzes butyrylcholine) also occur in red blood cells and plasma. While not functional in nervous

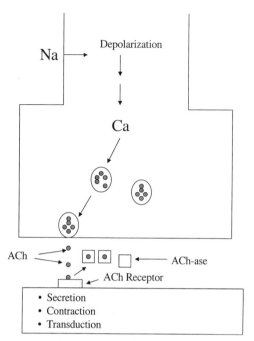

Figure 3. The neuromuscular junction, illustrating release of acethylcholine (ACh) and metabolism of ACh by acetylcholinesterase (AChE).

system physiology, these activities provide useful surrogate markers for nerve fiber AChE activity and can be measured nondestructively.

The measurement of AChE in brain tissue and of cholinesterase activity in blood has been employed to monitor exposures of fish and wildlife to OP and carbamate insecticides with good success (Mineau 1991). The perceived use of these measures is enhanced by proximity to the mode of toxic action and by the difficulty of measuring tissue levels of these compounds themselves, which is complicated by the number of compounds that may occur at a given field site and by their frequently rapid metabolism in animals. However, variability in enzyme activity measurements can be substantial, and quantitative associations between observed activities and health impacts are sometimes difficult. Also, while inhibition of cholinesterases by OPs is generally irreversible, inhibition by carbamates is reversible. That is, typically over a time frame of hours, the carbamate is released by the enzyme, thereby regenerating active enzyme, even in a moribund animal. Therefore, cholinesterases are considered more reliable markers for exposures to OPs than to carbamates. Fulton and Key (2001) provide a review of the use of AChE inhibition as a biomarker for OP exposure and effects in aquatic animals. Payne et al. (1996) describe the applicability of AChE inhibition for other pollution scenarios in aquatic systems, including urban runoff and paper mill effluents.

Acetylcholinesterase activity can be measured by using the procedures outlined in Ellman et al. (1961) and modified by Galgani and Bocquene (1991) to be used with a microplate reader. Rodriguez-Fuentes and Gold-Bouchot (2000) provide a similar method for determination of AChE activity in brains from tilapia. The method is again modified from Ellman et al. (1961) and involves monitoring the reaction of DTNB and acetylthiocholine iodide at an absorbance of 405 nm.

Case Studies

Clinch River, Watts Bar Reservoir

Perhaps one of the best examples of the uses of biochemical endpoints of fish in environmental assessment of a contaminated waterway is the study performed by Adams et al. (1999) on the Clinch River, Watts Bar Reservoir. The U.S. Department of Energy authorized a remedial investigation of the Clinch River, Poplar Creek system Superfund site in 1989. The site consists of 70 river km and 40 km² of surface area and is located adjacent to the Oak Ridge Reservation. The Clinch River and several tributary embayments in Melton Hill and Watts Bar reservoirs flow through this site. Contaminated tributary streams include McCoy Branch and Poplar Creek, which receive input from the Y-12 and Oak Ridge gaseous diffusion plants, as well as the Oak Ridge National Laboratory. Wastes included numerous radionucleides, metals (such as mercury), and organic contaminants (such as PCBs).

Two species of fish, largemouth bass *Micropterus salmoides* and bluegill *Lepomis macrochirus,* were collected for analyses at 9 sites and a reference location in this region. The purpose of the study was to measure a variety of individual fish health responses and assess fish community integrity, as well as determine the potential effects of contaminants on fish health. Three sample sites were located in the contaminated Poplar Creek embayment with additional sites representing a potential gradient of contaminant concentrations with higher levels projected near the Oak Ridge Reservation and lower values downstream in the Clinch River. The major changes observed included induction of EROD activity (CYP1A), organ dysfunction (alanine transaminase [ALT] and creatinine in plasma), increased frequency of histopathological lesions, impaired reproduction, and reduced fish community integrity. In Poplar Creek, where PCB and mercury concentrations were the highest, multivariate discriminate analyses using 20 responses at the individual level and below, including biochemical and physiological metrics, indicated that only 5 endpoints were statistically important in distinguishing among integrated site health responses for each fish species. These included: EROD, plasma ALT/creatinine, histopathological damage, and atretic oocytes (reproductive). Based on the relationship between higher-level effects and EROD, it was concluded that PCB exposure was likely a dominant factor in the adverse effects observed within individual bass and bluegill in this system. However, causal relationships between contaminant loading and fish community indices such as species richness and relative abundance were more difficult to establish because of confounding factors such as poor food and habitat availability.

Bayou Bartholomew, Arkansas

In a similar study on a much smaller scale, three biochemical endpoints were compared with animal health and fish community metrics using largemouth bass, white crappie *Pomoxis annularis*, common carp *Cyprinus carpio*, and bluegill collected from 13 sites on a low gradient waterway in southeastern Arkansas, Bayou Bartholomew (Schlenk et al. 1996). The stream originates in Pine Bluff Lake and drains approximately 2,239 square miles of, primarily, agricultural lands, eventually flowing into the Ouachita River. A total of 751 fish representing 44 species were collected with 134 individuals of the 4 species above used for biochemical endpoints. The three biochemical endpoints were CYP1A (measured as protein and catalytic activity-EROD), MT, and heme oxygenase (HSP 30, measured as protein). The original purpose of the study was to determine whether biochemical endpoints correlated with higher level biological responses in fish collected throughout the waterway. Results indicated that direct correlations were observed between CYP1A protein and individual health in largemouth bass, but the relationship was inversely related and did not correlate with CYP1A catalytic activity. This could be explained by at least two mechanisms: (1) catalytic activity was inhibited by a competing substrate, or (2) CYP1A was being degraded (see Cytochrome P450 1A section). To check the latter possibility, heme oxygenase expression was also measured in these individuals and was directly related to poor fish health, but inversely related to CYP1A protein. Thus, in unhealthy individuals, it was likely that degradation of CYP1A heme was occurring. This process can be initiated by a host of compounds, but usually occurs through oxidative stress (Hahn and Chandran 1996). Two suspected classes of compounds included planar halogenated aromatics and multivalence state metals. Each of these chemical classes can cause oxidative stress, heme degradation, and inhibit EROD activity (Albores et al. 1992; Stegeman and Hahn 1994). As described above, with the exception of Lake Pine Bluff where a pulp and paper plant is located, little industrial input occurs throughout the stream. Since the animals used for these studies were not collected from these upstream sites, halogenated aromatic hydrocarbons were unlikely candidates. In contrast, several concurrent studies carried out by the Arkansas Game and Fish Commission and Department of Environmental Quality indicated significant mercury contamination in largemouth bass from the areas where the animals used for biochemical analyses were also collected. In earlier studies performed in mercury-contaminated sites in south-western Arkansas, correlations were observed between hepatic and fillet mercury concentrations and hepatic MT expression in feral largemouth bass (Schlenk et al. 1995). Since no relationship between MT and fish health was observed in the Bayou Bartholomew study, mercury did not appear to be a candidate for the adverse effects observed in resident fish species. As the primary inputs of this stream appeared to be agricultural, arsenical biocides have been proposed recently as a primary stressor based upon their use in multiple formulations as herbicides in cotton culture for decades in this region. Arsenic has been shown to induce heme oxygenase and degrade CYP enzymes (Albores et al. 1992). Other metals capable of this mechanism are usually of industrial origin (i.e., Cd), and, given the lack of industrial input to this area, future efforts will be made to determine relative sediment concentrations and bioavailability of arsenicals in this area. As in the Adams et al. (1999) study described above, although strong correlations were observed between individual fish health and biochemical endpoints, no relationship was observed between fish community metrics and biochemical responses in the four species examined. Only when fish health indices of largemouth bass were compared with community indices were significant correlations present.

Summary

Numerous studies, including the case studies described above, demonstrate the advantages of using a suite of biochemical responses that are affected by specific chemical classes in a mechanism-based approach to identify potential stressors in aquatic systems. An important limitation to the use of biochemical responses in assessing chemical impacts in aquatic and other ecosystems pertains to the issue of biological levels of organization. Biochemical responses are by nature measures at lower levels of organization: molecular, biochemical, and cellular. Measurements at these levels provide important information concerning exposures and mechanisms of action, and can, oftentimes, be directly related to the health of individual organisms. However, the relationship of information at these lower levels to the effects of pollutants becomes more tenuous as one moves to higher levels of organization. Emergent properties (for example, competition and predator–prey relationships at the community level) that are not apparent at the lower levels come into play. Therefore, one cannot accurately predict higher-level effects of pollutants based solely upon lower level (e.g., biochemical) responses. This is an important limitation given that many assessments of aquatic systems are concerned primarily with conditions of populations, communities, and ecosystems. However, biochemical approaches have several advantages that motivate their use in environmental assessments. Their relative chemical specificity and proximity to modes of action underlie their ability to identify causative agents. Also, they generally have a rapid response time that allows

Table 1. Features of selected biochemical responses as biomarkers for pollutants in aquatic ecosystems.

Indicator	Advantages	Disadvantages	Application
CYP1A (cytochrome P4501A)	1. Relatively specific for planar aromatic compounds (PAH, PCB, etc.) 2. Potential relationships with hepatic lesions (PAH), reproduction (PCB/PAH), immune function (PCB/PAH)	1. Sensitive to hormonal impacts 2. Induction may be negatively impacted by coexposure to metals 3. Response may be diminished in nonmigratory species residing in chronically contaminated settings. 4. Although kits for expression are available; plate-reading spectro-photometer or fluorometer (EROD) is required	1. Assess exposure to planar aromatic hydrocarbons 2. Activity and expression should be evaluated in each sample
Phase II (conjugating enzymes)	1. Can be induced by a variety of organic compounds 2. Potential relationships with cancer susceptibility	1. Inductions highly variable among chemicals, species, and tissues 2. Includes several enzymes, each with its own method required	Best used in conjunction with CYP1A as index of susceptibility
MT (metallo-thionein)	1. Can be relatively specific for transition metals and oxidants 2. Indicator of available metal or oxidative stress within sampled cells 3. Simple spectrophotometric assays for measurement	May be induced by severe acute stress (i.e., part of general stress response)	1. Assess bioavailable exposure to transition metals or general stress. 2. Should be used with other indica-tor of oxidative damage (HSP, GSSG, Tbars) 3. Care should be taken to remove glutathione complexes by size exclusion prior to measurement
MDR (multidrug resistance)	1. Good indicator of stressor exposure and cellular transformation (i.e., precedes neoplastic transformation) 2. High degree of sequence identity (probes can be used in most species)	1. Nonspecific for metals or organics 2. Requires fluorometric analyses or immunoblotting	Used to determine exposure to nonspecific stressors and can be present in preneoplastic foci

Table 1. (continued.)

Indicator	Advantages	Disadvantages	Application
GSH and GSSG (reduced gluta-thione/oxidized glutathione)	1. Can provide an indication of oxidative stress from a variety of sources. 2. Both simple spectrophotomteric and more sensitive but difficult HPLC assays available	1. Relatively nonspecific with respect to chemical class 2. Not well characterized in aquatic models with most potential effectors	A potentially robust indicator of oxidative stress
FAC (fluorescent aromatic compounds)	Selective for PAH exposure	Requires fluorometric analyses and sophisticated dissection skills	Assesses PAH exposure; excellent in combination with CYP1A
Porphyrins	1. Relatively selective for exposure and adverse effects of halogenated aromatic compounds (PCBs, HCB, etc.) 2. Nondestructive	1. Baseline levels still uncharacterized 2. Effects may be complicated with metal coexposure	1. Assess exposure and potential adverse effects– of halogenated aromatic hydrocarbons 2. Can be useful with HSP proteins (i.e., HSP30) and CYP1A
Retinoids	Indicative of altered nutrient transformation	1. Although some relationships with planar aromatic compounds, response can be caused by nonspecific diminish-ment of animal health 2. Baseline levels still relatively uncharacterized	Should be used in concert with CYP1A or any of the other indicators of planar aromatic hydrocarbon exposure
AChE (acetyl-cholines-terase)	1. Highly specific indicator of cholinesterase-inhibiting compounds 2. Easily measured spectrophotometrically	Inhibition by some compounds (e.g., carbamates) reversible	Widely used marker for exposure and neurotoxic effects of organophosphate and carbamate insecticides

them to serve as early warning indicators of environmental harm. In many cases, they also serve as excellent markers of chemical exposure; this is particularly valuable for rapidly metabolized chemicals such as OPs and PAHs. Relatedly, these responses have the advantage over chemical residue analyses of indicating availability of the absorbed stressor in a cellular system. Finally, many biochemical measures are rapid and inexpensive to perform relative to many higher-level responses. In the context of assessments of aquatic systems, biochemical measures are generally of greatest use when integrated with other, more ecological, approaches.

References

Adams, S. M., M. S. Bevelhimer, M. S. Greeley Jr., D. A. Levine, and S. J. Teh. 1999. Ecological risk assessment in a large river reservoir: 6. Bioindicators of fish population health. Environmental Toxicology and Chemistry 18:628–640.

Albores, A., J. Koropatnick, M. G. Cherian, and A. J. Zelazowski. 1992. Arsenic induces and enhances rat hepatic metallothionein production in vivo. Chemico Biological Interactions 85:127–140.

Anderson, M. E. 1996. Glutathione. Pages 213-226 in N. A. Punchard and F. J. Kelly, editors. Free radicals: a practical approach. Oxford University Press, New York.

Andersson, T., L. Forlin, J. Hardig, and A. Larsson. 1988. Physiological disturbances in fish living in coastal water polluted with bleached kraft pulp mill effluents. Canadian Journal of Fisheries and Aquatic Sciences 45:1525–1536.

Arcand-Hoy, L. D., and C. D. Metcalfe. 1999. Biomarkers of exposure of brown bullheads (*Ameiurus Nebulosus*) to contaminants in the lower Great Lakes, North America. Environmental Toxicology and Chemistry 18:740–749.

Arfsten, D. P., D. J. Schaeffer, and D. C. Mulveny. 1996. The effects of near ultraviolet radiation on the toxic effects of polycyclic aromatic hydrocarbons in animals and plants. A review. Ecotoxicology and Environmental Safety 33:1–24

Armknecht, S. L., S. L. Kaattari, and P. A. Van Veld. 1998. An elevated glutathione S-transferase in creosote-resistant mummichog (*Fundulus heteroclitus*). Aquatic Toxicology 41:1–16.

Baer, K. N., and P. Thomas. 1990. Influence of capture stress, salinity and reproductive status on zinc associated with metallothionein-like proteins in the livers of three marine teleost species. Marine Environmental Research 29:277–287.

Baker, M. A., G. J. Cerniglia, and A. Zaman. 1990. Microtiter plate assay for the measurement of glutathione and glutathione disulfide in large numbers of biological samples. Analytical Biochemistry 190:360–365.

Besselink, H. T., and five coauthors. 1998. Alterations in plasma and hepatic retinoid levels in flounder (*Platichthys flesus*) after chronic exposure to contaminated harbour sludge in a mesocosm study. Aquatic Toxicology 42:271–285.

Beyer, J., and seven coauthors. 1996. Contaminant accumulation and biomarker responses in flounder (*Platichthys flesus*) and Atlantic cod (*Gadus morhua*) exposed by caging to polluted sediments in Soerfjorden, Norway. Aquatic Toxicology 36:75–98.

Boyer, P. M., A. Ndayibagira, and P. A. Spear. 2000. Dose-dependent stimulation of hepatic retinoic acid hydroxylation/oxidation and glucuronidation in brook trout,

Salvelinus fontinalis, after exposure to 3,3',4,4'-tetrachlorobiphenyl. Environmental Toxicology and Chemistry 19:700–705.

Branchaud, A., A. Gendron, R. Fortin, P. D. Anderson, and P. A. Spear. 1995. Vitamin A stores, teratogenesis, and EROD activity in white sucker, *Catostomus commersoni,* from Riviere Des Prairies near Montreal and a reference site. Canadian Journal of Fisheries and Aquatic Sciences 52:1703–1713.

Bucheli, T. D., and K. Fent. 1995. Induction of cytochrome P450 as a biomarker for environmental contamination in aquatic ecosystems. Critical Reviews in Environmental Science and Technology 25:201–268.

Collier, T. K., B. F. Anulacion, J. E. Stein, A. Goksoyr, and U. Varanasi. 1995a. A field evaluation of cytochrome P4501A as a biomarker of contaminant exposure in three species of flatfish. Environmental Toxicology and Chemistry 14:143–152.

Collier, T. K., and five coauthors. 1996. Incorporation of biomarkers into ecological risk assessments of contaminated nearshore marine habitats. Marine Environmental Research 42:274–275.

Collier, T. K., L. L. Johnson, C. M. Stehr, M. S. Myers, and J. E. Stein. 1998. A comprehensive assessment of the impacts of contaminants on fish from an urban waterway. Marine Environmental Research 46:243–247.

Collier, T. K., and six coauthors. 1995b. Biomarkers of PAH exposure in oyster toadfish (*Opsanis tau*) from the Elizabeth River, Virginia. Marine Environmental Research 39:348–349.

Cooper, P. S., W. K. Vogelbein, and P. A. Van Veld. 1999. Altered expression of the xenobiotic transporter P-glycoprotein in liver and liver tumours of mummichog (*Fundulus heteroclitus*) from a creosote-contaminated environment. Biomarkers 4:48–58.

Ecobichon, D. J. 1996. Toxic effects of pesticides. Page 1111 *in* C. D. Klaassen, editor. Casarett and Doull's toxicology: the basic science of poisons. McGraw-Hill, New York.

Ellman, G. L., K. D. Courtney, V. Andres, and R. M. Featherstone. 1961. A new and rapid colorimetric determination of acetylcholinesterase activity. Biochemical Pharmacology 7:88–95.

Elskus, A. A., E. Monosson, A. E. Mcelroy, J. J. Stegeman, and D. S. Woltering. 1999. Altered CYP1A expression in *Fundulus heteroclitus* adults and larvae: a sign of pollutant resistance? Aquatic Toxicology 45:99–113.

Eufemia, N. A., T. K. Collier, J. E. Stein, D. E. Watson, and R. T. Di Giulio. 1997. Biochemical responses to sediment-associated contaminants in brown bullhead (*Ameriurus nebulosus*) from the Niagara River ecosystem. Ecotoxicology 6:13–34.

Eufemia, N. A., and D. Epel. 2000. Induction of the multixenobiotic defense mechanism (Mxr), P-glycoprotein, in the mussel *Mytilus californianus* as a general cellular response to environmental stresses. Aquatic Toxicology 49:89–100.

Farag, A. M., M. A. Stansbury, C. Hogstrand, E. Macconnell, and H. L. Bergman. 1995. The physiological impairment of free-ranging brown trout exposed to metals in the Clark Fork River, Montana. Canadian Journal of Fisheries and Aquatic Sciences 52:2038–2050.

Farber, E. 1990. Clonal adaptation during carcinogenesis. Biochemical Pharmacology 39:1837–1846.

Fent, K., B. R. Woodin, and J. J. Stegeman. 1998. Effects of triphenyltin and other organotins on hepatic monooxygenase system in fish. Comparative Biochemistry and Physiology 121:277–288.

French, B. L., and five coauthors. 1996. Accumulation and dose–response of hepatic DNA adducts in English sole (*Pleuronectes vetulus*) exposed to a gradient of contaminated sediments. Aquatic Toxicology 36:1–16.

Fulton, M. H., and P. B. Key. 2001. Acetylcholinesterase inhibition in estuarine fish and invertebrates as an indicator of organophosphorous insecticide exposure and effects. Environmental Toxicology and Chemistry 20:37–45.

Galgani, F., and G. Bocquene. 1991. Semi-automated colorimetric and enzymatic assays for aquatic organisms using microplate readers. Water Research 25:147–150.

Gallagher, E. P., K. M. Sheehy, M. W. Lame, and H. J. Segall. 2000. In vitro kinetics of hepatic glutathione S-transferase conjugation in largemouth bass and brown bullheads. Environmental Toxicology and Chemistry 19:319–326.

George, S. C. 1994. Enzymology and molecular biology of phase II xenobiotic-conjugating enzymes in fish. CRC, Boca Raton, Florida.

Goksoyr, A., and A.-M. Husoy. 1992. The cytochrome P450 1A1 response in fish: application of immunodetection in environmental monitoring and toxicological testing. Marine Environmental Research 35:147–150.

Griffith, O. W. 1980. Determination of glutathione and glutathione disulfide using glutathione-reductase and 2-vinylpyridine. Analytical Biochemistry 106:207–212.

Habig, W. H., and W. B. Jakoby. 1981. Assays for differentiation of glutathione S-transferases. Methods in Enzymology 77:398–405.

Hahn, M. E. 1998. The aryl hydrocarbon receptor: a comparative perspective. Comparative Biochemistry and Physiology 121c:23–54.

Hahn, M. E., and K. Chandran. 1996. Uroporphyrin accumulation associated with cytochrome P4501A induction in fish hepatoma cells exposed to aryl hydrocarbon receptor agonists, including 2,3,7,8-tetrachlorodibenzo-p-dioxin and planar chlorobiphenyls. Archives of Biochemistry and Biophysics 329:163–174.

Hahn, M. E., B. L. Woodward, J. J. Stegeman, and S. W. Kennedy. 1996. Rapid assessment of induced cytochrome P4501A protein and catalytic activity in fish hepatoma cells grown in multiwell plates: response to TCDD, TCDF, and two planar PCBs. Environmental Toxicology and Chemistry 15:582–591.

Halliwell, B., and J. M. C. Gutteridge. 1999. Free radicals in biology and medicine. Oxford University Press, Oxford.

Hogstrand, C., G. Lithner, and C. Haux. 1991. The importance of metallothionein for the accumulation of copper, zinc and cadmium in environmentally exposed perch, *Perca fluviatilis*. Pharmacology and Toxicology 68:492–501.

Huuskonen, S., K. Koponen, O. Ritola, M. Hahn, and P. Lindstrom-Seppa. 1998. Induction of CYP1A and porphyrin accumulation in fish hepatoma cells (PLHC-1) exposed to sediment or water from a PCB-contaminated lake (Lake Kernaala, Finland). Marine Environmental Research 46:379–384.

Huuskonen, S. E., A. Tuvikene, M. Trapido, K. Fent, and M. E. Hahn. 2000. Cytochrome P4501A induction and porphyrin accumulation in PLHC-1 fish cells exposed to sediment and oil shale extracts. Archives of Environmental Contamination and Toxicology 38:59–69.

James, M. O., E. R. Bowen, P. M. Dansette, and J. R. Bend. 1979. Epoxide hydrolase and glutathione S-transferase activities with selected alkene and arene oxides in several marine species. Chemical-Biological Interactions 25:321–344.

Johnson, L., E. Casillas, S. Sol, T. K. Collier, J. E. Stein, and U. Varanasi. 1992. Contaminant effects on reproductive success in selected benthic fish. Marine Environmental Research 35:165–170.

Kagi, J. H. R., and A. Schaffer. 1988. Biochemistry of metallothionein. Biochemstry 27:8509–8515.

Kaplan, L. A. E., K. Van Cleef, and I. Wirgin. 1995a. Induction of metallothionein mRNA in environmentally exposed killifish (*Fundulus heteroclitus*). Marine Environmental Research 39:360.

Kaplan, L. A. E., K. Van Cleef, I. Wirgin, and J. F. Crivello. 1995b. A comparison of RT-PCR and Northern blot analysis in quantifying metallothionein mRNA levels in killifish exposed to waterborne cadmium. Marine Environmental Research 39:137–141.

Kille, P., and P. E. Olsson. 1994. Metallothionein and heavy metal poisoning. Biochemical Society Transactions 22:249.

Kille, P., P. E. Stephens, and J. Kay. 1991. Elucidation of cDNA sequences for metallothioneins from rainbow trout, stone loach and pike liver using the polymerase chain reaction. Biochimica Biophysica Acta 1089:407–410.

Klaassen, C. D., J. Liu, and S. Choudhuri. 1999. Metallothionein: an intracellular protein to protect against cadmium toxicity. Annual Reviews of Pharmacology and Toxicology 39:267–294.

Krahn, M. M., M. S. Myers, D. G. Burrows, and D. C. Malins. 1984. Determination of metabolites of xenobiotics in the bile of fish from polluted waterways. Xenobiotica 14:633–646.

Kurelec, B. 1992. The multixenobiotic resistance mechanism in aquatic organisms. Critical Reviews in Toxicology 22:23–43.

Kurelec, B., S. Krca, and D. Lucic. 1996. Expression of multixenobiotic resistance mechanism in a marine mussel *Mytilus galloprovincialis* as a biomarker of exposure to polluted environments. Comparative Biochemistry and Physiology 113c:283–289.

Lakritz, J., C. G. Plopper, and A. R. Buckpitt. 1997. Validated high-performance liquid chromatography electrochemical method for determination of glutathione and glutathione disulfide in small tissue samples. Analytical Biochemistry 247:63–68.

Lazo, J. S., S. M. Kuo, E. S. Woo, and B. R. Pitt. 1998. The protein thiol metallothionein as an antioxidant and protectant against antineoplastic drugs. Chemico Biological Interactions 111–112:255–262.

Lemaire, P., L. Forlin, and D. R. Livingstone. 1996. Responses of hepatic biotransformation and antioxidant enxymes to CYP1A-inducers (3-methylcholanthrene, betanaphthoflavone) in sea bass (*Dicentrarchus labrax*), dab (*Limanda limanda*) and rainbow trout (*Oncorhynchus mykiss*). Aquatic Toxicology 36:141–160.

Lenartova, V., K. Holovska, and P. Javorsky. 2000. The influence of environmental pollution on the SOD and GST-isoenzyme patterns. Water Science and Technology 42:209–214.

Lindstrom-Seppa, P., and A. Oikari. 1990. Biotransformation activities of feral fish in waters receiving bleached pulp mill effluents. Environmental Toxicology and Chemistry 9:1215–1424.

Livingstone, D. R., and eight coauthors. 1997. Apparent induction of a cytochrome P450 with immunochemical similarities to CYP1A in digestive gland of the common mussel (*Mytilus galloprovincialis* L.) with exposure to 2,2',3,4,4',5'-hexachlorobiphenyl and arochlor 1254. Aquatic Toxicology 38:205–224.

Melancon, M. J., and six coauthors. 1992. Metabolic products as biomarkers. Pages 87–123 *in* R. J. Huggett, R. A. Kimerle, P. M. Mehrle, and H. L. Bergman, editors.

Biomarkers: biochemical, physiological, and histological markers of anthropogenic stress. Lewis Publishers, Boca Raton, Florida.

Mineau, P., editor. 1991. Cholinesterase-inhibiting insecticides: their impact on wildlife and the environment. Elsevier, Amsterdam.

Minier, C., N. Eufemia, and D. Epel. 1999. The multi-xenobiotic resistance phenotype as a tool to biomonitor the environment. Biomarkers 4:442–454.

Myers, M. S., and six coauthors. 1998. Reductions in CYP1A expression and hydrophobic DNA adducts in liver neoplasms of English sole (*Pleuronectes vetulus*): further support for the "resistant hepatocyte" model of hepatocarcinogenesis. Marine Environmental Research 46:197–202.

Nishimoto, M., and five coauthors. 1995. Effects of a complex mixture of chemical contaminants on hepatic glutathione, L-cysteine and gamma-glutamylcysteine synthetase in English sole (*Pleuronectes vetulus*). Environmental Toxicology and Chemistry 14:461–469.

Novi, S., and five coauthors. 1998. Biotransformation enzymes and their induction by beta-napthoflavone in adult sea bass (*Dicentrarchus labrax*). Aquatic Toxicology 41:63–81.

Olsson, P. E. 1996. Metallothioneins in fish: induction and use in environmental monitoring. Pages 187–203 *in* E. W. Taylor, editor. Toxicology of aquatic pollution: physiological, molecular and cellular approaches. Cambridge University Press, UK.

Olsson, P. E., P. Kling, C. Petterson, and C. Silversand. 1995. Interaction of cadmium and oestradiol 17-beta on metallothionein and vitellogenin synthesis in rainbow trout (*Oncorhynchus mykiss*). Biochemistry Journal 307:197–203.

Payne, J. F., A. Mathieu, W. Melvin, and L. L. Fancey. 1996. Acetylcholinesterase, an old biomarker with a future? Field trials in association with two urban rivers and a paper mill in Newfoundland. Marine Pollution Bulletin 32:225–231.

Peakall, D. 1992. Animal biomarkers as pollution indicators. Ecotoxicology Series. Chapman and Hall, London.

Peters, L. D., J. P. Shaw, M. Nott, C. M. O'Hara, and D. R. Livingstone. 1999. Development of cytochrome P450 as a biomarker of organic pollution in *Mytilus* sp.: field studies in United Kingdom (*Sea Empress* oil spill) and the Mediterranean Sea. Biomarkers 4:425–441.

Petrivalsky, M., and five coauthors. 1997. Glutathione-dependent detoxifying enzymes in rainbow trout liver: search for specific biochemical markers of chemical stress. Environmental Toxicology and Chemistry 16:1417–1421.

Ramsdell, H. S., and D. L. Eaton. 1990. Mouse liver glutathione S-transferase isoenzyme activity toward aflatoxin B1-8,9-epoxide and benzo[a]pyrene7,8-dihydrodiol-9,10-epoxide. Toxicology and Applied Pharmacology 105:216–225.

Reichert, W. L., and seven coauthors. 1998. Molecular epizootiology of genotoxic events in marine fish: linking contaminant exposure, DNA damage, and tissue-level alterations. Mutation Research 411:215–225.

Rodriguez-Fuentes, G., and G. Gold-Bouchot. 2000. Environmental monitoring using acetylcholinesterase inhibition in vitro. A case study in two Mexican lagoons. Marine Environmental Research 50:357–360.

Roesijadi, G. 1992. Metallothionein in metal regulation and toxicology. Aquatic Toxicology 22:81–114.

Sarasquete, C., and H. Segner. 2000. Cytochrome P4501A (CYP1A) in teleostean fishes. A review of immunohistochemical studies. Science of the Total Environment 247:313–332.

Sato, M., and I. Bremner. 1993. Oxygen free radicals and metallothionein. Free Radical Biology and Medicine 14:325–337.

Schlenk, D., and M. Brouwer. 1991. Isolation of three copper metallothionein isoforms from the blue crab (*Callinectes sapidus*). Aquatic Toxicology 20:25–34.

Schlenk, D., M. Chelius, L. Wolford, S. Khan, and K. M. Chan. 1997a. Characterization of hepatic metallothionein expression in channel catfish (*Ictalurus punctatus*) by severse transcriptase polymerase chain reaction. Biomarkers 2:161–167.

Schlenk, D., W. C. Colley, A. El-Alfy, R. Kirby, and B. R. Griffin. 2000. Effects of the oxidant potassium permanganate on the expression of gill metallothionein mRNA and its relationship to sublethal whole animal endpoints in channel catfish. Toxicological Sciences 54:177–182.

Schlenk, D., E. J. Perkins, G. Hamilton, Y. S. Zhang, and W. Layher. 1996. Correlation of hepatic biomarkers with whole animal and population/community metrics. Canadian Journal of Fisheries and Aquatic Sciences 53:2299–2309.

Schlenk, D., and C. D. Rice. 1998. Effect of zinc and cadmium treatment on hydrogen peroxide-induced mortality and expression of gluathione and metallothionein in a teleost hepatoma cell line. Aquatic Toxicology 43:121–129.

Schlenk, D., L. Wolford, M. Chelius, J. Steevens, and K. M. Chan. 1997b. Effect of arsenite, arsenate, and the herbicide monosodium methyl arsenate (MSMA) on hepatic metallothionein expression and lipid peroxidation in channel catfish. Comparative Biochemistry and Physiology 118C:177–183.

Schlenk, D., Y. S. Zhang, and J. Nix. 1995. Expression of hepatic metallothionein messenger RNA in feral and caged fish species correlates with residual mercury levels. Ecotoxicology and Environmental Safety 31:282–286.

Senft, A. P., T. P. Dalton, and H. G. Shertzer. 2000. Determining glutathione and glutathione disulfide using the fluoresence probe o-phthalaldehyde. Analytical Biochemistry 280:80–86.

Spear, P. A., A. Y. Bilodeau, and A. Branchard. 1992. Retinoids: from metabolism to environmental monitoring. Chemosphere 25:1733–1738.

Stegeman, J. J. 1993. The cytochromes P450 in fish. Molecular biology frontiers. Elsevier Science Publishers, Amsterdam.

Stegeman, J. J., and M. E. Hahn. 1994. Biochemistry and molecular biology of monooxygenases: current perspectives on forms, functions, and regulation of cytochrome P450 in aquatic species. Pages 87–206 *in* D. C. Malins and G. K. Ostrander, editors. Aquatic toxicology: molecular, biochemical, and cellular perspectives. Lewis Publishers, Boca Raton, Florida.

Stephensen, E., and five coauthors. 2000. Biochemical indicators of pollution exposure in shorthorn sculpin (*Myoxocephalus scorpius*), caught in four harbors on the southwest coast of Iceland. Aquatic Toxicology 48:431–442.

Synder, M. J. 2000. Cytochrome P450 enzymes in aquatic invertebrates: recent advances and future directions. Aquatic Toxicology 48:529–547.

Thakker, D. R., and five coauthors. 1985. Polycyclin aromatic hydrocarbons: metabolic activation to ultimate carcinogens. Pages 178–242 *in* M. W. Anders, editor. Bioactivation of foreign compounds. Academic Press, Orlando, Florida.

Tom, M., E. Jakubov, B. Rinkevich, and B. Herut. 1999. Monitoring of hepatic metallothionein mRNA in the fish *Lithognathus mormyrus*: evaluation of transition metal pollution in a Mediterranean coast. Marine Pollution Bulletin 38:503–508.

Tong, Z., and M. O. James. 2000. Purification and characterization of hepatic and intestinal phenol sulfotransferases with high affinity for benzo[a]pyrene phenols from channel catfish, *Ictalurus punctatus*. Archives of Biochemistry and Biophysics 376:409–419.

Tuvikene, A., and five coauthors. 1999. Oil shale processing as a source of aquatic pollution: monitoring of the biologic effects in caged and feral freshwater fish. Environmental Health Perspectives 107:745–752.

van der Oost, R., A. Goksoyr, M. Celaner, H. Heida, and N. Vermeulen, P. E. 1996. Biomonitoring of aquatic pollution with feral eel (*Anguilla anguilla*) II. Biomarkers: pollution-induced biochemical responses. Aquatic Toxicology 36:189–222.

Van Veld, P. A., U. K. Ko, W. K. Vogelbein, and D. J. Westbrook. 1991. Glutathione S-transferase in intestine, liver and hepatic lesions of mummichog (*Fundulus heteroclitus*) from a creosote-contaminated environment. Fish Physiology and Biochemistry 9:361–369.

Van Veld, P. A., W. K. Vogelbein, R. Smolowitz, B. R. Woodin, and J. J. Stegeman. 1992. Cytochrome P4501A1 in hepatic lesions of a teleost fish (*Fundulus heteroclitus*) collected from a polycyclic aromatic hydrocarbon contaminated site. Carcinogenesis 13:505–507.

Viarengo, A., B. Burlando, F. Dondero, A. Marro, and R. Fabbri. 1999. Metallothionein as a tool in biomonitoring programs. Biomarkers 4:455–466.

Vigano, L., A. Arillo, F. Melodia, P. Arlati, and C. Monti. 1998. Biomarker responses in cyprinids of the middle stretch of the River Po, Italy. Environmental Toxicology and Chemistry 17:404–411.

Weinstein, J. E., and J. T. Oris. 1999. Humic acids reduce the bioaccumulation and photoinduced toxicity of fluoranthene in fish. Environmental Toxicology and Chemistry 18:2087–2094.

Williams, C. A. H. 1994. Toxicity resistance in mummichog (*Fundulus heteroclitus*) from a chemically contaminated environment. M.S. thesis, College of William and Mary, Gloucester, England.

Williams, D. E., and D. R. Buhler. 1984. Benzo(a)yyrene hydroxylase catalyzed by purified isozymes of cytochrome P-450 from beta-naphthoflavone-fed rainbow trout. Biochemical Pharmacology 33:4743–4754.

Wong, C. K. C., H. Y. Yeung, R. Y. H. Cheung, K. K. L. Yung, and M. H. Wong. 2000. Ecotoxicological assessment of persistent organic and heavy metal contamination in Hong Kong coastal sediment. Archives of Environmental Contamination and Toxicology 38:486–493.

Xu, H., S. Lesage, and K. R. Munkittrick. 1994. Suitability of carboxylated porphyrin profiles as a biochemical indicator in whitefish (*Coregonus clueaformis*) exposed to bleached kraft pulp mill effluent. Environmental Toxicology and Water Quality 9:223–230.

3

Molecular Biomarkers in Aquatic Organisms: DNA Damage and RNA Expression

ISAAC I. WIRGIN AND CHRISTOPHER W. THEODORAKIS

Introduction

Molecular biomarkers of environmental contamination can be defined as biological macromolecules—nucleic proteins, lipids, or carbohydrates—for which alteration of their structure or function can be used as an indication of xenobiotic exposure, effect, and susceptibility to toxicant-induced diseases. For the purposes of this chapter, these biomarkers will be limited to nucleic acid-based (RNA and DNA) endpoints. To be useful to resource managers, a biomarker should be indicative of exposure to environmentally relevant doses of xenobiotics, exhibit quantitative and predictive dose-response relationships, and reflect the overall health of the organism and, possibly, the vitality of the population or community. Evaluation of the biomarker response should be rapid, cost effective, and compatible with large sample sizes necessary for rigorous statistical analysis. Also, it would be advantageous if such biomarkers were nondestructive (i.e., nonlethal).

The problems encountered in the use of molecular biomarkers in sentinel species are unlike those in laboratory exposures of inbred animals to single chemicals. These include unknown or low levels of exposures, variation in inherent biological factors (age, sex, reproductive maturity), genetic variability (in terms of toxicokinetic rates, metabolism of, and susceptibility to xenobiotics), and exposure to complex mixtures of xenobiotic and natural chemicals. All of these factors serve to increase levels of variation in the biomarker response, thus requiring large sample sizes to achieve adequate statistical power. In turn, this necessitates technical approaches that are sufficiently streamlined and cost effective to process robust numbers of samples.

Recent advancements in mammalian molecular biology will undoubtedly be transferred to the field of ecotoxicology. This will result in a rapid rate of identification of sequence and expression patterns of many genes that are vulnerable to DNA damage and whose expression is often significantly altered (usually increased) by exposure to xenobiotics. Techniques to rapidly and sensitively assess these molecular alterations are currently available and are continuously being updated. Thus, the major and most intractable challenge impeding the use of molecular biomarkers in environmental toxicology is defining the cause and effect relationships between altered molecular responses, organismal health, and population performance. Thus, the discussion below will focus on methodology and application of molecular biomarkers, specifically DNA damage and mRNA-based gene expression assays, with special reference to the use of these biomarkers in biomonitoring programs and in determining mechanistic linkages between molecular responses and effects at higher levels of biological organization.

DNA Damage

The study of DNA damage and repair is important to ecotoxicology because genotoxins (chemicals that alter the structure and function of DNA) are ubiquitous and are often released into the environment in large quantities. Damage to DNA is of concern because it may lead to irreversible effects such as carcinogenesis or teratogenesis and can have deleterious effects on reproduction and fitness (Theodorakis et al. 1996) and so may be manifested at the population level. Unlike other xenobiotic effects, "subtle" DNA damage may be transmitted to subsequent generations and, thus be of evolutionary importance (see Chapter 5). For these reasons, DNA damage could also contribute to higher-level ecological effects at the community or ecosystem level (Shugart and Theodorakis 1996; Shugart 1998, 2000).

The amount of detectable DNA damage in contaminated organisms may be only slightly higher than that of reference organisms; this is because DNA damage is a steady-state process; it is constantly being formed and repaired (Freidberg 1985). Thus, observable DNA damage is that which is unrepaired at the time the tissue is collected. Also, a large amount of DNA damage is often lethal to cells. These factors limit the range of DNA damage that can be detected. Observable DNA damage is an indication that this steady state has been shifted toward the accumulation of DNA damage through an increase in the magnitude of DNA damage or a decrease in repair, which argues in favor of viewing DNA damage as a biomarker of contaminant effect and not simply of exposure.

DNA damage can have detrimental individual- or population-level effects on the health of aquatic organisms. For example, DNA damage has been associated with tumorigenesis (Baumann 1998; Stein et al. 1994; Wirgin et al. 1994), a shortened life span (Dey et al. 1993), cell lethality (Kampf and

Eichorn 1983), and programmed cell death or apoptosis (Payne et al. 1995). DNA lesions may detrimentally affect fecundity and embryonic development (Anderson and Wild 1994; Theodorakis et al. 1996) or immune function (Hurks et al. 1995) because cells that are rapidly dividing, such as reproductive and leukopoetic tissues, are particularly prone to genotoxicity. DNA damage has also been correlated with acute toxicity (Choi et al. 2000) and a reduction in growth rate (Steinert et al. 1998). Because DNA repair can deplete cellular energy stores (Pieper et al. 1999), chronic DNA damage could have impacts on bioenergetics status. These effects on fitness components (fecundity, growth, immune function, life span, and bioenergetics) may ultimately affect population growth and sustainability.

Genotoxic responses are induced by a wide variety of contaminants, including polyaromatic, nitroaromatic, and chlorinated hydrocarbons (Huang et al. 1995; Fu and Herreno-Saenz 1999; Fu et al. 1999; Pickering 1999), heavy metals (Snow 1992), and pesticides (Vijayaraghavan and Nagarajan 1994; Campana et al. 1999). Pollution often consists of complex chemical mixtures, so an assay that integrates responses to many chemicals is an asset in biomonitoring. Genotoxic responses are often correlated with a variety of other effects (tissue damage, enzyme inhibition, lipid peroxidation, etc.) due to multiple modes of action of these mixtures or their constituent genotoxins. Also, each individual genotoxin can elicit different DNA lesions, including adducts, base oxidation, DNA strand breaks, and cytogenetic effects. Different assays may be employed to assay each response, each with different limitations and advantages (Table 1).

DNA Adducts

Many genotoxins can form DNA adducts, exogenous molecules that form covalent bonds to the nucleotide bases. Exposure to some chemicals, for example diethylnitrosamine (DEN), leads to the transfer of an alkyl group (in this case, an ethyl) to the nucleotide base (Van Zeeland 1996). For example, waterborne exposure of Japanese medaka *Oryzias latipes* to DEN resulted in the accumulation of a variety of ethyl-DNA adducts in liver (Law et al. 1998). Many chemicals are not initially genotoxic but are metabolized to genotoxic intermediates that form covalent bonds with the nucleotide bases (e.g., aflatoxins and polycyclic aromatic hydrocarbons [PAH]; Shimada et al. 1996; Pickering 1999; Wang and Groopman 1999). There are several different analytical techniques available for the detection of DNA adducts. In aquatic organisms, the two most common being ^{32}P-postlabeling and high-pressure liquid chromatography (HPLC) with fluorescence detection.

The ^{32}P-postlabeling technique has been used in monitoring genotoxic exposure in a wide variety of aquatic taxa (e.g., Kurelec et al. 1989b; Harvey et al. 1997; Willett et al. 1997). The primary benefits to the ^{32}P-postlabeling assay include its extreme sensitivity (1 adduct in 10^9 to 10^{10} nucleotides),

Table 1. Comparison of various techniques for analysis of DNA damage.

Technology	Type of lesion	Throughput	Sensitivity[a]	Advantages	Limitations
[32]P post-labelling	Adducts Oxidative damage	Low	High	High sensitivity Capable of detecting a wide range of adducts	Costly Laborious, time consuming Requires radioactivity Identity unknown without GC-MS
HPLC	PAH adducts Oxidative damage	Medium	Medium-high	Lower cost and labor than [32]P (adducts) Specificity Unambiguous genotoxic effects	Specificity Identity unknown without GC-MS Costly instrumentation (HPLC)
Alkaline unwinding	Strand breaks	Medium	Low-medium	Very low cost; simple methodology	Indirect quantification of strand breaks
CF[b] gel electrophoresis	Strand breaks	High	Low-medium	Low cost of instrumentation Relatively little labor involved Assays many cells at once Direct quantification of strand breaks	Less sensitive than SCGE
PFGE[c]	Strand breaks	High	Medium	Efficient separation of DNA Increased sensitivity compared to CFGE	Costly equipment Long electrophoresis runs
SCGE[d]	Strand breaks	Low	High	High sensitivity; requires little tissue Highly dose-responsive	Laborious quantification Must examine many cells/sample

Table 1. continued.

Technology	Type of lesion	Throughput	Sensitivity[a]	Advantages	Limitations
PCR	Various lesions Point mutations	High	High	Rapid analysis; requires little tissue RAPD: many loci scored simultaneously Q-XLPCR: known loci Mutations: in many cases consequences of lesion are known (e.g., neoplasia)	RAPD: unknown lesions, unknown loci Q-XLPCR: unknown lesions, indirect quantification, examines limited number of loci Mutations: requires specialized techniques to quantify; only examines limited loci
Microscopy	Chromosomal abnormalities Micronuclei Apotosis	Low	Low-medium	Unambiguous detrimental effect Unambiguous indicator of genotoxicity Irreversible effect Multiple endpoints examined	Laborious quantification Must examine many cells/sample Chromosomes require special sample preparation
Flow cytometry	Clastogenesis Micronuclei Aneupolidy Cell cycle Apoptosis	Medium	Medium-high	Rapid analysis of large numbers of cells Multiple endpoints examined	Prohibitive cost of instrument Rigorous quality controls must be employed

[a]Relative ability to detect rare lesions or to use small amounts of DNA
[b]Constant field gel electrophoresis ("standard" agarose gel electrophoresis) or CFGE
[c]Pulse-field gel electrophoresis
[d]Single-cell gel electrophoresis or "comet assay"

realistic demands for quantity of DNA (low μg amounts), the ability to detect DNA adducts of unknown structure (Stein et al. 1994), linear dose–response functions for both PAH concentration and length of exposure (French et al. 1996), and resulting cumulative exposure histories (Reichert et al. 1998). The major disadvantages to ^{32}P-postlabeling are that the specific identity of the adducts may be unknown, it is a very laborious process, it can be somewhat expensive, and the use of radionuclides may inhibit its use in some laboratories. Furthermore, care must be taken to control for nonspecific artifacts and false positives (Harvey and Parry 1998).

In this technique, DNA is isolated and enzymatically digested into constituent nucleosides, the digest is enriched in DNA adducts by selective removal of normal nucleotides, nucleotides which are ^{32}P-phosphorylated, separated by 2-D thin-layer chromatography (TLC), and visualized by autoradiography on X-ray film or phosphor imaging. The labeled nucleotides appear as dark spots on the X-ray film or computer image (Figure 1) and are identified by comparing radiographs from reference and contaminated organisms or with the use of standard reference compounds. The adducted nucleotides can be recovered from the TLC plate and quantified by liquid scintillation counting or identified by gas chromatography-mass spectrophotometry (GC-MS) (Phillips et al. 1993).

A second technique (Shugart et al. 1987; McCarthy et al. 1989; Theodorakis et al. 1992) analyzes adducts by hydrolyzing them from the DNA using hy-

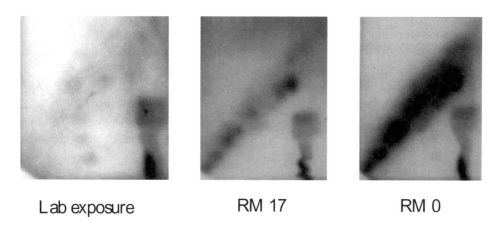

Lab exposure   RM 17   RM 0

Figure 1. Autoradiograms of ^{32}P-postlabeled hepatic DNA digests of juvenile Atlantic tomcod collected from two locales (River mile (RM) 17 and RM 0) on the Hudson River, New York, and from control juvenile tomcod that were laboratory spawned and reared. DNA adducts are represented by a diagonal radioactive zone with the origin at the bottom left hand corner. Hepatic DNAs were analyzed by the nuclease P1 version of the ^{32}-P postlabeling assay and thin-layer chromatography. DNA adduct levels are highest in tomcod from RM 0 and lowest in the laboratory spawned fish. Autoradiograms were kindly provided by William L. Reichert, Environmental Conservation Division, NMFS, Seattle, Washington.

drogen chloride and analyzing them via reverse-phase HPLC with fluorescence detection (PAHs and their metabolites are naturally fluorescent). The advantages of this technique are its simplicity (provided an appropriate HPLC is available) and no requirement of radioactivity. This method is specific to fluorescent DNA adducts (e.g., PAHs), which may be an advantage or disadvantage, depending upon the circumstances. Also, unless appropriate standards are available, the identity of the specific adducts may be unknown. Nonetheless, the majority of the fluorescent DNA-adducts are PAHs, and HPLC fractions can be collected and subsequently analyzed by GC-MS to identify their chemical composition.

DNA adducts that are commonly found in aquatic organisms include PAHs, nitroaromatic compounds (Shimada et al. 1996), and aflatoxins (Wang and Groopman 1999). They have been found in freshwater, estuarine, and marine fish (Chipman and Marsh 1991; Wirgin et al. 1994; Jones and Parry 1992), amphibians (Marty et al. 1998), and invertebrates (Harvey et al. 1999) exposed to contaminated waters and sediments. In most (Stein et al. 1994), although not all, cases (Kurelec et al. 1989a), levels of hepatic DNA adducts in fish have corresponded with known sediment or tissue contaminant concentrations. The liver is typically the major organ used for the determination of adduct levels because (1) large amounts of DNA can be easily extracted from this large, soft organ, (2) it typically accumulates lipophilic contaminants and heavy metals, and (3) it exhibits among the highest mixed-function oxidase activity of any organ, and many adducting chemicals must be bioactivated to become genotoxic. However, DNA adducts have also been found in blood, gill, brain, and kidney of fish (Theodorakis et al. 1992; Ericson et al. 1998, 1999). DNA adducts may accumulate over time, and this accumulation may be tissue dependent. For example, Theodorakis et al. (1992) detected adducts in fish blood and gill DNA after 1 week of exposure, but not in the liver until after 8 weeks. The amount of adducts in all tissues increased over time but more rapidly in the blood than in the gill or liver. This may reflect differences in toxicokinetics or metabolism of PAHs or relative DNA repair capacity in the different tissues.

Adduct formations in aquatic species are of special concern because they may be persistent lesions (French et al. 1996), giving an indication of cumulative exposure, and are often associated with more serious lesions. For example, PAH adducts were associated with hepatic carcinomas in populations of English sole *Pleuronectes vetulus* (Myers et al. 1998; Reichert et al. 1998), winter flounder *P. americanus* (Varanasi et al. 1989), and Atlantic tomcod *Microgadus tomcod* (Wirgin et al. 1994). In English sole, these adducts were not only associated with neoplastic lesions but also degenerative and preneoplastic lesions. Additionally, Ploch et al. (1998) found that brown bullhead catfish *Ameiurus nebulosus* were more susceptible to both PAH-induced adducts and tumor formation than were channel catfish *Ictalurus punctatus*, indicating species-specific susceptibility to adduct formation and providing further evidence that DNA adduct formation is involved in carcinogenesis in catfish. Thus, the advantage of using adducts as tools for monitoring is that they are unambiguous indicators of genotoxicity, the identity of the genotoxicants-causing effects can be clearly identified, and these DNA lesions have been clearly associated with health effects such as neoplasia or tissue damage (Stein et al. 1994; Kriek et al. 1998).

Base Oxidations

Base oxidations can be formed by the action of hydroxide radicals (OH·), singlet oxygen (O·), and superoxide anion (O2-), arising from xenobiotic exposure (see Chapter 2). These radicals may oxidize biomolecules (proteins, lipids, and DNA), altering their structure and function. This could also lead to production of reactive macromolecules (e.g., lipoperoxides) that could cause further DNA lesions (Kelly et al. 1998). One process by which DNA damage may be incurred is through nucleotide base oxidation, for example, by covalent attachment of hydroxyl radicals to these bases (Berger et al. 1991; Shigenaga and Ames 1991; Wagner et al. 1992). Reaction of oxyradicals with DNA bases may also lead to opening of the pyrimidine rings in guanine and adenine (Malins and Haimanot 1991).

There are several methods available for the detection and quantification of oxidized bases. One commonly used in aquatic toxicology is HPLC with electron capture detection (Shigenaga and Ames 1991). Alternatively, open ring structures may also be detected with GC-MS and selective ion detection (Malins and Haimanot 1991). The advantages of the HPLC technique are that specific types of DNA damage can be identified and quantified, and damage on different nucleotides can be differentiated. However, not all types of oxidative damage on all types of nucleotides may be determined.

Oxidative DNA damage has been observed in fish exposed to xenobiotic agents. For example, oxidized bases have been found in the liver and gills of fish exposed to single chemicals or complex mixtures (Malins and Haimanot 1991; Ploch et al. 1998; Rodriguez-Ariza et al. 1999). Oxidative damage to DNA is of concern because it may lead to mutagenesis and carcinogenesis (Sahu 1991). For example, Malins and Haimanot (1991) found that oxidative DNA damage was greatest in tumorous livers of Puget Sound English sole from contaminated sites, least in fish from references areas, and intermediate in tumor-free fish from contaminated sites. This approach also revealed positive associations between levels of oxidized guanine and adenine and severity of preneoplastic and nonneoplastic hepatic lesions in immature English sole from the same area (Malins et al. 1996; see Chapter 7 for further discussion).

DNA Strand Breaks

There are several mechanisms by which exposure to genotoxic agents can lead to DNA strand breaks (Eastman and Barry 1992). For example, strand breaks form during the process of nucleotide excision repair (NER) or base excision repair (BER) (de Laat et al. 1999; Memisoglu and Samson 2000) during which repair enzymes make a nick in the DNA adjacent to the damage, excise a short stretch of DNA containing the damaged base, fill in the gap, and ligate the new stretch of DNA to the old strand (Figure 2). Single-strand breaks (SSB) exist after the nicking and excision steps and before

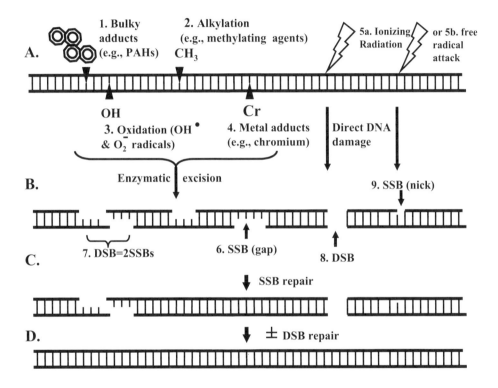

Figure 2. Schematic diagram of the types of DNA damage that can lead to formation of DNA single-strand breaks (SSB) and double-strand breaks (DSB). (A) DNA strand breakage is initiated by modification of the DNA nucleotide bases by (1) adduction to bulky molecules such as PAHs or aflatoxins, (2) alkylation (typically addition of methyl or ethyl groups, (3) oxidation via reaction with oxyradicals, or (4) adduction of certain metals. (B) Enzymatic excision of the damaged bases yields single-strand gaps (6), or, if two SSBs on alternate strands are close enough, DSBs (7). Alternatively, DNA strand breaks can be produced by interaction with radioactive particles (5a) or free radicals (5b), producing DSBs (8) or single-strand nicks (9). SSBs are rapidly repaired (C), while DSBs are repaired more slowly and may not be repaired at all (D).

ligation. DNA strand breaks may also occur during replication of DNA containing adducts and other base alterations that halt DNA polymerization catalyzed by DNA polymerase α (Wood and Shivji 1997). However, DNA polymerase ζ is able to bypass the damaged site, leaving a temporary gap in the newly synthesized DNA opposite the damage on the template strand (Wood and Shivji 1997). Strand breaks can also be directly formed as a result of an interaction of a radiation particle, oxyradical, or biomolecular radical with the sugar-phosphate backbone of the DNA molecule (Shulte-Frohlinde and Von Sonntag 1985; Figure 2). Single-strand breaks occur on only one of the DNA strands; thus, the DNA chromosome remains intact, but the integrity is compromised. Double-strand breaks (DSB) occur on two adjacent strands (e.g., by coincident formation of SSBs on adjacent DNA strands in close proximity or by interactions of DNA with high-energy radioactive particles; Shulte-Frolinde and Von Sonntag 1985). Although they may be formed by similar processes, SSBs and DSBs may differ in substantial ways. For example, repair of SSBs occurs via NER or BER, while repair of DSBs may involve direct rejoining of the broken ends or recombination between the damaged DNA molecule and its homologue (Chu 1997; Kanaar et al. 1998). Double-strand breaks are more difficult to repair than SSBs and have more dire consequences in terms of cell lethality and carcinogenic potential (Ward 1988).

There are several different methods with which to assess SSBs. In all protocols, the double-strand (DS) DNA must first be separated into its component strands, usually by treatment with sodium hydroxide. When DNA is exposed to alkaline conditions, the DNA begins to unwind into single strands at a rate proportional to the number of its SSBs (sites where unwinding is initiated; Shugart 1988). This property of DNA is the basis for the alkaline unwinding assay. In this technique, the DNA is allowed to partially unwind under alkaline conditions, after which the fraction of DNA remaining in the DS state (F) is then used as an index of the relative number of SSBs. There are two different methods of doing this: one based on differential fluorescence and one based on differential elution from hydroxylapatite. The differential fluorescence technique uses the fact that Hoescht dye, a fluorescent DNA dye, emits roughly twice the fluorescence when bound to DS DNA as when bound to single-strand (SS) DNA (Shugart 1988). For the differential elution technique ("batch elution technique"), the DNA is adsorbed onto hydroxylapatite and sequentially eluted with two different molarities of phosphate buffer, because SS DNA is eluted with a lower molarity buffer than is DS DNA. The amount of DNA in the two eluated fractions is then used to calculate the F value (Herbet and Hansen 1998).

A number of studies have used alkaline unwinding to investigate genotoxic exposure in both field and laboratory studies of fish and invertebrates (Shugart et al. 1989; Everaarts et al. 1991; Theodorakis et al. 1992; Fossi et al. 1996; Rao et al. 1996; Sugg et al. 1996). The amount of SSBs may vary, depending on the time of exposure or organs examined (Theodorakis et al. 1992, 1996; Sugg et al. 1996). Very high levels of genotoxins may elicit fewer strand breaks than lower concentrations, possibly due to induction of apoptosis (Theodorakis et al. 2000). Using alkaline unwinding, correlations were found between contaminants in the sediment and DNA strand breaks in freshwater and marine organisms or between SSBs in fish and the index of biotic integrity scores in contaminated streams (Adams et al. 1989a, 1989b; Everaarts 1995).

There are also other techniques that use agarose gel electrophoresis to quantify DNA strand breakage. Agarose gel electrophoresis is routinely used in most DNA- and RNA-based approaches to separate nucleic acids on the basis of their size and charge. For example, the single-cell gel electrophoresis (SCGE) or the "comet" assay (Singh et al. 1988; Singh 1996) is used to measure DNA strand breakage in individual cells. In this assay, individual cells or nuclei are isolated and then are suspended in low melting point agarose, coated onto a microscope slide, and subjected to electrophoresis. When stained with a fluorescent DNA-binding dye and observed with fluorescence microscopy, the electrophoresed nuclei appear in the shape of comets because smaller DNA fragments leave the nucleus (forming a "comet tail"), while large DNA fragments remain within. Strand breakage is quantified by calculating the "tail moment," the relative length of the tail compared with the size of the nucleus (Figure 3). Single-cell gel electrophoresis can be

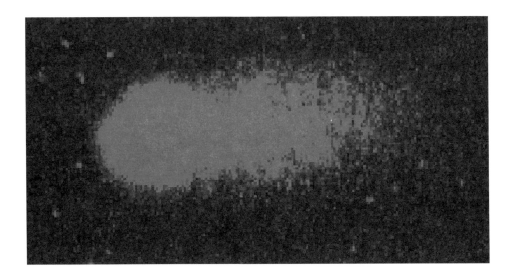

Figure 3. Photomicrograph of an ethidium bromide-stained "comet" from the single cell gel electrophoresis assay. The "head" of the comet contains high molecular weight DNA that did not migrate away from the original position of the nucleus, while the "tail" contains fragmented DNA. Photograph provided courtesy of Simazaki Dai, Engineering Research Institute, School of Engineering, University of Tokyo, Japan.

performed at alkaline or near-neutral pH for determination of SSBs or DSBs, respectively (Fairbairn et al. 1995). The advantages of the use of SCGE are its high sensitivity and dose-dependant responses (Mitchelmore et al. 1998). Also, the DNA does not have to be extracted from the nuclei, a process that can, in and of itself, result in induction of strand breakage (see below). Furthermore, this technique requires very little tissue, which may facilitate its use when collecting nondestructive samples. However, SCGE can be very time consuming, because many cells from each tissue are analyzed in order to get an estimation of strand breakage. It also requires the use of a fluorescent microscope, which may be much more expensive than standard electrophoresis gel image analysis equipment. A number of studies have used the comet assay for detecting genotoxicity in fish, aquatic invertebrates, and amphibians (Deventer 1996; Nacci et al. 1996; Mitchelmore and Chipman 1998; Mitchelmore et al. 1998; Ralph and Petras 1998). However, Belpaeme et al. (1998) found that care must be exercised to standardize assay conditions because such variables as preparation and storage of slides and tissues may affect the results.

Another electrophoresis technique used to detect DNA strand breaks uses standard agarose gel electrophoresis with DNA extracted from whole tissues (Theodorakis et al. 1994; Theodorakis et al. 1996, 2000). In this technique, the average molecular length (Ln) of the DNA is determined using alkaline gel electrophoresis (Freeman and Thompson 1990). Undamaged high molecular weight DNA (>50-100×10^3 base pairs) is compressed into a single band. When strand breakage occurs, there is "smearing" below this band, indicating the presence of smaller DNA fragments and a lower Ln (Figure 4). This technique can also be used with neutral (pH 8.0) electrophoresis buffers to obtain an indication of DSBs. However, assay conditions must be rigidly standardized and all samples must be treated equally in order to avoid bias (nonspecific degradation) due to factors other than contaminants, including freezing, mechanical shearing, or enzymatic degradation. For SSBs, this bias can be reduced by subtracting the DS Ln from the SS Ln in order to get the number of SSBs (Theodorakis et al. 2000). Alternatively, cells or nuclei may be embedded in low melting agarose plugs (Theodorakis et al. 1994) to avoid DNA shearing. When performing DSB analysis with this technique, high molecular weight DNA may not migrate out of the plug, so the amount of DNA migrating out of the plug with the amount remaining within is reflective of the number of DSBs (Theodorakis et al. 1994). This assay is typically performed with pulse-field gel electrophoresis, but comparable results can be obtained using standard constant field gel electrophoresis, which requires much less expensive equipment (Wlodek et al. 1991). The advantages of using electrophoresis techniques are that they are relatively simple, have high throughput, are inexpensive, and do not require potentially laborious microscopic evaluation. The disadvantages are that, if one is not careful, nonspecific DNA degradation may bias results, it requires more tissue

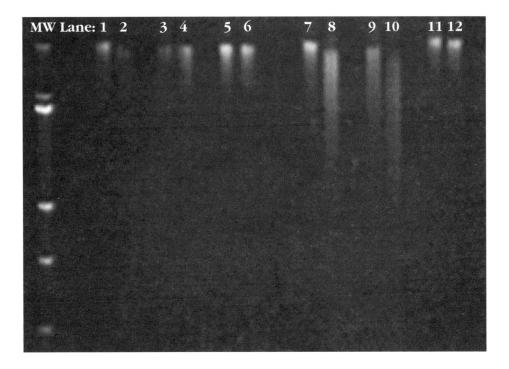

Figure 4. Photograph of an agarose gel containing DNA samples after electrophoresis. Each of the labeled lanes represents a separate DNA sample. MW = molecular weight marker. Note that, in some samples, the DNA is more smeared (e.g., lanes 8, 9, and 10) than it is in others (e.g., lanes 5, 6, 11, and 12), indicating more strand breaks.

and may be less sensitive than SCGE. These assays have been used to detect SBs in fish exposed to radionuclides or X-rays (Theodorakis et al. 1996; Theodorakis and Shugart 1998) and complex mixtures both in the laboratory (Theodorakis et al. 1992, 1994) and in the field (Theodorakis et al. 2000). In addition, Martin and Black (1998) and Black and Belin (1998) have successfully used this technique to detect genotoxic exposure in fish exposed to strip mine wastes and bivalves exposed to urban nonpoint runoff.

Polymerase-Chain-Reaction-Based Methods to Detect DNA Damage

The polymerase chain reaction (PCR) is at the center of the many new approaches that sensitively, rapidly, and, potentially, reproducibly assess alterations in DNA structure and expression (See Chapter 5 for a detailed description of PCR). Historically, one of the difficulties in applying molecular approaches to ecotoxicology was the inability to isolate a sufficient amount of target gene DNA or RNA to permit mutation or gene expression analyses. The polymerase chain reaction alleviates this problem by in vitro amplification of DNA or RNA. The use of PCR usually obviates the need for radio- or chemiluminescent-labeling to visualize, characterize, and quantify concentrations of target DNA or mRNA molecules. Polymerase chain reaction amplification usually exhibits sufficient fidelity and sensitivity to allow for the routine screening of PCR products for mutations and may allow for the detection of mutations that are present in only small numbers of cells within a population of primarily normal cells (Parsons and Heflich 1997; Roy et al. 1999).

However, the use of PCR is not without inherent difficulties and limitations. The PCR-based approaches (other than randomly amplified polymorphic DNA [RAPD] or arbitrarily primed PCR [AP-PCR]) require DNA sequence information to allow for the development of taxon-specific PCR primers. While in some instances, primers from closely related taxa may be used, it is not always the case. Species-specific PCR primers will almost always provide more reliable amplifications and greater yield of product than heterologous primers, and their development is not that technologically demanding, so it should not be avoided. Assuming the target gene has been characterized in related species, conserved sequences can be identified by computer alignments, and these can be used to design PCR primers to amplify a DNA fragment in the desired taxon. The DNA sequence data from this product can be used to design species-specific PCR primers.

Nonspecific DNA Damage

Several new PCR-based approaches are being tested to quantify nonspecific DNA damage in natural populations of plants and animals. Randomly amplified polymorphic DNA and AP-PCR are techniques that use single short oligonucleotide primers (10-12 mers) to amplify random short stretches of genomic DNA. If PCR conditions for RAPDs are optimized so that reproducible DNA fragments are obtained (Atienzar et al. 2000b), RAPDs can be used to compare the stability of specific DNA fragments (e.g., the intensities, number, or size of amplified DNA bands) before and after xenobiotic treatment. Genotoxic effects, as revealed with RAPDs, were found to be correlated to changes in fitness parameters such as growth and integrity of the photosynthetic apparatus in UV-B exposed marine microalgae *Palmaria palmata* or growth and reproduction in benozo[a]pyrene (B[a]P) exposed *Daphnia* (Atienzar et al. 1999, 2000a). The AP-PCR technique has been used as an index of mutagenesis in tumorous tissue (Navarro and Jorcano 1999). The sensitivity of RAPDs and AP-PCR in detecting DNA alterations in environmentally exposed aquatic populations has yet to be established. See Chapter 5 for further discussion and examples.

Locus-Specific DNA Damage

In most applications, PCR is used to amplify short stretches of DNA (up to 2,000 base pairs); however, protocols (long template PCR) have been developed to allow for PCR amplification of longer sequences (20-35 kilobase pairs [kb]) of DNA. Quantitative extra long PCR (Q-XLPCR) is such an approach for screening long sequences of DNA (10-18 kb) and detecting non-specific DNA damage with PCR products, provided stringent experimental precautions are observed and appropriate controls are introduced. The assumption behind Q-XLPCR is that many lesions (bulky adducts, modified bases, etc.) will "block" DNA polymerases and halt the extension step in PCR, reducing yields of PCR product. Thus, highly damaged DNA will provide a reduced amount of Q-XLPCR product compared to less damaged DNA. The longer the stretch of template DNA investigated, the greater the opportunity to detect DNA lesions. Unlike RAPDs and AP-PCR, Q-XLPCR requires locus and species-specific PCR primers, but, the tradeoff is the potential analysis of DNA damage at loci of known function (e.g., oncogenes, tumor suppressor genes) or that may be particularly sensitive to damage due to less efficient repair (e.g., mitrochondrial DNA [mtDNA]).

In studies with cell cultures, Q-XLPCR demonstrated that overall levels of DNA damage were 3-6-fold higher in mtDNA or nuclear DNA (nDNA) from genotoxin-exposed cells than in unexposed cells (Ballinger et al. 1996; Salazar and Van Houten 1997; Yakes and Van Houten 1997). Importantly, mtDNA damage was dose responsive to genotoxin levels and was associated with impaired mtDNA function and induction of apoptosis. The use of this approach in detecting mtDNA and nDNA damage is currently being assessed in PAH- and polychlorinated biphenyl (PCB)-treated and environmentally exposed Atlantic tomcod and mummichog *Fundulus heteroclitus* (Roy and Wirgin, unpublished data).

Somatic Mutations

There are currently several PCR-based methods for detecting point mutations (Cotton 1993). The most direct method of analyzing mutations is to amplify a section of a toxicologically significant locus, usually an oncogene or tumor suppressor gene, and determine DNA sequence polymorphisms by DNA sequencing or restriction enzyme digestion (RE) analysis (Tawata et al. 2000). Restriction enzyme analysis is more rapid and less costly than sequencing but is only able to reveal mutations at specific nucleotide sites, while sequencing examines the entire amplified DNA fragment. Therefore, sequencing is more informative, cost effective, and the preferred method to search for mutations in previously undescribed systems. However, once the mutational profile has been described within a PCR product, RE may provide an alternative to routinely screen for defined mutations.

Other methods, such as SS conformational polymorphism (SSCP) and denaturing gradient gel electrophoresis (DGGE), take advantage of the fact that DNA fragments of slightly different sequence either (1) assume different three-dimensional conformations when subjected to electrophoresis in DNA sequencing gels (SSCP; Nataraj et al. 1999) or (2) denature (i.e., the strands separate) at different concentrations of a denaturing agent (e.g., urea; DGGE; Fodde and Losekoot 1994) and, therefore, variant alleles exhibit different electrophoretic mobilities. These methods are rapid techniques that can quickly assess mutagenesis in a large number of samples (PCR products); however, the sequence of the mutant DNA is unknown and must be determined by DNA sequencing.

Additionally, even newer PCR-based approaches have been developed in which very short, labeled DNA probes are designed to hybridize to DNA sequences that span previously defined mutational hot spots, either additions or deletions or single base substitutions (Pals et al. 1999). After PCR and hybridization, the probe-target DNA complex is heated and melting curves (when the probe comes off the target DNA) are characterized. The probe will come off the mutated DNA at a lower temperature than the normal DNA because of single base mismatch. The advantages of this approach are that real time PCR-based technology is used, all reactions occur in a single tube or capillary, and mutants that do not contain diagnostic RE digestion sites can be detected quickly. The disadvantage is that the equipment to conduct this analysis is costly and development time to optimize the assay may be considerable.

The above techniques can only detect mutations that are relatively abundant in a DNA sample (e.g., DNA from tumor tissue), but, they are not sensitive enough to detect mutations when they are rare (e.g., in nontumorous tissue). More sensitive techniques can be used to selectively amplify mutated sequences or selectively destroy wild type sequences (Parsons and Heflich 1997). In one such assay, the amplified DNA is digested with REs that cut the DNA at specific sequences and then reamplify the DNA with PCR (Jenkins et al. 1999). If there is a mutation at the restriction site, the enzyme will not cut the DNA and will be amplified with PCR, otherwise, if there is no mutation at the restriction site, the DNA will be cut and will not be amplified with PCR (i.e., presence of amplification product indicates a mutation). Although this technique is highly sensitive, able to detect up to 1 mutated DNA sequence in 10,000 or more, it can only detect mutations at the restriction site. PCR-based 3' primer mismatch assays have also been used to detect mutated DNA with single base changes that are in low copy number (Roy et al. 1999). In this approach, alternative PCR primers are designed such that at their 3' ends are either perfect complements to the normal DNA sequence or are perfect complements of the mutated sequences. Primers with the normal sequence at their 3' end will only amplify DNA with the normal sequence, and mutant primers will only amplify DNA with the mutated sequence. This assay works because the PCR process is very sensitive to 3' mismatch in primer sequence. In all these techniques, it must be kept in mind that "mutations" can be formed in vitro by imperfect amplification during the PCR process, so that

DNA polymerases with "proofreading" ability must be used during the PCR process or strict controls must be employed to account for this.

Somatic (nonheritable) mutations are of concern because of their association with carcinogenesis. Hepatic and epithelial tumors have been observed in carcinogen-exposed laboratory fish and natural populations of fish (Baumann and Okihiro 2000). In mammalian systems, many such cancers are characterized by alterations in oncogenes and tumor suppressor genes (point mutations, additions/deletions, translocations, etc.), and probably represent the key events in the initiation of chemical carcinogenesis. In turn, DNA lesions in these genes lead to their aberrant expression (over-, under-, or nonexpression) in neoplastic tissue, and such effects are a seminal example of the interaction between DNA damage and gene expression. Approximately 200 oncogenes and 30 tumor suppressor genes have been identified in mammals, although only a small fraction have been characterized in fish (see Van Beneden and Ostrander 1994; and Baumann and Okihiro 2000). These genes often function in signal transduction, regulation of cell growth, differentiation, and DNA repair. Use of these alterations as biomarkers offers the advantage of relating molecular damage at single loci to aberrant cellular function and eventually to higher-level biological effects at the organismal level and perhaps at population levels (Anderson et al. 1994).

K-*ras* is the oncogene most characterized and studied to date in fish (Nemoto et al. 1986; Mangold et al. 1991; Rotchell et al. 1995; Peck-Miller et al. 1998; for review see Rotchell et al. 2001b), and oncogenic (tumor-initiating) mutations that commonly occur in mammalian tumors are often observed in fish tumors. Usually, although not always, these mutations in fish occur at the same sites (codons) in K-*ras* as in chemically induced mammalian tumors suggesting a common etiology. Laboratory exposures found K-*ras* mutations in carcinogen-induced liver, stomach, and swim bladder tumors in rainbow trout *Oncorhynchus mykiss* (Bailey et al. 1996) and in DEN-induced hepatic tumors in Japanese medaka *Oryzias latipes* (Torten et al. 1996; R. J. Van Beneden, personal communication), but an absence of K-*ras* and H-*ras* mutations in MNNG (Lee et al. 1999) and N-methyl-N-nitrosourea (MNU)-induced papillary thyroid tumors in *Rivulus marmoratus* (Lee et al. 2000). Because mutational hot spots and signatures in chemically treated fish, rodent models, and spontaneous human tumors are often identical, it suggests that molecular mechanisms of DNA damage and perhaps repair are frequently similar in the two taxa.

In natural populations, mutations in K-*ras* have been observed in tumors from some, but not all species investigated. K-*ras* activation was found to be a common event in liver lesions of winter flounder from Boston Harbor (McMahon et al. 1990) and in hepatocellular carcinomas in Atlantic tomcod from the Hudson River (Wirgin et al. 1989) but were not observed in necrotic, preneoplastic, or neoplastic hepatic lesions in English sole from Puget Sound, Washington (Peck-Miller et al. 1998) or in liver tumors of European flounder *Platichthys flesus* from the North Sea (Vincent-Hubert 2000). *Ras* activation in natural populations is assumed to be associated with exposure to high levels of environmental PAHs. *Ras* activation was also correlated with elevated levels of PCB and PAH contaminants and hepatic DNA adducts in hyperplastic liver cells in dragonets *Callionymus lyra* from the Seine River estuary (Vincent et al. 1998).

The most persistent and population-relevant molecular effects are those that are transmitted to future generations (germ line polymorphisms). For example, pink salmon *Oncorhynchus gorbuscha* embryos from some Exxon Valdez oil-exposed populations in Prince William Sound, Alaska, exhibited reduced survivorship compared with conspecifics from cleaner streams (Bue et al. 1996), and this persisted for several generations even after remediation. Laboratory studies found that descendants of oiled lineages, when reared under clean hatchery conditions, exhibited significantly reduced embryo survivorship compared with descendants of nonoiled streams suggesting that heritable damage had been incurred by environmental oiling (Bue et al. 1998). Further laboratory studies found that naïve pink salmon embryos exposed to weathered Prudehoe Bay crude oil under environmentally relevant conditions developed DNA damage in K-*ras*, but not at mitochondrial, microsatellite, or tumor suppressor gene loci (Roy et al. 1999). Mutations in K-*ras* occurred at the same codons that are most frequently mutated in chemically induced mammalian and rainbow trout tumors suggesting a common chemical etiology. Further studies demonstrated significant impacts on growth and survivorship of experimentally oiled embryos when released into the environment to complete their life cycles (Rice et al. 1999).

Tumor suppressor genes (e.g., p53 and Rb) have been characterized in several fish taxa and are highly conserved between fish and mammals (de Fromentel et al. 1992; Krause et al. 1997; Rotchell et al. 2001a). Mutations in p53 are the most frequent molecular alterations in most human and chemically induced rodent tumors. Unlike in mammals, mutations in p53 have not been detected in chemically induced fish tumors (Bailey et al. 1996; Krause et al. 1997). This may simply represent a failure to detect key sites involved in tumorigenesis or may reflect taxon-specific sequence differences between fish and mammals. However, the preliminary observation of decreased levels of basal expression of p53 protein in brown bullhead compared to channel catfish may help to explain the increased sensitivity of brown bullhead to PAH-induced carcinogenesis (Rau and Di Giulio, personal communication). In contrast, clams *Mya arenaria* from a population in PCB-polluted New Bedford Harbor, Massachusetts, exhibiting a high prevalence of leukemia, had a moderate level of p53 mutations, but, these variants may have been germ line polymorphisms rather than somatic mutations (Barker et al. 1997). The gene Rb has been characterized in Japanese medaka (Rotchell et al. 2001a), and Rb mutations have been found using PCR and SSCP analyses in

4/5 liver tumors of medaka cotreated with DEN and methylene chloride (Rotchell et al. 2001c). Further characterization of these alterations using DNA sequence analysis revealed 7 point mutations and a single deletion.

Cytogenetic Effects

Cytogenetic effects are alterations in chromosome structure (chromosome fragmentation, translocations, inversions, and acentric fragments) or chromosome number (e.g., anueploidy: one or more missing chromosomes). They are of concern because, unlike other DNA damage, they are irreversible effects that can affect fitness of exposed organisms via cell death or tumorigenesis (Simic et al. 1986). Traditional assays to measure such effects use condensed chromosomes in metaphase cells that are analyzed microscopically in order to determine the karyotype, (i.e., number and appearance of the chromosomes). This technique has been applied to a number of aquatic species (Jones and Parry 1992; Michailova et al. 1998; Das and John 1999; Hughes 1999; Jha et al. 2000). In order to obtain sufficient numbers of metaphase cells for analysis, cultured cells are incubated with colchicine (which blocks progression past metaphase) or live animals are dosed with the material. Karyological examination can be quite laborious and time consuming, but, there are several less laborious alternatives that have been used for determination of cytogenetic effects, including micronucleus and flow cytometric analyses.

Micronucleus Analysis

Micronuclei (MN) are extranuclear DNA fragments enclosed in a small envelope of nuclear membrane. They are formed when a piece of DNA is left behind after segregation of chromosome homologues during anaphase, either because they do not possess the centromere (e.g., chromosome fragments or acentric chromosomes) or because spindle formation or attachment is interrupted (Schmid 1976; Figure 5). Micronuclei can be detected by light or fluorescent microscopic examination of prepared slides stained with DNA-binding dyes (Figure 6). The advantages of this technique are that it provides an unambiguous index of genotoxicity, requires nothing more than a microscope, and is not as complex or laborious as karyotypic analysis. However, microscopic enumeration of many cells can still be time consuming. There are many examples that use MN as an indicator of genotoxic effects in aquatic organisms. The tissues used for this purpose include blood (Al-Sabti and Metcalfe 1995; Gauthier 1996), liver (Rao et al. 1997), and gills (Campana et al. 1999). Micronuclei can also give an indication of carcinogenic risk because Dopp et al. (1996) found that mussels from a contaminated habitat exhibited both an increased frequency of leukemia and an increase in number of micronuclei. The micronucleus test can also be used

A. Micronuclei

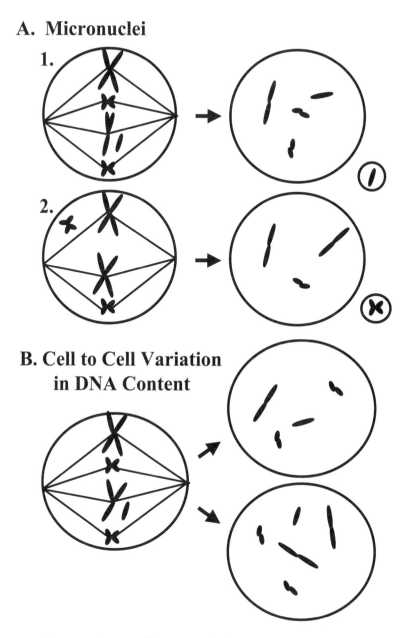

B. Cell to Cell Variation in DNA Content

Figure 5. Schematic diagram illustrating the formation of micronuclei (A) and increase cell-to-cell variation in DNA content (B) in cells exposed to genotoxins. Micronuclei may be formed when chromosomes are fragmented (A1) or when chemicals interfere with mitotic spindle formation (A2). In either case, the aberrant chromosome or chromosome fragment does not segregate properly, and part of the nuclear membrane can condense around it after mitosis. An increase in variation in DNA content is produced when chromosomes are fragmented (B) and the fragment cannot segregate properly, leading to unequal distribution of chromatin in the 2 daughter cells.

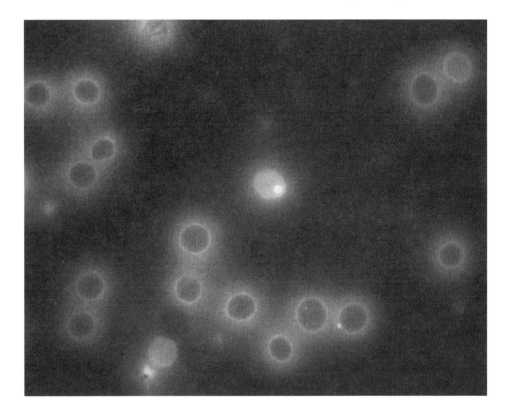

Figure 6. Red blood cell with micronucleus stained with the fluorescent, DNA binding stain acridine orange (arrow). Photograph provided courtesy of Brenda Rodgers, Texas Tech University.

to examine many types of genotoxic responses simultaneously (e.g., mitotic delay, apoptosis, and chromosome breakage, loss, and nondisjunction; Kirsch-Volders et al. 1997).

Flow Cytometry

Another method of analyzing tissues for cytogenetic damage involves flow cytometry. In this technique, cells or nuclei are isolated and stained with a fluorescent DNA-binding chemical (propidium iodide) and injected into a flow cytometer to determine the relative amount of DNA per nucleus. When analyzed in this way, different cell populations in different phases of the cell cycle (G1, S, and G2 + M) can be distinguished. Genotoxic effects are reflected as an increase in the cell-to-cell variation in DNA content (usually reported as half-peak coefficient of variation (CV) of the cells in G1 (Figure 7). When fragmented or acentric chromosomes are produced, cell division may result in an unequal distribution of DNA strands among daughter cells (Figure 5). Several workers have used this technique with aquatic wildlife exposed to complex mixtures of industrial pollutants, insecticides, or crude

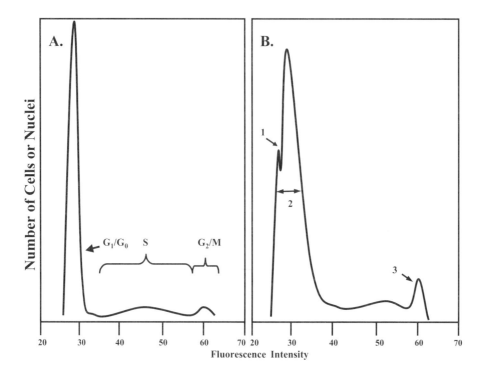

Figure 7. Schematic diagram of a hypothetical output from a flow cytometer, expressed in the form a graph of relative fluorescence intensity versus number of cells. Cells in the different phases of the cell cycle are indicated in (A). Note that fluorescence intensity is directly proportional to amount of DNA per nucleus. Thus, cells in the G_2/M phase have twice the DNA as cells in the G_1/G_0 phase and, hence, twice as much fluorescence. Panel (B) indicates several toxicological effects that can be determined by flow cytometry: (1) aneuploid cells may appear as a "shoulder" on the G_1/G_0 peak; (2) an increase in the half peak coefficient of variation in DNA content of G_1/G_0 cells (visualized here as an increase in the width of the G_1/G_0 peak) indicates chromosomal breakage; (3) alterations of cells in different phases of the cell cycle (e.g., an increase in the number of cells in G_1/M) can indicate contaminant-induced perturbations of the cell cycle.

oil (Custer et al. 1994; Lamb et al. 1995; Wickliffe and Bickham 1998) and fish exposed to complex mixtures of genotoxicants (Theodorakis et al. 1992; Bickham et al. 1998). Theodorakis et al. (2000) have also used this technique to show that the amount of chromosomal damage is correlated to changes in fish community composition in a contaminated stream.

In addition to chromosomal fragmentation, flow cytometry has been used for monitoring other genotoxic effects. For example, Dallas et al. (1998) found evidence of anueploidy in fish from a radionuclide-contaminated reservoir. Other workers have used flow cytometry for the detection of MN in fish cells exposed to various genotoxins in the laboratory (Kohlpoth et al.

1999). Exposure to genotoxicants has also induced an increase in the average size of the nuclei in liver of bluegill *Lepomis macrochirus* exposed to contaminated sediment (Theodorakis et al. 1992) and frogs exposed to radioactive contamination (Vinogradov and Chubinishvili 1999). However, the etiology and significance of changes in DNA content in exposed organisms is unclear.

Additionally, flow cytometric techniques can detect alterations in the cell cycle of aquatic organisms in contaminated habitats (Lamb et al. 1995). Both Dallas et al. (1998) and Whittier and McBee (1999) found alterations in the proportions of cells in G1 and G2 in radionuclide-exposed fish and pesticide-exposed ducks, respectively. However, such alterations may reflect nongenotoxic effects (e.g., tissue damage, metabolic inhibition, or perturbations in intracellular signaling). Alternatively, such effects may be due to "cell cycle checkpoints," arrests or delays in the cell cycle that occur if DNA damage is present (Schwartz and Rotter 1999). The most commonly studied cell cycle checkpoints are delays after G1 or G2 that prevent the cells from entering the DNA synthesis (S) phase or mitosis, respectively, before the DNA damage is repaired. Thus, alterations in the ratio of cells in G1/G2 might be a reflection of DNA damage.

Like all assays of genotoxic induction, flow cytometry has advantages and disadvantages. The advantages are that it is a very rapid method of assessing cytogenetic damage, providing multiple endpoints such as half peak CV, G1/G2 ratios, and quantification of micronuclei. One of the main disadvantages is the prohibitive cost of the instrument. Also, this technique is sensitive to slight variations in procedures, which may result in biases if strict quality control measures are not maintained (McCreedy et al. 1999).

Gene Expression Assays

Why Use Gene Expression Assays?

Exposure to xenobiotics can increase (induce) or decrease levels of expression of toxicant-sensitive genes. Indicators of gene expression diagnose exposure to causal stressors and quantify their effects on gene regulation. Recent evidence from the human genome project indicates that there are about 30,000 genes in the vertebrate genome, and it has been estimated that expression of about 2% of mammalian genes, or about 600 genes, is regulated by exposure to toxicants (Farr and Dunn 1999). Changes in the expression of genes, more often than not, result in cellular and tissue change and altered organismic performance. Furthermore, higher level toxic responses to all xenobiotics are believed to be mediated by altered expression of not one but many genes in signaling pathways. Thus, altered gene expression is an early event that precedes overt toxicities and should be associated with and predictive of higher level biological effects.

In laboratory waterborne exposures or intraperitoneal (i.p) treatments of fish to environmentally relevant concentrations of toxic chemicals, gene expression can be induced 500-fold or more compared with controls (Courtenay et al. 1999). Similarly, fish from highly contaminated locales can exhibit up to 25-35-fold higher levels of gene expression than those from reference sites (Wirgin et al. 1994; Yuan et al. 2001). Thus, induced gene expression can provide the large window of expression needed to achieve significance in statistical comparisons between or among exposure groups. However, there is often high interindividual variability in levels of gene expression due to different exposure histories or interindividual genetic variability in gene inducibility (Courtenay et al. 1994). Interpopulation variation in gene inducibility may occur and may reflect genetic adaptation, physiological acclimation to xenobiotics, or random genetic drift (Prince and Cooper 1995b; Nacci et al. 1999; Courtenay et al. 1999). Interspecific differences in gene inducibility also occur (i.e., some species are more sensitive than others; Wirgin et al. 1996).

Unlike measures of DNA damage, gene expression assays most often provide a transient picture of exposure history. Physiochemical properties of the xenobiotic and the tissue used for analysis will dictate the kinetics of mRNA induction and clearance. For example, a single i.p treatment of fishes with rapidly metabolized PAHs significantly induces hepatic cytochrome P4501A1 (CYP1A1) within hours and induced expression persists for hours or, at most, for days, but a single treatment with persistent xenobiotics such as 2,3,7,8 tetrachlorodibenzo-p-dioxin (TCDD) or coplanar PCBs can elevate levels of gene expression for weeks or months after exposure (Courtenay et al. 1999; Figure 8). Route of exposure can also influence the persistence of induced gene expression. For example, continuous flowthrough exposure of rainbow trout to β-NF (a PAH-like compound) resulted in induced CYP1A1 mRNA levels for extended periods (up to at least 21 days; Haasch et al. 1993b). Also, gene induction by lipophilic xenobiotics may be more prolonged in liver than in other tissues. In fact, the kinetics of induced gene expression in natural populations may provide information on the identity of the inducing agent (Kreamer et al. 1991; Courtenay et al. 1993).

Induced gene expression is often considered a marker of exposure rather than effect, but it almost certainly precedes and is predictive of damage at the histological, organismal, and population levels. For example, CYP1A1 catalyzes the conversion of procarcinogenic PAHs to metabolites, which adduct to DNA and mutate specific nucleotides in critical target genes, thus, presumably, initiating chemical carcinogenesis. Some studies have demonstrated correlations between levels of hepatic CYP1A1 expression, hepatic DNA adducts, activated oncogenes, and hepatic cancers in environmentally exposed fish (reviewed in Wirgin and Waldman 1998). Furthermore, an elevated prevalence of hepatic tumors in one population was associated with a truncated age structure (Dey et al. 1993). Additionally, persistent induction

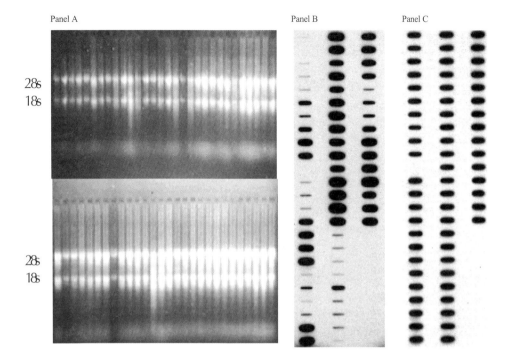

Figure 8. Panel A depicts an ethidium bromide stained two-tiered 1.0% agarose gel loaded with total hepatic RNAs from Atlantic tomcod from the Hudson River and Miramichi River that were i.p. injected with graded doses of PCB77 or PCB169. Presence of intact 28S rRNA and 18S rRNA bands is indicative of the quality of RNA isolations from each and their suitability for subsequent RNA hybridization analysis.

Panel B is an autoradiograph of the same Atlantic tomcod samples shown in Panel A that were subsequently slot blotted to a nylon membrane and hybridized to a ^{32}P radiolabeled, tomcod CYP1A1 cDNA probe.

Panel C is an autoradiograph of the same slot blot depicted in Panel B that was stripped of the tomcod CYP1A1 cDNA probe and rehybridized to a ^{32}P radiolabeled rat 18S rRNA housekeeping probe.

of CYP1A1 by PCBs has been associated with the formation of reactive oxygen species in scup *Stenotomus chrysops* (Schlezinger et al. 1999) and other vertebrate taxa (Schlezinger et al. 2000), inactivation of CYP1A catalytic activity, and perhaps overt cellular toxicity. Other studies have implicated induced CYP1A in apoptosis and cardiovascular dysfunction in early life stages of fishes exposed to PCBs and TCDD (Cantrell et al. 1998). Similarly, very recent studies have demonstrated correlations between induced vitellogenin expression in male fish, histological pathologies, depressed serum testosterone levels, reduced sperm viability, and compromised embryo survivorship. Thus, induction of gene expression may be predictive of perturbations at the organismal or even population levels in some fish.

Finally, interspecific, interpopulation, and interindividual variation in inducibility or activities of genes whose products activate and detoxify xenobiotics or repair DNA damage probably play a major role in differing susceptibilities to toxicant-induced disease. Thus, a major focus of human environmental research has been the identification of gene-environment interactions at toxicant-sensitive loci. Genetic polymorphisms have been identified at toxicant-sensitive loci in fish (Roy et al. 1995; Roy and Wirgin 1997), however, their functional significance in conferring susceptibility to disease or other higher level effects is unknown and currently under investigation.

Molecular Mechanisms of Gene Induction

Induction of gene expression by xenobiotics may occur via transcriptional activation or regulation at the posttranscriptional level. Among the toxicologically relevant genes, transcriptional activation of CYP1A1 is best understood. Activation of CYP1A1 transcription is mediated by the aryl hydrocarbon receptor (AhR) pathway. In unexposed animals, the AhR is located in the cytoplasm bound to two molecules of heat shock protein 90 (hsp90) and a single immunophilin molecule (Carver and Bradfield 1997). Following exposure, ligand (toxicant) diffuses through the cell membrane; it binds with AhR and displaces hsp90 and the immunophilin homologue. The AhR-ligand complex is then translocated by an unknown mechanism to the nucleus where it binds with ARNT (aryl hydrocarbon receptor nuclear translocator). This complex then binds to enhancer elements (termed "dioxin or xenobiotic responsive elements") in the promoter region of many inducible genes (e.g., CYP1A1) disrupting chromatin structure and increasing access of other transcription factors to regulatory elements just upstream of the gene's coding region and, thereby, initiating their transcription (Hahn 1998). (Note: promoter regions are DNA sequences where RNA polymerase binds to DNA at the "beginning" [5' upstream sequence] of a gene sequence, thus initiating transcription; enhancer elements are DNA sequences that facilitate this process. See Alberts et al. [1994] or Lewin [2000] for a more detailed description). Most importantly, it is believed that most AH-induced toxic responses (e.g., carcinogenicity, immunosuppression, early life mortality, and teratogenicity) result from activation of the AhR pathway (Hahn 1998). Thus, induction of CYP1A1 serves as a surrogate measure of most toxic responses resulting from exposure to AHs. Although endogenous ligands for AhR are largely unknown, AhR-deficient rodent models exhibit prolonged doubling times of liver cells (Ma and Whitlock 1996), bile duct fibrosis, liver abnormalities, and compromised immune system development (Fernandez-Salguero et al. 1995; Schmidt et al. 1996), suggesting that alterations in AhR function may present deterimental organismic consequences in the absence of toxicants. Similar receptor-ligand responsive element mediated mechanisms of transcriptional activation occur for many toxicologically important genes induced by endogenous (e.g., steroids) and exogenous (e.g., heavy metals) ligands (Olsson and Kille 1997; Kumar and Thompson 1999).

Other xenobiotically induced genes (e.g., the stress protein hsp70; Sanders 1993) are posttranscriptionally activated, either through stabilization of the mRNA (which results in accumulation of RNA of that gene) or by regulation of the rate of translation (in which the level of mRNA remains relatively constant). Another type of posttranscriptional induction may occur via stabilization of the protein (without concomitant increase in mRNA), as occurs with p53 via protein phosphorylation in response to DNA damage (Coleman et al. 2000).

Assays to Quantify Expression of mRNA

There are several different assays for examining gene expression in terms of mRNA levels in different tissues, each with it own advantages, disadvantages, and applications (Bader et al. 1992; Belin 1998). The three most common methods-blotting (northern, slot, and dot), reverse transcriptase polymerase chain reaction (RT-PCR), and ribonuclease protection assays-will be discussed here and their attributes are summarized in Table 2.

Northern, Slot, and Dot Blots

In northern blot hybridizations (Kroczek 1993), RNAs are separated using denaturing agarose gel electrophoresis, and 28S and 18S rRNA bands are visualized by staining of gels in ethidium bromide solution (Figure 8a). Integrity of the two rRNAs indicates that the RNAs are of suitable quality to continue with the experiment. RNAs are then transferred by capillary action onto membranes composed of nylon or nitrocellulose, RNAs are immobilized by heat or UV, and membranes are exposed to solutions containing a labeled DNA probe (a piece of DNA whose sequence is complementary to all or a portion of the mRNA under investigation so that the probe binds to this mRNA on the membrane). The sequence of the DNA probe is chosen according to the mRNA sequence of the gene of interest. A simplified example is given below:

mRNA: ....AAUCGGUUAUGCAUUAUCGGUACGAAGUACUAAGUGUAC....
Probe: TACGTAATAGCCATGCCTT

The DNA probe is labeled with a radioactive isotope or covalently bound to an enzyme that catalyzes a light-emitting reaction. The membrane with the bound, labeled probe is then exposed to X-ray film, and the radiation from the isotope or the light from the enzymatically catalyzed reaction darkens the film where the probe is bound to the mRNA of interest (this process is called "autoradiography"). The intensity of darkness of the band, slot, or blot on the X-ray film reflects the amount of labeled probe hybridized at that

Table 2. Comparative evaluation of the techniques for quantification of gene expression via mRNA.

Technique	Technical	Throughput[a]	Sensitivity[b]	Quantitation[c]	Applications (AP), advantages (AD), and limitations (L)
Northern blot	Low	S: low G: low	Low	Absolute	AP: known gene expression assays, separation of mRNA based on size AD: not technically difficult, absolute quantification of mRNA, absolute quantitation of mRNAL: low sensitivity, low throughput, uses radioactivity[d]
Dot or slot blot	Very low	S: medium G: low	Low	Absolute	AP: known gene expression assays AD: not technically difficult, higher throughput than (1)d. L: low sensitivity, uses radioactivity
Reverse transcriptase PCR (RT-PCR)	Medium	S: medium G: medium	Very high	Relative	AP: known gene expression assays AD: high sensitivity, higher throughput than (1) or (2)[f], uses very little mRNA, uses no radioactivity L: semiquantitative
Ribonuclease protection	High	S: medium G: low	Medium high	Absolute	AP: known gene expression assays AD: more sensitive than (1) or (2), absolute quantitation of assay mRNA L: construction of probes may be difficult, uses radioactivity[e]
Differential display	High	S: low G: high	Very high	Yes/No	AP: gene discovery[g]AD: No a priori DNA sequence information needed, can survey many genes at once L: false positives, uses radioactivity[e]
AFLP-DD	High	S: low G: high	Very high	Yes/No	AP: gene discovery[g], expression profiling[h] AD: No a priori DNA sequence information needed, can survey many genes at once, fewer false positives than (5) L: Uses radioactivity
Microarray	Very high	S: low G: very high	Medium	Relative	AP: Gene discovery[g], expression profiling[h] AD: Can survey very many genes at once, fewer false positives than (5) or (6). L: Specialized and costly equipment needed, requires a priori DNA sequence information.

[a]S = number of samples that can be assayed at one time; G = number of genes that can be assayed at one time.
[b]Ability to detect small amounts of RNA or rare transcripts.
[c]Absolute = gives a quantitative index of absolute differences in amount of RNA between 2 samples; Relative = relative difference in amount of RNA between 2 samples (includes semiquantitative); Yes/No = gives an indication if 2 samples are different, but not by how much.
[d]Dot or slot blot manifolds can assay up to 96 samples at once.
[e]Nonradioactive alternatives (e.g., chemiluminescence, fluorescence, silver staining) may be available and are cheaper and quicker than autoradiography. Although the sensitivity may approach or equal autoradiography in some cases, the detection limit may be less.
[f]Technology is available for PCR amplification of 96-360 samples at once, allowing multiple samples or genes to be simultaneously examined.
[g]Gene discovery, at least in this case, refers to expression assays for anonymous genes whose function is not known a priori; but may be identified a posteriori.
[h]Simultaneous examination of expression of many genes with known function and identity.

spot, which is proportional to the amount of membrane bound mRNA (i.e., gene expression; Figure 8b). Following autoradiography, the DNA probe for the gene of interest is washed off the membrane and the membrane is rehybridized as described above to a constitutive expressed, housekeeping DNA probe such as actin or rRNA (Figure 8c). This ensures that equal amounts of total RNA of each sample have been applied to the gel or membrane and permits normalization of the expression level of the gene of interest.

In dot and slot blotting, the RNA solution is applied directly to the membrane using a multiwelled (24-96; slots or dots) manifold. The RNA is immobilized on the membrane by heat or UV (ultraviolet) light and the mRNA of interest is visualized and its expression quantified as described above (Figure 8b,c).

Blotting techniques directly quantify native tissue concentrations of the mRNA species (gene) of interest. Thus, results of these analyses very accurately reflect levels of gene expression. The major disadvantage of blotting approaches is that they might not be sensitive enough to detect mRNA levels of rare transcripts (basal levels of expression or from genes that are not highly induced) or from very young life stages. Northern analysis offers several advantages not inherent with other blotting approaches. Northern gels allow an evaluation of the integrity of RNA preparations such that degraded RNAs can be deleted from subsequent analysis. Second, the molecular size of the target mRNA is determined, thus, confirming the identity of the hybridized mRNA band. Finally, because electrophoresis separates mRNAs of different sizes, northern blotting can reveal multiple mRNAs of similar sequence but different length that might hybridize with the same probe.

Dot and slot blotting are usually more quantitative than northern analysis because the target mRNA of interest is confined to a very small area. Currently, many labs employ a two-step process: (1) northern gels to evaluate the integrity of all RNA isolations and (2) slot or slot blotting to quantify mRNA expression.

Ribonuclease Protection Assay

In the ribonuclease protection assay (RPA; Sharma et al. 1998), cDNA is synthesized from the mRNA of interest (e.g., metallothioncin) and inserted into a plasmid (a small, circular, extra-chromosomal DNA from bacteria). Using an RNA polymerase, "antisense" RNA (i.e., RNA that has a sequence complementary to the original mRNA) is then made from this plasmid and labeled:

"sense" mRNA: AAUCGGUUAUGCA
"antisense" RNA: UUAGCCAAUACGU

Radioactive- or chemiluminescent-labeled antisense RNA probe is added to samples of mRNA from experimental organisms and allowed to hybridize to their "sense" mRNA in a liquid solution. A ribonuclease is then added that digests excess single-stranded RNA probe and unhybridized sample RNA, but the hybridized mRNA is protected because it is double-stranded. The hybridized RNA is then electrophoresed in denaturing polyacrylamide gels, and autoradiography is used to visualize and quantify mRNA species of interest. The advantage of RPA is that, like blotting approaches, it directly quantifies native tissue concentrations of the mRNA of interest but is 10-100 times more sensitive. The disadvantage is that initial construction of the insert-containing plasmid and the antisense RNA can be time consuming and technically demanding. RPA is almost never used in biomonitoring studies.

Reverse Transcriptase Polymerase Chain Reaction

Reverse transcriptase polymerase chain reaction is a modification of the PCR process in which the starting material is mRNA rather than DNA (Vu et al. 2000). Since mRNA cannot serve as a template for PCR, it first must be reverse transcribed into cDNA that, in turn, is quantitatively amplified in PCR, as outlined below.

(1) Start with mRNA: 5'-AAUCGGUUAUGCA-3'
(2) Using mRNA as a template, reverse transcriptase enzyme makes cDNA
    mRNA: 5'-AAUCGGUUAUGCA-3'
    cDNA: 3'- TTAGCCAATACGT-5'
(3) Taq DNA polymerase amplifies the cDNA to double-stranded DNA:
    5'-AATCGGTTATGCA-3'
      | | | | | | | | | | | | |
    3'-TTAGCCAATACGT-5'

In RT-PCR, the final yield of DNA product should be proportional to the amount of mRNA isolated from the tissue sample. Products of RT-PCR are usually visualized in agarose electrophoresis gels by ethidium bromide staining (Figure 9), but if more sensitivity is required, RT-PCR products can be radiolabeled during the PCR step. Levels of RT-PCR products are quantified by scanning or phosphor imaging, or in "real time" using more sophisticated thermal cyclers. In contrast to blotting approaches and RPA, RT-PCR provides only an indirect measure of mRNA content. Because of the multiplicity of steps in the RT-PCR process and the power of PCR in amplifying a template, in order to be quantitative and reproducible, it is crucial that internal mRNA standards are designed that will compensate for inter-experiment variation in both the reverse transcription and PCR processes (Stenman et al. 1999; Freeman et al. 1999). These standards should (1) have identical primer annealing sequences as the native mRNA target species and (2) be distinguishable from the native PCR product in gel electrophoresis (Figure 9). Often mRNA standards are synthesized that are truncated compared to the native mRNA (Figure 9) or that contain restriction enzyme digest sites or that

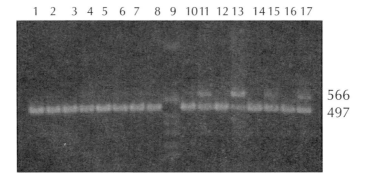

Figure 9. Competitive RT-PCR analysis of CYP1A1 mRNA expression of Atlantic tomcod larvae that were exposed to waterborne B[a]P (1 ppm) or corn oil vehicle. Lanes 1–8 contain RT-PCR reactions from four corn oil exposed larvae and lanes 10–17 contain RT-PCR reactions from four B[a]P exposed larvae. The top band is the native 566 base pair CYP1A1 PCR product and the lower band is the truncated 497 base pair CYP1A1 standard PCR product. Two RT-PCR reactions are depicted for each larva, each with a constant amount of competitive CYP1A1 mRNA standard (0.2 pg) and each with two different volumes of RNA sample. Lane 9 contains a 1 kb molecular size standard.

are absent in the native RT-PCR product. Reverse transcriptase polymerase chain reaction offers the advantages of its being a rapid and very sensitive assay that requires nanogram amounts of template RNA and that allows for the detection of relatively rare mRNA transcripts. Thus, RT-PCR can be applied to very small, young life stages in which the amount of mRNA is very limiting (Roy et al. 2001). However, before its application, the exact DNA sequence of the target gene in the species of interest must be known and considerable effort must be expended in the development of internal standard mRNAs.

Examples of the Use of Gene Expression Assays in Biomonitoring

Cytochrome P4501A1

Most gene expression studies used to evaluate xenobiotics exposures in fishes have focused on the cytochrome P4501A1 gene in the discussion below, as recommended by Stegeman and Hahn (1994), we use the designation CYP1A1 for those forms and species in which DNA or protein sequence information confirms this identity and CYP1A for those examples in which this information is lacking. Cytochrome P450s are a super family of phase I enzymes that oxidize both endogenous and exogenous substrates. CYP1A1 is conserved among all vertebrates and is highly inducible by exposure to some contaminants of greatest concern in aquatic ecosystems, including PCBs, dioxins, furans, and PAHs (Stegeman and Hahn 1994). In naïve fishes, CYP1A1 expression is proportional to dose at both the protein (Stegeman and Hahn 1994) and mRNA levels (Courtenay et al. 1999). In laboratory experiments, levels of CYP1A1 in fishes are usually significantly induced at environmentally relevant levels of xenobiotics, and, in field studies, levels of gene expression are frequently highly induced in fishes from contaminated sites (Collier et al. 1992; Stegeman and Hahn 1994; Wirgin and Waldman 1998). CYP1A1 is inducible in a variety of tissues, including liver, intestine, gill, heart, kidney, gonad, and sometimes spleen (Bello et al. 2001), and is detectable at the mRNA, protein, and catalytic (ethoxyresorufin O-deethylase [EROD], aryl hydrocarbon hydroxylase [AHH]) levels. Most studies have investigated hepatic expression because of the liver's role in biotransformation of xenobiotics, its size, and the ease with which its proteins and mRNA can be isolated. To date, most studies have evaluated CYP1A expression at the protein and enzyme levels; however, quantification of CYP1A1 mRNA may offer advantages over these other approaches. Because transcriptional activation is an early event in induced expression and it provides a large window of induction, CYP1A1 mRNA levels may provide the most dose-related response to aromatic hydrocarbon (AH) exposure. For example, substrate inhibition in AH-treated fish can decrease catalytic activity and levels of CYP1A1 protein, providing unrealistically low levels of gene expression (Gooch et al. 1989; Haasch et al. 1993a).

Many studies have investigated CYP1A expression in natural populations of bottom-dwelling fishes, so we restrict our discussion to those for which there is concurrent data on higher level effects (e.g., tumor prevalence, altered age structure, reproductive effects) or DNA damage. For example, in English sole from Puget Sound, Washington, elevated levels of hepatic CYP1A protein expression and activities were concomitant with high prevalences of hepatic neoplasms; preneoplastic, regenerative, and degenerative lesions; high levels of PAH metabolites in bile (an indication of PAH exposure); and hepatic DNA adducts (Myers et al. 1998). Levels of CYP1A expression corresponded well with sediment concentrations of PAHs, but not other AHs (Collier et al. 1992, 1995).

Winter flounder from Boston Harbor, Massachusetts, and other contaminated Atlantic coast estuaries exhibit an elevated prevalence of neoplastic and nonneoplastic hepatic lesions compared to flounder from cleaner locales (Murcelano and Wolke 1985). In most cases, flounder from sites with the highest prevalences of lesions or hepatic burdens of PCBs exhibited the greatest induction of hepatic CYP1A or EROD activity although fold induction (2-4-fold) between polluted and cleaner sites was not as great as might have been anticipated (Stegeman et al. 1987; Monosson and Stegeman 1994). Additionally, although CYP1A protein levels were significantly higher in flounder from PCB-polluted New Bedford Harbor, Massachusetts, than two cleaner estuaries, there was no difference in EROD activity among sites, perhaps due to inhibition of EROD activity by PCBs or other contaminants in fish from the polluted site (Elskus et al. 1989).

The mummichog is a very common, tolerant, estuarine species, found along most of the Atlantic coast of North America, which exhibits a very limited home range such that biomarker responses may closely reflect local environmental conditions. Although tumors are rare coastwide in mummichog from highly contaminated locales, there was a 90% prevalence of gross hepatic lesions in fish from the Elizabeth River, Virginia, of which 33% were hepatocellular carcinomas (Vogelbein et al. 1990). Levels of CYP1A protein and EROD activity were compared among hepatocellular carcinomas, foci of cellular alteration, and adjacent nonneoplastic hepatic tissue in mummichog from the Elizabeth River population. Levels of CYP1A protein and EROD expression were 15-85% lower in hepatic lesions than in adjacent normal tissue (Van Veld et al. 1992), a result that is consistent with that observed in rodents and another fish model. This suggests that depressed CYP1A expression may serve as a conserved mechanism whereby phenotypically altered hepatocytes are able to survive and proliferate in organisms from highly polluted environments. This also points to the need to avoid tissues containing lesions when using CYP1A as a biomarker in monitoring programs.

Levels of CYP1A expression in liver and gut were compared among mummichog from three sites in the lower Chesapeake Bay, including the highly contaminated site and a cleaner site in the Elizabeth River (Van Veld and Westbrook 1995). Levels of hepatic and intestinal CYP1A protein were elevated 6- and 56-fold, respectively, in fish from the two Elizabeth River locales compared to a reference site in the Chesapeake Bay. Surprisingly, there was no difference in hepatic EROD activity among the three sites, but significantly increased expression of intestinal EROD activity (up to 7-fold) was observed in fish from the two sites in the Elizabeth River. However, when mummichog from the Elizabeth River and the reference site were i.p. injected with 3-methylcholanthrene (a PAH), levels of hepatic CYP1A and EROD were induced 418- and 13-fold, respectively, in fish from the reference site and 4-fold or less in fish from the Elizabeth River. In total, these studies suggest that inhibition of hepatic CYP1A expression in liver may be occurring in mummichog from the most impacted site and that diet is a significant route of exposure to contaminants in fish from these two populations.

Another example can be found in Atlantic tomcod from the Hudson River, New York, that exhibited very high levels of hepatic DNA adducts, an activated K-*ras* oncogene in liver tumors, a 90% prevalence of hepatocellular carcinomas in 2-year-old fish, and a reduced life expectancy compared with tomcod from other populations (Wirgin et al. 1994, 1989; Dey et al. 1993). Levels of hepatic CYP1A1 mRNA differed 28-fold between tomcod from the Hudson River and cleaner Atlantic coast estuaries and generally correlated with levels of sediment contamination (Kreamer et al. 1991; Wirgin et al. 1994). In further studies, CYP1A1 mRNA levels in juvenile tomcod from 42 sites in the Hudson River estuary were used to evaluate the microgeographic distribution of bioavailable contaminants and the correspondence between hepatic concentrations of halogenated aromatic hydrocarbons (HAH; expressed as TCDD toxic equivalence quotients [TEQs]) and hepatic CYP1A1 expression (Yuan et al. 2001). Significant spatial heterogeneity in CYP1A1 mRNA levels (up to 23-34-fold) was found among sites within the Hudson River estuary. The CYP1A1 mRNA expression was highest in tomcod from the Newark Bay complex and lowest at the furthest upstream sites in the main stem of the Hudson River. Although CYP1A1 mRNA and TCDD TEQs were highest in tomcod from the Newark Bay complex, a lack of relationship between hepatic HAH TEQs and CYP1A1 mRNA in tomcod from the main stem of the Hudson River suggested that CYP1A1 may not always be reflective of levels of bioavailable HAH contaminants, particularly at sites highly polluted with mixtures of pollutants or that compounds other than those measured were contributing to CYP1A1 expression. This lack of correspondence between hepatic CYP1A1 mRNA and hepatic contaminant levels may have been due to acquired resistance (discussed below).

Endogenous and Exogenous Modulators of CYP1A1 Expression

Inducibility of CYP1A1 (in terms of CYP1A1 mRNA, protein, and enzyme activities) in many species of fishes is significantly impacted by endogenous factors such as gender and reproductive status. Seasonal variation in CYP1A enzyme activities between and within sexes-differential expression between mature females and males or immature females was reported in winter flounder and scup (Edwards et al. 1988; Gray et al. 1991). Hylland et al. (1998) reported that 12% (males) and 46% (females) of total annual variability in hepatic EROD activity in the Norwegian flatfish species *Platichthys flesus* could be explained by season, gender, and maturity. Other studies have suggested that this may be influenced by sex hormones because estradiol (E2)-mediated suppression was demonstrated to override or modulate regulation by CYP1A inducers and that inhibition occurred at the transcriptional level (Gray et al. 1991; Elskus et al. 1992). In environmentally exposed Atlantic tomcod from the Miramichi and Kouchibouguac rivers, New Brunswick, high levels of sex hormones in prespawning females, but not in males, were correlated with depressed levels of hepatic CYP1A1 mRNA (Williams et al. 1998). The implications of these results for biomonitoring programs are obvious and suggest that gender and reproductive status be considered in data interpretation and, better yet, that the prespawning and spawning periods be avoided in sample acquisitions.

Variation in natural and anthropogenically impacted exogenous factors, such as temperature or starvation, can also modulate CYP1A expression levels (Wall and Crivello 1999; Jorgensen et al. 1999; Kloepper-Sams and Stegeman

1992). Although protein concentrations and kinetics of protein accumulation were affected by temperature in mummichog, levels of CYP1A mRNA were not, although the half-life of the mRNA was much longer in the cold compared to warm acclimated fish (Kloepper-Sams and Stegeman 1992). Starvation is another factor, which has been demonstrated to impact cytochrome P450 activities in fish (Wall and Crivello 1999), perhaps through a starvation-mediated increase in hepatic PCB concentration (Jorgensen et al. 1999).

Resistance to CYP1A1 Induction

Several studies of CYP1A1 expression have reported probable acquired resistance of natural populations to xenobiotics. These results have precipitated investigations of the molecular mechanisms of selection and adaptation to xenobiotics and the associated fitness costs. Several populations of fishes from Atlantic coast estuaries highly polluted with dioxins, furans, PCBs, and PAHs exhibit an absence or reduced inducibility of CYP1A compared with reference populations. Mummichog from estuaries highly polluted with AHs exhibited impaired CYP1A inducibility (Prince and Cooper 1995b; Nacci et al. 1999; Elskus et al. 1999) that is heritable (Nacci et al. 1999; K. Cooper personal communication) and associated with resistance to AH (PCBs, TCDD, and PAH)-induced mortality in embryos and adults (Nacci et al. 1999; Prince and Cooper 1995a). Primary cultures of hepatocytes prepared from mummichog from the resistant New Bedford Harbor population were 14-fold less sensitive to TCDD-induced CYP1A1 activity than reference hepatocytes (Bello et al. 2001). Atlantic tomcod from the Hudson River, New York, also exhibited impaired hepatic CYP1A1 mRNA inducibility when treated with a variety of coplanar PCBs (PCBs 77, 81, 126, 169) or TCDD, but not PAHs (Wirgin et al. 1992; Courtenay et al. 1999; Wirgin, unpublished data). Interestingly, reduced inducibility of CY1A1 mRNA was also observed in heart, kidney, intestine, and spleen of tomcod from the Hudson River compared to tomcod from elsewhere (Wirgin, unpublished data) Competitive RT-PCR analysis of individual PCB-treated tomcod larvae suggested that modulation of CYP1A1 mRNA inducibility may involve a single generation physiological acclimation response (Roy et al. 2001). Interestingly, mummichog from the creosote (PAH)-resistant Elizabeth River, Virginia, population exhibited increased susceptibility to low oxygen concentration and fluoranthene mediated phototoxicity compared to reference fish suggesting that decreased CYP1A inducibility may be associated with increased sensitivity to other environmental stressors (R. Di Giulio personal communication).

Because AhR is required to activate CYP1A1 transcription and AH-induced toxicities, several laboratories are investigating the possibility that reduced AhR expression or function may be responsible for modulation of CYP1A inducibility in resistant fish (Roy and Wirgin 1997; Hahn 1998; Powell et al. 2000). ARNT and AhR mRNA expression were compared between F_1 offspring of mummichog from New Bedford Harbor and a reference population; expression of one of the two forms of AhR (AhR1) was higher in the resistant population, an unexpected result (Powell et al. 2000). However, the importance of AhR1 in binding dioxinlike compounds is questionable, and the importance of its upregulation in conferring resistance is unknown. High levels of genetic divergence in AhR cDNA have been observed between resistant and nonresistant tomcod populations, but, the functional significance of this variation requires further examination (Roy and Wirgin 1997; Yuan and Wirgin, authors' unpublished data). Because of the roles of AhR in normal hepatic and immunological function, downregulation of AhR function would imply detrimental organismic consequences in the absence of AH compounds. The molecular mechanisms of resistance in these fish are still open to investigation, but these results suggest that CYP1A1 inducibility should be rigorously evaluated with controlled laboratory experiments before data from field studies is interpreted.

Effects of Mixtures on CYP1A1 Expression

Aquatic organisms from highly industrialized and urban sites are usually exposed to complex mixtures of contaminants including individual AH congeners, AH compounds, and metals. Several studies have exposed fish under controlled laboratory conditions to cadmium (Cd) and mixtures of Cd and PAHs, and have demonstrated effects of this metal on CYP1A induction (Forlin et al. 1986; George 1989; van den Hurk et al. 1998a, 1998b). Additionally, copper (Cu), lead (Pb), and zinc (Zn) were shown to reduce PAH-induced CYP1A mRNA expression and enzyme activities, sometimes by more than 50%, in 3-methylcholanthrene-treated rainbow trout (Risso-de Faverney et al. 2000). Similarly, Cd, arsenic (As), nickel (Ni), and chromium (Cr) solutions were observed to dose responsively depress CYP1A1 mRNA expression in Atlantic tomcod compared to that of B[a]P or PCB77 alone (Sorrentino and Wirgin, unpublished data). The molecular mechanisms whereby CYP1A1 transcriptional expression is impacted in these metal-treated models have yet to be explored. Additionally, tributyltin and triphenyltin have been demonstrated to significantly inhibit levels of CYP1A gene expression in vitro and in vivo in a variety of fishes (Fent and Stegeman 1993; Fent and Bucheli 1994). It has yet to be determined if CYP1A inhibition by tributyltin and triphenyltin occurs at the transcriptional level, and the impacts of organotins or other metals on CYP1A expression on environmentally exposed fish from contaminated sites has yet to be evaluated.

Mixtures of individual AH compounds may interact other than additively in activating CYP1A1 expression. Recent studies with individual PCB congeners indicate that congeners that are low-intrinsic efficacy agonists or competitive antagonists may weakly bind with the AhR and, thereby, decrease CYP1A1 inducibility by strong inducers such as TCDD (Hestermann et al. 2000). Similarly, the noninducing PAH, fluoranthene, was shown to inhibit CYP1A protein induction in vivo in mummichog by B[a]P (Willett et al. 2000).

This is an area of investigation that is sure to receive much attention in the near future.

Metallothionein Induction

Metallothioneins (MT) are a class of highly conserved, small, cysteine-rich (30%) metalloproteins that efficiently bind divalent heavy metals (George et al. 1996). They are believed to play a major role in Zn and Cu homeostasis, in detoxification of metals such as Cd and mercury (Hg), and as scavengers of oxyradicals. It has been proposed that upregulation of MT can serve as an early warning signal of metal-induced toxicity, and depletion of MT has been correlated with metals-induced in vivo toxicity (Masters et al. 1996).

Metallothionein is inducible in fishes by environmentally relevant levels of metals such as Cd, Hg, Zn, Cu, and some, although not all, arsenical species (Eller-Jessen and Crivello 1998; Schlenk et al. 1997b, 2000). Gene inducibility in fish varies among tissues with high levels of MT expression often seen in liver, kidney, and gills; however, hepatic and renal MT mRNA expression was not always linear with Cd dosage, particularly at higher dosages (George et al. 1996). Additionally, the kinetics of gene induction differed among tissues and metals (George et al. 1996). Among the metals, Zn and Cd generally are stronger inducers (10-20-fold) than Cu of in vivo and in vitro MT expression (Schlenk et al. 1997a; Olsson and Kille 1997). Induction of MT mRNA in fish injected with Cd was rapid, resulting in significantly induced and peaked MT mRNA expression in gill, kidney, and liver, and with a half-life of 5-7 d in all three tissues (George et al. 1996). Significant interspecific variation in MT inducibility has been observed among fish species (Olsson 1996; Lam et al. 1998) but was not associated with differential toxicity (Olsson and Kille 1997). In laboratory experiments, MT expression is regulated at the transcriptional level. Cis-acting metal-responsive elements (Olsson and Kille 1997) and trans-acting metal-responsive transcription factors (Der Maur et al. 1999) have been characterized. However, posttranslational regulation of MT levels has been hypothesized for oysters *Crassostrea viginica* from Cd-contaminated Chesapeake Bay, in which MT mRNA levels did not correlate with tissue levels of Cd or MT protein (Roesijadi 1999).

Development of MT antibodies alleviated the difficulties encountered in tedious assays employed in the past to measure MT levels (e.g., nonspecific protein purification, metal saturation assays, atomic absorption analysis; Schlenk et al. 1997a); however, there is limited availability and cross-reactivity of these antibodies (Hogstrand and Haux 1990). Alternatively, advances in molecular biology allow characterization of fish MT cDNA sequences and offer the promise of relative ease of development of species-specific MT DNA hybridization probes or primers for RT-PCR. To date, the use of MT has not been as extensive as that of CYP1A1 in environmental biomonitoring programs. This may be due to high variability in MT expression, perhaps owing to (1) gene regulation by endogenous inducing factors other than metals (glucocorticoids, cytokines, oxyradicals; Olsson et al. 1995; Viarengo et al. 1999), (2) absence of a clear association between MT levels and higher-level toxic endpoints, and (3) interindividual variability in MT inducibility. Additionally, temperature (Tom et al. 1999), gender (Hamza-Chaffai et al. 1997), and maturity (Eller-Jessen and Crivello 1998) can modulate MT expression. Finally, interactive effects between metals and PAHs were reported. Induction of MT and MT mRNA by Cd, Zn, Cu, and lead (Pb) was reduced in rainbow trout hepatocytes concurrently treated with 0.5 mM 3-methylcholanthrene (Risso-de Faverney et al. 2000). Despite these difficulties, fish MT expression has been used in environmental monitoring programs in the Iberian peninsula (Bebianno and Machado 1997), Israel (Tom et al. 1999), and Hong Kong (Wong et al. 2000). In most cases, levels of MT expression correlated with suspected sediment or tissue concentrations of metal; however, levels of MT expression did not correlate with community or fish health indices in a low gradient stream in Arkansas (Schlenk et al. 1996).

Expression of the Multixenobiotic Defense Mechanism

A multidrug-resistance mechanism was first described in drug-resistant tumor cell lines because of their ability to pump out hydrophobic drugs in an ATP-dependent manner. A 170 kDa protein was identified, P-glycoprotein (P-gp), which was responsible for this phenomenon. Subsequently, this same efflux mechanism (termed multixenobiotic resistance) was found to protect the gill, mantle, and liver of aquatic invertebrates and fishes against the bioaccumulation of a variety of hydrophobic xenobiotics. Thus, it was proposed that P-gp expression or efflux activity could serve as biomarkers of xenobiotic exposure in aquatic taxa, but results have been mixed. For example, immunodetection revealed elevated P-gp protein in liver and tumors of mummichog from the population in the Elizabeth River, compared with a reference site (Cooper et al. 1999). This suggests that increased P-gp expression may play a role in genetic adaptation and neoplasia. However, laboratory exposure of fish to 3-methylcholanthrene failed to increase hepatic P-gp levels (Cooper et al. 1999). P-glycoprotein expression was often elevated in marine mussels and oysters from polluted sites (Minier et al. 1993; Kurelec et al. 1996). However, large interindividual, interspecific, and seasonal variation in basal and induced P-gp expression in mollusks (Smital et al. 2000) led to the recommendation that P-gp expression not be used as a biomarker of xenobiotic exposure (Kurelec et al. 1996).

Preliminary studies quantifying P-gp functional activity in aquatic invertebrates proved more promising as a biomarker. Basal levels of P-gp functional activity correlated well with the levels of pollution at locales from which test organisms were collected (Smital et al. 2000). However, it was demonstrated that induction of P-gp expression and functional activity in the California mussel *Mytilus californianus* was a general cellular response to environmental stressors including heat and non-P-gp substrates (Eufemia and Epel 2000). Thus, while P-gp expression may provide an important functional signal as to exposure to hydrophobic xenobiotics and organismal response, nonspecificity of the response and high levels of interindividual variation may introduce too much uncertainty for this biomarker to be useful in a management context.

Vitellogenin Expression

In recent years, considerable concern has been expressed regarding the potential effects of endocrine disrupting chemicals on reproduction and on development of the reproductive system of populations from impacted waters including an increased incidence of hermaphroditism, altered sex ratios, and impaired structure and, perhaps, function of reproductive organs. It has been proposed that decreased population size and recruitment may result from exposure of adults and sensitive early life stages to endocrine-disrupting chemicals (e.g., alkylphenol polyethoxylates, pharmaceuticals, HAHs, PAHs, PCBs, dioxins, furans, pesticides, some heavy metals, and others). The mechanism is probably the disruption of normal physiological processes by mimicking 17β estradiol (E_2).

Vitellogenin (VTG) is a yolk precursor glycophospholipoprotein that is normally synthesized in livers of female fish and transported by the blood to developing oocytes, where it is taken up and cleaved into phosvitin and lipovitellin, which serve as food supplies for developing embryos. Transcription of VTG is regulated by serum E_2 levels, binding of E_2 to estrogen receptors in the liver, and binding of activated estrogen receptors to estrogen receptor response elements in the promoter of E_2-responsive genes, such as VTG. Although not normally active, VTG can be induced in the livers of male and juvenile fish treated with E_2 or E_2-mimics and could serve as a biomarker of exposure to xenoestrogens.

Assays have been developed and implemented in laboratory and field experiments to quantify VTG expression at the protein and mRNA levels. The window of induced serum VTG expression in environmentally exposed fish is often very high, up to 60,000-fold compared with fish from reference sites. In biomonitoring programs, elevated VTG expression was observed in caged rainbow trout and wild populations of male flounder *Platichythys* in the United Kingdom and *Pleuronectes yokahame* in Japan (Sumpter and Jobling 1995; Hashimoto et al. 2000), carp *Cyrpinus carpio* just downstream from a sewage treatment plant in St. Paul, Minnesota (Folmar et al. 1996), and largemouth bass *Micropterus salmoides* from two rivers in northwestern Florida (Orlando et al. 1999). In some cases, elevated VTG was associated with decreased levels of plasma testosterone, increased liver mass, kidney pathologies, and elevated prevalences of testicular abnormalities. In controlled laboratory experiments, E_2-induced overexpression of VTG in male summer flounder *Paralichthys dentatus* resulted in VTG accumulation and pathological manifestations in liver, kidney, and testes (Folmar et al. 2001).

Similarly, 4-tert-octylphenol, and alkylphenol polyethoxylate induced over-expression of serum VTG in male Japanese medaka and resulted in significantly reduced fertilization success with sperm from treated fish and reduced survivorship of the embryonic offspring of treated males (Gronen et al. 1999). Thus, it is easy to envision how exposure to VTG-inducing chemicals could have potential effects at higher biological levels in impacted populations.

High Throughput Assays

A major limitation to the more extensive and effective use of altered gene expression and genetic damage at specific loci as biomarkers is the small number of genes identified to date at which these perturbations occur. Also, because overt toxicities are almost certainly preceded by alteration of expression or structure of multiple genes, the ability to simultaneously monitor batteries of genes would prove more informative in monitoring programs and predictive of perturbations. Recently, high throughput technologies have been developed to allow for the identification and characterization of heretofore undescribed genes that are sensitive to xenobiotic exposure.

Microarray Hybridizations: Gene Chips

Microarray hybridization is an approach that permits characterization and comparison of the expression levels of hundreds to thousands of genes simultaneously between toxicant-treated and control cells or tissues. This approach can be used to identify novel genes that are sensitive to toxicant exposure and to characterize the expression profile of batteries of genes whose expression may be coordinately regulated by xenobiotics. These gene expression profiles can then be associated with phenotypic manifestations of toxicity to begin to establish cause and effect relationships that can be used in risk management. Once identified and characterized, these functionally related batteries of sensitive genes can then be used to develop custom-designed toxicology arrays that can be routinely used in biomonitoring aquatic environments.

The microarray technique is essentially a reverse Northern blot. In Northern blotting, DNA probes in solution are hybridized to immobilized mRNAs; in microarrays, it is the reverse. The cDNA probes are deposited and immobilized on a support platform-nylon membranes, glass slides, or plastic chips. These probes may be full-length cDNAs, expressed sequence tags (EST; smaller fragments of cDNA that have been cloned), or oligonucleotides (very short pieces of cDNA) directly synthesized or deposited on the support (van Hal et al. 2000; Xiang and Chen 2000). These arrays consist of a rectangular matrix of a very large number (up to 20,000 or more) of very small dots of these probes. Total RNA or mRNA is then isolated from toxicant-exposed and reference cells or tissues, converted to cDNAs, labeled with fluorescent probes, and hybridized to the arrays. The relative expression of the genes in toxi-

cant-exposed and control samples is compared using fluorescence microscopy (Watson et al. 1998). This analysis will reveal which genes are
upregulated, downregulated, or unaffected by toxicant exposure (Figure 10).
By the use of cluster analysis, it will also identify groups of genes that are
coordinately regulated by toxicant exposure. This approach serves to initially identify xenobiotically sensitive genes whose specificity and respon-

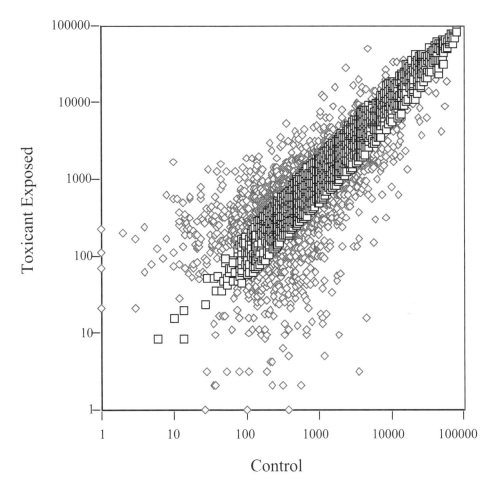

Figure 10. Microarray analysis of gene expression in toxicant-exposed and control
tissue. Each data point represents the relative expression of a single gene. Data is
presented for 6,8000 genes. X-axis depicts fluorescence from hybridization with
mRNA from the control tissue and Y-axis depicts fluorescence from hybridization
with mRNA from the toxicant-exposed tissue. Expression levels of genes in black did
not differ by more than two fold between the toxicant exposed and control cells and
are assumed not to be up- or downregulated by toxicant exposure. Genes whose
expression levels are in red differed by more than two fold between the toxicant-
exposed and control cells and are assumed to be up- or downregulated by toxicant
exposure.

siveness to graded doses of contaminants should then be evaluated by traditional methods of RNA analyses (Bartoseiwicz et al. 2000).

As most often applied in mammalian toxicology, cDNAs of known sequence and function are applied to the hybridization platform. This is facilitated by genome projects that identify all the genes and other noncoding sequences within a species' genome. Efforts are currently underway to conduct such analyses in the zebrafish *Danio rerio*, Japanese pufferfish *Fugu rubripes*, and perhaps other fish taxa. To date, microarrays are commercially available and are being used to evaluate the effects of contaminants on the expression of human, rat, mouse, and *Drosophila* genes. Unfortunately, because of the need for optimal hybridization signal strength and specificity, arrays from one species cannot be used in a second species, even if closely related. Thus, a major limitation to the widespread implementation of this approach to aquatic taxa is the lack of complete genomic sequence information. However, identification of sensitive target genes in one species such as zebrafish should lead to the easy development of PCR primers and oligonucleotides for that gene in other fish. Thus, it is easy to see how genomic DNA characterization in one species will open the door for expanded gene expression analyses in a wide variety of species.

Alternatively, it is also possible to use microarray technology with cDNA probes from anonymous genes (Oleksiak et al. 2001). These anonymous ESTs can be isolated by cloning random cDNAs from any taxon of interest into plasmids in order to form a cDNA library (a collection of plasmids, each with a different cDNA insert transfected into a bacterial host). The inserts in these random clones are then PCR-amplified and then systematically applied to microarray platforms and hybridized as described above. If a particular probe is found to be differentially expressed, the clone from which it came can be fully characterized using DNA sequencing in order to determine the identity of the gene from which the cDNA originated. Alternatively, cDNAs can be sequenced prior to their use in the microarrays to minimize redundancy. Efforts are currently underway to use this technology in estrogen treated and natural populations of fish (Denslow et al. 2001b).

Differential Display

Differential display (DD) is another technique that analyzes the expression of anonymous genes to identify those whose expression differs between toxicant-exposed and control cells or organisms. This is a second approach to identify novel genes that may serve as targets for gene expression studies. In this technique, cDNA is prepared from mRNA via reverse transcription using either a poly T (all mRNAs have a string of adenines ["poly A"] at their 3' end) or random hexamer (6 base pair) primers (Liang and Pardee 1995). Polymerase chain reaction is then used to amplify the cDNA using one poly T primer and one arbitrary primer. Theoretically, by using all 240 combinations of commercially available sets of random PCR primers, all the cDNAs within a vertebrate genome can be screened for differential expression of toxicant-sensitive genes. The PCR products are then separated on sequencing gels and visualized by autoradiography (Liang and Pardee 1995) or nonradioactively (Doss 1996; van der Houven van Oordt et al. 1999), producing a ladder-like "RNA fingerprint" similar to DNA fingerprints (see Chapter 5 for a discussion of DNA fingerprinting).

Differential expression between toxicant-exposed and control samples is determined by the presence or absence or intensity of the fingerprint bands. The PCR fragments that are differentially expressed are then isolated from the gel, their full-length cDNAs obtained, sequenced to determine their identity, and their expression patterns evaluated in RNA hybridizations. The advantage of this technique is that no DNA sequence information is needed, and it is a relatively rapid way to examine large numbers of anonymous genes. The disadvantage is that the rate of false positives or irreproducible results (due to PCR artifacts) can be high, so that replicate reactions need to be run and the differential expression patterns need to be verified with traditional RNA analyses. A recent modification to this technique is the incorporation of amplified fragment length polymorphism (AFLP) technologies into differential display techniques (AFLP-DD) (Ivashuta et al. 1999), which produces display patterns that are much more reproducible. This technique is essentially the same as the AFLP technique described in Chapter 5, except that cDNA is used instead of genomic DNA. The advantage of DD or AFLP-DD over microarray hybridizations is that it is less cost prohibitive and no complicated equipment other than a DNA sequencing device is needed.

To date, differential display has been used in aquatic taxa to explore the molecular mechanisms of gonadal tumorigenesis in soft-shell clams *Mya arenaria* from eastern Maine (Rhodes and Van Beneden 1996; Van Beneden et al. 1998; Sultan et al. 2000), to identify TCDD-inducible genes in soft-shell clams (Van Beneden et al. 1999), genes that are upregulated by Cu and permethrin treatment in staghorn coral *Acripora cervicornis* (Morgan et al. 2001) and novel genes that are differentially expressed in E_2, diethylstilbestrol (DES), or ethinylestradiol (EE_2)-treated sheepshead minnows *Cyprinodon variegatus* (Denslow et al. 2001a). For example, using 18% of the 240 potential primer pair combinations, Denslow et al. (2001a) reported the identification of 48 genes, including VTG, which were differentially expressed in E_2, DES, and EE_2-treated fish.

Which Assay to Use?

The assay system to use in aquatic biomonitoring depends upon a number of factors. First and foremost, the endpoint chosen depends upon the inducing agent. To this end, basic research in mechanisms of induction of molecular biomarkers must be pursued before applying any techniques. Also, some molecular biomarkers are indicative of exposure or effects of a broad class of chemicals, while others are very specific indicators of exposure to particular chemicals (Table 3). Thus, the use of particular biomarkers depends upon whether the objective is monitoring integrated exposure to complex mixtures or identification of specific chemicals that may be causing effects. In addition, some molecular biomarkers have been well characterized in terms of mechanism of induction and effects at higher levels of biological organization (CYP1A1 expression, DNA damage), while others have not (e.g., P-Gp). Hence, if the objective is the application of biomarkers to risk assessment, the former may be a better choice. If the objective is basic research into novel molecular responses, the latter may prove more interesting. The tissue that is to be examined may also determine which is the most appropriate biomarker to use and vice versa. A related factor is the amount of tissue that is available for analysis because some assays require less DNA or RNA than others and some are more sensitive than others (see Tables 1 and 2). This may depend on whether destructive or nondestructive sampling is employed. In addition, the method of regulation of gene expression should be taken into account when choosing which assay to employ, because some genes are regulated at the transcriptional, posttranscriptional, or posttranslational level. For example, using an mRNA-based assay to measure induction of p53 may be useless, because its expression is frequently induced via stabilization of the protein (posttranslational), without changes in the level of transcription or translation. Finally, when deciding upon which technique to use for a particular biomarker (e.g., alkaline unwinding versus SCGE for DNA strand breaks, RT-PCR versus Northern blotting for gene expression), logistical constraints may play a role, such as availability of suitable equipment, cost of analysis, or ability to use and dispose of radioactive materials. However, logistical constraints should not be the only factor in this decision because specificity, sensitivity, and appropriateness of the biomarker should be an equal or greater determining factor.

Future Studies

One avenue of promising future research is elucidation of molecular biomarkers, pathways, and mechanisms of sublethal toxicity that can affect components of fitness, such as bioenergetic status, immune function, growth, development, survival, and reproduction. Such studies can provide mechanistic linkages among molecular, organismal, and population-level effects and would be instrumental in establishing causality between pollution exposure and effects (Decaprio 1997). These questions can begin to be addressed by microarray and differential display technologies. Further, comparative studies examining expression of DNA damage or xenobiotic-inducible genes in multiple species could provide information as to the mechanism of community-level effects of contamination. Molecular biomarkers could also provide

Table 3. Comparison of various molecular biomarkers.

Biomarker	Inducing agent[a]	Specificity[a]	Kinetics[b]	Inducibility by nonanthropogenic factors	Associated with higher-level effects[c]	
					Correlation	Mechanism
Adducts	PAHs, organo-chlorines, aflo-toxins	Mid (general adducts) to very high (B[a]P adducts)	Delayed, cumulative Persistent	Very low	Highly	I: highly E: potentially
DNA base oxidations	Various chemicals, ionizing radiation	Mid	Rapid, transient (cumulative[d])	Low[e]	Highly	I: highly E: potentially
DNA strand breaks	Various chemicals, ionizing radiation	Mid	Rapid, transient (cumulative[d])	Low[e]	Highly	potentially
Point mutations	Various chemicals, ionizing radiation	Mid	Delayed, cumulative Persistent	Low[e]	I: highly E: unknown	I: highly E: potentially
Cytogenetic effects	Various chemicals, ionizing radiation	Mid	Delayed, cumulative Irreparable	Low	Highly	I: highly E: potentially
CYP1A1	AHs	High	Immediate, cumulative[d] Transient or persistent	Low	Highly	Highly
Metallothionein	Metals	High	Immediate, transient	High	Moderately	Poorly
P-glycoprotein	Various chemicals	Low	Unknown	Medium	Unknown	Unknown
Vitellogenin	Various endocrine disruptors	Mid	Rapid, transient	High(?) Low(?)	Moderately	Potentially
High-throughput assays	Potentially all chemicals	Potentially high to very high	Immediate, transient, or persistent[f]	Low-high[f]	Potentially	Potentially

[a]Low = not specific to any class of xenobiotics; Mid = specific to broad classes of xenobiotics (genotoxins, endocrine disruptors); High = specific to chemical classes (metals, Ahs); Very high = able to identify specific chemicals.

[b]Immediate = effects induced very quickly (within minutes to hours after exposure); Rapid = effects induced within hours or days of exposure; Delayed = effects apparent only after weeks or months of exposure; Transient = quickly repaired or subsides soon after cessation of exposure; Cumulative = effects increase with increasing exposure time; Persistent = effects may remain after cessation of exposure; Irreparable = effects not reversible except by cell death or cell turnover.

[c]Individual-level effects (I), including health and fitness components, or higher-level (population-community) ecological effects (E). Potentially = potential mechanisms of such relationships have been identified, but not yet well characterized.

[d]Depending on toxicokinetics of inducing agent.

[e]May be induced at low levels in nonexposed individuals due to endogenous metabolism or "mistakes" in DNA repair.

Simultaneous examination of multiple genes can give information on modes of action, and potentially be used as a molecular "signature" to identify exposure to specific chemicals.

[f]Depending on the genes examined.

mechanistic linkages between aquatic ecosystem and human health (Adams and Greeley 1996), a prospect that is facilitated by the fact that aquatic organisms are increasingly being used in molecular biomedical research (Morizot et al. 1998; Powell et al. 1996; see special issue of Marine Biotechnology Volume 3 [Supplement 1]). Another possible future avenue of research is on "epigenetic" effects of pollutant exposure-nongenotoxic alterations of DNA methylation and acetylation patterns. Such effects are found to reduce or silence gene expression in mammalian cells (Klein and Costa 1997), may be one mechanism of carcinogenesis (Robertson and Jones 2000), and are induced in fish exposed to xenobiotics (Shugart 1990).

Relationships between gene expression and DNA damage also merit further examination. This has been addressed for CYP1A1 expression and PAH adducts in the preceding text, but other pathways (e.g., estrogen signaling; Roy and Liehr 1999) or hypoxia and antioxidant response (Englander et al. 1999; Haddad et al. 2000), need further investigation. Interactions between pathways such as AhR and hypoxia response may warrant investigation given the interactive role of ARNT in both pathways. Furthermore, the relationship between DNA damage, mutagenesis, and aberrant expression of oncogenes or tumor suppressor genes in aquatic organisms is in its infancy and offers exciting possibilities for future research in biomedicine and ecotoxicology. Also, many mammalian DNA damage-inducible genes have been identified (Fornace 1992), and the use of homologous gene expression in aquatic organisms may provide more sensitive biomarkers of genotoxicant exposure and effects than DNA damage per se. Additionally, the role of DNA repair genes is only beginning to be explored in fish and warrants further investigation.

Further examination of molecular responses to contaminant exposure can offer additional insight into the mechanisms of resistance of aquatic organisms to xenobiotic stress. This aspect of toxicology has been underappreciated in the past, but is fast becoming an active area of research that has relevance for both environmental and biomedical toxicology (see Theodorakis and Shugart 1998 for discussion). Genetic adaptation, as well as induction of heritable mutations and population bottlenecks, can result in multigenerational effects on the population genetic structure of exposed organisms (see Chapter 5 and Bickham et al. 1998). Identification of novel toxicant-sensitive genes by microarrays and differential display should provide additional more-relevant target loci for population genetic analyses. Integration of genotoxic or other molecular biomarkers into population genetic analyses could provide valuable insight as to the etiology and consequences of population genetic alterations, as well as elucidate patterns of biomarker expression within and between populations (Theodorakis 2001; Theodorakis et al. 2001).

Other advances in biotechnology will undoubtedly be incorporated into aquatic toxicology. For example, transgenic fish or invertebrates can be used in mutagenesis studies and in elucidation of gene expression pathways and their toxicological relevance (Amanuma et al. 2000; Cioci et al. 2000; Winn et al. 2001). In this regard, transgenic zebrafish and Japanese medaka have been developed as in vivo tools to assess the presence of environmental mutagens. Advances in PCR (Vrana 1996), such as real-time monitoring of amplification (Gibson et al. 1996), can be applicable to PCR-based gene expression, mutagenesis, and DNA damage assays in aquatic organisms. Also, recent advances in capillary electrophoresis in detection of xenobiotics and biomolecules (Deyl et al. 1994; Sovocool et al. 1999; Ren 2000) could also allow significant advances in detection of adducts, nonspecific DNA damage, point mutations, and single-gene or high throughput expression assays in aquatic organisms. Finally, oligonucleotide arrays can be used to study the frequency of single-nucleotide polymorphisms at multiple sites simultaneously in sensitive loci (discussed in Chapter 5), such as practiced with p53 in mammals (Hacia 1999).

In the very near future, technical advances in mammalian molecular toxicology will undoubtedly permit the identification of batteries of new genes whose structures or expression are altered by toxicant exposure. Use of this information should be translated quickly into the aquatic arena and greatly increase the arsenal and specificity of target genes for molecular biomarker studies. Importantly, rigorous statistical analysis of expression data should identify clusters of genes whose expression profiles are coordinately altered by in vivo toxicant exposures. As a result, unlike current investigations that exclusively focus on single genes, it should soon be possible to investigate the genotoxic and epigenetic effects of toxicant exposure on entire molecular pathways using custom designed aquatic microarrays. This approach has already been applied to gene expression induced by PAHs and DNA damage in mammalian models (Amundson et al. 1999; Bartoseiwicz et al. 2000). Thus, we envision the development of custom-designed microarrays for aquatic organisms that will contain a battery of genes that have been demonstrated to be inducible by toxicant exposure and that will be used in routine biomarker monitoring programs. Assays that monitor patterns of induced gene expression in multiple coordinately regulated genes in aquatic organisms should provide increased reproducibility, confidence in the accuracy, and specificity of the results. More importantly, this development should aid in bridging the gap between molecular and organismal or population effects. As a result, alterations in the "right" molecular biomarkers should prove predictive of effects at higher levels of biological organization and, thus be viewed more favorably by resource managers.

To date, efforts to link molecular responses to higher-level biological effects have focused on cancer as the endpoint. However, with the aid of array hybridization analysis, it should soon be possible to identify novel genes whose altered expression is associated with, and perhaps predictive of, other toxic endpoints in aquatic organisms. For example, early life mortality and teratogenicity in fishes are sensitive to exposure to dioxins, furans, and dioxinlike PCBs in controlled laboratory experiments and sensitivity correlated with reduced recruitment to impacted Great Lakes populations (Elonen et al. 1998). Early life toxicity in fish is associated with dysfunction of the cardiovascular system (Cantrell et al 1998) and can serve as a model for the study of this effect in other organisms. Array hybridization analysis of dioxin- or PCB-exposed early life stages should identify clusters of novel target genes that are sensitive to these contaminants and alterations that are predictive of effects at the organismal and population levels. Similarly, this approach can be used to identify genes, in addition to vitellogenin, whose expression is regulated by exposure to endocrine disrupting chemicals; this information can be used in mechanistic studies to more fully understand the specificity and higher-level effects of these xenobiotics.

Conclusions

Judicious use of molecular biomarkers should include intensive species-specific investigations of those intrinsic and extrinsic factors, other than xenobiotic exposure that can modulate the biomarker response. These may include intrinsic biological parameters such as sex, nutritional status, or reproductive maturity and extrinsic factors such as temperature, salinity, exposure to complex mixtures, and prior exposure history, all of which have been shown to impact molecular responses. The use of molecular markers does not obviate the need to measure target tissue residues of xenobiotics in a subset of environmentally exposed samples. Although the molecular response may have been calibrated in naïve animals under controlled laboratory conditions in experiments with graded doses of chemical, it is still important to evaluate the dose-dependency of this response in natural populations.

Most importantly, molecular sensitivity must be defined on a taxon-specific basis in controlled laboratory experiments prior to the implementation of field monitoring programs and data interpretation. Many studies have demonstrated that species and populations within species differ significantly in gene inducibility and DNA damage. Genetic adaptation or physiological acclimation may result in diminished sensitivities to xenobiotic exposure. These effects may be particularly pronounced at those loci whose products convert xenobiotics to more or less reactive metabolites. It is likely that many populations that are chronically exposed to high levels of xenobiotics will exhibit compensatory responses to survive in toxicant-compromised environments. Finally, within populations, high levels of interindividual variability in gene expression are invariably observed, even among cohorts sharing similar exposure histories. This too probably reflects inherent genetic variation in inducibility. It is important that intrapopulation variation in molecular response be described and quantified in pilot experiments prior to the design of field studies, such that the potential statistical significance of results can be addressed.

Molecular approaches offer inherent advantages not seen with other biomarkers, including sensitivity, cost, time, and reproducibility. Molecular biomarkers can offer considerable specificity as to the identity of chemical-inducing or damaging agents. Specificity of the response is certain to increase in the near future as the battery of isolated xenobiotically sensitive genes increases. Most importantly, molecular biomarkers should provide direct quantitative mechanistic links between exposure and higher-level biological effects. Rather than merely biomarkers of exposure, they will become biomarkers of specific adverse biological effects.

Acknowledgment

IW acknowledges the support of NIEHS Center Grant ES00260.

References

Adams, S. M., W. D. Crumby, M. S. Greeley, L. R. Shugart, and C. F. Saylor. 1989a. Responses of fish populations and communities to pulp-mill effluents: a holistic assessment. Ecotoxicology and Environmental Safety 24:347–360.

Adams, S. M., and M. S. Greeley. 1996. Establishing possible links and relationships between aquatic ecosystem health and human health: an integrated approach. Pages 91–102 *in* R. T. Di Giulio and E. Monosson, editors. Interconnections between human and ecosystem health. Chapman and Hall, New York.

Adams, S. M., K. L. Shepard, M. S. Greeley, B. D. Jimenez, M. G. Ryon, and L. R. Shugart. 1989b. The use of bioindicators for assessing the effects of pollutant stress on fish. Marine Environmental Research 28:459–464.

Alberts, B., D. Bray, L. Lewis, K. Roberts, and J. D. Watson. 1994. Molecular biology of the cell, 3rd edition. Garland, New York.

Al-Sabti, K., and C. D. Metcalfe. 1995. Fish micronuclei for assessing genotoxicity in water. Mutation Research 343:121–135.

Amanuma, K., H. Takeda, H. Amanuma, and Y. Aoki. 2000. Transgenic zebrafish for detecting mutations caused by compounds in aquatic environments. Nature Biotechnology 18:62–65.

Amundson, S. A., M. Bittner, Y. Chen, J. Trent, P. Meltzer, and A. J. Fornace, Jr. 1999. Fluorescent cDNA microarray hybridization reveals complexity and heterogeneity of cellular genotoxic stress responses. Oncogene 18:3666–3672.

Anderson, S., W. Sadinski, L. Shugart, P. Brussard, M. Depledge, T. Ford, J. E. Hose, J. Stegeman, W. Suk, I. Wirgin, and G. Wogan. 1994. Genetic and molecular ecotoxicology: a research framework. Environmental Health Perspectives 102(Supplement 12):3–8.

Anderson, S. L., and G. C. Wild. 1994. Linking genotoxic responses and reproductive success in ecotoxicology. Environmental Health Perspectives 102(Supplement 12):9–12.

Atienzar, F. A., M. Conradi, A. J. Evenden, A. N. Jha, and M. H. Depledge. 1999. Qualitative assessment of genotoxicity using random amplified polymorphic DNA: comparison of genotoxic template stability with key fitness parameters in *Daphnia magna* exposed to benzo[a]pyrene. Environmental Toxicology and Chemistry 18:2275–2282.

Atienzar, F. A., B. Cordi, M. E. Donkin, A. J. Evenden, A. N. Jha, and M. H. Depledge. 2000a. Comparison of ultraviolet-induced genotoxicity detected by random amplified polymorphic DNA with chlorophyll fluorescence and growth in a marine macroalgae, *Palmaria palmata*. Aquatic Toxicology 50:1–12.

Atienzar, F, A. Evenden, A. Jha, D. Savva, and M. Depledge. 2000b. Optimized RAPD analysis generates high-quality genomic DNA profiles at high annealing temperature. BioTechniques 28:52–54.

Bader M., M. Kaling, R. Metzger, J. Peters, J. Wagner, and D. Ganten. 1992. Basic methodology in the molecular characterization of genes. Journal of Hypertension 10:9–16.

Bailey, G. S., D. E. Williams, and J. D. Hendricks. 1996. Fish models for environmental carcinogenesis: the rainbow trout. Environmental Health Perspectives 104(Supplement 1):5–21.

Ballinger, S. W., T. G. Bouder, G. S. Davis, S. A. Judice, J. A. Nicklas, and R. J. Albertini. 1996. Mitochondrial genome damage associated with cigarette smoking. Cancer Research 56:5692–5697.

Barker, C. M., R. J. Calvert, C. W. Walker, and C. L. Reinisch. 1997. Detection of mutant p53 in clam leukemia cells. Experimental Cell Research 232:240–245.

Bartoseiwicz, M., M. Trounstine, D. Barker, R. Joahston, and Alan Buckpitt. 2000. Development of a toxicological gene array and quantitative assessment of this technology. Archives of Biochemistry and Biophysics 376:66–73.

Baumann P. C. 1998. Epizootics of cancer in fish associated with genotoxins in sediment and water. Mutation Research/Reviews in Mutation Research 411:227–233.

Baumann, P. C., and M. S. Okihiro. 2000. Cancer. Pages 591–616 *in* G. K. Ostrander, editor. The laboratory fish. Academic Press, San Diego, California.

Bebianno, M. J., and L. M. Machado. 1997. Concentrations of metals and metallothioneins in *Mytilus galloprovincialis* along the south coast of Portugal. Marine Pollution Bulletin 34:666–671.

Belin D. 1998. The use of RNA probes for the analysis of gene expression. Northern blot hybridization and ribonuclease protection assay. Methods in Molecular Biology 86:87–102.

Bello, S. M., D. G. Franks, J. J. Stegeman, and M. E. Hahn. 2001. Acquired resistance to Ah receptor agonists in a population of Atlantic killifish (*Fundulus heteroclitus*) inhabiting a marine Superfund site: in vivo and in vitro studies on the inducibility of xenobiotic metabolizing enzymes. Toxicological Sciences 60:77–91.

Belpaeme, K., K. Cooreman, and M. K. Volders. 1998. Development and validation of the in vivo alkaline comet assay for detecting genomic damage in marine flatfish. Mutation Research 415:167–184.

Berger, M., C. Anselmino, J.-F. Mouret, and J. Cadet. 1991. Higher performance liquid chromatography-electrochemical assay for monitoring the formation of 8-oxo-7,8-dihydroadenine and its related 2'-deoxyribonucleoside. Journal of Liquid Chromatography 13:929–940.

Bickham, J. W., G. T. Rowe, G. Palatnikov, A. Mekhtiev, M. Mekhtiev, R. Y. Kasimov, D. W. Hauschultz, J. K. Wickliffe, and W. J. Rogers. 1998. Acute and genotoxic effects of Baku Harbor sediment on Russian sturgeon, *Acipenser gueldenstaedtii*. Bulletin of Environmental Contamination and Toxicology 61:51–518.

Black, M., and J. Belin. 1998. Evaluating sublethal indicators of stress in Asiatic clams (*Corbicula fluminea*) caged in an urban stream. ASTM Special Technical Publication 1333:76–91.

Bue, B. G., S. Sharr, S. D. Moffitt, and A. K. Craig. 1996. Effects of the Exxon Valdez oil on pink salmon embryos and preemergent fry. Pages 619–627 *in* S. D. Rice, R. B. Spies, D. A. Wolfe, and B. A. Wright, editors. Proceedings of the Exxon Valdez Oil Spill Symposium. American Fisheries Society, Symposium 18, Bethesda, Maryland.

Bue, B. G., S. Sharr, and J. E. Seeb. 1998. Evidence of damage to pink salmon populations inhabiting Prince William Sound, Alaska, two generations after the Exxon Valdez oil spill. Transactions of the American Fisheries Society 127:35–43.

Campana, M. A., A. M. Panzeri, V. J. Moreno, and F. N. Dulout. 1999. Genotoxic evaluation of the pyrethroid lambda-cyhalothrin using the micronucleus test in erythrocytes of the fish *Cheirodon interruptus interruptus*. Mutation Research/ Genetic Toxicology and Environmental Mutagenesis 438:155–161.

Cantrell, S. M., J. Joy-Schlezinger, J. J. Stegeman, D. E. Tillit, and M. Hannink. 1998. Correlation of 2,3,7,8-tetrachlorodibenzo-p-dioxin induced apoptotic cell death in the embryonic vasculature with embryotoxicity. Toxicology and Applied Pharmacology 148:24–34.

Carver, L. A., and C. A. Bradfield. 1997. Ligand-dependent interaction of the aryl hydrocarbon receptor with a novel immunophilon homologue in vivo. Journal of Biological Chemistry 272:11452–11456.

Chipman, J. K., and J. W. Marsh. 1991. Bio-techniques for the detection of genetic toxicity in the aquatic environment. Journal of Biotechnology 17:199–208.

Choi, K., M. Zong, and P. G. Meier. 2000. Application of a fish DNA damage assay as a biological toxicity screening tool for metal plating wastewater. Environmental Toxicology and Chemistry 19:242–247.

Chu, G. 1997. Double strand break repair. Journal of Biological Chemistry 272:24097–24100.

Cioci, L. K., L. Qiu, and J. H. Freedman. 2000. Transgenic strains of the nematode *Caenorhabditis elegans* as biomonitors of metal contamination. Environmental Toxicology and Chemistry 19:2122–2129.

Coleman, M. S., C. A. Afshari, and J. C. Barrett. 2000. Regulation of p53 stability and activity in response to genotoxic stress. Mutation Research 462:179–188.

Collier, T. K., B. F. Anulacion, J. E. Stein, A. Goksoyr, and U. Varanasi. 1995. A field evaluation of cytochrome P4501A as a biomarker of contaminant exposure in three species of flatfish. Environmental Toxicology and Chemistry 14:143–152.

Collier, T. K., S. D. Connor, B.-T. L. Eberhart, B. F. Anulacion, A. Goksoyr, and U. Varanasi. 1992. Using cytochrome P450 to monitor the aquatic environment: initial results from regional and national surveys. Marine Environmental Research 34:195–199.

Cooper, P. S., W. K. Vogelbein, and P. A. Van Veld. 1999. Altered expression of the xenobiotic transporter P-glycoprotein in liver and liver tumors of mummichog

(*Fundulus heteroclitus*) from a creosote-contaminated environment. Biomarkers 4:48–58.

Cotton R. G. 1993. Current methods of mutation detection. Mutation Research 285:125–144.

Courtenay, S., P. J. Williams, C. Grunwald, T.-L. Ong, B. Konkle, and I. I. Wirgin. 1994. An assessment of within group variation in CYP1A1 mRNA inducibility in Atlantic tomcod. Environmental Health Perspectives 102(Supplement 12):85–90.

Courtenay, S., C. Grunwald, G.-L. Kreamer, R. Alexander, and I. Wirgin. 1993. Induction and clearance of cytochrome P4501A mRNA in Atlantic tomcod caged in bleached kraft mill effluent in the Miramichi River. Aquatic Toxicology 27:225–244.

Courtenay, S. C., C. M. Grunwald, G.-L. Kreamer, W. L. Fairchild, J. T. Arsenault, M. Ikonomou, and I. I. Wirgin. 1999. A comparison of the dose and time response of CYP1A1 mRNA induction in chemically treated Atlantic tomcod from two populations. Aquatic Toxicology 47:43–69.

Custer, T. W., J. W. Bickham, T. B. Lyne, T. Lewis, L. A. Ruedas, C. M. Custer, and M. J. Melancon. 1994. Flow cytometry for monitoring contaminant exposure in black-crowned night herons. Archives of Environmental Contamination and Toxicology 27:176–179.

Dallas, C. E., S. F. Lingenfelser, J. T. Lingenfelser, K. Holloman, C. H. Jagoe, J. A. Kind, R. K. Chesser, and M. H. Smith. 1998. Flow cytometric analysis of erythrocyte and leukocyte DNA in fish from Chernobyl-contaminated ponds in the Ukraine. Ecotoxicology 7:211–219.

Das, P., and G. John. 1999. Induction of sister chromatid exchanges and chromosome aberrations in vivo in *Etroplus suratensis* (Bloch) following exposure to organophosphorus pesticides. Toxicology Letters 104:111–116.

Decaprio, A. P. 1997. Biomarkers: coming of age for environmental health and risk assessment. Environmental Science and Technology 31:1837–1848.

de Fromentel, C. C., F. Pakdel, A. Chapus, C. Baney, P. May, and T. Soussi. 1992. Rainbow trout p53: cDNA cloning and biochemical characterization. Gene 112:241–245.

de Laat, W. L., N. G. J. Jaspers, and J. H. J. Hoeijmakers. 1999. Molecular mechanism of nucleotide excision repair. Genes and Development 13:768–785.

Denslow, N. D., C. J. Bowman, R. J. Ferguson, H. S. Lee, M. J. Hemmer, and L. C. Folmar. 2001a. Induction of gene expression in sheepshead minows (*Cyprinodon variegatus*) treated with 17β-estradiol, diethylstilbestrol, or ethinylestradiol: the use of mRNA fingerprints as an indicator of gene regulation. General and Comparative Endocrinology 121:250–260.

Denslow, N. D., H. S. Lee, C. J. Bowman, M. J. Hemmer, and L. C. Folmar. 2001b. Multiple responses in gene expression in fish treated with estrogen. Comparative Biochemistry and Physiology. Part B: Biochemistry and Molecular Biology 129:277–282.

Der Maur, A. A., T. Belser, G. Elgar, O. Georgiev, and W. Schaffner. 1999. Characterization of the transcription factor MTF-1 from the Japanese pufferfish (*Fugu rubripes*) reveals evolutionary conservation of heavy metal stress response. Biological Chemistry 380:175–185.

Deventer, K. 1996. Detection of genotoxic effects on cells of liver and gills of *B. rerio* by means of single cell gel electrophoresis. Bulletin of Environmental

Contamination and Toxicology 56:911–918.

Dey, W. P., T. H. Peck, C. E. Smith, and G.-L. Kreamer. 1993. Epizoology of hepatic neoplasia in Atlantic tomcod (*Microgadus tomcod*) from the Hudson River estuary. Canadian Journal of Fisheries and Aquatic Sciences 50:1897–1907.

Deyl, Z., F. Tagliaro, and F. Miksik. 1994. Biomedical applications of capillary electrophoresis. Journal of Chromatography B: Biomedical Applications 656:3–27.

Dopp, E., C. M. Barker, D. Schiffmann, and C. L. Reinisch. 1996. Detection of micronuclei in hemocytes of *Mya arenaria*: association with leukemia and induction with an alkylating agent. Aquatic Toxicology 34:31–45.

Doss, R. P. 1996. Differential display without radioactivity, a modified procedure. BioTechniques 21:408–412.

Eastman, A., and M. A. Barry. 1992. The origins of DNA breaks: a consequence of DNA damage, DNA repair, or apoptosis? Cancer Investigations 10:229–240.

Edwards, A. J., R. F. Addison, D. E. Willis, and K. W. Renton. 1988. Seasonal variation of hepatic mixed function oxidases in winter flounder (*Pseudopleuronectes americanus*). Marine Environmental Research 4:299–309.

Eller-Jessen, K., and J. F. Crivello. 1998. Subcutaneous NaAs3+ exposure increases metallothionein mRNA and protein expression in juvenile winter flounder. Aquatic Toxicology 42:301–320.

Elonen, G. E., R. L. Spehar, G. W. Holcombe, R. D. Johnson, J. D. Fernandez, R. J. Erickson, J. E. Tietge, and P. M. Cook. 1998. Comparative toxicity of 2,3,7,8-tetrachlorodibenzo-p-dioxin to seven freshwater fish species during early life-stage development. Environmental Toxicology and Chemistry 17:472–483.

Elskus, A. A., E. Monosson, A. E. McElroy, J. J. Stegeman, and D. S. Woltering. 1999. Altered CYP1A expression in *Fundulus heteroclitus* adults and larvae: a sign of pollutant resistance? Aquatic Toxicology 45:99–113.

Elskus, A. A., R. Pruell, and J. J. Stegeman. 1992. Endogenously-mediated, pretranslational suppression of cytochrome P4501A in PCB-contaminated flounder. Marine Environmental Research 34:97–101.

Elskus, A. A., J. J. Stegeman, L. C. Susani, D. Black, R. J. Pruell, and S. J. Fluck. 1989. Polychlorinated biphenyls concentration and cytochrome P-450E expression in winter flounder from contaminated environments. Marine Environmental Research 28:25–30.

Englander, E. W., G. H. Greeley Jr., G. Wang, J. R. Perez-Polo, and H.-M. Lee. 1999. Hypoxia-induced mitochondrial and nuclear DNA damage in the rat brain. Journal of Neuroscience Research 58:262–269.

Ericson, G., B. Liewenborg, C. Naef, L. Balk, and A. Goksoeyr. 1998. DNA adducts in perch, *Perca fluviatilis*, from a creosote contaminated site and in perch exposed to an organic solvent extract of creosote contaminated sediment. Marine Environmental Research 46:341–344.

Ericson, G., E. Noaksson, and L. Balk. 1999. DNA adduct formation and persistence in liver and extrahepatic tissues of northern pike (*Esox lucius*) following oral exposure to benzo[a]pyrene, benzo[k]fluoranthene and 7H-dibenzo[c,g]carbazole. Mutation Research 427:135–145.

Eufemia, N. A., and D. Epel. 2000. Induction of the multixenobiotic defense mechanism (MXR), P-glycoprotein, in the mussel *Mytilus californianus* as a general cellular response to environmental stress. Aquatic Toxicology 49:89–100.

Everaarts, J. M. 1995. DNA integrity as a biomarker of marine pollution: strand breaks in seastar (*Asterias rubens*) and dab (*Limanda limanda*). Marine Pollution Bulletin 31:431–438.

Everaarts, J. M., L. R. Shugart, M. K. Gustin, W. E. Hawkins, and W. W. Walker. 1991. Biological markers in fish: DNA integrity, hematological parameters and liver somatic index. Marine Environmental Research 35:101–107.

Fairbairn, D. W., P. L. Olive, and K. L. O'Neill. 1995. The comet assay: a comprehensive review. Mutation Research 339:37–59.

Farr, S., and R. T. Dunn, II. 1999. Concise review: Gene expression applied to toxicology. Toxicological Sciences 50:1–9.

Fent, K., and T. D. Bucheli. 1994. Inhibition of hepatic microsomal monooxygenase system by organotins in vitro in freshwater fish. Aquatic Toxicology 28:107–126.

Fent, K., and J. J. Stegeman. 1993. Effects of tributyltin in vivo on hepatic cytochrome P450 forms in marine fish. Aquatic Toxicology 24:219–240.

Fernandez-Salguero, P., T. Pineau, D. M. Hilbert, T. McPhail, S. S. T. Lee, S. Kimura, D. W. Nebert, S. Rudikoff, J. M. Ward, and F. J. Gonzalez. 1995. Immune system impairment and hepatic fibrosis in mice lacking the dioxin-binding Ah receptor. Science 268:722–726.

Fodde R., and M. Losekoot. 1994. Mutation detection by denaturing gradient gel electrophoresis. Human Mutation 3:83–94.

Folmar, L. C., N. D. Denslow, V. Rao, M. Chow, D. A. Crain, J. Enblom, J. Marcino, and L. J. Guillete Jr. 1996. Vitellogenin induction and reduced serum testosterone concentrations in feral male carp (*Cyprinus carpio*) captured near a major metropolitan sewage treatment plant. Environmental Health Perspectives 104:1096–1101.

Folmar, L. C., G. R. Gardner, M. P. Schreibman, L. Magliulo-Cepriano, L. J. Mills, G. Zaroogian, R. Gutjahr-Gobell, R. Haebler, D. B. Horowitz, and N. D. Denslow. 2001. Vitellogenin-induced pathology in male summer flounder (*Paralichthys dentatus*). Aquatic Toxicology 51:431–441.

Forlin, L., C. Haux, L. Karlsson-Norrgren, P. Runn, and A. Larsson. 1986. Biotransformation enzyme activities and histopathology in rainbow trout, *Salmon gairdneri*, treated with cadmium. Aquatic Toxicology 8:51–64.

Fornace, A. J., Jr. 1992. Mammalian genes induced by radiation; activation of genes associated with growth control. Annual Review of Genetics 26:507–526.

Fossi, M. C., L. Lari, S. Casini, N. Mattei, C. Savelli, J. C. Sanchez-Hernandez, S. Castellani, M. Depledge, S. Bamber, C. Walker, D. Savva, and O. Sparagano. 1996. Biochemical and genotoxic biomarkers in the Mediterranean crab *Carcinus aestuarii* experimentally exposed to polychlorobiphenyls, benzopyrene and methyl-mercury. Marine Environmental Research 42:29–32.

Freeman, S. E., and B. D. Thompson. 1990. Quantitation of ultraviolet radiation-induced cyclobutyl pyrimidine dimers in DNA by video and photographic densitometry. Analytical Biochemistry 186:222–228.

Freeman W. M., S. J. Walker, and K. E. Vrana. 1999. Quantitative RT-PCR: pitfalls and potential. 26:112–122.

Freidberg, E. C. 1985. DNA repair. Plenum, New York.

French, B. L., W. L. Reichert, T. Hom, M. Nishimoto, H. R. Sanborn, and J. E. Stein. 1996. Accumulation and dose-response of hepatic DNA adducts in English sole (*Pleuronectes vetulus*) exposed to a gradient of contaminated sediments. Aquatic Toxicology 36:1–16.

Fu, P. P., and D. Herreno-Saenz. 1999. Nitro-polycyclic aromatic hydrocarbons: a class of genotoxic environmental pollutants. Journal of Environmental Science and Health, Part C: Environmental Carcinogenesis and Ecotoxicology Reviews C17:1–43.

Fu, P. P., L. S. Von Tungeln, L.-H. Chiu, Li-Hsueh, and Z. Y. Own. 1999. Halogenated-polycyclic aromatic hydrocarbons: a class of genotoxic environmental pollutants. Journal of Environmental Science and Health, Part C: Environmental Carcinogenesis and Ecotoxicology Reviews C17:71–109.

Gauthier, L. 1996. The amphibian micronucleus test, a model for in vivo monitoring of genotoxic aquatic pollution. Alytes 14:53–84.

George, S. G. 1989. Cadmium effects on plaice liver xenobiotic and metal detoxification systems: dose-response. Aquatic Toxicology 15:303–310.

George, S. G., K. Todd, and J. Wright. 1996. Regulation of metallothionein in teleosts: induction of MT mRNA and protein by cadmium in hepatic and extraheptic tissues of a marine flatfish, the turbot (*Scophthalmus maximus*). Comparative Biochemistry and Physiology 113C:109–115.

Gibson, U. E. M., C. A. Heid, and P. M. Williams. 1996. A novel method for real time quantitative RT-PCR. Genome Research 6:995–1001.

Gooch, J. W., A. A. Elskus, P. J. Kloepper-Sams, M. E. Hahn, and J. J. Stegeman. 1989. Effects of ortho- and non-ortho substituted polychlorinated biphenyl congeners on the hepatic monooxygenase system in scup (*Stenotomus chrysops*). Toxicology and Applied Pharmacology 98:422–433.

Gray, E. S., B. R. Woodin, and J. J. Stegeman. 1991. Sex differences in hepatic monooxygenases in winter flounder (*Pseudopleuronectes americanus*) and scup (*Stenotomus chrysops*) and regulation of P450 forms by estradiol. Journal of Experimental Zoology 259:330–342.

Gronen, S., N. Denslow, S. Manning, S. Barnes, D. Barnes, and M. Brower. 1999. Serum vitellogenin levels and reproductive impairment of male Japanese medaka (*Oryzias latipes*) exposed to 4-tert-octylphenol. Environmental Health Perspectives 107:385–390.

Haasch, M. L., R. Prince, P. J. Wejksnora, K. R. Cooper, and J. J. Lech. 1993a. Caged and wild fish: induction of hepatic cytochrome P-450 (CYP1A1) as an environmental biomonitor. Environmental Toxicology and Chemistry 12:885–895.

Haasch, M. L., E. M. Quardokus, L. A. Sutherland, M. S. Goodrich, and J. J. Lech. 1993b. Hepatic CYP1A1 induction in rainbow trout by continuous flow-through exposure to β-naphthoflavone. Fundamental and Applied Toxicology 20:72–82.

Hacia J. G. 1999. Resequencing and mutational analysis using oligonucleotide microarrays. Nature Genetics 21(1 Supplement):42–47.

Haddad, J. J. E., R. E. Olver, and S. C. Land. 2000. Antioxidant/pro-oxidant equilibrium regulates HIF-1 α, and NF-κ B redox sensitivity. Evidence for inhibition by glutathione oxidation in alveolar epithelial cells. Journal of Biological Chemistry 275:21130–21139.

Hahn, M. E., 1998. Mechanisms of innate and acquired resistance to dioxin-like compounds. Reviews in Toxicology. Series B, Environmental Toxicology 2:395–443.

Hamza-Chaffai, A., C. Amhard-Triquet, and A. El-Abed. 1997. Metallothionein-like protein: is it an efficient biomarker of metal contamination? A case study based on fish from the Tunisian coast. Archives of Environmental Contamination and Toxicology 33:53–62.

Harvey, J. S., B. P. Lyons, T. S. Page, C. Stewart, and J. M. Parry. 1999. An assessment of the genotoxic impact of the Sea Empress oil spill by the measurement of DNA adduct levels in selected invertebrate and vertebrate species. Mutation Research/ Genetic Toxicology and Environmental Mutagenesis 441:103–114.

Harvey, J. S., B. P. Lyons, M. Waldock, and J. M. Parry. 1997. The application of the ^{32}P-postlabelling assay to aquatic biomonitoring. Mutation Research 378:77–88.

Harvey, J. S., and J. M. Parry. 1998. Application of the ^{32}P-postlabelling assay for the detection of DNA adducts: false positives and artifacts and their implications for environmental biomonitoring. Aquatic Toxicology 40:293–308.

Hashimoto, S., H. Bessho, A. Hara, M. Nakamura, T. Iguchi, and K. Fujita. 2000. Elevated serum vitellogenin levels and gonadal abnormalities in wild male floun-der (*Pleuronectes yokohamae*) from Tokyo Bay, Japan. Marine Environmental Research 49:37–53.

Herbet, A., and P.-D. Hansen. 1998. Genotoxicity in fish embryos. Pages 491–505 *in* P. G. Wells, K. Lee and C. Blaise, editors. Microscale testing in aquatic toxicol-ogy: advances, techniques, and practice. CRC Press, Boca Raton, Florida.

Hestermann, E. V., J. J. Stegeman, and M. E. Hahn. 2000. Relative contributions of affinity and intrinsic efficacy to aryl hydrocarbon receptor ligand potency. Toxi-cology and Applied Pharmacology 168:160–172.

Hogstrand, C., and C. Haux. 1990. A radioimmunoassay for perch (*Perca fluviatilis*) metallothionein. Toxicology and Applied Pharmacology 103:56–65.

Huang, Q., X. Wang, Y. Liao, L. Kong, S. Han, and L. Wang. 1995. Discriminant analysis of the relationship between genotoxicity and molecular structure of organochlorine compounds. Bulletin of Environmental Contamination and Toxi-cology 55:796–801.

Hughes, J. B. 1999. Cytological-cytogenetic analyses of winter flounder embryos. Marine Pollution Bulletin 38:30–35.

Hurks, H. M. H., C. Out-Luiting, B. J. Vermeer, F. H. J. Claas, and A. M. Mommaas. 1995. The action spectra for UV-induced suppression of MLR and MECLR show that immunosuppression is mediated by DNA damage. Photochemistry and Pho-tobiology 62:449–453.

Hylland, K., T. Nissen-Lie, P. G. Christensen, and M. Sandvik. 1998. Natural modula-tion of hepatic metallothionein and cytochrome P4501A in flounder, *Platichthys flesus* L. Marine Environmental Research 46:1–5.

Ivashuta, S., R. Imai, K. Uchiyama, and M. Gau. 1999. The coupling of differential display and AFLP approaches for nonradioactive mRNA fingerprinting. Molecu-lar Biotechnology 12:137–141.

Jenkins G. J., H. S. Suzen, R. A. Sueiro, and J. M. Parry. 1999. The restriction site mutation assay: a review of the methodology development and the current status of the technique. Mutagenesis 14:439–448.

Jha, A. N., V. V. Cheung, M. E. Foulkes, S. J. Hill, and M. H. Depledge. 2000. Detec-tion of genotoxins in the marine environment: adoption and evaluation of an integrated approach using the embryo-larval stages of the marine mussel, *Mytilus edulis*. Mutation Research/Genetic Toxicology and Environmental Mutagenesis 464:213–228.

Jones, N. J., and J. M. Parry. 1992. The detection of DNA adducts, DNA based changes and chromosome damage for the assessment of exposure to genotoxic pollutants. Aquatic Toxicology 22:323–344.

Jorgensen, E. H., B. E. Bye, and M. Jobling. 1999. Influence of nutritional status on biomarker responses to PCB in the Arctic charr (*Salvelinus alpinus*). Aquatic Toxicology 44:233–244.

Kampf, G., and K. Eichhorn. 1983. DNA strand breakage by different radiation qualities and relations to cell killing. Studies in Biophysics 93:17–26.

Kanaar, R., J. H. J. Hoeijmakers, and D. C. van Gent. 1998. Molecular mechanisms of DNA double-strand break repair. Trends in Cell Biology 8:483–489.

Kelly, S. A., C. M. Havrilla, T. C. Brady, K. H. Abramo, and E. D. Levin. 1998. Oxidative stress in toxicology: established mammalian and emerging piscine model systems. Environmental Health Perspectives 106:375–384.

Kirsch-Volders, M., A. Elhajouji, E. Cundari, and P. Van Hummelen. 1997. The in vitro micronucleus test: a multi-endpoint assay to detect simultaneously mitotic delay, apoptosis, chromosome breakage, chromosome loss and non-disjunction. Mutation Research/Genetic Toxicology and Environmental Mutagenesis 392:19–30.

Klein, C. B., and M. Costa. 1997. DNA methylation, heterochromatin and epigenetic carcinogens. Mutation Research/Reviews in Genetic Toxicology 386:163–180.

Kloepper-Sams, P. J., and J. J. Stegeman. 1992. Effects of temperature acclimation on the expression of hepatic cytochrome P4501A mRNA and protein in the fish *Fundulus heteroclitus*. Archives of Biochemistry and Biophysics 299:38–46.

Kohlpoth, M., B. Rusche, and M. Nuesse. 1999. Flow cytometric measurement of micronuclei induced in a permanent fish cell line as a possible screening test for the genotoxicity of industrial waste waters. Mutagenesis 14:397–402.

Krause, M. K., L. D. Rhodes, and R. J. Van Beneden. 1997. Cloning of the *p53* tumor suppressor gene from the Japanese medaka (*Oryzias latipes*) and evaluation of mutational hotspots in MNNG-exposed fish. Gene 189:101–106.

Kreamer, G.-L., K. Squibb, D. Gioeli, S. J. Garte, and I. Wirgin. 1991. Cytochrome P450IA mRNA expression in feral Hudson River tomcod. Environmental Research 55:64–78.

Kriek, E., M. Rojas, K. Alexandrov, and H. Bartsch. 1998. Polycyclic aromatic hydrocarbon-DNA adducts in humans: relevance as biomarkers for exposure and cancer risk. Mutation Research 400:215–231.

Kroczek R. A. 1993. Southern and northern analysis. Journal of Chromatography 618:133–145.

Kumar R., and E. B. Thompson. 1999. The structure of the nuclear hormone receptors. Steroids 64:310–319.

Kurelec, B., A. Garg, M. Krca, M. Chacko, and R. C. Gupta. 1989a. Natural environment surpasses polluted environment in inducing DNA damage in fish. Carcinogenesis 10:1337–1339.

Kurelec, B., A. Garg, S. Krca, and R. C. Gupta. 1989b. DNA adducts as biomarkers in genotoxic risk assessment in the aquatic environment. Marine Environmental Research 28:317–321.

Kurelec, B., S. Krca, and D. Lucci. 1996. Expression of multixenobiotic resistance mechanism in a marine mussel *Mytilus galloprovincalis* as a biomarker of exposure to polluted environments. Comparative Biochemistry and Physiology 113C:283–289.

Lam, K., P. W. Ko, J. K.-Y. Wong, and K. M. Chan. 1998. Metal toxicity and metallothionein gene expression studies in common carp and tilapia. Marine Environmental Research 46:563–566.

Lamb, T., J. W. Bickham, T. B. Lyne, and J. W. Gibbons. 1995. The slider turtle as an environmental sentinel: multiple tissue assays using flow cytometric analysis. Ecotoxicology 4:5–13.

Law J. M., M. Bull, J. Nakamura, and J. A. Swenberg. 1998. Molecular dosimetry of DNA adducts in the medaka small fish model. Carcinogenesis 19:515–518.

Lee, J.-S., E.-H. Park, and J. Choe. 1999. Absence of the c-Ha-ras and c-Ki-ras oncogene mutations in the hermaphroditic fish *Rivulus marmoratus* papillary thyroid carcinomas induced by N-methyl-N'-nitro-N-nitrosoguandiine. Journal of Applied Ichthyology 15:93–96.

Lee, J.-S., E.-H. Park, J. Choe, and J. K. Chipman. 2000. N-methyl-N-nitrosourea (MNU) induces papillary thyroid tumours which lack ras gene mutations in the hermaphorditic fish *Rivulus marmoratus*. Teratogenesis, Carcinogenesis and Mutagenesis 20:1–9.

Lewin, B. 2000. Genes VII. Oxford University Press, Oxford, U.K.

Liang, P., and A. B. Pardee. 1995. Recent advances in differential display. Current Opinion in Immunology 7:274–280.

Ma, A., and J. P. Whitlock Jr. 1996. The aromatic hydrocarbon receptor modulates the Hepa 1c1c7 cell cycle and differentiated state independent of dioxin. Molecular and Cellular Biology 16:2144–2150.

Malins, D. C., and R. Haimanot. 1991. The etiology of cancer: hydroxyl radical-induced DNA lesions in histologically normal livers of fish from a population with liver tumors. Aquatic Toxicology 20:123–130.

Malins, D. C., N. L. Polisar, M. M. Garner, and S. J. Gunselman. 1996. Mutagenic DNA base modifications are correlated with lesions in nonneoplastic hepatic tissue of the English sole carcinogenesis model. Cancer Research 56:5563–5565.

Mangold, K., Y. J. Chang, C. Mathews, K. Marien, J. Hendricks, and G. Bailey. 1991. Expression of *ras* genes in rainbow trout liver. Molecular Carcinogenesis 4:97–102.

Martin, L. K., Jr., and M. C. Black. 1998. Biomarker assessment of the effects of coal strip-mine contamination on channel catfish. Ecotoxicology and Environmental Safety 41:307–320.

Marty, J., J. E. Djomo, C. Bekaert, and A. Pfohl-Leszkowicz. 1998. Relationships between formation of micronuclei and DNA adducts and EROD activity in newts following exposure to benzo(a)pyrene. Environmental and Molecular Mutagenesis 32:397–405.

Masters, J. R. W., R. Thomas, A. G. Hall, L. Hogarth, E. C. Matheson, A. R. Cattan, and H. Lohrer. 1996. Sensitivity of testis tumor cells to chemotherapeutic drugs: role of detoxifying pathways. European Journal of Cancer 32A:1248–1253.

McCarthy, J. F., D. N. Jacobson, L. R. Shugart, and B. D. Jimenez. 1989. Pre-exposure to 3-methylcholanthrene increases benzo(a)pyrene adducts on DNA of bluegill sunfish. Marine Environmental Research 28:1–4.

McCreedy, C. D., J. P. Robinson, C. E. Dallas, and C. H. Jagoe. 1999. Pages 401–412 *in* D. S. Henshel, M. C. Black, and M. C. Harrass, editors. Quality control in the application of flow cytometry to studies of environmentally induced genetic damage. ASTM Special Technical Publication No. 1364. ASTM, Conshohocken, Pennsylvania.

McMahon, G., L. J. Huber, M. J. Moore, J. J. Stegeman, and G. N. Wogan. 1990. Mutations in c-k-*ras* oncogenes in diseased livers of winter flounder from Boston Harbor. Proceedings of the National Academy of Sciences USA 87:841–845.

Memisoglu, A., and L. Samson. 2000. Base excision repair in yeast and mammals. Mutation Research 451:39–51.

Michailova, P., N. Petrova, G. Sella, L. Ramella, and S. Bovero. 1998. Structural-functional rearrangements in chromosome G in *Chironomus riparius* (Diptera, Chironomidae) collected from a heavy metal-polluted area near Turin, Italy. Environmental Pollution 103:127–134.

Minier, C., F. Akcha, and F. Galgani. 1993. P-glycoprotein expression in *Crassotrea gigas* and *Mytilus edulis* in polluted seawater. Comparative Biochemistry and Physiology 106B:1029–1036.

Mitchelmore, C. L., C. Birmelin, D. R. Livingston, and J. K. Chipman. 1998. Detection of DNA strand breaks in isolated mussel (*Mytilus edulis* L.) digestive gland cells using the "comet" assay. Ecotoxicology and Environmental Safety 41:51–58.

Mitchelmore, C. L., and J. K. Chipman. 1998. DNA strand breakage in aquatic organisms and the potential value of the comet assay in environmental monitoring. Mutation Research 399:135–147.

Monosson, E., and J. J. Stegeman. 1994. Induced cytochrome P4501A in winter flounder, *Pleuronectes americanus*, from offshore and coastal sites. Canadian Journal of Fisheries and Aquatic Sciences 51:933–941.

Morgan, M. B., T. W. Snell, and D. L. Vogelien. 2001. Assessing coral stress responses using molecular biomarkers of gene transcription. Environmental Toxicology and Chemistry 20:537–543.

Morizot D. C., B. B. McEntire, L. Della Coletta, S. Kazianis, M. Schartl, and R. S. Nairn. 1998. Mapping of tyrosine kinase gene family members in a *Xiphophorus* melanoma model. Molecular Carcinogenesis 22:150–157.

Murcelano, R. A., and R. E. Wolke. 1985. Epizootic carcinoma in the winter flounder, *Pseudopleuronectes americanus*. Science 228:587–589.

Myers, M. S., L. L. Johnson, T. Hom, T. K. Collier, J. E. Stein, and U. Varanasi. 1998. Toxicopathic hepatic lesions in subadult English sole (*Pleuronectes vetulus*) from Puget Sound, Washington, USA: relationships with other biomarkers of contaminant exposure. Marine Environmental Research 45:47–67.

Nacci, D., L. Corio, D. Champlin, S. Jayaraman, R. McKinney, T. R. Gleason, W. R. Munns, J. L. Specker, and K. R. Cooper. 1999. Adaptations of wild populations of the estuarine fish *Fundulus heteroclitus* to persistent environmental contaminants. Marine Biology 134:9–17.

Nacci, D. E., S. Cayula, and E. Jackim. 1996. Detection of DNA damage in individual cells from marine organisms using the single cell gel assay. Aquatic Toxicology 35:197–210.

Nataraj, A. J., I. Olivos-Glander, N. Kusukawa, and W. E. Highsmith Jr. 1999. Single-strand conformation polymorphism and heteroduplex analysis for gel-based mutation detection. Electrophoresis 20:1177–1185.

Navarro J. M., and J. L. Jorcano. 1999. The use of arbitrarily primed polymerase chain reaction in cancer research. 20:283–290.

Nemoto, N., K.-I. Kodama, A. Tazawa, P. Masahito, and T. Ishikawa. 1986. Extensive sequence homology of the goldfish *ras* gene to mammalian *ras* genes. Differentiation 32:17–23.

Oleksiak, M. F., K. J. Kolell, and D. L. Crawford. 2001. Utility of natural populations for microarray analyses: isolation of genes necessary for functional genomic studies. Marine Biotechnology 3(Supplement 1):S203–S211.

Olsson, P. E. 1996. Metallothioneins in fish: induction and use in environmental monitoring. Pages 187–203 *in* E. W. Taylor, editor. Toxicology of aquatic pollution: physiological, molecular, and cellular approaches. Cambridge University Press, U.K.

Olsson, P. E., and P. Kille. 1997. Functional comparison of the metal-regulated transcriptional control regions of metallothionein genes from cadmium-sensitive and tolerant fish species. Biochimica et Biophysica Acta 1350:325–334.

Olsson, P. E., P. Kling, L. J. Erkell, and P. Kille. 1995. Structural and functional analysis of the rainbow trout (*Oncorhynchus mykiss*) metallothionein-A gene. European Journal of Biochemistry 230:344–349.

Orlando, E. F., N. D. Denslow, L. C. Folmar, and L. J. Guillette Jr. 1999. A comparison of the reproductive physiology of largemouth bass, *Micropterus salmoides*, collected from the Escambia and Blackwater rivers in Florida. Environmental Health Perspectives 107:199–204.

Pals, G., K. Pindolia, and M. J. Worsham. 1999. A rapid and sensitive approach to mutation detection using real-time polymerase chain reaction and melting curve analyses, using BRCA1 as an example. Molecular Diagnosis 4:241–246.

Parsons B. L., and R. H. Heflich. 1997. Genotypic selection methods for the direct analysis of point mutations. Mutation Research/Reviews in Mutation Research 387:97–121.

Payne, C. M., C. Bernstein, and H. Bernstein. 1995. Apoptosis overview emphasizing the role of oxidative stress, DNA damage and signal-transduction pathways. Leukemia and Lymphoma 19:43–93.

Peck-Miller, K. A., M. Meyers, T. K. Collier, and J. E. Stein. 1998. Complete cDNA sequence of the Ki-*ras* proto-oncogene in the liver of wild English sole (*Pleuronectes vetulus*) and mutation analysis of hepatic neoplasms and other toxicopathic liver lesions. Molecular Carcinogenesis 23:207–216.

Phillips, D. H., M. Castegnaro, and H. Bartsch, editors. 1993. Postlabelling methods for detection of DNA adducts. IARC Scientific Publications No. 124. International Agency for Research on Cancer, Lyon, France.

Pickering, R. W. 1999. A toxicological review of polycyclic aromatic hydrocarbons. Journal of Toxicology: Cutaneous and Ocular Toxicology 18:101–135.

Pieper, A. A., A. Verma, J. Zhang, and S. H. Snyder. 1999. Poly(ADP-ribose) polymerase, nictric oxide and cell death. Trends in Phamacological Sciences 20:171–181.

Ploch, S. A., L. C. King, R. T. Di Giulio, and A. Goksoeyr. 1998. Comparative time-course of benzo[a]pyrene-DNA adduct formation, and its relationship to CYP1A activity in two species of catfish. Marine Environmental Research 46:345–349.

Powell J. F., S. L. Krueckl, P. M. Collins, and N. M. Sherwood. 1996. Molecular forms of GnRH in three model fishes: rockfish, medaka and zebrafish. Journal of Endocrinology 150:17–23.

Powell, W. H., R. Bright, S. M. Bello, and M. E. Hahn. 2000. Developmental, and tissue-specific expression of AHR1, AHR2, and ARNT2 in dioxin-sensitive, and -resistant populations of the marine fish *Fundulus heterolclitus*. Toxicological Sciences 57:229–239.

Prince, R., and K. R. Cooper. 1995a. Comparisons of the effects of 2,3,7,8-tetrachlordibenzo-p-dioxin on chemically impacted and nonimpacted subpopulations of *Fundulus heteroclitus*. I. TCDD toxicity. Environmental Toxicology and Chemistry 14:579–588.

Prince, R., and K. R. Cooper. 1995b. Comparisons of the effects of 2,3,7,8-tetrachlorodibenzo-p-dioxin on chemically impacted and nonimpacted subpopulations of *Fundulus heteroclitus*. II. Metabolic considerations. Environmental Toxicology and Chemistry 14:589–596.

Ralph, S., and M. Petras. 1998. Caged amphibian tadpoles and in situ genotoxicity monitoring of aquatic environments with the alkaline single cell gel electrophoresis (comet) assay. Mutation Research/Genetic Toxicology and Environmental Mutagenesis 313:235–250.

Rao, S. S., T. Neheli, J. H. Carey, and V. W. Cairns. 1997. Fish hepatic micronuclei as an indication of exposure to genotoxic environmental contaminants. Environmental Toxicology and Water Quality 12:217–222.

Rao, S. S., T. A. Neheli, J. H. Carey, A. Herbert, and P. D. Hansen. 1996. DNA alkaline unwinding assay for monitoring the impact of environmental genotoxins. Environmental Toxicology and Water Quality 11:351–354.

Reichert, W. L., M. S. Myers, K. Peck-Miller, B. French, B. F. Anulacion, T. K. Collier, J. E. Stein, and U. Varanasi. 1998. Molecular epizootiology of genotoxic events in marine fish: linking contaminant exposure, DNA damage, and tissue-level alterations. Mutation Research 411:215–225.

Ren, J. 2000. High-throughput single-strand conformation polymorphism analysis by capillary electrophoresis. Journal of Chromatography B: Biomedical Science Applications 741:115–128.

Rhodes, L. D., and R. T. J. Van Benden. 1996. Gene expression analysis in aquatic animals using differential display polymerase chain reaction. Pages 161–183 *in* G. Ostrander, editor. Techniques in aquatic toxicology. CRC Press, Boca Raton, Florida.

Rice, S. D., R. E. Thomas, R. Heintz, A. Moles, M. Carls, M. Murphy, J. W. Short, and A. Wertheimer. 1999. Synthesis of long term impacts to pink salmon following the Exxon Valdez oil spill: persistence, toxicity, sensitivity, and controversy. Final Report Project 99329 submitted to Exxon Valdez Trustee Council.

Risso-de Faverney, C., M. Lafaurie, J.-P. Girard, and R. Rahmani. 2000. Effects of heavy metals, and 3-methylcholanthrene on expression, and induction of CYP1A1 and metallothionein levels in trout (*Oncorhynchus mykiss*) hepatocyte cultures. Environmental Toxicology and Chemistry 19:2239–2248.

Robertson, K. D., and P. A. Jones. 2000. DNA methylation: past, present, and future directions. Carcinogenesis 21:462–467.

Rodriguez-Ariza, A., J. Alhama, F. M. Diaz-Mendez, and J. Lopez-Barea. 1999. Content of 8-oxodG in chromosomal DNA of *Sparus aurata* fish as a biomarker of oxidative stress and environmental pollution. Mutation Research/Genetic Toxicology and Environmental Mutagenesis 438:97–107.

Roesijadi, G. 1999. The basis for increased metallothionein in a natural population of *Crassostrea virginica*. Biomarkers 4:467–472.

Rotchell, J. M., J. B. Blair, J.-K. Shim, W. E. Hawkins, and G. K. Ostrander. 2001a. Cloning of the retinoblastoma cDNA from the Japanese medaka (*Oryzias latipes*) and preliminary evidence of mutational alterations in chemically induced retinoblastomas. Gene 263:231–237.

Rotchell, J. M., J. A. Craft, and R. M. Stagg. 1995. Chemically-induced genetic damage in fish: isolation and characterization of the dab (*Limanda limanda*) Ras gene. Marine Pollution Bulletin 31:457–459.

Rotchell, J. M., J.-S. Lee, J. K. Chipman, and G. K. Ostrander. 2001b. Structure, expression and activation of fish *ras* genes. Aquatic Toxicology 55:1–21.

Rotchell, J. M., E. Ulnal, R. J. Van Beneden, and G. K. Ostrander. 2001c. Retinoblastoma gene mutations in chemically induced liver tumor samples of Japanese medaka (*Oryzias latipes*). Marine Biotechnology 3(Supplement 1):S44–S49.

Roy, D., and J. G. Liehr. 1999. Estrogen, DNA damage, and mutations. Mutation Research 424:107–115.

Roy, N. K., S. Courtenay, Z. Yuan, M. Ikonomou, and I. Wirgin. 2001. An evaluation of the etiology of reduced CYP1A1 messenger RNA expression in the Atlantic tomcod from the Hudson River, New York, using reverse transcriptase polymerase chain reaction analysis. Environmental Toxicology and Chemistry 20:1022–1030.

Roy, N. K, G.-L. Kreamer, B. Konkle, C. Grunwald, and I. Wirgin. 1995. Characterization and prevalence of a polymorphism in the 3' untranslated region of cytochrome P4501A1 in cancer-prone Atlantic tomcod. Archives of Biochemistry and Biophysics 322:204–213.

Roy, N. K., J. Stabile, J. E. Seeb, C. Habicht, and I. Wirgin. 1999. High frequency of K-*ras* mutations in pink salmon embryos experimentally exposed to Exxon Valdez oil. Environmental Toxicology and Chemistry 18:1521–1528.

Roy, N. K., and I. Wirgin. 1997. Characterization of the aromatic hydrocarbon receptor gene and its expression in Atlantic tomcod. Archives of Biochemistry and Biophysics 344:373–386.

Sahu, S. C. 1991. Role of oxygen free radicals in the molecular mechanisms of carcinogenesis: a review. Journal of Environmental Science and Health, Part C: Environmental Carcinogenesis and Ecotoxicology Reviews C9:83–112.

Salazar, J. J., and B. Van Houten. 1997. Preferential mitochondrial DNA injury caused by glucose oxidase as a steady generator of hydrogen peroxide in fibroblasts. Mutation Research 385:139–149.

Sanders, B. M. 1993. Stress proteins in aquatic organisms: an environmental perspective. Critical Reviews in Toxicology 23:49–75.

Schlenk, D., M. Chelius, L. Wolford, S. Khan, and K. M. Chan. 1997a. Characterization of hepatic metallothionein expression in channel catfish (*Ictalurus punctatus*) by reverse-transcriptase polymerase chain reaction. Biomarkers 2:161–167.

Schlenk, D., W. C. Colley, A. El-Alfy, R. Kerby, and B. R. Griffin. 2000. Effects of the oxidant potassium permanganate on the expression of gill metallothionein mRNA and its relationship to sublethal whole animal endpoints in channel catfish. Toxicological Sciences 54:177–182.

Schlenk, D., E. J. Perkins, G. Hamilton, Y. S. Zhang, and W. Layher. 1996. Correlation of hepatic biomarkers with whole animal and population-community metrics. Canadian Journal of Fisheries and Aquatic Sciences 53:2299–2309.

Schlenk, E., L. Wolford, M. Chelius, J. Steevens, and K. M. Chan. 1997b. Effect of arsenite, arsenate, and the herbicide monosodium methyl arsonate (MSMA) on hepatic metallothionein expression and lipid peroxidation in channel catfish. Comparative Biochemistry and Physiology 118C:177–183.

Schlezinger, J. J., J. Keller, L. A. Verbrugge, and J. J. Stegeman. 2000. 3,3',4,4'-tetrachlorobiphenyl oxidation in fish, bird and reptile species: relationship to cytochrome P450 1A inactivation and reactive oxygen production. Comparative Biochemistry and Physiology Part C 125:273–286.

Schlezinger, J. J., R. D. White, and J. J. Stegeman. 1999. Oxidative inactivation of cytochrome P-450 1A (CYP1A) stimulated by 3,3',4,4'-tetrachlorobiphenyl: production of reactive oxygen by vertebrate CYP1As. Molecular Pharmacology 56:588–597.

Schmid W. 1976. The micronucleus test for cytogenetic analysis. Pages 31–53 *in* A. Hollander, editor. Chemical mutagens, principles and methods for their detection, volume 6. Plenum, New York.

Schmidt, J. V., G. H.-T. Su, J. K. Reddy, M. C. Simon, and C. A. Bradfield. 1996. Characterization of a murine AhR null allele: involvement of the Ah receptor in hepatic growth and development. Proceedings of the National Academy of Sciences USA 93:6731–6736.

Schwartz, D., and V. Rotter. 1999. p53-dependent cell cycle control: response to genotoxic stress. Seminar Cancer Biology 8:325–336.

Sharma, V., M. Xu, J. Vail, and R. Campbell. 1998. Comparative analysis of multiple techniques for semiquantitation. Biotechnology Techniques 12:521–524.

Shigenaga, M. K., and B. N. Ames. 1991. Assays for 8-hydroxy-2'-deoxyguanosine: a biomarker of in vivo oxidative DNA damage. Free Radical and Biology Medicine 10:211–216.

Shimada T., C. L. Hayes, H. Yamazaki, S. Amin, S. S. Hecht, F. P. Guengerich, and T. R. Sutter. 1996. Activation of chemically diverse procarcinogens by human cytochrome P-450 1B1. Cancer Research 56:2979–2984.

Shugart, L. R. 1988. Quantification of chemically induced damage to DNA of aquatic organisms by the alkaline unwinding assay. Aquatic Toxicology 13:43–52.

Shugart L. R. 1990. 5-methyl deoxycytidine content of DNA from bluegill sunfish *Lepomis macrochirus* exposed to benzo[a]pyrene. Environmental Toxicology and Chemistry 9:205-208.

Shugart, L. R. 1998. Structural damage to DNA in response to toxicant exposure. Pages 151–168 *in* V. E. Forbes, editor. Genetics and ecotoxicology. Taylor and Francis, Philadelphia.

Shugart, L. R. 2000. DNA as a biomarker of exposure. Ecotoxicology 9:329–340.

Shugart, L. R., M. K. Gustin, D. M. Laird, and D. A. Dean. 1989. Susceptibility of DNA in aquatic organisms to strand breakage: effect of X-rays and gamma radiation. Marine Environmental Research 28:339–343.

Shugart, L., J. McCarthy, B. Jimenez, and J. Daniels. 1987. Analysis of adduct formation in the bluegill sunfish (*Lepomis macrochirus*) between benzo(a)pyrene and DNA of the liver and hemoglobin of the erythrocyte. Aquatic Toxicology 9:319–325.

Shugart, L. R., and C. Theodorakis. 1996. Genetic ecotoxicology: the genotypic diversity approach. Comparative Biochemistry and Physiology 113C:273–276.

Shulte-Frohlinde, D., and C. Von Sonntag. 1985. Radiolysis of DNA and model systems in the presence of oxygen. Pages 11–40 *in* H. Sies, editor. Oxidative stress. Academic Press, London.

Simic, M. G., L. Grossman, and A. D. Upton. 1986. Mechanisms of DNA damage and repair: implications for carcinogenesis. Plenum, New York.

Singh, N. P. 1996. Microgel electrophoresis of DNA from individual cells: principles and methodology. Pages 3–24 *in* G. P. Pfiefer, editor. Technologies for detection of DNA damage and mutations. Plenum, New York.

Singh, N. P., M. T. McCoy, R. R. Tice, and E. L. Schneider. 1988. A simple technique for quantification of low levels of DNA damage in individual cells. Experimental Cell Research 175:184–191.

Smital, T., R. Sauerborn, B. Pivcevic, B. Krca, and B. Kurelec. 2000. Interspecific differences in P-glycoprotein mediated activity of multixenobiotic resistance mechanism in several marine and freshwater invertebrates. Comparative Biochemistry and Physiology Part C 126:175–186.

Snow, E. T. 1992. Metal carcinogenesis: mechanistic implications. Pharmacology Therapeutics Journal 53:31–65.

Sovocool G. W., W. C. Brumley, and J. R. Donnelly. 1999. Capillary electrophoresis and capillary electrochromatography of organic pollutants. Electrophoresis 20:3297–3310.

Stegeman, J. J., and M. E. Hahn. 1994. Biochemistry and molecular biology of monooxygenases: current perspectives on forms, functions, and regulation of cytochrome P450 in aquatic species. Pages 187–206 *in* D. C. Malins and G. K. Ostrander, editors. Aquatic toxicology: molecular, biochemical, and cellular perspectives. Lewis Publishers, Boca Raton, Florida.

Stegeman, J. J., F. Y. Teng, and E. A. Snowberger. 1987. Induced cytochrome P450 in winter flounder (*Pseudopleuronectes americanus*) from coastal Massachusetts evaluated by catalytic assay and monoclonal antibody probes. Canadian Journal of Fisheries and Aquatic Sciences 44:1270–1277.

Stein, J., W. Reichert, and U. Varanasi. 1994. Molecular epizootiology: assessment of exposure to genotoxic compounds in teleosts. Environmental Health Perspectives 102(Supplement 12):19–23.

Steinert, S. A., R. Streib-Montee, J. M. Leather, and D. B. Chadwick. 1998. DNA damage in mussels at sites in San Diego Bay. Mutation Research 399:65–85.

Stenman, J., P. Finne, A. Stahls, R. Grenman, U.-H. Stenman, A. Palotie, and A. Orpana. 1999. Accurate determination of relative messenger RNA levels by RT-PCR. Nature Biotechnology 17:720–722.

Sugg, D. W., J. W. Bickham, J. A. Brooks, M. D. Lomakin, C. H. Jagoe, C. E. Dallas, M. H. Smith, R. J. Baker, and R. K. Chesser. 1996. DNA damage and radiocesium. Environmental Toxicology and Chemistry 15:1057–1063.

Sultan, A., A. Abelson, V. Bresler, L. Fishelsom, and O. Mokady. 2000. Biomonitoring marine environmental quality at the level of gene-expression: testing the feasibility of a new approach. Water Science and Technology 42:269–274.

Sumpter, J. P., and S. Jobling. 1995. Vitellogenesis as a biomarker for estrogenic contamination of the aquatic environment. Environmental Health Perspectives 103(Supplement 7):173–178.

Tawata M., K. Aida, and T. Onaya. 2000. Screening for genetic mutations. A review. Combinatorial Chemistry and High Throughput Screening 3:1–9.

Theodorakis, C. W. 2001. Integration of genotoxic, and population genetic end-points in biomonitoring, and risk assessment. Ecotoxicology 10:227–236.

Theodorakis, C. W., J. W. Bickham, T. Lamb, P.A Medica, and T. B. Lyne. 2001. Integration of genotoxicity, and population genetic analysis in kangaroo rats (*Dipodomys merriami*) exposed to radionuclide contamination. Environmental Toxicology and Chemistry 20:317–326.

Theodorakis, C. W., B. G. Blaylock, and L. R. Shugart. 1996. Genetic ecotoxicology I.: DNA integrity and reproduction in mosquitofish exposed in situ to radionuclides. Ecotoxicology 5:1–14.

Theodorakis, C. W., S. J. D'Surey, J. W. Bickham, T. B. Lyne, B. P. Bradley, W. E. Hawkins, W. L. Farkas, J. F. McCarthy, and L. R. Shugart. 1992. Sequential expression of biomarkers in bluegill sunfish exposed to contaminated sediment. Ecotoxicology 1:45–73.

Theodorakis, C. W., S. J. D'Surey, and L. R. Shugart. 1994. Detection of genotoxic insult as DNA strand breaks in fish blood cells by agarose gel electrophoresis. Environmental Toxicology and Chemistry 13:1023–1031.

Theodorakis, C. W., and L. R. Shugart. 1998. Natural selection in contaminated habitats: a case study using RAPD genotypes. Pages 123–150 *in* V. E. Forbes, editor. Genetics and ecotoxicology. Taylor and Francis, Philadelphia.

Theodorakis, C. W., C. D. Swartz, W. J. Rogers, J. W. Bickham, K. C. Donnelly, and S. M. Adams. 2000. Relationship between genotoxicity, mutagenicity, and fish community structure in a contaminated stream. Journal of Aquatic Ecosystem Stress and Recovery 7:131–143.

Tom, M., E. Jakubov, B. Rinkevich, and B. Herut. 1999. Monitoring of hepatic metallothionein mRNA levels in the fish *Lithognathus mormyrus*: evaluation of transition metal pollution in a Mediterranean coast. Marine Pollution Bulletin 38:503–508.

Torten, M., Z. Liu, M. S. Okihiro, S. J. Teh, and D. E. Hinton. 1996. Induction of *ras* oncogene mutations and hepatocarcinogenesis in medaka (*Oryzias latipes*) exposed to diethylnitrosoamine. Marine Environmental Research 42:93–98.

Van Beneden, R. J., and G. K. Ostrander. 1994. Expression of oncogenes and tumor suppressor genes in teleost fish. Pages 295–326 *in* D. C. Malins and G. K. Ostrander, editors. Aquatic toxicology: molecular, biochemical, and cellular perspectives. Lewis Publishers, Boca Raton, Florida.

Van Beneden, R. J., L. D. Rhodes, G. R. Gardner, and A. Goksøyr. 1998. Studies of the molecular basis of gonadal tumors in the marine bivalve, *Mya arenaria*. Marine Environmental Research 46:209–213.

Van Beneden, R. J., L. D. Rhodes, and G. R. Gardner. 1999. Potential alterations in gene expression associated with carcinogen exposure in *Mya arenaria*. Biomarkers 4:485–491.

Van den Hurk, P., M. Faisal, and M. H. Roberts Jr. 1998a. Interaction of cadmium and benzo[a]pyrene in mummichog (*Fundulus heteroclitus*): effects on acute mortality. Marine Environmental Research 46:525–528.

Van den Hurk, P., M. H. Roberts Jr., and M. Faisal. 1998b. Interaction of cadmium and benzo[a]pyrene in mummichog (*Fundulus heteroclitus*): biotransformation in isolated hepatocytes. Marine Environmental Research 46:529–532.

van der Houven van Oordt, C. W., T. G. Schouten, A. J. vand der Eb, and M. L. Breuer. 1999. Differentially expressed transcripts in X-ray-induced lymphomas by dioxygenin-labeled differential display. Molecular Carcinogenesis 24:29–35.

van Hal, N. L. W., O. Vorst, A. M. M. L. van Houwelingen, E. J. Kok, A. Peijnenburg, A. Aharoni, A. J. van Tunen, and J. Keijer. 2000. The application of DNA microarrays in gene expression analysis. Journal of Biotechnology 78:271–280.

Van Veld, P. A., W. K. Vogelbein, R. Smolowitz, B. R. Woodin, and J. J. Stegeman. 1992. Cytochrome P450IAI in hepatic lesions of a teleost fish (*Fundulus heteroclitus*) collected from a polycyclic aromatic hydrocarbon-contaminated site. Carcinogenesis 13:505–507.

Van Veld, P. A., and D. J. Westbrook. 1995. Evidence for depression of cytochrome P4501A in a population of chemically resistant mummichog (*Fundulus heteroclitus*). Environmental Sciences 3:221–234.

Van Zeeland, A. A. 1996. Molecular dosimetry of chemical mutagens. Relationship between DNA adduct formation and genetic changes analyzed at the molecular level. Mutation Research 353:123–150.

Varanasi, U., W. L. Reichert, and J. E. Stein. 1989. ^{32}P-postlabeling analysis of DNA adducts in liver of wild English sole (*Parophrys vetulus*) and winter flounder (*Pseudopleuronectes americanus*). Cancer Research 49:1171–1177.

Viarengo, A., B. Burlando, F. Dondero, A. Marro, and R. Fabbri. 1999. Metallothionein as a tool in biomonitoring programmes. Biomarkers 6:455–466.

Vijayaraghavan, M., and B. Nagarajan. 1994. Mutagenic potential of acute exposure to organophosphorus and organochlorine compounds. Mutation Research 321:103–111.

Vincent, F., J. de Boer, A. Pfohl-Leszkowicz, Y. Cherrel, and F. Galgani. 1998. Two cases of *ras* mutation associated with liver hyperplasia in dragonets (*Callionymus lyra*) exposed to polychlorinated biphenyls and polycyclic aromatic hydrocarbons. Molecular Carcinogenesis 21:121–127.

Vincent-Hubert, F. 2000. cDNA cloning and expression of two Ki-*ras* genes in the flounder, *Platichythus flesus*, and analysis of hepatic neoplasms. Comparative Biochemistry and Physiology 126:17–27.

Vinogradov, A. E., and A. T. Chubinishvili. 1999. Genome reduction in a hemiclonal frog *Rana esculenta* from radioactively contaminated areas. Genetics 151:1123–1125.

Vogelbein, W. K., J. W. Fournie, P. A. Van Veld, and R. J. Huggett. 1990. Hepatic neoplasms in the mummichog *Fundulus heteroclitus* from a creosote-contaminated site. Cancer Research 50:5978–5986.

Vrana, K. E. 1996. Advancing technologies in gene amplification. Trends in Biotechnology 14:413–415.

Vu, H. L., S. Troubetzkoy, H. H. Nguyen, M. W. Russell, and J. Mestecky. 2000. A method for quantification of absolute amounts of nucleic acids by (RT)-PCR, and a new mathematical model for data analysis. Nucleic Acids Research 28:E18–E18.

Wagner, J. R., C. C. Hu, and B. N. Ames. 1992. Endogenous oxidative damage of deoxycytidine in DNA. Proceedings of the National Academy of Sciences USA 89:3380–3384.

Wall, K. L., and J. Crivello. 1999. Effects of starvation on liver microsomal P450 activity in juvenile *Pleuronectes americanus*. Comparative Biochemistry and Physiology Part C 123:273–277.

Wang, J. S., and J. D. Groopman. 1999. DNA damage by mycotoxins. Mutation Research 424:167–181.

Ward, J. F. 1988. DNA damage produced by ionizing radiation in mammalian cells: identities, mechanisms of formation, and repairability. Progress in Nucleic Acids Research and Molecular Biology 35:95–125.

Watson, A., A. Mazumder, M. Stewart, and S. Balasubramanian. 1998. Technology for microarray analysis of gene expression. Current Opinion in Biotechnology 9:609–614.

Whittier, J. B., and K. McBee. 1999. Use of flow cytometry to detect genetic damage in mallards dosed with mutagens. Environmental Toxicology and Chemistry 18:1557–1563.

Wickliffe, J. K., and J. W. Bickham. 1998. Flow cytometric analysis of hematocytes from brown pelicans (*Pelecanus occidentalis*) exposed to planar halogenated hydrocarbons and heavy metals. Bulletin of Environmental Contamination and Toxicology 61:239–246.

Willett, K. L., S. J. McDonald, M. A. Steinberg, K. B. Beatty, M. C. Kennicutt, and S. H. Safe. 1997. Biomarker sensitivity for polynuclear aromatic hydrocarbon contamination in two marine fish species collected in Galveston Bay, Texas. Environmental Toxicology and Chemistry 16:1472–1479.

Willett, K. L., D. M. Wassenberg, L. A. Lienesch, W. L. Reichert, and R. T. DiGiulio. 2000. In vivo inhibition of CYP1A by the PAH fluoranthene. SETAC, Nashville, Tennessee, November 12–16, 2000.

Williams, P. J., S. C. Courtenay, and C. E. Wilson. 1998. Annual sex steroid profiles and effects of gender and season on cytochrome P450 mRNA induction in Atlantic tomcod (*Microgadus tomcod*). Environmental Toxicology and Chemistry 17:1582–1588.

Winn, R. N., M. Norris, S. Muller, C. Torres, and K. Brayer. 2001. Bacterophage and plasmid pUR288 transgenic fish models for detecting in vivo mutations. Marine Biotechnology 3(Supplement 1):S185–S195.

Wirgin, I. I., D. C. Currie, and S. J. Garte. 1989. Activation of the K-*ras* oncogene in liver tumors of Hudson River tomcod. Carcinogenesis 10:2311–2315.

Wirgin, I. I., C. Grunwald, S. Courtenay, G.-L. Kreamer, W. L. Reichert, and J. E. Stein. 1994. A biomarker approach to assessing xenobiotic exposure in Atlantic tomcod from the North American Atlantic coast. Environmental Health Perspectives 102:764–770.

Wirgin, I. I., B. Konkle, M. Pedersen, C. Grunwald, P. J. Williams, and S. Courtenay. 1996. Interspecific differences in cytochrome P4501A mRNA inducibility in four species of Atlantic coast anadromous fishes. Estuaries 19:913–922.

Wirgin, I. I., G.-L. Kreamer, C. Grunwald, K. Squibb, S. J. Garte, and S. Courtenay. 1992. Effects of prior exposure history on cytochrome P450IA mRNA induction by PCB congener 77 in Atlantic tomcod. Marine Environmental Research 34:103–108.

Wirgin, I. I., and J. R. Waldman. 1998. Altered gene expression and genetic damage in North American fish populations. Mutation Research/Fundamental and Molecular Mechanisms of Mutagenesis 399:193–219.

Wlodek, D., J. Banath, and P. L. Olive. 1991. Comparison between pulsed-field and constant-field gel electrophoresis for measurement of DNA double-strand breaks in irradiated Chinese hamster ovary cells. International Journal of Radiation Biology 60:779–790.

Wong, C. K. C., H. Y. Yeung, R. Y. H. Cheung, K. K. L. Yung, and M. H. Wong. 2000. Ecotoxicological assessment of persistent organic and heavy metal contamination in Hong Kong coastal sediment. Archives of Environmental Contamination and Toxicology 38:486–493.

Wood, R. D., and M. K. K. Shivji. 1997. Which DNA polymerases are used for DNA-repair in eukaryotes? Carcinogenesis 18:605–610.

Xiang, C. C., and Y. Chen. 2000. cDNA microarray technology and its applications. Biotechnology Advances 18:35–46.

Yakes, F. M., and B. Van Houten. 1997. Mitochondrial DNA damage is more extensive and persists longer than nuclear DNA damage in human cells following oxidative stress. Proceeding of the National Academy of Sciences USA 94:514–519.

Yuan, Z., M. Wirgin, S. Courtenay, M. Ikonomou, and I. Wirgin. 2001. Is hepatic cytochrome P4501A1 expression predictive of dioxins, furans, and PCBs in Atlantic tomcod from the Hudson River estuary? Aquatic Toxicology 54:217–230.

4

Physiological and Condition-Related Indicators of Environmental Stress in Fish

Bruce A. Barton, John D. Morgan, and Mathilakath M. Vijayan

Introduction

Stress has been defined as "the nonspecific response of the body to any demand made upon it" (Selye 1973). Brett (1958) extended Selye's definition in a fisheries context to include that the response to stress causes an extension of a physiological condition beyond its normal resting state to the point that the chances of survival may be reduced. This is a useful working definition as it incorporates both the notion of a physiological change occurring within the organism in response to a stimulus (stressor) and the idea that, as a result, some aspect of fish performance may be compromised. Long-term exposure to environmental stressors is a concern to biologists and managers because of the possible detrimental effects on important fish performance features such as metabolism and growth, disease resistance, reproductive capacity, and, ultimately, the health, condition, and survival of fish populations. Despite general acceptance by biologists and managers that stress can affect fish adversely, the phenomenon of stress is still poorly understood. Nevertheless, a common theme exists among widely ranging perceptions about stress, that there is a suite of nonspecific biological responses to a stimulus at different levels of organization.

A common misconception is that stress, in itself, is detrimental to the fish. This is, however, not necessarily the case. The response to stress is an adaptive mechanism that allows the fish to cope with real or perceived stressors in order to maintain its normal or homeostatic state (Figure 1). Quite simply, stress can be considered as a state of threatened homeostasis that is reestablished by a complex suite of adaptive responses (Chrousos 1998).

111

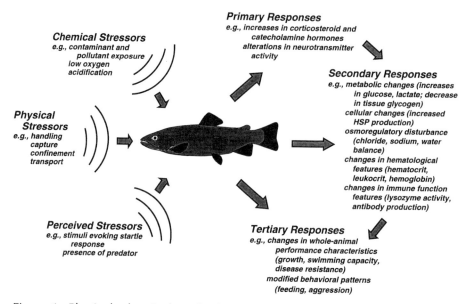

Figure 1. Physical, chemical, and other perceived stressors act on fish to evoke physiological and related effects, which are grouped as primary, secondary, and tertiary or whole-animal responses. In many instances, the primary and secondary responses, in turn, may directly affect secondary and tertiary responses, respectively, as indicated by the arrows (see text).

Also, studies have shown that repeated exposure to mild stressors can habituate fish and attenuate the neuroendocrine and metabolic responses to subsequent exposure to stressors (Iwama et al. 1998; Reid et al. 1998).

If the intensity of the stressor is overly severe or long-lasting, however, physiological response mechanisms may be compromised and can become detrimental to the fish's health and well-being, or maladaptive, a state associated with the term "distress" (Selye 1974; Barton and Iwama 1991). This view of an organism's ability or inability to adjust to a disturbance is consistent with the general adaptation syndrome (GAS) paradigm of Selye (1950), which considers that an organism passes through three stages in response to stress: (1) an alarm phase consisting of the organism's perception of the stimulus and recognition of it as a threat to homeostasis, (2) a stage of "resistance" during which the organism mobilizes its resources to adjust to the disturbance and maintain homeostasis, and (3) a stage of "exhaustion" that follows if the organism, in spite of mounting a stress response, is incapable of coping with the disturbance. The first two stages are usually manifested by measurable physiological changes at different levels of organization, particularly at the lower levels, whereas the final stage is the maladaptive phase normally associated with the development of pathological states, which can alter the health and condition of fish, eventually resulting in mortality.

Rand and Petrocelli (1985) defined acute as "having a sudden onset, lasting a short time; of a stimulus, severe enough to induce a response rapidly," whereas chronic was defined as that "involving a stimulus that is lingering or continues for a long time; often signifies periods from several weeks to years." In accordance with these definitions, acute aquatic toxicity tests are generally 4 d or less, but chronic tests are designed to study effects of continuous long-term exposures (Rand and Petrocelli 1985). Deciding seemingly arbitrary limits that differentiate acute and chronic stress at a sublethal level of response, however, is problematic. Researchers generally consider that acute exposures to stressors and subsequent responses to them are brief, ranging from seconds to minutes or possibly hours. Chronic disturbances are generally regarded as continuous or, at least, continual lasting from hours to weeks, months, or even throughout the fish's life, which is consistent with the view used in aquatic toxicology. However, it may be more useful to distinguish between acute and chronic in the context of stress responses from a functional perspective, such as described earlier. Thus, acute stress may be considered as that which usually occurs from brief or short-lived disturbances and results in physiologically adaptive responses to regain homeostasis. But responses that challenge compensatory mechanisms to the point that maladaptive or pathological conditions result are clearly chronic (Dhabhar and McEwan 1997) and can be either sublethal or eventually lethal depending on whether the fish's adaptive capacity is adequate to allow recovery from the stress.

Physiological responses to environmental stressors have been broadly grouped as primary and secondary (Figure 1). Primary responses, which involve the initial neuroendocrine responses, include the release of catecholamines from chromaffin cells (Randall and Perry 1992; Reid et al. 1998) and the stimulation of the hypothalamic-pituitary-interrenal (HPI) axis culminating in the release of corticosteroid hormones into circulation (Donaldson 1981; Wendelaar Bonga 1997; Mommsen et al. 1999). Secondary responses include changes in plasma and tissue ion and metabolite levels and hematological features, all of which relate to physiological adjustments such as in metabolism, respiration, acid–base status, hydromineral balance, and immune function (Pickering 1981; Iwama et al. 1997; Mommsen et al. 1999). Additionally, tertiary responses occur (Figure 1), which refer to aspects of whole-animal performance such as changes in growth, condition, overall resistance to disease, metabolic scope for activity, behavior, and ultimately survival (Wedemeyer and McLeay 1981; Wedemeyer et al. 1990). This grouping is simplistic, however, as stress, depending on its magnitude and duration, may affect fish at all levels of organization from molecular and biochemical to population and community (Adams 1990). Moreover, responses to stress at different levels of organization are not only interrelated functionally to each other, but often interregulated as well. It is useful to consider responses of fish in an integrated or holistic sense rather than simply observe isolated physiological phenomena in order to appreciate the animal's response at the population level to environmental stressors. This holistic view of stress becomes especially important for the development and validation of subcellular and molecular indicators as biomarkers for stress as they relate to biologically significant effects at higher levels of organization. Thus, including physiological and condition-based indices as components of more comprehensive bioassessment programs would be prudent.

Much of our present knowledge about physiological responses of fish to stress has been gained from studying the primary responses of the HPI axis and the brain-chromaffin axis to stressors and the subsequent or secondary effects associated with neuroendocrine stimulation on metabolism, reproduction, and the immune system (Randall and Perry 1992; Pickering 1993; Iwama et al. 1997; Reid et al. 1998; Mommsen et al. 1999). The investigation of heat-shock or stress proteins (HSP) in fish as a general indicator of the cellular response to various stressors is a recent and rapidly emerging field (Iwama et al. 1998, 1999). Also, studies have shown that HSP induction can be used as a sensitive and reliable indicator of cellular effects associated with acute and chronic exposure to stressors (Vijayan et al. 1997a, 1998). However, most research in this area is descriptive at this point and elucidation of possible functional relationships between the cellular responses to stress and neuroendocrine, immune, and other physiological systems may provide valuable information on the mechanisms involved in the stress tolerance process in fish. The majority of previous research on stress physiology in fish during the last few decades has focused on aquaculture. Many reviews have been published in that context (e.g., Barton and Iwama 1991; Iwama et al. 1997; Pickering 1998), and a few have dealt with the nature of stress responses and manipulations of stress-hormone levels in fish (Mazeaud et al. 1977; Pickering 1981; Gamperl et al. 1994; Wendelaar Bonga 1997; Mommsen et al. 1999; Schreck 2000). Comparatively less information is available on physiological and condition-related responses of fish to environmental perturbations associated with natural or anthropogenic stressors (e.g., Cairns et al. 1984; Adams 1990; Niimi 1990; Brown 1993; Hontela 1997).

Physiological Stress Indicators

Primary Indicators

When fish are exposed to a stressor, the physiological stress response is initiated by the recognition of a real or perceived threat by the central nervous system (CNS). The sympathetic nerve fibers, which innervate the chromaffin cells (adrenal medulla homologue), stimulate the release of catecholamines via cholinergic receptors (Reid et al. 1996, 1998). Because catecholamines, predominantly epinephrine in teleostean fishes, are stored in the chromaffin cells, their release is rapid and the circulating levels of these hormones increase immediately with stress (Mazeaud et al. 1977; Randall and Perry 1992; Reid et al. 1998). Therefore, it is very difficult to measure resting levels of catecholamines unless the fish are cannulated, resulting in little application of this approach to measure stress in fish (Gamperl et al. 1994).

The release of cortisol in teleostean and other bony fishes is delayed relative to catecholamine release. The pathway for cortisol release begins in the HPI axis with the release of corticotropin-releasing hormone (CRH), or factor (CRF), chiefly from the hypothalamus in the brain, which stimulates the corticotrophic cells of the anterior pituitary to secrete adrenocorticotropin (ACTH). Circulating ACTH, in turn, stimulates the interrenal tissue (adrenal cortex homologue) located in the kidney to synthesize and release corticosteroids into circulation for distribution to target tissues. Cortisol is the principle corticosteroid in teleostean, other neopterygian, and chondrostean fishes (Sangalang et al. 1971; Idler and Truscott 1972; Hanson and Fleming 1979; Barton et al. 1998) whereas 1α-hydroxycorticosterone is the major corticosteroid in elasmobranchs (Idler and Truscott 1966, 1967). Cortisol synthesis and release from interrenal cells has a lag time of several minutes, unlike chromaffin cells, and, therefore, proper sampling protocol can allow measurement of resting levels of this hormone in fish (Wedemeyer et al. 1990; Gamperl et al. 1994). As a result, the circulating level of cortisol is commonly used as an indicator of the degree of stress experienced by fish (Barton and Iwama 1991; Wendelaar Bonga 1997). Control of cortisol release is through negative feedback of the hormone at all levels of the HPI axis (Fryer and Peter 1977; Donaldson 1981; Bradford et al. 1992; Wendelaar Bonga 1997). Regulation of the HPI axis is far more complicated than this description implies, however, and is beyond the scope of this chapter. For additional details, Sumpter (1997), Hontela (1997), and Wendelaar Bonga (1997) have provided more complete descriptions of the endocrine stress axis in fish.

Other hormones can become either elevated or suppressed during stress, including, notably, thyroxine, prolactin, somatolactin, gonadotropins, and reproductive steroids in circulation (Barton 1997; Wendelaar Bonga 1997); however, these other hormones have not yet been demonstrated to be useful stress indicators and, therefore, are not discussed in this chapter.

Interest has focused recently on the responses of central brain monoamines, specifically the catecholamines (dopamine and derivatives) and indoleamines (serotonin and derivatives) in response to stress (Winberg and Nilsson 1993a). Serotonin, in particular, has been implicated in both epinephrine and cortisol regulation in fish during stress (Fritsche et al. 1993; Winberg and Nilsson 1993a; Winberg et al. 1997). These neurotransmitters are directly involved in behavioral changes associated with feeding activity, aggressive interactions among individuals, and establishment of social hierarchies (Winberg et al. 1992a; Winberg and Nilsson 1993b; Alanärä et al. 1998; Winberg and LePage 1998). Winberg et al. (1992b), for example, found that handling Arctic char *Salvelinus alpinus* daily for 4 weeks not only increased serotonin's metabolite, 5-hydroxyindoleacetic acid (5-HIAA), but also

5-HIAA/serotonin (5-HT) ratios, both of which are direct indicators of increased brain serotonergic activity. Increased levels of 5-HIAA and 5-HIAA/ 5-HT ratios are evident in subordinate low-ranking fish in dominance hierarchies (Winberg et al. 1992a; Winberg and Nilsson 1993b), which may result from social stress. Changes in brain monoamines, notably those in serotonergic activity, have also been reported in fish exposed to environmental stressors including ammonia (Atwood et al. 2000) and copper (DeBoeck et al. 1995). Moreover, such changes were correlated with growth suppression in a dose-dependent fashion following chronic ammonia exposure (Atwood et al. 2000). Taken together, these correlative results suggest the potential of using brain monoamine levels as indicators of tertiary changes associated with chronic stress in fish, such as growth and behavior; however, more study is needed to substantiate these relationships.

Secondary and Tertiary Indicators

Coping with stress is an energy-demanding process and, therefore, the physiological adjustments to stress are geared toward increased tissue oxygen availability and metabolic energy substrate mobilization and utilization in order to cope with the increased energy demand. The primary response, the release of corticosteroids and catecholamines, can directly or indirectly affect the secondary stress responses (Figure 1) including metabolism; hydromineral balance; cardiovascular, respiratory, and immune functions; and those related to hematology (e.g., circulating erythrocytes and leukocytes, differential leukocyte ratios, hemoglobin).

A major role of catecholamines is to modulate cardiovascular and respiratory functions, thereby maintaining adequate oxygen supply to the tissues (Randall and Perry 1992). This homeostatic mechanism includes increased blood flow to the gills, gill permeability, and lamellar perfusion to enhance oxygen uptake (Wendelaar Bonga 1997; Reid et al. 1998; Cech 2000). Accompanying the increased gill permeability, however, is an increase in water and ion transfer across the gill membrane, which can cause a temporary hydromineral imbalance in the fish. In freshwater fishes, this can be manifested during acute stress as a loss in major blood ions, such as sodium and chloride, and total osmolality, and an uptake of water (Eddy 1981; Wendelaar Bonga 1997; Cech 2000). Marine fishes during stress, however, may exhibit an increase in ion influx and osmolality and a loss of water across the gills (Eddy 1981; Wendelaar Bonga 1997; Cech 2000).

Catecholamines are also important for energy substrate mobilization, which is necessary to cope with the increased energy demand from metabolic activity associated with stress in fish (Barton and Schreck 1987a; Davis and Schreck 1997; Fabbri et al. 1998). One of the well characterized effects of catecholamines, especially epinephrine, is the stimulation of glycogenolysis, the conversion of stored tissue glycogen to glucose, in response to increased metabolic demands during stress (Mazeaud and Mazeaud 1981; Randall and Perry 1992; Gamperl et al. 1994; Wendelaar Bonga 1997). This effect of epinephrine is very rapid as it is mediated by the beta-receptors and includes the activation of cyclic adenosine monophosphate (cAMP) and phosphorylation of glycogen phosphorylase in fish hepatocytes (Randall and Perry 1992; Fabbri et al. 1998). As this pathway can be rapidly activated, the stressor-induced increase in circulating glucose levels, at least immediately after stress, is mediated primarily by catecholamine stimulation of glycogenolysis (Vijayan and Moon 1994; Vijayan et al. 1994a; Fabbri et al. 1998).

Cortisol has been shown to have several functions in fish including in metabolism, osmoregulation, and immune function (see Mommsen et al. 1999 for review). The metabolic role of cortisol has focused mainly on the gluconeogenic actions of the hormone (Vijayan et al. 1994b, 1996, 1997b). Cortisol treatment has been shown to increase plasma glucose concentration and liver gluconeogenic capacity including the activities of several key enzymes in intermediary metabolism (Mommsen et al. 1999). In addition, using RU486, a cortisol receptor antagonist, Vijayan et al. (1994b) showed that cortisol has a direct stimulatory effect on gluconeogenesis from alanine. As amino acids are preferred substrates for piscine gluconeogenesis (Suarez and Mommsen 1987), it is likely that cortisol's effect on gluconeogenesis is mediated via cortisol-induced peripheral proteolysis. Although no direct effect of cortisol on fish muscle proteolysis is apparent, studies have shown that cortisol treatment elevates plasma amino acid concentration and the activities of hepatic enzymes involved in amino acid catabolism (Vijayan et al. 1996, 1997b). These studies argue for a role for cortisol in the stress-induced elevation of plasma glucose concentration to cope with the increased energy demand (Mommsen et al. 1999). In addition to increasing plasma glucose concentration, cortisol may also be channeling carbon substrates for liver glycogen repletion after stress has occurred in fish (Vijayan et al. 1994a, 1997b). Therefore, it is thought that the immediate glucose elevation after stress is due to catecholamine stimulation, whereas the longer-term maintenance of glucose, in the absence of glycogen depletion after stress, is mediated by cortisol-induced gluconeogenesis (Vijayan et al. 1994a, 1997b; Mommsen et al. 1999). The effect of cortisol on in vivo hepatic glycogenolysis is less clear with studies showing an increase, decrease, or no change in liver glycogen content with cortisol treatment in fish (see Mommsen et al. 1999). Several factors, including species differences, nutritional status, and prior rearing history of the fish, may influence the differing liver glycogen response to cortisol treatment (Mommsen et al. 1999).

Circulating lactate, or lactic acid, concentrations often increase during stressful conditions as a result of increased muscular activity associated with the stress (Driedzic and Hochachka 1978). Plasma lactate increases in fish that follow physical disturbances have been well documented and were used extensively to monitor various types of physical stressors in fish before the advent of techniques to measure hormonal responses (Barton 1997). The role of stress hormones in the lactate response to stress is not clear, although studies have shown that cortisol increases gluconeogenic rates from lactate in isolated hepatocytes or liver slices (Mommsen et al. 1999).

Heat-shock or stress proteins are also being used increasingly as an indicator of cellular stress in fish (see Bradley 1993; Bradley et al. 1994; Iwama et al. 1998, 1999). There are several proteins that belong to the HSP family. The 70-kDa protein (HSP70) has been the most widely studied in fish, although HSP90 appears to be more directly involved in corticosteroid receptor activation (Pratt 1997). A recent study showed that cortisol modulates heat-shock-induced HSP90 transcription in fish hepatocytes, suggesting a possible link between the endocrine stress axis and the cellular stress response in fish (Sathiyaa et al. 2001). Heat-shock protein is expressed in response to stressors that seem to affect the protein machinery, and a recent study showed that a handling disturbance alone did not induce the expression of HSP70 in trout liver (Vijayan et al. 1997a). However, contaminants at sublethal concentrations, including bleached-kraft pulp mill effluent, induce the expression of HSP70 in salmonids (Vijayan et al. 1998) and white suckers *Catostomus commersoni* (Janz et al. 1997). Similarly, coho salmon *Oncorhynchus kisutch* chronically infected with bacterial kidney disease exhibited induction of HSP70 in their tissues (Forsyth et al. 1997). Therefore, these proteins may be sensitive indicators of adaptive cellular responses to stressors, especially as the onset of rainbow trout *O. mykiss* mortality with contaminant exposure was coincident with a lack of liver HSP70 expression (Vijayan et al. 1998). Studies on the role of stress hormones in the regulation of HSP expression are currently underway. Cortisol has been shown to modulate HSP70 expression (M. Vijayan, C. Pereira, and G. Iwama, University of Waterloo, personal communication) and HSP90-mRNA expression (Sathiyaa et al. 2001) in primary cultures of trout hepatocytes, whereas, the effect of cortisol on HSP expression in vivo is not clear (Deane et al. 1999). No information is available about the role of catecholamines on HSP expression in fish.

Both the corticosteroids and catecholamines can directly or indirectly affect aspects of tertiary responses or performances of particular concern to biologists including disease resistance (see Chapter 6), scope for growth, feeding and avoidance behavior (see Chapter 11), and reproductive capacity (Randall and Perry 1992; Iwama et al. 1997; Mommsen et al. 1999; see Chapter 9). Some of these tertiary responses, such as growth, condition, and general health, are reflected in changes in the various indices described later in this chapter. The secondary physiological changes that occur tend to take longer to manifest themselves in circulation than primary responses, from minutes to hours (e.g., glucose, lactate, chloride), but often remain altered for more extended periods. Timing of changes in tertiary or whole-animal performance characteristics may be variable; for example, swimming stamina may be affected relatively quickly, whereas, alterations in the immune system or reproductive function may not appear for hours or days or even weeks. Nevertheless, most research tends to support the notion that the magnitude and duration of the response reflect the severity and duration of the stressor. Thus, many of the primary and secondary physiological stress responses documented in fish have become well established as useful monitoring tools to assess the degree of stress experienced by fishes. It is emphasized, however, that functional relationships between sublethal physiological changes and mortality are not yet very well understood (Davis et al. 2001).

Factors Influencing Physiological Responses

The physiological responses of fish to stressful encounters are influenced by nonstress factors that affect both the magnitude of the response and recovery from it. These factors are primarily genetic, developmental, nutritional, and environmental. Fishes exhibit a wide variation in their responses to stress, particularly endocrine responses (Barton and Iwama 1991; Gamperl et al. 1994), ranging as much as two orders of magnitude in response to the same stressor (Table 1). The cause for such major differences between and among taxa is likely genetic. It is unknown, however, whether fishes that display relatively low corticosteroid stress responses are actually "less stressed"

Table 1. Examples of mean (± SE) plasma cortisol concentrations in selected juvenile freshwater fishes before and 1 h after being subjected to an identical 30-s aerial emersion (handling stressor). All species were acclimated to their respective environmental preferenda and are listed in increasing order of the magnitude of the 1-h poststress cortisol response (from Barton and Zitzow 1995; Barton and Dwyer 1997; Barton et al. 1998, 2000; Barton 2000; and personal communications for Arctic grayling [B. Barton and W. Dwyer], common carp [N. Ruane and J. Komen], and yellow perch [A. Haukenes and B. Barton]).

| | Cortisol (ng/mL) | |
Species	Prestress	Poststress
Pallid sturgeon *Scaphirhynchus albus*	2.3 ± 0.3	3.0 ± 0.3
Hybrid sturgeon *S. albus* × *platorynchus*	2.2 ± 0.4	3.2 ± 0.3
Paddlefish *Polyodon spathula*	2.2 ± 0.6	11 ± 1.8
Arctic grayling *Thymallus arcticus*	1.1 ± 0.3	26 ± 4.4
Rainbow trout *Oncorhynchus mykiss*	1.7 ± 0.5	43 ± 3.5
Common carp *Cyprinus carpio*	7.4 ± 2.9	79 ± 14
Brook trout *Salvelinus fontinalis*	4.0 ± 0.6	85 ± 11
Yellow perch *Perca flavescens*	3.4 ± 1.1	85 ± 12
Bull trout *Salvelinus confluentus*	8.1 ± 1.2	90 ± 11
Brown trout *Salmo trutta*	1.0 ± 0.3	94 ± 11
Lake trout *Salvelinus namaycush*	2.8 ± 0.4	129 ± 11
Walleye *Stizostedion vitreum*	11 ± 4.4	229 ± 16

than others or are as "stressed" but have a different capacity to respond to stress. Barton et al. (1998, 2000), for example, observed that low poststress changes also occurred among secondary physiological indicators measured in chondrosteans in addition to cortisol, suggesting a reduced response overall to the stressors. Consistent response differences to stress are not only evident among fish species (Barton and Iwama 1991; Vijayan and Moon 1994; Ruane et al. 1999; Barton 2000) but also among strains or stocks within the same species (Woodward and Strange 1987; Iwama et al. 1992) and even within the same population (Pottinger et al. 1992), a trend that appears to be at least partially heritable (Pottinger et al. 1994). Differences in physiological mechanisms that would account for these variations remain largely unexplored, but Pottinger et al. (2000) found that very high poststress cortisol levels in European chub *Leuciscus cephalus* were associated with low corticosteroid receptor affinity.

The developmental stage of the fish can also affect its responsiveness to stress. A fish's ability to respond to a disturbance develops very early in life, for example, as early as two weeks after hatching in some fishes (Barry et al. 1995; Stephens et al. 1997) and even earlier in cyprinids (Stouthart et al. 1998). Little evidence exists to suggest that fish show a consistent increase in stress responses as they develop, but they do appear to have heightened responses during periods of metamorphosis. Juvenile anadromous salmonids, for example, appear to be especially sensitive to certain stressors, particularly physical disturbances, during the period of parr–smolt transformation (Barton et al. 1985a; Maule et al. 1987). As fish mature, primary stress responses may actually decrease in magnitude, possibly as a result of a reduced threshold for regulatory feedback with the onset of maturity (Pottinger et al. 1995).

Almost all environmental nonstress factors examined to date can influence the degree to which fish respond to stress including acclimation temperature, external salinity, nutritional state, water quality, time of day, light, fish density, and even background color (Barton and Iwama 1991; Barton 1997). An awareness of the extent to which these nonstress factors modify responses is important to both researchers and managers wishing to interpret experimental results and compare them with published values. For example, the fish's acclimation temperature or nutritional state is likely to have an appreciable effect on the magnitude of poststress elevations of cortisol and glucose, particularly the latter (Davis et al. 1984; Barton and Schreck 1987b; Barton et al. 1988; Davis and Parker 1990; Vijayan and Moon 1992, 1994). In certain instances, stress-modifying factors that are themselves chronically stressful, such as poor water quality or toxicants, can actually exacerbate (Barton et al. 1985b) or attenuate (Pickering and Pottinger 1987; Hontela 1997; Wilson et al. 1998) the cortisol response to a second stressor.

The length of time between discrete stressors, the effect of multiple stressors, and the severity of continuous stressors are important factors that influence how fish will respond to stress. Schreck (2000) describes conceptually how fish can respond to (1) sequentially applied separate stressors, with and without sufficient recovery intervals for compensation to occur; and (2) single or multiple continuous concurrent stressors. If a stressor is persistent, the fish will either compensate eventually for the disturbance and become acclimated or habituated to it or die (Schreck 2000)—responses that are consistent with Selye's GAS model. Thus, it is important for investigators and managers to also know whether fish from experimental or monitored populations are naive or have been exposed to other stressors in their environment prior to study.

Condition-Based Stress Indicators

Condition Indices

Tertiary or whole-animal responses to stress, including changes in growth, condition, and health, indicate the extent to which stress may affect fish performance and provide a basis for understanding the effects of environmental perturbations on fish populations. While assessment of physiological stress indicators certainly provides useful data on the status of fish populations, some of the sophisticated methods employed for measurement are not practical in field bioassessment situations. Although sample kits are available for some physiological parameters (Iwama et al. 1995; Wells and Pankhurst 1999), most methods are used mainly in research and are often beyond the resources or expertise available to the monitoring agency. Thus, the use of physiological test kits for assessing stressed states in fish may be accompanied by recommendations for determining a variety of relatively simple indices related to higher level responses, such as the fish's health and condition, which occur following manifestation of primary and secondary responses (Morgan and Iwama 1997). These indices include a number of organosomatic indices, types of condition factors, and growth, and have been reviewed previously (Anderson and Gutreuter 1983; Busaker et al. 1990; Goede and Barton 1990; Anderson and Neumann 1996). Condition-based indicators, by nature, appear to be relatively insensitive to environmental stressors compared with most specific measurements of blood chemistry or other physiological indicators. Nevertheless, data on fish condition are easy to collect and can contribute appreciably toward understanding long-term trends in fish populations exposed to chronic environmental stressors.

Condition Factor and Length–Weight Relationship

Fulton's condition factor is a measure of robustness and is a function of fish weight divided by the cube of the length and approaches unity with a scaling constant. Condition factor (K, or often CF) is usually expressed in metric units (e.g., $K = g/mm^3 \times 10^5$; Anderson and Neumann 1996), but its calculation in English units (e.g., C = in/lb × 10^4) is still used in many aquacultural applications (Piper et al. 1982). Calculated values for K and C assume a constant length exponent of 3, but, in actuality, this exponent normally ranges between 2.5 and 3.5 (Carlander 1969) depending on the species' body shape and its life stage. For fusiform fishes, such as salmonids, the length exponent as determined from the length–weight relationship is usually within 2.8 and 3.2 (Barton 1996). Determination of length–weight relationships (i.e., log[weight] = a + b·log[length]) for specific populations or year-classes by linear regression of the two variables has also been used extensively to compare fish populations, particularly in ecological studies and for management use (Anderson and Neumann 1996). Published K values and length–weight relationships for many freshwater populations of the most important North American species are available in three volumes by Carlander (1969, 1977, 1997).

Changes in condition factor or the length–weight relationship can reflect the nutritional or energy status of the fish (Lambert and Dutil 1997; Grant et al. 1998; Grant and Brown 1999) as the fish's weight relative to length will change with storage or mobilization of energy reserves such as proteins, carbohydrates, or lipids. Declines in reserves, such as liver glycogen or visceral fat deposits, may indicate a change in feeding behavior and food consumption, whereas muscle wasting (proteolysis) will indicate a starvation response. Moreover, changes in condition factor resulting from changes in energy reserves and tissue biochemistry may be accompanied by concomitant changes in body water content (Adams et al. 1985; Cunjak and Power 1986; Lambert and Dutil 1997). These changes may be natural occurrences relating to season (e.g., overwintering) or life history stages (e.g., parr–smolt transformation in salmonids), or may result from the presence of disease or exposure to stressors.

Relative Weight Index

The length–weight relationship changes during a fish's life history, which complicates population assessments that compare different year-classes, or cohorts. Wege and Anderson (1978) developed the relative weight index (W_r) to accommodate these allometric growth differences over a wide range of sizes. The W_r is a comparison of the mean weight of the population in question (W) with a standard weight (W_s) and is calculated as follows: $W_r = W/W_s × 100$, after first determining a W_s value from: log(W_s) = a + b·log(L), where L is the mean length of the population with mean weight W and intercept a and slope b are empirically based, published standard values (Anderson and Neumann 1996).

The W_r index can, thus, be a very useful parameter for assessing fish population status, general fish health, fish stocking, and other management programs; these uses and the underlying theory are reviewed extensively by Murphy et al. (1990) and Blackwell et al. (2000). However, W_r has two drawbacks. First, W_s is calculated from the intercept (a) and slope (b) values provided from a compilation of as many populations as possible with the assumption that the more populations that are included in the calculation, the closer the regression parameters (and, thus, W_s) will be to a hypothetical ideal standard (Cone 1989). Second, W_s regression parameter values (a and b) are not published for all species, but mainly for freshwater fishes important to managers. Despite these shortcomings, W_r provides another quantitative tool for evaluating fish condition, at least for the approximately 56 freshwater species for which published regression parameters for W_s equations are available (Anderson and Neumann 1996; Bister et al. 2000; Blackwell et al. 2000; Hyatt and Hubert 2001).

Organosomatic Indices

Ratios of the mass of particular organs or tissues relative to total body mass can be used as indices of change in nutritional and energy status. Commonly used organosomatic indices include: hepatosomatic index (HSI, liver:body weight), gonadosomatic index (GSI, gonads:body weight), viscerosomatic index (VSI, entire viscera:body weight), and splenosomatic index (SSI, spleen:body weight) (Anderson and Gutreuter 1983; Goede and Barton 1990).

Necropsy-Based Indicators

If the stress experienced by fish is overly severe or long-lasting, then pathological conditions may ensue as compensatory mechanisms are exceeded (Moberg 1985). The necropsy-based condition assessment method described by Goede and Barton (1990) is based on that assumption. This approach was developed to meet the needs of field biologists and hatchery managers for a simple and inexpensive method to monitor and detect trends in their fish stocks. This system is useful for establishing long-term databases on the health and condition of fish populations, but, it is not designed to be a post hoc diagnostic tool. The system can be used alone or in combination with other physiological and hematological measurements (Iwama et al. 1995). As stated by Goede and Barton (1990), this method requires the following assumptions: (1) in fish under stress, tissue and organ function will change in order to maintain homeostasis; (2) if a change in function persists in response to continuous stress, there will be a gross change in structure of organs or tissues; (3) if the appearance of all organs and tissues is normal according to the necropsy criteria, there is a good likelihood that the fish is normal; and (4) if the appearance of an organ or tissue system departs from the normal or control condition, the fish is responding to changes brought about by the environmental stressor.

During the past decade, the necropsy-based assessment method or health condition profile (HCP) has been used in its original form mainly for monitoring hatchery stocks (Novotny and Beeman 1990; Wagner et al. 1996). This protocol has also been modified to provide a quantitative index that can be compared statistically. Adams et al. (1993) assigned numerical values to the necropsy categories, which could then be summarized in a health assessment index (HAI). The HAI has been used to evaluate wild fish populations subjected to various environmental perturbations (Adams et al. 1993; Barton 1994; Schlenk et al. 1996; Bergstedt and Bergerson 1997; Raymond and Shaw 1997; Steyermark et al. 1999; Sutton et al. 2000; McKinney et al. 2001) and, in some instances, modified or extended to incorporate additional site-specific observations (Steyermark et al. 1999; McKinney et al. 2001). Other external indices, such as skin ulceration (e.g., Barton 1994; Noga et al. 1998) or tumor incidence (e.g., Baumann 1992), can be used separately to evaluate fish populations or incorporated into a standardized examination system. Specific protocols for investigating fish kills are available (Meyer and Barclay 1990) and should be followed in such instances, especially where a legal chain of custody of data is required.

Measuring and Interpreting Stress Responses

Physiological Indicators

As judged by the prevalence in the literature and inferred earlier from numerous studies, the most popular approaches for evaluating physiological responses of fish to environmental disturbances are measurements of plasma cortisol, glucose, lactate, chloride (and other ions), and osmolality, and various hematological features. Typical ranges for resting and poststress-elevated values for commonly measured primary and secondary physiological responses are listed in Table 2. However, readers should note that these are approximate ranges of values that serve as a guideline only and have limited diagnostic value because stress responses are highly variable depending on genetic makeup, early life history, nutritional status, and the fish's environment. Detailed summaries of endocrine changes in fish following exposure to both physical and chemical disturbances are found in Barton and Iwama (1991), Brown (1993), Gamperl et al. (1994), and Hontela (1997). Folmar (1993) provides an extensive compilation of chemical contaminant effects on a number of blood chemistry features in fishes including hormone levels, secondary physiological parameters, and hematology. Species-specific blood chemistry summaries can also serve as useful guidelines for interpreting stress-induced physiological changes (e.g., Hille 1982; Roche and Bogé 1996; Noga et al. 1999). Extensive data indicating the point at which certain secondary physiological features may actually indicate tertiary changes or a life-threatening situation are not available, but, plasma chloride and osmolality concentrations less than 90 meq/L and 200 mOsm/kg, respectively, have been suggested as indicative of compromised osmoregulatory ability in salmonids (Wedemeyer 1996).

Table 2. Ranges of typical resting and stress-elevated values for primary and secondary physiological parameters used as indicators of stress in fish (compiled from Wedemeyer et al. 1990; Barton and Iwama 1991; Folmar 1993; Gamperl et al. 1994; and authors' unpublished data). However, considerable variation among these values and many exceptions outside of these ranges exist depending on species, genetic background, rearing history, and environmental conditions (see text and cited reviews).

Physiological parameter	Resting	Poststress
plasma epinephrine (nmoles/L)	1–6	5–200
plasma norepinephrine (nmoles/L)	1–14	10–100
plasma cortisol (ng/mL)	2–50	30–300
plasma glucose (mg/dL)	50–150	100–250
plasma lactate (mg/dL)	20–30	40–80
plasma chloride (meq/L)	100–130	≈10% ↑ or ↓ [a]
plasma sodium (meq/L)	140–170	≈10% ↑ or ↓ [a]
plasma potassium (meq/L)	2–6	≈10% ↑ or ↓ [a]
plasma osmolality (mOsm/kg)	290–320	≈10% ↑ or ↓ [a]
hemoglobin (g/dL)	5–9	< 4
hematocrit (% packed cell volume)	25–40	40–50+

[a] Blood ions and other features related to hydromineral status will fluctuate upward or downward depending on whether fish is marine or freshwater species, respectively.

Characteristic resting and stress-altered levels of important central monoamines in fishes are summarized in Table 3. Previous studies have demonstrated that these neurotransmitters and their metabolites appear to be concentrated in the telencephalon, the hypothalamus, and the brain stem. Analysis of whole brains would be expected to yield values lower than in specific brain areas because of the dilution effect of other tissues in the preparation. Most studies have examined teleost brains, but Sipiorski (2000) found that chondrostean brain tissue contained monoamine concentrations within the

Table 3. Ranges of characteristic levels of important brain monoamine neurotransmitters in bony fishes (summarized from published data compiled by Sipiorski 2000). All values are in pg/mg brain tissue and include both pre- and poststress values as considerable overlap exists among both species and experiments (5-HT: serotonin, or 5-hydroxytryptamine; 5-HIAA: 5-hydroxyindoleacetic acid; DA: dopamine; DOPAC: 3,4-dihydroxyphenylacetic acid).

Brain region	5-HT	5-HIAA	DA	DOPAC
Telencephalon	210–1,000	25–270	180–300	
Brain stem	200–1,800	33–200	125–180	
Hypothalamus	290–1,500	50–360	285–610	
Whole brain	30–500	40–210	31–280	10–109

same order of magnitude as those in teleosts but at the low end or somewhat lower than the ranges reported in Table 3.

Methods of physiological stress assessment in fish have been described (Wedemeyer et al. 1990; Iwama et al. 1995) and include simple assay kits (e.g., glucose, lactate) and easy-to-use meters (e.g., chloride, osmolality) for many of the physiological features of general interest. Measuring hormones such as cortisol is more complicated and usually involves a radioimmunoassay or enzyme-linked immunosorbent assay (ELISA) technique; a number of clinical kits for those assays are now available but should be calibrated and verified for the taxonomic group being studied. Detection methods for HSPs have been described previously and usually involve gel electrophoresis followed with Western blotting or ELISA, or both (Forsyth et al. 1997; Vijayan et al. 1997a; Iwama et al. 1998). Determination of neurotransmitters, such as the brain indoleamines and catecholamines, requires high performance liquid chromatography (HPLC). Inexpensive, readily available, portable meters, such as those used clinically for glucose, hemoglobin, plasma protein, and other features, have been tested for their efficacy in fish stress assessment (Iwama et al. 1995; Morgan and Iwama 1997; Wells and Pankhurst 1999) and such approaches show promise as future useful tools for field monitoring programs. Such kits should be compared with proven laboratory methods for accuracy and reproducibility before use.

Interpreting the physiological variables can be more problematic than measuring the actual responses for three major reasons. As discussed earlier, other nonstress factors such as the various genetic, developmental, and environmental factors can have a modifying effect on the magnitude and duration of the stress response. Without knowing the extent to which other nonstress factors may have altered the response, it is difficult to interpret the biological significance of that response in a relative context. A second factor complicating data interpretation is the variation and apparent inconsistency among fishes in the responses of different blood chemistry characteristics. For example, a species that shows the greatest endocrine response increase (e.g., plasma cortisol) compared with other taxa may not be the same species that elicits the greatest increase in a secondary response, such as glucose or lactate, when subjected to the identical stressor. Thus, a species or group that appears "most stressed" as indicated by one particular level of response may not necessarily reflect that same degree of stress if measured by another level of response (Barton 2000). Such discrepancies among different physiological indicators emphasize the importance of not relying on a single indicator but multiple indicators and also the need for appropriate controls in stress assessment. A third complicating factor is the nature of the stress response itself. The response to stress is a dynamic process and physiological measurements taken during a time course are only representative instantaneous "snapshots" of that process. A significant delay, depending on the level and type of response, can occur from initial perception of the stressor by the CNS to the time when the physiological feature of interest reaches a peak level of response. Thus, the measurement of a particular stress indicator may not necessarily reflect the degree of stress experienced by the fish at that instant but more likely be representative of the extent of the earlier or initial response. This time lag between perception of the stressor and manifestation of a measurable response can complicate interpretation of results.

Physiological measurements provide a useful approach to evaluate responses of fish to acute stressors but may not necessarily be so for monitoring fish experiencing sublethal chronic stress. Unless the stressors, singularly or in combination, are severe enough to challenge the fish's homeostatic mechanisms beyond their compensatory limits or permanently alter them, which ultimately may cause death, physiological processes generally adapt to compensate for the stress (Schreck 1981, 2000). In these cases, blood chemistry parameters, such as cortisol titer, may appear normal and other approaches may be needed to determine the fish's physiological status.

As one alternative approach for evaluating the stressed states in fish, increased continued activity of the HPI axis, which implies a continuous response to a stressor, can be determined from tissue histology. Continued synthetic and secretory activity of the interrenal tissue during chronic stress can result in hypertrophy (increase in size) and hyperplasia (increase in abundance) of the interrenal cells. Using standard tissue histological preparation and examination techniques, interrenal cells can be measured and counted, and, thus, quantified. This approach has been used to assess chronically stressed states in fish subjected to social stressors including rearing density (Noakes and Leatherland 1977; Fagerlund et al. 1981) and exposure to acidification (Brown et al. 1984; Tam et al. 1988), heavy metals (Norris et al. 1997), and other environmental pollutants (Ram and Singh 1988; Servizi et al. 1993). Continual interrenal activity, however, will also downregulate the HPI axis as a result of negative feedback by cortisol. Thus, when a second acute stressor subsequently challenges fish exposed to a chronic stressor, the corticosteroid response to the additional stress may be reduced considerably relative to controls (Hontela 1997). Impaired interrenal function from chronic stress has been demonstrated in vivo by subjecting fish to an acute physical disturbance after being exposed to various contaminants (Hontela et al. 1992; Wilson et al. 1998; Norris et al. 1999; Laflamme et al. 2000).

Another approach for determining the effect of chronic stress on the HPI axis in fish is by assessing the functional integrity of the interrenal tissue in vitro. Brodeur et al. (1997) developed a relatively simple perifusion bioassay protocol to measure the corticosteroidogenic capacity of ACTH-stimulated interrenal tissue removed from chronically stressed fish. More recently, Leblond and Hontela (1999) and Leblond et al. (2001) described a method of preparing and using interrenal cell suspensions for quantifying the extent of in vitro steroidogenic inhibition at the cellular level. These and similar approaches have been used by this group of investigators and others to evaluate the mechanisms involved in the depression of interrenal capacity following exposure to contaminants including heavy metals and organochlorine compounds (Brodeur et al. 1998; Girard et al. 1998; Wilson et al. 1998; Leblond and Hontela 1999; Benguira and Hontela 2000; Laflamme et al. 2000; Leblond et al. 2001).

A summary of advantages and disadvantages of the physiological indices commonly used for measuring stressed states in fish is presented in Table 4.

Condition-Based Indicators

Condition Factor and Length–Weight Relationship

Condition factors are often used in stress assessment studies as they are derived from easily obtained length and weight measurements (Anderson and Neumann 1996). As mentioned previously, declines in condition factor or changes in the length–weight relationship can indicate a change in nutritional or energy status, as reflected by depletions in liver glycogen and body fat deposits. These declines in energy reserves and, thus, condition factors may be caused by external stressors including high rearing densities in hatcheries, acidification, and other adverse environmental conditions (see Goede and Barton 1990). Such stressors may directly affect feeding behavior and food intake, alter metabolic rates, or divert energy away from storage reserves to cope with the stress. However, declines in condition factors in fish and changes in their length–weight relationships may also occur for reasons other than stress. These include seasonal and developmental changes such as natural fluctuations in food availability (Adams et al. 1982), sexual maturation and gonad development (Medford and Mackay 1978), and parr–smolt transformation in juvenile anadromous salmonids (Vanstone and Markert 1968). Furthermore, condition factors may not change in situations where other condition-based biomarkers indicate problems with fish health (Steyermark et al. 1999). Due to their ease of measurement, condition factors will undoubtedly continue to be used in stress assessment studies as indicators of general health. However, the results should be interpreted with caution for the above reasons.

Organosomatic Indices

Organosomatic indices such as HSI, GSI, VSI, and SSI are also relatively easy to measure in the laboratory and field and have been used in a number of stress assessment studies. The assumption generally made with some of these indices (i.e., HSI, GSI, VSI) is that lower than normal values indicate a diversion of energy away from organ or tissue growth in order to combat a stressor of some type. The HSI, sometimes referred to as the liver-somatic index (LSI), is the most frequently used organ mass ratio (Goede and Barton 1990).

Table 4. Physiological responses and their advantages and disadvantages as potential indicators of the degree of stress experienced by fish (see text for details).

Feature	Uses or functions	Advantages	Disadvantages
Heat-shock proteins (HSPs)	Indicator of the cellular stress response to various types of stressors	Sensitive indicator of the adaptive cellular response to acute and chronic stress Can be used as an indicator of both exposure and cellular effects (at the level of cell function) associated with stress	Still considered as a research area; the nonspecificity of the HSP70 response may limit its use as an indicator of stress Functional relationships with the endocrine responses not well established Most of the studies have used HSP70, and the response of other HSPs to different stressors needs to be established in order to validate their use as sensitive indicators Requires sophisticated methodology including Western blotting or ELISA
Brain neurotransmitters	Indicates central nervous system response to stress perception	May help explain mechanisms underlying changes in peripheral endocrine responses and certain behaviors associated with stress Can help establish neuroendocrine links between perception of the stressor and resultant physiological responses	Requires sophisticated HPLC analysis Requires rapid brain removal and instant freezing Whole brain analysis may yield lower values because of dilution effect Proper interpretation requires analysis of preparations from specific brain sites
Plasma catecholamines	Rapid endocrine response (primary) to most forms of disturbance Associated functionally with several physiological roles including tissue oxygen delivery and fuel mobilization for tissue metabolism	Very responsive to acute stress	Difficult to obtain samples from unstressed fish without cannulation because of the rapidity of response May require sophisticated analytical methods (HPLC)
Plasma cortisol	Primary endocrine response used commonly as an indicator of exposure to stressors in aquaculture systems Multiple functional roles including liver metabolism, immune function and osmo-	Predictable indicator of magnitude and duration of acute stress Useful endocrine indicator because of time lag between perception of the stressor and manifestation	Influenced by genetic, developmental and environmental nonstress factors Response may become desensitized or habituated with

Table 4. (continued.)

Feature	Uses or functions	Advantages	Disadvantages
	regulation	of the measurable response in circulation	chronic stress Requires RIA or ELISA for analysis An assay for 1α-hydro-droxycorticosterone in elasmobranchs is not readily available
Plasma glucose	Metabolic response to stress; indicative of catecholamine- and cortisol-stimulated mobilization of energy reserves	Useful indicator of acute and chronic stress in teleost fishes, especially with respect to chronic stress-induced food-deprivation; less responsive in chondrosteans Can be measured using portable inexpensive kits Lab method is simple with a spectrophotometer using commercial reagents	Fields kits require calibration with lab method to verify accuracy Readings influenced by species, rearing history and other environmental nonstress factors such as temperature and diet
Plasma lactate	Metabolic response to stress; by-product of anaerobic metabolism	Useful indicator of acute stress when associated with muscular activity Lab method is simple with a spectrophotometer using commercial reagents	Not a useful indicator when stressor does not involve or result in strenuous muscular activity
Tissue glycogen	Indicates metabolic reserves stored in liver and muscle	Depletion indicates mobilization and use of energy reserves possibly due to stress or muscular activity	Liver glycogen is very much dependent on the prior history of the animal; values may be high or low depending on when the animal fed and, as a result, any stress effect may be masked
Plasma chloride	Change indicates hydromineral imbalance suggesting possible osmoregulatory dysfunction	Easy to measure using chloridometer (chloride meter)	Not a particularly sensitive indicator in chondrosteans or euryhaline teleosts
Plasma sodium	Change indicates hydromineral imbalance suggesting possible osmoregulatory dysfunction	Standardized saltwater challenge test for salmonids has been developed for comparison of results	Not a particularly sensitive indicator in chondrosteans or euryhaline teleosts Sodium ion analyzer is relatively expensive
Plasma osmolality	Change indicates hydromineral imbalance suggesting possible osmoregulatory dysfunction	Easy to measure with osmometer (vapor pressure or freezing-point depression models)	Not a particularly sensitive indicator in chondrosteans or euryhaline teleosts Does not give indication of what ion is in imbalance

Table 4. (continued.)

Feature	Uses or functions	Advantages	Disadvantages
Plasma protein	Change indicates water imbalance suggesting possible osmoregulatory dysfunction	Very easy to measure using hand-held refractometer	Not a sensitive stress indicator
Hematocrit	A measure of the cellular fraction of blood; determined as packed cell volume. Increases may indicate stress-induced splenic release of red blood cells	Easy to measure in blood collected with capillary tubes	Not a very sensitive stress indicator. Difficult to ascertain whether differences result from changes in blood volume or red blood cell number. A low value may also indicate anemia and, thus, possible disease presence. Requires special hematocrit centrifuge to process samples
Leukocrit	An indicator of the fraction of white blood cells (i.e., all leukocytes)	Easy to measure in blood collected with capillary tubes; i.e., from hematocrit samples	Not a sensitive stress indicator. High value indicates possible pathogen infection; low value may result from stress
Hemoglobin	A possible indicator of the oxygen-binding capacity of the blood	Easy to measure with commercial hand-held meter or commercial test kit	Not a sensitive stress indicator (see hematocrit)

The liver serves as a major storage site for glycogen and the HSI can, therefore, provide an indication of the nutritional state of the fish, as well as reflect seasonal changes in growth in fish populations (Adams and McLean 1985). The HSI values have been shown to decrease in fish stressed by adverse changes in water quality (Lee et al. 1983), altered water flows (Barnes et al. 1984), and repeated handling (Barton et al. 1987). It has also been demonstrated that HSI values increase after exposure to certain types of contaminants, particularly petroleum hydrocarbons (Fletcher et al. 1982; Fabacher and Baumann 1985; Baumann et al. 1991). This increase in liver mass is presumably due to an increase in liver cell number (hyperplasia) and size (hypertrophy) as a consequence of the induction of the mixed-function-oxidase system in the liver to detoxify the contaminants (Poels et al. 1980). It is important to note that the various organosomatic indices may vary naturally with food availability, state of sexual maturation, and life history stage, often in concert with the season. The GSI, in particular, will exhibit a dramatic decrease after spawning has occurred because of the loss of gonadal products, especially in female fish. These factors should be taken into account when attempting to use organosomatic indices in stress-related studies, especially in the case of the HSI and GSI (see Goede and Barton 1990).

The presence of parasites in the various organs may also confound the interpretive value of organosomatic indices (e.g., Steyermark et al. 1999).

Necropsy-Based Assessment

As mentioned previously, the necropsy-based condition assessment method is a rapid, easy, and inexpensive procedure to detect changes in the health and condition of fish populations. The necropsy method is based on a systematic examination of the condition of external and internal tissues and organs (Table 5). A brief description of the procedure is given below, and more details can be found in Goede and Barton (1990) and Goede (1991).

The desired sample size to establish a health condition profile (HCP) in any given treatment is 20 fish. The fish should be examined soon after capture to prevent discoloration of the various organs. It is convenient to lay out the fish in a row and begin with an assessment of the external features. Observations of fin damage, opercles, eyes, gills, pseudobranchs, and the thymus tissue are made according to the classification scheme and coding system outlined in Table 5. After the external observations are completed, the fish is opened up using a pair of dissecting scissors to expose the internal organs. A ventral cut from the anal vent forward to the pectoral girdle is usually the most efficient method to open the fish. The physical appearance of the liver, kidney, and spleen is evaluated, as well as the relative amount of mesenteric fat, hindgut inflammation, and bile in the gallbladder (Table 5). The necropsy data can be recorded on a standardized data sheet and summarized to provide the following information: (1) percentage of fish with normal and abnormal eyes, gills, pseudobranchs, thymus, spleens, hindguts, kidneys, and livers; (2) mean index values of damage to fins, thymus hemorrhage, mesenteric fat deposition, hindgut inflammation, and bile color. Data entry and calculations can also be performed in the field using a laptop computer and spreadsheet program (Goede and Houghton 1987).

In its original form, the necropsy-based assessment method was used to determine HCPs mainly for hatchery fish populations (Novotny and Beeman 1990; Iwama et al. 1995; Wagner et al. 1996). Iwama et al. (1995) determined HCPs on a hatchery population of juvenile Atlantic salmon *Salmo salar* and, while all fish sampled exhibited normal swimming and feeding patterns, one tank contained fish that were known to be infected with *Aeromonas salmonicida*, the causative agent of furunculosis. They found a higher proportion of abnormal gills, kidneys, and livers, and a greater incidence of fin damage in fish that were infected compared with healthy fish that were held in different tanks. Thus, the HCP may be useful as an early warning system to detect departures from normal before more sophisticated and expensive diagnostic techniques, such as the previously discussed physiological indicators, are employed, although the necropsy method is not meant to serve as a diagnostic tool.

Table 5. Description of variables and numerical ranking system used in the health condition profile (HCP) and health assessment index (HAI) fish necropsy systems (modified from Goede and Barton 1990 and Adams et al. 1993, and adapted from Morgan and Iwama 1997).

Variable	Condition	HCP value	HAI value
Fins	No active erosion	0	0
	Light active erosion	1	10
	Moderate active erosion	2	20
	Severe active erosion with hemorrhaging	3	30
Opercles	No shortening	0	not
	Slight shortening	1	assigned
	Severe shortening, gills exposed	2	
Eyes	Normal	N	0
	Blind (one or both)	B	30
	Exophthlamic; swollen, protruding (one or both)	E	30
	Hemorrhaging (one or both)	H	30
	Missing one or both eyes	M	30
	Other	OT	30
Gills	Normal	N	0
	Frayed; erosion at tips of gills	F	30
	Clubbed; swelling at the end of gills	C	30
	Marginate; colorless margin along tips	M	30
	Pale in color	P	30
	Other	OT	30
Pseudobranchs	Normal; flat or concave in appearance	N	0
	Swollen and convex in aspect	S	30
	Lithic; white mineral deposits	L	30
	Swollen and lithic	S&L	30
	Inflamed; redness, hemorrhage	I	30
	Other	OT	30
Thymus	No hemorrhage	0	0
	Mild hemorrhage	1	10
	Moderate hemorrhage	2	20
	Severe hemorrhage	3	30
Mesenteric fat	No fat deposits	0	not
	Less than 50% coverage of pyloric caeca with fat	1	assigned
	50% of pyloric caeca covered with fat	2	
	More than 50% of caeca covered with fat	3	
	Pyloric caeca completely fat covered	4	

Table 5. (continued.)

Variable	Condition	HCP value	HAI value
Spleen	Normal; black, very dark red, or red	B	0
	Normal; granular, rough appearance of spleen	G	0
	Nodular; cysts or nodules in the spleen	D	30
	Enlarged; noticeably enlarged	E	30
	Other	OT	30
Hindgut	Normal; no inflammation	0	0
	Slight inflammation or reddening	1	10
	Moderate inflammation or reddening	2	20
	Severe inflammation or reddening	3	30
Kidney	Normal; firm, red color, lying flat against the backbone	N	0
	Swollen or enlarged	S	30
	Mottled; gray discoloration	M	30
	Granular; granular appearance and texture	G	30
	Urolithiasis; creamy white deposits in the kidney	U	30
	Other	OT	30
Liver	Normal; red or pink color	A	0
	Fatty or "coffee with cream" color	C	30
	Nodules or cysts in the liver	D	30
	Focal discoloration	E	30
	General discoloration in whole liver	F	30
	Other	OT	30
Bile	yellow color bile, gall bladder mostly empty	0	not assigned
	yellow bile, mostly full bladder	1	
	light green bile, full bladder	2	
	dark green to blue-green bile, full bladder	3	

Adams et al. (1993) modified the original field necropsy method to a health assessment index (HAI) system to provide a quantitative index so that statistical comparisons can be made between data sets. In this system, all variables are assigned numerical values to allow the calculation of a single HAI for each fish and, thus, mean values for each treatment (Table 5; Adams et al. 1993). During the past few years, the HAI and variations incorporating species-specific features have been used in several field studies involving contamination assessments (Barton 1994; Raymond and Shaw 1997; Steyermark et al. 1999) and other environmental disturbances (Sutton et al. 2000; McKinney 2001). These studies indicate that the HAI shows promise in detecting health problems in fish populations exposed to environmental stressors, particularly in bottom-feeding species that are in contact with contaminated sediments, such as the ictalurids (e.g., Steyermark et al. 1999). If the HAI approach is used, the HCP-ranking descriptions should also be recorded for possible future use.

A summary of advantages and disadvantages of the condition indices described in this chapter are presented in Table 6.

Summary

Knowledge and understanding of what constitutes stress in fish has increased immensely in the past few decades, notably in the area of physiological mechanisms and responses that lead to changes in metabolism and growth, immune functions, reproductive capacity, and normal behavior. Many of these changes are now used routinely for assessing stressed states in fish by measuring the primary and secondary physiological responses in individuals and tertiary or whole-animal changes that can relate to stress-induced alterations in fish populations. Stress is mediated through neuronal and endocrine pathways, known as the primary response, following initial perception of the stressor, which in turn influence secondary physiological features and whole-animal performance characteristics in the fish. Initially, the stress response is considered adaptive, one designed to help the fish overcome the disturbance and regain a homeostatic or normal state. If the stressor is overly severe or long-lasting, however, the fish may no longer be able to cope with it and enters a maladaptive or distressed state leading to a pathological condition or possibly death.

Typical primary responses used for evaluating stress in fish include determining circulating levels of cortisol and, to a lesser extent, catecholamines. Secondary responses, which may or may not be caused directly by the endocrine response, include measurable changes in blood glucose, lactate or lactic acid, and major ions (e.g., chloride, sodium, and potassium), and tissue levels of glycogen and HSPs. Many other apparent nonstress factors, however, influence characteristic physiological stress responses in fish that biologists need to be aware of for properly interpreting data, including genetic (e.g., species, strain), developmental (e.g., life history stage), and environmental (e.g., temperature, nutrition, water quality) factors.

Some of the whole-animal or tertiary changes, such as growth, condition, and general health, are reflected in various indices that can be useful to describe stressed states in fish. Some indices for this purpose include K (or CF), the length–weight relationship, W_r, and organosomatic indices (ratios of organ masses to total body mass) such as HSI, GSI, VSI, and SSI. Additionally, using a necropsy-based index system can help develop useful data on the normal status of fish populations for possible future detection of changes

Table 6. Uses, advantages, and disadvantages of selected condition-based indicators used to evaluate organismal changes in fish resulting from stress (see text for details).

Indicator	Uses	Advantages	Disadvantages
Condition factor (K or CF)	A measure of robustness of the fish Can be used to indicate changes in energy storage and metabolism and, possibly, feeding activity	Length and weight data easy to collect without invasive procedures	Relatively insensitive to acute stressors Assumes constant length exponent of 3 Varies considerably with sexual maturation and reproduction, especially in female fish because of egg mass Also varies with season and developmental state Changes occurring in body-water content can confound interpretation
Length–weight relationship	A linear regression equation used for comparing fish populations	Length and weight data easy to collect without invasive procedures A more useful comparative index for wild populations than K as length exponent is empirically based rather than fixed	Relatively insensitive to acute stressors Comparison of different life history stages problematic; relationship can change because of allometric growth
Relative weight index (Wr)	An empirically based index for comparing fish populations in management studies	Accounts for allometric growth differences over fish's life history	Published values needed for calculating index only available for limited number of species Assumes empirical data used for calculating index are representative of the ideal standard
Hepatosomatic index (HSI) Gonadosomatic index (GSI) Viscerosomatic index (VSI) Splenosomatic index (SSI)	HSI: indicator of nutritional or energetic status; possible indicator of chronic exposure to specific toxicants GSI: useful indicator of status of gonads or reproductive state of the fish VSI: indicator of nutritional or energetic status SSI: indicator of hematopoietic capacity of the fish and blood transfer into and out of circulation	Relatively simple to obtain data and calculate index	Requires sacrificing fish to remove and weigh organ Not particularly sensitive to acute stress Can vary with life history stage and season

Table 6. (continued.)

Indicator	Uses	Advantages	Disadvantages
Health condition profile (HCP)	Provides a long-term data set on the normal status of the population	Method is simple and inexpensive Changes in HCP over time may reflect changes in health of the population May serve as an early warning system to detect departures from normal in the population	Parameters used in the HCP may not be particularly sensitive to acute stress
Health assessment index (HAI)	Extends HCP information by providing a quantitative index for the population	Method is simple and inexpensive Changes in HAI over time may indicate changes in the population resulting from chronic environmental perturbation	Parameters used may not be particularly sensitive to acute stress Assigned numerical values are not empirically or experimentally based

resulting from environmental stressors. Index data on fish condition and health are relatively easy to collect and can contribute appreciably toward understanding long-term trends in fish populations subjected to perturbations. Understanding whole-animal changes is important to managers as these condition-based indices often provide clues that help relate the physiological stress responses of individual fish and changes manifested at the population level of organization.

Acknowledgments

We are grateful to Scott Lankford, Joe Cech, and Gary Wedemeyer for their constructive review comments on an earlier draft of this chapter. We also thank Neil Ruane, Wageningen University, and Alf Haukenes, University of South Dakota, for allowing us to use their unpublished data.

References

Adams, S. M., editor. 1990. Biological indicators of stress in fish. American Fisheries Society, Symposium 8, Bethesda, Maryland.

Adams, S. M., J. E. Breck, and R. B. McLean. 1985. Stress-induced mortality of gizzard shad in a southeastern U.S. reservoir. Environmental Biology of Fishes 13:103–112.

Adams, S. M., A. M. Brown, and R. W. Goede. 1993. A quantitative health assessment index for rapid evaluation of fish condition in the field. Transactions of the American Fisheries Society 122:63–73.

Adams, S. M., and R. B. McLean. 1985. Estimation of largemouth bass, *Micropterus salmoides* Lacépède, growth using the liver somatic index and physiological variables. Journal of Fish Biology 26:111–126.

Adams, S. M., R. B. McLean, and J. A. Parotta. 1982. Energy partitioning in large-mouth bass under conditions of seasonally fluctuating prey availability. Transactions of the American Fisheries Society 111:549–558.

Alanärä, A., S. Winberg, E. Brännäs, A. Kiessling, E. Höglund, and U. Elofsson. 1998. Feeding behavior, brain serotonergic activity levels, and energy reserves of Arctic char (*Salvelinus alpinus*) within a dominance hierarchy. Canadian Journal of Zoology 76:212–220.

Anderson, R. O., and S. J. Gutreuter. 1983. Length, weight, and associated structural indices. Pages 283–300 *in* L. A. Nielsen and D. L. Johnson, editors. Fisheries techniques. American Fisheries Society, Bethesda, Maryland.

Anderson, R. O., and R. M. Neumann. 1996. Length, weight, and associated structural indices. Pages 447–482 *in* B. R. Murphy and D. W. Willis, editors. Fisheries techniques, 2nd edition. American Fisheries Society, Bethesda, Maryland.

Atwood, H. L., J. R. Tomasso Jr., P. J. Ronan, B. A. Barton, and K. J. Renner. 2000. Brain monoamine concentrations as predictors of growth inhibition in channel catfish exposed to ammonia. Journal of Aquatic Animal Health 12:69–73.

Barnes, M. A., G. Power, and R. G. H. Downer. 1984. Stress-related changes in lake whitefish (*Coregonus clupeaformis*) associated with a hydroelectric control structure. Canadian Journal of Fisheries and Aquatic Sciences 41:141–150.

Barry, T. P., J. A. Malison, J. A. Held, and J. J. Parrish. 1995. Ontogeny of the cortisol stress response in larval trout. General and Comparative Endocrinology 97:57–65.

Barton, B. A. 1994. Monitoring the health and condition of mountain whitefish populations using the autopsy-based assessment method: a case study. Pages 459–463 *in* D. D. MacKinlay, editor. High performance fish: proceedings of the international fish physiology symposium. Vancouver, British Columbia.

Barton, B. A. 1996. General biology of salmonids. Pages 29–95 *in* W. Pennell and B. A. Barton, editors. Principles of salmonid culture. Developments in aquaculture and fisheries science 29, Elsevier B. V., Amsterdam.

Barton, B. A. 1997. Stress in finfish: past, present and future: a historical perspective. Pages 1–33 *in* G. K. Iwama, A. D. Pickering, J. P. Sumpter, and C. B. Schreck, editors. Fish stress and health in aquaculture. Society for experimental biology seminar series 62, Cambridge University Press, Cambridge, U.K.

Barton, B. A. 2000. Salmonid fishes differ in their cortisol and glucose responses to handling, and transport stress. North American Journal of Aquaculture 62:12–18.

Barton, B. A., H. Bollig, B. L. Hauskins, and C. R. Jansen. 2000. Juvenile pallid (*Scaphirhynchus albus*) and hybrid pallid × shovelnose (*S. albus* × *platorynchus*) sturgeons exhibit low physiological responses to acute handling, and severe confinement. Comparative Biochemistry and Physiology, Part A 126:125–134.

Barton, B. A., and W. P. Dwyer. 1997. Physiological stress effects of continuous- and pulsed-DC electroshock on juvenile bull trout. Journal of Fish Biology 51:998–1008.

Barton, B. A., and G. K. Iwama. 1991. Physiological changes in fish from stress in aquaculture with emphasis on the response and effects of corticosteroids. Annual Review of Fish Diseases 1:3–26.

Barton, B. A., A. B. Rahn, G. Feist, H. Bollig, and C. B. Schreck. 1998. Physiological stress responses of the freshwater chondrostean paddlefish (*Polyodon spathula*) to acute physical disturbances. Comparative Biochemistry and Physiology, Part A 120:355–363.

Barton, B. A., and C. B. Schreck. 1987a. Metabolic cost of acute physical stress in juvenile steelhead. Transactions of the American Fisheries Society 116:257–263.

Barton, B. A., and C. B. Schreck. 1987b. Influences of acclimation temperature on interrenal and carbohydrate stress responses in juvenile chinook salmon (*Oncorhynchus tshawytscha*). Aquaculture 62:299–310.

Barton, B. A., C. B. Schreck, and L. D. Barton. 1987. Effects of chronic cortisol administration and daily acute stress on growth, physiological conditions, and stress responses in juvenile rainbow trout. Diseases of Aquatic Organisms 2:173–185.

Barton, B. A., C. B. Schreck, R. D. Ewing, A. R. Hemmingsen, and R. Patiño. 1985a. Changes in plasma cortisol during stress and smoltification in coho salmon, *Oncorhynchus kisutch*. General and Comparative Endocrinology 59:468–471.

Barton, B. A., C. B. Schreck, and L. G. Fowler. 1988. Fasting and diet content affect stress-induced changes in plasma glucose and cortisol in juvenile chinook salmon. Progressive Fish-Culturist 50:16–22.

Barton, B. A., G. S. Weiner, and C. B. Schreck. 1985b. Effect of prior acid exposure on physiological responses of juvenile rainbow trout (*Salmo gairdneri*) to acute handling stress. Canadian Journal of Fisheries and Aquatic Sciences 42:710–717.

Barton, B. A., and R. E. Zitzow. 1995. Physiological responses of juvenile walleyes to handling stress with recovery in saline water. Progressive Fish-Culturist 57:267–276.

Baumann, P. C. 1992. The use of tumors in wild fish populations of fish to assess ecosystem health. Journal of Aquatic Ecosystem Health 1:135–146.

Baumann, P. C., M. J. Mac, S. B. Smith, and J. C. Harshbarger. 1991. Tumor frequencies in walleye (*Stizostedion vitreum*) and brown bullhead (*Ameiurus nebulosus*) and sediment contaminants in tributaries of the Laurentian Great Lakes. Canadian Journal of Fisheries and Aquatic Sciences 48:1804–1810.

Benguira, S., and A. Hontela. 2000. Adrenocorticotrophin- and cyclic adenosine 3',5'-monophosphate-stimulated cortisol secretion in interrenal tissue of rainbow trout exposed in vitro to DDT compounds. Environmental Toxicology and Chemistry 19:842–847.

Bergstedt, L. C., and E. P. Bergersen. 1997. Health and movements of fish in response to sediment sluicing in the Wind River, Wyoming. Canadian Journal of Fisheries and Aquatic Sciences 54:312–319.

Bister, T. J., D. W. Willis, M. L. Brown, S. M. Jordan, R. M. Neumann, M. C. Quist, and C. S. Guy. 2000. Proposed standard weight (*Ws*) equations, and standard length categories for 18 warmwater nongame and riverine fish species. North American Journal of Fisheries Management 20:570–574.

Blackwell, B. G., M. L. Brown, and D. W. Willis. 2000. Relative weight (*Wr*) status and current use in fisheries assessment and management. Reviews in Fisheries Science 8:1–44.

Bradford, C. S., M. S. Fitzpatrick, and C. B. Schreck. 1992. Evidence for ultra-short-loop feedback in ACTH-induced interrenal steroidogenesis in coho salmon: acute self expression of cortisol secretion in vitro. General and Comparative Endocrinology 87:292–299.

Bradley, B. P. 1993. Are the stress proteins indicators of exposure or effect? Marine Environmental Research 35:85–88.

Bradley, B. P., C. M. Gonzalez, J. A. Bond, and B. E. Tepper. 1994. Complex mixture analysis using protein expression as a qualitative and quantitative tool. Environmental Toxicology and Chemistry 13:1043–1050.

Brett, J. R. 1958. Implications and assessment of environmental stress. Pages 69–83 *in* P. A. Larkin, editor. The investigation of fish-power problems. H. R. MacMillan lectures in fisheries, University of British Columbia, Vancouver.

Brodeur, J. C., C. Daniel, A. C. Ricard, and A. Hontela. 1998. In vitro response to ACTH of the interrenal tissue of rainbow trout (*Oncorhynchus mykiss*) exposed to cadmium. Aquatic Toxicology 42:103–113.

Brodeur, J. C., C. Girard, and A. Hontela. 1997. Use of perifusion to assess in vitro the functional integrity of interrenal tissue in fish from polluted sites. Environmental Toxicology and Chemistry 16:2171–2178.

Brown, J. A. 1993. Endocrine responses to environmental pollutants. Pages 276–296 *in* J. C. Rankin and F. B. Jensen, editors. Fish ecophysiology. Fish and fisheries series 9, Chapman and Hall, London.

Brown, S. B., J. G. Eales, R. E. Evans, and T. J. Hara. 1984. Interrenal, thyroidal, and carbohydrate responses of rainbow trout (*Salmo gairdneri*) to environmental acidification. Canadian Journal of Fisheries and Aquatic Sciences 41:36–45.

Busaker, G. P., I. R. Adelman, and E. M. Goolish. 1990. Growth. Pages 363–387 *in* C. B. Schreck and P. B. Moyle, editors. Methods for fish biology. American Fisheries Society, Bethesda, Maryland.

Cairns, V. W., P. V. Hodson, and J. O. Nriagu, editors. 1984. Contaminant effects on fisheries. Wiley, New York.

Carlander, K. D. 1969. Handbook of freshwater fishery biology, volume 1. Iowa State University Press, Ames.

Carlander, K. D. 1977. Handbook of freshwater fishery biology, volume 2. Iowa State University Press, Ames.

Carlander, K. D. 1997. Handbook of freshwater fishery biology, volume 3. Iowa State University Press, Ames.

Cech, Jr., J. J. 2000. Osmoregulation in bony fishes. Pages 614–622 *in* R. R. Stickney, editor. Encyclopedia of aquaculture. John Wiley and Sons, New York.

Chrousos, G. P. 1998. Stressors, stress, and neuroendocrine integration of the adaptive response. Annals of the New York Academy of Sciences 851:311–335.

Cone, R. S. 1989. The need to reconsider the use of condition indices in fishery science. Transactions of the American Fisheries Society 118:510-514.

Cunjak, R. A., and G. Power. 1986. Seasonal changes in the physiology of brook trout, *Salvelinus fontinalis* (Mitchill), in a sub-Arctic river system. Journal of Fish Biology 29:279–288.

Davis, K. B., and N. C. Parker. 1990. Physiological stress in striped bass: effect of acclimation temperature. Aquaculture 91:349–358.

Davis, K. B., M. A. Suttle, and N. C. Parker. 1984. Biotic and abiotic influences on corticosteroid hormone rhythms in channel catfish. Transactions of the American Fisheries Society 113:414–421.

Davis, L. E., and C. B. Schreck. 1997. The energetic response to handling stress in juvenile coho salmon. Transactions of the American Fisheries Society 126:248–258.

Davis, M. W., B. L. Olla, and C. B. Schreck. 2001. Stress induced by hooking, net towing, elevated sea water temperature, and air in sablefish: lack of concordance between mortality and physiological measures of stress. Journal of Fish Biology 58:1–15.

Deane, E. E., S. P. Kelly, C. K. M. Lo, and N. Y. S. Woo. 1999. Effects of GH, prolactin and cortisol on hepatic heat shock protein 70 expression in a marine teleost *Sparus sarba*. Journal of Endocrinology 161:413–421.

DeBoeck, G., G. E. Nilsson, U. Elofsson, A. Vlaeminck, and R. Blust. 1995. Brain monoamine levels and energy status in common carp (*Cyprinus carpio*) after exposure to sublethal levels of copper. Aquatic Toxicology 33:265–277.

Dhabhar, F. S., and B. S. McEwan. 1997. Acute stress enhances while chronic stress suppresses cell-mediated immunity *in vivo*: a potential role for leukocyte trafficking. Brain, Behavior, and Immunity 11:286–306.

Donaldson, E. M. 1981. The pituitary-interrenal axis as an indicator of stress in fish. Pages 11–47 *in* A. D. Pickering, editor. Stress and fish. Academic Press, New York.

Driedzic, W. R., and P. W. Hochachka. 1978. Metabolism in fish during exercise. Pages 503–543 *in* W. S. Hoar and D. J. Randall, editors. Fish physiology, volume 7. Academic Press, New York.

Eddy, F. B. 1981. Effects of stress on osmotic and ionic regulation in fish. Pages 77–102 *in* A. D. Pickering, editor. Stress and fish. Academic Press, New York.

Fabacher, D. L., and P. C. Baumann. 1985. Enlarged livers and hepatic microsomal mixed function oxidase components in tumor-bearing brown bullheads from a chemically contaminated river. Environmental Toxicology and Chemistry 4:703–710.

Fabbri, E., A. Capuzzo, and T. W. Moon. 1998. The role of circulating catecholamines in the regulation of fish metabolism: an overview. Comparative Biochemistry and Physiology, Part C 120:177–192.

Fagerlund, U. H. M., J. R. McBride, and E. T. Stone. 1981. Stress-related effects of hatchery rearing density on coho salmon. Transactions of the American Fisheries Society 110:644–649.

Fletcher, G. L., M. J. King, J. W. Kiceniuk, and R. F. Addison. 1982. Liver hypertrophy in winter flounder following exposure to experimentally oiled sediments. Comparative Biochemistry and Physiology, Part C 73:457–462.

Folmar, L. C. 1993. Effects of chemical contaminants on blood chemistry of teleost fish: a bibliography and synopsis of selected effects. Environmental Toxicology and Chemistry 12:337–375.

Forsyth, R. B., E. P. M. Candido, S. L. Babich, and G. K. Iwama. 1997. Stress protein expression in coho salmon with bacterial kidney disease. Journal of Aquatic Animal Health 9:18–25.

Fritsche, R., S. G. Reid, S. Thomas, and S. F. Perry. 1993. Serotonin-mediated release of catecholamines in the rainbow trout, *Oncorhynchus mykiss*. Journal of Experimental Biology 178:191–204.

Fryer, J. N., and R. E. Peter. 1977. Hypothalamic control of ACTH secretion in goldfish. III. Hypothalamic cortisol implant studies. General and Comparative Endocrinology 33:215–225.

Gamperl, A. K., M. M. Vijayan, and R. G. Boutilier. 1994. Experimental control of stress hormone levels in fishes: techniques and applications. Reviews in Fish Biology and Fisheries 4:215–255.

Girard, C., J. C. Brodeur, and A. Hontela. 1998. Responsiveness of the interrenal tissue of yellow perch (*Perca flavescens*) from contaminated sites to an ACTH challenge test in vivo. Canadian Journal of Fisheries and Aquatic Sciences 55:438–450.

Goede, R. W. 1991. Fish health/condition assessment procedures. Utah Division of Wildlife Resources, Fisheries Experiment Station, Logan, Utah.

Goede, R. W., and B. A. Barton. 1990. Organismic indices and an autopsy-based assessment as indicators of health and condition in fish. Pages 93–108 *in* S. M. Adams, editor. Biological indicators of stress in fish. American Fisheries Society, Symposium 8, Bethesda, Maryland.

Goede, R. W., and S. Houghton. 1987. AUSUM: a computer program for the autopsy-based fish health/condition assessment system. Utah Division of Wildlife Resources, Fisheries Experiment Station, Logan, Utah.

Grant, S. M., and J. A. Brown. 1999. Variations in condition of coastal Newfoundland 0-group Atlantic cod (*Gadus morhua*): field and laboratory studies using simple condition indices. Marine Biology 133:611–620.

Grant, S. M., J. A. Brown, and D. L. Boyce. 1998. Enlarged fatty livers of small juvenile cod: a comparison of laboratory-cultured and wild juveniles. Journal of Fish Biology 52:1105–1114.

Hanson, R. C., and W. R. Fleming. 1979. Serum cortisol levels of juvenile bowfin, *Amia calva*: effects of hypophysectomy, hormone replacement and environmental salinity. Comparative Biochemistry and Physiology, Part A 63:499–502.

Hille, S. 1982. A literature review of the blood chemistry of rainbow trout, *Salmo gairdneri* Rich. Journal of Fish Biology 20:535–569.

Hontela, A. 1997. Endocrine and physiological responses of fish to xenobiotics: role of glucocorticosteroid hormones. Reviews in Toxicology 1:1–46.

Hontela, A., J. B. Rasmussen, C. Audet, and G. Chevalier. 1992. Impaired cortisol stress response in fish from environments polluted by PAHs, PCBs, and mercury. Archives of Environmental Contamination and Toxicology 22:278–283.

Hyatt, M. W., and W. A. Hubert. 2001. Proposed standard-weight equations for brook trout. North American Journal of Fisheries Management 21:253–254.

Idler, D. R., and B. Truscott. 1966. 1α-Hydroxycorticosterone from cartilaginous fish: a new adrenal steroid in blood. Journal of the Fisheries Research Board of Canada 23:615–619.

Idler, D. R., and B. Truscott. 1967. 1α-Hydroxycorticosterone: synthesis *in vitro* and properties of an interrenal steroid in the blood of cartilaginous fish (Genus *Raja*). Steroids 9:457–477.

Idler, D. R., and B. Truscott. 1972. Corticosteroids in fish. Pages 126–252 *in* Steroids in nonmammalian vertebrates. Academic Press, New York.

Iwama, G. K., J. C. McGeer, and N. J. Bernier. 1992. The effects of stock and rearing density on the stress response in juvenile coho salmon (*Oncorhynchus kisutch*). ICES Marine Science Symposium 194:67–83.

Iwama, G. K., J. D. Morgan, and B. A. Barton. 1995. Simple methods for monitoring stress and general condition of fish. Aquaculture Research 26:273–282.

Iwama, G. K., A. D. Pickering, J. P. Sumpter, and C. B. Schreck, editors. 1997. Fish stress and health in aquaculture. Society for experimental biology seminar series 62, Cambridge University Press, Cambridge, U.K.

Iwama, G. K., P. T. Thomas, R. B. Forsyth, and M. M. Vijayan. 1998. Heat shock protein expression in fish. Reviews in Fish Biology and Fisheries 8:35–56.

Iwama, G. K., M. M. Vijayan, R. B. Forsyth, and P. A. Ackerman. 1999. Heat shock proteins and physiological stress in fish. American Zoologist 39:901–909.

Janz, D. M., M. E. McMaster, K. R. Munkittrick, and G. Van Der Kraak. 1997. Elevated ovarian follicular apoptosis and heat shock protein-70 expression in white sucker exposed to bleached kraft pulp mill effluent. Toxicology and Applied Pharmacology 147:391–398.

Laflamme, J.-S., Y. Couillard, P. G. C. Campbell, and A. Hontela. 2000. Interrenal metallothionein and cortisol secretion in relation to Cd, Cu, and Zn exposure in yellow perch, *Perca flavescens*, from Abitibi lakes. Canadian Journal of Fisheries and Aquatic Sciences 57:1692–1700.

Lambert, Y., and J.-D. Dutil. 1997. Can simple condition indices be used to monitor and quantify seasonal changes in the energy reserves of Atlantic cod (*Gadus morhua*)? Canadian Journal of Fisheries and Aquatic Sciences 54(Supplement 1):104–112.

Leblond, V. S., M. Bisson, and A. Hontela. 2001. Inhibition of cortisol secretion in dispersed head kidney cells of rainbow trout (*Oncorhynchus mykiss*) by endosulfan, an organochlorine pesticide. General and Comparative Endocrinology 121:48–56.

Leblond, V. S., and A. Hontela. 1999. Effects of *in vitro* exposures to cadmium, mercury, zinc and 1-(2-chlorophenyl)-1-(4-chlorophenyl)-2,2-dichloroethane on steroidogenesis by dispersed interrenal cells of rainbow trout (*Oncorhynchus mykiss*). Toxicology and Applied Pharmacology 157:16–22.

Lee, R. M., S. B. Gerking, and B. Jezierska. 1983. Electrolyte balance and energy mobilization in acid-stressed rainbow trout, *Salmo gairdneri*, and their relation to reproductive success. Environmental Biology of Fishes 8:115–123.

Maule, A. G., C. B. Schreck, and S. L. Kaattari. 1987. Changes in the immune system of coho salmon (*Oncorhynchus kisutch*) during the parr-to-smolt transformation and after implantation of cortisol. Canadian Journal of Fisheries and Aquatic Sciences 44:161–166.

Mazeaud, M. M., and F. Mazeaud. 1981. Adrenergic responses to stress in fish. Pages 49–75 *in* A. D. Pickering, editor. Stress and fish. Academic Press, New York.

Mazeaud, M. M., F. Mazeaud, and E. M. Donaldson. 1977. Primary and secondary effects of stress in fish. Transactions of the American Fisheries Society 106:201–212.

McKinney, T., A. T. Robinson, D. W. Speas, and R. S. Rogers. 2001. Health assessment, associated metrics, and nematode parasitism of rainbow trout in the Colorado River below Glen Canyon Dam, Arizona. North American Journal of Fisheries Management 21:62–69.

Medford, B. A., and W. C. Mackay. 1978. Protein and lipid content of gonads, liver, and muscle of northern pike (*Esox lucius*) in relation to gonad growth. Journal of the Fisheries Research Board of Canada 35:213–219.

Meyer, F. P., and L. A. Barclay, editors. 1990. Field manual for the investigation of fish kills. U.S. Fish and Wildlife Service Resource Publication 177, Washington, D.C.

Moberg, G. P. 1985. Biological response to stress: key to assessment of animal well-being? Pages 27–49 *in* G. P. Moberg, editor. Animal stress. American Physiological Society, Bethesda, Maryland.

Mommsen, T. P., M. M. Vijayan, and T. W. Moon. 1999. Cortisol in teleosts: dynamics, mechanisms of action, and metabolic regulation. Reviews in Fish Biology and Fisheries 9:211–268.

Morgan, J. D., and G. K. Iwama. 1997. Measurements of stressed states in the field. Pages 247–268 *in* G. K. Iwama, A. D. Pickering, J. P. Sumpter, and C. B. Schreck, editors. Fish stress and health in aquaculture. Society for experimental biology seminar series 62, Cambridge University Press, Cambridge, U.K.

Murphy, B. R., M. L. Brown, and T. A. Springer. 1990. Evaluation of the relative weight (*Wr*) index, with new applications to walleye. North American Journal of Fisheries Management 10:85–97.

Niimi, A. J. 1990. Review of biochemical methods and other indicators to assess fish health in aquatic ecosystems containing toxic chemicals. Journal of Great Lakes Research 16:529–541.

Noakes, D. L. G., and J. F. Leatherland. 1977. Social dominance and interrenal activity in rainbow trout, *Salmo gairdneri* (Pisces, Salmonidae). Environmental Biology of Fishes 2:131–136.

Noga, E. J., S. Botts, M.-S. Yang, and R. Avtalion. 1998. Acute stress causes skin ulceration in striped bass and hybrid bass (*Morone*). Veterinary Pathology 35:102–107.

Noga, E. J., C. Wang, C. B. Grindem, and R. Avtalion. 1999. Comparative clinopathological responses of striped bass and palmetto bass to acute stress. Transactions of the American Fisheries Society 128:680–686.

Norris, D. O., S. Donahue, R. M. Dores, J. K. Lee, T. A. Maldonado, T. Ruth, and J. D. Woodling. 1999. Impaired adrenocortical response to stress by brown trout, *Salmo trutta*, living in metal-contaminated waters of the Eagle River, Colorado. General and Comparative Endocrinology 113:1–8.

Norris, D. O., S. B. Felt, J. D. Woodling, and R. M. Dores. 1997. Immunocytochemical and histological differences in the interrenal axis of feral brown trout, *Salmo trutta*, in metal-contaminated waters. General and Comparative Endocrinology 108:343–351.

Novotny, J. F., and J. W. Beeman. 1990. Use of a fish health condition profile in assessing the health and condition of juvenile chinook salmon. Progressive Fish-Culturist 52:162–170.

Pickering, A. D., editor. 1981. Stress and fish. Academic Press, London.

Pickering, A. D. 1993. Endocrine-induced pathology in stressed salmonid fish. Fisheries Research 17:35–50.

Pickering, A. D. 1998. Stress responses of farmed fish. Pages 222–255 *in* K. D. Black and A. D. Pickering, editors. Biology of farmed fish. Sheffield Academic Press, Sheffield, U.K.

Pickering, A. D., and T. G. Pottinger. 1987. Poor water quality suppresses the cortisol response of salmonid fish to handling and confinement. Journal of Fish Biology 30:363–374.

Piper, R. G., I. B. McElwain, L. E. Orme, J. P. McCraren, L. G. Fowler, and J. R. Leonard. 1982. Fish hatchery management. U.S. Fish and Wildlife Service, Washington, D.C.

Poels, C. L. M., M. A. van der Gaag, and J. F. J. van der Kerkhoff. 1980. An investigation into the longterm effects of Rhine water on rainbow trout. Water Research 14:1029–1035.

Pottinger, T. G., P. H. M. Balm, and A. D. Pickering. 1995. Sexual maturity modifies the responsiveness of the pituitary-interrenal axis to stress in male rainbow trout. General and Comparative Endocrinology 98:311–320.

Pottinger, T. G., T. R. Carrick, A. Appleby, and W. E. Yeomans. 2000. High blood cortisol levels and low cortisol receptor affinity: Is the chub, *Leuciscus cephalus*, a cortisol-resistant teleost? General and Comparative Endocrinology 120:108–117.

Pottinger, T. G., T. A. Moran, and J. A. W. Morgan. 1994. Primary and secondary indices of stress in the progeny of rainbow trout (*Oncorhynchus mykiss*) selected for high and low responsiveness to stress. Journal of Fish Biology 44:149–163.

Pottinger, T. G., A. D. Pickering, and M. A. Hurley. 1992. Consistency in the stress response of individuals of two strains of rainbow trout, *Oncorhynchus mykiss*. Aquaculture 103:275–289.

Pratt, W. B. 1997. The role of the hsp90-based chaperone system in signal transduction by nuclear receptors and receptor signalling via Mapkinase. Annual Review of Pharmacology and Toxicology 37:297–326.

Ram, R. J., and S. K. Singh. 1988. Long-term effect of ammonium sulfate fertilizer on histopathology of adrenal in the teleost, *Channa punctatus* (Bloch). Bulletin of Environmental Contamination and Toxicology 41:880–887.

Rand, G. M., and S. R. Petrocelli, editors. 1985. Fundamentals of aquatic toxicology. Hemisphere Publishing, Washington, D.C.

Randall, D. J., and S. F. Perry. 1992. Catecholamines. Pages 255–300 *in* W. S. Hoar and D. J. Randall, editors. Fish physiology, volume 12B. Academic Press, New York.

Raymond, B. A., and D. P. Shaw. 1997. Fraser River action plan resident fish condition and contaminants assessment. Water Science and Technology 35:389–395.

Reid, S. G., N. J. Bernier, and S. F. Perry. 1998. The adrenergic stress response in fish: control of catecholamine storage and release. Comparative Biochemistry and Physiology, Part C 120:1–27.

Reid, S. G., M. M. Vijayan, and S. F. Perry. 1996. Modulation of catecholamine storage and release by the pituitary-interrenal axis in the rainbow trout (*Oncorhynchus mykiss*). Journal of Comparative Physiology B 165:665–676.

Roche, H., and G. Bogé. 1996. Fish blood parameters as a potential tool for identification of stress caused by environmental factors and chemical intoxication. Marine Environmental Research 41:27–43.

Ruane, N. M., S. E. Wendelaar Bonga, and P. H. M. Balm. 1999. Differences between rainbow trout and brown trout in the regulation of the pituitary-interrenal axis and physiological performance during confinement. General and Comparative Endocrinology 113:210–219.

Sangalang, G. B., M. Weisbart, and D. R. Idler. 1971. Steroids of a chondrostean: corticosteroids and testosterone in the plasma of the American Atlantic sturgeon, *Acipenser oxyrhynchus* Mitchill. Journal of Endocrinology 50:413–421.

Sathiyaa, R., T. Campbell, and M. M. Vijayan. 2001. Cortisol modulates HSP90 mRNA expression in primary cultures of trout hepatocytes. Comparative Biochemistry and Physiology, Part B 129:679–685.

Schlenk, D., E. J. Perkins, W. G. Layher, and Y. S. Zhang. 1996. Correlating metrics of fish health with cellular indicators of stress in an Arkansas bayou. Marine Environmental Research 42:247–251.

Schreck, C. B. 1981. Stress and compensation in teleostean fishes: response to social and physical factors. Pages 295–321 *in* A. D. Pickering, editor. Stress and fish. Academic Press, New York.

Schreck, C. B. 2000. Accumulation and long-term effects of stress in fish. Pages 147–158 *in* G. P. Moberg and J. A. Mench, editors. The biology of animal stress. CABI Publishing, Wallingford, U. K.

Selye, H. 1950. Stress and the general adaptation syndrome. British Medical Journal 1(4667):1383–1392.

Selye, H. 1973. The evolution of the stress concept. American Scientist 61:692–699.

Selye, H. 1974. Stress without distress. McClelland Stewart, Toronto.

Servizi, J. A., R. W. Gordon, D. W. Martens, W. L. Lockhart, D. A. Metner, I. H. Rodgers, J. R. McBride, and R. J. Norstrom. 1993. Effects of biotreated bleached kraft mill effluent on fingerling chinook salmon (*Oncorhynchus tshawytscha*). Canadian Journal of Fisheries and Aquatic Sciences 50:846–857.

Sipiorski, J. T. 2000. Neurotransmitter activity in the fore- and hind-brain of the pallid sturgeon (*Scaphirhynchus albus*) following acute and chronic stress. Masters thesis, University of South Dakota, Vermillion.

Stephens, S. M., J. A. Brown, and S. C. Frankling. 1997. Stress responses of larval turbot, *Scophthalmus maximus* L., exposed to sub-lethal concentrations of petroleum hydrocarbons. Fish Physiology and Biochemistry 17:433–439.

Steyermark, A. C., J. R. Spotila, D. Gillette, and H. Isseroff. 1999. Biomarkers indicate health problems in brown bullheads from the industrialized Schuylkill River, Philadelphia. Transactions of the American Fisheries Society 128:328–338.

Stouthart, A. J. H. X., E. C. H. E. T. Lucassen, F. J. C. van Strien, P. H. M. Balm, R. A. C. Lock, and S. E. Wendelaar Bonga. 1998. Stress responsiveness of the pituitary-interrenal axis during early life stages of common carp (*Cyprinus carpio*). Journal of Endocrinology 157:127–137.

Suarez, R. K., and T. P. Mommsen. 1987. Gluconeogenesis in teleost fishes. Canadian Journal of Zoology 65:1869–1882.

Sumpter, J. P. 1997. The endocrinology of stress. Pages 95–118 *in* G. K. Iwama, A. D. Pickering, J. P. Sumpter, and C. B. Schreck, editors. Fish stress and health in aquaculture. Society for experimental biology seminar series 62, Cambridge University Press, Cambridge, U.K.

Sutton, R. J., C. C. Caldwell, and V. S. Blazer. 2000. Observations of health indices used to monitor a tailwater trout fishery. North American Journal of Fisheries Management 20:267–275.

Tam, W. H., J. N. Fryer, I. Ali, M. R. Dallaire, and B. Valentine. 1988. Growth inhibition, gluconeogenesis, and morphometric studies of the pituitary and interrenal cells of acid-stressed brook trout (*Salvelinus fontinalis*). Canadian Journal of Fisheries and Aquatic Sciences 45:1197–1211.

Vanstone, W. E., and J. R. Markert, J. R. 1968. Some morphological and biochemical changes in coho salmon, *Oncorhynchus kisutch*, during parr-smolt transformation. Journal of the Fisheries Research Board of Canada 25:2403–2418.

Vijayan, M. M., T. P. Mommsen, H. E. Glemet, and T. W. Moon. 1996. Metabolic effects of cortisol treatment in a marine teleost, the sea raven. Journal of Experimental Biology 199:1509–1514.

Vijayan, M. M., and T. W. Moon. 1992. Acute handling stress alters hepatic glycogen metabolism in food-deprived rainbow trout (*Oncorhynchus mykiss*). Canadian Journal of Fisheries and Aquatic Sciences 49:2260–2266.

Vijayan, M. M., and T. W. Moon. 1994. The stress-response and the plasma disappearance of corticosteroid and glucose in a marine teleost, the sea raven. Canadian Journal of Zoology 72:379–386.

Vijayan, M. M., C. Pereira, R. B. Forsyth, C. J. Kennedy, and G. K. Iwama. 1997a. Handling stress does not affect the expression of hepatic heat shock protein 70 and conjugation enzymes in rainbow trout treated with β-naphthoflavone. Life Sciences 61:117–127.

Vijayan, M. M., C. Pereira, E. G. Grau, and G. K. Iwama. 1997b. Metabolic responses to confinement stress in tilapia: the role of cortisol. Comparative Biochemistry and Physiology, Part C 116:89–95.

Vijayan, M. M., C. Pereira, G. Kruzynski, and G. K. Iwama. 1998. Sublethal concentrations of contaminants induce the expression of hepatic heat shock protein 70 in two salmonids. Aquatic Toxicology 40:101–108.

Vijayan, M. M., C. Pereira, and T. W. Moon. 1994a. Hormonal stimulation of hepato-cyte metabolism in rainbow trout following an acute handling stress. Compara-tive Biochemistry and Physiology, Part C 108:321–329.

Vijayan, M. M., P. K. Reddy, J. F. Leatherland, and T. W. Moon. 1994b. The effect of cortisol on hepatocyte metabolism in rainbow trout: a study using the steroid analogue RU486. General and Comparative Endocrinology 96:75–84.

Wagner, E. J., M. D. Routledge, and S. S. Intelmann. 1996. Fin condition and health profiles of albino rainbow trout reared in concrete raceways with and without a cobble substrate. Progressive Fish-Culturist 58:38–42.

Wedemeyer, G. A. 1996. Physiology of fish in intensive culture systems. Chapman and Hall, New York.

Wedemeyer, G. A., B. A. Barton, and D. J. McLeay. 1990. Stress and acclimation. Pages 451–489 in C. B. Schreck and P. B. Moyle, editors. Methods for fish biol-ogy. American Fisheries Society, Bethesda, Maryland.

Wedemeyer, G. A., and D. J. McLeay. 1981. Methods for determining the tolerance of fishes to environmental stressors. Pages 247–275 in A. D. Pickering, editor. Stress and fish. Academic Press, New York.

Wege, G. J., and R. O. Anderson. 1978. Relative weight (Wr): a new index of condi-tion for largemouth bass. Pages 79–91 in G. D. Novinger and J. G. Dillard, editors. New approaches to the management of small impoundments. American Fisheries Society, North Central Division Special Publication 5.

Wells, R. G. M., and N. W. Pankhurst. 1999. Evaluation of simple instruments for the measurement of blood glucose and lactate, and plasma protein as stress indica-tors in fish. Journal of the World Aquaculture Society 30:276–284.

Wendelaar Bonga, S. E. 1997. The stress response in fish. Physiological Reviews 77:591–625.

Wilson, J. M., M. M. Vijayan, C. J. Kennedy, G. K. Iwama, and T. W. Moon. 1998. β-naphthoflavone abolishes the interrenal sensitivity to ACTH stimulation in rain-bow trout. Journal of Endocrinology 157:63–70.

Winberg, S., and O. LePage. 1998. Elevation of brain 5-HT activity, POMC expres-sion, and plasma cortisol in socially subordinate rainbow trout. American Jour-nal of Physiology 274:R645–R654.

Winberg, S., A. Nilsson, P. Hylland, V. Soderstom, and G. E. Nilsson. 1997. Serotonin as a regulator of hypothalamic-pituitary-interrenal activity in teleost fish. Neuro-science Letters 230:113–116.

Winberg, S., and G. E. Nilsson. 1993a. Roles of brain monoamine neurotransmitters in agonistic behaviour and stress reactions, with particular reference to fish. Comparative Biochemistry and Physiology, Part C 106:597–614.

Winberg, S., and G. E. Nilsson. 1993b. Time course of changes in brain serotonergic activity and brain tryptophan levels in dominant and subordinate juvenile Arctic charr. Journal of Experimental Biology 179:181–195.

Winberg, S., G. E. Nilsson, and K. H. Olsén. 1992a. Changes in brain serotonergic activity during hierarchic behavior in Arctic charr (Salvelinus alpinus L.) are socially induced. Journal of Comparative Physiology A 170:93–99.

Winberg, S., G. E. Nilsson, and K. H. Olsén. 1992b. The effect of stress and starvation on brain serotonin utilization in Arctic charr (Salvelinus alpinus). Journal of Experimental Biology 165:229–239.

Woodward, C. C., and R. J. Strange. 1987. Physiological stress responses in wild and hatchery-reared rainbow trout. Transactions of the American Fisheries Society 116:574–579.

5

Genetic Responses as Population-Level Biomarkers of Stress in Aquatic Ecosystems

CHRISTOPHER W. THEODORAKIS AND ISAAC I. WIRGIN

Introduction

Population genetics is the branch of genetics that examines patterns of genetic variation within populations of organisms, the processes that lead to changes in genetic structure (e.g., gene, allele, and haplotype frequencies), and genetic variation within populations ("genetic diversity") and between populations ("genetic distance"). Many natural factors can cause changes in population genetic variation or allele frequencies. These natural processes either occur slowly relative to the age of the species (e.g., natural selection, geologic processes, etc.) or occur repeatedly over long periods of time (e.g., population crashes due to disturbances such as fires or floods). In either case, there has been sufficient time for natural organisms to evolve and adapt in response to these phenomena. Alternatively, there are many, shorter-term, anthropogenic factors that can lead to changes in genetic diversity or genetic structure of aquatic populations. These include habitat alteration and fragmentation, changes in physical characteristics of the water (temperature and pH), and changes in chemical composition of the water (nutrient load or xenobiotic chemicals). As opposed to long-term natural processes, these anthropogenic factors occur quickly relative to the age of the species or population—usually within decades, years, or even days—so that organisms may not have time to genetically adapt to these factors. These short-term environmental stressors may lead to population crashes or even local extinction of species.

Population crashes or local extinctions may occur as the result of any of the aforementioned anthropogenic activities. However, the focus of this chapter is primarily on changes in population genetic structure and diversity as a result of xenobiotic chemical exposure, a subdiscipline that has been termed "evolutionary toxicology" (Bickham and Smolen 1994). Such genetic changes are of concern because they may affect the viability of the population or alter the population's ability to evolve and adapt to natural environmental changes, and they may persist long after the chemical contamination has abated. Changes in population genetic diversity may be induced by both natural and anthropogenic factors.

Factors Influencing Genetic Diversity in Populations

Natural Factors

Genetic diversity within populations is usually measured in terms of heterozygosity (proportion of heterozygous loci relative to total number of loci screened) or, when using various DNA fingerprinting techniques, average similarity in fingerprint patterns between individuals. Genetic diversity between populations is usually reported as genetic distance and can be visually represented using dendrograms (Figure 1). Factors other than chemical contamination may confound determination of the effects of environmental pollutants, population genetic diversity, and population genetic structure. For example, if the population in question was recently colonized by a few individuals, then this population may have low levels of genetic variation, regardless of the relative amount of environmental contamination. The population in the colonized habitat may be genetically different from the source population not because of any selective constraints from natural or anthropogenic factors, but merely because of stochastic sampling of genetic variants from the founder population. This situation is termed the "founder effect." Stochastic sampling of genetic variants may also occur through a process of "genetic drift." In this situation, populations diverge genetically (the genetic distance increases) because random sampling may lead to stochastic changes in population genetic structure. Such random sampling may occur as a result of differences in recruitment or mortality among genotypes that occur merely by chance. However, the effects of genetic drift are greater in small populations than in large ones. Consequently, genetic bottlenecks and the founder effect may enhance the rate of genetic drift because they temporarily reduce population size (Hartl and Clark 1997).

However, if differences in recruitment or mortality between genotypes do not occur by chance, then natural selection may be responsible for differences between populations. Even if differences between contaminated and reference sites are due to natural selection, chemical contamination may not be the selective agent, because natural populations may experience many different selective pressures. In addition, the genetic structure of the population may not be directly affected by contaminant exposure, but by factors that are covariates of the level of environmental contamination. For example, an efflu-

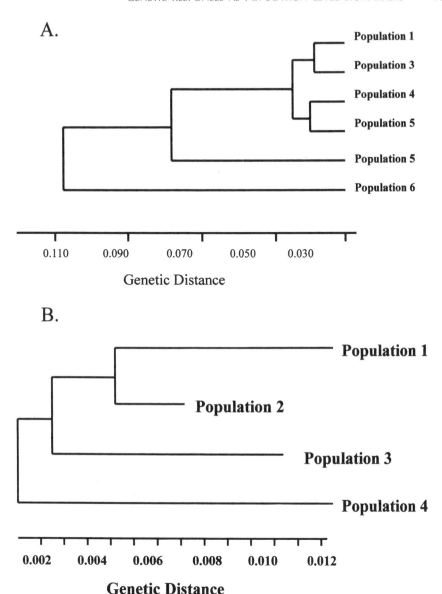

Figure 1. Examples of two different methods of displaying genetic distances between populations. The Unweighted Pair-Group Method Using Arithmetic Averages (UPGMA; A) assumes equal rates of evolution among all branches of the tree, while the Neighbor-Joining method (B) does not.

ent discharge may alter stream flow characteristics or the temperature of the water, so that genetic differences between contaminated and reference streams may be due to selective constraints related to flow or temperature regime rather than environmental contamination.

In addition to natural selection, other processes that can influence population genetic structure are migration and gene flow. In the context of population genetics, migration collectively refers to immigration and emigration. Gene flow is the exchange of genes or alleles between populations as a result of migration (Hartl and Clark 1997). The effect of migration is to reduce between-population variation (reduce genetic distance) and increase within-population variation, for example, by introducing novel alleles or genotypes into a population. Therefore, populations that are isolated tend to have less genetic variation than those that are more open to migration, all other things being equal. Migration may not be symmetrical; in other words, migration from population A to population B may be greater than the reverse. This is especially pronounced in populations that exhibit "source-sink" dynamics (Nunney 1999). In this situation, some populations may continuously exist at a density far below their carrying capacity or may experience periodic extinction or recolonization events ("sink" populations). Because emigration rate is directly related to population growth rate, sink populations may have a low emigration rate because of a low population growth rate.

In contrast, "source" populations may have high growth rates and concordantly high emigration rates. Alternatively, the densities of source populations may temporarily exceed their carrying capacities, thus necessitating emigration to new populations. The result is a net gene flow from source to sink populations. Also, there may be little or no genetic differentiation between source and sink populations, even though they are separated physically (Maruyama and Kamura 1980). Source-sink dynamics typically occur in patchy, heterogeneous habitats where some patches are less productive or more ephemeral than others (Nunney 1999). These types of interactions are pronounced in aquatic environments where water currents, floods, droughts, and seasonal anoxia may produce highly heterogeneous (both spatially and temporally) environments and local extinctions (Schlosser 1991).

In natural ecosystems, population-level processes (genetic drift, selection, migration, and mutation) do not act in isolation, but simultaneously. For example, natural selection or drift may act to increase interpopulation differentiation, but migration acts to reduce it. Also, genetic drift may tend to decrease genetic variation, whereas mutation would tend to increase it. Therefore, the amount of genetic variation within and among subpopulations, as well as within the population as a whole, depends on a balance between antagonistic (e.g., migration and drift) and synergistic forces (e.g., natural selection and drift; Hartl and Clark 1997).

Genetic variation within and between populations may not only be affected by these dynamic processes, but also by the properties of the populations themselves (Hartl and Clark 1997). For example, older populations tend to have greater genetic diversity than younger populations. This is because processes that tend to increase genetic diversity, such as mutation and immigration, are dictated by random chance (at least to some degree). Thus, the longer the population has existed, the greater the probability mutations will occur or immigrants will enter the population resulting in a higher genetic diversity. Larger populations also tend to have more genetic diversity than smaller ones, in part for similar reasons. Larger populations are also less affected by genetic drift and inbreeding. Inbreeding occurs less frequently in larger, randomly mating populations because there is a greater pool of potential mates.

Another characteristic of populations that contributes to their genetic diversity is the amount of population subdivision. In such an instance, individuals in one subpopulation are more similar to each other than they are to individuals in another subpopulation, but subpopulations within a population are more genetically similar to each other than they are to other populations. Population subdivision may act to increase the genetic diversity as a whole because of heterogeneous selective pressures or patterns of genetic drift. However, apparent heterozygosity may be less if a population is sampled without taking into account subdivision. Such a phenomenon is known as the "Wallund effect" (Hartl and Clark 1997).

Xenobiotic-Induced Factors

Xenobiotic-induced changes in population genetic structure can include reductions in genetic diversity, increases in mutation rates, and alterations of selection pressure. Xenobiotic-induced reductions in genetic diversity are thought to occur as a result of temporary ("population bottleneck") or permanent reduction in population size (Figure 2). This reduction may occur either through an increase in adult mortality or a decrease in reproductive and recruitment success (fecundity, egg, larval, juvenile survival,

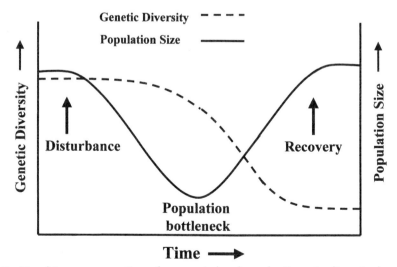

Figure 2. Graphic representation of a genetic bottleneck. Genetic diversity begins to decline after the population starts to crash, but the diversity remains low even though the population recovers, because of loss of alleles/genotypes during the bottleneck.

etc.) as a result of a variety of environmental stressors including acute or chronic toxicity. If sufficiently severe, a reduction in population size may lead to a decrease in genetic variability both within individuals and within populations. A decrease in heterozygosity within individuals is of concern because it has often been associated with decreased resistance to disease (Wolfarth 1985; Ferguson and Drahushchak 1990), decreased growth rates (Wolfarth 1985), fecundity (Leberg 1990), and, in general, increased vulnerability to a variety of environmental stressors. Hence, decreases in individual heterozygosity may affect population growth and recruitment. Reduction in the genetic diversity of the population as a whole is also of concern for a number of reasons. First, the evolutionary plasticity and ability for a population to adapt to changing environmental conditions require a certain amount of genetic variability in the population. A reduction in genetic variation due to environmental contaminant exposure may compromise this ability. Second, it has been found that populations with lower genetic variability have slower growth rates (Leberg 1990; Albert 1993). Finally, recent evidence supports the long-held supposition that the amount of inbreeding and genetic diversity are correlated with the probability of extinction (Saccheri et al. 1998). Consequently, changes in genetic diversity at the individual or population levels may ultimately affect long-term viability and sustainability of the population.

Pollution exposure may also affect the genetic diversity and genetic structure of natural populations via alterations in the patterns of natural selection by selecting for organisms that are more tolerant of the mode of action of the toxicant. By selecting for more tolerant organisms, genetic variation may be reduced in the population. This situation occurs because certain genotypes in the population may confer a selective advantage in contaminated habitats. Usually, these genotypes are present as a small fraction of the total gene pool and represent only a small amount of the genetic variability. When exposed to environmental contamination, these genotypes increase in frequency relative to all other genotypes, leading to a reduction in overall genetic variability of the population. Selection for pollution-tolerant organisms may also result in a situation in which organisms that are adapted to anthropogenic chemicals may be less well adapted to natural stressors (Weis et al. 1982). Also, such natural selection tolerance has associated costs in terms of reduction in fitness of individuals and recruitment into populations. For example, Schlueter et al. (1995) found that fish with genotypes that were associated with copper tolerance were also associated with a smaller body size. Hence, copper tolerant individuals may have a slower growth rate than copper-sensitive individuals. Thus, pollution-induced changes in population genetic structure (the distribution of gene or allele frequencies) may alter the population's ability to adapt to novel selective pressures, increase their vulnerability to stressors, or reduce the average fitness of the individuals in the population.

Changes in population genetic structure as a result of population bottle-necks or contaminant exposure may also ultimately be manifested effects at the population or community levels. For example, Krane et al. (1999) found that genetic diversity was correlated with changes in the index of biotic integrity (IBI)—an index of pollutant effects that measures changes in com-munity status based on several integrated metrics of community structure and organism health. However, Fore et al. (1995) found no such correla-tions, perhaps because of low variation in IBI scores (Fore et al. 1995). Alternatively, the lack of correspondence between the findings of Fore et al. (1995) and Krane et al. (1999) may be due to differences in life history or vagility between the species studied by Fore et al. (1995) versus those stud-ied by Krane et al. (1999). It has also been suggested that such changes in population genetic structure may take place sooner than the loss of species from the community (Benton and Guttman 1992), thus, providing an early warning indicator of community-level effects. Model simulations have also suggested that, given that the system under study is properly understood, changes in allele frequencies may reflect population-level effects of pollu-tion (Newman and Jagoe 1998).

In addition to acting via genetic bottlenecks or selection, environmental contamination may also change population genetic structure by inducing heritable mutations. There are numerous genotoxic pollutants that can cause an increase in mutation rates, including radiation, polycyclic aromatic hydro-carbons, and heavy metals. However, mutations, even pollutant-induced ones, are rare events, so any effect would be relatively slight. Therefore, detecting a slight increase in mutation rates against a background of natural genetic variation would be difficult. Also, any deleterious mutations would be quickly eliminated from the gene pool, particularly if they were dominant. Accord-ingly, most of the induced mutations that would be detectable would be neutral mutations, those that have no adverse effect on the biochemistry or physiology of the organisms.

Deleterious mutations, however, may persist in populations at low fre-quencies if they are recessive. The relative frequency of deleterious muta-tions in a population is referred to as the "genetic load." The genetic load of small populations may be greater than that of large populations, because the effects of inbreeding are more pronounced in small populations, particularly when migration is limited. Inbreeding often leads to an increase in the fre-quency of deleterious recessive mutations, a situation that may be exacer-bated if the mutation rate is elevated by exposure to genotoxic agents. Ge-netic drift may facilitate accumulation of deleterious mutations in a population (Gabriel and Bürger 1994), and effects of genetic drift are more pronounced in smaller populations. An increase in the genetic load may drive the popu-lation to extinction, a situation termed "mutational meltdown," but this, gen-erally, is of concern to asexual species or small populations of sexual species (Gabriel and Bürger 1994). An increased genetic load may decrease individual and mean population fitness, thereby reducing population size. Because accumulation of deleterious mutations occurs more quickly in smaller populations (due to inbreeding and genetic drift), a positive feedback loop between increased genetic load and decreased population size can develop and subsequently drive the population to extinction. An increased genetic load may also act synergistically with stochastic processes to increase extinction risk in small populations (Gabriel and Bürger 1994, and references therein). Such a situation may be exacerbated by chemicals that are both toxic (reducing population size) and mutagenic.

An increase in the genetic load of a population may reduce fitness of the constituent individuals and ultimately result in a reduction of population growth rate or population size. For example, pink salmon *Oncorhynchus gorbuscha* embryos that were exposed to crude oil from the *Exxon Valdez* oil spill in tributaries of Prince William Sound, Alaska, exhibited lower survivorship than those from tributaries that were not contaminated with crude oil. This effect persisted after the toxic effects of the oil had abated (Bue et al. 1998). Thus, it is possible that this reduced embryo viability may have been due to an increase in the number of deleterious mutations in these populations. In fact, the ability of weathered Prudhoe Bay crude oil to elicit K-*ras* somatic mutations in pink salmon embryos was demonstrated in subsequent controlled laboratory experiments (Roy et al. 1999). Additionally, population level perturbations in the F_1 progeny of experimentally oiled pink salmon embryos have recently been observed (J. Rice, North Carolina State University, personal communication). However, population genetic model simulations indicated that there was an "inconsistency between observed egg mortality and that expected if lethal heritable mutations had been induced by exposure to crude oil" (Cronin and Bickham 1998). In another study, mosquitofish in a radionuclide-contaminated reservoir displayed a higher number of abnormal embryos than fish sampled at a reference site (Blaylock and Frank 1980). When these fish were reared in the laboratory, subsequent generations also showed an increased frequency of embryonic abnormalities, suggesting that there was an increased genetic load of deleterious mutations in this population (Trabalka and Allen 1977).

Because there are numerous factors other than environmental contamination that can influence population genetic structure, care needs to be taken when evaluating the contribution of toxicants alone on genetic structure. For example, multiple reference populations and, if possible, multiple contaminated populations experiencing similar types of contamination need to be sampled. If only one contaminated and one reference population are compared, the only conclusion that can be drawn is that the two populations are genetically different. No inferences about the effects of contamination can be made, because any two populations are likely to be genetically different. In this case, it cannot be determined if the differences between one contaminated and one reference site are greater than would be expected on average. The use of multiple reference populations allows a more accurate estimation of the natural variability within and between populations, which permits determination if the differences between contaminated and reference sites are greater than would be expected without contaminant effects.

Also, a "weight of evidence" approach that employs multiple lines of evidence to assess contaminant effects should be used. For example, trends in genetic variability should be compared not only with trends in contaminant concentrations, but also with other biomarker responses that reflect effects on organisms at the biochemical and physiological levels. When differences in gene or allele frequencies are found between populations, claims that these are due to contaminant-mediated selection should be supported with other lines of evidence, such as differences in fecundity and survival between genotypes when exposed to contamination (Guttman 1994; Diamond et al. 1991; Theodorakis and Shugart 1997; Theodorakis et al. 1998, 1999). Additional evidence such as differences in biomarker responses between genotypes (Theodorakis and Shugart 1998b; Theodorakis et al. 1999) or biochemical differences between alleles (Changon and Guttman 1989b) would provide further evidence of contaminant-induced selection.

Approaches to Quantify Genetic Diversity

Allozymes

Allozymes are allelic forms of enzymes. Structurally, variant allozymes differ in their constituent amino acids, and alleles are distinguished on the basis of variation in their isoelectric points. However, only those alleles whose substituted amino acid result in a change in charge are distinguishable. When dissolved in appropriate buffers, each allele has a unique electrical charge and, therefore, migrates at a different rate when analyzed by electrophoresis. Starch gels or cellulose acetate films are commonly used for this purpose (Murphy et al. 1996). Starch gels can be cut into horizontal slices so that multiple enzymes can be analyzed from the same gel. On cellulose acetate films, only one enzyme can be examined at a time, but the amount of tissue and time needed to analyze the samples is much less.

Enzymes are visualized by incubating a gel slice or cellulose acetate film with the appropriate substrate plus a reagent that forms a colored precipitate when the enzyme-catalyzed reaction takes place. The identity of the enzyme is determined by the substrate used for the staining process; for example, incubating the gel with isocitrate will stain the enzyme isocitrate dehydrogenase. The enzymes appear as bands on the gel or the film (Figure 3). Alleles are identified by their relative electrophoretic mobility, which correspond with the relative magnitude of the electrical charge at the pH of the buffer. For example, in Figure 3A, the "fast" allele has a greater net negative charge, so it migrates to the positive electrode at a faster rate. Allozyme loci may

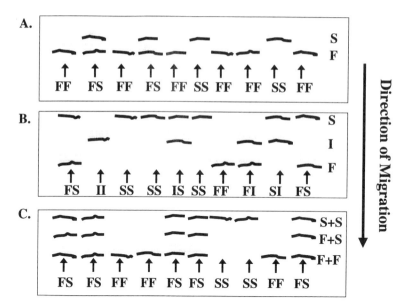

Figure 3. Schematic representation of an allozyme gel. A. An allozyme locus with 2 alleles, the "fast" allele (F) and the "slow" allele (S), so called because of their electrophoretic mobility (determined by net charge). Each lane represents a different individual (indicated by arrows). There are 3 genotypes, the fast homozygote (FF), the slow homozygote (SS), and the heterozygote (FS) B. A locus with 3 alleles, "fast," "slow," and "intermediate." C. An enzyme composed of a dimer. The heterozygote possesses 3 bands, the fast/fast dimer, the slow/slow dimer, and the fast/slow dimer.

consist of more than two alleles, as depicted in Figure 3B as "fast," "slow," and "intermediate." These enzymes can also be composed of dimers (two proteins attached by noncovalent bonds), trimers (three proteins), or tetramers (four proteins) of the same peptide. A hypothetical example is illustrated in Figure 3C, where the heterozygote displays three bands: the "fast/fast" dimer (F + F), the "slow/slow" dimer (S + S), and the "fast/slow" dimer (F + S). Heterozygotes for enzymes that are trimers and tetramers display more complex patterns.

There are several advantages and disadvantages of using allozymes to examine population genetic structure. One advantage is that multiple loci can be scored from the same electrophoresis gel. Also, each enzyme has a known specific function (the identity of the enzyme is determined by the substrate used in the staining reaction), so that functional significance of the locus is known. Furthermore, allozymes are expressed as codominant markers; that is, both alleles are visualized in individuals that are heterozygotes, which facilitates determination of heterozygosity and allele frequencies. The disadvantages of using allozymes are that only variation at specific loci is measured and only variants that result in changes in isoelectric points can be detected. Also, because of their functional significance, these loci are much less variable than some of the other genetic markers discussed in this chapter.

To date, most of the research on effects of pollution on population genetic structure has used allozyme markers. Some of these studies have used allozymes to examine reductions in genetic variation due to environmental contamination (Kopp et al. 1992; Fore et al. 1995; Hummel et al. 1995). However, caution should be exercised when examining genetic variation in natural populations because there is the possibility that populations exhibit subdivision or other types of nonrandom structure. For instance, Woodward et al. (1996) found heterozygote deficiencies in populations of chironomids from a mercury-polluted lake, but they concluded that this was due to fine-scale substructuring of the population. When samples were collected without regard to this population heterogeneity, apparent loss of heterozygosity was attributable to the Wallund effect (defined earlier).

Allozyme markers have also been used to provide evidence of contaminant-induced selection. Initial evidence of such selection may be inferred by differences in allele frequencies between contaminated and reference sites. Such differences were found to occur in both fish (Changon and Guttman 1989a, 1989b; Fore et al. 1995; Schlueter et al. 1997) and invertebrates (Benton and Guttman 1992; Nicola et al. 1992; Tanguy et al. 1999). However, differences in allozyme allele frequencies are not definitive evidence of contaminant-induced selection because other factors such as neutral mutation and genetic drift, founder effects, or gene flow can also contribute to such differences. One way to test for effects due to contaminants versus other factors would be to expose organisms to contaminants in the laboratory and then determine if some component of fitness (fecundity, survival, growth, etc.) was dependent on genotype (e.g., Changon and Guttman 1989a). In a related study, Tanguy et al. (1999) found that oyster populations from metal contaminated sites possessed alleles that were present at a higher frequency than in reference sites. These differences in allele frequencies corresponded to genotype-dependant metal resistance in the laboratory with individuals possessing alleles that were more common at contaminated sites also being more resistant to metal toxicity in the laboratory. However, this pattern of correlation between laboratory and field studies may not always be so clear-cut. For example, survival time of eastern mosquitofish *Gambusia holbrooki* exposed to heavy metals was correlated with allozyme genotype, but this pattern was not consistent for fish sampled from different populations or during different years (Diamond et al. 1991). In fact, Lee et al. (1992) suggested that differences between polluted and reference populations may also be due to correlations between broods or other subunits of a structured population. Hence, the efficacy of selection of contaminant-resistant genotypes using allele frequencies may depend on characteristics of the life history or behavior of the particular species in question, as well as local population structure.

In order to demonstrate contaminant-mediated selection using allozyme markers, additional evidence could be garnered from demonstration of a biochemical basis for differential chemical sensitivity among genotypes. For example, Gillespie and Guttman (1988) reported that certain alleles were present at a higher frequency in contaminated populations of central stoneroller *Campostoma anomalum* than in reference populations. These genotypes also had longer survival times when exposed to copper in the laboratory (Changon and Guttman 1989a). They further demonstrated that the enzymatic activity of these particular alleles was less inhibited by copper in vitro than were the alleles that were less prevalent in noncontaminated populations (Changon and Guttman 1989b). This study was significant because it not only linked genotype frequencies with a component of selection (survival), but also demonstrated a biochemical basis for this phenomenon. However, selection may not necessarily act directly on the allozyme loci themselves, instead these loci may be closely linked to other genes (e.g., detoxification enzymes, etc.) upon which selection acts.

In addition to affecting mortality and survival of organisms (viability selection), contaminant-mediated selection may also act via fecundity selection. For example, Mulvey et al. (1995) found that reproductive performance (number of gravid females and number of developing embryos per female) in mosquitofish exposed to mercury differed between glucose-phosphate isomerase genotypes. These differences in reproductive performance were also consistent with differences in survival.

Randomly Amplified Polymorphic DNA

Randomly amplified polymorphic DNA (RAPD) is a technique for measuring genetic diversity using the polymerase chain reaction (PCR) with short (10 base pair) oligonucleotide primers (small pieces of DNA) to amplify DNA fragments. The specific mechanism of PCR amplification is illustrated in Figure 4. The DNA to be amplified (template DNA) is mixed with primers, free nucleotides, and a DNA polymerase enzyme. When the DNA is heated to around 95°C, the DNA "denatures" or separates into single strands (Figure 4A). The DNA solution is then cooled to 35–45°C (Figure 4B; Note: for other types of PCR reactions, e.g., amplifying sections of mitochondrial DNA or microsatellites [see below], the mixture is cooled to 50–60°C.) At this point, the primer binds (anneals) with any section of the template with a sequence that is homologous to the primer (e.g., a primer with the sequence AATGGCTAGG will bind with the template DNA wherever there is a sequence of TTACCGATCC). A DNA polymerase then catalyzes the polymerization of a new DNA strand, starting at the end of the primer (Figure 4C). This process is then repeated 30–40 times, resulting in the amplification of DNA fragments that are bounded by two priming sites. When analyzed by gel electrophoresis, RAPD fragments produce banding patterns similar to other types of DNA fingerprints (Figure 5). These bands have often been found to exhibit Mendelian inheritance typically acting as dominant markers

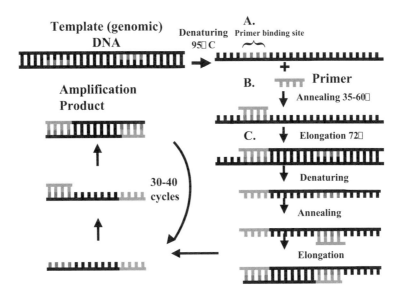

Figure 4. Representation of the basic components of a PCR reaction. The reaction constituents consist of template DNA (purified genomic DNA), free nucleotides and "primers"—short pieces of DNA. After denaturation (A), the primers bind to the DNA wherever there is a complimentary sequence (e.g., a primer with a sequence of AACTGGTCGG will bind to a DNA with a sequence of TTGACCAGCC; (B). DNA polymerases then add additional nucleotides, using the genomic DNA as a template (C). In the RAPD and AP-PCR techniques, only 1 primer is used and it is 10 (RAPD) or 15–25 (AP-PCR) nucleotides long. In all other applications (e.g., amplifying mtDNA or microsatellites), 2 different primers are used.

(Williams et al. 1990). Genetic variation between individuals is visualized as the presence or absence of some of these bands because genetic polymorphisms (point mutations, deletions, inversions, etc.) within or between primer binding sites may prevent the primer from binding and, thus, the DNA fragment from being amplified. The RAPD technique has been employed in such areas as population genetics, taxonomy, species identification, and parentage assessment (Hadrys et al. 1992; Dinesh et al. 1993b). In addition to DNA fingerprinting, RAPD markers have also been used to isolate microsatellite markers in *Daphnia spp.* (Ender et al. 1996).

The RAPD technique has several advantages over some other currently used techniques. First, because it is PCR-based, it requires little template DNA and does not require the use of radioactive probes as do other methods of fingerprinting. Second, the RAPD technique investigates genetic differences that occur at multiple loci, unlike allozyme studies that only examine a few specific loci. Thus, it is possible to characterize previously unidentified loci that are under selective pressure from contaminant exposure. Third, the RAPD technique does not require a priori knowledge of the DNA sequence of the organisms in question because the sequence of the primers is arbitrary (with a few restrictions).

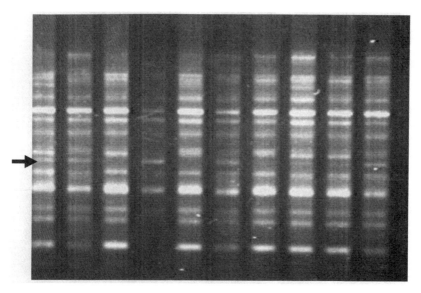

Figure 5. Picture of a RAPD gel. Each lane of bands represents a separate individual. The arrow indicates a variable band (present in some, absent in others).

There are, however, some disadvantages to the RAPD technique. The genomic identity and possible biological significance of RAPD bands are currently unknown. These bands can be amplified from various locations in the genome, from coding or noncoding regions, or from repetitive or single-copy DNA. Therefore, assigning adaptive significance to particular bands should be viewed with caution, at least until rigorous investigation of the genomic locations and possible functions of the amplification sites have been thoroughly characterized. A second drawback of RAPDs is that the banding patterns are sensitive to reaction conditions; variation in template DNA quality or concentration, reagent composition, or PCR thermal cycle profiles may lead to changes in banding patterns. As a result, in some cases, bands are produced that are not reproducible or are difficult to score. However, if reactions are rigorously standardized and appropriate quality control procedures are maintained, reproducible results may be attained (Dinesh et al. 1993a). Furthermore, the molecular basis for band absence is usually unknown, thus, two different individuals that do not display a particular band may not necessarily possess the same genotype. These markers are also inherited in a dominant fashion rather than codominant. That is, heterozygotes may be indistinguishable from the dominant homozygote. Consequently, determination of heterozygosity and allele frequencies from RAPD markers may be biased (Lynch and Milligan 1994). Finally, the RAPD bands are usually identified by molecular weight and not by nucleotide sequence. Therefore, two DNA fragments with similar molecular weights, but different sequences may be identified as a single band. These drawbacks are further discussed and reviewed in Hadrys et al. (1992).

In ecotoxicological research, RAPD markers have been used to infer loss of genetic diversity in crayfish populations from contaminated sites (Krane et al. 1999) and altered genetic structure in fish populations in a contaminated stream (Nadig et al. 1998). In another set of studies, populations of western mosquitofish *Gambusia affinis* from radionuclide-contaminated and reference sites were examined using the RAPD technique (Theodorakis and Shugart 1997). Surprisingly, the genetic diversity in this case was actually greater in the contaminated sites than in the reference sites. This is in concordance with other studies that found evidence of increased genetic diversity in contaminated sites (Prus-Glowacki et al. 1999). In addition, Theodorakis and Shugart (1997) found several RAPD bands that were present at an increased frequency in radionuclide-contaminated populations. Furthermore, they discovered that these bands may be markers of relative radioresistance (Theodorakis and Shugart 1997; Theodorakis et al. 1998, 1999).

The RAPD technique and related assays can also be used to investigate mutations and related phenomena. For example, Kubota et al. (1992) irradiated male Japanese medaka *Oryzias latipes* with X-rays and found that some of the progeny developed abnormally. Using arbitrarily primed PCR (AP-PCR), a technique very similar to RAPD, they found that these fish lost some of the parental bands. They suggested that this was due to mutations or other DNA lesions. Several recent investigators have also suggested that RAPD or AP-PCR banding patterns may change as a result of contaminant exposure (Savva 1998). These results have been found for *Daphnia* (Atienzar et al. 1999), cultured fish cells (Becerril et al. 1999), and the plant *Arabadobsis thaliana* (Conte et al. 1998). It has been suggested that loss of these bands may be attributable to induction of mutations or other DNA lesions such as bulky adducts (Atienzar et al. 1998). An alternative explanation may be that loss of these bands represents induced DNA rearrangements (e.g., transposition) or genomic instability, as suggested by Atienzar et al. (1999). For example, Keshava et al. (1999) used the RAPD technique to analyze genomic instability and genomic rearrangements in human and mouse tumors and tumor cell lines. They found that the RAPD banding patterns differed between tumorous tissue or tumor cell lines and the normal tissue from which they were derived. To date, however, this type of analysis has yet to be applied to environmental analysis.

Amplified Fragment Length Polymorphism

Like RAPD, amplified fragment length polymorphisms (AFLP) is another method of generating DNA fingerprints from anonymous loci (Savelkoul et al. 1999). One of the advantages of AFLP, relative to RAPD, is that more marker bands are produced, which may allow a greater amount of genetic diversity to be uncovered with AFLP than with RAPD. This may allow greater discrimination of genetic differences between populations—a property important if comparing populations that are closely related or that have the potential for gene flow. One disadvantage compared to RAPD is that the methodologies for production of AFLP fingerprint patterns are slightly more complex and time consuming, and, as a result, this technique is more expensive than RAPD.

In the AFLP technique, genomic DNA is first digested with various restriction enzymes, which cut the DNA into short pieces. Short pieces of DNA of known sequences, termed "adaptors," are then attached (or ligated) onto the ends of these restriction enzyme fragments with ligation enzymes. PCR primers are then used to amplify these DNA fragments. The PCR primers are designed to be homologous to sequences of both the adaptor DNA and to the ends of the restriction fragment; the sequence of ends of the restriction fragments is known because each restriction enzyme will only cut DNA at specific sequences. These DNA fragments are then separated by gel electrophoresis, typically polyacrylamide because of its greater resolving power. This results in a banding pattern similar to RAPD, but typically with many more bands (50–100 per reaction, as compared to an average of 10–30 with RAPD). The reproducibility of this technique has been reported to be good.

Representative applications of the AFLP technique include genomic mapping of plant and animal species, medical diagnostics, phylogenetic and systematic studies, and population genetics (Mueller and Wolfenbarger 1999; Savelkoul et al. 1999). To date, however, this technique has not yet been applied to the determination of the effects of environmental contamination on population genetic diversity in fish or other aquatic species. This is unfortunate because this technique holds much promise for these type of studies.

Microsatellite Analysis

Microsatellites are loci in the nuclear genome that contain di- (e.g., AC), tri- (e.g., ACC), or tetranucleotide (e.g., GATA) repeats arrayed in tandem (Figure 6). The number of repeats per microsatellite locus generally varies between 5 and 50 copies. Microsatellite loci are abundant in all eukaryotic genomes. It has been estimated that there are 10^3 to 10^5 microsatellite loci dispersed at 7 to 10^{100} kilobase pair (kb) intervals in the eukaryotic genome (Wright and Bentzen 1994). The number of copies of repeats at a microsatellite locus often differs among individuals within a population and serves as the basis for microsatellite allelic variation (Figure 6). The functional significance of variation at microsatellite loci is largely unknown. However, microsatellite instability has been shown to be associated with progression of various types of human cancers (Thomas et al. 1996). The use of microsatellites can make a significant contribution to ecotoxicological research because they are highly variable (allowing greater discrimination between populations), are associated with environmentally induced dis-

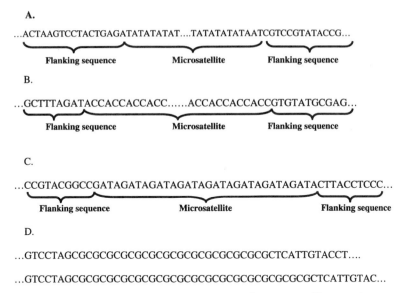

A.

...ACTAAGTCCTACTGAGATATATATAT....TATATATATAATCGTCCGTATACCG...

Flanking sequence   Microsatellite   Flanking sequence

B.

...GCTTTAGATACCACCACCACC.......ACCACCACCACCGTGTATGCGAG...

Flanking sequence   Microsatellite   Flanking sequence

C.

...CCGTACGGCCGATAGATAGATAGATAGATAGATAGATAGATACTTACCTCCC...

Flanking sequence   Microsatellite   Flanking sequence

D.

...GTCCTAGCGCGCGCGCGCGCGCGCGCGCGCGCGCGCTCATTGTACCT....

...GTCCTAGCGCGCGCGCGCGCGCGCGCGCGCGCGCGCGCGCGCGCGCTCATTGTAC...

Figure 6. Representation of the DNA sequences of microsatellites (only 1 strand of the DNA is represented, the other strand is complementary). The microsatellites shown are 2 (A), 3 (B), and 4 (C) base pair repeats. The PCR primers used to amplify the microsatellites are complementary to the flanking sequences, which are often conserved between closely related species. (D) an example of 2 microsatellite alleles that differ in the number of repeats.

ease in humans (i.e., cancer), and have a high mutation rate (facilitating detection of environmentally induced mutations).

Analysis of genetic diversity via microsatellites is a PCR-based approach in which oligonucleotide primers (fragments of DNA usually 20–30 nucleotides long) are designed based on the DNA sequences on either side (flanking regions) of the microsatellite repeats (Figure 6). PCR primer pairs are selected such that PCR products are of small molecular size (<200 bp), providing relative ease in distinguishing small differences in the molecular size of PCR products among individuals by using polyacrylamide gel electrophoresis. Each individual shows a one or two band pattern, one band inherited from each parent. The PCR products are labeled with fluorescent or radioactive tags and visualized using fluorescence detection or autoradiography. An example of such an autoradiograph is shown in Figure 7.

One of the major drawbacks in the use of microsatellites is the need to develop PCR primers for each new taxonomic group investigated. Usually 4–8 months are required to develop a genomic DNA library, identify, isolate, and characterize a battery of 10–20 microsatellite loci; determine their single copy flanking sequences; optimize PCR parameters for each locus; and identify a subset of 4–8 loci that reveal sufficient levels of genetic variation for population studies. Of course, in many instances, microsatellite primers developed for one species will amplify diagnostic loci in closely related species. Most importantly,

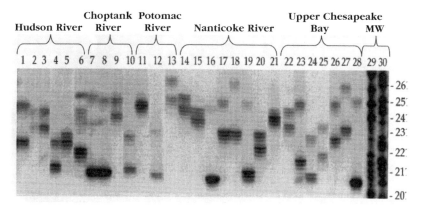

Figure 7. Autoradiograph showing high levels of microsatellite allelic diversity in Atlantic coast striped bass at the SB113 locus. Samples in lanes 1–6 are from the Hudson River, NY; lanes 7–10 the Choptank River, MD; lanes 11–13 the Potomac River, MD-VA; lanes 14–21 the Nanticoke River, MD; and lanes 22–28 the Upper Chesapeake Bay, MD. Lanes 29–30 contain DNA sequencing ladders used as molecular weight markers (MW). The fainter bands below each band are "shadow" or "stutter" bands, and are artifacts of the PCR amplification.

within the past five years, microsatellite analysis has developed into a powerful tool in the arsenal of population geneticists and fishery managers, and, thus, microsatellite loci have been identified and PCR primers developed for a surprisingly large number and diversity of North American fish taxa.

The sensitivity of approaches in detecting subtle alterations in overall genetic diversity is, in large part, dependent on levels of genetic variation encountered in natural populations. The higher the levels of variation, the greater the likelihood in detecting differences in levels of diversity among populations. The popularity of microsatellite analysis stems in large part from the exceedingly high levels of allelic diversity observed at individual microsatellite loci in almost all species investigated to date. Mean numbers of alleles at individual microsatellite loci in populations of freshwater, anadromous, and marine fish were 7.5, 11.3, and 20.6, respectively (DeWoody and Avise 2000). For example, striped bass *Morone saxatilis* allozymes proved almost monomorphic in many Atlantic and Pacific coast populations (Waldman et al. 1988), levels of mitochondrial DNA diversity were exceedingly low (Wirgin et al. 1989), single copy anonymous DNA loci exhibited a maximum of 2 alleles/locus (Wirgin and Maceda 1991; Wirgin et al. 1997), and band sharing in multilocus DNA fingerprints was so high that single fragments were used in distinguishing some populations (Wirgin et al. 1991). In contrast, up to 28 alleles were observed at single microsatellite loci, and the mean number of alleles detected at 8 loci in striped bass was 9 (Roy et al. 2000). Thus, microsatellites appear to have a much higher level of allelic diversity, making this a more desirable technique for detecting effects of environmental stressors than many other genetic techniques.

Microsatellite analysis also offers the major advantage of investigating codominant DNA markers that are inherited in Mendelian fashion. This is not the case with dominant or recessive RAPD or haploid mtDNA markers. In population studies, the extent to which genotypes conform to Hardy-Weinberg equilibrium confers additional information about population structure and those processes (migration or selective pressure) that may have contributed to deviations from expected genotype frequencies. Most importantly, microsatellite analysis provides the high degree of reproducibility often not seen with RAPDs analysis. It is our opinion that microsatellite analysis is a very powerful tool available for assessing overall genetic diversity in natural populations.

However, to date, because of the need to develop taxon-specific PCR primers, microsatellites have received little attention in ecotoxicology. In one such study, Dimsoski and Toth (1998) investigated the possibility of population bottleneck events in six populations of central stonerollers from tributaries of the Great Miami River basin. They reported that four of the six populations had heterozygosity deficiencies for at least 5/6 microsatellite loci investigated, probably resulting from a recent population bottleneck, possibly as a result of pollution stress. Microsatellite loci have also been used to investigate pollution-induced increases in mutation rates by comparing genotypes in parents and offspring (Ellegren et al. 1997).

Mitochondrial DNA

Mitochondrial DNA (mtDNA) is a closed, circular molecule of about 17,000 base pairs in fish that exists in multiple copies within individual cells. Studies have demonstrated that mtDNA is almost always maternally inherited and does not undergo recombination. Hence, in the absence of additional mutations, all descendants of a single ancestral female will exhibit identical mtDNA haplotypes. Mitochondrial DNA contains 13 protein-encoding genes, 2 ribosomal RNAs (rRNA), and 22 transfer RNAs (tRNA) with little intergenic space (Figure 8).

Despite its economical size and conserved gene synteny, it has been demonstrated in many taxa, including fishes, that mtDNA evolves (changes in DNA sequence) approximately 5–10 times more rapidly than protein-encoding nuclear DNA (nDNA) genes. This may be due to relaxed functional constraints on individual mtDNA molecules because of copy multiplicity within individual mitochondrion or mitochondrion multiplicity within individual tissues. Not all areas of mtDNA evolve at equal rates—many studies have demonstrated that the mtDNA control region (d-loop) evolves 4–5 times more rapidly than the entire the mitochondrial genome (Brown et al. 1996). Additionally, the mitochondrial genome has no proofreading repair mechanism in the DNA replication process (Matson et al. 2000). Furthermore, it has been shown that mitochondria are preferential targets for lipophilic contaminants and reactive oxygen species and, as a result, levels of DNA adducts and overall DNA damage may be higher in mtDNA than nDNA.

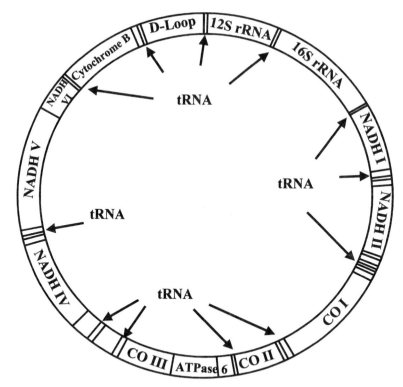

Figure 8. Example of the orientation of coding genes, rRNA and tRNA genes on a mtDNA molecule.

Initially, the study of mtDNA variation necessitated the isolation of pure mitochondria, organelle lysis, and purification of mtDNA by lengthy ultra-centrifugation steps. However, with the advent of PCR, analysis of mtDNA variation has become far less demanding and processing the large number of samples needed for statistical precision has become possible. Today, total DNA is rapidly and inexpensively isolated from virtually any tissue and PCR is used to amplify a specific 0.5–2.0 kb area of mtDNA of interest. Highly conserved areas in mtDNA have been identified and these have been used to design "universal" PCR primers. These can then be used to amplify a species-specific mtDNA fragment from which species-specific PCR primers can be designed. Two alternatives are then available to screen the PCR prod-uct for polymorphisms: direct DNA sequencing or restriction fragment length polymorphism (RFLP) analysis. DNA sequencing provides definitive infor-mation on most, if not all, nucleotide characters within the PCR product; whereas, RFLP analysis provides information on only a small subset of these characters. With the advent and increased availability of automated technol-ogy, DNA sequencing is, by far, the method of choice.

The DNA sequences of mtDNA (or any DNA sequence, for that matter) can also be used to construct phylogenies or evolutionary histories of the haplotypes. These phylogenies are usually based on DNA sequence similarities between haplotypes and can be visualized as a so-called "phylogenetic tree." A hypothetical example is presented in Figure 9. The distribution of the haplotypes (or genotypes, if nuclear DNA is used) and the evolutionary relationships between them is termed "phylogeography" (Avise 1998). Phylogenetic analysis can contribute much to the study of population genetics, such as age of haplotypes, their evolutionary relationships, and possible gene flow and migration events.

Compared with allozyme analysis, very few studies have evaluated mtDNA variation in impacted versus reference populations. Murdoch and Hebert (1994) compared mtDNA diversity in brown bullhead *Ameiurus nebulosus* from industrialized and matched cleaner sites in Lake Ontario and Lake Erie. Using RFLP analysis of the entire mtDNA molecule, they found significantly decreased mtDNA haplotype diversity in fish from the industrialized sites and concluded that this resulted from population bottlenecks due to chronic historical exposure to contaminants. Conversely, other studies have found increases in mtDNA diversity at contaminated sites compared with reference sites (Matson et al. 2000) and suggested that this may be due to recolonization of the contaminated locales by immigrating voles from several genetically divergent populations or an increased mutation rate in the impacted populations.

Invertebrate species have also been the subject of mtDNA studies. In one study, individuals from five invertebrate species living near (<50 m) offshore oil and gas production platforms in the Gulf of Mexico exhibited significantly lower levels of haplotype diversity than populations at stations greater than 3 km from the platforms (Street and Montagna 1996). Haplotype diversity was inversely correlated with multivariate measurements of concentrations of sediment-borne contaminants. Even though haplotype diversity was lower at the more contaminated sites, effects could not be distinguished between contamination exposure, disturbance due to drilling, or some other nonanthropogenic cause. Therefore, Street et al. (1998) performed controlled laboratory experiments, using three generations of an estuarine species of copepod *Nitocra lacustris*. These experimental populations exhibited a significant 60% reduction in mtDNA haplotype diversity following exposure to phenanthrene-contaminated sediment. Changes in genetic diversity were associated with increased adult mortality and reduced survival of offspring, which resulted from an increase in the frequency of the most common mtDNA haplotype and a decrease in the frequency of rarer haplotypes. This study provides evidence that exposure to toxic chemicals leads to the reduction in genetic diversity seen in the field and that reduced viability may have been the etiological agent.

Chen and Hebert (1999a, 1999b) also examined the possibility of a pollution-induced increase in mutation rate using direct sequencing of the mtDNA control region to identify recent mutational derivatives (terminal branch haplotypes [TBH]; see Figure 9) and compare their frequencies in bullhead

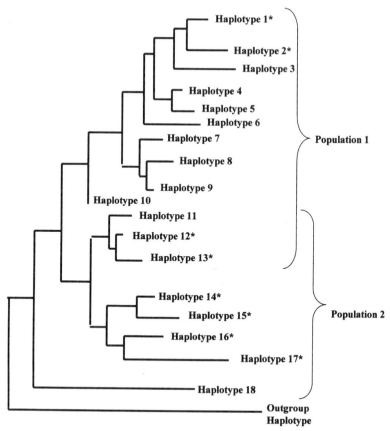

Figure 9. An example of a phylogenetic tree of 17 mitochondrial haplotypes from 2 hypothetical populations. Several things are inferred from this tree: (1) haplotypes 11–13 are found in both populations, suggesting that there has been gene flow between them; (2) the age of the haplotypes can be inferred from the position on the tree: for example, haplotypes 1 & 2 are terminal branch haplotypes, while haplotype 18 is deeply rooted in the tree, so 18 is much older than 1 & 2; (3) haplotype 1 is more similar to 2 than it is to 3; (4) haplotype 10 is probably the ancestral haplotype to all other haplotypes unique to population 1; (5) haplotypes 8 and 9 share a common ancestor, and 9 is more similar to this common ancestor than 8 (8 is more "derived"). The "Outgroup" is a closely related species used to "root" the tree. Terminal branch haplotypes are indicated by an asterisk (*).

from these two sets of Great Lake sites. The hypothesis was that this TBH would be more frequent in fish from the contaminated sites because of an increase in their mutation rate. However, no difference in the frequencies of TBHs was observed between fish from the heavily contaminated sites and the cleaner sites. Instead, variation in the frequencies of TBHs among sites probably reflected demographic factors. Fish from bay habitats exhibited higher levels of TBH than those from tributaries, probably reflecting the larger population sizes in the bays.

A pollution-induced increase in mutation rate may also be reflected as an increase in heteroplasmy. Usually all copies of mtDNA within a single tissue or organism are identical (homoplasmy); however, several taxa of lower vertebrates and invertebrates often exhibit more than one form of mtDNA within individual tissues or organisms (heteroplasmy). Heteroplasmy may arise from the presence of newly developed mutations, paternal leakage, or maternally derived transmission of heteroplasmic oocytes. Heteroplasmy may occur for single base substitutions or for variation in the overall size of the mtDNA molecule, usually resulting from the presence of differing numbers of tandem repeats in the mtDNA control region. The induction of an increased level of heteroplasmy as a result of environmental contamination has been investigated by Baker et al. (1999), and they concluded that mtDNA heteroplasmy merits more study to determine its effectiveness as an index of pollution-induced increases in mutation rate.

An additional method of determining mutation rates is by comparison of the mitochondrial DNA sequences of mother and offspring. This is because mtDNA is maternally inherited, so that the mtDNA sequences of the offspring should be genetic clones of the mother. This technique was used by Johnson et al. (1999) to directly sequence a 433 bp portion of cytochrome b (cyt b) to compare the mtDNA cyt b mutation rate in 18 female wood ducks *Aix spónsa* and their offspring from three radiocesium-contaminated reservoirs on the Savannah River site with that from 2 females and their offspring from a captive breeding population (the control). DNA from females was noninvasively isolated from feathers, and, for embryos, liver and muscle served as the source of DNA. In total, four haplotypes were detected among the 40 individuals tested and maximum divergence between any two individuals was at two base positions (0.46% sequence divergence). No mutations were observed between females and their offspring suggesting that chronic exposure to radiocesium had not increased the mtDNA mutation rate in these wood duck populations.

A final use for mitochondrial DNA is in determination of genotype-specific toxicant resistance. Sturmbauer et al. (1999) identified five genetically distinct mtDNA lineages based on 376 bp of rRNA sequences among *Tubifex tubifex* worms from central and eastern European rivers. In controlled laboratory exposures, tolerance to cadmium was found to vary by a factor of 8 among representatives of these lineages, and, at the same time, the frequencies of these mtDNA lineages differed significantly among rivers, and the relative abundance of the resistant lineages was higher in populations that were more tolerant of environmental contamination.

Multilocus DNA Fingerprinting

DNA fingerprinting examines allelic diversity at nDNA loci that contain much longer tandem repeats (9–65 bp; minisatellites) than those at microsatellite loci (2–4 bp). Unlike microsatellites that are distributed throughout the entire genome at regular 10 kbp intervals, minisatellites are frequently clustered on chromosomes. Similar to microsatellites, DNA fingerprinting detects very high levels of interindividual variation (hypervariability) in DNA banding patterns. Hypervariability at microsatellite loci results from a high germinal mutation rate producing novel length alleles resulting from changes in the number of repeated units. Minisatellite mutations probably result from a combination of processes including unequal recombination at meiosis, gene conversion, and slippage at replication forks (Yauk 1998).

Unlike microsatellite analysis, DNA fingerprinting examines genetic variation at multiple nDNA loci simultaneously. Because the target tandem repeat sequences are fairly long and conserved, universal DNA probes (M13mp9 phage, *Drosophila* Per, pUCG containing 25 tandem copies of Jeffrey's core sequence, etc.) can be used to detect multilocus fingerprints in a variety of disparate animal taxa including fishes (Castelli et al. 1990). Because of the large number of bands and loci represented on DNA fingerprinting gels, there may be uncertainty regarding assignment of bands to individual loci. Therefore, it is inadvisable to use the Hardy–Weinberg model in analyzing allelic frequencies. Instead, the extent of band sharing between individuals is calculated and then the band sharing index is used as a measure of dissimilarity among individuals and diversity within populations (Lynch 1990; Figure 10). Additionally, DNA fingerprinting is not a PCR-based approach and, therefore, demands on tissue size and condition are much more stringent than those needed for microsatellite analysis. If one family of bands proves particularly informative in DNA fingerprinting gels, the locus that encodes these bands can be cloned, flanking sequences characterized, and PCR primers can be designed. In this way, a PCR-based assay for that one minisatellite locus can be developed.

In DNA fingerprinting, relatively large amounts of high molecular weight (intact) DNAs are isolated from tissues or blood, and DNA is then fragmented by digestion with one or more restriction enzymes. The digested DNA fragments are then electrophoretically separated on agarose gels and transferred, usually by capillary action, to hybridization membranes and fixed by baking or UV light. The DNA fragments on the membranes are marked by hybridization with labeled DNA fingerprinting probes and visualized by exposure to X-ray film. The amount of DNA fragment sharing among individuals is then determined within a population; these data are then used to calculate intrapopulation genetic diversity. Alternatively, the fragment sharing is compared between parents and their offspring. In this case, any "novel" fragments that are present in the offspring and absent from either of their two parents must have originated from de novo germ line mutations (Figure 10).

Yauk and Quinn (1996) used multilocus DNA fingerprinting to compare mutation rates between the populations of herring gulls *Lárus argentátus* in highly chemically contaminated Hamilton Harbor, Ontario,

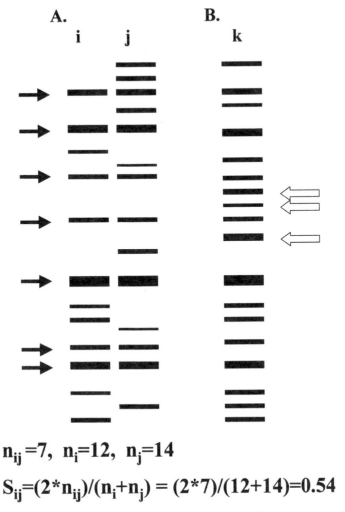

$$n_{ij} = 7, \quad n_i = 12, \quad n_j = 14$$

$$S_{ij} = (2 * n_{ij})/(n_i + n_j) = (2*7)/(12+14) = 0.54$$

Figure 10. A simulated example of a DNA fingerprint. (A) The figure indicates the method used to calculate a similarity index (S) between individuals within populations. The similarity between individual i and j is calculated from the number of shared bands (n_{ij}; indicated by dark arrows) and the number of bands in i (n_i) and j (n_j). For within population diversity, the average S is calculated for all ij pairs within the population. For between population diversity (genetic distance) i & j are from different populations. (B) Individual k represents an offspring. If i & j are known to be the parents of k, any bands in k should be found in i or j. The bands found in k and not found in i or j (light arrows) may be mutational derivatives. If i is known to be the mother and the father is unknown, the results indicate that j is probably not the father (which must be an individual that possesses the bands indicated by the light arrows).

two cleaner sites in the Great Lakes, and a pristine site in the Bay of Fundy. Using four DNA fingerprinting probes, they found that the number of "novel" fragments present in nestlings from Hamilton Harbor was 2–8 times higher than those in nestlings from the three control sites indicating that the mutation rate in the Hamilton Harbor population far exceeded that at the clean sites.

Several studies have used a modification of DNA fingerprinting to examine the relationship between exposure to pollutants and levels of genetic diversity in aquatic and nonaquatic plants. Keane et al. (1999) used three synthetic DNA probes composed of tandemly repeated "core" sequences (GACA, GATA, GCAC) in multilocus DNA fingerprints to quantify genetic diversity in cattails *Typha latifolia* from five sites on the Wurtsmith Air Force Base near Oscoda, Michigan. Soils, sediments, and groundwater at two of these sites were contaminated with fuels, solvents, and organic mixtures. Cattails from the two most polluted sites showed significantly higher levels of allelic diversity as reflected by higher heterozygosity, higher proportion of polymorphic loci, and lower probability of two specimens within a site sharing all their bands. In total, 9 private alleles (those unique to each population) were detected among specimens from the five sites, all of which were found in cattails from the two most polluted sites. The investigators suggested that either a higher mutation rate at the impacted sites or increased mortality and subsequent recolonization could be invoked to explain these findings. These same investigators (Keane et al. 1998) used the same DNA fingerprinting techniques to quantify genetic diversity in wild red raspberries *Rubus idaeus* L. growing in plumes of water at 16 sites at the Wurtsmith Air Force Base, of which half were considered contaminated with six organic pollutants and iron. In contrast to results in cattails, number of population bands, number of private alleles, and alleles per individual were significantly higher at the uncontaminated sites. The conflicting results between cattails and raspberries may be due to differences in levels of contamination or contamination uptake in aquatic versus terrestrial systems, differences in physiological or molecular responses to toxicants between the two species, differences in life history characteristics and dispersal mechanisms (wind-propagated cattail versus animal-propagated raspberry seeds), or some other unknown mechanism.

Comparison Among Techniques

There are a wide array of techniques and approaches for studying the effects of environmental contaminants on genetic diversity in aquatic organisms. Each technique has its advantages and disadvantages, depending upon the study situation and the objective of the experiment (Table 1). Methods that potentially provide a large number of variable markers, such as RAPD, AFLP, or microsatellites, may be good for studies that require a high degree of discrimination between populations (e.g., if populations have been isolated or exposed to environmental contaminants for only short periods of time). Markers that evolve relatively rapidly (e.g., mitochondrial DNA) would also provide a high degree of discrimination. If the amount of tissue is limited or the quality of the DNA is compromised, then PCR-based assays would be preferred over those requiring whole genomic or mitochondrial DNA. If the use of radioisotopes is a constraint, this would preclude use of such techniques as radioactively labeled sequencing microsatellite gels or Southern blots. Also, some markers may not be variable enough in some species to distinguish between populations. For instance, one species may have a highly variable mtDNA, but low variation in allozymes, while another species may show the opposite patterns. Also, the specific marker to use may depend on whether one wishes to use markers that are selectively neutral (e.g., microsatellites) or those that are adaptive (e.g., certain allozymes). A comparison of the properties of each of these markers described herein is summarized in Table 2.

Many of the DNA-based molecular methods offer both logistical and analytical advantages over previously used protein-based techniques. Logistically, all PCR-based approaches (RAPD, AFLP, mtDNA, and microsatellites) offer the luxury of analysis of very small, noninvasively acquired, archived, or alcohol-preserved tissue samples. Processing of small amounts of tissue presents the opportunity to quantify genetic diversity in young life stages (embryos or larvae) that probably are particularly vulnerable to toxic effects of xenobiotics. Analysis of noninvasively acquired tissues offers the potential for repeated sampling of the same individuals over time and the conservation of threatened resources. PCR-based approaches also offer the potential to conduct retrospective analysis on archived samples and compare genetic diversity in extirpated or historically pristine populations to that in extant populations. Obviously, relaxed demands on tissue preservation offer far greater flexibility in field collection efforts.

Analytically, most of these approaches provide the opportunity to focus on DNA sequences that are noncoding (are not used to produce proteins), probably have little or no functional significance, and, as a result, are often rapidly evolving (i.e., change rapidly in DNA sequence [mtDNA] or number of repeats [microsatellites] over time). Moderate to high levels of allelic or haplotype variation are needed to sensitively compare genetic diversity among natural populations and, thus, all of the approaches that use noncoding loci provide that ability. Examination of population genetics using noncoding loci provides the potential to focus on loci that are not subject to selective pressure. Thus, in most cases, levels of genetic variation will be a function of either mutation rate or historical population size—two variables that can be affected by contaminant exposure. In the event that unusually high levels of genetic diversity are observed in some populations, mtDNA heteroplasmy may provide a novel and sensitive empirical approach to distinguish newly generated from older mutations and, thus, the potential for an evaluation of the likelihood of an increased mutation rate in these populations.

Of all these approaches, microsatellites may provide the most accurate and reproducible estimates of genetic diversity. First, microsatellites exhibit

Table 1. Summary of advantages, disadvantages, and common applications of the various population genetic techniques in aquatic ecotoxicology.

Technique	Advantages	Disadvantages	Application
Allozymes	Relatively simple methodology. Function of loci known. Codominant inheritance facilitates determination of allele frequencies. No DNA sequence information or complex instrumentation needed.	Often low or no genetic variability. Slow rates of evolution: may not be suitable for closely related or recently separated populations. Degraded issues cannot be used.	Adaptation and resistance to toxicants. Genetic variability and gene flow if high variability is not needed.
mtDNA	Restriction enzyme analysis not complex. Often high variability. Evolves relatively quickly. Degraded DNA can be used for PCR.	Functional significance of some regions may not be known. DNA sequencing capabilities may not be available for some laboratories. A priori knowledge of DNA sequence information needed for PCR.	Genetic diversity and gene flow. Phylogenetic analysis. Distinguishing between closely related or recently separated populations. Analysis of changes in mutation rate.
RAPDs	High variability. No a priori knowledge of DNA sequence needed. Relatively simple methodology. Degraded DNA can be used. No specialized equipment. Multiple loci scored simultaneously.	Problems with reproducibility. Rigid standardization and QA/QC required. Functional significance of variability. unknown. Strict dominant inheritance complicates determination of allele frequencies.	Genetic diversity and gene flow. Adaptation to xenobiotic stressors. Useful for species without a priori DNA sequence information. Relative ease/ low cost make it attractive for laboratories not equipped for advanced molecular genetics.
AFLPs	High variability. No a priori knowledge of DNA sequence needed. Relatively simple methodology. Degraded DNA can be used. More reproducible than RAPDs. Multiple loci scored simultaneously.	Functional significance of variability unknown. Strict dominant inheritance complicates determination of allele frequencies. Additional steps during analysis make it more laborious and time consuming than RAPDs.	Genetic diversity and gene flow. Useful for species without a priori DNA sequence information. Relative ease, low cost make it attractive for laboratories not equipped for advanced molecular genetics.
DNA fingerprinting	High variability. Conserved hybridization probes often available. Numerous loci scored simultaneously. No a priori knowledge of DNA sequence.	Sequencing equipment and autoradiography may be prohibitive for some laboratories. Relatively large quantities of high quality DNA required.	Genetic diversity and gene flow. Distinguishing between closely related or recently separated populations. Useful for species without a priori DNA sequence information. Analysis of changes in mutation rate.
Microsatellites	Very high variability. Conserved PCR primers may be available. Codominant inheritance facilitates determination of allele frequencies. Evolves relatively quickly. Degraded DNA can be used.	Sequencing equipment or autoradiography may be prohibitive for some laboratories. Development of PCR primers may be laborious and time consuming.	Genetic diversity and gene flow. Distinguishing between closely related or recently separated populations. Analysis of changes in mutation rate.

among the highest levels of variability of all loci investigated with heterozygosities often approaching 1.0 (i.e., 100% of the loci examined are heterozygous), even in taxa that exhibited depauperate levels of diversity with allozyme or mtDNA analyses. Second, microsatellites offer the reproducibility sometimes not seen with RAPD analysis. Third, because microsatellites are codominantly expressed, they provide the ability to unequivocally identify allelic variants at individual loci, thereby, allowing Hardy Weinberg analysis of genotype frequencies. Microsatellite primers are currently available for a wide variety of commercially, recreationally, and ecologically important fish taxa and are becoming increasingly more common for invertebrate and plant species.

Whatever the genetic biomarker used, it should be kept in mind that each marker only reveals information about a very small portion of the genome. Thus, patterns that emerge from one set of markers may be different than patterns revealed by a different set of markers. These discrepancies may be determined by the evolutionary history of the organism or the specific locus in question. The discrepancies between loci in patterns of genetic diversity also may be influenced by constraints (or lack thereof) on evolutionary rates; functional loci (e.g., protein-coding genes or structural or regulatory elements), which may be less variable because changes in DNA sequence would be selected against. Such differences between markers may also be determined by life history of the organisms. For example, mtDNA is maternally inherited, while nuclear DNA is biparentally inherited. Therefore, if male dispersal is much greater than female dispersal, populations may appear to be subdivided with mtDNA markers, but not with nuclear markers (Bickham et al. 1996, 1998; Theodorakis et al. 2000).

Conclusions and Future Directions

Population genetics is an effective tool that can contribute to evaluating the effects and risks of environmental stressors on the health of aquatic organisms. This approach can provide an indication of chronic effects of environmental pollution on populations of affected organisms. Changes in patterns of genetic variation within and between populations may affect a population's long-term viability and adaptability or have detrimental effects on fitness components (fecundity, growth, and long-term survival) of individuals, which could ultimately affect population growth or ecosystem productivity. Also, changes in patterns of genetic diversity may provide an early warning of effects at the community or ecosystem level. There is currently a need to determine the relationship between population genetic effects and those at the community or ecosystem level. Within this context, examination of population genetic effects has some advantages compared to assessing community level effects such as the index of biotic integrity. In contrast to the community-level investigations, collecting samples of small fish for popula-

Table 2. A comparison of methods used to quantify genetic diversity and structure in natural populations.

Technique	Genetic variability	Function	Data interpretation
Allozymes	low	known	easy
mtDNA			
Whole genome	moderate	unknown	easy
Selected regions	low-high	known	easy
RAPDs	high	unknown	difficult
AFLPs	high	unknown	difficult
DNA fingerprinting	high	unknown	difficult
Microsatellites	very high	unknown	easy

tion genetic analysis can be accomplished faster and with less cost than community-level surveys. Also, once in the laboratory, PCR-based assays can be automated, eliminating operator subjectivity and allowing low-cost analysis of large numbers of samples necessary for environmental monitoring. The most useful approach to using such information would be a tiered strategy, in which the lower cost and less labor-intensive genetic assays could be used to help to prioritize sites for cleanup or more extensive monitoring. In evaluating the effects of pollutants on ecosystem health, it may be most informative to quantify and compare genetic diversity at various trophic levels within the food chains of impacted and nonimpacted ecosystems, including plants, invertebrates, and predatory fish species.

There will also, undoubtedly, be advances in the future that will facilitate monitoring population genetic diversity or allow more of the genome to be examined. For example, hybridization of PCR-amplified DNA to arrays of oligonucleotides immobilized on glass slides or silicon microchips could allow simultaneous screening and genotyping of thousands of loci containing single-nucleotide polymorphisms (Lemieux et al. 1998; Ramsay 1998; Sapolsky et al. 1999). Although these techniques are becoming more common in biomedical research, they have yet to be exploited in population genetic studies. Such advances will serve to enhance our ability to detect chronic effects of environmental contamination on natural populations.

Thus, the study of population genetics in aquatic ecosystems can provide a meaningful contribution to the fields of ecotoxicology and ecological risk assessment. We believe that population genetics can be used as a population-level biomarker of contaminant exposure and effects. Because population genetic effects are not specific to any one particular chemical and are "emergent properties" of population-level perturbations, (Bickham and Smolen 1994), they can be widely used in many situations as indicators of mixtures of complex chemicals and also, potentially, be applied to distinguishing between different sources of environmental stressors in aquatic systems.

Table 2. (continued).

Tissue demands	PCR-based	Relative reproducibility	Cost
stringent	no	high	low
stringent	no	high	high
relaxed	yes	high	low or high[a]
relaxed	yes	low	low
relaxed	yes	moderate	low
stringent	no	moderate	high
relaxed	yes	high	low or high[b]

[a] Low for RFLP analysis; high for sequence analysis
[b] Low if taxon-specific PCR primers have already been developed; high if taxon-specific PCR primers need to be developed

Acknowledgments

IW acknowledges the support of NIEHS Center Grant ES00260.

References

Albert, P. L. 1993. Strategies for population reintroduction: effects of genetic variability on population growth and size. Conservation Biology 7:194–199.

Atienzar, F., and seven coauthors. 1998. Application of the arbitrarily primed polymerase chain reaction for the detection of DNA damage. Marine Environmental Research 46:331–335.

Atienzar, F. A., M. Conradi, A. J. Evenden, A. N. Jha, and M. H. Depledge. 1999. Qualitative assessment of genotoxicity using random amplified polymorphic DNA: comparison of genotoxic template stability with key fitness parameters in *Daphnia magna* exposed to benzo[a]pyrene. Environmental Toxicology and Chemistry 18:2275–2282.

Avise, J. C. 1998. The history and purview of phylogeography: a personal reflection. Molecular Ecology 7:371–379.

Baker, R. J., J. A. DeWoody, A. J. Wright, and R. K. Chesser. 1999. On the utility of heteroplasmy in genotoxicity studies: an example from Chornobyl. Ecotoxicology 8:301–309.

Becerril, C., M. Ferrero, F. Sanz, and A. Castano, A. 1999. Detection of mitomycin C-induced genetic damage in fish cells by use of RAPD. Mutagenesis 14:449–456.

Benton, M. J., and S. I. Guttman. 1992. Allozyme genotype and differential resistance to mercury pollution in the caddisfly, *Nectopsyche albida*. 1. Single-locus genotypes. Canadian Journal of Fisheries and Aquatic Science 49:142–146.

Bickham J. W., T. R. Loughlin, D. G. Calkins, J. K. Wickliffe, and J. C. Patton. 1998. Genetic variability and population decline in Steller sea lions from the Gulf of Alaska. Journal of Mammalogy 79:1390–1395.

Bickham J. W., J. C. Patton, and T. R. Loughlin. 1996. High variability for control-region sequences in a marine mammal: implications for conservation and biogeography of Steller sea lions (*Eumetopias jubatus*). Journal of Mammalogy 77:95–108.

Bickham, J. W., and M. Smolen. 1994. Somatic and heritable effects of environmental genotoxins and the emergence of evolutionary toxicology. Environmental Health Perspectives 102:25–28.

Blaylock, B. G., and M. L. Frank. 1980. Effects of chronic low-level irradiation on *Gambusia affinis*. Pages 81–90 *in* N. Egami, editor. Radiation effects on aquatic organisms. University Park Press, Baltimore, Maryland.

Brown, J. R., K. Beckenbach, A. T. Beckenbach, and M. J. Smith. 1996. Length variation, heteroplasmy and sequence divergence in the mitochondrial DNA of four species of sturgeon (*Acipenser*). Genetics 142:525–535.

Bue, B. G., S. Sharr, and J. E. Seeb. 1998. Evidence of damage to pink salmon populations inhabiting Prince William Sound, Alaska, two generations after the *Exxon Valdez* oil spill. Transactions of the American Fisheries Society 127:35–43.

Castelli, M., J.-C. Philipart, G. Vassart, and M. Georges. 1990. DNA fingerprinting in fish: a new generation of genetic markers. Pages 514–520 *in* N. C. Parker et al. Fish-marking techniques. American Fisheries Society, Symposium 7, Bethesda, Maryland.

Changon, N. L., and S. I. Guttman. 1989a. Differential survivorship of allozyme genotypes in mosquitofish populations exposed to copper or cadmium. Environmental Toxicology and Chemistry 8:319–326.

Changon, N. L., and S. I. Guttman. 1989b. Biochemical analysis of allozyme copper and cadmium tolerance in fish using starch gel electrophoresis. Environmental Toxicology and Chemistry 8:1141–1147.

Chen, J. Z., and P. D. N. Hebert. 1999a. Intraindividual sequence diversity and a hierarchical approach to the study of mitochondrial DNA mutations. Mutation Research 434:205–217.

Chen, J. Z., and P. D. N. Hebert. 1999b. Terminal branch haplotype analysis: a novel approach to investigate newly arisen variants of mitochondrial DNA in natural populations. Mutation Research 434:219–231.

Conte, C., and six coauthors. 1998. DNA fingerprinting analysis by a PCR based method for monitoring the genotoxic effects of heavy metals pollution. Chemosphere 37:2739–2749.

Cronin, M. A., and J. W. Bickham. 1998. A population genetic analysis of the potential for a crude oil spill to induce heritable mutations and impact natural populations. Ecotoxicology 7:259–278.

DeWoody, J. A., and J. C. Avise. 2000. Microsatellite variation in marine, freshwater, and anadromous fishes compared with other animals. Journal of Fish Biology 56:461–473.

Diamond, S. A., M. C. Newman, M. Mulvey, and S. I. Guttman. 1991. Allozyme genotype and time-to-death of mosquitofish, *Gambusia holbrooki*, during acute inorganic mercury exposure: a comparison of populations. Aquatic Toxicology 21:119–134.

Dimsoski, P. and G. P. Toth. 1998. Development of DNA-based microsatellite marker technology for studies of genetic diversity in central stoneroller (*Campostoma anomalum*) populations. Page 56 *in* Society of Environmental Toxicology and Chemistry Abstract Book, SETAC 19th Annual Meeting, Charlotte, North Carolina. Society of Environmental Toxicology and Chemistry, Pensacola, Florida.

Dinesh, K. R., W. K. Chan, T. M. Lim, and V. P. E. Phang. 1993a. RAPD markers in fishes: an evaluation of resolution and reproducibility. Asia Pacific Journal of Molecular Biology and Biotechnology 3:112–118.

Dinesh, K. R., T. M. Lim, K. L. Chua, W. K. Chan, and V. P. E. Phang. 1993b. RAPD analysis: an efficient method of DNA fingerprinting in fishes. Zoological Science 10:849–854.

Ellegren, H., G. Lindgren, C. R. Primmer, and A. P. Moeller. 1997. Fitness loss and germline mutations in barn swallows breeding in Chernobyl. Nature (London) 389:593–596.

Ender, A., K. Schwenk, T. Staedler, B. Streit, and B. Schierwater. 1996. RAPD identification of microsatellites in *Daphnia*. Molecular Ecology 5:437–441.

Ferguson, M. M., and L. R. Drahushchak. 1990. Disease resistance and enzyme heterozygosity in rainbow trout. Heredity 64:413–417.

Fore, S. A., S. I. Guttman, A. J. Bailer, D. J. Altfater, and B. V. Counts. 1995. Exploratory analysis of population genetic assessment as a water quality indicator. 1995. 1. *Pimephales notatus*. Ecotoxicology and Environmental Safety 30:24–35.

Gabriel, W., and R. Bürger. 1994. Extinction risk by mutational meltdown: synergistic effects between population regulation and genetic drift. Pages 69–86 *in* V. Loeschcke, J. Tomiuk and S. K. Jain, editors. Conservation genetics. Birkhäuser Verlag, Basel, Switzerland.

Gillespie, R. B., and S. I. Guttman. 1988. Effects of contaminants on the frequencies of allozymes in populations of central stonerollers. Environmental Toxicology and Chemistry 8:309–317.

Guttman, S. I. 1994. Population genetic structure and ecotoxicology. Environmental Health Perspectives 102(Supplement 12): 97–100.

Hadrys, H., M. Balick, and B. Schierwater. 1992. Applications of random amplified polymorphic DNA (RAPD) in molecular ecology. Molecular Ecology 1:55–63.

Hartl, D. L., and A. G. Clark. 1997. Principles of population genetics, 3rd edition. Sinauer Associates, Sunderland, Massachusetts.

Hummel, H., and nine coauthors. 1995. Uniform variation in genetic traits of a marine bivalve related to starvation, pollution and geographic clines. Journal of Experimental Marine Biology and Ecology 191:133–150.

Johnson, K. P., J. Stout, I. L. Brisbin Jr., R. M. Zink, and J. Burger. 1999. Lack of demonstrable effects of pollutants on cyt b sequences in wood ducks from a contaminated nuclear reactor cooling pond. Environmental Research 81:146–150.

Keane, B., S. Pelikan, G. P. Toth, M. K. Smith, and S. H. Rogstand. 1999. Genetic diversity of *Typha latifolia* (Typhaceae) and the impact of pollutants examined with tandem-repetitive DNA probes. American Journal of Botany 86:1226–1238.

Keane, B., M. K. Smith, and S. H. Rogstand. 1998. Genetic variation in red raspberries (*Rubus Idaeus* L.; Rosaceae) from sites differing in organic pollutants compared with synthetic tandem repeat DNA probes. Environmental Toxicology and Chemistry 17:2027–2034.

Keshava, C., N. Keshava, G. Zhou, W. Z. Whong, and T. M. Ong. 1999. Genomic instability in silica- and cadmium chloride-transformed BALB/c-3T3 and tumor cell lines by random amplified polymorphic DNA analysis. Mutation Research 425:117–123.

Kopp, R. L., S. I. Guttman, and T. E. Wissing. 1992. Genetic indicators or environmental stress in central mudminnow (*Umbra limi*) populations exposed to acid deposition in the Adirondack Mountains. Environmental Toxicology and Chemistry 11:665–676.

Krane, D. E., D. C. Sternberg, and G. A. Burton. 1999. Randomly amplified polymorphic DNA profile-based measures of genetic diversity in crayfish correlated with environmental impacts. Environmental Toxicology and Chemistry 18:504–508.

Kubota, Y, A. Shimada, and A. Shima. 1992. Detection of x-ray-induced DNA damages in malformed dominant lethal embryos of the Japanese medaka (*Oryzias latipes*) using AP-PCR fingerprinting. Mutation Research 283:263–270.

Leberg, P. L. 1990. Influence of genetic variability on population growth: implications for conservation. Journal of Fish Biology 37 (Supplement A): 193–196.

Lce, C. J., M. C. Newman, and M. Mulvey. 1992. Time to death of mosquitofish (*Gambusia holbrooki*) during acute inorganic mercury exposure: population structure effects. Archives of Environmental Contamination and Toxicology 22:284–287.

Lemieux, B., A. Aharoni, and M. Schena. 1998. Overview of DNA chip technology. Molecular Breeding 4:277–289.

Lynch, M. 1990. The similarity index and DNA fingerprinting. Molecular Biology and Evolution 7:478–489.

Lynch, M, and B. G. Milligan. 1994. Analysis of population genetic structure with RAPD markers. Molecular Ecology 3:91–99.

Macnair, M. R. 1991. Why the evolution of resistance to anthropogenic toxins normally involves major gene changes: the limits to natural selection. Genetica 84:213–219.

Maruyama T, and M. Kamura. 1980. Genetic variation and effective population size when local extinction and recolonization of subpopulations are frequent. Proceedings of the National Academy of Sciences USA 77:6710–6714.

Matson, C. W., B. E. Rodgers, R. K. Chesser, and R. J. Baker. In press. Genetic diversity of *Clethrionomys glareolus* populations from highly contaminated sites in the Chernobyl region, Ukraine. Environmental Toxicology and Chemistry.

Mueller, U. G., and L. L. Wolfenbarger. 1999. AFLP genotyping and fingerprinting. Trends in Ecology and Evolution 14:389–394.

Mulvey, M., and five coauthors. 1995. Genetic and demographic responses of mosquitofish (*Gambusia holbrooki* Girard 1859) populations stressed by mercury. Environmental Toxicology and Chemistry 14:1995.

Murdoch, M. H., and P. D. N. Hebert. 1994. Mitochondrial DNA diversity of brown bullhead from contaminated and relatively pristine sites in the Great Lakes. Environmental Toxicology and Chemistry 13:1281–1289.

Murphy, R. W., J. W. Sites Jr., D. G. Buth, and C. H. Haufler. 1996. Proteins: isozyme electrophoresis. Pages 51–120 *in* D. M. Hillis, C. Moritz and B. K. Mable, editors. Molecular systematics, 2nd edition. Sinauer Associates, Sunderland, Massachusetts.

Nadig, S. G., K. L. Lee, and S. M. Adams. 1998. Evaluating alterations of genetic diversity in sunfish populations exposed to contaminants using RAPD assay. Aquatic Toxicology 43:163–178.

Newman, M. C., and R. H. Jagoe. 1998. Allozymes reflect the population-level effect of mercury: simulations of the mosquitofish (*Gambusia holbrooki* Girard) GPI-2. Ecotoxicology 7:141–150.

Nicola, M. D. E., C. Gambardella, and S. M. Guarino. 1992. Interactive effects of cadmium and zinc pollution on PGI and PGM polymorphisms in *Idotea baltica*. Marine Pollution Bulletin 24:619–621.

Nunney, L. 1999. The effective size of a hierarchically structured population. Evolution 53:1–10.

Prus-Glowacki, W., A. Wojnicka-Poltorak, J. Oleksyn, and P. Reich. 1999. Industrial pollutants tend to increase genetic diversity: evidence from field-grown European Scots pine populations. Water, Air, and Soil Pollution 116:395–402.

Ramsay, G. 1998. DNA chips: state-of-the art. Nature Biotechnology 16:40–44.

Roy, N. K., J. Stabile, J. E. Seeb, C. Habicht, and I. Wirgin. 1999. High frequency of K-*ras* mutations in pink salmon embryos experimentally exposed to *Exxon Valdez* oil. Environmental Toxicology and Chemistry 18:1521–1528.

Saccheri, I, and five coauthors. 1998. Inbreeding and extinction in a butterfly metapopulation. Nature (London) 392:491–494.

Sapolsky, R. J., and five coauthors. 1999. High-throughput polymorphism screening and genotyping with high-density oligonucleotide arrays. Genetic Analysis: Biomolecular Engineering 14:187–192.

Savelkoul, P. H. M., and eight coauthors. 1999. Amplified-fragment length polymorphism analysis: the state of an art. Journal of Clinical Microbiology 37:3083–3091.

Savva, D. 1998. Use of DNA fingerprinting to detect genotoxic effects. Ecotoxicology and Environmental Safety 41:103–106.

Schlosser, I. J. 1991. Stream fish ecology: a landscape perspective. Bioscience 41:704–712.

Schlueter, M. A., S. I. Guttman, J. T. Oris, and A. J. Baile. 1995. Survival of copper-exposed juvenile fathead minnows (*Pimephales promelas*) differs among allozyme genotypes. Environmental Toxicology and Chemistry 14:1727–1734.

Schlueter, M. A., S. I. Guttman, J. T. Oris, and A. J. Baile. 1997. Differential survival of fathead minnows, *Pimephales promelas*, as affected by copper exposure, prior population stress, and allozyme genotypes. Environmental Toxicology and Chemistry 16:939–947.

Street, G. T., G. R. Lotufo, P. A. Montagna, and J. W. Fleeger. 1998. Reduced genetic diversity in a meiobenthic copepod exposed to a xenobiotic. Journal of Experimental Marine Biology and Ecology 222:93–111.

Street, G. T., and P. A. Montagna. 1996. Loss of genetic diversity in Harpacticoida near offshore platforms. Marine Biology 126:271–282.

Sturmbauer, C., G. B. Opadiya, H. Niederstaetter, A. Riedmann, and R. Dallinger. 1999. Mitochondrial DNA reveals cryptic oligochaete species differing in cadmium resistance. Molecular Biology and Evolution 16:967–974.

Tanguy, A., N. F. Castro, A. Marhic, and D. Moraga. 1999. Effects of an organic pollutant (tributyltin) on genetic structure in the Pacific Oyster *Crassostrea gigas*. Marine Pollution Bulletin 38:550–559.

Theodorakis, C. W., J. W. Bickham, T. Elbl, L. R. Shugart, and R. K. Chesser. 1998. Genetics of radionuclide-contaminated mosquitofish populations and homology between *Gambusia affinis* and *G. holbrooki*. Environmental Toxicology and Chemistry 10: 1992–1998.

Theodorakis, C. W., J. W. Bickham, T. Lamb, P. A. Medica, and T. B. Lyne. 2000. Integration of genotoxicity, and population genetic analysis in kangaroo rats *(Dipodomys merriami)* exposed to radionuclide contamination. Environmental Toxicology and Chemistry 20:317–326.

Theodorakis, C. W., T. Elbl, and L. R. Shugart. 1999. Genetic ecotoxicology IV: survival and DNA strand breakage is dependant on genotype in radionuclide-exposed mosquitofish. Aquatic Toxicology 45:279–291.

Theodorakis, C. W., and L. R. Shugart. 1997. Genetic ecotoxicology: II. Population genetic structure in radionuclide-contaminated mosquitofish (*Gambusia affinis*). Ecotoxicology 6:335–354.

Theodorakis, C. W., and L. R. Shugart. 1998a. Natural selection in contaminated habitats: A case study using RAPD genotypes. Pages 123–150 *in* V.E. Forbes, editor. Genetics and ecotoxicology. Taylor & Francis, Philadelphia.

Theodorakis, C. W., and L. R. Shugart. 1998b. Genetic ecotoxicology III: the relationship between DNA strand breaks and genotype in mosquitofish exposed to radiation. Ecotoxicology 7:227–236.

Thomas, D. C., A. Umar, and T. A. Kunkel. 1996. Microsatellite instability and mismatch repair defects in cancer cells. Mutation Research-Fundamental and Molecular Mechanisms of Mutagenesis 350:201–205.

Trabalka, J. R., and C. P. Allen. 1977. Aspects of fitness of a mosquitofish *Gambusia affinis* exposed to chronic low-level environmental radiation. Radiation Research 70:198–211.

Waldman, J. R., J. Grossfield, and I. Wirgin. 1988. Review of stock discrimination techniques for striped bass. North American Journal of Fisheries Management 8:410–425.

Weis, J. S., P. Weis, and M. Heber. 1982. Variation in response to methylmercury by killifish (*Fundulus heteroclitus*) embryos. In J. G. Pearson, R. Foster and W.E. Bishop, editors. Aquatic toxicology and risk assessment: 5th Conference (ASTM S.T.P. 776). American Society for Testing and Materials, Philadelphia.

Williams, J. G. K., A. R. Kubelik, K. J. Livak, J. A. Rafaski, and S. V. Tingey. 1990. DNA polymorphisms amplified by arbitrary primers are useful as genetic markers. Nucleic Acids Research 18:6531–6535.

Wirgin, I. I., C. Gunwald, S. J. Garte, and C. Mesing. 1991. Use of DNA fingerprinting in the identification and management of a striped bass population in the southeastern United States. Transactions of the American Fisheries Society 120:273–282.

Wirgin, I. I., and L. Maceda. 1991. Development and use of striped bass-specific RFLP probes. Journal of Fish Biology 39 (Supplement A):159–167.

Wirgin, I. I., R. Proenca, and J. Grossfield. 1989. Mitochondrial DNA diversity among populations of striped bass in the southeastern United States. Canadian Journal of Zoology 67:891–907.

Wirgin, I. I., J. R. Waldman, L. Maceda, J. Stabile, and V. J. Vecchio. 1997. Mixed-stock analysis of Atlantic coast striped bass *(Morone saxatilis)* using nuclear DNA and mitochondrial DNA markers. Canadian Journal of Fisheries and Aquatic Sciences 54:2814–2826.

Wolfarth, G. W. 1985. Selective breeding of the common carp. *In* R. Billard and J. Marcel, editors. Aquaculture of cyprinids. Institut National de la Recherche Agronomie Jouy en Josas, France.

Woodward, L. A., M. Mulvey, and M. C. Newman. 1996. Mercury contamination and population-level responses in chironomids: can allozyme polymorphism indicate exposure? Environmental Toxicology and Chemistry 15:1309–1316.

Wright, J. M., and P. Bentzen. 1994. Microsatellites: genetic markers for the future. Reviews in Fish Biology and Fisheries. 4:384–388.

Yauk, C. 1998. Monitoring for induced heritable mutations in natural populations: application of minisatellite DNA screening. Mutation Research 411:1–10.

Yauk, C. L., and J. S. Quinn. 1996. Multilocus DNA fingerprinting reveals high rate of heritable genetic mutation in herring gulls nesting in an industrialized urban site. Proceedings of the National Academy of Sciences USA 93:12137–12141.

6

Immunological Indicators of Environmental Stress and Disease Susceptibility in Fishes

CHARLES D. RICE AND MARY R. ARKOOSH

Introduction

The impact of stress on the immune system of fishes is perhaps one of the most pondered, yet least understood aspects of fish biology. Professionals in the aquaculture industry and in fish biology research are well aware that stressful situations increase the incidence of infectious disease, often resulting in significant losses to both enterprises. The term stressor, however, is difficult to define in terms of its impact on fish health. Normal density-dependent factors in the environment (i.e., biotic factors) drive natural selection toward populations that are not overly sensitive to such "eustress." The same may be said of density-independent factors (i.e., abiotic factors). An excess of either of these factors, however, can overwhelm or "distress" most, if not all, of the individuals in a given population. When the physiological integrity of an individual is overwhelmed, immune suppression often follows. This condition often leads to outbreaks of diseases from pathogens already present in the environment or those pathogens held in check by the host under nonstressed situations. In reality, it is difficult to define a stress beyond, "we know it when we see it." Overcrowding; rapid changes in temperature, salinity, dissolved oxygen; and handling are common stressors affecting immunocompetence in fish (Norris 2000). Because immune function can be altered so quickly by stressors, particularly acute stressors, one must question why such deleterious physiological reactions have not been selected against throughout vertebrate evolution. The stress response favors survival of the whole animal at a time when altered immunity is a secondary concern. The negative effects of handling stress or predation, for example, on immune function are secondary to the animal when the primary need is to mobilize resources via glycolysis or gluconeogenesis in order to escape the source of the stress. Hormones and steroids that involve glycolysis and gluconeogenesis (catecholamines, glucocorticoids) are potent immunomodulators not only of immune responses but of the anatomical structure of the immune system as well. A balance between the need for energy and any disadvantageous immune suppression usually occurs as an integrated organismal response.

Immune function and the effects of stressors on immune function are best understood within the context of how the immune system is integrated within the whole organism; this is the basis for the discipline known as immmunophysiology. An understanding of the anatomy and physiology of the immune system in fishes allows us to understand how the immune function is affected by physiological stress. As in the case for other vertebrate physiological systems, the immune system of fishes comprises discrete cell types organized into specific tissues and organs, which are innervated and contain a rich blood supply. The primary role of the immune system is to distinguish (sense) foreign or nonself and altered self. Another role of the immune system is to sense loss of tissue integrity or wound formation and respond by remodeling tissue and eliminating pathogens introduced in the wound. Infectious disease agents are typical examples of nonself, while neoplasia (tumors) may be an example of altered self. Wounds caused by predation or handling may be examples of loss of tissue integrity. As discussed elsewhere in this chapter, an immune response is integrated within the neuroendocrine system, and, by "sensing" nonself and altered self, the immune system becomes an extension of other sensory systems. The existence and nature of foreign materials within the organism must be communicated to other cells, tissues, and organs of both the immune system and the whole animal. Finally, the whole animal responds to this information using the cells, tissues, and organs of the immune system. Throughout vertebrate evolution, the selective pressures for "immunoefficiency" have been intense, resulting in a finely tuned and well-orchestrated system. As with the other vertebrate physiological systems, the very basic aspects of immune function are conserved throughout the vertebrate taxa. There are, however, over 20,000 different species of fishes ranging from the primitive agnathan cyclostomes (e.g., hagfish), to the gnathan fishes including the chondrosteans (sturgeons, paddlefish), the elasmobranchs (sharks, skates, and rays), and the teleosts or modern bony fishes (cod, trout, catfish, tuna, killifish, zebrafish, bluegill, grouper, etc.). This chapter focuses on teleosts, the primary concern for environmental scientists and the majority of those involved in aquaculture. The chapter provides a general overview of the immune system of teleosts, the nature and effects of common environmental stressors on these systems, and current methods and approaches for detecting changes in immunological function in teleost. The major emphasis is placed on species commonly

used in aquaculture and in environmental sciences. For more detailed reviews on the subject of comparative immunobiology and fish immunology,
the reader is referred to Iwama and Nakanishi (1996).

Fish Immunophysiology

As is the case with higher vertebrates, specific cells and tissues of the immune system of teleosts are located in primary, secondary, and tertiary
lymphomyeloid organs. Primary lymphoid organs are specialized for hematopoiesis and the generation of predetermined lineages of specialized
cells (Figure 1). In teleosts, these organs are the thymus and anterior kidney.
The secondary organs are those responsible for the generation of immune
responses involving the interaction of multiple cell types, a process programmed to mount specific immune responses against specific antigens. In
higher vertebrates, this takes place in the spleen and lymph nodes, however,
teleosts have undeveloped lymphatic systems and rely on the spleen for
filtration of body fluids. As with the bone marrow of higher vertebrates,
specific immune responses also occur in the anterior kidney of fishes. Tertiary organs are primarily mucosal structures involving loose patches of lymphoid cells associated with the surface areas of the alimentary canal. Tertiary
tissues are distributed along the buccal cavity, nasopharyngeal cavity, and

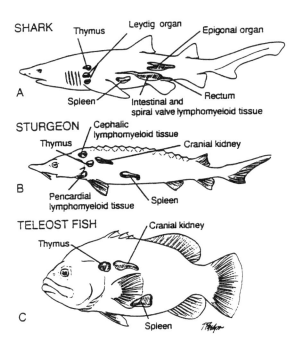

Figure 1. The locations of lymphopoietic tissues in shark (A), sturgeons (B), and
most teleosts (C). (Source: Kennedy-Stoskopf 1993).

intestinal cavity of higher vertebrates, but seem to be restricted to the intestinal regions of teleosts.

The most effective and first line of defense against intruding pathogens is a structural barrier. The primary roles of this structural integument are secretion of mucus and antimicrobial compounds and maintaining a low pH in the stomach. These functions require little effort on the part of the organism when energy mobilization to the cells and organs of the immune system are not required. Once the surface of the animal is breached, the immune system deals with the invading organisms through two responses. The innate immune response involves circulating and fixed tissue granulocytic phagocytes (neutrophils, eosinophilic granular cells) and monocytes and macrophages. In the case of tissue damage that may result from numerous events such as predation, skin abrasions, and parasitism, phagocytes migrate (traffic) to the site of injury and serve as the first line of active defense and wound repair. Phagocytes, namely granulocytes and monocytes, not only actively engulf invading microorganisms and tissue debris, they also release a myriad of enzymes and free radicals that retard or eliminate the growth of pathogens.

Monocytes also migrate to fixed tissues throughout the organism where they differentiate into tissue-specific macrophages. Brain macrophages are called microglia, liver macrophages are Kupffer's cells, serosal macrophages in the peritoneal cavity, and, lymphomyeloid macrophages in the spleen and anterior kidney. In the connective tissues, they are referred to as histocytes and become part of the functioning connective tissue. Along with the stromal tissues and vascular endothelial cells, phagocytes form the reticuloendothelial system (RES), a network of tissues specialized to trap, engulf, and process foreign material. These phagocytic processes can easily be demonstrated by administering particulates such as sheep red blood cells, bacteria, or iron particles intravenously and examining the spleen, anterior kidney, or liver at various times following administration. These particulates become trapped in the phagocytes of the tissues and, in the case of iron particles, the tissues become darkened with the agent.

Nonspecific cytotoxic cells (NCC) are involved in the initial immune response to virally infected cells. Nonlymphomyeloid tissues such as the liver also perform a role in the response to invading pathogens through the production of acute phase proteins that are released in the bloodstream to facilitate the functions of phagocytes, NCC, and endothelial cells, and to augment other humoral factors, including lectins, lytic enzymes, transferrin, and components of the complement system. Complement is the collective name for a group of serum proteins that participate in an enzymatic cascade, ultimately leading to the cytolytic destruction of target membranes via the membrane attack complex. Each step in the cascade forms two products: a split product and the next component in the pathway. Complement split products are critical to enhancing cell-mediated components of innate immune responses.

Once released from the cascade, the split products bind to both pathogen and specific receptors on monocytes, macrophages, and granulocytes, enhancing phagocytosis of pathogens, thereby increasing the overall effectiveness of phagocytosis. The complement cascade is activated by binding immunoglobulin to pathogens, directly by pathogens bearing receptors for split products or by mannose-binding lectin on the cell wall of microbial pathogens. This aspect of innate immunity seems to be highly conserved among vertebrates.

Complement in teleosts is similar in composition and function to that of higher vertebrates (Yano 1996). From an evolutionary view, it is interesting to note that the components of complement are produced in a concerted effort by both the liver and macrophages and some endothelial and epithelial cells. One role of the liver in the immunophysiology of the organism probably reflects its ontogenic function as a primary site for hematopoiesis during development. Taken together, the natural barriers of the skin, mucosal secretions, secreted antimicrobial compounds, the phagocytes and NCC, the RES, and the numerous secreted humoral factors comprise the innate resistance to invading pathogens. A defining feature of the innate immune system and its responses is that prior exposure to a particular pathogen does not improve or enhance its efficiency. Innate responses are "designed" to eliminate the pathogen or to slow the disease progression long enough for the specific acquired immune response to become activated.

Teleosts have the capacity to mount acquired antigen-specific responses to invading pathogens in a manner similar to that of higher vertebrates, including mammals. These responses involve both humoral and cell-mediated components, namely B cells and T cells, respectively. The specificity is due to specific B cell receptors (membrane-bound immunoglobulin) and T cell receptors. Rearrangement of heavy and light chain variable region genes during B cell development and the variable regions of the T cell receptor genes during T cell development provide the animal with the ability to recognize a wide range of specific antigens in the environment. Similar to the systems of higher vertebrates, specific immune responses of teleosts are under the influence of genetic restriction via the major histocompatibility complex (MHC) gene loci. Antigens are processed into cleaved peptide fragments and presented by macrophages and B cells to T cells in the context of MHC-II gene products just as they are in mammals. Recent evidence of cytotoxic T cells in fish suggests that endogenous antigens (altered self proteins and viral particles) are endogenously processed by proteosomes and presented to cytotoxic T cells in the context of MHC-1 gene products, as they are in mammals.

The predominant immunoglobulin of fishes is an immunoglobulin type M (IgM)-like molecule. Unlike the pentameric IgM of mammals, IgM of teleosts is tetrameric. Also unlike mammals, teleosts lack the ability to undergo immunoglobulin class switching upon subsequent exposure to specific antigens. Structurally, antibodies have an antigen-binding end consisting of light and heavy chain variable regions and a constant region at the other end. The variable region is able to recognize a specific configuration of the antigen (epitope) while the constant end, also known as the fragmented constant portion (Fc portion), provides a functional role to the entire molecule. Phagocytes and NCC have receptors for the Fc regions, at least in higher vertebrates, thus, providing a cell the ability to direct its actions when, otherwise, it lacks the ability to recognize specific targets. In general, antibody-coated (opsonized) pathogens are more easily phagocytized and destroyed than those left unopsonized. In the presence of multiple IgM molecules, the complement cascade is activated, thereby forming both the membrane attack complex end products and the split products that further activate components of the innate immune response. Antibody molecules can also neutralize certain pathogens by blocking their ability to bind or activate their specific molecular receptors.

During the generation of a specific immune response, undifferentiated B cells and T cells are activated, proliferate, and become differentiated plasma cells and T-helper or T-cytotoxic cells, respectively. During proliferation and differentiation, a portion of the B cells and T cells are committed to becoming memory cells, thereby providing the animal with the ability to mount a stronger, faster response upon subsequent exposure. Although there is no evidence for immunoglobulin class switching in teleosts, there is evidence for affinity maturation (Kaattari 1994).

Host–Pathogen Interactions

As defined earlier, the immune system is designed to "sense" nonself and altered-self or the loss of tissue integrity in the form of wounds and to respond to this information. Immunologists often divide immune responses into afferent and efferent phases. Detection of nonself, usually an infectious disease agent, begins with entry. The most common portals of entry or penetration are the gills, the gastrointestinal tract, the skin, and by other nonclassical means such as the olfactory sac epithelium, the thymus, the pseudobranch, and possibly the ovum. The reader is referred to an excellent review on the subject by Evelyn (1996).

The marine pathogen *Vibrio anguillarum* represents a typical microbe to which most fish are immunologically responsive. Once barriers are breached and the nonspecific immune response is initiated, the specific response is summoned. This particular pathogen is actively phagocytized by the epithelial cells of the gill (Nelson et al. 1985; Baudin-Laurencin and Germon 1987) and the skin (Kanno et al. 1989), as well as the gastrointestinal tract, under experimental conditions (Grisez et al. 1996). Fish are also highly responsive to this pathogen when administered intraperiotoneally (Regala et al. 2001). It is believed that *V. anguillarum* particles are taken up by gill epithelial cells, passed on to blood-borne monocytes in the gill, and finally transferred to the anterior kidney and spleen. Although fish do not generate lymphoid germinal centers as seen in higher vertebrates, they do have melanomacrophage centers or macrophage aggregations in the spleen and anterior kidney. Particles become trapped in these centers following uptake, therefore, it is speculated that these areas may be primitive germinal centers. As discussed below, these macrophage aggregations are also associated with other conditions including aging, oxidative stress, pollution, and disease resistance (Camp et al. 2000). The pathogen is processed in the lymphoid tissues and is presented to T cells resulting in the proliferation and differentiation of antigen-specific B cells into antigen-specific antibody-secreting plasma cells. The antibody is secreted into the bloodstream where it is easily detected in serum or plasma. Specific antibodies are also secreted in the mucus, indicating an active process by which B cells, possibly those in circulation, are directed toward the integument. Skin mucosal antibodies may be an indication of skin-related uptake mechanisms and responses to the antigen (Anderson 1990). Circulating antibodies then opsonize these particles via specific interactions with heavy and light chain variable regions in the antigen-binding regions of the molecule. The heavy chain nonvariable (constant) regions of the Ig molecule are effective in giving specificity to macrophages and other phagocytes via their receptors for the constant regions (Fc receptors). Activated T cells, macrophages, other phagocytes, and B cells secrete proinflammatory cytokines that then potentiate and regulate immune responses.

Neuroendocrine-Immune Connections

As discussed above, the anterior kidney tissue of teleosts is specialized for hematopoiesis, being a source for both mature and immature cells and a site for immune responses, including antibody production. Anterior kidney tissues are also the site for chromaffin tissues and the interrenal glands. Chromaffin tissues secrete catecholamines (norepinephrine and epinephrine) while the interrenal glands secrete glucocorticoids, namely cortisol. As pointed out in other sections of this book, stress can be acute, chronic, or both. Catecholamines are released during an acute stress response, resulting in a burst of epinephrine and norepinephrine release that drives the "flight or fight" response. Glucocorticoids are typically elevated during prolonged chronic stress. Prolonged stress can deplete the animal of energy reserves and eventually lead to death due to ion loss in individual cells.

A typical stress response in fish begins the release of central nervous system-derived corticotropin-releasing hormone. Corticotropin-releasing hormone stimulates the release of adrenocorticotropic hormone (ACTH) in the pituitary gland, which then travels to the distal location of the cortex of the adrenal glands, at least in higher vertebrates, where it stimulates the release of cortisol or corticosterone. Both of these hormones are potent suppressors of immune functions. In higher vertebrates, the adrenal glands are not located in or on lymphoid tissues. Thus, physiological stress has an immediate impact on the primary lymphoid organ in teleosts in the form of catecholamine release by the chromaffin tissues and glucocorticoid release by the interrenal tissues. In terms of anatomy and physiology, it would seem that teleosts may be especially sensitive to stress because of the immunosuppressive activity of these hormones. Apparently, fish leukocytes can also produce hormones classically associated with the neuroendocrine system. Arnold and Rice (2000) show that channel catfish leukocytes secrete ACTH.

Cortisol is strongly associated with immune suppression in fish. Generally speaking, elevated cortisol levels resulting from handling stress reduce peripheral blood lymphocytes and monocytes and increase neutrophils (Bonga 1997). In channel catfish *Ictalurus punctatus,* a single injection of cortisol that elevates circulating levels to 100 ng/mL also increases circulating neutrophils, but lowers lymphocyte counts (Ellsaesser and Clem 1987). Lymphocytes from these same fish were less responsive to mitogen-stimulated proliferation in vitro compared to controls. In contrast, cortisol administration in Atlantic salmon *Salmo salar* via injection did not alter circulating neutrophil and lymphocyte counts, but it did reduce the in vitro mitogenic response (Espelid et al. 1996). Earlier studies by Pickering and Pottinger (1989) show that even moderately elevated levels of cortisol are associated with increased susceptibility to diseases in brown trout *Salmo trutta.*

Maule et al. (1987, 1993) demonstrate that smolting coho salmon *Oncorhynchus kisutch* have high levels of cortisol, and these levels were not only associated with increased susceptibility to pathogens, but their lymphocytes had a reduced capacity to secrete antibodies. In general, nonspecific proinflammatory functions are highly susceptible to glucocorticoids. For example, phorbol ester-induced respiratory burst activity and phagocytosis are suppressed by cortisol in the dab *Limanda limanda* while basal levels are elevated. An elevated basal (nonstimulated) level of respiratory burst activity may be due to the gluconeogenic actions of cortisol. Cortisol also inhibits nitric oxide synthase activity in a goldfish *Carassius auratus* macrophage cell line (Wang and Belosevic 1995). From studies in mammalian systems, it is now clear that proinflammatory functions are under the control of NFκ-β, a highly conserved promoter for a variety of proinflammatory gene products. Translocation of NFκ-β from the cytosol to the nucleus is inhibited by glucocorticoids, therefore, reduced nonspecific defenses in fish by cortisol is expected. It is important to recognize that in fish cortisol is both a glucocorticoid (gluconeogenic) and a mineralocorticoid (ion balance) (Norris 2000). Moreover, since gill integrity is essential to maintaining ion balance, any stressor affecting gill function would be expected to initiate a rise in cortisol levels because of its mineralocorticoid role. In this case, immunomodulation as a result of cortisol elevation is an indirect effect of the environmental stressor on gill function. This point is often overlooked by fish biologists.

The literature supporting the immunomodulating effect of cortisol is extensive. The following is a summary of recent reviews by Bonga (1997), Barton and Iwama (1991), and Ellis (1981). In the natural environment and in captivity, however, there are many situations associated with elevated cortisol in fish. As reviewed by Norris (2000), anesthetics, confinement, feeding, handling, and light all increase cortisol levels in fish, as does temperature and water quality changes, and reproductive and social status. It should be noted that, like higher vertebrates, including humans, there are genetic strains and individuals exhibiting different cortisol responses to stress. At the population level, a given species of fish would, therefore, be expected to show a great deal of variation in stress-related cortisol effects on immune functions.

Much less is known about the influence of catecholamines on immune function in fish compared to cortisol. Flory (1989) demonstrated that beta-adrenergic agonists suppress, while alpha-2 adrenergic agonists, as well as cholinergic agonists, enhance in vitro antibody response in rainbow trout *Oncorhynchus mykiss*. Catecholamines also stimulate the respiratory burst of rainbow trout phagocytes (Bayne and Levy 1991). These investigators showed that phenylephrine, an alpha-2-adrenergic agonist, enhances this activity. In mummichogs *Fundulus heteroclitus*, epinephrine causes a rapid decrease in lymphocytes and increase in granulocytes, suggesting that catecholamines are responsible for this effect of acute stress.

Why would highly immunomodulating stress-related hormones be released directly near a major lymphoid organ? We are just now beginning to realize that in fish, like mammals, leukocytes produce and secrete neuroendocrine hormones (Arnold and Rice 2000). Some of these, namely ACTH, are fairly potent activators of leukocyte function (Bayne and Levy 1991) and may be the counteracting force that controls the effects of local neuroendocrine effects in the lymphoid tissues. The production of potentially immunomodulating hormones by leukocytes may be a physiological feedback mechanism to cope with high levels of stress hormones released in hematopoietic compartments that are highly vascularized and innervated (Flory 1989; Flory and Bayne 1991).

Effects of Biotic Stress on Immune Function in Fishes

The expression of disease in fish is dependent upon several factors, including the quality of the environment, differential susceptibility of individual fish to the infectious agent as a result of their genetic predisposition or physiological health, and the presence and virulence of the infectious agents (Snieszko 1973). A delicate balance exists between these variables in their contributions to disease expression as part of a normal functioning environment and disease expression as part of an environment experiencing pathology (Rapport et al. 1985). The interaction of these factors, as well as the element of immunosuppression can create what has been referred to as "a web of causation for disease" (Hedrick 1998). Environmental quality, disease susceptibility, and parasite presence and virulence are controlled by abiotic, biotic, and genetic factors and have the potential to influence this delicate balance and increase the potential for disease expression in fish. The increased potential for disease, in turn, can regulate the population (Anderson and May 1979), influence the structure of the community (Freeland 1983), and alter social behavior (Loehle 1995). Host–parasite interactions are considered to be of even greater evolutionary significance than predator–prey interactions (Price 1980; Kennedy 1984). Therefore, it is critical that our assessment of the effects of biotic stressors on the immune system also examine the influence of parasites upon the ecology and behavior of the host. Although numerous biotic factors have the potential to influence immune function, we will concentrate primarily on social behavior, predation and predator avoidance, handling and crowding stress, and pathogens.

Social Behavior

Hamilton and Zuk (1982) suggest that secondary sexual characteristics may provide a signal of the presence of parasite resistance genes in a potential mate. Paradoxically, the expression of these secondary sexual characteristics may cause suppression of immune function. The maintenance of secondary sexual characteristics may be a handicap considering the potential problems incurred by a suppressed immune system. The immunocompetence-handicap hypothesis suggests that the increased production of hormones needed for expression of the secondary sexual characteristics may be responsible for suppressing the immune response (Folstad and Karter 1992). Another hypothesis suggests that the expression of the sexual characteristics uses energy resources that normally would be allocated for immune function (Zahavi 1975; Gustafsson et al. 1994; Sheldon and Verhulst 1996).

Few fish studies exist that correlate the male's ability to mount an immune response with the expression of its secondary sexual characteristics. However, studies have examined pathogen prevalence and expression of secondary sexual characteristics and have demonstrated contrasting results. For example, it was found that the more sexually mature male rainbow trout and brown trout possess an increased intensity and prevalence of parasite and fungal infection (Richards and Pickering 1978; Robertson 1979; Pickering and Christie 1980) relative to their less mature counterparts. Yet other experiments have demonstrated a negative relationship between secondary sexual characteristics and parasite load. For example, a negative relationship has been demonstrated between a male's parasite load and exaggerated secondary sexual characteristics in guppies *Poecilia reticulata* and threespine sticklebacks *Gasterosteus aculeatus* (Milinski and Bakker 1990; Houde and Torio

1992). It has also been demonstrated that nonparasitized guppy males are chosen for mating by female guppies over males parasitized with an ecto-parasitic monogenean parasite, (*Gyrodactylus* sp.*) that were displaying re-duced mating behavior (Kennedy et al. 1987). Interestingly, in a species of fish where the sex roles are reversed and females compete for males, it was demonstrated that pipefish *Syngnathus typhle* parasitized with the trematode *Cryptocotyle* sp. produce fewer eggs and appear to be less attractive to the males (Rosenqvist and Johansson 1995). Therefore, these studies with fish demonstrate that the correlation between parasite load and secondary sexual characteristics may be positive in some cases and negative in others.

Social Hierarchy and Stress

Numerous studies have demonstrated that social hierarchy in fish appears to correlate with stress responses (Ejike and Schreck 1980; Peters et al. 1980; Scott and Currie 1980). Typically, the most dominant fish appear to be less stressed than the fish lower in the hierarchy. Studies with rainbow trout and tilapia *Tilapia mossambica* × *T. honorum* have demonstrated that the subor-dinate (β) fish in the hierarchy have more morphological alterations in their hematopoietic tissue (Peters and Schwarzer 1985; Cooper et al. 1989), al-tered immune function (Ghoneum et al. 1988; Peters et al. 1991) and are more susceptible to disease (Peters et al. 1988) relative to the dominant (α) fish. Tilapia have been used in a number of these studies because they are very aggressive when kept together in a small group, as they are for artificial spawning.

Studies in rainbow trout have shown that phagocytes from the subordi-nate fish show signs of degeneration (Peters et al. 1991). Necrotic tissue formed in the areas where the phagocytes disintegrated. In tilapia, melanomacrophage centers from the pronephros had a distinct vacuolar degeneration and the nuclei were vesicular in shape in β fish while the pronephric cells in α fish appeared normal (Cooper et al. 1989). Subordinate fish were found to have altered natural cytotoxicity activity, phagocyte func-tion, suppressed mitogenic response, and marked leukopenia (Ghoneum et al. 1988; Faisal et al. 1989; Peters et al. 1991). Subordinate fish have been found to be more susceptible to the opportunistic pathogen *Aeromonas hydrophilia* (Peters et al. 1988).

Predation and Predator Avoidance

Data of the influence of predator avoidance on the immune system in fish are limited. However, researchers have examined various aspects of preda-tor avoidance once infection has occurred. They found that infected fish were more susceptible to predation (Herting and Witt 1967; Milinski 1984; Mesa et al. 1998). Juvenile chinook salmon *Oncorhynchus tshawytscha* infected with either moderate or high levels of *Renibacterium salmoninarum* were more vulnerable to predation by either northern pike minnow *Ptychocheilus oregonensis* or smallmouth bass *Micropterus dolomieu*. Bluegill *Lepomis macrochirus* infected with *Flexibacter columnaris* were also more vulnerable to predation. Threespine sticklebacks infested by plerocercoids of the cestode *Schistocephalus solidus* feed much more closely to predators than uninfected fish (Milinski 1984). In contrast, studies examining predator avoidance behavior in fish infected with ectoparasites found no alteration in behavior (Coble 1970; Vaughan and Coble 1975). For some parasites, predation allows for a greater chance for the parasite to be transmitted to its next host. However, not all parasites that influence predator avoidance require another host for completion of their life cycle. Therefore, this increased vulnerability to predation can be at a high cost to the pathogen as well as the host.

Handling and Crowding Stress

Handling of fish and crowded stocking conditions are often associated with intensive aquaculture. In some cases, fish raised under these crowded conditions are released to supplement declining wild stocks. However, it appears that attempts to restock wild runs with fish from intensive aquaculture have not succeeded. In particular, salmon stocks are still on the decline. It has even been demonstrated that wild stocks have a greater survival rate than their hatchery raised cousins (Felton et al. 1990), suggesting that hatchery rearing conditions may be influencing survival through some means related to stress early in their development.

A number of studies demonstrate that crowding and handling stressors can cause immunosuppression. For example, in channel catfish, handling stress was associated with a suppressed mitogenic and antibody response (Ellsaesser and Clem 1986). Both T- and B-leukocytes appeared to be affected. The number of B-leukocytes and the surface immunoglobulin decreased as did the mitogenic response of cells to a T cell mitogen. Maule et al. (1989) were able to demonstrate that after handling, chinook salmon were immunosuppressed as demonstrated by a reduced plaque forming cell (PFC) response of the anterior kidney lymphocytes. These fish were also more susceptible to mortality caused by the pathogen *Vibrio anguillarum*. Mazur and Iwama (1993) demonstrated that juvenile Atlantic salmon reared under crowded conditions (64 kg/m^3) also had a reduced humoral immune response. Common carp *Cyprinus carpio* under crowding stress experienced a significant drop in the nonspecific immune parameters of lysozyme activity, complement levels, and phagocytosis (Yin et al. 1995). Whether immunosuppression as a result of these stressors is due to increases in plasma cortisol is debatable (Strange and Schreck 1978; Ellsaesser and Clem 1986; Salonius and Iwama 1993). It is apparent that most studies examining handling and crowding stressors are focused upon freshwater species (Waring et al. 1992) and little information is available on the affect of these stressors on the immune response of marine teleosts.

Pathogens

There is some debate as to whether pathogens themselves activate the hypothalamic-pituitary-adrenocortical system. Mesa et al. (1998) found that cortisol is significantly increased during the later stages of bacterial kidney disease (BKD). *Renibacterium salmoninarum*, the causative agent of BKD, appears to have the ability to suppress the immune response in fish (Turaga et al. 1987; Sakai et al. 1991). Leukocytes from salmon exposed to either the whole bacteria or to the major surface protein of the bacteria, p57, were immunosuppressed as determined by their altered chemiluminescence response (Sakai et al. 1991). Juvenile salmon exposed in vitro and in vivo to p57 or the pathogen, respectively, produced a suppressed B cell response (Turaga, et al. 1987; Fredriksen et al. 1997). Additionally, in studies with fish infected with *Cryptobia salmositica*, *Ichthyophonus hoferi*, and erythrocytic necrosis virus, plasma cortisol levels were not elevated (Laidley et al. 1988; Rand and Cone 1990; Haney et al. 1992). However, the immune function of animals infected with these pathogens has yet to be determined.

The Effects of Pollution Stress

Considering the anatomical and physiological features of the immune system in vertebrates, it is not surprising that the immune system is a sensitive target for environmental pollutants (Rice 2001). The mechanisms behind this immunomodulation can be either direct or indirect and sometimes both (Figure 2). While a significant percentage of leukocytes are located in the fixed organs and tissues of the immune system, most are found in the vascular compartment. More important, cells are continually trafficking between the peripheral blood compartment and these fixed tissues (primary, secondary, and tertiary organs). As emphasized earlier, the nonspecific barriers (skin, gut pH, mucus secretions, etc.) to pathogen entry are also considered to have a role in the immune response, even though their functions may be passive. Contaminants may also disrupt these barriers, leading to increased susceptibility to pathogen invasions.

Contaminants are either absorbed or actively taken up by fish via the gills, skin, buccal cavity, or gastrointestinal tract. As a result, the leukocytes are directly exposed to the contaminants as they enter the bloodstream. For those compounds that are metabolically activated by the liver and other tissues to more toxic metabolites (e.g., polycyclic aromatic hydrocarbons [PAH]), circulating leukocytes may be directly exposed to these reactive intermediates. In addition, it is now known that leukocytes express the aryl-

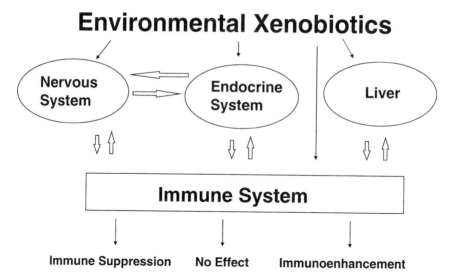

Figure 2. General scheme for direct and indirect effects of foreign compounds on immunity and potential outcomes. (Source: Karras and Holsapple 1995; Aaltonen 2000).

hydrocarbon receptor (Ahr) (Crawford et al. 1997) and have endogenous cytochrome P450 enzymes capable of metabolizing xenobiotics in a manner similar to hepatic parenchymal cells (Ladics et al. 1992). Lymphoid organs are designed as filtering devices, therefore, these organs are potential targets for compounds in circulation. Another feature that makes the immune system a vulnerable target for contaminants is the dependency upon proliferation and differentiation of T cells, B cells, and monocytes to become effector cells. Compounds or active metabolites of some compounds that are highly mutagenic, carcinogenic, or have a strong oxidative property can easily target these cells. The immune system depends heavily upon inter- and intra-cellular communication via both cognate (cell-to-cell contact) and noncognate (via hormones and cytokines) mechanisms. Several contaminants, especially metals, alter the normal signaling between cells (Burnett 1997). Finally, the immune system is integrated within the neuroendocrine system. Therefore, contaminants that alter neuroendocrine function or that induce the stress response will indirectly affect the immune system. For example, metals may alter gill membrane integrity resulting in a critical ion imbalance. Cortisol, acting as a mineralocorticoid, is then released, which may indirectly affect immune functions. It is critical, therefore, that before immunotoxicity is ascribed to a particular contaminant or mixtures of contaminants, we must first rule out the possibility that other organ systems are the primary target. In essence, truly immunotoxic agents target the immune system before other systems are altered (Figure 3).

Research into the immunotoxicity of environmental contaminants in fish has increased dramatically over the last two decades, resulting in a large

Immunotoxic Agents Target the Immune System

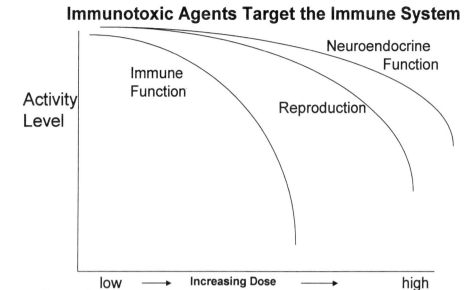

Figure 3. In a dose-dependent manner, directly immunotoxic agents target the immune system before other organ systems are affected.

amount of literature on the subject. The reader is referred to a partial listing of key reviews from the past 25 years: Snieszko (1974), Zeeman and Brindley (1981), Sinderman (1980), Zeeman (1986), Weeks et al. (1992), Dunier and Siwicki (1993), Anderson (1990), Anderson (1994), Zelikoff (1994), Anderson and Zeeman (1995), Rice (2001). The following is a brief overview of selected citations representing the many papers that demonstrate the immunotoxicity of environmental contaminants and is not intended to be exclusive. The above reviews contain many more examples and are, by design, much more inclusive.

Immunotoxicity of Select Contaminants

Polycyclic Aromatic Hydrocarbons

Polycyclic aromatic hydrocarbons are by-products of fossil fuels and their combustion products. They are ubiquitous in distribution, particularly in aquatic habitats near heavily populated areas where atmospheric deposition, terrestrial runoff, and boating activity is high. Of the PAHs, benzo-*a*-pyrene (BAP) has received the most attention in terms of immunotoxicity. As the prototype experimental PAH, BAP is mutagenic, carcinogenic, and immunotoxic, following metabolic activation by hepatic xenobiotic metabolizing enzymes, namely CYP1A. Protein and nucleic acid adducts and reactive intermediates (oxidative stress) seem to account for its wide range of toxicities. In fish, dimethylbenzanthracene and BAP are very potent immunotoxic (Faisal et al. 1991a; Seeley and Weeks-Perkins 1991; Hart et al. 1998) and carcinogenic agents (Vogelbein et al. 1990). In recent years, it has become clear that certain populations of fish inhabiting systems heavily contaminated with PAHs, such as Virginia's Elizabeth River system, have become resistant to overt toxicity, although the population harbors neoplastic lesions (Armknecht et al. 1998; Vogelbein et al. 1999). Earlier work in the Elizabeth River system by Faisal et al. (1991a) demonstrated reduced NCC activity. Kelly-Reay and Weeks-Perkins (1994) show that macrophage respiratory burst activity was higher in mummichogs from a creosote-contaminated station in the Elizabeth River system than in fish from the York River, Virginia. To date, antigen specific antibody responses have not been examined in any species of fish from the Elizabeth River system. Arkoosh et al. (1991, 1996) demonstrated that salmon from Puget Sound, Washington, another system heavily contaminated with PAHs, have reduced immunological memory and reduced peripheral blood mitogenic responses. Faisal et al. (1998) and Rose et al. (2000) demonstrated that lymphoid tissues and isolated leukocytes of the mummichog metabolize BAP to structures associated with DNA adducts in those tissues. Macrophage function in several species of fish seems to be particularly sensitive to PAHs (Weeks and Warinner et al. 1984, 1986; Seeley and Weeks-Perkins 1992; Tahir et al. 1993; Tahir and Secombes 1995). Exposure to PAHs (as creosote) lowers the mitogen-induced lymphocyte proliferation in spot *Leiostomus xanthurus* (Faisal et al. 1991b), as well as phagocyte activity and mitogen-induced lymphocyte proliferation in rainbow trout (Karrow et al. 1999).

Halogenated Aromatic Hydrocarbons

Halogenated aromatic hydrocarbons (HAH) include the polychlorinated biphenyls, polychlorinated dioxins, furans, and naphthalenes. The prototype congener of this class and the most immunotoxic, at least in rodents, is 2,3,7,8-tetrachlorodibenzo-*p*-dioxin (TCDD) (Holsapple et al. 1991). The more toxic HAHs are those processing a planar, TCDD-like structure that binds with high affinity to the cytosolic Ah receptor. The nonplanar HAHs are also toxic, but the mechanisms of action may be due to their effects on biological membranes, cell signaling, and possibly by acting as hormone-mimics (Soto et al. 1992; Ganey et al. 1993; Tithof et al. 1995). Some nonplanar polychlorinated biphenyls (PCB) can activate human neutrophils, thereby initiating the oxidive burst response. To date, such properties in fish immune systems have not been explored.

While TCDD is very immunotoxic in mice, especially for primary antibody responses, it seems to be much less so in fish (Rice 2001). Spitzbergen et al. reported that TCDD did not significantly affect the antibody response in rainbow trout, but exposure did enhance susceptibility to Infectious hematopoietic necrosis virus. Perhaps nonspecific aspects of immunity, such as NCC activity, are targeted. In support of this interpretation, Rice and Schlenk

(1995) reported that PCB-126, a TCDD-like coplanar PCB, did not affect the antibody response to *Edwardsiella ictaluri*, however, NCC and phagocyte activity were suppressed. Dietary aroclor 1254 suppresses the antibody response in rainbow trout and increases disease susceptibility (Cleland 1988; Thuvander and Carlstein 1991), but it increases anterior kidney B cell populations and enhances phagocyte activity in channel catfish, while having no effect on antibody responses (Rice et al. 1998).

In the pulp and paper industry, bleached kraft mill effluents are mixtures of halogenated organics, insoluble wood fibers, and particles (Jokinen et al. 1995). Even after treatment, significant amounts of halogenated phenols are released to the environment. This source of HAHs into the environment has been investigated by several researchers for its potential immunotoxicity. Subchronic exposure to these effluents lowered both the basal and antigen-induced antibody response in roach *Rutilus rutilus* (Jokinen et al. 1995; Aaltonen et al. 1997, 2000).

Heavy Metals

The immunotoxicity of metals in fish has been reviewed extensively by Zelikoff (1993, 1994), Zelikoff et al. (1995), and Anderson (1996). In one particular study, Burnett (1997) demonstrates that low levels of mercury chloride induce an influx of calcium into red drum *Sciaenops ocellatus* lymphocytes, stimulate tyrosine-kinase activity, and induce cellular proliferation. A similar property is associated with tributyltin (TBT) immunotoxicity (Rice and Weeks 1989), although this compound is used as antifoulant biocide, not a heavy metal. Copper is also a potent immunomodulatory metal in fish (reviewed by Anderson 1996). Following exposure to copper, rainbow trout are more susceptible to *Yersinia ruckeri* (Knittel 1981). Anderson et al. (1989) later shows that copper inhibits the in vitro antibody response in rainbow trout, suggesting that this may be the target response in this species. For example, rainbow trout exposed to copper tailings in a river system draining a mining operation had reduced lymphocyte mitogenic responses. Depending upon the dose, cadmium is known to reduce immune responses (Thuvander 1989) and increase protection against pathogens (MacFarlane et al. 1986). This reduced response may be true for most metals because of their ability to mimic or interfere with intracellular signals involving calcium (Burnett 1997). Nickel and zinc exposure is associated with reduced serum antibody levels in salmonids (O'Neill 1981). In some cases, particularly with high concentrations, the metal may affect the viability of pathogens, thus, their numbers, which then may reduce the overall vigor of the host immune response.

Pesticides

Most pesticides are neurotoxicants, therefore, their effects on the immune systems of fishes is probably through indirect mechanisms. Several studies show reduced immune functions in fishes by pesticides (Dunier and Siwicki

1993; Anderson 1996), but the doses are usually high and the overt neuro-toxicity was not monitored. Depending on the concentration, the biocide TBT can both activate and inhibit reactive oxygen formation in toadfish *Opsanus tau* macrophages (Rice and Weeks 1989, 1990, 1991). This activation is associated with an influx of calcium and seems to be consistent with other reports related to TBT-induced lymphoid toxicity, including thymic atrophy (Chow et al. 1992; Raffray et al. 1993). Tributyltin is generally regarded as a classical immunotoxic agent in fish (Rice and Weeks 1989; Rice et al. 1995; Regala et al. 2001). Tributyltin seems to target B cells, therefore, the antibody response (O'Halloran et al. 1998), as well as phagocyte function at high doses (Rice et al. 1995).

Environmental Case Study

Puget Sound

Research has been performed over a 10-year period by the Northwest Fisheries Science Center in Seattle, Washington, on the effects of contaminants on the immune system in fish. Their primary studies have focused on salmon. Studies by these researchers demonstrate that juvenile fall chinook salmon bioaccumulate significant concentrations of chemical contaminants during their relatively short residence time in Puget Sound, primarily through exposure from their diet, as they migrate to the ocean. High levels of PAHs and PCBs or their metabolites were detected in stomach contents and tissues of these fish (McCain et al. 1990; Stein et al. 1995). These findings revealed that salmon in Puget Sound are exposed to contaminants that may suppress their immune function.

Field studies focusing on the effects of environmental exposure to chemical contaminants on the primary and secondary B-leukocyte responses of fish had not been previously performed. Immunocompetence of salmon from the Puget Sound study sites was evaluated by analyzing the functional ability of B-leukocytes to produce an in vitro primary and secondary PFC response to a T-independent antigen, trinitrophenyl-lipopolysaccharide, and to a T-dependent antigen, TNP-keyhole-limpet hemocyanin. Leukocytes of juvenile salmon collected from hatcheries and from a nonurban estuary were able to generate a significantly higher secondary PFC response to a foreign antigen than that produced during the primary PFC response, which is the normal and expected response (Arkoosh et al. 1991). However, an enhanced secondary PFC response did not occur with leukocytes of juvenile salmon exposed to pollutants from Puget Sound.

The suppressed secondary immune response in juvenile salmon from contaminated areas within Puget Sound suggest that these fish might be more susceptible to disease than those from nonpolluted environments. To examine this possibility, juvenile fall chinook salmon were collected from urban and nonurban estuaries and from the respective releasing hatcheries, and exposed in the laboratory to the marine pathogen *Vibrio anguillarum*. Juvenile chinook salmon from the contaminated estuary were more susceptible to *V. anguillarum* induced mortality than fish from the corresponding hatchery. In contrast, juvenile fall chinook salmon from the nonurban estuary showed no greater susceptibility to *V. anguillarum*-induced mortality than the fish from the corresponding hatchery (Arkoosh et al. in 1998). The results of these initial disease challenges indicate that juvenile chinook salmon from contaminated waterways within Puget Sound may be immunosuppressed and more susceptible to disease than are chinook from noncontaminated waterways.

There is also evidence of altered immune function in English sole *Pleuronectes vetulus* exposed to contaminants found in Puget Sound. Researchers from the Northwest Fisheries Science Center examined the immune response of English sole exposed to either PAHs in the sound, PAH-contaminated sediment, or to PAH-contaminated sediment extracts from Eagle Harbor (Arkoosh et al. 1996; Clemons et al. 1999). Eagle Harbor is located at Bainbridge Island in Puget Sound and has high levels of PAHs. The leukoproliferative (mitogenic) and the macrophage production of cytotoxic reactive oxygen intermediates (ROI) response of fish exposed to PAHs were found to be increased over controls. Augmentation of the production of macrophage ROIs has concomitantly been associated with an increase in peroxidative damage of kidney and gill tissues (Fatima et al. 2000). However, the direct effect of augmented ROIs and leukoproliferative response on English sole's immune function is unknown.

Assessment of Immunological Functions in Fishes

In order to determine how stress affects immune function, there must be a balance between the complexity of assays and, thus, the inherent training required of personnel, the expense, and how much detailed information is required. The most sensitive assays of immune function, such as gene expression, radioimmunoassay, cytometry, and complex cellular functions are expensive and require advance training. On the other hand, the nitroblue tetrazolium reduction assay with phagocytes or serum agglutination assays can be performed by aquaculturists or field biologists with limited training and equipment (Rice et al. 1996). In the end, however, no single assay, regardless of expense or complexity, is able to assess total immune function in the host. Like other physiological systems, there is redundancy in immune defenses. For example, neutrophils, eosinophilic granular cells, and macrophages are actively phagocytic, and both macrophages and NCC can lyse tumor targets (Seeley and Weeks-Perkins 1991).

Unless a new assay is being designed and validated that measures immune function, most researchers focus on established assays. Most of these assays fall within the broadly defined categories of hematological, nonspecific defense mechanisms and specific immune response assays (Table 1). It is extremely expensive to conduct a large array of assays, therefore, particular endpoints are usually selected for measurement. The ultimate goal for researchers attempting to understand the effects of stress on immune function is to be systematic in their approach. To that end, mammalian immunotoxicologists developed a tier I and II battery of assays (Burns et al. 1996) (Table 2), with tier I being simpler, less expensive, and requiring less personnel training. In the case of immunotoxicity testing, tier I tests are designed to screen compounds for their immunotoxicity relative to general toxicity. If a compound is shown to be immunotoxic then tier II tests are employed to determine the specific target and provide information about possible mechanisms of action. For the tier I group of assays, the standard tests include lymphoid organ weights, cellularity of these organs, and hematological parameters. General pathology is determined using vital organs in

Table 1. Hematological, nonspecific defense mechanisms, and specific immune response assays.

1. Hematological and physiological assays—blood assays
    Hematocrit: Percent of red blood cell pack
    Leukocrit: Percent of white blood cell pack
    Cell counts and differentials: Numbers of cells and types
    Lysozyme levels: Enzyme level in blood
    Serum immunoglobulin level: Specific and non-specific
    Serum protein level: Total protein in serum
2. Nonspecific defense mechanism or specific immune response assays
    Phagocytosis: Percents and indexes: engulfment by phagocytic cells
    Glass or plastic adherence: Stickiness of phagocytic cells
    Pinocytosis: Engulfment of fluids by phagocytes
    Macrophage and neutrophil activation:
        Nitroblue tetrazolium (NBT) assay: detects intracellular superoxide anion
        Cytochrome C reductase assay: detects extracellular superoxide anion
        "Greiss reaction: measures nitric oxide production, indicator of iNOS activity"
        Dihydrorhodamine dye: fluourescence detection of intracelluar hydrogen peroxide
        Chemiluminescence: detects superoxide anion production
    Chemotaxis: movement of cells toward a stimulus
    Blastogenesis (mitogenesis): proliferation of lymphocytes in response to mitogens
    Non-specific cytotoxic cell activity: natural killer-like activity (target cell lysis)
3. Specific immune response assays
    Scale rejection: cell-mediated immunity
    Passive hemolytic plaque assay (Jerne assay): Antibody-secreting cells
    Elispot Assay: antibody secreting cells
    Serum antibody levels: antigen specific titers or concentration
    Cytotoxic T-cell activity: specific immunity and cell-mediated

Table 2. Tier approach to evaluating the effects of immunomodulators in fish.

Tier I Immunoassays

"General toxicity (behavior, body wts, organ wts, organ cellularity, hematology, leukocyte differentials)"

"Nonspecific immune function assays (natural killer activity, phagocytosis, oxidative burst activity, mitogenesis)"

"Primary antibody response (hemolytic plaque assay, ELISpot, antibody titers and concentration)"

Tier II Immunoassays

Primary and secondary antibody responses
Cytotoxic T cell-like activity
Cytokine profiles (activity and type)
Immunoglobulin switching (in higher vertebrates)
Leukocyte marker profiles
Tumor or pathogen challenge
Others (National Toxicology Program guidelines)

an effort to determine if lymphoid organs are targeted relevant to other systems. Basic immune functions are measured as well, including granulocyte or macrophage phagocytosis, respiratory burst activity, or phagocyte chemotaxis. The ability of the test animal to mount a primary antibody response to a model antigen can also be followed to determine if the humoral response is altered. The cell-mediated immune response is measured using mitogen-induced lymphocyte proliferation tests and a test for natural killer cell activity. Tier II tests are usually initiated following these tier I assays. Tier II assays involve more probing into the regulatory components of immune function, are more involved, expensive, and time consuming, and also require significantly more training and equipment than tier I tests. Leukocyte cell surface markers, secondary antibody responses (the memory response), cytotoxic T cell function, and delayed type hypersensitivity are evaluated during or after the tier II tests. Although designed by the National Toxicology Program for rodent models, the framework for evaluating immune function is a logical approach and is applicable to fish and for almost any stressor. More importantly, the tier approach provides a methodical means for systematically testing critical immune parameters.

Fish immunologists are hampered by the simple fact that the availability of fish-specific reagents is far less than that for rodent models. For example, antibody probes for cell surface markers are commercially available for rodent studies. In addition, many of the reagents used by mammalian physiologists, immunologists, and toxicologists are highly cross-reactive. Since the different rodent models are merely strains of a given species (mouse, rat), there is no need to redevelop reagents. In contrast, reagents developed for one species of fish rarely cross-react with other species. For example, antibodies specific for channel catfish immunoglobulin (a B cell marker) do not recognize striped bass *Morone saxatilis* immunoglobulin and vice versa. The same is true, however, for different species of mammals. As a result, fish biologists are forced to either develop their own reagents or choose assays of immune function status that consistently measure the same endpoint in all species.

Despite these limitations, fish biologists are quite capable of conducting tier I tests for stress-induced immunomodulation in most species of fish. Most of the common fish immunological indicators can be divided into non-specific assays, nonspecific or specific assays used with antigen stimulation, and specific immune response assays (Table 3). Most of the fish immunology literature consists of studies where only one or a select few assays of immune function are used. From Table 3, the reader can easily see how these assays can be regrouped within the National Toxicology Program's tier I and II approach described above (Table 2). Indeed, virtually all of the assays listed in the first two categories of Table 3 fit well within the tier I battery of assays. The passive hemolytic plaque assay for evaluating the primary antibody response by enumerating the numbers of antigen-specific producing cells (B cells) has now been replaced by simply measuring circulating serum antibody concentration (Burns et al. 1996). The most common means for this is the enzyme-linked immunosorbent assay.

It is generally agreed among comparative immunologists that fish rely on innate, nonspecific immune defenses more so than antigen-specific defenses. Many aspects of highly advanced mammalian immune systems are associated with regulating antigen-specific immune responses such as antibody class switching that requires T-helper cells and a network of cytokines. Since antibody class switching does not occur in fish, much of the energetics of the immune system are dedicated to innate immunity. Capitalizing on this, field biologists have used macrophage or granulocyte function (phagocytosis, oxidative burst activity, chemotaxis) as a reliable marker of immune status for the past several decades (see review by Fournier et al. 2000). Leukocyte proliferation (blastogenesis, mitogenesis) in response to lectins or bacterial lipopolysaccharides is easily performed by field biologists with training in immunology and cell culture.

As investigators attempt to understand the effects of environmental stressors on fish, they are pressed with a critical question. Can immunological tests predict or indicate specific stressors? In natural situations where a myriad of interactions between pathogens, the host, and the environment exist, it is difficult, if not impossible, to determine which of the many stressors is responsible for disease outbreaks. Broad sampling of populations over a wide geographical range may yield trends, but environmental conditions are constantly changing. Furthermore, many species of fish, especially those of commercial interest, have a wide range of mobility during their life cycle. An immunological indicator may not distinguish between past and present exposures to stress. Anderson (1990) presented a discussion on the biological levels for investigat-

Table 3. Immunological indicator assays that can be used to determine immunomodulation in fish.

Assay	Sensitivity	Statistical variation	Materials needed	Example reference
Tier I				
1. Hematocrit (red blood cell level)	low	high	few	Rice and Schlenk (1995)
2. Leukocrit (white blood cell level)	low	high	few	Pickering and Pottinger (1987)
3. Blood cell differentials	fair	high	microscope	Rice et al. (1995)
4. Lysozyme levels	good	good	spectrometer	Karrow et al. (1998)
5. Serum immunoglobulin	very good	low	specific antibodies (and plate reader)	Ottinger and Kaattari (1998)
6. Serum total protein	very good	low	spectrometer	Rice and Weeks (1989)
7. Phagocytic index	fair	good	microscope	Zelikoff et al (1995)
8. Bacteriacidal activity	good	low	plate reader	Roszell and Anderson (1996)
9. Pinocytosis	fair	good	microscope	Weeks et al (1987)
10. Oxidative burst activity				
NBT assay	very good	low	96 well plate reader	Rice et al (1996)
Griess Assay	very good	low	96 well plate reader	Neumann et al (1995)
DHR assay	very good	low	fluorescence plate reader	Roszell and Rice (1998)
Chemiluminescence	very good	low	luminometer or scintillation counter	Roszell and Anderson (1993)
11. Chemotaxis	fair	moderate	specialized chambers and membranes	Weeks et al (1986)
12. Blastogenesis (mitogenesis)	good	good	plate reader or scintillation counter	Burnett (1997)
13. NCC assay	good	good	plate reader or beta-counter	Faisal et al (1991)
Tier II				
14. Passive hemolytic plaque assay (Jerne plaque assay	very good	low	dissecting microscope (cell culture incubator)	Arkoosh et al (1991)
15. Elispot assay (plaque assay modified for ELISA	very good	low	dissecting micrscope (cell culture incubator)	Aalotnen et al (2000)
16. Serum antibody titers using agglutination assays	low	high	minimal	Robohm (1986)
17. Serum antibody titers using ELISA	very good	low	96 well plate reader specific antibodies	Regala et al (2001)

ing the effects of anthropogenic stress on fish health (Figure 4). More information on the effects of environmental conditions may be gained by monitoring cage exposures in the wild. However, this technique has limitations because the caging effect alone may be stressful, especially if natural movements and activity (feeding) are restricted. Controlled laboratory exposures to a particular stressor yields still more information about the influence of potential environmental conditions and the investigator can control for more variables (e.g.,

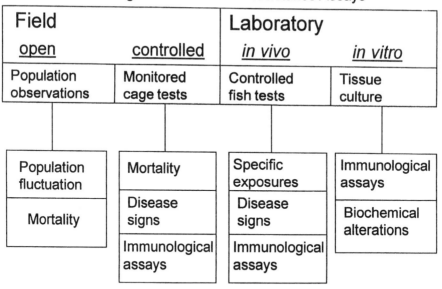

Figure 4. Biological levels for the investigation of the effects of pollutants, drugs, or stress on fish by immunological and disease-resistance assays (Source: Anderson 1990).

water quality). Mechanisms behind immunomodulation can be elucidated from in vitro tests, yet this approach is far removed from realistic field situations. Ultimately, investigators want to know if a particular stressor lowers resistance to pathogens leading to disease outbreaks. Disease resistance (pathogen challenge) trials may provide answers, but this approach needs controls as well, most of which do not mimic field conditions. In the end, predictive testing using the tier I and tier II approach may be the best approach to determining whether or not a particular stressor lowers disease resistance. Mammalian immunotoxicologists, via the tier approach to toxicity testing, have determined that the primary antibody response to sheep red blood cells during or after exposure is highly predictive of increased disease susceptibility (Burns et al. 1996). For each species of fish in question, investigators may find similar results. However, this too must be empirically validated.

In summary, we now know that the immune system of fishes is sensitive to environmental stress, and, in many cases, we also know the mechanisms behind this modulation. Fish biologists are left with the reality that the immune system is integrated within the neuroendocrine-immune system and the effects of stress on immune function are merely physiological responses at the organismal level. Immunology is a complicated discipline requiring significant education and training to understand the implications of immunomodulation by environmental stress. While almost any field or laboratory technique can be "learned," the need for specific training comes into play when troubleshooting is required and when data must be interpreted.

References

Aalotnen, T. 2000. Effects of pulp and paper mill effluents on fish immune defence. Ph.D. thesis. University of Jyvaskyla, Finland.

Aaltonen, T. M., E. I. Jokinen, H. M. Salo, S. E. Markkula, and R. Lammi. 2000. Modulation of immune parameters of roach, *Rutilus rutilis,* exposed to untreated ECF and TCF bleached pulp effluents. Aquatic Toxicology 47:277–289.

Aaltonen, T. M., E. T. Valtonen, and E. I. Jokinen. 1997. Immunoreactivity of roach, *Rutilus rutilis*, following laboratory exposure to bleached pulp and paper mill effluents. Ecotoxicology and Environmental Safety 38:266–271.

Anderson, D. P. 1990. Immunological indicators: effects of environmental stress on immune protection and disease outbreaks. Pages 38–50 *in* S. M. Adams, editor. Biological indicators of stress in fish. American Fisheries Society, Symposium 8, Bethesda, Maryland.

Anderson, D. P. 1996. Environmental factors in fish health: immunological aspects. Pages 289–310 *in* W. S. Hoar, D. J. Randall, and A. P. Farrell, editors. The fish immune system: organism, pathogen, and environment, volume 15. Academic Press, New York.

Anderson, D. P., O. W. Dixon, J. E. Bodammer, and E. F. Lizzio. 1989. Suppression of antibody-producing cells in rainbow trout spleen sections exposed to copper in vitro. Journal of Aquatic Animal Health 1:57–61.

Anderson, D. P., and M. G. Zeeman. 1995. Immuntoxicology in fish. Pages 371–404 *in* G. M. Rand, editor. Fundamentals of aquatic toxicology, 2nd edition. Taylor and Francis, Washington, D.C.

Anderson, R. M., and R. M. May. 1979. Population biology of infectious diseases: part I. Nature(London) 280:361–367.

Anderson, R. S. 1994. Modulation of blood cell-mediated oxyradical production in aquatic species: implications and applications. Pages 241–265 *in* D. C. Malins and G. K. Ostrander, editors. Aquatic toxicology: molecular, biochemical, and cellular perspectives. CRC Press, Boca Raton, Florida.

Arkoosh, M. R., E. Casillas, E. Clemons, B. B. McCain, and U. Varanasi. 1991. Suppression of immunological memory in juvenile chinook salmon (*Oncorhynchus tshawytscha*) from an urban estuary. Fish and Shellfish Immunology 1:261–277.

Arkoosh, M. R., E. Casillas, P. Huffman, E. Clemons, J. Evered, J. E. Stein, and U. Varanasi. 1998. Increased susceptibility of juvenile chinook salmon from a contaminated estuary to *Vibrio anguillarum*. Transactions of the American Fisheries Society 127:360–374.

Arkoosh, M. R., E. Clemons, P. Huffman, H. R. Sanborn, E. Casillas, and J. E. Stein. 1996. Leukoproliferative response of splenocytes from English sole (*Pleuronectes vetulus*) exposed to chemical contaminants. Environmental Toxicology and Chemistry 15(7):1154–1162.

Armknecht, S. L., S. L. Kaattari, and P. A. Van Veld. 1998. An elevated glutathione S-transferase in creosote-resistant mummichog, *Fundulus heteroclitus*. Aquatic

Toxicology 41:1–16.

Arnold, R. E., and C. D. Rice. 2000. Channel catfish, *Ictalurus punctatus*, leukocytes secrete adrenal corticotropin hormone (ACTH). Fish Physiology and Biochemistry 22:303–310.

Barton, B. A., and G. K. Iwama. 1991. Physiological changes in fish from stress in aquaculture with emphasis on the responses and effects of corticosteroids. Annual Review of Fish Diseases 1:3–26.

Baudin-Laurencin, F., and E. Germon. 1987. Experimental infection in rainbow trout, *Salmo gairdneri* R., by dipping in suspensions of *Vibrio anguillarum*: ways of bacterial penetration; influence of temperature and salinity. Aquaculture. 67:203–205.

Bayne, C. J., and S. Levy. 1991. Modulation of the oxidative burst in trout myeloid cells by adrenocorticotropic hormone and catecholamines: mechanisms of action. Journal of Leukocyte Biology 50:554–560.

Bonga, S. E. W. 1997. The stress response in fish. Physiological Reviews 77:591–625.

Burnett, K. G. 1997. Evaluating intracellular signaling pathways as biomarkers for environmental contaminant exposures. American Zoologist 37:585–594.

Burns, L. A., B. J. Meade, and A. E. Munson. 1996. Toxic responses of the immune system. Pages 355–402 *in* C. D. Klaassen, editor. Casarett and Doull's toxicology: the basic science of poisons, 5th edition. McGraw-Hill, New York.

Camp, K. C., W. R. Wolters, and C. D. Rice. 2000. Survivability and immune responses after challenge with *Edwardsiella ictaluri* in susceptible and resistant families of channel catfish, *Ictalurus punctatus*. Fish and Shellfish Immunology 10:475–487.

Chow, S. C., G. E. Kass, M. J. McCabe, and S. Orrenius. 1992. Tributyltin increases cytosolic free Ca^{++} concentration in thymocytes by mobilizing intracellular Ca^{++}, activating a Ca^{++} entry pathway, and inhibiting Ca^{++} efflux. Analytical Biochemistry and Biophysics 298:143–149.

Cleland, G. B., P. J. McElroy, and R. A. Sonstegard. 1988. The effect of dietary exposure to Aroclor 1254 and/or mirex on humoral immune expression of rainbow trout, *Salmo gairdneri*. Aquatic Toxicology 12:141–146.

Clemons, E., M. R. Arkoosh, and E. Casillas. 1999. Enhanced superoxide anion production in activated peritoneal macrophages from English sole (*Pleuronectes vetulus*) exposed to polycyclic aromatic compounds. Marine Environmental Research 47:71–87.

Coble, D. W. 1970. Vulnerability of fathead minnows infected with yellow grub to largemouth bass predation. Journal of Parasitology 56:395–396.

Cooper, E. L., G. Peters, I. I. Ahmed, M. Faisal, and M. Ghoneum. 1989. Aggression in *Tilapia* affects immunocompetent leukocytes. Aggressive Behavior 15:13–22.

Crawford, R. B., M. P. Holsapple, and N. R. Kaminski. 1997. Leukocyte activation induces aryl hydrocarbon receptor upregulation, DNA binding, and increased CYP1A1 expression in the absence of exogenous ligand. Molecular Pharmacology 52:921–927.

Dunier, M., and A. K. Siwicki. 1993. Effects of pesticides and other organic pollutants in the aquatic environment on immunity in fish: a review. Fish and Shellfish Immunology 3:423–438.

Ejike, C., and C. B. Schreck. 1980. Stress and social hierarchy rank in coho salmon.

Transactions of the American Fisheries Society 104:423–426.

Ellis, A. E. 1981. Stress and the modulation of defense mechanisms in fish. Pages 147–170 in A. D. Pickering. Stress and fish. Academic Press, London.

Ellsaesser, C. F., and L. W. Clem. 1986. Hematological and immunological changes in channel catfish stressed by handling and transport. Journal of Fish Biology 28:511–521.

Ellsaesser, C. F., and L. W. Clem. 1987. Cortisol-induced hematologic and immunologic changes in channel catfish, *Ictalurus punctatus*. Comparative Biochemistry and Physiology Part A. Comparative Physiology 87:405–408.

Espelid, S., G. B. Lokken, K. Steiro, and J. Bogwald. 1996. Effects of cortisol and stress on the immune system in Atlantic salmon, *Salmo salar* L. Fish and Shellfish Immunology 6:95–110.

Evelyn, T. 1996. Infection and disease. Pages 339–359 in W. S. Hoar, D. J. Randall, and A. P. Farrell, editors. The fish immune system: organism, pathogen, and environment, volume 15. Academic Press, New York.

Faisal, M., F. Chiappelli, I. I. Ahmed, E. L. Cooper, and H. Weiner. 1991a. Social confrontation "stress" is associated with an endogenous opioid mediated suppression of proliferative responses to mitogens and nonspecific cytotoxicity. Brain Behavior and Immunity 3:223–333.

Faisal, M., E. E. Elsayed, and W. L. Rose. 1998. Formation of DNA adducts in hemopoetic organs and isolated leukocytes of the mummichog *Fundulus heteroclitus* following exposure to benzoy-a-pyrene. Marine Environmental Research 46:359–360.

Faisal, M., R. J. Huggett, W. K. Vogelbein, and B. A. Weeks. 1991b. Evidence of aberration of the natural cytotoxic cell activity in *Fundulus heteroclitus* (Pisces: *Cytprinodontidae*) from the Elizabeth River, Virginia. Veterinary Immunology and Immunopathology 29:339–351.

Faisal, M., M. S. M. Marzouk, C. L. Smith, and R. J. Huggett. 1991b. Mitogen induced proliferative responses of lymphocytes from spot, *Leiostomus xanthurus,* exposed to polycyclic aromatic hydrocarbon contaminated environments. Immunopharmacology and Immunotoxicology 13:311–327.

Fatima, F., I. Ahmad, I. Sayeed, M. Athar, and S. Raisuddin. 2000. Pollutant-induced over-activation of phagocytes in concomitantly associated with peroxidative damage in fish tissues. Aquatic Toxicology 49:243–250.

Felton, S. P., J. Wenjuan, and S. B. Mathews. 1990. Selenium concentrations in coho salmon outmigrants smolts and returning adults: a comparison of wild versus hatchery reared fish. Diseases of Aquatic Organisms 9:157–161.

Flory, C. M. 1989. Autonomic innervation of the spleen of the coho salmon, *Oncorhynchus kisutch:* a histochemical demonstration and preliminary assessment of its immunoregulatory role. Brain Behavior Immunity 3:331–344.

Flory, C. M., and C. J. Bayne. 1991. The influence of adrenergic and cholinergic agents on the chemiluminescent and mitogenic responses of leukcoytes from the rainbow trout, *Oncorhynchus mykiss*. Developmental and Comparative Immunology 15:135–142.

Folstad, I., and A. J. Karter. 1992. Parasites, bright males, and the immunocompetence handicap. American Naturalist 139:603–622.

Fournier, M., D. Cyr, B. Blakley, H. Boermans, and P. Brousseau. 2000. Phagocytosis as a biomarker of immunotoxicity in wildlife species exposed to environmental xenobiotics. American Zoologist 40:412–420.

Fredriksen, A., C. Endresen, and H. I. Wergeland. 1997. Immunosuppressive effect of a low molecular weight surface protein from *Renibacterium salmoninarum* on lymphocytes from Atlantic salmon (*Salmo salar* L.). Fish and Shellfish Immunology 7:273–282.

Freeland, W. J. 1983. Parasites and the coexistence of animal host species. American Naturalist 121:223–236.

Ganey, P. E., J. E. Sirois, M. Denison, J. P. Robinson, and R. A. Roth. 1993. Neutrophil function after exposure to polychlorinated biphenyls *in vitro*. Environmental Health Perspectives 101:430–434.

Ghoneum, M., M. Faisal, G. Peters, I. I. Ahmed, E. L. Cooper. 1988. Suppression of natural cytotoxic cell activity by social aggressiveness in *Tilapia*. Developmental and Comparative Immunology 12:595–602.

Grisez, L., M. Chair, P. Sorgeloose, and F. Ollevier. 1996. Mode of infection and spread of *Vibrio anguillarum* in turbot, *Scophthalmus maximus*, larvae after oral challenge through live feed. Diseases of Aquatic Organisms 26:181–187.

Gustafsson, L., D. Nordling, M. S. Andersson, B. C. Sheldon, and A. Qvarnstrom. 1994. Infectious disease, reproductive effort, and the cost of reproduction in birds. Philosophical Transactions of the Royal Society of London. Series B. Biological Sciences 346:323–331.

Hamilton, W. D., and M. Zuk. 1982. Heritable true fitness and bright birds: a role for parasites? Science 218:384–386.

Haney, D. C., D. A. Hursh, M. C. Mix, and J. R. Winton. 1992. Physiological and hematological changes in chum salmon artificially infected with erythrocytic necrosis virus. Journal of Aquatic Animal Health 4:48–57.

Hart, L. J., S. A. Smith, B. J. Smith, J. Robertson, E. G. Besteman, and S. D. Holladay. 1998. Subacute immunotoxic effects of the polycyclic aromatic hydrocarbon 7,12-dimethylbenzanthracene (DMBA) on spleen and pronephros leukocytic cell counts and phagocytic cell activity in tilapia, *Oreochromis niloticus*. Aquatic Toxicology 41:17–29.

Hedrick, R. P. 1998. Relationships of the host, pathogen, and environment: implications for diseases of cultured and wild populations. Journal of Aquatic Animal Health 10(2):107–111.

Herting, G. E., and A. Witt Jr. 1967. The role of physical fitness of forage fishes in relation to their vulnerability to predation by bowfin (*Amia calva*). Transactions of the American Fisheries Society 96:427–430.

Holsapple, M. P., N. K. Snyder, S. C. Wood, and D. L. Morris. 1991. A review of 2,3,7,8-tetrachlorodibenzo-p-dioxin-induced changes in immunocompetence: 1991 update. Toxicology 69:219–255.

Houde, A., and A. J. Torio. 1992. Effect of parasitic infection on male color pattern and female choice in guppies. Behavior Ecology 3:346–351.

Iwama, G., and T. Nakanishi. 1996. The fish immune system: organism, pathogen, and environment. Academic Press, New York.

Jokinen, E. I., T. M. Aaltonen, and E. T. Valtonen. 1995. Subchronic effects of pulp and paper mill effluents on the immunoglobulin synthesis of roach, *Rutilus rutilus*. Ecotoxicology and Environmental Safety 32:219–225.

Kaattari, S. L. 1994. Development of a piscine paradigm of immunological memory.

Fish and Shellfish Immunology 4:447–457.

Kanno, T., T. Nakai, and K. Muroga. 1989. Mode of transmission of vibriosis among ayu, *Plecoglossus altivelis*. Journal of Aquatic Animal Health 1:2–6.

Karras, J. G., and M. P. Holsapple. 1995. Structure and function of the immune system. Pages 3–12 *in* R. J. Smialowicz and M. P. Hosapple, editors. Experimental immunology. CRC Press, Boca Raton, Florida.

Karrow, N. A., H. J. Boermans, D. G. Dixon, A. Hontella, K. R. Solomon, J. J. Whyte, and N. C. Bols. 1999. Characterizing the immunotoxicity of creosote to rainbow trout, *Oncorhynchus mykiss*: a microcosm study. Aquatic Toxicology 45:223–239.

Kelly-Reay, K., and B. A. Weeks-Perkins. 1994. Determination of the macrophage chemiluminescent response in *Fundulus heteroclitus* as a function of pollution stress. Fish and Shellfish Immunology. 4:95–105.

Kennedy, C. E. J., J. A. Endler, S. L. Poynton, and H. McMinn. 1987. Parasite load predicts mate choice in guppies. Behavior Ecology and Sociobiology 21:291–295.

Kennedy, C. R. 1984. Host-parasite interrelationships: strategies of coexistence and coevolution. Pages 34–59 *in* C. J. Barnard, editor. Producers and scroungers. Croom Helm, London.

Kennedy-Stoskopf, S. 1993. Immunology. Pages 149–159 *in* M. K. Stoskopf, editor. Fish medicine. Saunders, Philadelphia.

Knittel, M. D. 1981. Susceptibility of steelhead trout, *Salmo gairdneri* Richardson, to red mouth infection, *Yersinia ruckeri* following exposure to copper. Journal of Fish Diseases 4:33–40.

Ladics G. S., T. T. Kawabata, A. E. Munson, and K. L. White. 1992. Metabolism of Benzo(a)pyrene by murine splenic cell types. Toxicology and Applied Pharmacology 116:248.

Laidley, C. W., P. T. K. Woo, and J. F. Leatherland. 1988. The stress-response of rainbow trout to experimental infection with the blood parasite *Cryptobia salmositica* Katz, 1951. Journal of Fish Biology 32:253–261.

Loehle, C. 1995. Social barriers to pathogen transmission in wild animal populations. Ecology 76(2):326–335.

MacFarlane, R. D., G. L. Bullock, and J. J. A. McLaughlin. 1986. Effects of five metals on susceptibility of striped bass to *Flexibacter columnaris*. Transactions of the American Fisheries Society 115:227–231.

Maule, A. G., C. B. Schreck, and S. L. Kaattari. 1987. Changes in the immune system of coho salmon, *Oncorhynchus kisutch*, during the parr-to-smolt transformation and after implantation of cortisol. Canadian Journal of Fisheries and Aquatic Science 44:161–166.

Maule, A. G., C. B. Schreck, and C. Sharpe. 1993. Seasonal changes in cortisol sensitivity and glucocorticoid receptor affinity and number in leukocytes of coho salmon. Fish Physiology and Biochemistry 10:497–506.

Maule, A. G., R. A. Tripp, S. L. Kaattari, and C. B. Schreck. 1989. Stress alters immune function and disease resistance in chinook salmon (*Oncorhynchus tshawytscha*). Journal of Endocrinology 120:135–142.

Mazur, C. F., and G. K. Iwama. 1993. Handling and crowding stress reduces number of plaque-forming cells in Atlantic salmon. Journal of Aquatic Animal Health 5:98–101.

McCain, B. B., D. C. Malins, M. M. Krahn, D. W. Brown, W. D. Gronlund, L. K.

Moore, and S.-L. Chan. 1990. Uptake of aromatic and chlorinated hydrocarbons by juvenile chinook salmon (*Oncorhynchus tshawytscha*) in an urban estuary. Archives of Environmental Contamination and Toxicology 11:143–162.

Mesa, M. G., T. P. Poe, A. G. Maule, and C. B. Schreck. 1998. Vulnerability to predation and physiological stress responses in juvenile chinook salmon (*Oncorhynchus tshawytscha*) experimentally infected with *Renibacterium salmoninarum*. Canadian Journal of Fisheries and Aquatic Sciences 55:1599–1606.

Milinski, M. 1984. Risk of predation of parasitized sticklebacks (*Gasterosteus aculeatus* L.) under competition for food. Behavior 93:203–216.

Milinski, M., and T. C. M. Bakker. 1990. Female sticklebacks use male coloration in mate choice and hence avoid parasitized males. Nature(London) 344(6264):331–333.

Nelson, J. S., J. S. Rohovec, and J. L. Fryer. 1985. Location of *Vibrio anguillarum* in tissues of infected rainbow trout, *Salmo gairdneri,* using fluorescent antibody technique. Fish Pathology 20:229–235.

Neumann, N. F., D. Fagan, and M. Belosevic. 1995. Macrophage activating factor(s) secreted by mitogen stimulated goldfish kidney leukocytes synergize with bacterial lipopolysaccharide to induce nitric oxide production in teleost macrophages. Developmental and Comparative Immunology 19:473–482.

Norris, D. O. 2000. Endocrine disruptors of the stress axis in natural populations: how can we tell? American Zoologist 40:393–401.

O'Halloran, K., J. T. Ahokas, and P. F. A. Wright. 1998. Response of fish immune cells to in vitro organotin exposures. Aquatic Toxicology 40:141–156.

O'Neill, J. G. 1981. The humoral immune response of *Salmo trutta* L. and *Cyprinus carpio* L. exposed to heavy metals. Journal of Fish Biology 19:297–306.

Ottinger, C. A., and S. L. Kaattari. 1998. Sensitivity of rainbow trout leucocytes to aflatoxin B_1. Fish and Shellfish Immunology 8:515–530.

Peters, G., H. Delventhal, and H. Klinger. 1980. Physiological and morphological effects of social stress on the eel, *Anguilla anguilla* L. Archiv fuer Fischereiwissenschaft 307:157–180.

Peters, G., M. Faisal, T. Lang, and I. Ahmed. 1988. Stress caused by social interaction and its effect on susceptibility to *Aeromonas hydrophilia* infection in rainbow trout, *Salmo gairdneri*. Diseases of Aquatic Organisms 4:83–89.

Peters, G., A. Nubgen, A. Raabe, and A. Mock. 1991. Social stress induces structural and functional alterations of phagocytes in rainbow trout (*Oncorhynchus mykiss*). Fish and Shellfish Immunology 1:17–31.

Peters, G., and R. Schwarzer. 1985. Changes in hemopoietic tissue of rainbow trout under influence of social stress on the eel, *Anguilla anguilla* L. Archiv Fuer Fischereiwissenschaft 30:157–180.

Pickering, A. D., and P. Christie. 1980. Sexual differences in the incidence and severity of ectoparasitic infestation of the brown trout, *Salmo trutta* L. Journal of Fish Biology 16: 669–683.

Pickering, A. D., and T. G. Pottinger. 1987. Crowding causes prolonged leucopenia in salmonid fish, despite interrenal acclimation. Journal of Fish Biology 30:701–712.

Pickering, A. D., and T. G. Pottinger. 1989. Stress response and disease resistance in salmonid fish: effects of chronic elevation of cortisol. Fish Physiology and Biochemistry 7:253–258.

Price, P. W. 1980. Evolutionary biology of parasites. Princeton University Press, New Jersey.

Raffray, M., D. McCarthy, R. T. Snowden, and G. M. Cohen. 1993. Apoptosis as a mechanism of tributyltin cytotoxicity to thymocytes: relationship of apoptotic markers to biochemical and cellular effects. Toxicology and Applied Pharmacology 119:122–130.

Rand, T. G., and D. K. Cone. 1990. Effects of *Ichthyophonus hoferi* on condition indices and blood chemistry of experimentally infected rainbow trout (*Oncorhynchus mykiss*). Journal of Wildlife Diseases 26:323–328.

Rapport, D. J., H. A. Regier, T. C. Hutchinson. 1985. Ecosystem behavior under stress. American Naturalist 125:617–640.

Regala, R. P., T. Schwedler, I. R. Dorociak, and C. D. Rice. 2001. The effects of tributyltin (TBT) and 3,3',4,4',5-pentachlorbiphenyl (PCB-126) mixtures on antibody responses and phagocyte oxidative burst activity in channel catfish, *Ictalurus punctatus*. Archives of Environmental Contamination and Toxicology 40:386–391.

Rice, C. D. 2001. Fish immunotoxicology: understanding mechanisms of action. Pages 96–138 *in* D. Schlenk and W. H. Benson, editors. Target organ toxicity in marine and freshwater teleosts, volume 2: systems. Taylor and Francis Publishers, London.

Rice, C. D., M. M. Banes, and T. C. Ardelt. 1995. Immunotoxicity in channel catfish, *Ictalurus punctatus*, following acute exposure to tributyltin. Archives of Environmental Contamination and Toxicology 28:464–470.

Rice, C. D., D. H. Kergosien, and S. M. Adams. 1996. Innate immune function as a bioindicator of pollution stress in fish. Ecotoxicology and Environmental Health Safety 33:186–192.

Rice, C. D., L. E. Roszell, M. M. Banes, and R. E. Arnold. 1998. Effects of dietary PCBs and nonyl-phenol on immune function and CYP1A activity in channel catfish, *Ictalurus punctatus*. Marine Environmental Research 46:351–354.

Rice, C. D, and D. Schlenk. 1995. Immune function and cytochrome P4501A activity after acute exposure to 3,3',4,4',5-pentachlorobiphenyl (PCB 126) in channel catfish. Journal of Aquatic Animal Health. 7(3):195–204.

Rice C. D. and B. A. Weeks. 1989. The influence of tributyltin on *in vitro* activation of toadfish macrophages. Journal of Aquatic Animal Health 1:62–68.

Rice C. D. and B. A. Weeks. 1990. The influence of *in vivo* tributyltin exposure on reactive oxygen species formation in macrophages from the toadfish. Archives of Environmental Contamination and Toxicology 19:854–857.

Rice C. D. and B. A. Weeks. 1991. Tributyltin stimulates reactive oxygen formation in toadfish macrophages. Developmental and Comparative Immunology 15:431–436.

Richards, R. H., and A. D. Pickering. 1978. Frequency and distribution patterns of *Saprolegnia* infection in wild and hatchery-reared brown trout *Salmo trutta* L. and char *Salvelinus alpinus* L. Journal of Fish Diseases 1:69–82.

Robertson, D. 1979. Host parasite interactions between *Ichtyobodo necator* (Henneguy 1883) and farmed salmonids. Journal of Fish Diseases 2:481–493.

Robohm, R. A. 1986. Paradoxical effects of cadmium exposure on antibacterial antibody responses in two fish species: inhibition in cunners *(Tautogolabrus adspersus)* and enhancement in striped bass *(Morone saxatilis)*. Veterinary Immunology and Immunopathology 12:251–262.

Rose, W. L., B. L. French, W. L. Reichert, and M. Faisal. 2000. DNA adducts in hematopoietic tissues and blood of the mummichog, *Fundulus heteroclitus,* from a creosote-contaminated site in the Elizabeth River, Virginia. Marine Environmental Research 50:581–589.

Rosenqvist, G., and K. Johansson. 1995. Male avoidance of parasitized females explained by direct benefits in a pipefish. Animal Behavior 49:1039–1045.

Roszell, L. E., and R. S. Anderson. 1993. Immunosuppressive effect of pentachlorophenol on phagocytes from an estuarine teleost, *Fundulus heteroclitus,* as measured by chemiluminescence activity. Archives of Environmental Contamination and Toxicology 25:489–492.

Roszell, L. E., and R. S. Anderson. 1996. Effect of in vivo pentachlorophenol exposure on *Fundulus heteroclitus* phagocytes: modulation of bactericidal activity. Diseases of Aquatic Organisms 26:205–211.

Roszell, L. E., and C. D. Rice. 1998. Innate cellular immune function of anterior kidney leucocytes in the gulf killifish, *Fundulus grandis*. Fish and Shellfish Immunology 8:129–142.

Sakai, M. M. Konishi, S. Atsuta, and M. Kobayashi. 1991. The chemiluminescent response of leukocytes from the anterior kidney of rainbow trout *Oncorhynchus mykiss* vaccinated with *Vibrio anguillarum, Streptococcus* sp. or *Renibacterium salmoninarum.* Nippon Suisan Gakkaishi 57:237–241.

Salonius, K., and G. K. Iwama. 1993. Effects of early rearing environment on stress response, immune function, and disease resistance in juvenile coho (*Oncorhynchus kisutch*) and chinook salmon (*O. tshawytscha*). Canadian Journal of Fisheries and Aquatic Sciences 50:759–766.

Scott, D. B. C., and C. E. Currie. 1980. Social hierarchy in relation to adrenocortical activity in *Xiphophorus helleri* Heckel. Journal of Fish Biology 16:265–277.

Seeley, K. R., and B. A. Weeks-Perkins. 1991. Altered phagocytic activity of macrophages in oyster toadfish from a highly polluted subestuary. Journal Aquatic Animal Health 3:224–227.

Sheldon, B. C., and S. Verhulst. 1996. Ecological immunology: costly parasite defenses and trade-offs in evolutionary ecology. Trends in Ecology and Evolution 11(8):317–321.

Sinderman, C. J. 1980. Pollution effects on fisheries: potential management activities. Helgolander Meeresunters 33:674–686.

Snieszko, S. F. 1973. Recent advances in scientific knowledge and development pertaining to diseases of fishes. Advances in Veterinary Science and Comparative Medicine 17:291–314.

Snieszko, S. F. 1974. The effects of environmental stress on outbreaks of infectious diseases of fishes. Journal of Fish Biology 6:374–383.

Soto, A. M., T. M. Lin, H. Justicia, R. M. Slivia, and C. Sonnenschein. 1992. An "in culture" bioassay to access the estrogenicity of xenobiotics (E-screen). Pages 295–309 *in* T. Colborn, and C. Clement, editors. Chemically induced alterations in sexual and functional development: the wildlife/human connection. Princeton Scientific Publishing, New Jersey.

Stein, J. E., T. Hom, T. K. Collier, D. W. Brown, and U. Varanasi. 1995. Contaminant exposure and biochemical effects in outmigrant juvenile chinook salmon from urban and nonurban estuaries of Puget Sound, Washington. Environmental Toxicology and Chemistry 14:1019–1029.

Strange, R. J., and C. B. Schreck. 1978. Anesthetic and handling stress on survival and cortisol concentration in yearling chinook salmon (*Oncorhynchus tshawytscha*). Journal of the Fisheries Research Board of Canada 35:345–349.

Tahir, A., T. C. Fletcher, D. F. Houlihan, and C. J. Secombes. 1993. Effect of short-term exposure to oil-contaminated sediments on the immune response of dab, *Limanda limanda* L. Aquatic Toxicology 27:71–82.

Tahir, A., and C. J. Secombes. 1995. The effect of diesel oil-based drilling mud extracts on immune responses of rainbow trout. Archives of Environmental Contamination and Toxicology 29:27–32.

Thuvander, A. 1989. Cadmium exposure of rainbow trout, *Salmo gairdneri* R: effects on immune functions. Journal of Fish Biology 35:521–529.

Thuvander, A., and M. Carlstein. 1991. Sublethal exposure or rainbow trout, *Oncorhynchus mykiss*, to polychlorinated biphenyls: effects on the humoral immune response to *Vibrio anguillarum*. Fish and Shellfish Immunology 1:77–86.

Tithof, P. K, M. L. Contreras, P. E. Ganey. 1995. Aroclor 1242 stimulates the production of inositol phosphates in polymorphonuclear neutrophils. Toxicology and Applied Pharmacology 131:136–143.

Turaga, P., G. Wiens, and S. Kaattari. 1987. Bacterial kidney disease: the potential role of soluble protein antigen(s). Journal of Fish Biology 31(Supplement A):191–194.

Vaughan, G. E., and D. W. Coble. 1975. Sublethal effects of three ectoparasites on fish. Journal of Fish Biology 7:283–294.

Vogelbein, W. K., J. W. Fournie, P. S. Cooper, and P. A. Van Veld. 1999. Hepatoblastomas in the mummichog, *Fundulus heteroclitus* (L.), from a creosote-contaminated environment: a histologic, ultrastructural and immunohistochemical study. Journal of Fish Diseases 22:419–431.

Vogelbein, W. K., J. W. Fournie, P. A. Van Veld, and R. J. Huggett. 1990. Hepatic neoplasms in the mummichog, *Fundulus heteroclitus,* from a creosote-contaminated site. Cancer Research 50:5978–5986.

Wang, R., and M. Belosevic. 1995. The in vitro effects of estradiol and cortisol on the function of a long-term goldfish macrophage cell line. Developmental and Comparative Immunology 19:327–336.

Waring, C. P., R. M. Stagg, and M. G. Poxton. 1992. The effects of handling on flounder (*Platichthys flesus* L.) and Atlantic salmon (*Salmo salar* L.). Journal of Fish Biology 41: 131–144.

Weeks, B. A., D. P. Anderson, A. P. DuFour, A. Fairbrother, A. J. Goven., G. P. Lahvis, and G. Peters. 1992. Immunological biomarkers to assess environmental stress. Pages 211–234 in R. J. Hugget, R. A. Kimerle, P. M. Mehrle, and H. L. Bergman, editors. Biomarkers: biochemical, physiological, and histological markers of anthropogenic stress. Lewis, Boca Raton, Florida.

Weeks, B. A., and J. E. Warinner. 1984. Effects of toxic chemicals on macrophage phagocytosis in two estuarine fishes. Marine Environmental Research 14:327–335.

Weeks, B. A., and J. E. Warinner. 1986. Functional evaluation of macrophages in fish from a polluted estuary. Veterinary Immunology Immunopathology 12:313–320.

Yano, T. 1996. The non-specific immune system: humoral defense. Pages 106–140 *in* W. S. Hoar, D. J. Randall, and A. P. Farrell, editors. The fish immune system: organism, pathogen, and environment, volume 15. Academic Press, New York.

Yin, Z., T. J. Lam, and Y. M. Sin. 1995. The effects of crowding stress on the non-specific immune response in fancy carp (*Cyprinus carpio* L.). Fish and Shellfish Immunology 5:519–529.

Zahavi, A. 1975. Mate selection: a selection for a handicap. Journal of Theoretical Biology 53:205–214.

Zeeman, M. 1986. Modulation of the immune response in fish. Veterinary Immunology and Immunopathology 12:235–241.

Zeeman, M. G., and W. A. Brindley. 1981. Effects of toxic agents upon fish immune systems: a review. Pages 1–60 *in* R. P. Sharma, editor. Immunologic considerations in toxicology, volume 2. CRC Press, Boca Raton, Florida.

Zelikoff, J. T. 1993. Metal pollution-induced immunomodulation in fish. Annual Review of Fish Diseases 3:305–325.

Zelikoff, J. T. 1994. Fish immunotoxicology. Pages 71–95 *in* J. H. Dean, M. I. Luster, A. E. Munson, and I. Kimber, editors. Immuntoxicology and immunopharmacology, 2nd edition. Raven Press, New York.

Zelikoff, J. T., D. Bowser, K. S. Squibb, and K. Frenkel. 1995. Immunotoxicology of low level cadmium exposure in fish: an alternative animal model for immuntoxicological studies. Journal of Toxicology and Environmental Health 45:235–248.

7

Histopathological Biomarkers as Integrators of Anthropogenic and Environmental Stressors

MARK S. MYERS AND JOHN W. FOURNIE

Introduction

Routine histopathology is a useful tool that can assess effects of prior or current exposure to stressors at the level of the individual, representing the intermediate level of biological organization. The lesions detected in cells, tissues, or organs represent an integration of cumulative effects of physiological and biochemical stressors (lower level, earlier responses) and, therefore, can be linked to exposure, subsequent metabolism of chemical contaminants, earlier biochemical, and physiological responses (e.g., cytochrome P450-1A [CYP1A] induction, DNA adduct formation), and potentially to upper or higher levels of biological organization (Figure 1). Many of the lesions detected in field studies are identical to those induced in laboratory studies using fish and other vertebrates and, therefore, have the advantage of experimental validation. When effectively used in integrated studies involving multiple disciplines such as analytical chemistry, biochemistry, immunology, reproductive biology, and fish behavior, one can often identify the general stressors (e.g., classes of chemical contaminants) that are the likely etiologic agents for the lesions detected. In fact, application of the histopathological biomarker approach in field studies is most useful in multidisciplinary studies where linkages to both lower and upper levels of biological organization are possible. A growing body of literature from well-designed and implemented studies that use histopathological biomarkers of environmental stress provides a very strong basis for their use as indicators of adverse, chronic, sublethal effects and injury to aquatic vertebrates (especially fish) in environmental assessment studies. The emphasis of this chapter will be on reli-

Linkages between contaminants and liver diseases in fish

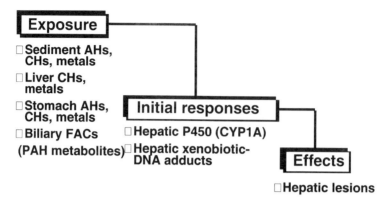

Figure 1. General approach to establishing linkages in field studies with demersal fishes (e.g., English sole, winter flounder, white croaker *Genyonemus lineatus*, starry flounder) between indicators of potential contaminant exposure (sediment contaminants), actual exposure (contaminants in stomach contents, liver, and bile), early responses to exposure (hepatic CYP1A, hepatic xenobiotic-DNA adducts), and more chronic effects in the form of toxicopathic liver lesions.

able, commonly used lesions, as well as lesions with potential as histopathological biomarkers of environmental stressors in fish, especially those associated with chemical contaminant exposure.

Advantages of Histopathology

Histopathological changes in tissues are cellular and subcellular morphologic manifestations and, as such, serve as integrators of the cumulative effects of alterations in physiological and biochemical systems in an organism. When exposure to a toxicant or other stressor is sublethal to the organism, damage may be detected in cells and tissues of that organism. These changes can be detected in situ in native animals in the environment being investigated, eliminating the less ecologically relevant alternative of using surrogate species in laboratory studies and extrapolating results to a different species in the natural environment. Many histopathological biomarkers in vertebrates are well validated experimentally by exposure to relatively specific stressors (e.g., liver neoplasms and other lesions related to neoplasia induced by chemical carcinogens), and standardized diagnostic criteria are available for these and other lesions.

The primary advantage of histopathology is that it permits the visual localization of injury to unique cells and tissues in multiple organs as it existed just prior to sacrifice and fixation of the tissue, as well as its in situ relationship within organs in an individual animal. In contrast to tissue homogenates, this approach provides a window to understanding the organization of cells, tissues, and organs, and the spatial and functional relationships between them. Organs are complex structures exhibiting associations of heterogeneous cell types that have differing susceptibilities to toxicants or other environmental stressors. By using routine histopathology and other methods such as immunohistochemistry and histochemistry, the morphological approach can identify alterations in specific target cells and also precisely localize unique cell products and enzymes (e.g., vitellogenin, CYP1A, and proliferating cell nuclear antigen [PCNA]) within cells and tissues. The routine histopathological approach is also relatively inexpensive, provides results rapidly, and lesions detected can be subjected to quantitative analyses (e.g., morphometry). By virtue of their intermediate position in terms of levels of biological organization, lesions in individual fish can be linked statistically to stressor exposure, early biochemical responses to this exposure, and, potentially, to effects at upper levels of biological organization. In this way, causal relationships can be established between exposure to stressors and effects at the individual and upper levels of biological organization. The use of histopathology as a research tool also permits documentation of essential biological characteristics of the animal being investigated for sexual, nutritional, and reproductive status, all variables that may also influence the histological evaluation of a specimen. Histopathological examination can also be used as an adjunct method to diagnose certain infectious diseases that may be directly or indirectly linked (e.g., via immunosuppression) to environmental or anthropogenic stressors, such as in the *Exxon Valdez* oil spill in Prince William Sound, Alaska (Marty et al. 1998), and juvenile salmon from urbanized estuaries on the Pacific Coast (Arkoosh et al. 1998). Lesions in native fish species can and have been used as manifestations of actual injury to biota in natural resource damage assessment studies (Collier et al. 1998a, 1998b) and in federally mandated environmental remediation investigations (Adams et al. 1999). Documentation of lesions in organisms sampled in the field is especially useful in multidisciplinary studies of recovery of biological resources in engineered environmental remediation of Superfund sites (Myers et al. 2000a, author's unpublished data; Baumann and Harshbarger 1998), in the tracking of temporal trends in lesion prevalences at multiple sites (Moore et al. 1996; Puget Sound Water Quality Action Team 2000; O'Neill personal communication), and evaluation of recovery following closure or modification of industrial or municipal waste treatment facilities (Baumann and Harshbarger 1995; Moore et al. 1996).

Limitations of Histopathology

The ability of the investigator to accurately detect alterations from the "range of normality" depends on a number of factors. First, the tissue must be properly fixed, processed, and stained. Second, accurate interpretation of histologic sections requires extensive prior experience in comparative histo-pathology, and the examiner must be familiar with the range of normal morphologic variations for the particular target species, such as variations due to sex, reproductive and nutritional status, and season (Timashova 1981). The examiner also must be aware of the extensive interspecies variability in the appearance of normal tissues in teleosts, as well as with common arti-facts of tissue collection, fixation, processing, and staining. It must also be realized that few if any lesions in wild fish are pathognomonic (uniquely distinctive for a disease) for exposure to particular chemical or environmen-tal stressors, since cellular and tissue responses to injury caused by these agents are most often very stereotyped and nonspecific. Also, some lesions known to be caused by exposure to groups of chemical contaminants are mimicked by exposure to natural toxins. For example, hepatic megalocytosis in fish can be caused by exposure to chemicals such as polycyclic aromatic hydrocarbons (PAH) (Schiewe et al. 1991) but can also result from exposure to certain naturally occurring algal toxins in salmon (Kent 1990; Anderson et al. 1993) and pyrrolizidine alkaloids in rainbow trout *Oncorhynchus mykiss* (Hendricks et al. 1981).

An additional challenge of histopathological studies is to distinguish toxi-copathic or stress-related lesions in tissue sections from those caused by infectious or parasitic agents. This is necessary to exclude the possibility that the detected lesion may have been caused by an infectious agent before considering a toxicopathic etiology. In most situations, it is possible to view the infectious agent and the associated inflammatory response in a tissue section. If a viral agent is suspected by virtue of suspicious nuclear or cyto-plasmic inclusions, a portion of the tissue can be extracted from the tissue block and reprocessed for examination by transmission electron microscopy (TEM) to view the virus. However, presence or absence of viral particles by TEM does not prove or disprove a viral etiology. A separate issue relating to diagnostic uncertainty is the fact that several infectious conditions exist in fish that grossly mimic neoplasia, but are most appropriately termed xenomas, pseudotumors, or pseudoneoplasms (Harshbarger 1984). Examples include the viral disease lymphocystis (Yasutake 1975) and X-cell pseudotumors caused by an, as yet, undefined protistan parasite such as an amoeba (Dawe 1981; Myers 1981; Myers, author's unpublished data), which previously and erro-neously have been referred to as true pseudobranchial tumors and epider-mal papillomas, even in the more recent literature (Syasina et al. 1999). Certain histopathological biomarkers in skin, fin, and gill (e.g., respiratory epithelial hyperplasia), while capable of being caused by chemical expo-sure, can also be the result of ongoing or past infections or infestations by external ciliates or metazoan parasites. Therefore, the absence of infectious or parasitic agents in tissue sections does not eliminate them as the causative agent for these lesions, and such lesions with highly nonspecific causes are of limited use as biomarkers of environmental stressors such as chemical contaminants.

Some histopathological lesions in certain species of fish cannot be applied universally in field studies, even in sympatric species, due to the wide variations in species susceptibility to stressors such as chemical contaminants. For example, while liver neoplasms, preneoplastic focal lesions, and other neoplasia-related lesions are considered highly effective biomarkers of PAH exposure in English sole *Pleuronectes vetulus* (Myers et al. 1999), similar lesions are rarely detected in a closely related and sympatric species, starry flounder *Platichthys stellatus* (Myers and Rhodes 1988; Myers et al. 1992, 1999; Collier et al. 1992). On the other hand, starry flounder are affected by a hepatic lesion termed hydropic vacuolation of biliary epithelium and hepatocytes that has been shown to be a useful biomarker for chemical contaminant exposure in this and other species (Myers et al. 1994; Moore et al. 1997; Stehr et al. 1998); this condition does not occur in English sole.

The use of histopathological biomarkers of environmental stress has been criticized because of the inherently subjective nature of this discipline. It has often been difficult to quantitatively correlate the results of fish histopathology studies with other more quantitative physiological and biochemical approaches, especially if lesion severity is incorporated into correlational analyses. There have been a few attempts to use morphometric or stereological methods (Weibel 1979, 1980) to quantify severity of injury in an organ to environmental stressors (Schwaiger et al. 1997; Adams et al. 1999) with relative degrees of success. It is these authors' and others' (Moore et al. 1996) opinion that the additional information gained by such time-consuming and tedious methods rarely justifies using these methods in large scale environmental monitoring studies. Notable exceptions are to validate results of a semiquantitative nature, as was done by Moore et al. (1996), or in the measurement of various parameters relating to macrophage aggregates in liver, kidney, and spleen (Blazer et al. 1994, 1997). To assure accuracy in histopathological diagnosis, it is advisable to consult and follow established standardized diagnostic criteria for lesions observed, such as for toxicopathic and neoplasia-related liver lesions in fish (Hendricks et al. 1984; Myers et al. 1987; Bunton 1990; Vogelbein et al. 1990; Bucke and Feist 1993; Moore and Myers 1994; Moore et al. 1997), and gill lesions (reviewed in Mallatt 1985). A number of general fish histology and histopathology texts also provides general guidance and criteria for histopathological diagnosis (Grizzle and Rogers 1976; Groman 1982; Yasutake and Wales 1983; Meyers and Hendricks 1985; Ferguson 1989; Roberts 1989; Takashima and Hibiya 1995; Bruno and Poppe 1996). General pathology (Cotran et al. 1999) and toxicologic pathology (Haschek and Rousseaux 1991) texts are also useful resources for establishing standardized criteria for histopathological diagnosis.

Practical Guidelines for Field Assessment

Field Protocols

Target or Sentinel Species Selection

For most histopathological studies in wild fish, it is preferable to select bottom-dwelling, bottom-feeding species in frequent or continuous contact with bottom sediments, where contaminant levels are usually highest due to adsorption to particulates. A species should also have a broad geographic distribution among sites to be sampled and should reside for the majority of its life history in the area of capture. The mummichog *Fundulus heteroclitus* is an excellent sentinel species with a restricted summer home range of 30–40 m (Vogelbein et al. 1990). Significant migration between contaminated and uncontaminated sites can seriously confound interpretation of lesion prevalence data, especially in the absence of chemical markers of chronic exposure (e.g., measurement of hepatic PCB concentrations). Problems of migration of the target species with respect to lesion data interpretation can be minimized by the concurrent collection and analysis of liver tissue for bioaccumulated compounds such as PCBs, DDTs, and xenobiotic-DNA adducts, which act as dosimeters of exposure to genotoxic compounds such as PAHs (Reichert et al. 1998), and are reliable indicators of chronic exposure to these compounds. DNA adducts have been shown to be significant risk factors for lesions occurring early in the histogenesis of neoplasia in English sole, especially hepatocellular nuclear pleomorphism, hepatic megalocytosis, and altered hepatocellular foci (Reichert et al. 1998; Myers et al. 1999). For more detailed information on effects of contaminants on DNA, see Chapter 5.

Method of Fish Capture

A capture technique that minimizes stress on the fish and that does not have a selection bias for either diseased or healthy fish should be used. In our experience, in marine and estuarine studies targeting demersal fishes, capture by otter trawl with 5–10 min tows is an optimal method; capture by fyke nets (Arcand-Hoy and Metcalfe 1999), electroshocking in freshwater habitats (Leadley et al. 1998), cast nets and trammel nets, or use of baited minnow traps (Vogelbein et al. 1990) have also been used effectively. In cases where a broad size range of fish is to be sampled, proper random subsampling techniques should be employed (for more detail on statistical sampling techniques, see Chapter 15). If fish cannot be necropsied immediately, they must be held alive, preferably in a flow-through system.

Sample Size

Sample size is a very important factor in any histopathological survey, and the required sample size will vary somewhat according to several factors and the objectives of the study. Generally, the higher the prevalence of a condition in a population, the lower the required sample size needed to detect this condition in a sample population. Tables are available in statistical texts and in specific guidelines for this type of study (e.g., Tetra Tech 1987) to approximate the required sample size. Certain parameters that will guide the researcher in selecting the proper sample size will have to be established by conducting smaller pilot studies (e.g., to determine approximate prevalence of a condition in a sample population). In determining whether the prevalence of a condition at multiple treatment (e.g., polluted) sites differs significantly from that in a reference (unpolluted site), the smaller the detectable difference in prevalence between two sites, the larger the required sample size needed to statistically verify this difference. For example, if the prevalence of hepatic neoplasms at the reference site is 0% and the prevalence at a test site is 25%, fewer samples will be needed to verify this difference than if the difference between the sites were, for example, only 5%. Generally, the minimum detectable prevalence at a test site decreases with increasing sample size. However, if reference sites can be located that have a demonstrable prevalence for a condition that approaches 0%, a general guideline of 30–60 specimens per site will probably provide the necessary statistical power to detect differences in lesion prevalences among the study sites, provided the prevalence of the same condition in a test area is high enough (10–20%). The chapter on statistical methods in this volume (Chapter 15) addresses these issues in more detail.

Supporting Biological Data

Any study using fish histopathology as a research tool should attempt to collect fish representing a range of ages and similar sex ratios among the sites or at least be able to adjust for the effects of age and sex on the risk of lesion occurrence. This is especially important for lesions such as liver neoplasms, preneoplastic focal lesions, and other neoplasia-related lesions that have a higher probability of occurrence in older fish (Rhodes et al. 1987; Baumann et al. 1990; Johnson et al. 1993; Myers et al. 1994; reviewed in Moore and Myers 1994; Moore et al. 1996). In general, if it is not possible to collect a homogeneous age distribution among the sites, the investigator will need to present lesion data by age-class or be able to account for the influence of age on risk of lesion occurrence by statistical methods such as stepwise logistic regression. Collection of anatomical structures such as otoliths, opercula, interopercular bones, scales, or pectoral spines from each specimen examined histologically, so age can be individually determined, should be considered obligatory; fish length is usually not a reliable estimate of age. In addition to basic length, weight, and gutted weight measurement, the sex and stage of sexual maturation should be noted for each fish, as these variables can influence the histologic appearance of the liver and may influence the probability of lesion occurrence. For example, hepatic neoplasms are more prevalent in female European flounder *Platichthys flesus* (Vethaak and Wester 1996).

Sampling Site Characterization

If sediment contaminant levels or water quality parameters are already known or can be reasonably predicted, sampling sites should be selected along a gradient of chemical contamination or other stressors in order to make the case more convincing when testing the hypothesis that higher lesion prevalences will be present in more contaminated areas. Included within this selection scheme should be severely polluted sites (if present) and a relatively uncontaminated or "reference" site. In a preliminary pilot study, a "worst case" scenario should be designed, in which the chosen test site is severely contaminated, with a relatively clean reference site used for comparison of the chosen endpoints. If an hypothesis holds up in this pilot study, then for the definitive study, multiple sampling sites should be chosen to represent a broad gradient of contaminant exposure. The precise location and depth of all sampling sites should always be documented, along with ambient physical conditions in the habitat area of the target species, including basic water quality parameters such as temperature, pH, salinity, and dissolved O_2 (DO).

Temporal Aspects of Sampling

Relatively little information is available regarding the seasonal effects on hepatic or other lesion prevalence in most species, but, as an example, significant seasonal differences are not seen in the prevalences of the important toxicopathic lesion types in Puget Sound English sole (Rhodes et al. 1987). However, seasonal spawning migrations can dramatically affect the distribution of diseased fish within a geographical region. It is recommended that sampling for all sites be done within the same season, when the target species is on its primary resident feeding grounds and not during periods of annual spawning migration. Because the nutritional status of fish can change dramatically according to season and related food availability and changes in feeding behavior, it is preferable to conduct all sampling during the same season to be able to control for influences of nutritional status on the histopathological endpoints being investigated.

Necropsy and Laboratory Protocols

Necropsy Protocol

Only recently killed (within several minutes) animals should be necropsied. All internal and external gross anomalies and other parameters such as those used in the necropsy-based health assessment index (Goede and Barton 1990; Adams et al. 1993) and the Biomonitoring of Environmental Status and Trends Program (Schmitt et al. 1999) should be described and recorded. Collect appropriate structures to determine fish age. If possible, the entire excised liver and gonad, along with the gutted carcass should be weighed for computation of liver somatic index, gonadosomatic index (GSI), and condition index. A histological sample is taken from all organs in which effects of stressors are predicted and from all tissues and organs in which grossly visible anomalies occur. Our laboratory typically collects sections of liver, head (anterior) and trunk (posterior) kidney, spleen, and gonad, unless grossly visible lesions are noted in other organs. In the absence of gross lesions, the pathologist should consistently collect tissues from the same region of each organ. If neurotoxicity is hypothesized, samples of brain, lateral line, and olfactory rosettes should be taken early in the sequence of tissue collection. Several texts exist on necropsy technique in fishes and should be consulted by new investigators (Yasutake and Wales 1983; Reimschuessel 1993; Kent and Poppe 1998; Schmitt et al. 1999; Fisher and Myers 2000).

Tissue Fixation and Processing

For routine fixation of specimens for paraffin embedment, tissues should be placed in an appropriate fixative immediately after excision from the freshly dead fish (<2–3 min after sacrifice) in a volume equal to or greater than 20 times the volume of the tissue collected. Protocols in our field studies (Stehr et al. 1993) call for collection of tissue sections (~3–4 mm thickness) into specialized cassettes that are then placed into 2 L jugs of fixative, 30 cassettes per jug. Fixation time should be 24–48 h, with storage of tissues preferably below 20°C. The choice of fixative should be based on study objectives and potential future uses of the specimens; a number of simple fixatives and fixative mixtures are available (Fournie et al. 2000). Our laboratories prefer Dietrich's or Davidson's fixatives for field studies because of their inherent decalcifying properties (acetic acid), ease of handling, and good nuclear fixation. For laboratory studies, especially those involving use of small fish species such as Japanese medaka *Oryzias latipes,* Bouin's or 10% neutral buffered formalin are the preferred and superior fixatives (Fournie et al. 1996a). If tissues are to be examined for ultrastructural changes by TEM, protocols for tissue collection and fixation are substantially different and specialized texts should be consulted (Hawkes 1974; Stehr and Myers 1990). For a detailed treatment of fixation and other histological procedures, see Hinton et al. (1984), Hinton (1990), and Fournie et al. (2000). Our laboratories process, infiltrate, and embed fish tissues for histology in Paraplast-extra, using an automated Shandon Hypercenter XP tissue processor.

Tissue Sectioning and Staining

It is usually sufficient in histopathological studies to use routine, standard histological techniques in preparing slides for microscopic examination (Luna 1968, 1992; Preece 1972). Sections are cut at 4–5 microns so that the full area of a sampled organ can be viewed, with one or two sections typically examined. Sections are routinely stained with Harris' hematoxylin and eosin or

Gill's hematoxylin and eosin-phloxine (similar to the AFIP method of Luna [1968]), with good results in terms of tissue definition and contrast. Many special stains are available to demonstrate specific tissues, storage products, pigments, and pathogens. Several of the most commonly used include the Perl's Prussian blue reaction for iron and hemosiderin (Luna 1968, 1992), which has been used very effectively to demonstrate resistance to iron up-take in hepatic foci of cellular alteration and neoplasms (Myers et al. 1987) and hemosiderin content in macrophage aggregates (Blazer et al. 1994, 1997); periodic acid-Schiff (for hepatocellular glycogen content estimations and mycotic infections); Masson's trichrome for connective tissue components; and Brown and Brenn's tissue Gram stain.

Tissue Examination, Diagnosis, and Quality Control

Slides should be read by an experienced fish histopathologist with formal training in human and veterinary or comparative pathology. To minimize interpretive bias, it is recommended to use a "blind" system in which the examiner is not aware of the site of capture of the specimens. This is accomplished by using a "pathology" number on the slide label generated from a random number table (e.g., P3073) matched with the actual specimen number (e.g., 95–2502). Standardized, concise, and consistent terminology for lesion descriptions should be followed. Examples include Myers et al. (1987) for a description of toxicopathic liver lesions in English sole; Vogelbein et al. (1990) for lesions in mummichog; and Murchelano and Wolke (1985, 1991), Moore and Stegeman (1994), and Moore et al. (1997) for hepatic lesions in winter flounder *Pleuronectes americanus.* In situations where more than one histopathologist is examining slides, it is essential to have a training program to establish standardized and consistent diagnostic criteria. Our laboratory usually has 2–3 slide examiners working on a particular project at one time. Each slide reader must complete a training period of 3–9 months under the teaching and supervision of the chief histopathologist before the diagnostic data from that slide reader begins to be incorporated into the database. Such a training period helps to ensure the consistency and accuracy of the diagnostic data. All unusual lesions are confirmed by consultation with the chief histopathologist.

Morphometric Analysis and Image Analysis

This approach has the advantage of taking much of the subjectivity out of histopathology and is assuming increasing importance as an adjunct tool, especially in quantifying alterations in proportions of normal and abnormal cells in an organ (Adams et al. 1999), cell proliferation indices such as PCNA (Ortego et al. 1995), cellular localization of cytochrome P450 enzymes (e.g., Myers et al. 1998a), quantitation of parameters relating to macrophage aggregates in tissues (Blazer et al. 1994, 1997; Couillard et al. 1999), and changes in nuclear and cytoplasmic diameters of cells. Morphometric analysis is especially useful when changes are subtle and cannot be clearly discerned by subjective evaluation. However, morphometry by image analysis is very labor intensive and impractical in large scale biomonitoring studies and, with the exception of its use to quantify parameters of macrophage aggregates, such an approach is rarely used in studies appearing in the peer-reviewed literature.

Use of Histochemistry, Immunohistochemistry

Histopathology also offers the opportunity to identify and localize specific cellular and macromolecular components by histochemistry and immunohistochemistry. Several pertinent examples of how these methods may be used in morphological studies of fish include identification of (1) xenobiotic metabolizing enzymes such as CYP1A both as an indicator of PAH exposure (Van Veld et al. 1990, 1997; Stegeman et al. 1991; Stegeman and Hahn 1994; Husøy et al. 1994) and in neoplasia-related hepatic lesions (Smolowitz et al. 1989; Van Veld et al. 1992; Myers et al. 1995, 1998a); (2) glutathione-S-transferase expression in hepatic lesions (Kirby et al. 1990; Stalker et al. 1991; Van Veld et al. 1991); (3) enzyme histochemical markers in liver such as alkaline phosphatase, gamma-glutamyltranspeptidase, and glucose-6-phosphate dehydrogenase as diagnostic aids for early preneoplastic changes (Teh and Hinton 1993; Kohler and Van Noorden 1998; reviewed in Moore and Myers 1994); (4) cytoplasmic filaments (cytokeratins) for cell identification in problematic tumor types (Bunton 1993); and (5) localization of PCNA for assessment of cell proliferation in tissue sections (Ortego et al. 1995; Kohler and Van Noorden 1998). Any of these more specific methods that give insight into the mechanisms of pathogenesis or characterization of a lesion can be used on a selective basis to better characterize the lesions detected in smaller, mechanistic studies, but, they have limited practicality or use as biomarkers in large scale biomonitoring programs involving examination of hundreds to thousands of fish.

Linking Histopathology to Other Response Levels

In studies investigating potential effects of environmental contaminants on fish, it is essential to collect the necessary samples to document potential exposure (e.g., sediment for chemical analyses), actual exposure and metabolism (e.g., stomach contents and liver for chemical analyses, bile for analysis of PAH metabolites as fluorescent aromatic compounds [FACs] [Krahn et al. 1987]), early biochemical responses (e.g., hepatic CYP1A, glutathione-S-transferase; see Chapter 2), and early biomarkers of molecular injury (e.g., hepatic DNA adducts; see Chapter 3). If necessary, because of tissue weight limitations in particular assays or because of financial constraints, tissue samples should be collected from individuals that can be composited later in the laboratory. Collection of liver tissue from individual fish for CYP1A quantitation (as aryl hydrocarbon hydroxylase [AHH] or ethoxyresorufin-O-deethylase [EROD] activities) and xenobiotic-DNA adducts is important in order to provide biomarkers that link exposure to early biochemical responses and, thence, to histopathological responses in individual fish.

Statistical Analyses

A statistical approach, such as stepwise logistic regression to identify relationships between potential biological and chemical risk factors and lesion occurrence, and a software package capable of performing multivariate analyses are recommended. Stepwise logistic regression is a robust statistical method for epidemiological and epizootiological studies that is commonly used on binomial (e.g., presence or absence of lesion) or proportional (e.g., lesion prevalence) data to examine the influence of multiple risk factors on the probability of disease occurrence, as well as exposure–response relationships in retrospective epidemiological studies (Breslow and Day 1980). This method allows for the simultaneous adjustment for biological risk factors (e.g., fish age, sex) included in the regression. The odds ratio, representing the degree of association between a risk factor and lesion occurrence (Fleiss 1981), as determined from variable coefficients of the logistic regression equations (Anderson et al. 1980; Schlesselman 1982), can be calculated as an estimate of relative risk for lesions in individual fish as related to risk variables such as site, sex, and age, as well as biomarker data and chemical data measured in individual fish, in a stepwise fashion. The relative risk of lesion occurrence at unique sites can also be estimated as compared with reference sites by this method. The most common application of this method in our laboratory's research (Johnson et al. 1993; Myers et al. 1994; Stehr et al. 1997, 1998; O'Neill et al. 1998) has been to determine the significance of relationships between lesion prevalence at sampling sites to discrete risk factors, such as levels of contaminants in sediments and fish tissues, while simultaneously adjusting for mean fish age and sex ratio. Results are typically expressed as the proportion of variation in lesion prevalence that can be attributed to significant risk factors (Figure 2). For these analyses, our laboratory uses the PECAN module (parameter estimation through conditional probability analysis) of the EGRET (epidemiological graphics, estimation, and testing) statistical package (Cytel Software Corporation).

Histopathological Biomarkers in Teleosts

Histopathological biomarkers of stress occurring in the main organs of fishes will be described. Because this chapter is not intended to be a comprehensive review of all lesions in all organs, only examples of lesions with established use or with good potential for use as biomarkers in field-caught fish will be described.

Optimal Risk Model for
Lesions in English Sole

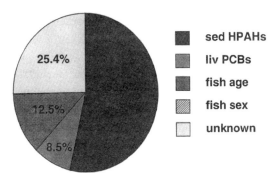

- sed HPAHs
- liv PCBs
- fish age
- fish sex
- unknown

Total variance explained by model: 74.6%
N = 49

Figure 2. Optimal risk model for risk factors significantly associated with toxicopathic liver lesions in English sole from the Puget Sound Ambient Monitoring Program (PSAMP), 1989–1995. This logistic regression model describes the proportion of variation in site-and-year-specific lesion prevalence explained by levels of high molecular weight PAHs (HPAHs) in sediment, PCBs in liver tissue, mean fish age and sex ratio. Total number of sites with full data sets included in this analysis is 49. Sex ratio was not a significant risk factor for toxicopathic lesion occurrence. Toxicopathic liver lesions included in this analysis were (1) hepatocellular nuclear pleomorphism/hepatic megalocytosis, (2) preneoplastic foci of cellular alteration, and (3) hepatic neoplasms. Any fish affected with any of these lesions was considered to be affected with a toxicopathic lesion. These significant risk factors account for approximately 75% of the variation in lesion prevalence among the sites, with sediment (HPAHs) accounting for the majority (53.6%) of this variation.

Liver

The liver is the organ most commonly examined and exhibits the most lesions with specific and strong associations to chemical contaminant exposure in fish. The liver is the primary target organ for effects of lipophilic xenobiotic chemicals, primarily due to its following functions (reviewed in Hinton 1993): (1) it is the major site of production of xenobiotic metabolizing enzymes of the cytochrome P450 system, which inactivates some toxicants and activates others to more toxic intermediates that can bind to cellular components such as DNA and initiate hepatotoxicity and hepatocarcinogenesis; (2) it stores lipids and carbohydrates absorbed by the gastrointestinal tract for transfer to further breakdown by other tissues; (3) bile produced in liver carries many conjugated metabolites of xenobiotics that can be released via the common bile duct into the intestine and returned back to the liver by enterohepatic recirculation; and (4) it is the major site of vitellogenin production in fish, mediated through estradiol binding. Lesions in the liver will not be described here in detail, but are presented below in tabular format and include comments on associated environmental stressors, advantages and limitations, and supportive literature references. Lesions listed below have been identified in field studies as being associated with chemical contaminant exposure and, except where noted, have also been validated in laboratory exposure experiments (reviewed in Metcalfe 1989). The majority of these lesions are involved in the stepwise, temporal histogenesis of neoplasia in the liver as demonstrated experimentally in rats (Farber and Sarma 1987) and fish, representing the phases of cytotoxicity, compensatory regeneration and proliferation, preneoplastic foci of cellular alteration, and neoplasms.

Lesions	Associated environmental stressors	Advantages/ limitations	Selected references
Loss of hepatocellular glycogen/increased basophilia	Nutritional stress, some toxicants, including TCDD	Dependent on season, reproductive, nutritional status; nonspecific change	Meyers and Hendricks 1985; Teh et al. 1997; Hinton and Lauren 1990; Walter et al. 2000; Spitsbergen et al. 1991
Hepatocellular coagulative necrosis, hypertrophy, hydropic degeneration, hepatocellular hyalinization	Toxicants, ischemia, infectious agents; plant toxins, infectious agents; toxicant classes include PAHs, PCBs, DDTs, chlordanes, TCDD	Must distinguish from autolysis; can be due to natural (algal) toxins; relatively specific for toxicants, if other etiologies ruled out	Meyers and Hendricks 1985; Hendricks et al. 1981; Kent 1990; Anderson et al. 1993; Hinton et al. 1992; Johnson et al. 1993; Myers et al. 1994; Spitsbergen et al. 1991
Hepatocellular/biliary epithelial cell regeneration and hyperplasia; also oval cell proliferation and cholangiofibrosis	Toxicants, infectious agents, natural toxins, as sequelae to necrosis; toxicant classes include PAHs, PCBs, DDTs, chlordanes; biliary lesions may be associated with infectious agents in some species	If infectious agents ruled out, are good biomarkers of toxicant exposure; often associated with hepatocellular and biliary neoplasms	Hayes et al. 1990; Baumann et al. 1996; Vogelbein et al. 1997; Johnson et al. 1993; Myers et al. 1987, 1994; Kent et al. 1988; Moore and Myers 1994; Moore et al. 1997
Hepatocellular nuclear pleomorphism, megalocytosis; also referred	Toxicants, plant and algal toxins; toxicant classes include PAHs, DDTs,	Specific for toxicant exposure in some fish (e.g., English sole);	Myers et al. 1987, 1994, 1999; Hendricks et al.

Lesions	Associated environmental stressors	Advantages/ limitations	Selected references
to as hepatocellular anisokaryosis 1992; Stephen et al.	PCBs; etiology may be unknown (Groff et al. species with high 1993)	other species not susceptible; normal in et al. 1992; hepatocellular polyploidy; often associated with hep. necrosis; induction in English sole by PAH exposure, in medaka by DENA; affects both juvenile and adult fish	1981; Kent et al. 1988, 1990; Groff et al. 1992; Hinton et al. 1988, 1992, 1993; Schiewe et al. 1991; Walter et al. 2000
Hydropic vacuolation of biliary epithelial cells/hepatocytes	Toxicants; includes PAHs, mainly assoc. with DDTs, chlordane; prevalence reduction in winter flounder from Boston. Hbr. associated with reduced chemical inputs (Moore et al. 1996)	Unique lesion, occurs at high prevalences in winter flounder, starry flounder, white croaker; other species not susceptible; assoc. with biliary tumors and neoplasia; no experimental model in fish or mammals; strongly age-related	Murchelano and Wolke 1985, 1991; Camus and Wolke 1991; Bodammer and Murchelano 1990; Moore et al. 1996, 1997; Stehr et al. 1998; Johnson et al. 1993; Myers et al. 1994, 1999; Augspurger et al. 1994
Foci of cellular alteration (FCA), or altered hepatocellular foci (AHF); includes clear cell, vacuolated, eosinophilic, and basophilic foci; amphophilic and mixed cell foci described in some species; based on staining changes in routine H & E-stained sections	Toxicants, including PAHs, chlorinated hydrocarbons such as PCBs, DDTs, many other initiators and promoters of carcinogenesis (rev. in Myers et al. 1987)	Specific for chemical exposure; validated in many fish and mammalian models; all AHF have neoplastic *potential*, but not all develop into tumors; basophilic focus most proximate to hep. neoplasms; as a group, tend to co-occur with hep. neoplasms; many species resistant; strongly age-related	Hendricks et al. 1984; Myers et al. 1987, 1991, 1994, 1999; Hawkins et al. 1990; Hoover 1984; Lauren et al. 1990; Murchelano and Wolke 1991; Johnson et al. 1993; Vethaak and Wester 1996; Perkins 1995; Baumann et al. 1990, 1996; Bucke and Feist 1993; Kranz and Dethlefsen 1990; Vogelbein et al. 1990, 1997; Hayes et al. 1990; Kohler et al.

Lesions	Associated environmental stressors	Advantages/ limitations	Selected references
			1990; Moore and Myers 1994; Barron et al. 2000; Farber and Sarma 1987 (rats)
Enzyme-altered foci; histochemical evaluation of FCA/AHF above; changes may precede lesions seen in H & E-stained sections	Toxicants, including PAHs, chlorinated hydrocarbons, other initiators and promoters of carcinogenesis (rev. in Myers et al., 1987; also in Teh and Hinton 1993)	Specific for chemical exposure; validated in fish and mammalian models; not practical for large field surveys; main advantage is enzyme changes may precede lesions seen in H&E sections; requires sections of frozen/freeze-dried tissue	Teh and Hinton 1993; Kohler and Van Noorden 1998; Stalker et al. 1991; Hinton et al. 1988
Hepatocellular adenoma and carcinoma; cholangioma, cholangiocarcinoma; mixed hepatobiliary carcinoma	Toxicants, including PAHs, chlorinated hydrocarbons, other initiators and promoters of carcinogenesis (rev. in Myers et al. 1987, also Metcalfe 1989)	Specific for chemical exposure; experimentally validated in numerous fish and mammalian models; some species highly resistant; strongly age-related	See references for AHF above; also Harshbarger and Clarke 1990; Smith et al. 1979; Dey et al. 1993; Moore and Stegeman 1994

The following case studies represent examples of field studies with fish where liver lesions have been successfully applied as histopathological biomarkers of xenobiotic contaminant exposure.

Brown Bullhead Catfish

The brown bullhead *Ameiurus nebulosus* is a benthic and philopathic freshwater species common to all of the lower Great Lakes and is affected with hepatic neoplasms and other neoplasia-related liver lesions. This species has been used in multiple field studies as a sentinel species for monitoring environmental health effects since the early 1970s (Brown et al. 1973; Baumann et al. 1982, 1987, 1990, 1996; Baumann and Harshbarger 1985; Black 1983). The results from most of these studies, especially those investigating the Black River, Ohio, have implicated high molecular weight PAHs in sediments as the probable etiologic agents for the hepatic tumors observed. Brown bullhead from areas with high tumor prevalence also have significantly elevated hepatic AHH activity and hepatic DNA adduct levels (Balch et al. 1995). Biliary and hepatocellular neoplasms have also been induced experimentally in this species via dietary exposure to PAH-contaminated extracts added to commercial feed (Black et al. 1985). Geographically broader studies have been conducted using multiple histopathological and biochemical biomarkers in brown bullhead from the lower Great Lakes (Baumann et al. 1996; Arcand-Hoy and Metcalfe 1999). Recent studies (Arcand-Hoy and Metcalfe 1999) found significant prevalences of hepatic neoplasms in the polluted sites of the Detroit River in Michigan (15%), Black River in Ohio (9.5%), and Hamilton Harbor, Ontario, Canada (6%), with no tumors found in fish from relatively clean reference sites. The liver lesion data were consistent with results of biochemical biomarkers such as bile FACs, hepatic EROD activity, and hepatic retinoid activity. Similar field and caging studies in the Detroit River in Michigan (Maccubbin and Ersing 1991; Leadley et al. 1998, 1999) found higher hepatic neoplasm prevalences at sites with elevated concentrations of sediment PAHs and demonstrated exposure to PAHs as measured by biliary FACs levels (Leadley et al. 1999). Results from these more recent studies essentially confirm the presence of hepatic neoplasms at contaminated sites as shown in previous studies examining more fish per site (Baumann and Harshbarger 1985; Baumann et al. 1987, 1990, 1996).

Additional strong evidence linking hepatic tumors in this species to PAHs is provided by a clear pattern of initial reduction in age-specific liver neoplasm prevalence after the closing of a coking plant on the Black River and subsequent reduction in PAH exposure (Baumann and Harshbarger 1995), followed by a brief increase in tumor prevalence three years after dredging activities and resuspension of PAHs in 1990, and a subsequent decrease in age-specific tumor prevalence in 1994 and 1998 (Baumann and Harshbarger 1998; Baumann and Okihiro 2000). In this case, reduction in the hypothesized causal variable (PAH exposure) preceded the reduction in the effect, satisfying the required temporal order for cause and effect in epidemiological studies (Susser 1986).

Mummichog

The use of hepatic lesions as biomarkers of chemical contaminant exposure is also represented by field studies with the highly territorial, estuarine mummichog. These studies are primarily from a site with sediments severely contaminated with creosote-derived PAHs from a wood treatment facility on the Elizabeth River, Virginia (Vogelbein et al. 1990, 1997). In samplings done adjacent to the creosoting facility, multiple types of neoplasms were detected in the liver at prevalences up to 35%, including hepatocellular adenomas and carcinomas, hepatoblastomas (Vogelbein et al. 1999), and pancreatic acinar cell adenomas and carcinomas (Fournie and Vogelbein 1994; Vogelbein and Fournie 1994). Altered hepatocellular foci (eosinophilic, clear cell, and basophilic foci) were also detected at high prevalences (73%) at this site. No significant hepatic lesions were detected at a relatively uncontaminated reference site or at a less contaminated site immediately across the river from the creosoting facility. Studies at less severely contaminated sites in the Elizabeth River and James River estuary (Vogelbein and Zwerner 2000) showed intermediate prevalences of neoplastic and preneoplastic lesions.

Immunohistochemical studies on CYP1A and glutathione-S-transferase in these focal lesions, as well as neoplasms in mummichog (Van Veld et al. 1991, 1992) indicate that the hepatocytes in these lesions have undergone an adaptive response that confers upon these hepatocytes a resistance to the cytotoxicity and genotoxicity of agents such as PAHs that require metabolic activation to exert their effects. This expression pattern is consistent with liver neoplasia induced in the laboratory by genotoxic chemical carcinogens in mammals (Buchmann et al. 1985; Roomi et al. 1985) and fish (Lorenzana et al. 1989) and in similar lesions in native English sole from PAH-contaminated environments (Myers et al. 1995, 1998a).

The most convincing evidence of the causal role of PAHs in the etiology of preneoplastic and neoplastic lesions in this species comes from an experiment (Vogelbein et al. 1998) in which mummichog were exposed for 12 months to a creosote-contaminated sediment or to sediment and diet amended with eight PAHs. Fish were sampled at 3, 9, and 12 months after initiation of exposure. No lesions were observed in any treatment at 3 months, but high prevalences of foci of cellular alteration were detected at 9 months (up to 23%). At 12 months up to 40% of fish exhibited preneoplastic focal lesions with up to 6% bearing frank hepatic neoplasms. These experiments are the first to demonstrate convincingly in fish, via the ecologically relevant route of sediment and dietary exposure, that environmental PAHs play a direct role in the etiology of liver neoplasia in wild fish populations.

English Sole

The collective research on the occurrence of hepatocellular and biliary neoplasms and a spectrum of other toxicopathic lesions involved in the stepwise pathogenesis of liver neoplasia in English sole (Myers et al. 1987) from Puget Sound and other Pacific Coast estuaries provides especially strong evidence of a causal relationship with exposure to environmental PAHs and certain chlorinated hydrocarbons (CHs) (Pierce et al. 1978; Malins et al. 1984, 1986, 1987; Myers et al. 1987, 1990, 1991, 1994; O'Neill et al. 1998) (Figure 3). These studies are based on the hypothesis that these liver lesions in English sole are the result of exposure to PAHs (as hepatoxicants and initiators of carcinogenesis) and CHs such as PCBs and DDTs (as hepatotoxicants and nongenotoxic promoters of carcinogenesis) (Williams and Weisburger 1986). In fact, evidence from the large set of many field studies and fewer laboratory studies that have validated the field findings, fulfills all of the classic criteria for causation in epidemiological studies of cancer and other diseases (Mausner and Bahn 1974; Colton and Greenberg 1982). These criteria include (1) strength of the association (Malins et al. 1984, 1987; Myers et al. 1990, 1991, 1994, 1998b, 1999; O'Neill et al. 1998); (2) consistency of the association (Malins et al. 1984, 1987; Myers et al. 1990, 1991, 1994, 1998b, 1999; O'Neill et al. 1998; Collier et al. 1998a); (3) toxicological and biological plausibility (Varanasi et al. 1986, 1989a, 1989b; Myers et al. 1987; Stein et al.

Known Risk Factors for Liver Disease, English Sole

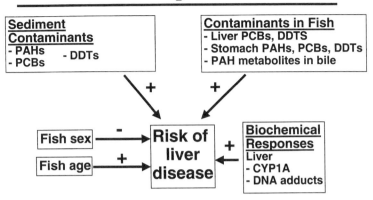

Figure 3. Generalized risk model for toxicopathic liver lesions in English sole, describing the risk factors that have been significantly associated with lesion occurrence in multiple field studies.

1992, 1994; French et al. 1996); (4) temporal sequence (i.e., exposure precedes disease) (Rhodes et al. 1987; Myers et al. 1987, 1998b), (5) dose–response gradient (Horness et al. 1998; Reichert et al. 1998; Myers et al. 2000, author's unpublished data); and (6) specificity of the association (Myers et al. 1993, 1994). Recent efforts to identify a threshold level of PAHs in sediments associated with occurrence of toxicopathic lesions in this species (Horness et al. 1998) showed that statistically significant sediment threshold levels ranged from 54 to 2,800 ppb (dry weight) total PAHs. Studies monitoring toxicopathic liver lesion occurrence in English sole subsequent to capping of PAH-contaminated sediments in Eagle Harbor, Washington (Myers et al. 2000, author's unpublished data), have also demonstrated a significant reduction in the relative risk of these lesions over time (1993–2001).

Among the case studies discussed above, two of them (brown bullhead and English sole) illustrate situations in which reduced exposure to the putatively causative chemical or chemical class was followed, after a variable lag time, by a reduction in the effect, in the form of neoplasia related liver lesions. This same pattern of reduction in prevalence has also been demonstrated for hepatic neoplasms and particularly for hydropic vacuolation of biliary epithelial cells and hepatocytes in winter flounder from a site in Boston Harbor. Reduction in prevalence of centrotubular hydropic vacuolation (from 77% in 1987 to ~30% in 1999) has been linked convincingly by Moore et al. (1996, personal communication) to reductions in chemical content of the Deer Island sewage effluent, as a result of active source reduction programs, passive reduction of industrial output related to economic recession, sludge removal from the Deer Island sewage effluent subsequent to December 1991, and finally, a switching of sewage effluent deposition from the

Deer Island site to an offshore site in 1998. Epizootiological or epidemiological studies cannot by themselves prove a causal relationship. However, the reduction in the frequency or intensity of effect in a wild population, brought about by reduction in exposure to the purported agent by natural causes or engineered remediation, can provide very strong support for a causal hypothesis (Fox 1991).

Posterior or Excretory Kidney

The posterior or excretory kidney in freshwater fish functions primarily to produce large amounts of dilute urine with little function in ionic regulation or acid–base balance. In marine fishes where the osmotic gradients are reversed, urine production is quite low and the main function of the kidney is excretion of divalent cations (Pritchard and Renfro 1982). The posterior kidney receives a major flow of blood from both the renal portal venous system and the renal arteries, and, at least in humans, the magnitude of blood flow per weight of the organ is higher in the kidney than any other organ (Finn 1983). The fish excretory kidney has been used as a model system for the effects of a number of environmental toxicants in laboratory studies such as mercury (Trump et al. 1975), copper (Baker 1969), hexachlorobutadiene (Reimschuessel et al. 1990a, 1990b), and others (reviewed in Pritchard and Renfro 1982). However, surprisingly few if any reliable biomarkers associated with xenobiotic chemical stress have been identified in the excretory portion of the fish kidney. Although we have encountered many types of renal lesions over the many years that our laboratories have conducted field studies, their prevalences are low and variable from year to year at the same site, and their etiologies are evidently quite nonspecific. Early field studies on English sole in Puget Sound (Rhodes et al. 1987) identified a number of renal lesions in the general categories of depositional disorders (e.g., mesangiosclerosis), degenerative and necrotic lesions (e.g., tubular necrosis, mesangiolysis), and proliferative conditions (e.g., glomerular hypercellularity) that occurred at higher estimated relative risks in sole from urbanized sites such as the Duwamish Waterway. Age was also a significant risk factor for these lesion categories. However, subsequent studies in English sole and other marine and estuarine species (Myers, author's unpublished data) have shown no consistent links between any of these lesions and chemical contaminant exposure, especially in comparison to those documented for the liver lesions described above. This result is consistent with the experience of others assessing potential effects of xenobiotics on liver and kidney in wild fish (Vogelbein and Zwerner 2000).

Lesions with potential use as biomarkers, based on their induction in laboratory exposure studies or occurrence in field-captured fish, are listed below.

Tubular Epithelial Degeneration, Necrosis, Vacuolation, Hyalinization, and Exfoliation

These lesions have been experimentally induced by heavy metals such as mercury (Trump et al. 1975), cadmium (Eisler and Gardner 1973; Hawkins et al. 1980), copper (Baker 1969), uranium (Cooley et al. 2000); insecticides such as DDT, lindane, and parathion (reviewed in Walsh and Ribelin 1975, and Meyers and Hendricks 1982); herbicides such as acrolein, amitrole-t, dinoseb, diquat, and paraquat (Hendricks 1979); and chlorinated hydrocarbons such as PCBs (Nestel and Budd 1975) and hexachlorobutadiene (Reimschuessel et al. 1989, 1990a). Similar lesions are observed in wild fish (McCain et al. 1982; Vogelbein and Zwerner 2000) but with no consistent associations to contaminant exposure.

Glomerular Lesions Such as Mesangiolysis and Mesangiosclerosis

A lesion called mesangiolysis, characterized by a ballooning dilatation and edema of the glomerular mesangium, is observed in marine flatfish (McCain et al. 1982; Rhodes et al. 1987). This lesion has some basis as a potentially toxicopathic renal lesion, since a similar lesion is inducible in mammals by exposure to Habu snake venom (Morita et al. 1978). Mesangiosclerosis, characterized by an increase in density of mesangial matrix of the glomerular tuft, is also observed in wild fish (McCain et al. 1982; Vogelbein and Zwerner 2000), is often found in conjunction with mesangiolysis (McCain et al. 1982), and is a common response to injury in the glomerulus in mammals (Schillings and Stekhoven 1980). This lesion is also found more commonly in older fish (Rhodes et al. 1987). A similar lesion, glomerular hyalinosis, was induced in medaka (Wester and Canton 1986) by an isomer (B-hexachlorocyclohexane) of the insecticide lindane and, therefore, may have use as a biomarker.

Anterior Kidney and Spleen: Hemopoietic Tissues

Hemopoietic tissues (cells of lymphoid and erythrocytic lineage) are concentrated in the anterior kidney and spleen of fish and may, because of their high turnover rate, be the target of certain toxicants. Although not commonly observed in wild fish, lesions such as necrosis, hypoplasia, hypocellularity, or depletion of cells in both the erthropoietic and especially the lymphoid lineage in anterior kidney and spleen have been reported in experimental exposures of fish to compounds as diverse as copper sulfate (Baker 1969); tributyltins (Schwaiger et al. 1994); DDT, dieldrin, endosulfan, and other pesticides (Walsh and Ribelin 1975); PAHs such as 7,12-dimethylbenzanthracene (Hart et al. 1998); TCDD (Spitsbergen et al. 1991; Walter et al. 2000); and PCBs (Nestel and Budd 1975; Hendricks et al. 1977; Spitsbergen et al. 1988; Thuvander et al. 1993). Splenic lymphoid depletion only occurred in redbreast sunfish *Lepomis auratus* and largemouth bass *Micropterus salmoides* exposed to various types of contaminant stress (Teh et al. 1997), suggesting a contaminant-related etiology. However, considering the number of other environmental stressors that may potentially cause erythroid or lymphoid depletion, necrosis, hypoplasia, and atrophy, these lesions can currently be regarded only as nonspecific biomarkers of stress in fish.

Spleen: Macrophage Aggregates

Macrophage aggregates (MA) are focal accumulations of macrophages commonly present in fish liver, kidney, and spleen (Wolke 1992; Blazer et al. 1994). They are believed to be functional equivalents of germinal centers, active in (1) the immune response, functioning as part of the reticuloendothelial system and in antigen presentation to immunocompetent cells; (2) storage, destruction, or detoxification of exogenous and endogenous substances such as necrotic debris and other products of tissue catabolism; (3) iron storage and recycling (Wolke et al. 1992; Blazer et al. 1997); and (4) in chronic inflammation as a special type of granuloma (Vogelbein et al. 1987), in which phagocytosis constitutes a defensive reaction to an injurious agent (Wolke et al. 1992). Macrophage aggregates contain three types of pigments: (1) ceroid or lipofuscin, (2) melanin, and (3) hemosiderin, all discernible with a Perl's Prussian blue stain for iron (Luna 1968; Blazer et al. 1994). These structures have long been proposed as histological and immunological biomarkers of fish health and environmental degradation and have been used to this end with differential success in a number of field studies (reviewed in Wolke et al. 1992; Blazer et al. 1994, 1997). Parameters of MAs typically measured by image analysis are number, size, and percent area occupied (Blazer et al. 1994, 1997).

Considering the multiple functions of MAs and the number of factors that influence their formation, they can be best regarded as sensitive, nonspecific biomarkers of exposure to multiple environmental stressors. As such, they can be effectively used as a type of screening assay for assessing effects of multiple environmental stressors in fish. Macrophage aggregates are known to increase with fish age, nutritional status, heat stress, exposure to chemical contaminants, infectious diseases (Wolke 1992), and stage of sexual maturation (Couillard et al. 1999). Low DO has also been suggested as a risk factor for increased MA parameters (Blazer et al. 1994). Increased numbers have been interpreted as indicative of an activated immune system, whether by increased exposure to antigens (e.g., infectious agents like bacteria) or chemical contaminants (Blazer et al. 1997). Because MAs probably do not disappear after removal of the stressor, they can be used as an integrator of past exposure to these multiple stressors.

Even though a small number of studies indicate that MAs are not a reliable biomarker of pollution exposure and effect (e.g., Haaparanta et al. 1996; Couillard et al. 1999), there are well over fifty literature references that report an increase in number, size, percent area occupied, or pigment content of these structures in fish from degraded sites. Examples include recent field studies in Canada targeting white sucker *Catostomus commersoni* showing age-adjusted increases in MA density in spleen to be consistently associated with exposure to bleached kraft pulp mill effluent (Couillard and Hodson 1996). Assessment of MAs was also successfully used in mummichog from a PAH-contaminated site in the Elizabeth River (increase in number/unit area in spleen), and in Atlantic croaker *Micropogonias undulatus* from Pensacola Bay (decrease in number/unit area) (Blazer et al. 1997). Macrophage aggregates were successfully used in brown bullhead from Hamilton Harbor, Ontario, where they were present at higher prevalences than in fish from an uncontaminated reference site (Baumann et al. 1996). Spleen of brown bullhead from a contaminated site in Lake Champlain also showed altered MA parameters as compared with the reference site (Blazer et al. 1994). Individuals of most fish species from chemically contaminated estuaries of the Virginian and Appalachian provinces in U.S. EPA's Environmental Monitoring and Assessment Program (EMAP) showed increases in MA numbers and the percentage area of the spleen composed of MAs (Blazer et al. 1994). Recently, Fournie et al. (2001) provided a critical value for the number of MAs/mm^2 for certain groups of estuarine fishes that discriminates between exposure to degraded and nondegraded environments.

The use of MAs as a biomarker in field studies has been validated, at least partially, by laboratory exposure studies, including (1) exposure of juvenile striped bass *Morone saxatilis* to arsenic, where an increase in number/unit area, size, and increase in hemosiderin levels within the MAs (possibly from hemolysis or as a mechanism to sequester arsenic in a less toxic, insoluble form) was noted (Blazer et al. 1997); (2) chromium exposure in plaice *Pleuronectes platessa* where an increase in number and decrease in size of MAs was noted (Kranz and Gerken 1987); and (3) winter flounder exposed to a range of PAH concentrations for 4 months, where reduced numbers of hepatic MAs were reported (Payne and Fancey 1989).

In summary, splenic MAs (less so in liver) have a practical application as a sensitive but nonspecific tissue response to multiple environmental stressors, and this application is best regarded as an initial screening tool or "early warning system" in fish. Sufficient field and laboratory studies exist to support the hypothesis that increased density of MAs is linked to contaminant exposure. Although exact mechanisms are not known, the formation of MAs and the accumulation of pigments increase in spleen and liver of fish from contaminated environments and, therefore, serve as reliable histopathological biomarkers. However, in assessment of MAs, it is important to account for other influencing variables such as fish age, nutritional status, water temperature, stage of sexual maturation, and presence of infectious diseases.

Ovary

Damage to reproductive organs is probably the most critical biological impact of environmental stress (including toxicant exposure) because of the potential effects at the population level. In addition to the lesions listed below, it is essential to examine gonadal tissues for confirmation of sex and to document the stage of sexual maturation, which may affect the morphology of other organs examined. Histological classification of gonads is viewed as being superior to all other methods of assessment of reproductive status (Hunter and Macewicz 1985). A strong familiarity with the pattern of gametogenesis and the seasonal variations in gonadal maturation in the target species is essential in distinguishing between normal gametocyte turnover and effects potentially due to environmental stressors. Numerous fish species normally undergo sex reversal during their life history; protogyny describes species whose gonads begin as ovaries and later develop into testes, and protandry refers to species that originally have testes that later develop into ovaries. These normal patterns are particularly common in marine fish (Atz 1964; Breeder and Rosen 1966) and must be accounted for in a histological assessment. Listed below are several examples of ovarian lesions or maturational indices with proven use as histopathologic biomarkers in fish.

Oocyte Atresia

The occurrence of atretic follicles in the fish ovary is normal. Atresia can affect oocytes at all stages of development and usually occurs as a result of resorption of ova in response to unfavorable environmental conditions such as altered temperature, reduced food availability, low DO, or altered salinity. Atresia can also represent simple resorption of residual, unspawned ova. However, marked increases in oocyte atresia can also result from chemical contaminant exposure, oil spills, lowered water pH, and other adverse environmental conditions that would be detrimental to larval survival (Stott et al. 1983; Cross and Hose 1988; Johnson et al. 1988, 1997; McCormick et al. 1989). Atretic oocyte frequency in both largemouth bass and bluegill *Lepomis macrochirus* was correlated with sediment levels of mercury (Adams et al. 1999). In a controlled field experiment, a significant depression of reproductive success was clearly linked to high rates of atresia (>20% of ovarian volume occupied by atretic oocytes) and reduced water pH (McCormick et al. 1989). Increased frequencies of oocyte atresia have also been demonstrated in experimental exposures of female fish to the natural estrogen, 17β-estradiol (Miles-Richardson et al. 1999), and a model environmental estrogen, ethinyl estradiol (Papoulias et al. 1999); hence, monitoring of oocyte atresia may have application in field studies of endocrine disruptors. In contrast, caged adult female fathead minnow *Pimephales promelas* exposed to wastewater treatment effluent failed to demonstrate an increase in oocyte atresia compared with fish caged at a reference site (Nichols et al. 1999). Morphologically, oocyte atresia appears as degeneration and necrosis of all components of the developing ovum, with activation and infiltration of the surrounding thecal cells and other macrophages that phagocytose this material. The presence of large proportions of atretic oocytes (as compared with normal oocytes) can be used as a biomarker of environmental stress with the understanding that such changes are not specific to particular stressors.

Hermaphroditism or Intersex Condition

This condition represents the presence of testicular tissue in ovary, which is exceedingly rare in wild fish but has potential use in endocrine disruption studies where environmental androgens are postulated as stressors. Experiments in medaka exposed to testosterone as newly hatched fry or at one week posthatch showed this intersex condition, with production of both mature ova and sperm (Koger et al. 2000). Newly hatched fry were the life stage most susceptible to endocrine-disrupting chemicals.

Atrophy, Inhibited Development

Ovarian atrophy is the underdevelopment or lack of development of ovarian tissue, which may be diagnosed grossly but is more definitively diagnosed histologically. For example, in laboratory experiments, newly hatched larvae of rainbow trout exposed to 1, 5, and 20 mg/L solution of a PCB mixture (Aroclor 1260) and subsequently examined grossly and histologically for gonadal abnormalities at approximately 6 months postfertilization, showed a significantly higher proportion of grossly abnormal females than in the controls. Histologically, the gross abnormalities were characterized by incomplete or inconsistent development of oocytes, a potentially serious anomaly indicating delay or prevention of sexual maturity in females (Matta et al., 1998). An example of field studies showing this effect is in the limited ovarian development seen grossly (assessed only by gonad weight) in lake whitefish *Coregonus clupeaformis* exposed to bleached kraft mill effluent in Jackfish Bay, Lake Superior (Munkittrick et al. 1992a, 1992b). Although this condition is rarely observed in wild fish, it is included here as a potential biomarker in endocrine disruptor studies.

Alterations in Ovarian Maturation

Histological evaluation of the reproductive stage is especially useful in field studies in which other reproductive parameters are also measured (e.g., plasma estradiol and vitellogenin levels, GSI, fecundity, etc.). For example, based on histological staging of ovaries, inhibited ovarian development and vitellogenesis have been significantly associated with PAH exposure in English sole (Johnson et al. 1988), and precocious maturation of female English sole has also been associated with chemical exposure, primarily with hexachlorobutadiene, hexachlorobenzene, and other chlorinated hydrocarbons in the contaminated Hylebos Waterway in Puget Sound, Washington (Johnson et al. 1999).

Testis

The testis includes a number of cellular components that may be targeted by toxicants or affected by other environmental stressors. These include Sertoli cells, which establish the blood–testis barrier; Leydig cells, in the interstitium of the testis involved in male hormone production; and the various stages of germ cells, from spermatogonia, primary and secondary spermatocytes, spermatids, and spermatozoa. An excellent review of the histological features of developmental stages in the testis is contained in Takashima and Hibiya (1995). Lesions with potential as biomarkers in the testis are limited and are represented by degenerative, necrotic, and atrophic lesions of the germinal epithelium, as observed in laboratory exposure studies; male hermaphroditism, intersex, and male feminization have been induced experimentally and have been reported in wild male fish associated with exposure to environmental estrogens.

Germinal Epithelial Degeneration, Necrosis, Atrophy

Degeneration and necrosis of the germinal epithelium, testicular atrophy, disorganization of spermatogenic lobules, and lobular fibrosis of the testis have been induced in fish by exposure to cadmium (Tafanelli and Summerfelt 1975) or Aroclor 1254 (Sangalang et al. 1981). Decreases in primordial germ cells and reduced spermatogenesis have also been induced in young male common carp *Cyprinus carpio* exposed to the natural estrogen, 17β-estradiol, or the environmental pseudoestrogen, 4-tert-pentylphenol (TPP) (Gimeno et al. 1998a), while adults exposed to TPP also showed reduced spermatozoa and spermatogenic cysts, reduced diameters of seminiferous lobules, lobular disorganization and fibrosis, and vacuolation and atrophy of the germinal epithelium (Gimeno et al. 1998b). In fertilized medaka eggs microinjected with ethinyl estradiol (Papoulias et al. 1999), testes showed reduced spermiogenesis and less advanced stages of maturity. Mature fathead minnows exposed to 17β-estradiol also exhibited loss of germinal cells, necrosis and mineralization of spermatozoa, and germ cell syncytia (Miles-Richardson et al. 1999). Accompanying these changes were hyperplasia and hypertrophy of Sertoli cells that had phagocytized the degenerate germ cells. However, caging studies with adult fathead minnows exposed to municipal wastewater treatment effluent (Nichols et al. 1999; see below) failed to demonstrate any testicular lesions of the types mentioned above that could be associated with exposure to environmental estrogens.

Male Hermaphroditism, Intersex Condition, Male Feminization

Another potential biomarker in endocrine disruptor studies is referred to as male hermaphroditism, intersex condition, or male feminization, all represented by the presence of scattered nests of ovarian follicles in the testis, usually as primary or secondary oocytes. This condition has been induced experimentally in juvenile and adult male fish of several species by exposure to β-hexachlorocyclohexane (Wester and Canton 1986), 17β-estradiol (Gimeno et al. 1998a; Koger et al. 2000), TPP (Gimeno et al. 1998b), and the xenoestrogen, nonylphenol (Gray and Metcalfe 1997). Posthatch exposure to TPP also induced oviduct development in young male carp (Gimeno et al.1998a). In field studies, the presence of ovarian tissue (usually primary oocytes) in testis, together with elevated levels of plasma vitellogenin in males, has also been observed and associated with exposure to municipal wastewater effluent containing environmental estrogens. The first studies showed high prevalences (up to 100%) of intersexuality of male roach *Rutilus rutilus* in several rivers of the United Kingdom, with claimed association with discharges from sewage treatment works known to contain estrogens and xenoestrogens (Jobling et al. 1998). Intersexuality also occurred in up to 17% of male European flounder from certain estuaries in the United Kingdom (Allen et al. 1999; Simpson et al. 2000a), in 8% of male European flounder from the highly contaminated Seine Bay (Minier et al. 2000), and in marbled flounder *Pleuronectes yokahamae* from Haneda in Tokyo Bay (Hashimoto et al. 2000). Intersex condition was also found in 2 of 7 male shovelnose sturgeon *Scaphirhynchus platorynchus* from the Mississippi River and was associated with exposure to chlordanes, PCBs, and DDEs (Harshbarger et al. 2000). The sensitivity of this condition as a biomarker of environmental estrogen exposure and effect in wild fish remains to be definitively established; caging studies in which adult fathead minnows were exposed for three weeks to effluent from seven different wastewater treatment plants in Michigan known to contain xenoestrogens (Nichols et al. 1999), failed to demonstrate any testicular or ovarian lesions attributable to endocrine disruptor exposure. Additional information related to the effects of environmental stressors on reproductive endpoints can be found in Chapter 9.

Gill

The teleost gill functions in respiration, ionic homeostasis, and excretion of nitrogenous wastes, and is in continuous exposure to the surrounding water. The gill is, therefore, a sensitive target for a variety of irritants and toxicants, including pH (McDonald 1983) and temperature extremes; transition (Baker 1969) and heavy metals (Hughes et al 1979; Verbost et al. 1987; Randi et al. 1996), including mercury (Pereira 1988); tributyltins (Grinwis et al. 1998); chlorine dioxide (Yonkos et al. 2000); detergents (Abel 1976); suspended solids; ammonia (Klontz et al. 1985; Klontz, 1993); nutritional imbalances (Ferguson 1989); whole crude oil and its water soluble fractions (Solangi and Overstreet 1982); organic pesticides; and numerous other chemical contaminants (see reviews by Mallatt 1985 and Eller 1975 for specific references). In a comprehensive review and statistical analysis of gill lesion types occurring under conditions of exposure to a number of irritants and toxicants, Mallatt (1985) made the overall conclusion that most gill lesions are stereotyped, nonspecific reactions and physiological responses to injury caused by multiple stressors. Overall, none are pathognomonic for unique stressors, and gill lesions are, therefore, difficult, if not impossible, to link to specific causes. In these authors' and others' (e.g., Vogelbein and Zwerner 2000) experience, this generalization is also true for gill lesions detected in field studies, where the issue of ongoing or past infections or infestations with pathogens or parasites further clouds the issue of lesion etiology.

If gills are collected, it is important that they consistently be collected first, within 1 min after sacrifice and preferably in neutral buffered formalin or especially Bouin's fixative (Speare and Ferguson 1989) to prevent formation of postmortem artifacts that can mimic toxicant-induced pathological lesions (Walsh and Ribelin 1975; Ferguson 1989; Speare and Ferguson 1989). A synthesis of data from Speare and Ferguson (1989) suggests that rapid fixation in Bouin's fixative following sacrifice produces the fewest postmortem artifactual changes as compared with neutral buffered formalin.

In general, the temporal histogenesis of gill lesions due to environmental stressors is probably quite similar to that described for "environmental gill disease syndrome" in cultured salmonids (Klontz 1993) and reinforced by detailed descriptions of acute and chronic gill lesions in salmonids by Ferguson (1989). Initial acute lesions are hypertrophy of the lamellar respiratory epithelium, followed by apparent lifting or separation of the epithelium from the underlying basal lamina and capillaries, with the resultant space becoming filled with an edematous, serous exudate and mononuclear inflammatory cells. The respiratory epithelium may undergo frank necrosis. Formation of lamellar synechiae or junctions between respiratory epithelial cells in adjacent lamellae may then occur, resulting in fusion of adjacent lamellae. These synechiae are common in toxicity with heavy and transition metals such as cadmium, mercury, and copper (Ferguson 1989). Other acute responses to injury are lamellar aneurysms or telangiectasia, due to rupture of supportive pillar cells between lamellar capillaries and pooling of blood. These vascular lesions can also simply be an artifactual result of trauma during capture or necropsy and should be used very cautiously as a biomarker (Speare and Ferguson 1989). A biomarker specific for exposure to chemicals whose metabolism is mediated by the P450-IA isozyme of the mixed function oxidase system is localization of its expression in the pillar or endothelial cells of the branchial vasculature (Van Veld et al. 1997). However, this is strictly a biomarker of exposure and not of effect.

Chronic responses to injury in the gill are also nonspecific reactions to multiple stressors including infectious agents. These are respiratory epithelial hyperplasia, compensatory to respiratory epithelial necrosis occurring at the lamellar tips or at the lamellar bases in the interlamellar region, where proliferation of the respiratory epithelium typically occurs in the normal cell turnover process (Ferguson 1989). This thickening of the respiratory epithelium reduces respiratory efficiency. In severe cases, hyperplasia results in fusion of adjacent lamellae with a further reduction in surface area available for respiration. Hyperplasia of chloride cells may also occur along the entire length of the lamella under altered external conditions such as increased water acidification (Leino and McCormick 1984). This is a compensatory response to ion loss and may represent a biomarker of adaptation to ionoregulatory stress (Hinton et al. 1992). Mucous cell hyperplasia is also a common chronic response to injury by a number of toxicants, irritants, and infectious agents, and typically co-occurs with respiratory epithelial hyperplasia. A common chronic sequela to lamellar telangiectasis or microaneuysms is the formation of thrombi and associated fibrosis; the presence of these types of conditions in association with the more acute lesions suggests a pathological process rather than an artifact.

In summary, the gill is an organ highly sensitive to a number of environmental stressors, but the suite of pathological responses to injury are quite stereotyped and nonspecific, making it very difficult in field studies to attribute alterations to specific etiologies. Several lesion types (epithelial lifting or epithelial capillary separation, epithelial hypertrophy, and lamellar telangiectasis) are common artifacts of delayed fixation or method of killing and, therefore, should be used very cautiously as field biomarkers of environmental stress.

Skin

As the main interface between the fish and its external environment, the skin and oral surfaces are susceptible to injury by numerous environmental stressors and exhibit lesions that have some potential as biomarkers. Because the entire depth of the fish epidermis consists of living cells with mitotic capacity, it is a frequent target of pathogens, irritants, and toxicants. Certain skin lesions have been classified as "pollution-associated diseases" (Sindermann 1990). However, the primary etiology of many of these skin diseases is extremely difficult to establish because of the presence of bacteria, viruses, and parasites associated with these lesions. The major complicating factor in determining the causality of an environmental stressor for skin lesions is the inability to rule out the role of infectious organisms in their etiology, whether as primary agents or secondary invaders of previously damaged epidermis, dermis, and underlying tissues. This remains a major obstacle in their application as biomarkers of specific environmental stressors in field studies. In our experience, it is sufficient to sample only grossly visible anomalies that are clearly not the result of collection trauma such as net damage. The following are a number of skin lesions that should be considered as potential biomarkers.

Skin Ulcers

Skin ulcers are common in a wide variety of fish species (Noga 2000), including European flounder (reviewed in Wiklund and Bylund 1993; Wiklund 1994), winter flounder (Levin et al. 1972), red hake *Urophycis chuss*

(Murchelano and Ziskowski 1979), Atlantic cod *Gadus morhua* (Dethlefsen 1980), Pacific cod *Gadus macrocephalus* (McCain et al. 1979), and Atlantic menhaden *Brevoortia tyrannus* and other species from U.S. Atlantic coast estuaries (e.g., Noga et al. 1996; Blazer et al. 1999; Noga 2000). Research on most of these enzootic and epizootic ulcerative conditions of the skin have implicated infectious organisms such as *Vibrio anguillarum*, *Aeromonas salmonicida* (Wiklund and Bylund 1993; Wiklund 1994), viruses (Jensen and Larsen 1982), and pathogenic fungi such as *Aphanomyces* spp. from skin ulcers (ulcerative mycosis) in menhaden (Blazer et al. 1999), Asian tropical fish (Willoughby et al. 1995; Vishwanath et al. 1998), and other members of the Oomycetes (reviewed in Dykstra and Kane 2000) as the *proximate* etiologic agents for these chronic, ulcerative conditions. These lesions may develop secondary to earlier acute injury to the epidermis by a multitude of stressors that can be considered as the *ultimate* cause of these lesions (Noga 2000). These stressors include fishing activities; spawning activity (e.g., as proposed in European flounder ulcers by Wiklund 1994) or other trauma; ectoparasitism; chemical contaminants; immunosuppression; stress through induction of stress hormones such as cortisol; drastic water temperature, pH, or salinity fluctuations; and hypoxia. Other adverse factors or stressors include exposure to blooms of the toxic dinoflagellate, *Pfiesteria piscicida*, possibly related to nutrient enrichment from municipal and agricultural sources (Noga et al. 1996; Noga 2000). However, there is still no convincing evidence that exposure to the *Pfiesteria* toxin is the primary initiating event in the development of this chronic, ulcerative mycotic condition in Atlantic menhaden in all or any particular epizootic event. Similarly, there is no convincing evidence directly associating ulcerative skin lesions in wild fish specifically to chemical contaminant exposure. As stated by Noga (2000), "The multifactorial pathways that operate at both the ecological and organismal levels, as well as the nonspecific response of the skin to insults make it very challenging to link epidemic skin ulcers to any single cause in natural aquatic populations. Consequently, using pathology to unequivocally identify the specific cause of a lesion (e.g., *Pfiesteria* exposure) is not a valid approach."

Epidermal Neoplasms

Several teleost species, primarily from chemically contaminated sites, are affected by papillomas and squamous cell carcinomas of the epidermis and oral mucosa. Although viruses have been implicated in the etiology of some of these tumors (McAllister et al. 1977), there is considerable evidence linking their occurrence to exposure to sediment contaminants (Harshbarger and Clark 1990).

Brown bullhead from the Great Lakes area exhibit oral and epidermal lesions ranging from hyperplastic plaques to papillomas to invasive squamous cell carcinomas (Smith et al. 1989a; Baumann et al. 1996). Fish 4 years of age from a severely PAH-contaminated site in the Black River were affected by up to 32% for lip papillomas and 18% for skin papillomas, while prevalences at a reference site were 0% for skin neoplasms and 1.5% for oral papillomas (Baumann et al. 1987). Later studies in the contaminated Hamilton Harbor in Lake Ontario and a reference site on Lake Erie (Smith et al. 1989a) showed prevalences of 55% for oral or skin papillomas at Hamilton Harbor as compared with 15% at the reference site. As reviewed in Baumann et al. (1996), although there is some experimental evidence to support a contaminant etiology for these skin and oral neoplasms in brown bullhead (Black et al. 1985) and no viruses have been identified in these tumors, a viral role in their etiology cannot be ruled out. A multifactorial etiology for these tumors is likely, which includes the involvement of chemical contaminants (Baumann et al. 1996).

A high prevalence (over 70%) of oral papillomas have also been observed in black bullhead *Ameiurus melas* from a chlorinated sewage treatment pond in Alabama (Grizzle et al. 1984). Bullhead caged in this pond also developed papillomas and exhibited elevated AHH activities in liver as compared with reference fish, indicating induction of CYP1A (Tan et al. 1981). Following a 90% reduction in residual chlorine in this pond over about 4 years, the prevalence of this condition dropped to 23%, suggesting that exposure to chlorine compounds was a factor in the etiology of these tumors.

White suckers *Catostomus commersoni* from the Great Lakes region also exhibit similar oral and skin hyperplastic and neoplastic lesions (Smith et al. 1989a, 1989b; Premdas et al. 1995). Their association with chemical exposure is not at all clear, since relatively high prevalences are detected in suckers from reference sites, and the prevalence data are difficult to interpret due to lack of data comparability among studies and the unknown migratory patterns of white suckers (Baumann et al. 1996). Experiments have shown that a high proportion of these tumors can regress and new tumors do not develop in fish held in "uncrowded" conditions, and few tumors regress while new tumors develop in previously unaffected fish held in "crowded" conditions (Smith and Zajdlik 1987; Premdas and Metcalfe 1994). These findings suggested an infectious etiology, and injection of cell-free filtrates from papilloma tissues into the lip induced papillomas in 50% of the fish (Premdas and Metcalfe 1996). However, horizontal transmission experiments were inconclusive, and suckers injected with a mixture of PCBs, organochlorine pesticides, polychlorinated dibenzodioxins, and dibenzofurans extracted from Lake Ontario rainbow trout muscle failed to develop tumors at the injection site. Although these studies are not definitive, chemicals are clearly not the primary etiological agent for lip papillomas in white suckers, and a primary viral etiology for these tumors is highly likely (Premdas and Metcalfe 1996). These tumors generally occur at higher prevalences in suckers from contaminated habitats, and, therefore, one cannot eliminate chemical contaminants and other environmental stressors as indirectly acting factors in this multifactorial disease (Premdas and Metcalfe 1996).

Epidermal neoplasms in the above species have been detected at relatively high prevalences and are probably related, in part, to chemical contaminant exposure. However, in our experience with other fish species examined in multiple, large field surveys such as NOAA's Status and Trends Program, and EPA's EMAP (Fournie et al. 1996b), the occurrence of neoplasms of the skin and oral mucosa are extremely rare. Therefore, their use as biomarkers in field applications is limited and highly dependent on the target species selected.

Chromatophoromas and Pigment Cell Hyperplasia

Dermal pigment cell neoplasms have been reported in multiple species, and associations with exposure to chemical contaminants in the environment have been claimed, with verification by laboratory exposure experiments (reviewed in Kimura et al. 1989; Kinae et al. 1990). In field studies in Japan, chromatophoromas occur in two species of croaker (*Nibea mitsukurii* and *Nibea albiflora*) at prevalences as high as 50% at sites receiving effluents from a kraft process pulp and paper mill and lowest prevalences at several reference sites. A related lesion, skin melanosis, or pigment cell hyperplasia, was also reported in up to 13% of sea catfish *Plotosus anguillaris* near zones of discharge of pulp and paper effluent. Extracts of these effluents contained compounds that were mutagenic in the Ames and Rec tests, including chloroacetones, tetrachlorocyclopentene-1,3-dione, and alpha-dicarbonyl compounds. Validation of the field findings in croaker was achieved by chromatophoroma induction in exposure to carcinogens such as 7,12-dimethylbenz[a]anthracene, N-methyl-N'-nitro-N-nitrosoguanidine, and nifurpirinol, as well as by exposure to 10% dilutions of two types of kraft mill effluents. In a similar exposure of sea catfish, pigment cell hyperplasia was induced in 70–100% of the exposed fish. Field data supporting a chemical etiology for these lesions comes from reductions in chromatophoroma prevalence in croaker at the most contaminated site, from 47% in 1973–1983 to approximately 20% in 1984–1987. This reduction followed installation of a waste water purification system at the pulp and paper mill and removal of contaminated sediments nearby (Kimura et al. 1989; Kinae et al. 1990). Because of these relatively convincing data, pigment cell neoplasms should be considered a potential biomarker of chemical contaminant exposure.

Musculoskeletal System

Gross deformities of the spine, including lordosis, scoliosis, and kyphosis are common in native fish species (Valentine 1975; Slooff 1982; Baumann and Hamilton 1984; Ferguson 1989). Other common anomalies of the fish skeletal system include vertebral fusion and dysplasia (Middaugh et al. 1990), opercular, jaw, and skull deformities (Slooff 1982; Lindesjoo and Thulin 1992, 1994), and fin erosion (Murchelano 1975; Wellings et al. 1976; Lindesjoo and

Thulin 1990; reviewed in Bodammer 2000). Chemical toxicants known to produce skeletal anomalies in fish include zinc, cadmium, lead, mercury, and organic compounds such as formalin, toxaphene, malathion, parathion, dylox, guthion, dursban, akton, methylparathion, trichlorphon, phosalone, demeton, Kepone, DDTs, PCBs, and TCDD (specific references cited in Slooff 1982). Exposure to complex effluents from a pulp mill and an ore smelter also cause vertebral and other skeletal abnormalities in fish (Bengtsson and Larsson 1986; Bengtsson et al. 1988a, 1988b; Mayer et al. 1988), as do creosote (Vines et al. 2000), low DO (Bengtsson 1979), genetic factors, unsuitable temperature, salinity fluctuations, ionizing radiation, dietary deficiencies, electric shock, parasites, and mechanical injuries (Slooff 1982). More frequently observed examples of these anomalies as potential biomarkers of exposure to environmental stressors are listed below.

Vertebral Deformities, Fractures

Two mechanisms for vertebral anomalies resulting in lordosis, scoliosis, kyphosis, and vertebral fractures have been proposed for these types of gross lesions. The first relates to vitamin C deficiencies, alterations in vitamin C metabolism, and vitamin C-dependent production of hydroxyproline and consequent effects on collagen matrix formation, or alterations in mineralization of bone. The second mechanism is muscular tetany (Hinton 1993). In the first, the causes are multifactorial, but include increased water temperature, low DO, simple vitamin C or B12 deficiency, radiation, heavy metals, phthalates, and kraft pulp mill effluent (Bengtsson 1974, 1975; Mayer et al. 1977; Hinton 1993), and organochlorine compounds like toxaphene, Kepone, mirex, and PCBs (Mehrle and Mayer 1975; Couch et al. 1977; Mayer et al. 1977). The backbone of fathead minnow chronically exposed to toxaphene showed decrease collagen deposition and hydroxyproline and an increase in calcium, resulting in a demonstrable increase in bone fragility (Mehrle and Mayer 1975). Cadmium has also been shown to cause vitamin C depletion in fish by increasing its utilization or decreasing its absorption (Thomas et al. 1982). Decreased bone mineralization by calcium and subsequent skeletal weakening can also be caused by cadmium-induced inhibition of calcium uptake across the gills and reduced availability for deposition in bone (Roch and Maly 1979; Reid and McDonald 1988), and displacement of calcium from binding sites by exposure to cadmium (Muramoto 1981), lead (Varanasi and Gmur 1978), and zinc (Sauer and Watabe 1984).

In tetany-mediated vertebral lesions, there are also many toxicant-related causes shown experimentally, including heavy metals, organochlorine pesticides such as toxaphene (Mehrle and Mayer 1975; Mayer et al. 1978) and Kepone or chlordecone (Couch et al. 1977), trifluralin (Couch et al. 1979), and organophosphate pesticides such as parathion and malathion (Weis and Weis 1976b, 1989; Kumar and Ansari 1984). Organophosphates appear to cause vertebral fractures by inhibition of acetylcholinesterase, causing increases in acetylcholine at nerve endings leading to muscular tetany (Ferguson 1989). Organochlorine pesticides appear to cause both muscular tetany and vitamin C depletion (Hinton 1993). Other causes of muscular tetany include electric shock, water temperature fluctuations, and parasitic infections (see Hinton 1993 for specific references). Gross lesions are best confirmed by histopathology to exclude parasitic infections and to visualize the microscopic appearance of bone matrix and bone articulations.

Vertebral Fusion, Dysplasia

This unique lesion has been induced by the aniline herbicide, trifluralin (Couch et al. 1979; Wells and Cowan 1982) and validated in field observations in brown trout *Salmo trutta* from a stream in which trifluralin was spilled (Wells and Cowan 1982). Similar vertebral lesions in juvenile inland silversides *Menidia beryllina* were induced by the organophosphorus pesticide, terbufos (Hemmer et al. 1990; Middaugh et al. 1990). Bone lesions ranged from small hyperostoses to nearly complete fusion of adjacent vertebrae and were characterized by exuberant bone growth with bone hypertrophy and proliferation of fibroblasts, osteoblasts, and osteocytes. The mechanism proposed was alteration of calcium metabolism in the corpuscles of Stannius and ultimobranchial glands (Couch et al. 1979), leading to altered bone deposition and stimulation of osteogenic cells (Middaugh et al. 1990). It is important to note that these vertebral anomalies were not visible grossly; analysis of X-ray radiographs and parallel histopathological specimens was necessary to detect these lesions.

Skull and Jaw Deformities

A skull or jaw abnormality in northern pike *Esox lucius*, characterized by distinct upward bending of the jaws, has been associated with bleached kraft pulp mill effluent exposure (Thulin et al. 1988; Lindesjoo and Thulin 1992); this association is strengthened by the reduction in its prevalence from 35% to 0% four years after improvement of the pulp mill effluent treatment. However, oxygen deficiency arising from effluent discharged into the receiving waters could not be discounted as a causative factor. Craniofacial malformations in zebra danio *Danio rerio* and rainbow trout larvae exposed in the laboratory to retene (associated with pulp and paper mill effluents) have also been demonstrated (Billiard et al. 1999).

Fin Erosion

This lesion complex has been commonly used in the past as a biomarker of general environmental degradation in bottom-dwelling species and is best described as a syndrome affecting skin, subcutaneous tissues, and the fin rays (reviewed in Bodammer 2000). Its etiology is probably multifactorial as a generalized reaction of the skin, dermis, and fin rays to a wide variety of toxicants and irritants, including low DO (Bodammer 2000). However, fin erosion has been associated with chemical contaminants (Wellings et al. 1976; Sherwood and Mearns 1977; Murchelano and Ziskowski 1982; Ziskowski et al. 1987; McCain et al. 1988; Overstreet 1988) including sewer outfalls (Cross 1985), bleached kraft pulp mill effluent (Lindesjoo and Thulin 1990; Khan et al. 1992; Sharples et al. 1994; Sharples and Evans 1996), and crude oil (Minchew and Yarbrough 1977). This chronic lesion is characterized by a complex set of lesions ranging from epidermal and mucous cell hyperplasia; epidermal cell necrosis; epidermal spongiosis; increased epidermal eosinophilic granular cells, subcutaneous fibrosis, fibroplasia, hyperemia, and inflammatory cell infiltrates; increases in dermal and epidermal melanophores; and fin ray necrosis, retraction, resorption, and occasional fusion (Murchelano 1975; Wellings et al. 1976; Bodammer 2000). A chemical contaminant-related etiology is suggested by reductions in prevalence following cessation of the bleaching process in Swedish pulp mills (Lindesjoo and Thulin 1994) and reduction of chemical releases in sewage outfalls, such as in Boston Harbor (Moore et al. 1996). However, the available evidence linking fin erosion to chemical contaminants is mainly circumstantial, and the highest prevalence of fin erosion found in one field survey of winter flounder was in the relatively uncontaminated area of the Gulf of Maine (Ziskowski et al. 1987). In our experience, fin erosion is not a particularly useful biomarker of chemical contaminant exposure because of these confounding factors and its lack of specificity.

Central and Peripheral Nervous Systems, Neurosensory Organs

Field studies in which components of the nervous system have been carefully examined in fish are extremely rare, which is surprising considering the high metabolic activity and consequent sensitivity of the central and peripheral nervous systems in fish to toxicants and the large body of information on pathological aspects of neurotoxicity of industrial and agricultural chemicals in mammals, including experimental models. The component cells of most interest toxicologically are neurons, glial cells (oligodedrogial cells, astrocytes, Schwann cells), and endothelial cells of the vasculature (Hinton 1993). Based on laboratory studies and a few field studies, lesions affecting these components have been observed in fish after exposure to toxicants and other stressors and could be useful as field biomarkers if assessed more frequently (Hinton 1993; Simpson et al. 2000b). Several examples of field and laboratory studies in which neuropathological effects have been documented in fish or amphibians follow.

Fish larvae from the Salton Sea, California, grossly exhibited reduced brain or eye development, with histological lesions including pyknosis of individual neurons, necrotic foci, and severe degeneration of all neural tissue (Matsui et al. 1992); lesions were circumstantially associated with undefined environmental contaminants. Retinal and brain cell necrosis also occurred in larval Pacific herring *Clupea pallasi* exposed to Prudhoe Bay crude oil (Marty et al. 1997), as compared with controls. In a review of experimental studies on the neurotoxicity of rubber tire leachate in larval sheepshead minnow *Cyprinodon variegatus* (Evans 1997), a number of histopathological effects were described, including retinopathy, vacuolated encephalopathy, and brain necrosis. However, these studies were not able to isolate the actual chemicals responsible for the neurotoxicity. Tadpoles *Physalaemus biligonigerus* exposed to the neurotoxic pyrethroid insecticide, cypermethrin, also showed massive cell death of neural cells in the developing telencephalon, probably representing apoptosis (Izaguirre et al. 2000), along with edema. Trimethyltin exposure also caused neuronal degeneration and necrosis in the telencephalon, diencephalon, and retina of urodele larvae and adults (*Triturus carnifex*) (Gozzo et al. 1994), which were linked to impaired swimming activity. Lake trout *Salvelinus namaycush* embryos and larvae exposed as fertilized eggs to TCDD exhibited degeneration and necrosis of the neuroepithelium of brain, spinal cord, and retina (Spitsbergen et al. 1991). Edema of the neuropile and hemorrhage and congestion of the meninges and brain parenchyma accompanied these lesions. These lesions were interpreted as sequelae of major circulatory derangements and hypoxia. Neuropathological effects of aluminum have also been observed in rainbow trout (Exley 1996). Histochemically localized aluminum in and around the cerebrovascular endothelium was associated with neuronal cell body membranes and with several distinct types of neuronal cellular debris, interpreted as neuronal degeneration. Extracellular aluminum deposits surrounded by an apparent proteinaceous matrix were most common. This study provides the first evidence that chronic environmental exposure to aluminum was associated with distinct degenerative neuropathology. Lesions were similar to those in human dementia cases and the amyloid deposition and senile plaques seen in Alzheimer's disease.

Lesions associated with chemosensory organs and tissues show that neurosensory systems are vulnerable to water pollutants, especially metals. These organs have not been extensively investigated in fish but deserve more attention considering that lesions in any of these organs could lead to marked effects on behaviors such as migratory homing, prey capture, swimming performance, and predator avoidance, with potentially serious population-level impacts. Early studies on the lateral line and olfactory organs (Gardner and LaRoche 1973; Gardner 1975) show that exposure of mummichog to copper, mercury, and silver produced degeneration and necrosis of the neurosensory cells of the lateral line and olfactory organ, while exposure to the pesticide methoxychlor induced only necrosis of the epithelium lining the lateral line organ. Studies on chronic copper exposure in rainbow trout embryos and alevins showed sequential changes in the olfactory epithelium; the initial lesion was mucous cell hyperplasia, followed by various stages of degeneration, necrosis, and eventual exfoliation of the olfactory epithelium and associated cells in the rosette (Saucier et al. 1991). These lesions were partially reversible after cessation of exposure, owing to the strong regenerative capacity of olfactory epithelium in fish (Stewart and Brunjes 1990). In the Atlantic silversides *Menidia menidia* (Gardner 1975) and inland silverside *M. beryllina* (Solangi and Overstreet 1982), exposure to crude oil and its saltwater soluble and insoluble fractions produced necrosis of the mucosal and sustentacular cells of the olfactory lamellae, and hyperplasia of the sustentacular cells. Crude oil exposure in the hogchoker *Trinectes maculatus* produced only necrosis of neurosensory and sustentacular cells of the olfactory organ (Solangi and Overstreet 1982). Fathead minnow exposed to the water-soluble fraction of aviation fuel also showed necrosis of the olfactory rosette (Latendresse and Fisher 1983). Cadmium exposure also causes necrosis of the olfactory epithelium (Stromberg et al. 1983), and ionic zinc also causes degeneration of the olfactory receptor cells (Cancalon 1982). Experimental studies in chinook salmon *Oncorhynchus tshawytscha* and rainbow trout exposed to copper documented loss of specific olfactory receptor cells, probably via cellular necrosis and apoptosis. These morphologic changes were strongly associated with neurophysiological impairment (electroencephalogram response to serine) that could lead to reduced survival and reproductive potential (Hansen et al. 1999). Other stressors causing various pathological effects on neurosensory tissue include detergents and low or high pH (reviewed in Klaprat et al. 1992).

Studies on pathological responses in the fish eye to environmental stressors are also relatively rare, especially field studies. Lesions of the cornea (ulcerations, edema, corneal epithelial loss and hyperplasia, keratitis), and lens (hydropic swelling, lysis and regeneration of lens fibers, lens epithelial hyperplasia, cataracts) as pathological responses to toxicants and other stressors are common but lack specificity (Wilcock and Dukes 1989). Cataract is the most common lesion in the fish eye that, unfortunately, has multiple etiologies, thereby, limiting its value as a biomarker. These causes include dietary imbalances, environmental pollutants, excessive sunlight, cold temperatures, and digenetic trematode metacercariae *Diplostomum* spp. The pathologic response in the lens to a variety of stressors follows a stereotyped sequence of vacuolization, fiber lysis, and unsuccessful fiber regeneration (Wilcock and Dukes 1989). Retinal lesions in fish have also received little attention but can be manifest as retinal pigment epithelial hypertrophy, hyperplasia, fibrous metaplasia, and phagocytic activity, while the neural retina itself may undergo necrosis associated with metazoan parasites (Wilcock and Dukes 1989). Several laboratory studies have examined pathological responses to toxicant exposure in the eye and especially the retina. Rainbow trout alevins exposed to benzo(a)pyrene exhibited microphthalmia, optic fissures, reduced mitotic rates in the sensory retina, retinal folding, and poor retinal differentiation (Hose et al. 1984). Lake trout embryos and larvae exposed as fertilized eggs to TCDD also exhibited retinal cell degeneration and necrosis (Spitsbergen et al. 1991). Neoplasms originating from cells that correspond to primitive medullary epithelium (medulloepitheliomas) have been induced in medaka by exposure to methylazoxymethanol acetate (Hawkins et al. 1986); but, ocular tumors in wild fish are rare (Fournie and Overstreet 1985) and are not likely to be useful histological biomarkers. Laboratory studies with embryonic medaka (Hamm et al. 1998) exposed to the organophosphate pesticides, diazinon and diisopropylphosphorofluoridate, showed delayed differentiation of the retinal cell layers and foci of necrosis in the inner nuclear and ganglion cell layers of the retina. Inhibition of acetylcholinesterase was also demonstrated histochemically to be spatially linked to the necrotic foci. Grossly detected eye malformations (microphthalmia and anophthalmia) have been induced in fish embryos by another organophosphate pesticide, malathion (Weis and Weis 1976a), and methylmercury exposure produced retinal cell necrosis in medaka (Dial 1978). Overall, based on the number of lesions in the eye, and especially the retina, that have been induced experimentally by chemical toxicants, much more attention should be paid to its examination in field studies.

Histopathological Biomarkers in Marine Mammals

Although it is far beyond the scope of this chapter to discuss this topic in any detail, the use of histopathological biomarkers in marine mammals is attracting increased attention, and several exemplary case studies exist. Such biomarkers have been difficult to establish in marine mammals because of the typically long interval between death and necropsy in stranded animals and subsequent inability to get quality specimens for histology. The most reliable information on this topic comes from studies of animals killed and quickly necropsied in sanctioned hunts or in studies where moribund animals are captured and brought to a rehabilitation facility for observation and prompt collection of tissues if death occurs or the animals are euthanatized.

Multiple types of neoplasms, nonneoplastic lesions, and infectious diseases have been observed in diverse organs in multiple studies of beluga whales *Delphinapterus leucas* from the St. Lawrence Waterway since the early 1980s (Martineau et al. 1988; Girard et al. 1991; Beland et al. 1993; De Guise et al. 1994a, 1994b, 1995a, 1995b; Lair et al. 1997, 1998). Occurrence of these lesions has been circumstantially associated with massive tissue levels of PCBs, other lipophilic chlorinated hydrocarbons like DDTs, and also with exposure to PAHs. Adducts of BaP-DNA have been detected in brains and livers of belugas from this region (Beland et al. 1993), at levels associated with carcinogenesis in laboratory animals. As of 1994, of the 75 individual cases of tumors reported worldwide in cetaceans, 37% of these come from a small (~500) population of beluga whales from the St. Lawrence Waterway (De Guise et al. 1994). The proposed mechanism of tumor induction could be direct, via exposure to carcinogenic initiators and promoters, or indirect through suppressed immunosurveillance of tumor cells (De Guise et al. 1995). Immunosuppression by exposure to environmental contaminants has also been proposed as the mechanism underlying the increased prevalence of infectious diseases in these belugas (De Guise et al. 1995).

A more recent case study clearly documents neuropathology in the brain hippocampus from California sea lions *Zalophus californianus* exhibiting neurological seizures as a result of acute domoic acid poisoning, which could be linked unambiguously to a toxic diatom bloom (*Pseudo-nitzchia australis*) along the central California coast (Scholin et al. 2000). This opportunistic study clearly showed trophic transfer of domoic acid from diatoms to planktivorous northern anchovy *Engraulis mordax* to sea lions, and the initial algal bloom was linked to increased nutrient levels in the seawater. Brain lesions were characterized by zonal vacuolation and necrosis of the anterior ventral hippocampal neuropile. This lesion is identical to those observed in mice, macaques, and humans exposed to domoic acid either accidentally or in experimental studies.

Summary

Histopathology is an extremely useful tool for assessing effects of exposure to stressors at the level of the individual. Even though the histopathological approach is somewhat qualitative, it is very valuable because the observed lesions represent an integration of cumulative effects of biochemical and physiological changes, as well as representing actual injury to the organism. This approach also allows identification of specific cells, tissues, and organs that have been affected. We have presented information in this chapter on a number of lesions in various tissues and organs that may at some point in time have the potential as reliable histopathological biomarkers of specific environmental stressors, but require further validation by epizootiological and laboratory studies. Examples include lesions of the excretory kidney, gonad, gills, central nervous system, neurosensory organs, and hemopoietic tissues in head kidney. In our opinion, there are only a small number of histopathological biomarkers that have been shown via rigorous epizootiological methods and experimental validation to be reliable indicators of exposure to various stressors. These include liver lesions, splenic macrophage aggregates, some skin lesions, and certain musculoskeletal abnormalities. Some are relatively specific, while others are more nonspecific, and all require the investigator to account for seasonal, physiological, age-, and sex-related variation. For example, an epizootic of liver neoplasms and other liver lesions involved in the stepwise histogenesis of liver neoplasia in a population of fish indicates that they have been exposed to hepatotoxic and hepatocarcinogenic contaminants. An elevated number of splenic macrophage aggregates is a less specific indicator, but still indicates that the affected fish probably have been exposed to contaminated sediments, low DO, or

Table 1. Summary of selected histopathologic biomarkers of environmental stressors in various organs of fish.

Organ and lesion	Specificity to unique stressors	Assessed in field studies?	Experimentally induced?	Advantages, limitations, other comments
Liver				
1) Hepatic neoplasms; foci of cellular alteration	High; PAHs, CHs	Yes; numerous studies, species	Yes; numerous studies, species	Specific for chemical exposure; many species resistant; tend to occur at highly contaminated sites; age-related
2) Hepatocellular nuclear pleomorphism; hepatic megalocytosis; aniso-karyosis	Moderate; PAHs, CHs; also plant and algal toxins	Yes, numerous studies, species	Yes; numerous studies, species	Specific for toxicant exposure in some species (English sole); many species resistant; may be normal in species with high degree of hepatocellular polyploidy
3) Hydropic vacuolation of biliary epithelial cells, hepatocytes	Moderate; PAHs, CHs, chlordanes, DDTs, possibly metals such as tributyltins	Yes, numerous studies, species	No	High prevalences in winter and starry flounders, white croaker; associated with neoplasia; other species resistant; prevalences not correlated with contaminant exposure (e.g., rock sole); age-related
Excretory kidney				
1) Tubular epithelial degeneration, necrosis, vacuolation, hyalini-	Low; contaminants of many classes; also infectious etiologies; nonspecific	Yes; few studies, species; no associations with contaminant exposure	Yes; many studies, species mainly with metals, pesticides, herbicides	Prevalences in field studies low, no consistent associations with con-

Organ and lesion	Specificity to unique stressors	Assessed in field studies?	Experimentally induced?	Advantages, limitations, other comments
zation				taminants; easily confused with autolytic changes
Spleen				
1) Macrophage aggregates	Low; associated with many stressors, from chemicals, to temperature, age, nutrition, past infections, sexual maturation stage	Yes; many studies, species, with relative degrees of success	Yes; several studies, species	Highly sensitive to stressors, changes are persistent, easily measured; very nonspecific; must control for fish age, can use as screening tool
Ovary				
1) Oocyte atresia	Low; associated with many stressors, from chemicals, to temperature, salinity, DO, food availability, pH	Yes; many studies, species, correlates with many chemicals, including PAHs, pH, mercury	Yes, several studies, species; includes 17B-estradiol, ethinyl estradiol	Easily assessed, but very nonspecific; occurs normally in ovary, data should be expressed by ordinal severity or frequency/mm^2
Testis				
1) Germinal epithelial degeneration, necrosis, atrophy	Low; multiple chemicals, including Cd, PCBs, 17β-estradiol, 4-tertpentylphenol, ethinyl estradiol	No; single caging study in fish exposed to municipal wastewater effluent showed no effects	Yes; all studies inducing these effects experimental	High potential for population level effect; insufficient field application to be useful biomarker
2) Intersex condition; hermaphroditism; male feminization	Moderate; several chemicals, including 17β-estradiol, environmental estrogens like nonylphenol,	Yes; several studies, several species; associations with synthetic estrogens, surfactants, sewage effluents;	Yes; 17β-estradiol, environmental estrogens like nonylphenol, β-hexachlorocyclohexane,	High potential for population level effect; relatively specific effect of exposure to estrogenic

Organ and lesion	Specificity to unique stressors	Assessed in field studies?	Experimentally induced?	Advantages, limitations, other comments
	β- hexachloroxy-clohex-ane, 4-tert- pentylphenol	marine, estuarine and freshwater fishes	4-tert-pentylphenol	or anti-androgenic chemicals; must know normal maturation pattern of species
Gill				
1) Respiratory epithelial hyperplasia	Low; associated with multiple stressors, biotic, physical, and chemical	Yes; many studies and species, few consistent associated with chemicals	Yes; many studies, species, and stressors	Highly nonspecific response to multiple stressors; often response to external parasitism, past or current
Skin				
1) Skin ulcerations	Low; associated with multiple stressors, biotic, chemical and physical, including trauma, spawning activities	Yes; many studies and species; no consistent association with chemicals	Yes; few studies, most involve pathogens such as bacteria, fungi	May show high prevalences in field; almost impossible to separate proximate cause (e.g., infectious agents) from ultimate cause (e.g., skin irritation or necrosis by trauma, chemicals, poor water quality). Cause not determinable by histopathology alone
2) Epidermal, oral neoplasms	Moderate; mainly assoc. with chemical contaminants, (e.g., PAHs, CHs); viral etiology not ruled out	Yes; many studies in several species, mainly brown and black bullhead, white sucker	Yes; very few studies, data on chemical induction not consistent	Prevalences higher in areas with higher chemical contaminants (PAHs, CHs), but transmission studies strongly suggest a

Organ and lesion	Specificity to unique stressors	Assessed in field studies?	Experimentally induced?	Advantages, limitations, other comments
				primary viral etiology, but viruses not isolated; probable multifactorial etiology
Musculoskeletal system				
1) Vertebral deformities	Low; associated with multiple stressors; biotic, chemical (BKME, toxaphene) physical, including posttrauma; many lesions, different mechanisms	Yes; many studies, species; associated with many different chemical classes (e.g., BKME, pesticides, etc.)	Yes; few studies and species; mainly with toxaphene and trifluralin	Represents a number of different types of lesions including vertebral fractures, fusion, with different pathological mechanisms; other than vertebral fusion with trifluralin, "broken back" syndrome with toxaphene, lesions relatively nonspecific
Central nervous system				
1) Brain necrosis, vacuolated encephalopathy, neuronal and neuroepithelial degeneration, atrophy; similar degenerative, necrotic lesions in spinal cord	Unknown; too few field studies done to assess specificity; lab studies suggest some specificity	Very few; no strong association with particular stressors	Yes; few studies, few stressors (e.g., aluminum, rubber tire leachate, TCDD	Experimental studies in fish and mammals suggest this and other CNS lesions could be good biomarkers of neurotoxicant exposure; much more research necessary.

some other stressor. Because histopathology is a biologically meaningful method of evaluating the effects of stressors on animals, it should be an essential component of all environmental assessments. In order to assess the health status of a body of water, it is necessary to evaluate the health of the organisms inhabiting the particular body of water, which requires a histopathological examination of representative tissues and organs from those organisms.

Acknowledgments

The assistance of Sandie O'Neill of the Washington Department of Fish and Wildlife in the preparation of Figure 2 is gratefully acknowledged. This document represents manuscript #01–09 of the Environmental Conservation Division, Northwest Fisheries Science Center, NMFS/NOAA in Seattle, Washington, and contribution #1139 of the National Health and Environmental Effects Research Laboratory, U.S. Environmental Protection Agency, Gulf Breeze, Florida.

References

Abel, P. D. 1976. Toxic action of several lethal concentrations of an anionic detergent on the gills of the brown trout (*Salmo trutta* L.). Journal of Fish Biology 9:441–446.

Adams, S. M., M. S. Bevelhimer, M. S. Greeley Jr., D. A. Levine, and S. J. Teh. 1999. Ecological risk assessment in a large river-reservoir: bioindicators of fish population health. Environmental Toxicology and Chemistry 18:628–640.

Adams, S. M., A. M. Brown, and R. W. Goede. 1993. A quantitative health assessment index for rapid evaluation of fish condition in the field. Transactions of the American Fisheries Society 122:63–73.

Allen, Y., A. P. Scott, P. Mattheissen, S. Haworth, J. E. Thain, and S. Feist. 1999. Survey of estrogenic activity in United Kingdom estuarine and coastal waters and its effect on gonadal development of the flounder *Platichthys flesus*. Environmental Toxicology and Chemistry 18:1791–1800.

Anderson, R. J., H. A. Luu, D. Z. X. Chen, C. F. B. Holmes, M. L. Kent, M. Le Blanc, F. J. R. Taylor, and D. E. Williams. 1993. Chemical and biological evidence links microcystin-LR to salmon "netpen liver disease." Toxicon 31:1315–1323.

Anderson, S., A. Augier, W. W. Hauck, D. Oakes, W. Vandaele, and H. I. Weisburg. 1980. Statistical methods for comparative studies. Wiley, New York.

Arcand-Hoy, L. D., and C. D. Metcalfe. 1999. Biomarkers of exposure of brown bullheads (*Ameiurus nebulosus*) to contaminants in the lower Great Lakes, North America. Environmental Toxicology and Chemistry 18:740–749.

Arkoosh, M. R., E. Casillas, E. Clemons, A. N. Kagley, R. Olson, P. Reno, and J. E. Stein. 1998. Effect of pollution on fish diseases: potential impacts on salmonid populations. Journal of Aquatic Animal Health 10:182–190.

Atz, J. W. 1964. Intersexuality in fishes. Pages 145–323 *in* C. N. Armstrong and A. J. Marshall, editors. Intersexuality in vertebrates including man. Academic Press, New York.

Augspurger, T. P., R. L. Herman, J. T. Tanacredi, and J. S. Hatfield. 1994. Liver lesions in winter flounder (*Pseudopleuronectes americanus*) from Jamaica Bay, New York: indications of environmental degradation. Estuaries 17:172–180.

Baker, J. T. P. 1969. Histological and electron microscopical observations on copper poisoning in the winter flounder (*Pseudopleuronectes americanus*). Journal of the Fisheries Research Board of Canada 26:2785–2793.

Balch, G. C., C. D. Metcalfe, W. L. Reichert, and J. E. Stein. 1995. Biomarkers of exposure of brown bullheads to contaminants in Hamilton Harbor, Ontario. Pages 249–273 *in* F. M. Butterworth, editor. Biomonitors and biomarkers as indicators of environmental change. Plenum, New York.

Barron, M. G., M. J. Anderson, D. Cacela, J. Lipton, S. W. Teh, D. E. Hinton, J. T. Zelikoff, A. L. Dikkeboom, D. E. Tillitt, M. Holey, and N. Denslow. 2000. PCBs, liver lesions, and biomarker responses in adult walleye (*Stizostedium vitreum vitreum*) collected from Green Bay, Wisconsin. Journal of Great Lakes Research 26:250–271.

Baumann, P. C., and S. J. Hamilton. 1984. Vertebral abnormalities in white crappies, *Pomoxis annularis* Rafinesque, from Lake Decatur, Illinois, and an investigation of possible causes. Journal of Fish Biology 25:25–33.

Baumann, P. C., and J. C. Harshbarger. 1985. Frequencies of liver neoplasia in a feral fish population and associated carcinogens. Marine Environmental Research 17:324–327.

Baumann, P. C., and J. C. Harshbarger. 1995. Decline in liver neoplasms in wild brown bullhead catfish after coking plant closes and environmental PAHs plummet. Environmental Health Perspectives 103:168–170.

Baumann, P. C., and J. C. Harshbarger. 1998. Long term trends in liver neoplasm epizootics of brown bullhead in the Black River, Ohio. Environmental Monitoring and Assessment 53:213–223.

Baumann, P. C., J. C. Harshbarger, and K. J. Hartman. 1990. Relationship between liver tumors and age in brown bullhead populations from two Lake Erie tributaries. Science of the Total Environment 94:71–87.

Baumann, P. C., and M. S. Ohihiro. 2000. Cancer. Pages 591-616 *in* G. K. Ostrander, editor. The handbook of experimental animals: the laboratory fish. Academic Press, London.

Baumann, P. C., I. R. Smith, and C. D. Metcalfe. 1996. Linkages between chemical contaminants and tumors in benthic Great Lakes fish. Journal of Great Lakes Research 22:131–152.

Baumann, P. C., W. D. Smith, and W. K. Parland. 1987. Tumor frequencies and contaminant concentrations in brown bullheads from an industrialized river and a recreational lake. Transactions of the American Fisheries Society 116:79–86.

Baumann, P. C., W. D. Smith, and M. Ribick. 1982. Hepatic tumor rates and polynuclear aromatic hydrocarbon levels in tow populations of brown bullhead (*Ictalurus nebulosus*). Pages 93–102 *in* M. W. Cooke, A. J. Dennis, and G. L. Fisher, editors. Polynuclear aromatic hydrocarbons. Sixth International Symposium on Physical and Biological Chemistry. Battelle Press, Columbus, Ohio.

Beland, P., S. DeGuise, C. Girard, A. Lagace, D. Martineau, R. Michaud, D. C. G. Muir, R. Norstrom, E. Pelletier, S. Ray, and L. R. Shugart. 1993. Toxic compounds and health and reproductive effects in St. Lawrence beluga whales. Journal of Great Lakes Research 19:766–775.

Bengtsson, A., B. E. Bengtsson, and G. Lithner. 1988a. Vertebral defects in fourhorn sculpin, *Myoxocephalus quadricornis* L., exposed to heavy metal pollution in the Gulf of Bothnia. Journal of Fish Biology 33:517–529.

Bengtsson, B. E. 1974. Vertebral damage to minnows (*Phoxinus phoxinus*) exposed to zinc. Oikos 25:134–139.

Bengtsson, B. E. 1975. Vertebral damage in fish induced by pollutants. Pages 23–30 *in* J. H. Koeman and J. J. T. W. A. Strik, editors. Sublethal effects of toxical chemicals on aquatic animals. Academic Press, New York.

Bengtsson, B. E. 1979. Biological variables, especially skeletal deformations in fish, for monitoring marine pollution. Philosophical Transactions of the Royal Society of London Series B 286:457–464.

Bengtsson, B. E., and A. Larsson. 1986. Vertebral deformities and physiological effects in fourhorn sculpin *Myoxocephalus quadricornis* after long-term exposure to a simulated heavy metal-containing effluent. Aquatic Toxicology 9:215–229.

Bengtsson, B. E., A. Larsson, A. Bengtsson, and L. Renberg. 1988b. Sublethal effects of tetrachloro-1,2,-benzoquinone, a component in bleachery effluents from pulp mills, on vertebral quality and physiological parameters in fourhorn sculpin. Ecotoxicology and Environmental Safety 15:62–71.

Billiard, S. M., K. Querbach, and P. V. Hodson. 1999. Toxicity of retene to early life stages of two freshwater fish species. Environmental Toxicology and Chemistry 18:2070–2077.

Black, J. J. 1983. Field and laboratory studies on environmental carcinogenesis in Niagara River fish. Journal of Great Lakes Research 9:326–334.

Black, J. J., H. Fox, P. Black, and F. Bock. 1985. Carcinogenic effects of river sediment extracts in fish and mice. Pages 415–427 *in* R. L. Jolley, R. J. Bull, W. P. Davis, S. Katz, M. H. Roberts Jr., and V. A. Jacobs, editors. Water chlorination chemistry: environmental impact and health effects. Lewis Publishers, Chelsea, Michigan.

Blazer, V. S., D. E. Facey, J. W. Fournie, L. A. Courtney, and J. K. Summers. 1994. Macrophage aggregates as indicators of environmental stress. Pages169–185 *in* J. S. Stolen and T. C. Fletcher, editors. Modulators of fish immune responses: models for environmental toxicology/biomarkers, immunostimulators, volume 1. SOS Publications, Fair Haven, New Jersey.

Blazer, V. S., J. W. Fournie, and B. A. Weeks. 1997. Macrophage aggregates: biomarker for immune function in fishes? Pages 360–375 *in* F. J. Dwyer, T. T. Doane, and M. L. Hinman, editors. Environmental toxicology and risk assessment: modeling and risk assessment, 6th volume, ASTM STP 1317. American Society for Testing and Materials, West Conshohocken, Pennsylvania.

Blazer, V. S., W. K. Vogelbein, C. L. Densmore, E. B. May, J. H. Lilley, and D. E. Zwerner. 1999. *Aphanomyces* as a cause of ulcerative skin lesions of menhaden from Chesapeake Bay tributaries. Journal of Aquatic Animal Health 11:340–349.

Bodammer, J. E. 2000. Some new observations on the cytopathology of fin erosion disease in winter flounder *Pseudopleuronectes americanus*. Diseases of Aquatic Organisms 40:51–65.

Bodammer, J. E., and R. A. Murchelano. 1990. Cytological study of vacuolated cells and other aberrant hepatocytes in winter flounder from Boston Harbor. Cancer Research 50:6744–6756.

Breeder, C. M., and D. E. Rosen. 1966. Modes of reproduction in fishes. Natural History Press, Garden City, New York.

Breslow, N. E., and N. E. Day. 1980. Statistical methods in cancer research, volume 1. The analysis of case-control studies. International Agency for Research on Cancer, Lyon, France.

Brown, R. E., J. J. Hazdra, L. Keith, I. Greenspan, and J. B. G. Kwapinski. 1973. Frequency of fish tumors in a polluted watershed as compared to non-polluted Canadian waters. Cancer Research 33:189–198.

Bruno, D. W., and T. T. Poppe. 1996. A color atlas of salmonid diseases. Academic Press, London.

Buchmann, A., W. Kuhlmann, M. Schwarz, W. Kunz, C. R. Wolf, E. Moll, T. Friedberg, and F. Oesch. 1985. Regulation and expression of four cytochrome P-450 isoenzymes, NADPH-cytochrome P-450 reductase, the glutathione transferases B and C and microsomal epoxide hydrolase in preneoplastic and neoplastic lesions in rat liver. Carcinogenesis 6:513–521.

Bucke, D., and S. W. Feist. 1993. Histopathological changes in the livers of dab, *Limanda limanda* (L.). 1993. Journal of Fish Diseases 16:281–296.

Bunton, T. E. 1990. Hepatopathology of diethylnitrosamine in the medaka (*Oryzias latipes*) following short-term exposure. Toxicologic Pathology 18:313–327.

Bunton, T. E. 1993. The immuncytochemistry of cytokeratin in fish tissues. Veterinary Pathology 30:418–425.

Camus, A. C., and R. E. Wolke. 1991. Atypical hepatic vacuolated cell lesion in white perch *Morone americana*. Diseases of Aquatic Organisms 11:225–228.

Cancalon, P. 1982. Degeneration and regeneration of the olfactory cells induced by ZnSo4 and other chemicals. Tissue and Cell 14:717–733.

Collier, T. K., L. L. Johnson, M. S. Myers, C. M. Stehr, M. M. Krahn, and J. E. Stein. 1998a. Fish injury in the Hylebos Waterway in Commencement Bay, Washington. U.S. NOAA Technical Memorandum NMFS-NWFSC-36, Department of Commerce, Seattle, Washington.

Collier, T. K., L. L. Johnson, C. M. Stehr, M. S. Myers, and J. E. Stein. 1998b. A comprehensive assessment of the impacts of contaminants on fish from an urban estuary. Marine Environmental Research 46:243–247.

Collier, T. K., S. V. Singh, Y. C. Awasthi, and U. Varanasi. 1992. Hepatic xenobiotic metabolizing enzymes in two species of benthic fish showing different prevalences of contaminant-associated liver neoplasms. Toxicology and Applied Pharmacology 113:319–324.

Colton, T., and E. R. Greenberg. 1982. Cancer epidemiology. Pages 23–70 *in* V. Mike and K. E. Stanley, editors. Statistics in medical research: methods and issues, with applications in cancer research. John Wiley and Sons, New York.

Cooley, H. M., R. E. Evans, and J. F. Klaverkamp. 2000. Toxicology of dietary uranium in lake whitefish (*Coregonus clupeaformis*). Aquatic Toxicology 48:495–515.

Cotran, R. S., V. Kumar, T. Collins, and S. L. Robbins. 1999. Robbins' pathologic basis of disease. 6th edition. Saunders, Philadelphia, Pennsylvania.

Couch, J. A., J. T. Winstead, and L. R. Goodman. 1977. Kepone-induced scoliosis and its histological consequences in fish. Science 197:585–587.

Couch, J. A., J. T. Winstead, D. J. Hansen, and L. R. Goodman. 1979. Vertebral dysplasia in young fish exposed to the herbicide trifluralin. Journal of Fish Diseases 2:35–42.

Couillard, C. M., and P. V. Hodson. 1996. Pigmented macrophage aggregates: a toxic response in fish exposed to bleached-kraft mill effluent. Environmental Toxicology and Chemistry 15:1844–1854.

Couillard, C. M., P. J. Williams, S. C. Courtenay, and G. P. Rawn. 1999. Histopathological evaluation of Atlantic tomcod (*Microgadus tomcod*) collected at estuarine sites receiving pulp and paper mill effluent. Aquatic Toxicology 44:263–278.

Cross, J. N. 1985. Fin erosion among fishes collected near a southern California municipal wastewater outfall (1971–82). Fishery Bulletin 83:195–206.

Cross, J. N., and J. E. Hose. 1988. Evidence for impaired reproduction in white croaker (*Genyonemus lineatus*) from contaminated areas off southern California. Marine Environmental Research 24:185–188.

Dawe, C. J. 1981. Polyoma tumors in mice and X-cell tumors in fish, viewed through telescope and microscope. Pages 19–49 *in* C. J. Dawe, J. C. Harshbarger, and S. Kondo, editors. Phyletic approaches to cancer. Japan Scientific Society Press, Tokyo.

De Guise, S., A. Lagace, and P. Beland. 1994a. Tumors in St. Lawrence beluga whales. Veterinary Pathology 31:444–449.

De Guise, S., A. Lagace, and P. Beland. 1994b. Gastric papillomas in eight St. Lawrence beluga whales (*Delphinapterus leucas*). Journal of Veterinary Diagnostic Investigation 6:385–388.

De Guise, S., A. Lagace, P. Beland, C. Girard, and R. Higgins. 1995b. Non-neoplastic lesions in beluga whales (*Delphinapterus leucas*) and other marine mammals from the St. Lawrence Estuary. Journal of Comparative Pathology 112:257–271.

De Guise, S., D. Martineau, P. Beland, and M. Fournier. 1995a. Possible mechanisms of action of environmental contaminants on St. Lawrence beluga whales. Environmental Health Perspectives 103(Supplement 4):73–77.

Dethlefsen, V. 1980. Observations on fish diseases in the German Bight and their possible relation to pollution. Rapports et Proces-Verbaux des Reunions Commission Internationale pour l'Exploration Scientifique de la Mer Mediterranee Monaco 179:110–117.

Dey, W. P., T. H. Peck, C. E. Smith, and G.-L. Kreamer. 1993. Epizoology of hepatic neoplasia in Atlantic tomcod (*Microgadus tomcod*) from the Hudson River estuary. Canadian Journal of Fisheries and Aquatic Sciences 50:1897–1907.

Dial, N. A. 1978. Some effects of methylmercury on development of the eye in medaka fish. Growth 42:309–318.

Dykstra, M. J., and A. S. Kane. 2000. *Pfiesteria piscicida* and ulcerative mycosis of Atlantic menhaden: current status of understanding. Journal of Aquatic Animal Health 12:18–25.

Eisler, R. E., and G. R. Gardner. 1973. Acute toxicology to an estuarine teleost of mixtures of cadmium, copper and zinc salts. Journal of Fish Biology 5:131–142.

Eller, L. L. 1975. Gill lesions in freshwater teleosts. Pages 305–330 *in* W. E. Ribelin and G. Migaki, editors. The pathology of fishes. University of Wisconsin Press, Madison.

Evans, J. J. 1997. Rubber tire leachates in the aquatic environment. Reviews in Environmental Contamination and Toxicology 151:67–115.

Exley, C. 1996. Aluminum in the brain and heart of the rainbow trout. Journal of Fish Biology 48:706–713.

Farber, E., and D. S. R. Sarma. 1987. Biology of disease—hepatocarcinogenesis: a dynamic cellular perspective. Laboratory Investigations 56:4–22.

Ferguson, H. W. 1989. Systemic pathology of fish. Iowa State University Press, Ames.

Finn, W. F. 1983. Environmental toxins and renal disease. Journal of Clinical Pharmacology 23:461–472.

Fisher J. P., and M. S. Myers 2000. Fish necropsy. Pages 543–556 *in* G. K. Ostrander, editor. The handbook of experimental animals: the laboratory fish. Academic Press, London.

Fleiss, F. L. 1981. Statistical methods for rates and proportions, 2nd edition. Wiley, New York.

Fournie, J. W., W. E. Hawkins, R. M. Krol, and M. J. Wolfe. 1996a. Preparation of whole small fish for histological evaluation. Pages 577–587 *in* G. K. Ostrander, editor. Techniques in aquatic toxicology. CRC Press, Boca Raton, Florida.

Fournie, J. W., R. M. Krol, and W. E. Hawkins. 2000. Fixation of fish tissues. Pages 569–578 *in* G. K. Ostrander, editor. The handbook of experimental animals: the laboratory fish. Academic Press, London.

Fournie, J. W., and R. M. Overstreet. 1985. Retinoblastoma in the spring cavefish, *Chologaster agassizi* Putnam. Journal of Fish Diseases 8:377–381.

Fournie, J. W., J. K. Summers, L. A. Courtney, V. D. Engle, and V. S. Blazer. 2001. Utility of splenic macrophage aggregates as an indicator of fish exposure to degraded environments. Journal of Aquatic Animal Health 13:105–116.

Fournie, J. W., J. K. Summers, and S. B. Weisberg. 1996b. Prevalence of gross pathological abnormalities in estuarine fishes. Transactions of the American Fisheries Society 125:581–590.

Fournie, J. W., and W. K. Vogelbein. 1994. Exocrine pancreatic neoplasms in the mummichog (*Fundulus heteroclitus*). Toxicologic Pathology 22:237–247.

Fox, G. A. 1991. Practical causal inference for ecoepidemiologists. Journal of Toxicology and Environmental Health 33:359–373.

French, B. L., W. L. Reichert, T. Hom, M. Nishimoto, H. R. Sanborn, and J. E. Stein. 1996. Accumulation and dose–response of hepatic DNA adducts in English sole (*Pleuronectes vetulus*) exposed to a gradient of contaminated sediments. Aquatic Toxicology 36:1–16.

Gardner, G. R. 1975. Chemically induced lesions in estuarine or marine teleosts. Pages 657–695 *in* W. C. Ribelin and G. E. Migaki, editors. The pathology of fishes. University of Wisconsin Press, Madison.

Gardner, G. R., and G. LaRoche. 1973. Copper-induced lesions in estuarine teleosts. Journal of the Fisheries Research Board of Canada 30.363–368.

Gimeno, S., H. Komen, A. G. M. Gerritsen, and T. Bowmer. 1998a. Feminization of young males of the common carp, *Cyprinus carpio*, exposed to 4-*tert*-pentylphenol during sexual differentiation. Aquatic Toxicology 43:77–92.

Gimeno, G., H. Komen, S. Jobling, J. Sumpter, and T. Bowmer. 1998b. Demasculinization of sexually mature male common carp, *Cyprinus carpio*, exposed to 4-*tert*-pentylphenol during spermatogenesis. Aquatic Toxicology 43:93–109.

Girard, C., A. Lagace, R. Higgins, and P. Beland. 1991. Adenocarcinoma of the salivary gland in a beluga whale (*Delphinapterus leucas*). Journal of Veterinary Diagnostic Investigation 3:264–265.

Goede, R. W., and B. A. Barton. 1990. Organismic indices and an autopsy-based assessment as indicators of health and condition of fish. Pages 93–108 *in* S. M.

Adams, editor. Biological indicators of stress in fish. American Fisheries Society, Symposium 8, Bethesda, Maryland.

Gozzo, S., G. Peretta, U. Andreozzi, V. Monaco, and E. Rossiello. 1994. Neuropathology induced by trimethytin in the central nervous system of the urodele *Triturus carnifex*. Aquatic Toxicology 30:1–11.

Gray, M. A., and C. D. Metcalfe. 1997. Induction of testis–ova in Japanese medaka (*Oryzia latipes*) exposed to nonylphenol. Environmental Toxicology and Chemistry 16:1082–1086.

Grinwis, G. C. M., A. Boonstra, E. J. van den Brandhof, J. A. M. A. Dormans, E. Engelsma, R. V. Kuiper, H. van Loveren, P. W. Wester, M. A. Vaal, A. D. Vethaak, and J. G. Vos. 1998. Short-term toxicity of bis(tri-*n*-butyltin) oxide in flounder (*Platichthys flesus*): pathology and immune function. Aquatic Toxicology 42:15–36.

Grizzle, J. M., P. Melius, and D. R. Strength. 1984. Papilloma on fish exposed to chlorinated wastewater effluent. Journal of the National Cancer Institute 73:1133–1142.

Grizzle, J. M., and W. A. Rogers. 1976. Anatomy and histology of the channel catfish. Auburn Printing, Auburn, Alabama.

Groff, J. M., D. E. Hinton, T. S. McDowell, and R. P. Hedrick. 1992. Progression and resolution of megalocytic hepatopathy with exocrine pancreatic metaplasia in a population of cultured juvenile striped bass *Morone saxatilis*. Diseases of Aquatic Organisms 13:189–202.

Groman, D. B. 1982. Histology of the striped bass. American Fisheries Society, Monograph 3, Bethesda, Maryland.

Haaparanta, A., E. T. Valtonen, R. Hoffmann, J. Holmes. 1996. Do macrophage centres in freshwater fishes reflect the differences in water quality? Aquatic Toxicology 34:253–272.

Hamm, J. T., B. W. Wilson, and D. E. Hinton. 1998. Organophosphate-induced acetylcholinesterase inhibition and embryonic retinal cell necrosis *in vivo* in the teleost (*Oryzias latipes*). Neurotoxicology 19:853–870.

Hansen, J. A., J. D. Rose, R. A. Jenkins, K. G. Gerow, and H. L. Bergman. 1999. Chinook salmon (*Oncorhynchus tshawytscha*) and rainbow trout (*Oncorhynchus mykiss*) exposed to copper: neurophysiological and histological effects on the olfactory system. Environmental Toxicology and Chemistry 18:1979–1991.

Harshbarger, J. C. 1984. Pseudoneoplasms in ectothermic animals. Pages 251–273 *in* K. L. Hoover, editor. Use of small fish species in carcinogenicity testing. National Cancer Institute Monograph 65, National Institutes of Health, Bethesda, Maryland.

Harshbarger, J. C., and J. B. Clark. 1990. Epizootiology of neoplasms in bony fish of North America. Science of the Total Environment 94:1–32.

Harshbarger, J. C., M. J. Coffey, and M. Y. Young. 2000. Intersexes in Mississippi shovelnose sturgeon sampled below Saint Louis, Missouri, USA. Marine Environmental Research 50:247–250.

Hart, L. J., S. A. Smith, B. J. Smith, J. Robertson, E. G. Besteman, and S. D. Holladay. 1998. Subacute immunotoxic effects of the polycyclic aromatic hydrocarbon 7,12-dimethylbenzathracene on spleen and pronephros leucocytic cell counts and phagocytic cell activity in tilapia (*Oreochromis niloticus*). Aquatic Toxicology 41:17–29.

Haschek, W. G., and C. G. Rousseaux, editors. 1991. Handbook of toxicologic pathology. Academic Press, San Diego, California.

Hashimoto, S., H. Bessho, A. Hara, M. Nakamura, T. Iguchi, and K. Fujita. 2000. Elevated serum vitellogenin levels and gonadal abnormalities in wild male flounder (*Pleuronectes yokohamae*) from Tokyo Bay, Japan. Marine Environmental Research 49:37–53.

Hawkes, J. W. 1974. The structure of fish skin. I. General organization. Cell and Tissue Research 149:147–158.

Hawkins, W. E., J. W. Fournie, R. M. Overstreet, and W. W. Walker. 1986. Intraocular neoplasms induced by methylazoxymethanol acetate in Japanese medaka (*Oryzias latipes*). Journal of the National Cancer Institute 76:453–465.

Hawkins, W. E., L. G. Tate, and T. G. Sarphie. 1980. Acute effects of cadmium on the spot, *Leiostomus xanthurus* (Teleostei): tissue distribution and renal ultrastructure. Journal of Toxicology and Environmental Health 6:283–295.

Hawkins, W. E., W. W. Walker, J. S. Lytle, T. F. Lytle, and R. M. Overstreet. 1989. Carcinogenic effects of 7,12-dimethylbenz[a]anthracene on the guppy *Poecilia reticulata*. Aquatic Toxicology 15:63–82.

Hawkins, W. E., W. W. Walker, R. M. Overstreet, J. S. Lytle, and T. F. Lytle. 1990. Carcinogenic effects of some polycyclic aromatic hydrocarbons on the Japanese medaka and guppy in waterborne exposure. Science of the Total Environment 94:155–167.

Hayes, M. A., I. R. Smith, T. H. Rushmore, T. L. Crane, C. Thorn, T. E. Kocal, and H. W. Ferguson. 1990. Pathogenesis of skin and liver neoplasms in white suckers from industrially polluted areas in Lake Ontario. Science of the Total Environment 94:105–123.

Hemmer, M. J., D. P. Middaugh, and J. C. Moore. 1990. The effects of temperature and salinity on *Menidia beryllina* embryos exposed to terbufos. Diseases of Aquatic Organisms 8:127–136.

Hendricks, J. D. 1979. Appendix II. Effect of various herbicides on histology of yearling coho salmon. Pages 90–93 *in* H. W. Lorz, S. W. Glenn, R. H. Williams, C. M. Kunkel, L. A. Norris, and R. R. Loper, editors. Effects of selected herbicides on smolting of coho salmon. Corvallis Environmental Research Laboratory, U.S. Environmental Protection Agency, 600/3-79-071, Corvallis, Oregon.

Hendricks, J. D., T. R. Meyers, and D. W. Shelton. 1984. Histological progression of hepatic neoplasia in rainbow trout (*Salmo gairdneri*). National Cancer Institute Monograph 65:321–336.

Hendricks, J. D., T. P Putnam, D. D. Bills, and R. O. Sinnhuber. 1977. Inhibitory effect of a polychlorinated biphenyl (Aroclor 1254) on aflatoxin B1 carcinogenesis in rainbow trout (*Salmo gairdneri*). Journal of the National Cancer Institute 59:1545–1551.

Hendricks, J. D., R. O. Sinnhuber, M. C. Henderson, and D. R. Buhler. 1981. Liver and kidney pathology in rainbow trout (*Salmo gairdneri*) exposed to dietary pyrrolizidine (*Senecio*) alkaloids. Experimental Molecular Pathology 35:170–183.

Hinton, D. E. 1990. Histological techniques. Pages 191–211 *in* C. B. Shreck and P. B. Moyle, editors. Methods for fish biology. American Fisheries Society, Bethesda, Maryland.

Hinton, D. E. 1993. Toxicologic histopathology of fishes: a systemic approach and overview. Pages 177–215 *in* J. A. Couch and J. W. Fournie, editors. Pathobiology of marine and estuarine organisms. CRC Press, Boca Raton, Florida.

Hinton, D. E. 1994. Cells, cellular responses, and their markers in chronic toxicity of fishes. Pages 207–231 *in* D. C. Malins and G. K. Ostrander, editors. Aquatic toxicology: molecular, biochemical, and cellular perspectives. Lewis Publishers, Boca Raton, Florida.

Hinton, D. E., P. C. Baumann, G. R. Gardner, W. E. Hawkins, J. D. Hendricks, R. A. Murchelano, and M. S. Okihiro. 1992. Histopathologic biomarkers. Pages 155–209 *in* R. J. Huggett, R. A. Kimerle, P. M. Mehrle Jr., and H. L. Bergman, editors. Biomarkers: biochemical, physiological, and histological markers of anthropogenic stress. Lewis Publishers, Chelsea, Michigan.

Hinton, D. E., J. A. Couch, S. J. Teh, and L. A. Courtney. 1988. Cytological changes during progression of neoplasia in selected fish species. Aquatic Toxicology 11:77–112.

Hinton, D. E., and D. J. Lauren. 1990. Integrative histopathological approaches to detecting effects of environmental stressors on fishes. Pages 51–66 *in* S. M. Adams, editor. Biological indicators of stress in fish. American Fisheries Society, Symposium 8, Bethesda, Maryland.

Hinton, D. E., E. R. Walker, C. A. Pinkstaff, and E. M. Zuchelkowski. 1984. Morphological survey of teleost organs important in carcinogenesis with attention to fixation. National Cancer Institute Monograph 65:291–320.

Hoover, K. L., editor. 1984. Use of small fish species in carcinogenicity testing. National Cancer Institute Monograph 65, National Institutes of Health, Bethesda, Maryland.

Horness, B. H., D. P Lomax, L. L. Johnson, M. S. Myers, S. M. Pierce, and T. K. Collier. 1998. Sediment quality thresholds: estimates from hockey stick regression of liver lesion prevalence in English sole (*Pleuronectes vetulus*). Environmental Toxicology and Chemistry 17:872–882.

Hose, J. E., J. B. Hannah, H. W. Puffer, and M. L. Landolt. 1984. Histologic and skeletal abnormalities in benzo(a)pyrene-treated rainbow trout alevins. Archives of Environmental Contamination and Toxicology 13:675–684.

Hughes, G. M., S. F. Perry, and V. M. Brown. 1979. A morphometric study of effects of nickel, chromium, and cadmium on the secondary lamellae of rainbow trout gills. Water Research 13:665–679.

Hunter, J. R., and B. J. Macewicz. 1985. Rates of atresia in the ovary of captive and wild northern anchovy, *Engraulis mordax*. Fisheries Bulletin 83:119–136.

Husøy, A. M., M. S. Myers, M. L. Willis, T. K. Collier, M. Celander, and A. Goksøyr. 1994. Immunohistochemical localization of CYP1A and CYP3A-like isozymes in hepatic and extrahepatic tissues of Atlantic cod (*Gadus morhua* L.), a marine fish. Toxicology and Applied Pharmacology 129:294–308.

Izaguirre, M. F., R. C. Lajmanovich, P. M. Peltzer, A. P. Soler, and V. H. Casco. 2000. Cypermethrin-induced apoptosis in the telencephalon of *Physalaemus biligonigerus* tadpoles (Anura: Leptodactylidae). Bulletin of Environmental Contamination and Toxicology 65:501–507.

Jensen, N. J., and J. L. Larsen. 1982. The ulcus-syndrome in cod (*Gadus morhua*) IV. Transmission experiments with two viruses isolated from cod and *Vibrio anguillarum*. Nordisk Veterinaermedicin 34:136–142.

Jobling, S., M. Nolan, C. R. Tyler, G. Brighty, and J. P. Sumpter. 1998. Widespread sexual disruption in wild fish. Environmental Science and Technology 32:2498–2506.

Johnson, L. L., E. Casillas, T. K. Collier, B. B. McCain, and U. Varanasi. 1988. Contaminant effects on ovarian development in English sole (*Parophrys vetulus*) from Puget Sound, Washington. Canadian Journal of Fisheries and Aquatic Sciences 45:2133–2146.

Johnson, L. L., S. Y. Sol, D. P. Lomax, G. M. Nelson, D. A. Sloan, and E. Casillas. 1997. Fecundity and egg weight in English sole *Pleuronectes vetulus* from Puget Sound: influence of nutritional status and chemical contaminants. Fisheries Bulletin 95:231–249.

Johnson, L. L., S. Sol, G. M. Ylitalo, T. Hom, F. French, O. P. Olson, and T. K. Collier. 1999. Reproductive injury in English sole (*Pleuronectes vetulus*) from the Hylebos Waterway, Commencement Bay, Washington. Journal of Aquatic Ecosystem Stress and Recovery 6:289–310.

Johnson, L. L., C. M. Stehr, O. P. Olson, M. S. Myers, S. M. Pierce, C. A. Wigren, B. B. McCain, and U. Varanasi. 1993. Chemical contaminants and hepatic lesions in winter flounder (*Pleuronectes americanus*) from the northeast coast of the United States. Environmental Science and Technology 27:2759–2771.

Kent, M. L. 1990. Netpen liver disease (NLD) of salmonid fishes reared in sea water: species susceptibility, recovery, and probable cause. Diseases of Aquatic Organisms 8:21–28.

Kent, M. L., and T. T. Poppe. 1998. Diseases of seawater netpen-reared salmonid fishes. Pacific Biological Station, Fisheries and Oceans, Canada. Nanaimo, British Columbia.

Khan, R. A., D. Barker, R. Hooper, and E. M. Lee. 1992. Effect of pulp and paper effluent on a marine fish, *Pseudopleuronectes americanus*. Bulletin of Environmental Contamination and Toxicology 48:449–456.

Kimura, I., N. Kinae, H. Kumai, M. Yamashita, G. Nakamura, M. Ando, H. Ishida, and I. Tomita. 1989. Environment: peculiar pigment cell neoplasm in fish. Journal of Investigative Dermatology 92:248S–254S.

Kinae, N., M. Yamashita, I. Tomita, I. Kimura, H. Ishida, H. Kumai, and G. Nakamura. 1990. A possible correlation between environmental chemicals and pigment cell neoplasia in fish. Science of the Total Environment 94:143–153.

Kirby, G. M., M. Stalker, C. Metcalfe, T. Kocal, H. Ferguson, and M. A. Hayes. 1990. Expression of immunoreactive glutathione-S-transfer in hepatic lesions induced by aflatoxin B1 or 1,2 dimethylbenzanthracene in rainbow trout (*Oncorhynchus mykiss*). Carcinogenesis 12:2255–2257.

Klaprat, D. A., R. E. Evans, and T. J. Hara. 1992. Environmental contaminants and chemoreception in fishes. Pages 321–341 *in* T. J. Hara, editor. Fish chemoreception. Chapman and Hall, London.

Klontz, G. W. 1993. Environmental requirements and environmental diseases of salmonids. Pages 333–342 *in* M. K. Stoskopf, editor. Fish medicine. Saunders, Philadelphia.

Klontz, G. W., B. C. Stewart, and D. W. Eib. 1985. On the etiology and pathophysiology of environmental gill disease in juvenile salmonids. Pages 199–211 *in* A. E. Ellis, editor. Fish and shellfish pathology. Academic Press, New York.

Koger, C. S., S. J. Teh, and D. E. Hinton. 2000. Determining the sensitive developmental stages of intersex induction in medaka (*Oryzias latipes*) exposed to 17β-estradiol or testosterone. Marine Environmental Research 50:201–206.

Kohler, A. 1990. Identification of contaminant-induced cellular and subcellular lesions in the liver of flounder (*Platichthys flesus* L.) caught at differently polluted estuaries. Aquatic Toxicology 16:271–294.

Kohler, A., and C. J. F. Van Noorden. 1998. Initial velocities *in situ* of G6PDH and PGDH and expression of proliferating cell nuclear antigen (PCNA): sensitive diagnostic markers of environmentally induced hepatocellular carcinogenesis in a marine flatfish (*Platichthys flesus* L.). Aquatic Toxicology 40:233–252.

Krahn, M. M., D. G. Burrows, W. D. MacLeod Jr., and D. C. Malins. 1987. Determination of individual metabolites of aromatic compounds in hydrolyzed bile of English sole (*Parophrys vetulus*) from polluted sites in Puget Sound, Washington. Archives of Environmental Contamination and Toxicology 16:511–522.

Kranz, H., and V. Dethlefsen. 1990. Liver anomalies in dab *Limanda limanda* from the southern North Sea with special consideration given to neoplastic lesions. Diseases of Aquatic Organisms 9:171–185.

Kranz, H., and W. H. Gerken. 1987. Effects of sublethal concentrations of potassium dichromate on the occurrence of splenic melanomacrophage centres in juvenile plaice, *Pleuronectes platessa*, L. Journal of Fish Biology 31 (Supplement A):75–80.

Kumar, K., and B. A. Ansari. 1984. Malathion toxicity: skeletal deformities in zebrafish (*Brachydanio rerio*, Cyprinidae). Pesticide Science 15:107–111.

Lair, S., P. Beland, S. De Guise, and D. Martineau. 1997. Adrenal hyperplastic and degenerative changes in beluga whales. Journal of Wildlife Diseases 33:430–437.

Lair, S., S. De Guise, and D. Martineau. 1998. Uterine adenocarcinoma with abdominal carcinomatosis in a beluga whale. Journal of Wildlife Diseases 34:373–376.

Latendresse, J. R. II, and J. W. Fisher. 1983. Histopathologic effects of JP-4 aviation fuel on fathead minnows (*Pimephales promelas*). Bulletin of Environmental Contamination and Toxicology 30:536–543.

Lauren, D. J., S. J. Teh, and D. E. Hinton. 1990. Cytotoxicity phase of diethylnitrosamine-induced hepatic neoplasia in medaka. Cancer Research 50:5504–5514.

Leadley, T. A., L. S. Arcand-Hoy, G. D. Haffner, and C. D. Metcalfe. 1999. Fluorescent aromatic hydrocarbons in bile as a biomarker of exposure of brown bullheads (*Ameiurus nebulosus*) to contaminated sediments. Environmental Toxicology and Chemistry 18:750–755.

Leadley, T. A., G. Balch, C. D. Metcalfe, R. Lazar, E. Mazak, J. Habovsky, and G. D. Haffner. 1998. Chemical accumulation and toxicological stress in three brown bullhead (*Ameiurus nebulosus*) populations of the Detroit River, Michigan, USA. Environmental Toxicology and Chemistry 17:1756–1766.

Leino, R. L., and J. H. McCormick. 1984. Morphological and morphometrical changes in chloride cells in the gills of *Pimephales promelas* after chronic exposure to acid water. Cell and Tissue Research 36:121–128.

Levin, M. A., R. E. Wolke, and V. J. Cabelli. 1972. *Vibrio anguillarum* as a cause of disease in winter flounder (*Pseudopleuronectes americanus*). Canadian Journal of Microbiology 118:1585–1592.

Lindesjoo, E., and J. Thulin. 1990. Fin erosion of perch *Perca fluviatilis* and ruffe *Gymnocephalus cernua* in a pulp mill effluent area. Diseases of Aquatic Organisms 8:119–126.

Lindesjoo, E., and J. Thulin. 1992. A skeletal deformity of northern pike (*Esox lucius*) related to pulp mill effluents. Canadian Journal of Fisheries and Aquatic Sciences 49:166–172.

Lindesjoo, E., and J. Thulin. 1994. Histopathology of skin and gills of fish in pulp mill effluents. Diseases of Aquatic Organisms 18:81–93.

Lorenzana, R. M., O. R. Hedstrom, J. A. Gallagher, and D. R. Buhler. 1989. Cytochrome P450 isozyme distribution in normal and tumor-bearing hepatic tissue from rainbow trout (*Salmo gairdneri*). Experimental and Molecular Pathology 50:348–361.

Luna, L. G., editor. 1968. Armed Forces Institute of Pathology manual of histologic staining methods, 3rd edition. McGraw-Hill, New York.

Luna, L. G. 1992. Histopathological methods and color atlas of special stains and tissue artifacts. American Histolabs, Gaithersburg, Maryland.

Maccubbin, A. E., and N. Ersing. 1991. Tumors in fish from the Detroit River. Hydrobiologia 219:301–306.

Malins, D. C., M. M. Krahn, M. S. Myers, L. D. Rhodes, D. W. Brown, C. A. Krone, B. B. McCain, and S.-L. Chan. 1985. Toxic chemicals in sediments and biota from a creosote-polluted harbor: relationships with hepatic neoplasms and other hepatic lesions in English sole (*Parophrys vetulus*). Carcinogenesis 6:1463–1469.

Malins, D. C., B. B. McCain, D. W. Brown, S.-L. Chan, M. S. Myers, J. T. Landahl, P. G. Prohaska, A. J. Friedman, L. D. Rhodes, D. G. Burrows, W. D. Gronlund, and H. O. Hodgins. 1984. Chemical pollutants in sediments and diseases of bottom-dwelling fish in Puget Sound, Washington. Environmental Science and Technology 18:705–713.

Malins, D. C., B. B. McCain, M. S. Myers, D. W. Brown, M. M. Krahn, W. T. Roubal, M. H. Schiewe, J. T. Landahl, and S-L. Chan. 1987. Field and laboratory studies of the etiology of liver neoplasms in marine fish from Puget Sound. Environmental Health Perspectives 71:5–16.

Mallatt, J. 1985. Fish gill structural changes induced by toxicants and other irritants: a statistical review. Canadian Journal of Fisheries and Aquatic Sciences 42:630–648.

Martineau, D., A. Lagace, P. Beland, R. Higgins, D. Armstrong, and L. R. Shugart. 1988. Pathology of stranded beluga whales (*Delphinapterus leucas*) from the St. Lawrence Estuary, Quebec, Canada. Journal of Comparative Pathology 98:287–311.

Marty, G. D., E. F. Freiberg, T. R. Meyers, J. Wilcock, T. B. Farver, and D. E. Hinton. 1998. Viral hemorrhagic septicemia virus, *Ichthyophonus hoferi*, and other causes of morbidity in Pacific herring *Clupea pallasi* spawning in Prince William Sound, Alaska, USA. Diseases of Aquatic Organisms 32:15–40.

Marty, G. D., J. E. Hose, M. D. McGurk, E. D. Brown, and D. E. Hinton. 1997. Histopathology and cytogenetic evaluation of Pacific herring larvae exposed to petroleum hydrocarbons in the laboratory or in Prince William Sound, Alaska, after the *Exxon Valdez* oil spill. Canadian Journal of Fisheries and Aquatic Sciences 54:1846–1857.

Matsui, M., J. E. Hose, P. Garrahan, and G. A. Jordan. 1992. Developmental defects in fish embryos from Salton Sea, California. Bulletin of Environmental Contamination and Toxicology 48:914–920.

Matta, M. B., C. Cairncross, and R. M. Kocan. 1998. Possible effects of polychlorinated biphenyls on sex determination in rainbow trout. Environmental Toxicology and Chemistry 17:26–29.

Mausner, J. S., and A. K. Bahn. 1974. Epidemiology: an introductory text. Saunders, Philadelphia.

Mayer, F. L., Jr., B. E. Bengtsson, S. J. Hamilton, and A. Bengtsson. 1988. Effects of pulp mill and ore smelter effluents on vertebrae of fourhorn sculpin: laboratory and field comparisons. American Society of Testing and Materials, Special Technical Publication 971:406–419.

Mayer, F. L., P. M. Mehrle, and P. L. Crutcher. 1978. Interactions of toxaphene and vitamin C in channel catfish. Transactions of the American Fisheries Society 107:326–333.

Mayer, F. L., P. M. Mehrle, and R. A. Schoettger. 1977. Collagen metabolism in fish exposed to organic chemicals. U.S. Environmental Protection Agency, Ecological Research Series, EPA/600/3-77-085. Corvallis, Oregon.

McAllister, P. E., T. Nagabayashi, and K. Wolf. 1977. Viruses of eels with and without stomatopapillomas. Annals of the New York Academy of Sciences 298:233–244.

McCain, B. B., D. W. Brown, M. M. Krahn, M. S. Myers, R. C. Clark Jr., S.-L. Chan, and D. C. Malins. 1988. Marine pollution problems, North American West Coast. Aquatic Toxicology 11:143–162.

McCain, B. B., W. D. Gronlund, and M. S. Myers. 1979. Tumors and microbial diseases in Alaskan waters. Journal of Fish Diseases 2:111–130.

McCain, B. B., M. S. Myers, U. Varanasi, D. W. Brown, L. D. Rhodes, W. D. Gronlund, D. G. Elliott, W. A. Palsson, H. O. Hodgins, and D. C. Malins. 1982. Pathology of two species of flatfish from urban estuaries of Puget Sound. Interagency Energy/Environment R and D Program Report. EPA-600/7-82-001. U.S. Environmental Protection Agency, Center for Environmental Research Information, Cincinnati, Ohio.

McCormick, J. H., G. N. Stokes, and R. O. Hermanutz. 1989. Oocyte atresia and reproductive success in fathead minnows (*Pimephales promelas*) exposed to acidified hard water environments. Archives of Environmental Contamination and Toxicology 18:207–214.

McDonald, D. G. 1983. The effects of H^+ upon the gills of freshwater fish. Canadian Journal of Zoology 61:691–703.

Mehrle, P. M., and F. L. Mayer Jr. 1975. Toxaphene effects on growth and bone composition of fathead minnows, *Pimephales promelas*. Journal of the Fisheries Research Board of Canada 32:593–598.

Metcalfe, C. D. 1989. Tests for predicting carcinogenicity in fish. CRC Critical Reviews in Aquatic Sciences 1:111–129.

Meyers, T. R., and J. D. Hendricks. 1982. A summary of tissue lesions in aquatic animals induced by controlled exposures to environmental contaminants, chemotherapeutic agents, and potential carcinogens. Marine Fisheries Review 44:1–17.

Meyers, T. R., and J. D. Hendricks. 1985. Histopathology. Pages 283–331 *in* G. M. Rand and S. R. Petrocelli, editors. Fundamentals of aquatic toxicology. Hemisphere, Washington D.C.

Middaugh, D. P., J. W. Fournie, and M. J. Hemmer. 1990. Vertebral abnormalities in juvenile inland silversides *Menidia beryllina* exposed to terbufos during embryogenesis. Diseases of Aquatic Organisms 9:109–116.

Miles-Richardson, S. R., V. J. Kramer, S. D. Fitzgerald, J. A. Render, B. Yamini, S. J. Barbee, and J. P. Geisy. 1999. Effects of waterborne exposure of 17β-estradiol on secondary sex characteristics and gonads of fathead minnows (*Pimephales promelas*). Aquatic Toxicology 47:129–145.

Minchew, C. D., and J. D. Yarbrough. 1977. The occurrence of fin rot in mullet (*Mugil cephalus*) associated with crude oil contamination of an estuarine pond-ecosystem. Journal of Fish Biology 10:319–323.

Minier, C., F. Levy, D. Rabel, G. Bocquen'e, D. Godefroy, T. Bougeot, and F. Leboulenger. 2000. Flounder health status in the Seine Bay. A multibiomarker study. Marine Environmental Research 50:373–377.

Moore M. J., and M. S. Myers. 1994. Pathobiology of chemical-associated neoplasia in fish. Pages 327–386 *in* D. C. Malins and G. K. Ostrander, editors. Aquatic toxicology: molecular, biochemical, and cellular approaches. Lewis Publishers, London.

Moore, M. J., D. Shea, R. E. Hillman, and J. J. Stegeman. 1996. Trends in hepatic tumors and hydropic vacuolation, fin erosion, organic chemicals and stable isotope ratios in winter flounder from Massachusetts, USA. Marine Pollution Bulletin 32:458–470.

Moore, M. J., R. M. Smolowitz, and J. J. Stegeman. 1997. Stages of hydropic vacuolation in the liver of winter flounder *Pleuronectes americanus* from a chemically contaminated site. Diseases of Aquatic Organisms 31:19–28.

Moore, M. J., and J. J. Stegeman. 1994. Hepatic neoplasms winter flounder *Pleuronectes americanus* from Boston Harbor, Massachusetts, USA. Diseases of Aquatic Organisms 20:33–48.

Morita, I., I. Kihara, T. Oite, T. Yamamoto, and Y. Suzuki. 1978. Mesangiolysis: sequential ultrastructural study of Habu venom-induced glomerular lesions. Laboratory Investigation 38:94–102.

Munkittrick, K. R., M. E. McMaster, C. B. Portt, G. Van Der Kraak, I. R. Smith, and D. G. Dixon. 1992a. Changes in maturity, plasma sex steroid levels, hepatic MFO activity and the presence of external lesions in lake whitefish exposed to bleached kraft mill effluent (BKME). Canadian Journal of Fisheries and Aquatic Sciences 49:1560–1569.

Munkittrick, K. R., G. J. Van Der Kraak, M. E. McMaster, and C. B. Portt. 1992b. Reproductive dysfunction and MFO activity in three species of fish exposed to bleached kraft mill effluent at Jackfish Bay, Lake Superior. Water Pollution Research Journal of Canada 27:439–446.

Muramoto, S. 1981. Vertebral column damage and decrease of calcium concentration in fish exposed experimentally to cadmium. Environmental Pollution 24:125–133.

Murchelano, R. A. 1975. The histopathology of fin rot disease in winter flounder from the New York Bight. Journal of Wildlife Diseases 11:263–268.

Murchelano, R. A., and R. E. Wolke. 1985. Epizootic carcinoma in the winter flounder (*Pseudopleuronectes americanus*). Science 228:587–589.

Murchelano, R. A., and R. E. Wolke. 1991. Neoplasms and nonneoplastic liver lesions in winter flounder, (*Pseudopleuronectes americanus*), from Boston Harbor, Massachusetts. Environmental Health Perspectives 90:17–26.

Murchelano, R. A., and J. Ziskowski. 1979. Some observations on an ulcer disease of red hake *Urophycis chuss*, from the New York Bight. International Council for Exploration of the Sea. C. M. 1979/E:23. Copenhagen, Denmark.

Murchelano, R. A., and J. Ziskowski. 1982. Fin rot disease in the New York Bight 1973–1977. Pages 347–358 *in* G. F. Mayer, editor. Ecological stress and the New York Bight: science and management. Estuarine Research Federation, Columbia, South Carolina.

Myers M., B. Anulacion, B. French, T. Hom, W. Reichert, L. Hufnagle, and T. Collier. 2000. Biomarker and histopathologic responses in flatfish following site remediation in Eagle Harbor, WA. Marine Environmental Research 50(1–5):435–436.

Myers, M. S. 1981. Pathological anatomy of papilloma-like tumors in the Pacific Ocean perch *Sebastes alutus* from the Gulf of Alaska. M.S. thesis. University of Washington, Seattle.

Myers, M. S., B. L. French, W. L. Reichert, M. W. Willis, B. F. Anulacion, T. K. Collier, and J. E. Stein. 1998a. Reductions in CYP1A expression and hydrophobic DNA adducts in liver neoplasms of English sole (*Pleuronectes vetulus*): further support for the "resistant hepatocyte" model of hepatocarcinogenesis. Marine Environmental Research 46:197–202.

Myers, M. S., L. L. Johnson, T. Hom, T. K. Collier, J. E. Stein, and U. Varanasi. 1998b. Toxicopathic hepatic lesions in subadult English sole (*Pleuronectes vetulus*) from Puget Sound, Washington, USA: relationships with other biomarkers of contaminant exposure. Marine Environmental Research 45:47–67.

Myers, M. S., L. L Johnson, O. P. Olson, C. M. Stehr, B. H. Horness, T. K. Collier, and B. B McCain. 1999. Toxicopathic hepatic lesions as biomarkers of chemical contaminant exposure and effects in marine bottomfish species from the Northeast and Pacific coasts, USA. Marine Pollution Bulletin 37 (1–2):92–113.

Myers, M. S., J. T. Landahl, M. M. Krahn, L. L. Johnson, and B. B. McCain. 1990. Overview of studies on liver carcinogenesis in English sole from Puget Sound; evidence for a xenobiotic chemical etiology I: pathology and epizootiology. Science of the Total Environment 94:33–50.

Myers, M. S., J. T. Landahl, M. M. Krahn, and B. B. McCain. 1991. Relationships between hepatic neoplasms and related lesions and exposure to toxic chemicals in marine fish from the U.S. West Coast. Environmental Health Perspectives 90:7–15.

Myers, M. S., O. P. Olson, L. L. Johnson, C. S. Stehr, T. Hom, and U. Varanasi. 1992. Hepatic lesions other than neoplasms in subadult flatfish from Puget Sound, WA: relationships with indices of contaminant exposure. Marine Environmental Research 34:45–51.

Myers, M. S., and L. D. Rhodes. 1988. Morphologic similarities and parallels in geographic distribution of suspected toxicopathic liver lesions in rock sole (*Lepidopsetta bilineata*), starry flounder (*Platichthys stellatus*), Pacific staghorn sculpin (*Leptocottus armatus*), and Dover sole (*Microstomus pacificus*) as compared to English sole (*Parophrys vetulus*) from urban and non-urban embayments of Puget Sound, Washington. Aquatic Toxicology 11:410–411.

Myers, M. S., L. D. Rhodes, and B. B. McCain. 1987. Pathologic anatomy and patterns of occurrence of hepatic neoplasms, putative preneoplastic lesions and other idiopathic hepatic conditions in English sole (*Parophrys vetulus*) from Puget Sound, Washington. Journal of the National Cancer Institute 78:333–363.

Myers, M. S., C. M. Stehr, O. P. Olson, L. L. Johnson, B. B. McCain, S.-L. Chan, and U. Varanasi. 1993. National Status and Trends Program, National Benthic Surveillance Project: Pacific Coast, fish histopathology and relationships between toxicopathic lesions and exposure to chemical contaminants for Cycles I to V (1984–88). NOAA Technical Memorandum NMFS-NWFSC-6, U.S. Department of Commerce, Seattle, Washington.

Myers, M. S., C. M. Stehr, O. P. Olson, L. L. Johnson, B. B. McCain, S.-L. Chan, and U Varanasi. 1994. Relationships between toxicopathic hepatic lesions and exposure to chemical contaminants in English sole (*Pleuronectes vetulus*), starry flounder (*Platichthys stellatus*), and white croaker (*Genyonemus lineatus*) from selected marine sites on the Pacific Coast, USA. Environmental Health Perspectives 102(2):200–215.

Myers, M. S., M. L. Willis, A.-M. Husøy, A. Goksøyr, and T. K. Collier. 1995. Immunohistochemical localization of cytochrome P4501A in multiple types of contaminant-associated hepatic lesions in English sole (*Pleuronectes vetulus*). Marine Environmental Research 39:283–288.

Nestel, H., and J. Budd. 1975. Chronic oral exposure of rainbow trout to a polychlorinated biphenyl (Aroclor 1254): pathological effects. Canadian Journal of Comparative Medicine 39:208–215.

Nichols, K. M., S. R. Miles-Richardson, E. M. Snyder, and J. P. Giesy. 1999. Effects of exposure to municipal wastewater in situ on the reproductive physiology of the fathead minnow (*Pimephales promelas*). Environmental Toxicology and Chemistry 18:2001–2012.

Noga, E. J. 2000. Skin ulcers in fish: *Pfiesteria* and other etiologies. Toxicologic Pathology 28:807–823.

Noga, E. J., L. Khoo, J. B. Stevens, Z. Fan, and J. M. Burkholder. 1996. Novel toxic dinoflagellate causes epidemic diseases in estuarine fish. Marine Pollution Bulletin 32:219–224.

Ortego, L. S., W. E. Hawkins, W. W. Walker, R. M. Krol, and W. H. Benson. 1995. Immunocytochemical detection of proliferating cell nuclear antigen (PCNA) in tissues of marine animals used in toxicity bioassays. Marine Environmental Research 39:271–273.

Overstreet, R. M. 1988. Aquatic pollution problems, southeastern U.S. coasts: histopathological indicators. Aquatic Toxicology 11:213–239.

Papoulias, D. M., D. B. Noltie, and D. E. Tillitt. 1999. An in vivo model fish system to test chemical effects on sexual differentiation and development: exposure to ethinyl estradiol. Aquatic Toxicology 48:37–50.

Payne, J. F., and L. F. Fancey. 1989. Effect of polycyclic aromatic hydrocarbons on immune responses in fish: changes in melano-macrophage centers in flounder (*Pseudopleuronectes americanus*) exposed to hydrocarbon-contaminated sediments. Marine Environmental Research 28:431–435.

Pereira, J. J. 1988. Morphological effects of mercury exposure on windowpane flounder gills as observed by scanning electron microscopy. Journal of Fish Biology 33:571–580.

Perkins, E. M. 1995. An overview of hepatic neoplasms, putatively preneoplastic lesions, and associated conditions in fish sampled during the County Sanitation Districts of Orange County's 1986–1992 Ocean Monitoring Program. Bulletin of the Southern California Academy of Sciences 94:75–91.

Pierce, K. V., B. B. McCain, and S. R. Wellings. 1978. Pathology of hepatomas and other liver abnormalities in English sole (*Parophrys vetulus*) from the Duwamish River estuary, Seattle, Washington. Journal of the National Cancer Institute 60:1445–1453.

Preece, A. 1972. A manual for histologic technicians, 3rd edition. Little, Brown, Boston.

Premdas, P. D., and C. D. Metcalfe. 1994. Regression, proliferation, and development of lip papilloma in wild white suckers, *Catostomus commersoni*, held in the laboratory. Environmental Biology of Fishes 40:263–269.

Premdas, P. D., and C. D. Metcalfe. 1996. Experimental transmission of epidermal lip papillomas in white sucker, *Catostomus commersoni*. Canadian Journal of Fisheries and Aquatic Sciences 53:1018–1029.

Premdas, P. D., T. L. Metcalfe, M. E. Bailey, and C. D. Metcalfe. 1995. The prevalence and histological appearance of lip papilloma in white suckers (*Catostomus commersoni*) from two sites in central Ontario, Canada. Journal of Great Lakes Research 21:207–218.

Pritchard, J. B., and J. L. Renfro. 1982. Interactions of xenobiotics with teleost renal function. Pages 51–106 *in* L. J. Weber, editor. Aquatic toxicology. Raven Press, New York.

Puget Sound Water Quality Action Team. 2000. 2000 Puget Sound update, seventh report of the Puget Sound Ambient Monitoring Program. Puget Sound Water Quality Action Team, P. O. Box 40900, Olympia, Washington 98504-0900.

Randi, A. S., J. M. Montserrat, E. M. Rodriguez, and L. A. Romano. 1996. Histopathological effects of cadmium on the gills of the freshwater fish, *Macropsobrycon uruguayanae* Eigenmann (Pisces, Atherinidaie). Journal of Fish Diseases 19:311–322.

Reichert, W. L., M. S. Myers, K. Peck-Miller, B. L. French, B. F. Anulacion, T. K. Collier, J. E. Stein, and U. Varanasi. 1998. Genotoxic events in marine fish from exposure to complex mixtures of environmental contaminants. Mutation Research/Reviews in Mutation Research 411:215–225.

Reid, S. D., and D. G. McDonald. 1988. Effects of cadmium, copper, and low pH on ion fluxes in rainbow trout, *Salmo gairdneri*. Canadian Journal of Fisheries and Aquatic Sciences 45:244–253.

Reimschuessel, R. 1993. Postmortem examination. Pages 160–165 *in* M. K. Stoskopf, editor. Fish medicine. Saunders, Philadelphia.

Reimschuessel, R., R. O. Bennett, E. B. May, and M. M. Lipsky. 1989. Renal histopathological changes in the goldfish (*Carassius auratus*) after sublethal exposure to hexachlorobutadiene. Aquatic Toxicology 15:169–180.

Reimschuessel, R., R. O. Bennett, E. B. May, and M. M. Lipsky. 1990a. Renal tubular cell regeneration, cell proliferation and chronic nephrotoxicity in the goldfish *Carassius auratus* following exposure to a single sublethal dose of hexachlorobutadiene. Diseases of Aquatic Organisms 8:211–224.

Reimschuessel, R., R. O. Bennett, E. B. May, and M. M. Lipsky. 1990b. Development of newly formed nephrons in the goldfish kidney following hexachlorobutadiene-induced nephrotoxicity. Toxicologic Pathology 18:32–38.

Rhodes, L. D., M. S. Myers, W. D. Gronlund, and B. B. McCain. 1987. Epizootic characteristics of hepatic and renal lesions in English sole, *Parophrys vetulus*, from Puget Sound. Journal of Fish Biology 32:395–407.

Roberts, R. J., editor. 1989. Fish pathology, 2nd edition. Bailliere Tindall, Saunders, London.

Roch, M., and E. J. Maly. 1979. Relationship of cadmium-induced hypocalcemia with mortality in rainbow trout (*Salmo gairdneri*) and the influence of temperature on toxicity. Journal of the Fisheries Research Board of Canada 36:1297–1303.

Roomi, M. W., R. K. Ho, D. S. R. Sarma, and E. Farber. 1985. A common biochemical pattern preneoplastic hepatocyte nodules generated in four different models in the rat. Cancer Research 45:564–571.

Sangalang, G. B., H. C. Freeman, and R. Crowell. 1981. Testicular abnormalities in cod (*Gadus morhua*) fed Arochlor 1254. Archives of Environmental Contamination and Toxicology 10:617–626.

Saucier, D., L. Astic, P. Rioux, and F. Godinot. 1991. Histopathological changes in the olfactory organ of rainbow trout (*Oncorhynchus mykiss*) induced by early chronic exposure to a sublethal copper concentration. Canadian Journal of Zoology 69:2239–2245.

Sauer, G. R., and N. Watabe. 1984. Zinc uptake and its effect on calcification in the scales of the mummichog, *Fundulus heteroclitus*. Aquatic Toxicology 5:51–66.

Schiewe M. H, D. D. Weber, M. S. Myers, F. J. Jacques, W. L. Reichert, C. A. Krone, D. C. Malins, B. B. McCain, S.-L. Chan, and U. Varanasi. 1991. Induction of foci of cellular alteration and other hepatic lesions in English sole (*Parophrys vetulus*) exposed to an extract of an urban marine sediment. Canadian Journal of Fisheries and Aquatic Sciences 48:1750–1760.

Schillings, P. H. M, and J. H. S. Stekhoven. 1980. Atlas of glomerular histopathology. S. Karger, New York.

Schlesselman, J. 1982. Case-control studies: design, conduct and analysis. Oxford University Press, New York.

Schmitt, C. J., V. S. Blazer, G. M. Dethloff, D. E. Tillitt, T. S. Gross, W. L. Bryant Jr., L. R. DeWeese, S. B. Smith, R. W. Goede, T. M. Bartish, and T. J. Kubiak. 1999. Biomonitoring of Environmental Status and Trends (BEST) Program: field procedures for assessing the exposure of fish to environmental contaminants. Information and Technology Report USGS/BRD-1999–0007. U.S. Geological Survey, Biological Resources Division, Columbia, Missouri.

Scholin, C. A., F. Gulland, G. J. Doucette, S. Benson, M. Busman, F. P. Chavez, J. Cordaro, R. DeLong, A. De Vogelaere, J. Harvey, M. Haulena, K. Lefebvre, T. Lipscomb, S. Loscutoff, L. J. Lowenstine, R. Marin III, P. E. Miller, W. A. McLellan, P. D. R. Moeller, C. L. Powell, T. Rowles, P. Silvagni, M. Silver, T. Spraker, V. Trainer, and F. M. Van Dolah. 2000. Mortality of sea lions along the central California coast linked to a toxic diatom bloom. Nature (London)403:80–84.

Schwaiger, J., H. F. Falk, F. Bucher, G. Orthuber, R. Hoffmann, and R. D. Negele. 1994. Prolonged exposure of rainbow trout *(Oncorhynchus mykiss)* to sublethal concentrations of bis(tri-n-butyltin) oxide: effects on leucocytes, lymphatic tissues and phagocytosis activity. Pages 113–123 *in* R. Müler and R. Lloyd, editors. Sublethal and chronic effects of pollutants on freshwater fish. Blackwell, Cambridge, Massachusetts.

Schwaiger, J., R. Wanke, S. Adam, M. Pavert, W. Honnen, and R. Triebskorn. 1997. The use of histopathological indicators to evaluate contaminant-related stress in fish. Journal of Aquatic Ecosystem Stress and Recovery 6:75–96.

Sharples, A. D., D. Campin, and C. W. Evans. 1994. Fin erosion in a feral population of goldfish (*Carassius auratus* L.) exposed to bleached kraft mill effluent. Journal of Fish Diseases 17:483–493.

Sharples, A. D., and C. W. Evans. 1996. Pathology of fin erosion in goldfish *Carassius auratus*. Diseases of Aquatic Organisms 24:81–91.

Sherwood, M. J., and A. J. Mearns. 1977. Environmental significance of fin erosion in southern California demersal fishes. Annals of the New York Academy of Sciences 298:177–189.

Simpson, M. G., M. Parry, A. Kleinkauf, D. Swarbreck, P. Walker, and R. T. Leah. 2000a. Pathology of the liver, kidney, and gonad of flounder (*Platichthys flesus*) from a UK estuary impacted by endocrine disrupting chemicals. Marine Environmental Research 50:283–287.

Simpson, M. G., P. S. Widdowson, P. Walker, and R. T. Leah. 2000b. Mammalian neuropathology: a paradigm for the study of teleost neurotoxicity? Marine Environmental Research 50:126.

Sindermann, C. J. 1990. Principal diseases of marine fish and shellfish, volume 1, 2nd edition. Academic Press, New York.

Slooff, W. 1982. Skeletal anomalies in fish from polluted surface waters. Aquatic Toxicology 2:157–173.

Smith, C. E., T. H. Peck, R. J. Klauda, and J. B. McLaren. 1979. Hepatomas in Atlantic tomcod *Microgadus tomcod* (Walbaum) collected in the Hudson River estuary in New York. Journal of Fish Diseases 2:313–319.

Smith, I. R., K. W. Baker, M. A. Hayes, and H. W. Ferguson. 1989b. Ultrastructure of malpighian and inflammatory cells in epidermal papillomas of white suckers *Catostomus commersoni*. Diseases of Aquatic Organisms 6:17–26.

Smith, I. R., H. W. Ferguson, and M. A. Hayes. 1989a. Histopathology and prevalence of epidermal papillomas epidemic in brown bullhead, *Ictalurus nebulosus* (Lesueur), and white sucker, *Catostomus commersoni* (Lacepède) populations from Ontario, Canada. Journal of Fish Diseases 12:373–388.

Smith, I. R., and B. A. Zajdlik. 1987. Regression and development of epidermal papillomas affecting white suckers *Catostomus commersoni* (Lacepède), from Lake Ontario, Canada. Journal of Fish Diseases 10:487–494.

Smolowitz, R. M., M. J. Moore, and J. J. Stegeman. 1989. Cellular distribution of cytochrome P450E in winter flounder liver with degenerative and neoplastic disease. Marine Environmental Research 28:441–446.

Solangi, M. A., and R. M. Overstreet. 1982. Histopathological changes in two estuarine fishes, *Menidia beryllina* (Cope) and *Trinectes maculatus* (Bloch and Schneider), exposed to crude oil and its water-soluble fractions. Journal of Fish Diseases 5:13–35.

Speare, D. J., and H. W. Ferguson. 1989. Fixation artifacts in rainbow trout *(Salmo gairdneri)* gills: a morphometric evaluation. Canadian Journal of Fisheries and Aquatic Sciences 46:780–785.

Spitsbergen, J. M., K. A. Schat, J. M. Kleeman, and R. E. Peterson. 1988. Effects of 2,3,7,8-tetrachlorobenzo-p-dioxin or Aroclor 1254 on the resistance of rainbow trout, *Salmo gairdneri* Richardson, to infectious haematopoietic necrosis virus. Journal of Fish Diseases 11:73–83.

Spitsbergen, J. M., M. K. Walker, J. R. Olson, and R. E. Peterson. 1991. Pathologic alterations in early life stage of lake trout, *Salvelinus namaycush*, exposed to 2,3,7,8-tetrachlorodibenzo-*p*-dioxin as fertilized eggs. Aquatic Toxicology 19:41–72.

Stalker, M. J., B. M. Kirby, T. E. Kocal, I. R. Smith, and M. A. Hayes. 1991. Loss of glutathione-S-transferases in pollution-associated liver neoplasms in white suckers (*Catostomus commersoni*) from Lake Ontario. Carcinogenesis 12:2221–2226.

Stegeman, J. J., and M. E. Hahn. 1994. Biochemistry and molecular biology of monooxygenases: current perspectives on forms, functions, and regulation of cytochrome P450 in aquatic species. Pages 87–206 *in* D. C. Malins and G. K. Ostrander, editors. Aquatic toxicology: molecular, biochemical, and cellular perspectives. Lewis Publishers, Boca Raton, Florida.

Stegeman, J. J., R. M. Smolowitz, and M. E. Hahn. 1991. Immunohistochemical local-ization of environmentally induced cytochrome P450IA1 in multiple organs of the marine teleost *Stenotomus chrysops* (Scup). Toxicology and Applied Phar-macology 110:486–504.

Stehr, C. M., L. L. Johnson, and M. S. Myers. 1998. Hydropic vacuolation in the liver of three species of fish from the U.S. West Coast: lesion description and risk assessment associated with contaminant exposure. Diseases of Aquatic Organ-isms 32:119–135.

Stehr, C. M., and M. S. Myers. 1990. The ultrastructure and histology of cholangiocellular carcinomas in English sole (*Parophrys vetulus*) from Puget Sound, Washington. Toxicologic Pathology 18:362–372.

Stehr, C. M., M. S. Myers, D. G. Burrows, M. M. Krahn, J. P. Meador, B. B. McCain, and U. Varanasi. 1997. Chemical contamination of associated liver diseases in two species of fish from San Francisco Bay and Bodega Bay. Ecotoxicology 6:35–65.

Stehr, C. M., M. S. Myers, and M. L. Willis. 1993. Collection of fish tissues for the National Benthic Surveillance Project: necropsy procedure, tissue processing, and diagnostic procedure for histopathology. Pages 63–69 *in* G. G. Lauenstein and A. Y. Cantillo, editors. Sampling and analytical methods of the National Status and Trends Program National Benthic Surveillance and Mussel Watch Projects, 1984–1992, volume II. Comprehensive descriptions of complementary measurements. NOAA Technical Memorandum NOS ORCA 71. National Oce-anic and Atmospheric Administration, Silver Spring, Maryland.

Stein, J. E., T. K. Collier, W. L. Reichert, E. Casillas, T. Hom, and U. Varanasi. 1992. Bioindicators of contaminant exposure and sublethal effects: studies with benthic fish in Puget Sound, Washington. Environmental Toxicology and Chemistry 11:701–714.

Stein, J. E., W. L. Reichert, and U. Varanasi. 1994. Molecular epizootiology: assess-ment of exposure to genotoxic compounds in teleosts. Environmental Health Perspectives 102 (Supplement 12):19–23.

Stephen, C., M. L. Kent, and S. C. Dawe. 1993. Hepatic megalocytosis in wild and farmed chinook salmon *Oncorhynchus tshawytscha* in British Columbia, Canada. Diseases of Aquatic Organisms 16:35–39.

Stewart, J. S., and P. C. Brunjes. 1990. Olfactory bulb and sensory epithelium in goldfish: morphological alterations accompanying growth. Developmental Brain Research 54:187–193.

Stott, G. G., W. E. Haensley, J. M. Neff, and J. R. Sharp. 1983. Histopathologic survey of ovaries of plaice, *Pleuronectes platessa* L., from Aber Wrac'h, and Aber Benoit, Brittany, France: long-term effects of the Amoco Cadiz crude oil spill. Journal of Fish Diseases 6:429–437.

Stromberg, P. C., J. G. Ferrante, and S. Carter. 1983. Pathology of lethal and sublethal exposure of fathead minnows, *Pimephales promelas*, to cadmium: a model for aquatic toxicity assessment. Journal of Toxicology and Environmental Health 11:247–259.

Susser, M. 1986. Rules of inference in epidemiology. Regulatory Toxicology and Pharmacology 6:116–128.

Syasina, I. G., A. S. Sokolowsky, and M. Phedorova. 1999. Skin tumors in *Pleuronectes obscurus* (Pleuronectidae) represent a complex combination of epidermal pap-illoma and rhabdomyosarcoma. Diseases of Aquatic Organisms 39:49–57.

Tafanelli, R., and R. C. Summerfelt. 1975. Cadmium-induced histopathological changes in goldfish. Pages 613–645 *in* W. E. Ribelin and G. Migaki, editors. The pathology of fishes. University of Wisconsin Press, Madison.

Takashima, F., and T. Hibiya, editors. 1995. An atlas of fish histology, normal and pathological features, 2nd edition. Kodansha, Tokyo.

Tan, B., P. Melius, and J. Grizzle. 1981. Hepatic enzymes and tumor histopathology of black bullhead with papillomas. Pages 377–386 *in* M. Cooke and A. J. Dennis, editors. Polynuclear aromatic hydrocarbons: chemical analysis and biological fate. Battelle Press, Columbus, Ohio.

Teh, S. J., S. M. Adams, and D. E. Hinton. 1997. Histopathologic biomarkers in feral freshwater fish populations exposed to different types of contaminant stress. Aquatic Toxicology 37:51–70.

Teh, S. J., and D. E. Hinton. 1993. Detection of enzyme histochemical markers of hepatic preneoplasia and neoplasia in medaka (*Oryzias latipes*). Aquatic Toxicology 24:163–182.

Tetra Tech. 1987. Recommended protocols for fish pathology studies in Puget Sound. In Recommended protocols for measuring selected environmental variables in Puget Sound. Final report TC-3338-04, USEPA, Region X, Office of Puget Sound. Tetra Tech, Bellevue, Washington.

Thomas, P., M. Bally, and J. M. Neff. 1982. Ascorbic acid status of mullet *Mugil cephalus* Linn., exposed to cadmium. Journal of Fish Biology 20:183–196.

Thulin, J., J. Hoglund, and E. Lindesjoo. 1988. Diseases and parasites of fish in a bleached kraft mill effluent. Water Science and Technology 2:179–180.

Thuvander, A., E. Wiss, and L. Norrgren. 1993. Sublethal exposure of rainbow trout (*Oncorhynchus mykiss*) to Clophen A50: effects on cellular immunity. Fish and Shellfish Immunology 3:107–117

Timashova, L. V. 1981. Seasonal changes in the structure of the plaice, *Pleuronectes platessa*. Journal of Ichthyology 21:145–151.

Trump, B. F., R. T. Jones, and S. Sahaphong. 1975. Cellular effects of mercury on fish kidney tubules. Pages 585–612 *in* W. D. Ribelin and G. Migaki, editors. The pathology of fishes. University of Wisconsin Press, Madison.

Valentine, D. W. 1975. Skeletal anomalies in marine teleosts. Pages 695–716 *in* W. R. Ribelin and G. Migaki, editors. The pathology of fishes. University of Wisconsin Press, Madison.

Van Veld, P. A., U. Ko, W. K. Vogelbein, and D. J. Westbrook. 1991. Glutathione-S-transferase in intestine, liver and hepatic lesions of mummichog (*Fundulus heteroclitus*) from a creosote-contaminated environment. Fish Physiology and Biochemistry 9:369–376.

Van Veld, P. A., W. K. Vogelbein, M. K. Cochran, A. Goksøyr, and J. J. Stegeman. 1997. Route-specific cellular expression of cytochrome P4501A (CYP1A) in fish (*Fundulus heteroclitus*) following exposure to aqueous and dietary benzo[a]pyrene. Toxicology and Applied Pharmacology 142:348–359.

Van Veld, P. A., W. K. Vogelbein, R. Smolowitz, B. R. Woodin, and J. J. Stegeman. 1992. Cytochrome P450IA1 in hepatic lesions of a teleost fish (*Fundulus heteroclitus*) collected from a polycyclic aromatic hydrocarbon-contaminated site. Carcinogenesis 13:505–507.

Van Veld, P. A., D. J. Westbrook, B. R. Woodin, R. C. Hale, C. L. Smith, R. J. Huggett, and J. J. Stegeman. 1990. Induced cytochrome P450 in intestine and liver of spot

(*Leiostomus xanthurus*) from a polycyclic aromatic hydrocarbon contaminated environment. Aquatic Toxicology 17:119–132.

Varanasi, U., and D. J. Gmur. 1978. Influence of water-borne and dietary calcium on uptake and retention of lead by coho salmon *(Oncorhynchus kisutch)*. Toxicology and Applied Pharmacology 46:65–75.

Varanasi, U., M. Nishimoto, W. L. Reichert, and B.-T. L. Eberhart. 1986. Comparative metabolism of benzo[a]pyrene and covalent binding to hepatic DNA in English sole, starry flounder, and rat. Cancer Research 46:3817–3824.

Varanasi, U., W. L. Reichert, B.-T. Eberhart, and J. E. Stein. 1989a. Formation and persistence of benzo(a)pyrene-diolepoxide-DNA adducts in liver of English sole (*Parophrys vetulus*). Chemical-Biological Interactions 69:203–216.

Varanasi, U., W. D. Reichert, and J. E. Stein. 1989b. 32P-postlabelling analysis of DNA adducts in liver of wild English sole (*Parophrys vetulus*) and winter flounder (*Pseudopleuronectes americanus*). Cancer Research 49:1171–1177.

Verbost, P. M., G. Flik, R. A. C. Lock, and S. E. Wendelaar Bonga. 1987. Cadmium inhibition of Ca^{2+} uptake in rainbow trout gills. American Journal of Physiology 253:R216–R221.

Vethaak, A. D., and P. W. Wester. 1996. Diseases of flounder *Platichthys flesus* in Dutch coastal and estuarine waters, with particular reference to environmental stress factors. II. Liver histopathology. Diseases of Aquatic Organisms 26:99–116.

Vines, C. A., T. Robbins, F. J. Griffin, and G. N. Cherr. 2000. The effects of diffusible creosote-derived compounds on development in Pacific herring. Aquatic Toxicology 51:225–239.

Vishwanath, T. S., C. V. Mohan, and K. M. Shankar. 1998. Epizootic ulcerative syndrome (EUS), associated with a fungal pathogen, in Indian fishes: histopathology—"a case for invasiveness." Aquaculture 165:1–9.

Vogelbein, W. K., and J. W. Fournie. 1994. Ultrastructure of normal and neoplastic exocrine pancreas in the mummichog, *Fundulus heteroclitus*. Toxicologic Pathology 22:248–260.

Vogelbein, W. K., J. W. Fournie, P. S. Cooper, and P. A. Van Veld. 1999. Hepatoblastomas in the mummichog, *Fundulus heteroclitus* (L.), from a creosote-contaminated environment: a histologic, ultrastructural and immunohistochemical study. Journal of Fish Diseases 22:419–431.

Vogelbein, W. K., J. W. Fournie, and R. M. Overstreet. 1987. Sequential development and morphology of experimentally induced hepatic melano-macrophage centres in *Rivulus marmoratus*. Journal of Fish Biology 31 (Supplement A):145–153.

Vogelbein, W. K., J. W. Fournie, P. A. Van Veld, and R. J. Huggett. 1990. Hepatic neoplasms in the mummichog (*Fundulus heteroclitus*) from a creosote-contaminated site. Cancer Research 50:5978–5986.

Vogelbein, W. K., M. Unger, and D. Zwerner. 1998. Induction of liver neoplasms in *Fundulus heteroclitus* following laboratory exposure to contaminated sediment and diet. Marine Environmental Research 46:222–223.

Vogelbein, W. K., and D. W. Zwerner. 2000. The Elizabeth River Monitoring Program. 1998–99: mummichog liver histopathology as an indicator of environmental quality. Final report to the Virginia Department of Environmental Quality, Tidewater Regional Office, 5636 Southern Boulevard, Virginia Beach, Virginia 23462.

Vogelbein, W. K., D. E. Zwerner, M. A. Unger, C. L. Smith, and J. W. Fournie. 1997. Hepatic and extrahepatic neoplasms in a teleost fish from a polycyclic aromatic hydrocarbon-contaminated habitat in Chesapeake Bay, USA. Pages 55–63 *in* L. Rossi, R. Richardson, and J. Harshbarger, editors. Spontaneous animal tumors: a survey. (Proceedings of the First World Conference on Spontaneous Animal Tumors, Genoa, Italy, April 28–30, 1995). Press Point di Abbiategrasso, Milan, Italy.

Walsh, A. H., and W. E. Ribelin. 1975. The pathology of insecticide poisoning. Pages 515–557 *in* W. E. Ribelin and G. Migaki, editors. Pathology of fishes. University of Wisconsin Press, Madison.

Walter, G. L., P. D. Jones, and J. P. Geisy. 2000. Pathologic alterations in adult rainbow trout, *Oncorhynchus mykiss*, exposed to dietary 2,3,7,8-tetrachlorodibenzo-*p*-dioxin. Aquatic Toxicology 50:287–299.

Weibel, E. R. 1979. Stereological methods, volume I. Academic Press, New York.

Weibel, E. R. 1980. Stereological methods, volume II. Academic Press, New York.

Weis, J. S., and P. Weis. 1976a. Optical malformations induced by insecticides in embryos of the Atlantic silverside, *Menidia menidia*. Fisheries Bulletin 74:208–211.

Weis, J. S., and P. Weis. 1989. Effects of environmental pollutants on early fish development. Aquatic Science 1:45–73.

Weis, P., and J. S. Weis. 1976b. Abnormal locomotion associated with skeletal malformations in the sheepshead minnow, *Cyprinodon variegatus*, exposed to malathion. Environmental Research 12:196–200.

Wellings, S. R., C. E. Alpers, B. B. McCain, and B. S. Miller. 1976. Fin erosion disease of starry flounder (*Platichthys stellatus*) and English sole (*Parophrys vetulus*) in the estuary of the Duwamish River, Seattle, Washington. Journal of the Fisheries Research Board of Canada 33:2577–2586.

Wells, D. E., and A. A. Cowan. 1982. Vertebral dysplasia in salmonids caused by the herbicide trifluralin. Environmental Pollution (Series A)29:249–260.

Wester, P. W., and J. H. Canton. 1986. Histopathological study of *Oryzias latipes* (medaka) after long-term β-hexachlorocyclohexane exposure. Aquatic Toxicology 9:21–45.

Wiklund, T. 1994. Skin ulcer disease of flounder (*Platichthys flesus*): disease patterns and characterization of an etiological agent. Ph.D. dissertation. Department of Biology, Abo Academi University, Abo, Finland.

Wiklund, T., and G. Bylund. 1993. Skin ulcer disease of flounder *Platichthys flesus* in the northern Baltic Sea. Diseases of Aquatic Organisms 17:165–174.

Wilcock, B. P., and T. W. Dukes. 1989. The eye. Pages 156–173 *in* H. W. Ferguson, editor. Systemic pathology of fish. Iowa State University Press, Ames.

Williams, G. M., and J. H. Weisburger. 1986. Pages 65–82 *in* C. D. Klaasen, M. O. Amdur, and J. Doull, editors. Toxicology, the basic science of poisons, 3rd edition. Macmillan, New York.

Willoughby, L. G., R. J. Roberts, and S. Chinabut. 1995. *Aphanomyces invaderis* sp. nov., the fungal pathogen of freshwater tropical fish affected by epizootic ulcerative syndrome. Journal of Fish Diseases 18:273–275.

Wolke, R. E. 1992. Piscine macrophage aggregates: a review. Annual Reviews of Fish Diseases 2:91–108.

Yasutake, W. T. 1975. Fish viral diseases: clinical, histopathological, and comparative aspects. Pages 247–271 *in* W. E. Ribelin and G. Migaki, editors. The pathology of fishes. University of Wisconsin Press, Madison.

Yasutake, W. T., and J. H. Wales. 1983. Microscopic anatomy of salmonids: an atlas. U.S. Fish and Wildlife Service Resource Publication 150. U.S. Department of Interior, Fish and Wildlife Service, Washington, D.C.

Yonkos, L. T., D. J. Fisher, D. A. Wright, and A. S. Kane. 2000. Pathology of fathead minnows (*Pimephales promelas*) exposed to chlorine dioxide and chlorite. Marine Environmental Research 50:267–271.

Ziskowski, J. J., L. Despres-Patanjo, R. A. Murchelano, A. B. Howe, D. Ralph, and S. Atran. 1987. Disease in commercially valuable fish stocks in the northwest Atlantic. Marine Pollution Bulletin 18:496–504.

8

Evaluating Stress in Fish Using Bioenergetics-Based Stressor-Response Models

Daniel W. Beyers and James A. Rice

Foundations

An Energetic Perspective on Stress

Quantifying the influence of human-induced stress on energetics of aquatic organisms has been a recurrent theme in environmental science (Fry 1947; Warren 1971; Rice 1990; Beyers et al. 1999a, 1999b). Stressors may be natural, such as daily or seasonal temperature fluctuations, or anthropogenic such as contaminants, thermal effluents, or physical habitat modifications. Rice (1990) described applications of bioenergetics modeling for evaluating stress in fish. Since publication of Rice's (1990) chapter, applications of bioenergetics models have expanded, but there have been few attempts to use the approach for evaluating effects of stressors on organisms. The lack of application to stressors is partially due to different interpretations of how stressors are defined. In traditional applications of bioenergetics models, environmental conditions such as water temperature and prey density are regarded as external factors (driving variables) that regulate food consumption and fish physiology. Alternatively, these external factors can be regarded as natural sources of physiological stress to fish.

Selye (1956, 1973) developed a stress concept that describes how external factors influence the physiological condition of organisms. Selye (1973) defined stress as "the nonspecific response of the body to any demand made upon it" and defined stressor as the environmental factor that elicits stress. These definitions distinguish between specific and nonspecific effects of a stressor. A specific effect is the direct biological response to the stressor; for example, enzyme inhibition after exposure to an acetylcholinesterase-inhibiting pesticide. Nonspecific effects are all the accompanying changes that occur as an organism compensates the effects of enzyme inhibition (e.g., increased respiratory or heart rates). In theory, the distinction between specific and nonspecific effects is useful, but in practice it is difficult to separate the relative contribution of each in the response of an organism to a stressor. Consequently, it is common for investigators to refrain from distinguishing the relative contribution of specific and nonspecific responses to a stressor and to measure the cumulative effect from the two sources (see definitions in Adams 1990 and Newman 1995).

When organisms are exposed to stressors, their physiological response follows a pattern known as the general adaptation syndrome (Selye 1956, 1973; Beyers et al. 1999b), which is characterized by a three-phase time-dependent process (Figure 1). The first phase of the general adaptation syndrome is the stage of physiological alarm. During this stage, there is a stressor-induced loss, then recovery of compensatory ability. In fish, this stage may be associated with a loss of appetite, loss of equilibrium, and behavioral changes. The second phase of the general adaptation syndrome is the stage of resistance. This stage occurs when physiological adaptation is achieved and compensating for stressor effects becomes part of the normal cost of living for an exposed animal. This stage may be associated with increased metabolic and food consumption rates. If the stressor has sufficient magnitude and duration, the syndrome progresses to the stage of exhaustion. During this stage, cumulative effects of exposure cause premature death of the individual. Mortality occurs when compensating mechanisms break down because they are unable to sustain the level of action required to offset stressor effects. Despite some disagreement about the underlying mecha-

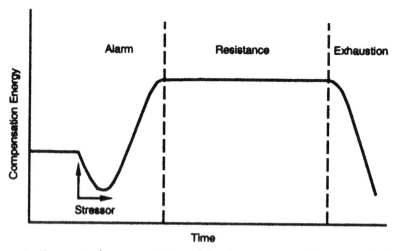

Figure 1. Conceptual representation of the three stages of the general adaptation syndrome illustrating how energy expenditure for compensating stressor effects changes with time. (Reproduced with modification from Selye 1956.)

nisms of the general adaptation syndrome (see Barton 1997), predictions based on it are consistent with empirical data that reveal energetic responses of fish to chemical stressors (O'Hara 1971; De Boeck et al. 1995, 1997; Beyers et al. 1999b).

Selye measured stress and the progression of the general adaptation syndrome by quantifying physical and biochemical changes in individuals, but stress can also be quantified by measuring energy intake and expenditures. For this energy-based perspective, stress can be defined as "the change in the energy budget resulting from the combined effects of specific and nonspecific responses to a stressor" (Beyers et al. 1999b). This broad definition of stress emphasizes that living organisms constantly expend energy to compensate effects of environmental change in order to survive, grow, and reproduce (Wedemeyer et al. 1984; Beyers et al. 1999b). The energetic cost of offsetting stressor effects can be represented by a compensation continuum. Stressors such as diel temperature fluctuations represent low-magnitude, short-duration environmental changes that organisms compensate routinely. Stressors such as chemical spills represent high-magnitude, long-duration environmental changes that may influence bioenergetics of exposed animals for weeks or even years. The compensation continuum concept avoids having to distinguish normal environmental variation from disturbance and emphasizes that effects of all environmental changes must be offset with respect to stressor magnitude and duration.

The Need for Integration

Ecological theory suggests that energy is a critical factor regulating the distribution and abundance of organisms (Hall et al. 1992). An energetics-based approach to evaluating stress is appealing because energy is a common currency of ecology, and it can be used to quantify effects of exposure regardless of whether the stressor is natural or anthropogenic. Common currency provides a basis for direct comparison of effects of different stressors and allows estimation of the cumulative cost of exposure to multiple stressors. Common currency also facilitates translation of effects to higher levels of biological organization. However, traditional approaches for quantifying effects of anthropogenic stressors have been dominated by toxicological methods that have not emphasized an energetics-based approach. Traditional toxicological analyses involve measurement of the response of growth, survival, or biochemical measures as functions of exposure to a stressor. Comparisons between exposed and unexposed test organisms are the basis for evaluating effects. In contrast, energetics-based analyses are well developed for applications in fisheries management and fish ecology. Bioenergetics models for fish represent an energy budget that balances energy intake from prey consumption against expenditures for metabolism, waste production, and storage of surplus energy in the form of growth. The basic bioenergetics model has the form

$$C = R + A + S + F + U + \Delta B \tag{1}$$

where C is food consumption, R is metabolism, A is activity, S is specific dynamic action (the cost of processing food), F is egestion (feces), U is excretion (urine), and ΔB is somatic growth (and reproductive growth in mature individuals). Food consumption is estimated using size- and temperature-dependent functions and a proportionality constant that adjusts for the fraction of maximum daily ration consumed by fish. Metabolic rate is estimated using size-, temperature-, and activity-dependent functions. The remaining components of the energy budget are primarily dependent on the amount and type of food eaten, but may also be influenced by temperature. Detailed presentations of fish bioenergetics models and their applications are available elsewhere (Brett and Groves 1979; Kitchell 1983; Adams and Breck 1990; Hansen et al. 1993; Ney 1993; Hanson et al. 1997).

Fish bioenergetics models have been intensively used for management and scientific purposes for over 20 years. Model-evaluation studies have demonstrated that when inputs (e.g., temperature regime and food consumption rate) are well known, there is good agreement between output predictions (e.g., growth) and observed responses (Rice and Cochran 1984; Whitledge et al. 1998; Beyers et al. 1999a; Madenjian et al. 2000). In addition, sensitivity analysis has been used to evaluate model behavior and predictions (Kitchell et al. 1977; Stewart et al. 1983; Bartell et al. 1986; Beyers et al. 1999a). Consequently, bioenergetics models are an excellent foundation for analyses of effects of stressors on fish. By merging traditional toxicological methods with bioenergetics modeling, an integration of two well-developed scientific fields can be achieved, with great potential for advancing the rigor, accuracy, and generality of environmental assessments (Table 1).

This chapter demonstrates the advantages of linking bioenergetics models with stressor-response relationships. The integrated bioenergetics-based stressor-response model (SRM) can be used to predict effects of anthropogenic and natural stressors on fish living in a fluctuating environment (Figure 2). Within this framework, the basic bioenergetics element of a SRM accounts for individual characteristics of fish and adjusts their energy budget in response to natural environmental stressors. The stressor-response element integrates these natural affects with those from anthropogenic sources according to stressor magnitude and duration. The SRM reallocates energy expenditures so that costs of maintenance are accounted for and effects of exposure are expressed as changes in growth or reproduction.

In the sections that follow, we describe strategies and approaches for general application of SRMs and review recent investigations of the energetic responses of fish to chemical, thermal, and physical stressors. We also present example applications that demonstrate how SRMs can be used to isolate or

Table 1. Comparisons of key characteristics of the scientific disciplines of fish bioenergetics and aquatic toxicology.

Fish bioenergetics	Aquatic toxicology
1. Energy is a fundamental universal currency; a variety of useful transformations are available.	1. Qualitatively different currencies or endpoints routinely used (e.g., survival, growth, enzymatic activity).
2. Effects of natural stressors can be easily accumulated and integrated with energetic cost of anthropogenic stressors.	2. No mechanism for superimposing effects of natural stressors onto laboratory-derived exposure–response relationships.
3. Strong foundation for quantitative prediction of effects of exposure to stressors using computer modeling.	3. Strong foundation for quantitative description of stressor–response relationships.
4. Emphasis on predicting effects in the field based on mechanistic relationships derived from laboratory experiments.	4. Emphasis on laboratory experiments that describe effects of stressor exposure with minimal uncertainty.
5. Emphasis on ecologically relevant endpoints (e.g., growth rates, population characteristics).	5. Emphasis on easily measured endpoints of limited ecological relevance (e.g., enzyme activity, lethal concentrations).
6. Rigor of model evaluation varies; no widely accepted procedures exist.	6. Quality assurance procedures have been established to ensure reliable data.
7. Not widely applied for risk assessment, but consistent and easily adapted for probabilistic methods.	7. Methodology for risk assessment is well developed.

combine effects of multiple stressors, and we conclude with a discussion about advantages and limitations of investigating effects of stressors from a bioenergetics perspective.

General Application of Stressor-Response Models

Null Versus Direct Modeling Approaches

Rice (1990) described two general strategies for using bioenergetics modeling to evaluate stress in fish: a null model approach and a direct approach. Selection of the appropriate strategy depends on objectives of the investigation and quality of available data. The null model approach uses an existing bioenergetics model to evaluate alternative hypotheses. A "healthy fish" model is used as a null hypothesis. Predictions from the basic or healthy fish bioenergetics model are compared to observed growth patterns to determine if fluctuations in natural environmental factors alone are sufficient to explain an observed outcome. Rice et al. (1983) used this approach to test the hypothesis that changes in water temperature could account for seasonal variation in the condition of largemouth bass *Micropterus salmoides* in Par Pond, South Carolina. Alternative hypotheses included stress effects due to infec-

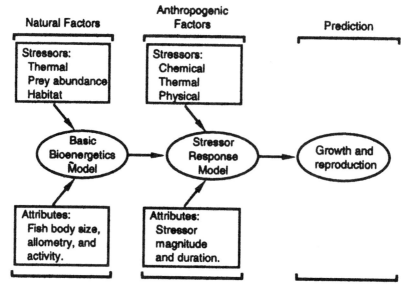

Figure 2. Illustration of how effects of natural and anthropogenic stressors are integrated using a bioenergetics-based stressor-response model (SRM). The basic bioenergetics model accounts for individual attributes of fish and adjusts their energy budget in response to natural stressors. The SRM integrates these effects with those from anthropogenic sources according to stressor magnitude and duration. The integrated model reallocates energy expenditures so that costs of maintenance are met and resulting effects are manifested as changes in growth or reproduction. (Reproduced with modification from Beyers et al. 1999a.)

tion, high contaminant concentrations, variation in activity, decline in condition from spawning, and seasonal changes in food availability. Bioenergetics analysis, using the null-model approach, suggested that the change in fish condition could be explained by a combination of seasonal variation in prey availability and water temperature and that more complicated interactions involving infection, contaminants, or activity were not required. This conclusion was subsequently confirmed by field investigations of factors that influence the condition of largemouth bass in Par Pond (Janssen and Giesy 1984). In this case, the null model approach was successfully used to eliminate several less parsimonious explanations and identify a likely hypothesis that was not apparent from the available data.

The second modeling approach is based on direct evaluation of effects of a stressor by explicitly modeling changes in energy acquisition and expenditure that result from exposure. Beyers et al. (1999a, 1999b) used this direct approach to assess the potential energetic costs incurred by largemouth bass living in a lake contaminated with sublethal concentrations of the persistent organochlorine pesticide, dieldrin. Traditional toxicological exposure methods were used to estimate concentration-response relationships for feeding and metabolic rates of largemouth bass. The relationships described the influence of dieldrin exposure on energy acquisition and expenditure and were integrated with a basic bioenergetics model for largemouth bass. The integrated SRM was used to predict the energetic cost of chemical exposure for fish living in a fluctuating natural environment. This analysis allowed direct comparison of effects of thermal and chemical stressors, and supported evaluation of potential costs and benefits of alternative management actions.

Essential Components of a Stressor-Response Model

Once a general strategy has been selected for assessing the bioenergetic effects of stressors, there are three main requirements for a successful application: (1) a basic bioenergetics model for the target species, (2) energetics-based exposure-response relationships for the stressor of interest, and (3) environmental data describing relevant natural and anthropogenic stressors in the field. For the first requirement, bioenergetics models are available for over 27 warm- and coldwater species, and life stage specific models have been developed for some fishes (Hanson et al. 1997). Most models are for species that have ecological, economic, or recreational value. In cases where a specific model is not available, a surrogate model can be constructed using parameters of closely related species (but see Hansen et al. 1993; Ney 1993). For the second requirement, exposure–response relationships can be quantified using standard toxicological methods that measure effects in terms of exposure- and time-dependent changes in food consumption, metabolic rate, activity, and growth (Beyers et al. 1999b). Other components of the basic bioenergetics budget (e.g., excretion) may also be influenced by stressor exposure, but are relatively unimportant because they do not have a strong influence on bioenergetics model predictions (Bartell et al. 1986). For the last requirement, attention should be given to describing the temporal and spatial variation in anthropogenic stressors and natural factors (e.g., water temperature) that influence fish bioenergetics. To ensure that simulated conditions accurately represent exposure conditions, environmental data should be collected with respect to life stage and the spatial and temporal distribution of the target species.

When dealing with chemical contaminants, incorporation of a bioaccumulation submodel is one potentially useful modification of a SRM. A bioaccumulation submodel allows the investigator to estimate the chemical dose and body burden associated with various exposure scenarios. Both measures should be related to the mechanism of toxic action within an organism and are expected to be better predictors of effects than waterborne or dietary exposure concentrations (McCarty and Mackay 1993). Linking bioaccumulation and bioenergetics models has produced some valuable insights about toxic effects and chemical dynamics, but none of the applications have translated chemical dose into energetic cost (Borgman and Whittle

1992; Post et al. 1996; Rodgers 1996; Muller and Nisbet 1997; Penttinen and Kukkonen 1998).

Investigations of Bioenergetic Responses to Stressors

Chemical Stressors

The components of the basic bioenergetics budget (equation 1) represent the physiological systems that may be influenced by chemical stressors. Several reviews of toxicological literature have been conducted using an energetics perspective, and they provide detailed examples and discussions of potential physiological responses to chemical stressors (Warren 1971; Adams and Breck 1990; Rice 1990; Widdows and Donkin 1991; Heath 1995; Beyers et al. 1999b). In general, the response of each physiological system varies with the state of progression of the general adaptation syndrome. Typically, food consumption and metabolism are reduced at the initiation of sublethal chemical exposure. As physiological adaptation occurs and compensatory mechanisms are induced, an increase in metabolic rate may be observed, as well as a corresponding increase in food consumption. Activity may increase or decrease depending on mode of action. Specific dynamic action, egestion, and excretion may also be influenced by chemical exposure, but are typically considered to be relatively constant proportions of energy consumed. Fish growth is correlated with the other components of the bioenergetics budget and represents the cumulative, integrated response of the organism to all stressor effects.

Recent investigations of energy-related responses to chemical exposure have emphasized the study of food consumption, metabolism, food conversion efficiency, and growth (De Boeck et al. 1995, 1997; Christiansen et al. 1998; Beyers et al. 1999b). These studies provide empirical data about the variety of ways chemical exposure may influence fish bioenergetics and also support the time-dependent predictions of the general adaptation syndrome. Other studies emphasize the combined effects of chemical and environmental stressors on bioenergetics. Lemly (1993) studied the influence of winter water temperature and photoperiod on toxicity of selenium to juvenile bluegill *Lepomis macrochirus*. Winter conditions reduced food consumption, while selenium exposure increased metabolic rate. The combined effects of reduced energy acquisition and increased energy expenditure were reflected by temperature- and selenium concentration-dependent responses for body condition, lipid content, and mortality. Lipid depletion during winter is a common phenomenon for bluegill and other centrarchids, but the effect of selenium exposure on metabolism accelerated the process and reduced energetic reserves required for overwinter survival. Dockray et al. (1996) and Linton et al. (1997) evaluated effects on rainbow trout *Oncorhynchus mykiss* of acidification or increased ammonia exposure in combination with increased summer temperatures (+2°C). Acidification and increased ammonia represent degraded environmental conditions that currently exist, but may become more important if ambient temperatures increase. Summer temperatures may be influenced by thermal effluents from industrial sources, water diversion and return for irrigation, canopy removal from forest clear-cutting, and global warming. Acidification increased food consumption, growth, and conversion efficiency of rainbow trout fed to satiation, whereas warming to temperatures near the upper thermal tolerance reduced appetite, growth, and metabolism. Combined effects of both stressors appeared to offset each other and produced an intermediate outcome (Dockray et al. 1996). Similarly, exposure to ammonia increased growth and conversion efficiency of rainbow trout, whereas warming reduced consumption and growth (Linton et al. 1997). However, in contrast to Dockray et al. (1996), the combined effects of ammonia exposure and warming did not offset each other and were reflected by slow growth and high metabolic rate. These examples illustrate the complexities involved in evaluating human-induced change within the context of a variable natural environment and the need for an approach that can predict combined effects of qualitatively different stressors.

Computer models that are based on well-established principles can contribute greatly to solving complex environmental problems. An assumption for a successful modeling application is that the fundamental principles that the computer model is based on (e.g., exposure–response relationships) are relevant to the simulated environmental conditions. When this is the case, several relatively simple submodels can be linked together to permit detailed understanding of the outcome of interacting processes. Although the resulting integrated model is based on simple principles, the outcome of the interactions can be too complex to be predicted intuitively. Computer programs make excellent bookkeeping tools for these applications because they eliminate guesswork about the predicted effects from multiple stressors. Rose et al. (1993) used this approach to link a bioenergetics model for striped bass *Morone saxatilis* with an individual-based population model. The integrated model allowed each individual in a population to have different bioenergetic characteristics. A variety of potential stressors were evaluated, including episodic and chronic exposure to toxics, temperature changes, and long-term reduction in livable habitat for various life stages. Model simulations began with the reproductive output from 100 adult female striped bass and followed the development and growth of eggs, larvae, and juveniles. Model predictions included the number of survivors to age-1, life stage duration, growth rates, survival, and mortality rates. The simulations suggest that chronic chemical exposures generally have proportionally greater effects than do temperature and episodic chemical exposures and that reduced habitat has the least influence on striped bass early life stages.

Thermal Stressors

In aquatic systems, water temperature has a dual role as a regulator of physiological processes and as a potential source of disturbance (damage). Water temperature acts as a regulator when thermal conditions are within the range of physiological adaptation of resident organisms. Within this range, physiological rates of poikilothermic animals increase or decrease with water temperature. When temperatures exceed the range of adaptation, normal function cannot be maintained because physiological rates approach zero at low temperatures, or denaturation of vital proteins and loss of integrity occur at high temperatures. From an energetics perspective, all thermal influences are sources of stress that must be compensated using resources represented in the bioenergetics budget. The influence of temperature on fish is so important that most bioenergetics models incorporate temperature-dependent relationships for food consumption, metabolism, and activity. Thus, empirical relationships do not need to be described to assess effects of thermal stressors on fish unless the investigation has an emphasis on atypical conditions such as rapid thermal changes or temperature extremes.

Inclusion of temperature dependence in fish bioenergetics models has made them a natural tool for assessing potential stress associated with global climate warming. Climate warming is anticipated to have a variety of effects on freshwater environments (Shuter and Meisner 1992). In lakes, climate warming will produce higher surface water temperatures, longer ice-free periods, and longer periods of thermal stratification. In rivers, warming will produce higher groundwater temperatures and corresponding increases in summer and winter water temperatures from headwaters to mouth. Thermal stress and the corresponding positive or negative energy flux associated with compensating climate-induced changes in water temperature will likely affect growth, reproduction, and survival of fish. Under most circumstances, these changes are likely to have positive effects for warm- and coolwater species, and negative effects for coldwater species. Many environmental assessments make the important assumption that fish thermoregulate behaviorally; that is, they seek out habitats where water temperature is close to their optimal temperature for growth. Consequently, predicted effects of climate warming on fish are strongly dependent on species' thermal optima and temperature availability. Shuter and Post (1990) used temperature dependence in fish bioenergetics to explain the historical geographic distribution of yellow perch *Perca flavescens*, Eurasian perch *P. fluviatilis*, and smallmouth bass *Micropterus dolomieu* and predicted that a warmer climate may allow these species to thrive north of their present distributions. A similar approach was used to assess the affect of climate warming on growth and prey consumption of warm-, cool-, and coldwater fishes in the Great Lakes (Hill and Magnuson 1990). This analysis emphasized the link between food-web dynamics and fish growth and suggested that climate warming will increase growth if there is a corresponding increase in prey consumption. This analysis also illustrated spatial links between site-specific physical habitat and access to temperatures that allow behavioral thermoregulation by fish and predicted that at some localities, lake stratification will constrain growth of coldwater fish because dissolved oxygen concentrations will not be sufficient at depths with optimal temperature.

Similar analyses of climate change have been conducted for stream fish. Ries and Perry (1995) used a bioenergetics model to assess the affect of global warming on growth and food consumption rates of brook trout *Salvelinus fontinalis* in central Appalachian streams. They examined the hypothesis that faster growth during winter (due to warming) will offset the opposing effects of above-optimal temperatures during summer, and concluded that (1) a 2°C warming will probably increase fish growth rates if a 15–20% increase in prey consumption also occurs, and (2) a 4°C increase will require a 30–40% increase in prey consumption to maintain current growth rates and may reduce the amount of habitat with suitable temperature regime. Van Winkle et al. (1997) used bioenergetics modeling to study effects of −2, +2, and +4°C climate change on reproductively active female rainbow trout in a California stream. The model tracked effects of temperature change on energy acquisition and energy allocation to respiration, growth, and reproduction of individual fish. The effects were complex because the model was designed to allow shifts in the balance between energy acquisition, energetic costs, and energy allocation at various times during an annual reproductive cycle. Predictions were that temperature increases will reduce growth and reproduction, but may improve fish condition, whereas temperature decreases will reduce growth, condition, gonad index, and reproduction. In addition, the analysis showed that small differences in physiological characteristics of individual fish can have a relatively large influence on survival, growth, and reproductive ability.

Temperature dependence in bioenergetics models has also made them a useful tool for assessing nonclimate-related reasons for shifts in geographical distribution of fish. Bioenergetics analysis and laboratory feeding trials were used to confirm conclusions of a field study that suggested human-induced changes in maximum annual water temperature and percent pool area were responsible for the decline of smallmouth bass and corresponding range expansion of largemouth bass in Ozark streams (Zweifel et al. 1999). Results showed that smallmouth bass have a lower optimum temperature for food consumption and a higher mass-specific demand for prey than largemouth bass. Human activities have increased the percentage of pool area in Ozark streams, which is associated with reduced food production and warmer water temperatures. Both these conditions favor largemouth bass since the species has a higher thermal optimum and lower demand for prey than smallmouth bass. Thus, the bioenergetics analysis provided a potential mechanistic explanation for correlative habitat–fish associations that were observed in the field.

In a more direct assessment of temperature-induced stress, Bevelhimer and Bennett (2000) integrated fish bioenergetics with a damage–repair model to investigate effects of thermal effluents on fish. Their model explicitly accounts for the influence of fluctuating environmental and effluent temperatures. The bioenergetics component of the model quantifies the influence of temperature on fish growth, while the damage–repair model estimates potentially lethal stress that accumulates when temperatures exceed established thresholds for adverse effects. The method for translating stress accumulated by the damage–repair model into energetic cost has not yet been developed, but the combined modeling framework is a useful general extension of bioenergetics models for stressors that have both sublethal and lethal effects.

Interactions of temperature effects with other environmental factors can be investigated by comparing model outputs to field data. Railsback and Rose (1999) investigated the effects of changing water temperature on stream-dwelling rainbow trout in California. Field and laboratory data were integrated with bioenergetics modeling to compare the relative effects of temperature and food consumption on predicted growth and to correlate model output to physical habitat, stream flow, temperature, and trout density. Growth was found to be affected more by factors controlling food consumption than by direct effects of temperature. Correlation analysis suggested that food consumption during fall and spring is influenced by temperature, stream flow, and trout density, and food consumption during summer is influenced by gradient, which may be correlated with food production.

Physical Stressors

Physical habitat of lakes and streams represents a three-dimensional space where fish potentially occur. Fish are constrained to a fraction of the total space by physical and environmental characteristics such as water velocity, temperature, dissolved oxygen, and by interactions with co-occurring organisms such as prey, predators, and competitors. Human-induced changes in physical habitat (physical stressors) can have broad energetic consequences; one of the most important consequences is an indirect effect on the abundance of food. Food consumption is the only energy input into a bioenergetics budget, and the quantity of food consumed influences all expenditures (equation 1). From a modeling perspective, it is difficult to predict absolutely how various physical changes will affect food consumption, but models can be used to estimate outcomes of a range of realistic scenarios to aid evaluation of potential costs and benefits. This approach was used to assess effects of alternative vegetation removal strategies on bluegill growth rates in lakes (Trebitz and Nibbelink 1996). It was hypothesized that edge habitat created by vegetation removal can increase per capita food availability and be reflected in faster bluegill growth rates. A computer model was used to combine bioenergetics, a functional feeding response, and vegetation geometry. Food consumption and growth of bluegill were dependent on plant-removal geometry, and the analysis suggested several potentially useful vegetation management strategies that should increase fish growth.

Other investigations of the influence of physical stressors on food consumption and bioenergetics of fish have been conducted in streams. Human modification and natural variation in stream discharge affect depth, velocity, substrate, cover, and water quality. These physical changes are sources of stress to fish and they respond by avoiding or minimizing stressor effects (Schreck et al. 1997). Avoiding stressors usually involves movement and associated changes in energy expenditures for activity. Accounting for activity costs of fish that are responding to physical stressors is challenging because it is difficult to observe and measure movement of fish in the field, but a variety of techniques have been used. Rincón and Lobón-Cerviá (1993) related seasonal shifts in microhabitat use of brown trout *Salmo trutta* to changing habitat characteristics in a river. Behavioral observations, temperature records, and bioenergetics relationships were used to estimate costs of swimming and standard metabolism in specific habitats. Seasonal changes in swimming cost were found to be more strongly related to water velocity than temperature. Other investigators have used physiological telemetry to obtain in situ measurements of activity (Lucas et al. 1993; Demers et al. 1996; Hinch and Rand 1998). Physiological telemetry provides real-time measurements of heart rate, ventilation rates, tailbeat frequency, or locomotor muscle contractions. These measures can be related to swim speed and used to estimate activity costs of fish in the field. For example, Hinch and Rand (1998) used physiological telemetry to assess swim speeds, behavior, and migration speeds of sockeye salmon *Oncorhynchus nerka* in different river reaches. A bioenergetics model was used to estimate reach-specific energy use per meter traveled based on average swim speed, migration time, body size, and river temperature. The affects of river features such as parallel banks, bends, and constrictions (physical stressors) on energy use were detected and estimated using this bioenergetics-based approach.

Many stream fishes incur energetic demands associated with maintaining position in flowing water because they prey on drifting invertebrates. Hill and Grossman (1993) showed that drift-feeding fish select habitats where net energy intake is maximized. They derived a model for net energy intake as a function of water velocity, based on the metabolic cost of swimming at velocity and the energetic gain associated with feeding on drift at that location. Predicted optimal velocities closely matched actual velocity use of rainbow trout and rosyside dace *Clinostomus funduloides*. This approach has been used to evaluate fish responses to changing discharge regimes. Braaten et al. (1997) compared predictions of net energy intake models to habitat use of trout in a large regulated river and a small headwater stream. Predictions of the model matched observed habitat use for fish in the large river, but not in the small stream where there may have been a mismatch between optimal velocity and habitats of sufficient depth to be occupied by fish. Spatially explicit analyses of net energy intake in streams can be achieved by linking fish bioenergetics with a hydraulic (physical) model (Ludlow and Hardy 1996). The hydraulic model component provides a spatial representation of water depths and velocities over a range of stream discharge magnitudes. The integrated bioenergetic and hydraulic model estimates the quantity and quality of fish habitat based on net energy intake. Application of the model suggested a reduction in habitat quality for rainbow trout and brown trout between summer and winter due to reduced velocities, temperature, and food availability. It also showed a different spatial distribution of optimal habitat between summer and winter that paralleled known habitat shifts of resident fish. This modeling approach is unique because it integrates a spatially explicit representation of a physical stressor (e.g., velocity) with bioenergetics constraints. Thus, it accounts for the interactions between changing discharge, water temperature, energetic cost of swimming, and velocity-dependent food availability. This example illustrates that in many aquatic systems, the physical environment is a dynamic background of stressors that have strong daily and seasonal influences on biological and chemical processes.

Incorporating a broad range of physical influences into assessments can provide valuable insights into system dynamics because simulated environmental conditions are controlled by the investigator. For example, Clark and Rose (1997) linked a hydraulic model of a hypothetical Appalachian stream to an individual-based model of resident rainbow and brook trout. The model simulated daily growth, mortality, movement, and spawning for both species for 100 years. Daily growth of fry, juveniles, and adults was calculated based on bioenergetics and consumption of invertebrate drift. The model simulated population dynamics and competition (biological stressors) between the two co-occurring species in a naturally fluctuating environment (thermal and physical stressors). Model predictions were in general agreement with field observations and, because mechanisms of environmental and species interactions were known, the results suggested several hypotheses that may explain competitive exclusion of brook trout in southern Appalachian streams.

Example Applications

Simulation Conditions

To demonstrate how a bioenergetics-based SRM can be used to analyze effects of stressors, we conducted six simulations (Table 2) based on the model described by Beyers et al. (1999a). The example simulations emphasize the flexibility of SRMs for assessing a variety of stressors, as well as the

Table 2. Conditions and outcomes of simulations using a bioenergetics-based stressor–response model to evaluate effects of natural and anthropogenic stressors on largemouth bass growth. B = baseline; T = thermal stressor; C = chemical stressor; D = dietary stressor.

Example simulation	Temperature regime	% maximum consumption	Chemical concentration (µg/L)	Final fish mass (g)
1. *(B)*	20°C	50	0	165
2. *(T)*	natural	50	0	158
3. *(T+C$_{continuous}$)*	natural	50	1	104
4. *(T+C$_{continuous}$)*	natural	50	$\bar{x}$ = 1; SD = 0.5	$\bar{x}$ = 100; SD = 2.71
5. *(T+C$_{episodic}$)*	natural	50	0, 1, 0	150
6. *(T+C$_{episodic}$+D)*	natural	50[a]	0, 1, 0	84.6

[a] Food consumption reduced by 40% on day 50 to 1.34 g/d or 50% of maximum consumption, whichever was less.

utility of the approach for unraveling complex interactions and revealing effects that are difficult to detect with other methods. Each simulation started with a 50-g largemouth bass feeding at 50% of its maximum ration for 160 d.

In the first example, the SRM was used to predict growth of control fish in a laboratory experiment where water temperature was a constant 20°C. This simulation represents application of the SRM to conditions commonly used in toxicological investigations (i.e., laboratory exposures). It also represents a baseline for evaluating effects of other natural and anthropogenic stressors.

In the second example, fish had the same initial size and ration used in the previous example, but growth was simulated using a natural thermal regime measured from a pond in Colorado (Beyers et al. 1999a). The regime started at 17°C on 9 May, increased to a maximum of about 25°C in midsummer, then declined to temperatures ranging from 13 to 15°C near the end of the simulation on 16 October (Figure 3A). Thus, the second simulation quantifies the energetic cost of exposure to a fairly typical fluctuating seasonal thermal regime.

The third example demonstrates the influence of continuous, chronic exposure to a waterborne chemical using conditions based on those likely to be observed in a contaminated pond. This simulation was based on example 2 with the addition of concentration- and time-dependent exposure–response relationships for the effects of dieldrin on largemouth bass metabolic and feeding rates (Beyers et al. 1999a). We assumed fish were continuously exposed to 1 mg/L dieldrin and constrained fish to respond as if they were fully acclimated to the chemical. It should be noted that this hypothetical dieldrin exposure scenario and the examples that follow are used to demon-

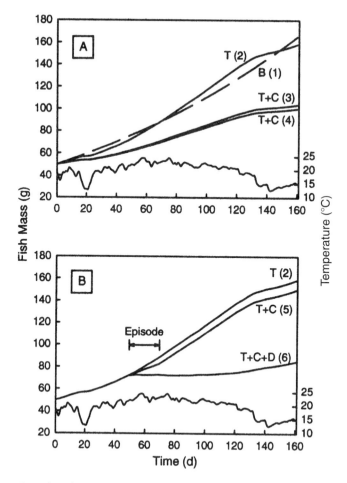

Figure 3. Simulated cumulative growth of largemouth bass exposed to natural and anthropogenic stressors. Predicted growth of fish in the laboratory at 20°C (dashed line) represents a baseline (B) for evaluating the effects of exposure to a thermal stressor (T), thermal and chemical stressors (T + C), and thermal, chemical, and dietary stressors (T + C+ D). The thermal stressor was the depicted natural temperature regime; the chemical stressor was simulated as a continuous exposure (upper panel) or 20-d episodic exposure that started on day 50 (lower panel); the dietary stressor was a simulated reduction in abundance of prey caused by the episodic chemical stressor. Numbers in parentheses refer to Table 2 for specific conditions and outcomes of each simulation.

strate applications of a SRM and are not intended to reflect existing conditions at any specific locality.

The fourth example illustrates how a SRM can be modified to incorporate stochastic variation in order to estimate probabilities of effects. The simulation was identical to example 3 except on each day of the 160-d exposure period, a new dieldrin concentration was selected from a normal distribution having a mean = 1 μg/L and SD = 0.5. This distribution was based on spatial and temporal variation of dieldrin concentrations in a pond (coefficient of variation = 0.5) sampled a total of 31 times at different locations on 13 occasions from June to November (Beyers et al. 1999a). Thus, we approximated realistic variation in dieldrin concentration based on observed data, but the average magnitude of concentration in the simulation was much higher than that observed in the field. One thousand solutions of the model were obtained in order to estimate mean and variance for the distribution of fish weights at the end of 160-d simulations.

The fifth example also evaluates the influence of dieldrin exposure, but is based on an episodic exposure using conditions patterned after those likely to be observed from a chemical spill in a riverine system. This simulation was based on example 2, with the addition of exposure to 1 µg/L dieldrin for 20 d starting on day 50. Dieldrin concentrations before and after the episode were 0 µg/L. We assumed that fish were not acclimated to dieldrin when the exposure began and that their metabolic rates changed as a function of time to a maximum limit approximated by the value for 16 d of exposure (Beyers et al. 1999b). We also made the conservative assumption that fish recovered instantaneously when the chemical episode subsided. In reality, at least several days are probably required for fish to recover from chemical exposure and for metabolism and food consumption rates to return to normal. However, data are not available to describe these recovery processes.

The last simulation incorporated the same thermal and chemical effects used in example 5, as well as an indirect dietary effect from the chemical exposure. We assumed that starting on the first day of the chemical episode, largemouth bass food consumption declined by 40% because the abundance of prey was reduced by chemical exposure, and additional prey (e.g., young-of-year forage fish) did not recruit to the population for the remainder of the summer. Thus, as in the previous example, fish recovered instantaneously from the physiological effects of exposure when the chemical episode subsided, but the dietary effects of reduced prey availability lingered. The simulated conditions are hypothetical, but they reflect what might occur in a system contaminated by a chemical spill where prey organisms are more strongly affected by exposure than the predator.

Simulation Results

Under laboratory conditions (example 1), largemouth bass grew to 165 g by the end of the 160-d simulation (Figure 3A; Table 2). This outcome was dependent on the feeding and temperature regimes investigated. In general, fish growth rates increase with water temperature to an optimum, then decline. The optimum temperature for feeding of largemouth bass is about 27.5°C (Hanson et al. 1997), therefore, growth rates could have increased or decreased had simulated conditions been different.

Final size of fish exposed to the natural thermal regime (example 2; Figure 3A) was 158 g. Inspection of the growth trajectory for this simulation reveals the direct effect of temperature on fish growth. Decreased growth associated with temperatures less than 20°C resulted in smaller fish size at the end of the simulation despite all other conditions being equivalent to those in example 1.

The final size of the fish subjected to the natural thermal regime and continuous chemical exposure (example 3; Figure 3A) was 104 g. The difference between final masses of fish in this simulation and example 2 is 54 g or about a 34% reduction in size. This difference in growth represents the cumulative energetic cost of exposure to the chemical stressor, since all other conditions were identical with the previous simulation. Similarly, the combined cost of chemical and thermal stress is represented by the difference between this and the baseline simulation (example 1).

Reparameterization of the chemical stressor as a stochastic variable (example 4) produced two interesting results. First, the average size of the fish at the end of the simulation was 100 g, which is 4 g less than the deterministic solution estimated in example 3. This difference is due to interactions of model parameters and nonlinear relationships within the model such that the relative influence of chemical exposure is not constant for all concentrations. The ability to accumulate and quantify subtle effects like these, which might otherwise go undetected, is a major advantage of modeling-based applications. The second interesting outcome is that the magnitude of variation in the final fish size is less than the variation in chemical concentration (coefficients of variation = 0.027 and 0.5, respectively). The variation in fish growth is a function of parameter variability as translated by the model equations. Thus, predicted variation may be influenced by interactions and nonlinear relationships between variables, as well as simulated conditions. For example, all fish were the same size at the beginning of the simulation and the magnitude of variation in size increased with time. The exact behavior of variation in final fish size can be described by thorough analysis of simulation output, but it is important to note that the outcome can change if simulation conditions are modified (Bartell et al. 1986). Output from a stochastic simulation is often interpreted using probability functions for the response variables (Suter and Barnthouse 1993). The probability density function for the final size in this simulation depicts the probability of observing a fish in a specific size category (Figure 4). The cumulative probability function represents the probability of observing a fish with a mass less than or equal to a specific value (e.g., P [fish $\leq$ 98 g] = 0.22). These analyses are useful for estimating the fraction of a population that may fall below or above a specified threshold value that has been established based on management or regulatory objectives.

Episodic chemical exposure (example 5) reduced the final size of the fish to 150 g or about 5% less than unexposed fish living in a natural thermal

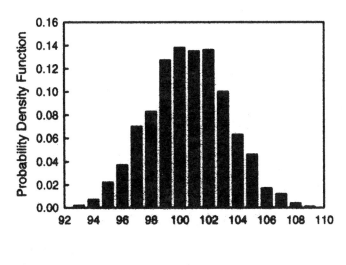

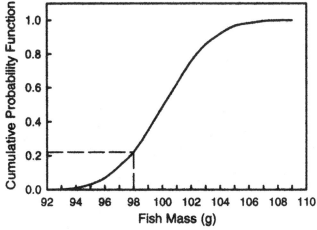

Figure 4. Probability functions for 1,000 solutions of a model predicting final size of largemouth bass after 160-d exposure to natural temperature regime and stochastically varying chemical concentrations. Refer to Table 2 (simulation 4) for simulation details and outcomes. Histograms represent probability of observing a fish having a mass less than or equal to the label value in the interval between each 1-g division. Dashed lines in lower panel show that the probability of observing a fish #98 g is about 0.22.

regime (example 2; Figure 3B). The SRM mechanistically accounted for the influence of chemical exposure on metabolic and food consumption rates. The combined effects of both of these changes caused a reduction in fish growth as long as chemical exposure continued. Once the chemical episode subsided, growth of exposed largemouth bass matched that of unexposed fish, but the size and energetic setback associated with exposure persisted to the end of the simulation.

The last simulation (example 6; Figure 3B) illustrates the importance of indirect effects of stressors, which can be particularly dramatic if they influence prey abundance or food consumption rates. In this example, the effect on largemouth bass of episodic chemical exposure is identical to that in the previous simulation but is small compared to the influence of chemical-induced reduction in prey availability. Food consumption rates in the two simulations were identical until day 50 when prey availability was reduced 40% by the chemical episode (Figure 5). The direct effects of chemical exposure subsided after 20 d, but the influence of prey reduction persisted until day 134 when declining water temperature decreased physiological demand for food to values less than availability.

Example 6 (Figure 3B) shows the ability of a SRM to partition effects from different sources of stress. The final size of the fish exposed to the combined effects of the chemical episode was 84.6 g or about 46% smaller than unexposed fish living in a natural thermal regime (example 2). About 11% of this difference was caused by direct exposure of largemouth bass to the chemical; the remaining 89% was caused by the indirect effect of chemical-induced reduction of prey abundance, which constrained largemouth bass feeding. This example illustrates that indirect effects of stressors can be more important than direct effects and that both influences can be evaluated simultaneously using an energetics-based approach. This example also shows the value of using a model that can account for the influence of natural stressors such as temperature. For example, if the influence of declining water temperature on demand for food had not been taken into account, the

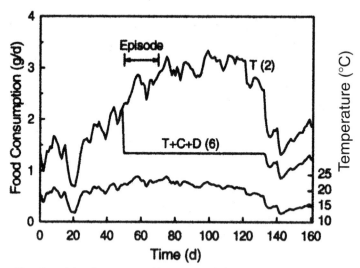

Figure 5. Simulated feeding rates of largemouth bass exposed to a thermal stressor (T), and thermal, chemical, and dietary stressors (T + C+ D). Feeding rates were identical until day 50 of the simulation when availability of prey was reduced 40% by the 20-d episodic chemical exposure. The influence of chemical-induced prey reduction persisted until day 134 when declining water temperature decreased physiological demand for food to quantities less than availability.

magnitude of the effect of the chemical exposure would have been overestimated.

Discussion

Advantages of Bioenergetics Models

Bioenergetics-based SRMs are founded on fundamental principles of toxicology and fish physiology. By linking laboratory-derived exposure–response relationships with bioenergetics models, the energetic cost of anthropogenic stress can be quantified for organisms living in a naturally fluctuating environment. The approach has numerous advantages stemming from its energetic foundation (Table 3).

The universal applicability of energy as a common currency allows direct comparisons of effects of natural and anthropogenic stressors. The effect of exposure to a single stressor or the cumulative cost of multiple stressors can be estimated simultaneously. Comparisons of relative effects of each stressor can be used to assess potential costs and benefits of alternative management strategies. For example, reducing contaminant concentrations may reduce

Table 3. Advantages and limitations of bioenergetics-based approaches for studying effects of stressors on fish.

Advantages

- Can account for simultaneous effects of natural and anthropogenic stressors.
- Can incorporate chronic or intermittent stressor effects and indirect (trophic-level) effects.
- Can measure and accumulate subtle effects of exposure to stressors.
- Growth reflects the integrated response of whole organism to stressors.
- Output can be related to fitness of fish in the natural environment.
- Changes in individual energetics can be used to predict population-level responses.
- Flexibility of computer modeling makes this approach generally applicable and valuable for prospective or retrospective environmental assessments.
- Bioenergetics models are available for many species; model structure and behavior has been thoroughly evaluated; accuracy of predictions is generally robust.

Limitations

- Application requires a bioenergetic model for the target species in the target environment.
- Data describing the influence of anthropogenic stressors on the energy-budgeting process are required, but traditional methods have not emphasized measurement of relevant endpoints.
- Animal behavior can be unpredictable and may be a significant source of uncertainty.
- Accuracy of the model should be evaluated before predictions are used to guide management actions.
- Importance of model parameters is dependent on conditions being simulated.

physiological stress from direct chemical exposure in fish, but if stress from natural variability in environmental conditions is relatively large, then the response to remediation may be undetectable in the field even though a very real effect was produced. Under these circumstances, an SRM can quantify the response to remediation so that the benefits of the management activity can be assessed. Alternatively, if more than one environmental factor can be managed, an SRM can quantify the magnitude of change associated with each potential management activity so that costs and benefits can be compared and priorities can be assigned for remediation.

The energetic foundation also allows bioenergetics-based SRM results to be translated into various commodities using energetic equivalents. For example, a stressor-induced increase in metabolic rate can be quantified in terms of joules, oxygen consumption, fish growth, fish mass, or increased prey consumption required to compensate effects of exposure. These various transformations are alternative metrics that can be used to place energetic effects into contexts that have intuitive meaning to public and scientific audiences.

Stressor-response models can also be used to assess the influence of stressors on individual fitness. Energetic expenditures of fish are distributed in a hierarchical fashion such that costs of maintenance (and stress) must be met first; surplus energy is conserved as growth. Fitness of fish depends on survival and reproductive success. Growth is linked to survival because fish size influences the outcome of ecological processes such as predation and competition. Growth is also linked to reproductive potential because larger females are usually more fecund and produce higher quality eggs than smaller fish (Wootton 1991). Direct estimates of fitness of fish in their natural environment are difficult to obtain, but there are two energetic correlates that can be measured. Net energy intake has been used as a surrogate for fitness based on the assumption that surplus energy is allocated to growth or reproductive output (Ware 1982; Hill and Grossman 1993). Other authors have extended this application and converted net energy intake into growth rate potential (Brandt et al. 1992; Brandt and Kirsch 1993; Luecke and Teuscher 1994) based on fish characteristics and spatially explicit environmental conditions. Net energy intake and growth rate potential are useful means of translating effects of chemical, physical, and biological stressors into the ecologically relevant response of fitness.

The energetic response of fish to stressors is consistent with the general adaptation syndrome, which describes physiological response to stressors as a function of time. The complex response of organisms to stressors can produce conflicting results unless the stage of progression of the general adaptation syndrome is considered. Metabolic and feeding rates of exposed animals may decrease initially, then increase to magnitudes above those of control organisms if the investigation is of long enough duration for physiological adaptation to occur. Conflicting interpretations of stressor-response data can be avoided by associating observed outcomes to the respective stage of the general adaptation syndrome. The transition from the stage of adaptation to the stage of exhaustion represents the translation of the *functional* response of energy reallocation to the *structural* response of mortality. This transition implies a mechanistic basis for linking stressor effects on physiology of individuals to changes in population structure. Although the transition cannot be quantified energetically, it can be related to probability of survival and explicitly incorporated into a simulation model of population-level effects.

The flexibility of a simulation-based approach allows evaluation of scenarios that cannot be studied in the laboratory. Typical laboratory exposure conditions include constant temperature, abundant food, and absence of predators or competitors. Laboratory tests are designed with these constraints in order to eliminate all sources of variation except those due to the experimental treatment. Consequently, laboratory tests support a strong case for cause and effect and allow estimation of the magnitude of effect with minimal uncertainty, but they may not accurately predict the response in the field. In the natural environment, stressor magnitudes change in time and space, and organisms may modify their exposure via attraction or avoidance. Stressor fluctuations and fish behavior in the field can be characterized with carefully designed investigations to obtain estimates of frequency and magnitude of exposure to stressors; but, mimicking these fluctuations in a laboratory is challenging. In contrast, computer models can easily simulate spatial and temporal distributions of stressors and organisms. Once estimates of stressor extent and fish behavior have been obtained, they can be explicitly integrated with bioenergetics simulations to predict effects under realistic conditions. Modeling can also track subtle effects that are difficult to detect with controlled studies. For example, in laboratory studies, sublethal effects on growth must accumulate for weeks or months before they have sufficient magnitude to be detected from the background noise of individual variability. Computer models can precisely follow each individual so that the cost of stressor exposure can be quantified instantaneously. Thus, SRMs provide a mechanistic means of superimposing environmental variability onto laboratory data. Laboratory investigations are a critical link in the process of model development because they describe a response as a function of exposure magnitude and time and provide empirical data for model parameterization and evaluation.

Bioenergetics-based methods are applicable to a diversity of biological systems and environmental issues. Fishery investigations have used bioenergetics as a basis for studies of food web dynamics, nutrient cycling, bioaccumulation, individual-based modeling, whole-system estimates of fish production, foraging theory, effects of spatial arrangement of prey and habitat on fish growth, and predator–prey interactions (Brandt and Hartman 1993; Hanson et al. 1997). Bioenergetics has also been applied to study chemical effects on aquatic invertebrates (Muller and Nisbet 1997; Penttinen and Kukkonen 1998; Maltby 1999), assessments of ecological effects of introduced species (Madenjian 1995; Young et al. 1996), and estimating waterbird predation on fish (Furness 1994; Madenjian and Gabrey 1995; Derby and Lovvorn 1997). Most of these applications have emphasized ecological questions, but the diversity of approaches attests to the flexibility and success of bioenergetics-based applications. All bioenergetics models are subject to the thermodynamic constraint that an organism's energy budget must balance. This constraint limits the magnitude of error that can be inadvertently introduced into bioenergetics models and probably contributes to successful applications. Thus, when data quantifying energetic effects of exposure are available, SRMs represent an efficient, cost-effective, and reliable means for predicting the outcome of exposure to a variety of stressors.

Limitations of Bioenergetics Models

There are drawbacks associated with the development and use of computer models (Hansen et al. 1993; Ney 1993). The application of any analytical technique requires understanding of implicit assumptions. Fish bioenergetics models are based on simple principles such as mass–balance equations that balance energy from food consumption against costs of metabolism, growth, and waste loss, and size- and temperature-dependent functions that adjust energy allocation based on individual characteristics and environmental conditions. These models reflect our perception of how energy flows through organisms, but they are simplifications of reality. Also, because the energy budget must balance, errors or uncertainties in model parameters and structure accumulate in the predicted response variable. Consequently, the accuracy of predicted effects should be evaluated using laboratory or field data. These evaluations serve as "reality checks" that, if successful, give confidence that the model is accurately predicting behavior of the study system. Unfortunately, as the level of ecological realism increases, it becomes more difficult to obtain data for comparing predicted and observed responses. General strategies have been described for evaluating complex, management-oriented models (Bart 1995).

Sensitivity analysis is an important component of model evaluation that involves quantifying the relative change in predicted outcome resulting from a change in parameter magnitude (Bartell et al. 1986). Sensitivity analysis aids interpretation of model output and can help guide selection of future topics for investigation. Several sensitivity analyses of bioenergetics models have been conducted, but conclusions of these investigations have limited scope because sensitivity may change when the respective models are modified or applied to a new system (Bartell et al. 1986). Beyers et al. (1999a) conducted sensitivity analysis of a SRM and demonstrated that as food consumption declined as a result of chemical exposure, the relative ranking in importance of many related parameters also declined and there was a corresponding increase in importance of respiration parameters. This result illustrates the need for restraint when relying on conclusions of previous analyses and the value of thorough analysis over the range of potential environmental conditions.

Applications of computer models to environmental problems are often limited by a shortage of biological data. Incomplete knowledge about animal behavior in a complex environment can have a strong influence on bioenergetics-based predictions. Organisms may minimize stress through trade-offs involving growth, feeding activity, or other behaviors (Gilliam and Fraser 1987; Wildhaber and Crowder 1995; Grand and Dill 1997; Whitledge et al. 1998). Behavior-mediated trade-offs can often be detected using visual observation, telemetry, or recapture of marked individuals (Little, this volume), but in some cases such intensive data-collection efforts may not be feasible due to time or monetary constraints. In these cases, there may be significant uncertainty in model predictions. The null model approach (*sensu* Rice 1990) can be used to evaluate potential sources of uncertainty by identifying a mismatch between model predictions and field observations, but the analysis can only provide hypotheses about potential trade-offs and cannot resolve uncertainties arising from a lack of data.

It is not clear how energetic effects of stressors that act in a probabilistic fashion on survival of organisms (e.g., mutagens) can be quantified in an ecologically meaningful way at the individual level of biological organization. For fish with cancer, the energetic implications of chemical exposure are trivial compared to the implications of premature death. For these types of stressors, individual energetics may be of little importance compared to loss of fish in the population and corresponding reductions in number of adults, reproduction, and recruitment (Power, Chapter 10).

Extrapolating from Individuals to Higher Levels of Biological Organization

Stressors act on individuals. The importance of individual variability is an emerging paradigm in ecology, and individual-based models can be used to extrapolate stressor effects to higher levels of organization (DeAngelis and Gross 1992; Uchmanski et al. 1999). Incorporating individual variability allows each member of a simulated population to have unique physiological or behavioral characteristics. It also allows individuals to independently encounter different environmental characteristics that may affect their survival and reproduction. Models that focus on the responses of stressed individuals can simulate mortality or the reallocation of energy away from reproduction, and the corresponding reduction in reproductive output can be evaluated using existing population modeling approaches (Clark and Rose 1997; Nisbet et al. 1997; Maltby 1999; Power, this volume). There is empirical evidence suggesting that individual characteristics influence the outcome of exposure to stressors and are consistent over a range of conditions. For example, toxicity tests routinely show that not all animals have the same sensitivity to a chemical (Newman and McCloskey 2000), and individual differences in critical swimming speeds of fish are consistent at various temperatures (Kolok 1992). Variation in condition and physiological rates probably contributes to individual differences. The benefits of attempting to incorporate this level of ecological realism into bioenergetics models are uncertain because there are costs associated with increased model complexity. In addition, variability in individual characteristics may be trivial compared to natural variation in environmental conditions (Hansen et al. 1993). Regardless of these uncertainties, this area of research warrants investigation because incorporating real individual variability into bioenergetics models will allow generation of prediction distributions that reflect the varying responses of members of a natural population.

Individual variability also affects the outcome of community-level processes such as predation and competition. These interactions have been studied using individual-based models that combine bioenergetics and predator–prey encounter models (reviewed by Hansen et al. 1993 and Hanson et al. 1997). These models simulate complex processes that may have outcomes that are difficult to predict or interpret without the aid of computer simulations. Model-based analysis may also suggest hypotheses that are outside a current paradigm (e.g., Clark and Rose 1997).

Changes in individual energetics ripple through the population and community and may, ultimately, be realized at the ecosystem level of organization in the form of increased export of energy, increased maintenance:biomass ratios, reduced internal cycling of essential elements, reduced efficiency of resource use, and reversal of successional trends (Odum 1985; Rapport and Whitford 1999). These ecosystem changes are correlated responses that are produced by the action of stressors on individual organisms. Typical ecosystem measures of stress are akin to indices because they represent the cumulative response of the biological components of a system. When a change is detected, indices do not reveal the underlying mechanism. Bioenergetics analyses can, potentially, untangle these complex problems by revealing mechanisms that explain the influence of suspected stressors on individuals and the corresponding changes in populations and communities.

Conclusion

Bioenergetics-based stressor-response models facilitate a management philosophy based on an interdisciplinary and ecosystem perspective because they inherently acknowledge the link between an organism's physiology and its constantly changing environment. They can also explicitly incorporate time-, magnitude-, and space-dependent effects of exposure to stressors. These properties can greatly enhance the ecological realism of environmental assessments.

Fish are excellent integrators of effects of chemical, thermal, physical, or biological stressors. However, different stressors can produce similar responses in fish and that complicates assessment of effects produced by a particular event or management activity. Bioenergetics-based SRMs separate and quantify the influence of different stressors using energy as a common currency. Effects of multiple stressors can be estimated regardless of whether the sources are anthropogenic or natural. Energetic cost of subtle or intense stressors can be accumulated and translated into ecologically meaningful response metrics such as growth rates or fish mass, and these characteristics can be used as a basis for predicting population-level effects.

The goal of investigations into effects of anthropogenic stressors on aquatic organisms should be to understand how exposure influences the ecology of target species, not just to describe how exposure affects an organism's physiology or growth. By integrating ecological energetics, stressor-response relationships, and environmental influences using bioenergetics-based stressor-response models, investigators can assess ecological effects of human-induced change at multiple levels of biological organization.

Acknowledgments

We thank Marshall Adams, Kevin Bestgen, Mark Bevelhimer, Charles Gowan, and Jeremy Monroe for discussions and comments that improved the manuscript. Manuscript preparation and editorial assistance was provided by Emily Plampin. This is contribution number 115 of the Larval Fish Laboratory.

References

Adams, S. M. 1990. Status and use of biological indicators for evaluating the effects of stress on fish. Pages 1–8 *in* S. M. Adams, editor. Biological indicators of stress in fish. American Fisheries Society, Symposium 8, Bethesda, Maryland.

Adams, S. M., and J. E. Breck. 1990. Bioenergetics. Pages 389–415 *in* C. B. Schreck and P. B. Moyle, editors. Methods for fish biology. American Fisheries Society, Bethesda, Maryland.

Bart, J. 1995. Acceptance criteria for using individual-based models to make management decisions. Ecological Applications 5:411–420.

Bartell, S. M., J. E. Breck, R. H. Gardner, and A. L. Brenkert. 1986. Individual parameter perturbation and error analysis of fish bioenergetics models. Canadian Journal of Fisheries and Aquatic Sciences 43:160–168.

Barton, B. A. 1997. Stress in finfish: past, present and future - a historical perspective. Pages 1–33 *in* G. K. Iwama, A. D. Pickering, J. P. Sumpter, and C. B. Schreck, editors. Fish stress and health in aquaculture. Cambridge University Press, New York.

Bevelhimer, M. S., and W. A. Bennett. 2000. Assessing cumulative thermal stress in fish during chronic intermittent exposure to high temperatures. Environmental Science and Policy 3:211–216.

Beyers, D. W., J. A. Rice, and W. H. Clements. 1999a. Evaluating biological significance of chemical exposure to fish using a bioenergetics-based stressor-response model. Canadian Journal of Fisheries and Aquatic Sciences 56:823–829.

Beyers, D. W., J. A. Rice, W. H. Clements, and C. J. Henry. 1999b. Estimating physiological cost of chemical exposure: integrating energetics and stress to quantify toxic effects in fish. Canadian Journal of Fisheries and Aquatic Sciences 56:814–822.

Borgman, U., and D. M. Whittle. 1992. Bioenergetics and PCB, DDE, and mercury dynamics in Lake Ontario lake trout (*Salvelinus namaycush*): a model based on surveillance data. Canadian Journal of Fisheries and Aquatic Sciences 49:1086–1096.

Braaten, P. J., P. D. Dey, and T. C. Annear. 1997. Development and evaluation of bioenergetic-based habitat suitability criteria for trout. Regulated Rivers: Research & Management 13:345–356.

Brandt, S. B., and K. J. Hartman. 1993. Innovative approaches with bioenergetics models. Transactions of the American Fisheries Society 122:731–735.

Brandt, S. B., and J. Kirsch. 1993. Spatially explicit models of striped bass growth potential in Chesapeake Bay. Transactions of the American Fisheries Society 122:845–869.

Brandt, S. B., D. M. Mason, and E. V. Patrick. 1992. Spatially-explicit models of fish growth rate. Fisheries 17:23–33.

Brett, J. R., and T. D. D. Groves. 1979. Physiological energetics. Pages 279–352 *in* W. S. Hoar, D. J. Randall, and J. R. Brett, editors. Fish physiology, volume 8. Academic Press, New York.

Christiansen, P. D., M. Brozek, and B. W. Hansen. 1998. Energetic and behavioral responses by the common goby, *Pomatoschistus microps* (Krøyer), exposed to linear alkylbenzene sulfonate. Environmental Toxicology and Chemistry 17:2051–2057.

Clark, M. E., and K. A. Rose. 1997. Individual-based model of stream-resident rainbow trout and brook char: model description, corroboration, and effects of sympatry and spawning season duration. Ecological Modelling 94:157–175.

DeAngelis, D. L., and L. J. Gross, editors. 1992. Individual-based models and approaches in ecology. Chapman and Hall, New York.

De Boeck, G., H. De Smet, and R. Blust. 1995. The effect of sublethal levels of copper on oxygen consumption and ammonia excretion in the common carp, *Cyprinus carpio*. Aquatic Toxicology 32:127–141.

De Boeck, G., A. Vlaeminck, and R. Blust. 1997. Effects of sublethal copper exposure on copper accumulation, food consumption, growth, energy stores, and nucleic acid content in common carp. Archives of Environmental Contamination and Toxicology 33:415–422.

Demers, E., R. S. McKinley, A. H. Weatherley, and D. J. McQueen. 1996. Activity patterns of largemouth and smallmouth bass determined with electromyogram biotelemetry. Transactions of the American Fisheries Society 125:434–439.

Derby, C. E., and J. R. Lovvorn. 1997. Predation on fish by cormorants and pelicans in a cold-water river: a field and modeling study. Canadian Journal of Fisheries and Aquatic Sciences 54:1480–1493.

Dockray, J. J., S. D. Reid, and C. M. Wood. 1996. Effects of elevated summer temperatures and reduced pH on metabolism and growth of juvenile rainbow trout (*Oncorhynchus mykiss*) on unlimited ration. Canadian Journal of Fisheries and Aquatic Sciences 53:2752–2763.

Fry, F. E. J. 1947. Effects of the environment on animal activity. University of Toronto Studies. Biological series No. 55. University of Toronto Press, Toronto.

Furness, R. W. 1994. An estimate of the quantity of squid consumed by seabirds in the eastern North Atlantic and adjoining seas. Fisheries Research 21:165–177.

Gilliam, J. F., and D. F. Fraser. 1987. Habitat selection under predation hazard: test of a model with foraging minnows. Ecology 68:1856–1862.

Grand, T. C., and L. M. Dill. 1997. The energetic equivalence of cover to juvenile coho salmon (*Oncorhynchus kisutch*): ideal free distribution theory applied. Behavioral Ecology 8:437–447.

Hall, C. A. S., J. A., Stanford, and F. R. Hauer. 1992. The distribution and abundance of organisms as a consequence of energy balances along multiple environmental gradients. Oikos 65:377–390.

Hansen, M. J., and six coauthors. 1993. Applications of bioenergetics models to fish ecology and management: where do we go from here? Transactions of the American Fisheries Society 122:1019–1030.

Hanson, P. C., T. B. Johnson, D. E. Schindler, and J. F. Kitchell. 1997. Fish bioenergetics 3.0. University of Wisconsin Sea Grant Institute, Report WISCU-T-97-001. Madison, Wisconsin.

Heath, A. G. 1995. Water pollution and fish physiology, 2nd edition. Lewis Publishers, New York.

Hill, D. K., and J. J. Magnuson. 1990. Potential effects of global climate warming on the growth and prey consumption of Great Lakes fish. Transactions of the American Fisheries Society 119:265–275.

Hill, J., and G. D. Grossman. 1993. An energetic model of microhabitat use for rainbow trout and rosyside dace. Ecology 74:685–698.

Hinch, S. G., and P. S. Rand. 1998. Swim speeds and energy use of upriver-migrating sockeye salmon (*Oncorhynchus nerka*): role of local environment and fish characteristics. Canadian Journal of Fisheries and Aquatic Sciences 55:1821–1831.

Janssen, J., and J. P. Giesy. 1984. A thermal effluent as a sporadic cornucopia: effects on fish and zooplankton. Environmental Biology of Fishes 11:191–203.

Kitchell, J. F. 1983. Energetics. Pages 312–338 *in* P. W. Webb and D. Weihs, editors. Fish biomechanics. Praeger, New York.

Kitchell, J. F., D. J. Stewart, and D. Weininger. 1977. Applications of a bioenergetics model to yellow perch (*Perca flavescens*) and walleye (*Stizostedion vitreum vitreum*). Journal of the Fisheries Research Board of Canada 34:1922–1935.

Kolok, A. S. 1992. The swimming performances of individual largemouth bass (*Micropterus salmoides*) are repeatable. Journal of Experimental Biology 170:265–270.

Lemly, A. D. 1993. Metabolic stress during winter increases the toxicity of selenium to fish. Aquatic Toxicology 27:133–158.

Linton, T. K., S. D. Reid, and C. M. Wood. 1997. The metabolic costs and physiological consequences to juvenile rainbow trout of a simulated summer warming scenario in the presence and absence of sublethal ammonia. Transactions of the American Fisheries Society 126:259–272.

Lucas, M. C., A. D. F. Johnstone, and I. G. Priede. 1993. Use of physiological telemetry as a method of estimating metabolism of fish in the natural environment. Transactions of the American Fisheries Society 122:822–833.

Ludlow, J., and T. B. Hardy. 1996. Comparative evaluation of suitability curve based habitat modeling and a mechanistic based bioenergetic model using two-dimensional hydraulic simulations in a natural river system. Pages 519–530 *in* Leclerc et al., editors. Ecohydraulics 2000: 2nd international symposium on habitat hydraulics, volume B. INRS-Eau, Quebec.

Luecke, C., and D. Teuscher. 1994. Habitat selection by lacustrine rainbow trout within gradients of temperature, oxygen, and food availability. Pages 133–149 *in* D. J. Strouder, K. L. Fresh, and R. J. Feller, editors. Theory and application in fish feeding ecology. University of South Carolina Press, Columbia.

Madenjian, C. P. 1995. Removal of algae by the zebra mussel (*Dreissena polymorpha*) population in western Lake Erie: a bioenergetics approach. Canadian Journal of Fisheries and Aquatic Sciences 52:381–390.

Madenjian, C. P., and S. W. Gabrey. 1995. Waterbird predation on fish in western Lake Erie: a bioenergetics model application. Condor 97:141–153.

Madenjian, C. P., D. V. O'Connor, and D. A. Nortrup. 2000. A new approach toward evaluation of fish bioenergetics models. Canadian Journal of Fisheries and Aquatic Sciences 57:1025–1032.

Maltby, L. 1999. Studying stress: the importance of organism-level responses. Ecological Applications 9:431–440.

McCarty, L. S., and D. Mackay. 1993. Enhancing ecotoxicological modeling and assessment. Environmental Science and Technology 27:1719–1728.

Muller, E. B., and R. M. Nisbet. 1997. Modeling the effect of toxicants on the parameters of dynamic energy budget models. Pages 71–81 *in* Dwyer, F. J., T. R. Doane, and M. L. Hinman, editors. Environmental toxicology and risk assessment: modeling and risk assessment, volume 6, ASTM STP 1317. American Society for Testing and Materials, West Conshohocken, Pennsylvania.

Newman, M. C. 1995. Quantitative methods in aquatic ecotoxicology. Lewis Publishers, Boca Raton, Florida.

Newman, M. C., and J. T. McCloskey. 2000. The individual tolerance concept is not the sole explanation for the probit dose-effect model. Environmental Toxicology and Chemistry 19:520–526.

Ney, J. J. 1993. Bioenergetics modeling today: growing pains on the cutting edge. Transactions of the American Fisheries Society 122:736–748.

Nisbet, R. M., E. B. Muller, A. J. Brooks, and P. Hosseini. 1997. Models relating individual and population response to contaminants. Environmental Modeling and Assessment 2:7–12.

Odum, E. P. 1985. Trends expected in stressed ecosystems. BioScience 35:419–422.

O'Hara, J. 1971. Alterations in oxygen consumption by bluegills exposed to sublethal treatment with copper. Water Research 5:321–327.

Penttinen, O. P., and J. Kukkonen. 1998. Chemical stress and metabolic rate in aquatic invertebrates: threshold, dose-response relationships, and mode of toxic action. Environmental Toxicology and Chemistry 17:883–890.

Post, J. R., R. Vandenbos, and D. J. McQueen. 1996. Uptake rates of food-chain and waterborne mercury by fish: field measurements, a mechanistic model, and an assessment of uncertainties. Canadian Journal of Fisheries and Aquatic Sciences 53:395–407.

Railsback, S. F., and K. A. Rose. 1999. Bioenergetics modeling of stream trout growth: temperature and food consumption effects. Transactions of the American Fisheries Society 128:241–256.

Rapport, J. D., and W. G. Whitford. 1999. How ecosystems respond to stress. BioScience 49:193–203.

Rice, J. A. 1990. Bioenergetics modeling approaches to evaluation of stress in fishes. Pages 80–92 in S. M. Adams, editor. Biological indicators of stress in fish. American Fisheries Society, Symposium 8, Bethesda, Maryland.

Rice, J. A., J. E. Breck, S. M. Bartell, and J. F. Kitchell. 1983. Evaluating the constraints of temperature, activity and consumption on growth of largemouth bass. Environmental Biology of Fishes 9:263–275.

Rice, J. A., and P. A. Cochran. 1984. Independent evaluation of a bioenergetics model for largemouth bass. Ecology 65:732–739.

Ries, R. D., and S. A. Perry. 1995. Potential effects of global climate warming on brook trout growth and prey consumption in central Appalachian streams, USA. Climate Research 5:197–206.

Rincón, P. A., and J. Lobón-Cerviá. 1993. Microhabitat use by stream-resident brown trout: bioenergetic consequences. Transactions of the American Fisheries Society 122:575–587.

Rodgers, D. W. 1996. Methylmercury accumulation by reservoir fish: bioenergetic and trophic effects. Pages 107–118 in L. E. Miranda and D. R. DeVries, editors. Multidimensional approaches to reservoir fisheries management. American Fisheries Society, Symposium 16, Bethesda, Maryland.

Rose, K. A., J. H. Cowan Jr., E. D. Houde, and C. C. Coutant. 1993. Individual-based modelling of environmental quality effects on early life stages of fishes: a case study using striped bass. Pages 125–145 in L. A. Fuiman, editor. Water quality and the early life stages of fishes. American Fisheries Society, Symposium 14, Bethesda, Maryland.

Schreck, C. B., B. L. Olla, and M. W. Davis. 1997. Behavioral responses to stress. Pages 145–170 in G. K. Iwama, A. D. Pickering, J. P. Sumpter, and C. B. Schreck, editors. Fish stress and health in aquaculture. Cambridge University Press, New York.

Selye, H. 1956. The stress of life. McGraw-Hill, New York.

Selye, H. 1973. The evolution of the stress concept. American Scientist 61:692–699.

Shuter, B. J., and J. D. Meisner. 1992. Tools for assessing the impact of climate change on freshwater fish populations. GeoJournal 28:7–20.

Shuter, B. J., and J. R. Post. 1990. Climate, population viability, and the zoogeography of temperate fishes. Transactions of the American Fisheries Society 119:314–336.

Stewart, D. J., D. Weininger, D. V. Rottiers, and T. A. Edsall. 1983. An energetics model for lake trout, *Salvelinus namaycush*: application to the Lake Michigan population. Canadian Journal of Fisheries and Aquatic Sciences 40:681–698.

Suter, G. W., and L. W. Barnthouse. 1993. Assessment concepts. Pages 21–88 in G. W. Suter, editor. Ecological Risk Assessment. Lewis Publishers, Boca Raton, Florida.

Trebitz, A. S., and N. Nibbelink. 1996. Effect of pattern of vegetation removal on growth of bluegill: a simple model. Canadian Journal of Fisheries and Aquatic Sciences 53:1844–1851.

Uchmanski, J., D. Aikman, T. Wyszomirski, and V. Grimm. 1999. Individual-based modelling in ecology. Ecological Modelling 115:109–110.

Van Winkle, W., K. A. Rose, B. J. Shuter, H. I. Jager, and B. D. Holcomb. 1997. Effects of climatic temperature change on growth, survival, and reproduction of rainbow trout: predictions from a simulation model. Canadian Journal of Fisheries and Aquatic Sciences 54:2526–2542.

Ware, D. M. 1982. Power and evolutionary fitness of teleosts. Canadian Journal of Fisheries and Aquatic Sciences 39:3–13.

Warren, C. E. 1971. Biology and water pollution control. Saunders, Philadelphia.

Wedemeyer, G. A., D. J. McLeay, and C. P. Goodyear. 1984. Assessing the tolerance of fish and fish populations to environmental stress: the problems and methods of monitoring. Pages 163–278 in V. W. Cairns, P. V. Hodson, and J. O. Nriagu, editors. Contaminant effects on fisheries. John Wiley, New York.

Whitledge, G. W., R. S. Hayward, D. B. Noltie, and N. Wang. 1998. Testing bioenergetics models under feeding regimes that elicit compensatory growth. Transactions of the American Fisheries Society 127:740–746.

Widdows, J., and P. Donkin. 1991. Role of physiological energetics in ecotoxicology. Comparative Biochemistry and Physiology C 100:69–75.

Wildhaber, M. L., and L. B. Crowder. 1995. Bluegill sunfish (Lepomis macrochirus) foraging behavior under temporally varying food conditions. Copeia 4:891–899.

Wootton, R. J. 1991. Ecology of teleost fishes. Chapman and Hall, New York.

Young, B. L., D. K. Padilla, D. W. Schneider, and S. W. Hewett. 1996. The importance of size-frequency relationships for predicting ecological impact of zebra mussel populations. Hydrobiologia 332:151–158.

Zweifel, R. D., R. S. Hayward, and C. F. Rabeni. 1999. Bioenergetics insight into black bass distribution shifts in Ozark border region streams. North American Journal of Fisheries Management 19:192–197.

9

Reproductive Indicators of Environmental Stress in Fish

MARK S. GREELEY, JR.

Introduction

Reproduction is arguably the most significant life function of fish affected by environmental stress, providing a crucial linkage between the effects of xenobiotics and other stressors on individual fish (through altered fecundity, increased mortality) and consequences at the population and community levels (population crashes, degraded communities). The reproductive potential of an individual fish depends primarily on survival from one age to another and on the number of offspring produced (Barnthouse 1993). From a broad perspective, therefore, any stressor that alters the reproductive life span, impedes expected reproductive output, or interferes with the development of juveniles to sexually active adults could be considered a reproductive hazard.

Reproductive potential and reproductive success are difficult concepts to measure at the population level; therefore, reproductive parameters of individual fish are often used as surrogate indicators of the reproductive health of the population. Fish reproduction is a complicated process whose success is dependent on a variety of processes and functions that include, among others, gonadogenesis, gametogenesis, mating, fertilization, early development, sexual differentiation, and sexual maturation. The complexity of the reproductive process in fish provides numerous targets for potential stressors but also yields a variety of potential biological indicators of environmental stress.

No discussion of reproductive indicators of environmental stress can be complete without mention of the rapidly expanding field of endocrine disruption. Endocrine-disrupting chemicals have been defined by a U.S. Environmental Protection Agency (USEPA) panel as substances that interfere with the production, release, transport, metabolism, binding, action, or elimination of natural hormones in the body responsible for the maintenance of homeostasis and the regulation of developmental processes (Kavlock et al. 1996). The potential for endocrine-disrupting chemicals to have serious impacts on fish, other wildlife, and humans led the USEPA to list this topic as one of six high priority research issues (USEPA 1996). Because reproduction and development are under the ultimate control of the endocrine system, almost any endocrine-disrupting effect could eventually prove to have reproductive or developmental consequences. However, stressors could also have direct or indirect effects on reproductive or developmental processes that are mediated through nonendocrine mechanisms. Therefore, although most endocrine-disrupting chemicals may ultimately prove to be reproductive hazards to some extent, it should be recognized that not all reproductive hazards are necessarily endocrine disruptors.

This chapter briefly summarizes the processes of reproduction and development in fish, describes various classes of potential biological indicators of reproductive and developmental dysfunction for use in field and laboratory investigations, and discusses the advantages and disadvantages of their use in assessing and monitoring environmental stress in fish. The working definition of environmental stress employed in this discussion is consistent with that employed elsewhere in this book, encompassing the potential effects of a wide variety of, generally, anthropomorphic influences on fish reproduction, including temperature extremes, acidification, and environmental pollutants. This concept is broader than and is not meant to be equated to the classical compensatory physiological stress responses of Selye (1950).

Reproduction in Fish

Detailed reviews of various aspects of the reproductive process in fish can be found elsewhere (Potts and Wootton 1984; Norris and Jones 1987; Munro et al. 1990; Redding and Patino 1993; Specker and Sullivan 1994; Goetz and Garczynski 1997; Van Der Kraak et al. 1998; McNabb et al. 1999). The purpose of the present brief and selective summary is to provide a framework for the discussion of reproductive indicators of environmental stress that follows, with a focus on likely stressor targets. Fish easily have the most diverse range of reproductive mechanisms and strategies to be found among the vertebrates and include oviparous species with yolk-laden eggs that develop externally following fertilization, and ovoviviparous and viviparous species that produce eggs that develop internally but differ in the degree to which the young depend on yolk for nourishment during early development. This summary focuses primarily on oviparous teleosts, the largest and most diverse group of fishes, but much of the information is applicable to other fish as well.

Endocrine and Paracrine Regulation of Reproduction

Reproduction is under the control of the brain–hypothalamus–pituitary–gonadal axis in fish (Van Der Kraak et al. 1998; Figure 1). Gonadal recrudescence and reproductive behavior are regulated by external cues such as temperature and photoperiod (Munro et al. 1990), which cause the hypothalamus to release gonadotropin-releasing hormone (GnRH) and various other neuroendocrine stimulatory factors, as well as inhibitory factors such as dopamine (Chang et al. 1984).

Gonadotropin-releasing hormone stimulates the pituitary to release gonadotropins (GtH) from the anterior pituitary, which in turn regulate gonadal development and function, as well as the production of reproductive steroids. The majority of fish appear to produce two gonadotropins, GtH-I and GtH-II, that are roughly analogous to follicle-stimulating hormone (FSH) and luteinizing hormone (LH), respectively, in mammals, although their functions in fish vary widely from their mammalian roles (Van Der Kraak et al. 1998). In species where two gonadotropin variants are present (some species may have only one gonadotropin), GtH-I appears to regulate gonadal developmental and gamete development through vitellogenesis in females

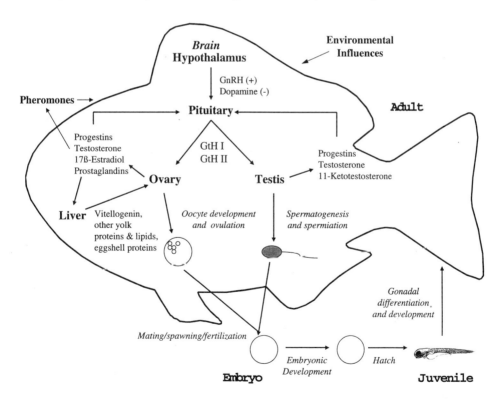

Figure 1. Summary of the major pathways and target receptors in fish reproduction.

and spermatogenesis in males. Gonadotropin-II, on the other hand, is involved in the regulation of later stages of gamete development including oocyte maturation and spermiogenesis.

Both GtH-I and GtH-II stimulate gonadal tissues to produce and release reproductive steroid hormones, including various estrogens, androgens, and progestins. The major estrogen in fish is 17β-estradiol, which in females has a prominent role of stimulating the liver to produce vitellogenin and other yolk precursors. Vitellogenin is a lipoglycoprotein precursor to the yolk proteins of egg-laying vertebrates, which is produced by the liver and transported to the ovaries where it is cleaved and incorporated into the yolk of vitellogenic oocytes (Wallace 1985; Guraya 1986; Mommsen and Walsh 1988; Specker and Sullivan 1994). Although other proteins and lipids are involved in yolk formation, vitellogenin is the frequent focus of ongoing research because of recent interest in the issue of endocrine disruption. Vitellogenin synthesis requires the interaction of estrogen with a receptor, present in target hepatocytes (Tata et al. 1987), which is normally found only in sexually mature females (Copeland et al. 1986). Male fish, which carry the gene for vitellogenin but do not typically produce the compound in detectable amounts, can be induced to produce vitellogenin if injected with estradiol (Emmersen and Petersen 1976). Male fish also produce vitellogenin if exposed to estrogens or estrogen agonists in the environment (Folmar et al. 1996; Harries et al. 1996, 1997) and tend to maintain detectable concentrations in the circulation for an extended duration due to poor metabolic clearance. Thus, the detection of circulating vitellogenin in male fish has proven to be a sensitive and reliable indicator of environmental exposure to endocrine-disrupting chemicals that act through estrogenic pathways.

Fish ovaries also produce other steroids involved in reproduction, including androgens such as testosterone and progestins such as 17α-20β-dihydroxyprogesterone (Nagahama 1987) and 17,20-α-trihydroxy-4-pregene-3-one (Thomas 1994). The former are primarily intermediates in the metabolic pathways for other steroids, although they may be present at circulating levels equal to or greater than the estrogens (Greeley et al. 1988); the latter are involved in the local regulation of oocyte growth, maturation and ovulation (expulsion of the oocyte from the surrounding follicular tissue), and prostaglandin synthesis by the ovary (Goetz 1997). All steroid hormones act through receptor-mediated mechanisms, typically using intracellular nuclear receptors but occasionally membrane receptors as well (Tsai and O'Malley 1994).

Prostaglandins are involved in ovulation and, with steroid hormones, may act as pheromones in at least some species (Stacey and Caldwell 1995). There is evidence that activin and inhibin, proteins belonging to the transforming growth-factor superfamily, and follistatin, a glycoprotein that binds activins and inhibins, may have roles in the paracrine regulation of oocyte maturation (Wu et al. 2000) and also in gonadotropin secretion by the pituitary (Ge et al. 1997).

Fish testes produce androgens, primarily 11-ketotestosterone, which stimulate spermatogenesis and induce and maintain the development of secondary sexual characteristics and reproductive behavior. Gonadal steroids also act through both endocrine and paracrine pathways, including positive and negative feedback loops, to affect the production and functioning of gonadotropins and other steroids. Additional hormones also have less well defined or specialized roles in regulating reproduction or early development in various fish species, including thyroid hormones, prolactin, and growth hormone among others (Norris and Jones 1987).

Gametogenesis

Fish gonads are composed of the gametes and various somatic tissues that have endocrine, support, and structural roles in fish reproduction. Fish ovaries are often paired structures that may produce from one to many clutches of eggs a year, with similarly wide variations in the associated patterns of oogenesis. Teleost oocytes, surrounded by supporting follicle cells and arrested in meiotic prophase I, arise from oogonia in germinal regions of the ovarian lumen, then undergo a series of developmental phases that include primary growth, cortical alveoli development, vitellogenesis, and maturation (Wallace et al. 1987). It is during the vitellogenic stage of development that greater than 90% of oocyte growth can occur, especially in freshwater species, from the incorporation of vitellogenin and other yolk precursor proteins and lipids. However, in other fish, particularly marine species with buoyant eggs, an additional major increment of growth occurs by hydration during postvitellogenic oocyte maturation (Craik and Harvey 1984; Greeley et al. 1991). During maturation, the oocyte resumes meiosis and proceeds into the second meiotic metaphase, the nucleus breaks down, and the cytoplasm undergoes a number of other changes, including further proteolytic cleavage of yolk proteins (Greeley et al. 1986), that prepare the oocyte for fertilization. The duration of vitellogenic oocyte growth can range from as little as two days in repeat spawners such as the Japanese medaka *Oryzias latipes* (Wallace and Selman 1981) to several months in fish such as salmonids that spawn only once (Tyler and Sumpter 1996). In contrast, oocyte maturation generally occurs in 24 hours or less in most teleosts (Wallace et al. 1987). After maturation, the now mature egg ovulates from the enveloping follicle cells into the ovarian lumen, surrounded by an eggshell (vitelline envelope; zona radiata) marked by a micropyle canal through which fertilization occurs.

The ovarian production of oocytes and eggs in fish is widely considered to follow one of three patterns: (1) synchronous development, (2) group-synchronous development, or (3) asynchronous development (Wallace and Selman 1981). In synchronous ovaries, oocytes develop and ovulate in a single cohort with no further replenishment, such as in salmonids that spawn once and die. In group-synchronous ovaries, oocytes develop in distinct cohorts or clutches during a reproductively active period, with replenishment occurring from an asynchronous collection of smaller oocytes. This pattern is prevalent among species that either spawn once annually over a period of years or several times during a year. In asynchronous ovaries, oocytes develop continuously and in no discernable pattern, a mode generally associated with tropical species that spawn more or less continuously. However, the latter may represent merely an extreme case of the group-synchronous pattern of development.

Teleost male germ cells consist of spermatogonia, spermatocytes, spermatids, and spermatozoa. In the first of two types of the teleost testis, spermatogonia occur along the entire length of testicular tubules (Grier 1981); in the second type, spermatogonia are restricted to the distal terminus of the tubules. In both testis types, cysts of synchronously developing germ cells form when spermatogenesis is initiated. Spermatogenesis proceeds within the cysts through the stages of primary spermatocytes (meiosis I), secondary spermatocytes (meiosis II), and spermatid development. Mature sperm formed from spermatids during the process of spermiogenesis are released into the tubular lumen in the first testis type; in the second, cysts move down the tubules during development, and sperm is released into a complex system of efferent ducts. Sperm may require additional maturation (activation) within the tubules, efferent ducts, in the receiving water, or even within the female reproductive tract in the case of internal fertilization before becoming able to successfully fertilize. Sperm morphology varies widely within the teleosts and includes aflagellated, single flagellated, and double-flagellated forms. In most fish, sperm are released free swimming into the environment and fertilization is external. In other species with internal fertilization, sperm may be packaged in clusters and delivered into the female reproductive tract via a specialized anal fin or gonopodium. The duration of spermatogenesis can vary from days to months depending on species.

Reproductive Behavior and Spawning

The term reproductive behavior encompasses a wide array of species-specific activities of fish prior to and during breeding, including migration to spawning sites, site selection, territorial defense, nest building, and courtship behaviors. Reproductive behavior is regulated by and, in turn, influences the circulating levels of, hormones ranging from the thyroid hormones involved in salmonid migratory activity (Liley 1969) to the androgens involved in fighting and territorial defense behaviors (Oliveira et al. 2001a, 2001b).

In many fish, reproductive behavior and physiology and spawning are synchronized between the genders by the release of pheromones into the water by the female (Stacey and Caldwell 1995). In goldfish *Carassius auratus,* these pheromones are also hormones, $17\alpha,20\beta$-dihydroxy-4-pregnen-3-one and prostaglandin $F_{2\alpha}$, which have internal endocrine roles, as well as an external synchronizing role (Stacey 1987). Atlantic salmon *Salmo salar* priming pheromone, which has a similar role in synchronizing spawning activity, is also thought to be prostaglandin $F_{2\alpha}$ (Olsén and Liley 1993; Moore and Waring 1996b).

Reproductive styles vary widely among fish (Balon 1975a), ranging from pelagic spawners with eggs that are fertilized externally and dispersed for wide distances by ocean currents to viviparous species that bear their developing young internally. Factors such as egg size and abundance (fecundity) can be affected by environmental conditions such as temperature, salinity, and nutrition (Tanasichuk and Ware 1987), with temperature stress generally resulting in a shift to higher fecundity but lower egg size. Fecundity is very dependent on nutritional status (Penzak 1985; Chappaz et al. 1987). Egg size varies considerably between fish species, ranging from less than 0.3 mm in the viviparous shiner perch *Cymatogaster aggregata* to nearly 90 mm in diameter in the coelacanth *Latimeria chalumnae* (reviewed in Wallace and Selman 1981). Furthermore, within a single species, egg size can vary according to the size of the female parent, nutritional status, salinity, season, and the latitude of the spawning site (Solemdal 1967; Bagenal 1969, 1971; Ware 1977; Knutsen and Tilseth 1985; Marteinsdottir and Able 1988; Fleming and Gross 1990). Within species, egg size has been inversely related with fecundity (Blaxter 1969), and positively correlated with size of larvae (Chambers et al. 1989). Egg size and number can vary seasonally and with parental size (Buckley et al. 1991).

Early Development and Hatching

Fish development is very diverse (reviewed in Blaxter 1969, 1988), exhibiting most developmental patterns found among the other vertebrates, as well as some unique specialized developmental processes. Development begins roughly with the process of fertilization (reviewed by Gilkey 1981), during which the male and female gametes unite and pronuclei fuse to form the zygote nucleus. In oviparous fish with external fertilization, sperm and eggs are generally shed in close proximity to one another. After the previously immotile sperm becomes activated either immediately prior to or during exposure to water, a now free-swimming sperm enters the egg through the micropylar canal; this event coincides with the activation of the egg and the resumption of development.

The terminology used to describe the early development of fishes varies considerably from species to species and from author to author (reviewed in

Blaxter 1988). Balon (1975b) summarizes fish life intervals as five periods beginning with the embryonic period that terminates at exogenous feeding (i.e., following yolk sac absorption in many fish), followed by the larval, juvenile, adult, and senescent periods. Other common usage has the embryo period ending at hatching, with a variety of terms used to describe subsequent developmental stages (i.e., protolarva, prelarva, postlarva, and larva) depending on the state of complexity of the young at the time of hatch or release (in the case of live-bearers). Regardless of the terminology employed, there are a number of significant developmental events occurring after fertilization (or after activation, since parthenogenetic species can develop in the absence of fertilization) that are relevant to the potential effects of environmental stressors on fish development and reproduction. Activation initiates the process of cleavage, a series of mitotic divisions that redistributes the relatively enormous amount of cytoplasm in the original single egg cell to the progressively smaller cells of the developing zygote. Development of the zygote to the embryo, and then on to the juvenile and adult stages, requires the exquisite choreography of simultaneous processes proceeding at specific rates and in particular orders and sequences. Organs and tissues are differentiated at various species-specific intervals during early development, which may make the embryo more-or-less sensitive to environmental stressors (i.e., environmental pollutants whose toxicity may be modulated by a newly developed liver). Blaxter (1988) recognizes five critical periods in fish development, including (1) hatching, (2) first feeding, (3) respiration, (4) swim-up, and (5) metamorphosis. Each of these critical periods, all not necessarily present in each species, occurs at species-specific times and stages during development and can be influenced by environmental variables, such as temperature, to various degrees.

Gonadal Differentiation

Gender determination in teleost fish is primarily genotypically derived (Yamamoto 1969), although endocrine and environmental factors have prominent roles in some species in determining phenotypic gender (Baroiller et al. 1999). All types of gonadal differentiation are found in teleosts (McNabb et al. 1999), including gonochorism (gonad either testis or ovary), hermaphroditism (both male and female germ tissue present simultaneously), protogyny (gonad changes from ovary to testis during the life cycle), and protandry (gonad changes from testis to ovary). The critical period for gender differentiation occurs early in development in many fish, around hatch in both medaka (Satoh and Egami 1972) and salmonids (Piferrer and Donaldson 1989). Gender differentiation based on the activity of the primordial germ cells can be detected as early as 24-hour posthatch in the medaka (Papoulias et al. 1999). However, in some species differentiation can occur much later, as in the European sea bass *Dicentrarchus labrax* whose critical period of androgen-inducible masculinization is located between 96 and 126 days postfertilization during a period of rapid primordial germ cell proliferation in the previously undifferentiated gonad (Blazquez et al. 2001). Estrogen receptors have been identified to be very active during the period of gonadal differentiation in transgenic zebrafish *Danio rerio* (Legler et al. 2000), suggesting an active involvement of receptor systems in sexual development.

Reproductive Indicators of Environmental Stress

Indicators of reproductive stress in fish can be measures of responses to specific stressors, such as a particular class of pollutants, or measures that integrate the effects of multiple stressors on individual fish or populations (also see Chapter 10). An example of the former is the appearance of vitellogenin in the plasma of male fish, which is a relatively specific response resulting from exposure to estrogenic compounds. Examples of the latter include classic early life history parameters (fecundity or egg production, egg size, and age at maturity) and even circulating levels of steroid hormones, although these endpoints can also represent specific targets of reproductive stressors as well. It should be emphasized that nonspecific reproductive responses are generally neutral as to direction of response; that is, alterations from normal in either positive or negative directions can represent a true stress-related response and could have adverse implications at either the individual or population levels of biological organization.

For organizational purposes, reproductive indicators can be loosely grouped into three arbitrary and overlapping classes of indicator: (1) endocrine-related indicators, (2) gonadal- and gamete-related indicators, and (3) developmental indicators. The emphasis in this discussion will be on the first two of these classes or those indicators that can be directly measured in adult fish or their gametes. Endocrine-related indicators include levels or activities of the gonadotropic hormones (Table 1), reproductive steroid hormones, aromatase and other steroidogenic enzymes, and hormone receptor analyses. Since vitellogenin and zona radiata protein induction, sex reversal/sex ratio, sexual characteristics, and reproductive behavior are so heavily dependent on endocrine regulation, they will be grouped into this category as well for the purposes of this chapter. Gonadal- and gamete-related indicators include gonadal condition measures such as the gonadosomatic index (GSI), gonadal histology, oocyte atresia, fecundity, fertility, and age/size at sexual maturity. Developmental indicators include, among many others, mortality, hatching success, hatching time, size at hatch, developmental abnormalities, and cytogenetic abnormalities. Following is a discussion of current or proposed reproductive indicators of reproductive stress, with examples of their use, as available, and consideration of the advantages and disadvantages of each indicator (summarized in Table 1). It is acknowledged, however, that, in reality, these indicators are rarely used singly as presented

Table 1. Characteristics of the major reproductive indicators of environmental stress in fish

Indicator	Assay techniques	Primary advantages	Primary disadvantages
Reproductive peptide hormones (gonadotropic hormones: GtH I, GtH-II; GnRH)	Radioimmuno-assay (RIA); enzyme-linked immunoassay (ELISA); mRNA assays	May be measured without sacrificing larger fish; relative ease of measurement in a few well-characterized species; provide information of mechanistic value	High species specificity; low stressor specificity; standardization problem in multiple spawners; diel and seasonal variation
Reproductive steroid hormones (17β-estradiol, testosterone, 11-keto-testosterone, various progestins)	RIA; ELISA; mRNA assays	May be measured without sacrificing larger fish; little species specificity, so relative ease of measurement; provide information of mechanistic value	Low stressor specificity; in some cases, specialized reagents needed for assay; standardization problem in multiple spawners; diel and seasonal variation; affected by capture and confinement stress; circulating and gonadal concentrations may not coincide
Steroid synthetic enzymes (aromatase, others)	Gene expression assays (northern blots, southern blots, RT-PCR)	Insight into mechanisms of stress effects	Little stressor specificity; limited ecological relevance without other reproductive measurements
Steroid hormone receptors	Competitive binding assays; gene expression assays (northern blots, southern blots, RT-PCR)	Insight into mechanisms of stress effects	Little stressor specificity; limited ecological relevance without other reproductive measurements; potential standardization problem in multiple spawners
Vitellogenin	RIA; ELISA; single radial immuno-diffusion assay; chemiluminescent immunoassays (CLIA); ribonuclease protection assay; SDS-PAGE densitometry; mRNA assays	When detected in males, high specificity for exposure to estrogenic compounds; associated with additional reproductive abnormalities in	High species specificity with most assay methods; ecological relevance uncertain; seasonal and possibly diel variation could be confounding factors

Table 1. continued.

Indicator	Assay techniques	Primary advantages	Primary disadvantages
Zona radiata proteins	ELISA	When detected in males, high specificity for exposure to estrogenic compounds	Ecological relevance uncertain
Sex ratio/sex reversal	Population sampling followed by phenotyping	Potentially high ecological relevance; can be related to population effects; relative ease of measurement; can be noninvasive; may be indicative of endocrine disruption	Low stressor specificity (although possibly indicative of endocrine disruption)
Sexual characteristics	Manual measurements	Medium to high ecological relevance; endocrine-related; potentially high sensitivity	Low stressor specificity (although possibly indicative of endocrine disruption); species specific
Sexual behavior	Manual and automated visual analysis	Medium to high ecological relevance; can be related to population effects	Stressor nonspecificity; can be difficult to measure in field
Gonadal somatic index	Manual measurement	Ease of measurement	Invasive, nonspecific; high within-species and within-individual variability; standardization problem in multiple spawners
Fecundity	Population sampling with manual enumeration of eggs; volumetric determination; extrapolation from developing clutch of oocytes (prespawn) or abundance of shed follicles (postspawn)	High ecological relevance, can be related to population effects; can be noninvasive; integrative response measure	Nonspecific (if specificity desired); high within-species and within-individual variability; standardization problem in multiple spawners; occasionally invasive
Egg weight or size	Manual or automated measurements	Medium to high ecological relevance; noninvasive	Nonspecific; high within-species and within-individual variability

Table 1. continued.

Indicator	Assay techniques	Primary advantages	Primary disadvantages
Fertility/gamete viability	Fertilization trials; sperm motility assays	Medium to high ecological relevance	Nonspecific; can be difficult to measure in field situation
Gonad abnormalities	Histopathology	Medium ecological relevance; can be indicative of endocrine-disruption	Low stressor specificity generally
Oocyte atresia	Histopathology; morphometric analysis; biochemical measures	Medium ecological relevance; good integrator of stress effects	Low stressor specificity; generally invasive
Age-at-maturity	Histopathology; morphometric analysis	High ecological relevance; can be related to population effects	Nonspecific; can be difficult to measure in the field; may be invasive
Developmental abnormalities	Histopathology; morphometric measurements; cytogenetic analysis; physiological and biochemical measures	Medium to high ecological relevance; high stressor specificity in some instances	Low stressor specificity in many instances

but are more often measured along with a larger suite of related or diverse indicators of stressor exposure or effects.

Endocrine-Related Indicators

Gonadotropins

Fish gonadotropin levels have been shown to be affected by in vivo exposure to phytoestrogens (MacLatchy and Van Der Kraak 1995), metals (Ma et al. 1995; Bieniarz et al. 1997), and pesticides (Singh and Singh 1980). Alterations to the in vitro secretion of gonadotropins following exposure to cadmium, lead, polychlorinated biphenyls (PCB), and paper and pulp mill effluent also have been reported (Van Der Kraak et al. 1992; Thomas 1993; Thomas and Khan 1997; Khan and Thomas 2000).

Considering the significant regulatory role of the gonadotropins to all aspects of fish reproduction, it is somewhat surprising that circulating levels of either GtH I or GtH II are not used more frequently as reproductive indicators of environmental stress. However, the structures of these protein hormones are not well conserved across the phyla, so their use as indicators is, of necessity, generally limited to those few fish species in which they are relatively well characterized and most easily measured. Also, since many of the actions of gonadotropins are mediated by steroid hormones, measurement of the latter endpoints may suffice for many purposes. The measurement of gonadotropin levels and the expression of the genes coding for gonadotropins should receive increased use as environmental stress indicators in the future, as these hormones and their mechanisms of action are more widely characterized and understood.

Steroid Hormones

Unlike the gonadotropins, the structure and functions of steroid hormones are relatively well conserved across the phyla, thus facilitating the quantitation of steroids such as 17β-estradiol and testosterone through the use of standard, commercially available assays. However, antisera and other reagents needed for assaying the more atypical steroids found in fish, such as 11-ketotestosterone, are generally not available commercially and thus must be synthesized in-house or obtained from colleagues. Nevertheless, the effort required to establish or acquire such assays is much less than with protein hormones such as the gonadotropins that must first be identified and characterized for each species studied.

Stressors can affect reproductive steroid hormone concentrations at the level of the hypothalamus or pituitary by impairing the supply of intermediaries such as cholesterol, by interfering with liver metabolism of steroids, or by decreasing production through effects on the activity of steroidogenic enzymes (Van Der Kraak et al. 1992; Monteiro et al. 2000). A number of environmental pollutants have been shown to alter circulating levels of estrogenic and androgenic steroid hormones following laboratory or environmental exposures, including cadmium (Sangalang and Freemen 1974), PCBs (Sivarajah et al. 1978; Freemen et al. 1984; Thomas 1989; Spies et al. 1996), aromatic hydrocarbons (Johnson et al. 1988), paper and pulp mill effluents (McMaster et al. 1991; Munkittrick et al. 1992, 1994; Karels et al. 1998), agrochemical and municipal runoff (Gallagher et al. 2001), cyanide (Ruby et al. 1993), organochlorine or organophosphorus pesticides (Singh and Singh 1987; Spies et al. 1996), alkylphenols (Arukwe et al. 1997a), and crude petroleum (Truscott et al. 1983). Alterations in testosterone concentrations in male fish appear, in particular, to be reliable indicators of exposure to estrogenic compounds in laboratory tests (Jobling et al. 1996; Mills et al. 2001) and in fish environmentally exposed to sewage treatment effluent (Folmar et al. 1996; Routledge et al. 1998). An example of the use of steroid hormones to examine the effects of paper and pulp mill effluent on female fish is shown in Figure 2 (modified from Sharp 1994). In this case, both plasma 17β-estradiol and testosterone were reduced immediately downstream of the effluent source as compared with an upstream reference population.

Environmental pollutants have also been shown to alter production or metabolism of steroids during incubations of fish gonadal tissue (Gagnon et al. 1994; Singh and Kime 1995; Loomis and Thomas 2000). The use of 17β-estradiol and testosterone production by gonadal tissues to evaluate fish

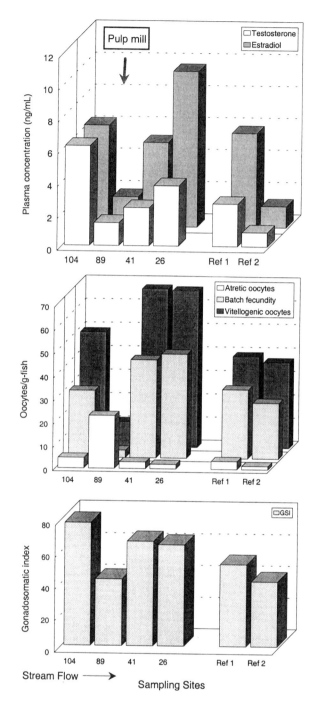

Figure 2. Reproductive parameters measured in redbreast sunfish females at sites along a river upstream (km 104) and downstream (km 89, 41, 26) of a paper and pulp mill and at two reference rivers. Note increase in atresia and decreases in fecundity and steroid hormones at the site (km 89) located immediately downstream of the mill. Modified from Sharp 1994.

environmentally exposed to endocrine disruptors was recently validated by an interlaboratory study of white suckers *Catostomus commersoni* exposed to pulp mill effluent (McMaster et al. 2001).

In a nationwide assessment of endocrine disruption in freshwaters of the United States, steroid hormone measurements have been used alongside other reproductive parameters as indicators of reproductive dysfunction and exposure to endocrine-disrupting chemicals (Goodbred et al. 1996; USGS 1998). Results of this study indicate widespread elevation of androgens, estradiol, and vitellogenin in carp from areas of known contamination, providing evidence of the usefulness of circulating steroid hormone measurements as indicators of environmental stress.

Care must be taken in the use and interpretation of steroid data because of potentially significant confounding factors. For instance, capture and confinement stress can suppress sex steroid levels, thus, the methods of obtaining samples may bias the results of such data (Jardine et al. 1996; Coward et al. 1997; Clearwater and Pankhurst 1997). Furthermore, patterns of circulating steroid content may not always adequately reflect patterns in ovarian (Hobby and Pankhurst 1997) or testicular (Cavaco et al. 1997) steroid content, particularly in fish with short-term reproductive cycles, and thus may not accurately reflect true reproductive status. Finally, natural variation in steroid hormones levels or production may overwhelm any stressor-related effect. For example, in many fish, the circulating concentrations of reproductive hormones can vary significantly throughout the reproductive cycle, especially in fish with multiple spawning habits; in the gulf killifish *Fundulus grandis*, plasma levels of estradiol and testosterone exhibit two-week cycles associated with semilunar tidal- and moon phase-related spawning cycles (Greeley et al. 1988). Steroid levels can also show distinct diel cycles or fluctuations as well (Matsuyama et al. 1990; Jardine et al. 1996; Haddy and Pankhurst 1998; Nash et al. 2000). Therefore, the use of circulating levels of reproductive steroid hormones or steroid production by gonadal tissues as reproductive indicators of environmental stress must somehow account for this natural variability, for instance, by taking samples from different study sites at the same time of the day and at a similar point in the breeding cycle.

Aromatase Activity

Cytochrome P450 aromatase, the steroidogenic enzyme responsible for the conversion of aromatizable androgens to estrogens, appears to be a key component in the mechanism of sex determination in fish. Its potential as a reproductive bioindicator of environmental stress in fish was demonstrated in a recent study correlating the expression of the aromatase gene and genes for two other steroidogenic enzymes in the channel catfish *Ictalurus punctatus* ovary with reproductive steroid levels and with ovarian developmental stages such as the first appearance of vitellogenic oocytes (Kumar et al. 2000).

Aromatase activity was previously used as a reproductive indicator along with steroid hormone levels and gonadal histology in a field study of alligators in a contaminated lake (Crain et al. 1997). In fish, measurement of aromatization activity in laboratory studies helped to explain the mechanisms underlying alterations in steroid hormone concentrations following fish exposure to paper and pulp mill effluents (McMaster et al. 1995) and the environmental estrogen 17α-ethinylestradiol (Scholz and Gutzeit 2000). As measures of steroid synthetic capacity and activity, aromatase and other steroidogenic enzymes are of specific value to studies that also include measurements of steroid hormone levels as indicators of environmental stress.

Hormone Receptors

The effects of steroid hormones on their target tissues are mediated through specific cellular receptors acting as ligand-dependent transcription factors. Although the effects of pollutants on endocrine function are thought to be largely mediated through interference with the genomic actions of steroid hormones at the level of nuclear steroid receptors, interference with steroid membrane receptors may also figure into the overall reproductive effect (Das and Thomas 1999; Loomis and Thomas 2000; Thomas 2000). Pollutants with endocrine-disrupting properties may affect receptor characteristics such as affinity and capacity through direct binding competition or effects on feedback mechanisms. Largely because of the ongoing emphasis on vitellogenin induction in male fish as an indicator of endocrine disruption, attention has naturally focused on the estrogen receptor (ER) that mediates this induction. Estrogen receptor capacity has been correlated with seasonal cycles of vitellogenesis in fish (Smith and Thomas 1990, Campbell et al. 1994a; Todo et al. 1995) and with estrogen induction of vitellogenin mRNA (Pakdel et al. 1991).

Estrogen receptor agonists also have been a recent focus of research in the area of endocrine disruptors (Purdom et al. 1994; Folmar et al. 1996; Harries et al. 1996, 1997, 1999; Janssen et al. 1997; Routledge et al. 1998; Tyler et al. 1999). Estrogenic compounds such as 4-nonylphenol and octylphenol have been shown to cause a significant upregulation of ERs in rainbow trout *Oncorhynchus mykiss* in vitro and in juvenile rainbow trout (Knudsen et al. 1998) and in Atlantic salmon (Yadetie et al. 1999) in vivo. Other environmental contaminants can have the opposite effect: Exposure of fish to PCBs in the laboratory (Thomas 1989) and in the environment (Garcia et al. 1997) was shown to result in decreases in ER-binding capacity. Cadmium was observed to exert an inhibitory influence on vitellogenesis through a negative effect on rainbow trout estradiol receptor transcriptional activity (Le Guével et al. 2000).

In addition to the vitellogenin connection, the current emphasis on the ER is partially due to the relatively well-characterized nature of the ER system in fish. Although androgens and other steroids are also important for fish reproduction (for instance, see Borg 1994), much less is known of these receptor systems in fish. Sperry and Thomas (1999) demonstrated distinct forms of androgen receptors in the kelp bass *Paralabrax clathratus* having differential sensitivities to both steroids and environmental chemical contaminants. Several known mammalian antiandrogenic compounds, including DDT, have been shown to bind to fish androgen receptors, but only in specific tissues (testis) and specific species (goldfish; Wells and Van Der Kraak 2000). Another known mammalian antiandrogen, vinclozolin, did not undergo discernable binding to high affinity, low capacity testosterone binding sites in fathead minnow *Pimephales promelas* brain and ovary cytosolic preparations (Makynen et al 2000) and had only limited effects on reproduction or development in this species. Thomas et al. (1998) showed that several estrogenic compounds could affect the binding of a progestin to its binding receptor on the Atlantic croaker *Micropogonias undulatus* sperm membrane. However, when several compounds known to affect the ER were tested in rainbow trout for their ability to displace the native ligand from the ER, the testosterone receptor, or the cortisol receptor, these compounds appeared specific for the ER (Knudsen and Pottinger 1999).

Because steroid receptor capacity varies seasonally, and in the case of the ER with the status of vitellogenesis, care must be taken in using these measures as indicators of reproductive competence or dysfunction. Standardization of sample collection to minimize natural variation as suggested for the steroid hormones is recommended. Also, more work is needed on the basic biology of receptor systems other than the ER in fish before these systems are readily adaptable to routine use in environmental monitoring or assessment situations.

Vitellogenin

The protocol recommended for Tier 1 testing of chemicals for endocrine-disrupting properties specifically recommends the testing of plasma vitellogenin in male fish as an indicator of endocrine disruption (U.S. EPA 1998). Vitellogenin measured in male fish has been used as an indicator of environmental exposure to estrogenic compounds in numerous studies (Purdom et al. 1994; Folmar et al. 1996; Harries et al. 1996, 1997, 1999; Janssen et al. 1997; Routledge et al. 1998; Tremblay and Van Der Kraak 1999; Rodgers-Gray et al. 2001). Figure 3 shows the vitellogenin induction that occurs in male rainbow trout during exposure to differing concentrations of effluent from two U.K. sewage treatment plants (modified from Harries et al 1999).

The induction by endocrine-disrupting compounds of vitellogenin in male fish may be accompanied by significant damage to the affected individuals, particularly at the histopathological level (see Chapter 7). Octylphenol has been shown to induce vitellogenin production in cultured brown bullhead *Ameiurus nebulosus* hepatocytes at concentrations below the toxic thresh-

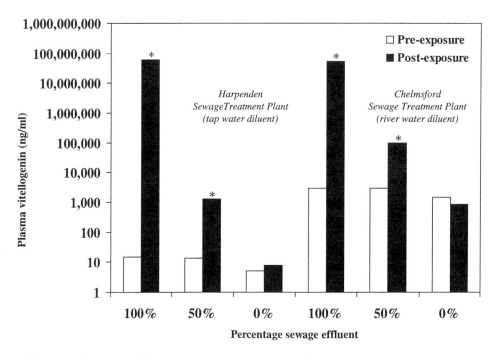

Figure 3. Plasma vitellogenin concentrations in rainbow trout *Oncorhynchus mykiss* held in various effluent concentrations from Chelmsford and Harpenden (U.K.) Sewage Treatment Works. Fish in Harpenden study were adult males, while fish in Chelmsford study were immatures of both sexes. Tap water was used as diluent in Harpenden study; river water was diluent in Chelmsford study. Modified from Harries et al. 1999.

old; but, at higher concentrations of octylphenol, vitellogenin production decreased and apoptic cell death was induced (Toomey et al. 1999). Disruption of spermatogenesis has been documented in male summer flounder *Paralichthys dentatus* and associated with vitellogenin induction, although the alterations could have been due to direct estrogen effects on androgen steroidogenesis in the testes (Folmar et al. 2001). Vitellogenin induction was also associated with renal and liver histopathologies that appear to have been directly attributable to vitellogenin accumulation in those tissues. Vitellogenin induction in male common carp *Cyprinus carpio* exposed to 4-tert-pentylphenol at much higher doses than necessary for causing histopathological damage led to significant reductions in primordial germ cells (PGC), an inhibition of spermatogenesis, the induction of an oviduct instead of the male vas deferens, and the development of oocytes (Gimeno et al. 1998).

In female fish, vitellogenin has been implicated in the downregulation of estradiol by feedback mechanisms (Reis-Henriques et al. 1997). Interestingly, in adult fathead minnows exposed to waterborne concentrations of 4-nonylphenol, it was the females and not the males who exhibited dose-dependent increases in vitellogenin with exposure (Giesy et al. 2000).

Although elevated vitellogenin is widespread in male European flounder *Platichthys flesus* collected in U.K. estuaries downstream of wastewater treatment facilities (Allen et al. 1999), only a few instances of testes with gross morphological abnormalities have been found to accompany these elevations in vitellogenin, nor have abnormal sex ratios been observed. However, in wild roach *Rutilis rutilis* living in rivers that receive treated sewage effluents, increases in plasma vitellogenin have been seen to be associated with both high incidences of intersex fish and significant levels of testicular abnormalities (Jobling et al. 2002). Thus, the evidence is somewhat contradictory whether vitellogenin induction in male fish as a result of exposure to estrogenic compounds in the environment has significant consequences at the population level and beyond (Kramer et al. 1998).

Vitellogenin can be measured by a variety of methods, including radio-immunoassay (RIA), enzyme-linked immunoassay (ELISA), and single radial immunodiffusion (Specker and Sullivan 1994). ELISA (for example, see Parks et al. 1999 for fathead minnow plasma vitellogenin ELISA) is considered the standard, but like the other immunological-based assays, requires specialized species-specific antisera. A ribonuclease protection assay (Korte et al. 2000) may be more sensitive than the ELISA, but again is species specific. A highly sensitive and specific chemiluminescent immunoassay (CLIA) has recently been developed for salmonids (Fukada et al. 2001).

The vitellogenin sequence is poorly conserved among the various fish groups (Lee et al. 1992). To make these assays more applicable across fish species, monoclonal and polyclonal antibodies with less specificity and greater cross-reactivity are being developed (Nilsen et al. 1998). However, because of the great amount of vitellogenin generally present in the serum of responding males (Figure 3), less sensitive but also less species-specific assays might be effective in some instances. An example of such an assay is the use of densitometry measurements with sodium dodecylsulphate-polyacrylamide gel electrophoresis (SDS-PAGE) as has been employed for assessing vitellogenin in European flounder exposed to contaminated sediments (Janssen et al. 1997). In addition, some researchers now measure vitellogenin mRNA expression rather than vitellogenin per se (Karels et al. 1998, Soimasuo et al. 1998).

As are most other reproductive processes, vitellogenesis is typically a very cyclic process, with pronounced annual variation in all fish species and even shorter cycles in multiple spawning species. When used solely as an indicator of exposure to endocrine disruptors in male fish in controlled laboratory studies, the inherent cyclicity of vitellogenesis should not be a significant confounding factor. However, the potential effects of natural cyclicity should be controlled through study design in field studies of male fish and in field or laboratory studies of female fish that seek to use vitellogenin levels as indicators of environmental stress.

Zona Radiata Proteins

Similar to vitellogenin, eggshell proteins (zona radiata proteins) are produced in the fish liver under the influence of estradiol and are transported to the developing oocyte to be incorporated into the surrounding zona radiata layer (Oppen-Berntsen et al. 1992). Also, like vitellogenin, zona radiata proteins have been shown to be induced in juvenile and male fish in response to treatment with both 17β-estradiol and estrogen-mimicking compounds (Arukwe et al. 1997b; Celius et al. 1999; Arukwe et al. 2000) and, in fact, may be a more sensitive responder than vitellogenin (Celius and Walther 1998). Zona radiata proteins can be assayed by ELISA (Oppen-Berntsen et al. 1994). Cautions and recommendations for use of the zona radiata proteins as indicators of environmental stress are as already indicated for vitellogenin. Little is known of the potential ecological relevance of such measurements.

Sex Ratio/Sex Reversal

Exposure of fish to steroids, endocrine-disrupting compounds, or even temperature or pressure extremes at the critical period of gender differentiation can alter the phenotypic sex from the genotypic sex. Using a medaka small fish model, newly hatched fry were determined to be the life stage most sensitive to hormone exposure and the most appropriate for use in investigating the potential effects of known endocrine-disrupting compounds on gender differentiation (Koger et al. 2000).

Measurements of sex ratio in the field or observations of sex reversal in the laboratory can be ecologically relevant measures of exposure to a variety of potential environmental stressors. For instance, intersex fish have been induced by laboratory exposure to xenobiotics (Gimeno et al. 1998; Gray et al. 1999). Intersex fish with both male and female germ cells have been found in the environment in close proximity to sewage treatment plants (Allen et al. 1999; Jobling et al. 1998, 2002), with the assumption being that estrogenic compounds in the effluents affected the fish during development and led to the "feminization" of the male fish. Populations of fish exposed to thermal stress also have produced intersex fish (Lukšieng et al. 2000). An example of such an "ovotestis" is shown in Figure 4.

In whitefish *Coregonus spp.,* population stress has been reported to alter sex ratio (George 1977). Exposure of juvenile roach to treated sewage effluent induced dose-dependent and persistent disruption in gonadal duct development, although no effect was seen on germ cell development (Rodgers-Gray et al. 2001). Environmental exposure to effluent from a pulp mill resulted in a sex ratio significantly biased toward males in an eelpout *Zoarces viviparus* population (Larsson et al. 2000), while a laboratory exposure to pulp mill effluent caused a similar male bias in fathead minnows (Kovacs et al. 1995). A single injection of ethinyl estradiol to medaka can cause phenotypic sex reversal of genetic males to females (Papoulias et al. 1999), with no instances of gonadal intersex reported. Male carp exposed to 4-tert-pentylphenol dur-

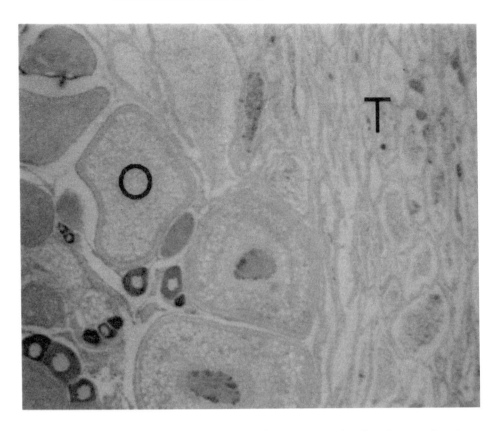

Figure 4. Redbreast sunfish ovo-testis in phenotypic male, showing ovarian tissue on the left (O) and testicular tissue on the right (T). Ovarian tissue appears normal, but testicular ducts contain only degenerating sperm, and no active spermatogenesis is evident.

ing sexual differentiation led to feminization and development of extensive histopathologies, including significant reductions in PGCs, inhibition of spermatogenesis, oviduct induction instead of the male vas deferens, plus the development of oocytes (Gimeno et al. 1998).

Sexual Characteristics

The appearance or absence of tubercles, an androgen-dependent secondary sexual characteristic in many male fish (Borg 1994), is a potential indicator of exposure to environmental stressors that affect or mimic normal androgen functioning. In a recent laboratory assessment of a fathead minnow reproductive performance test for endocrine-disrupting chemicals, reductions in the secondary sexual characteristics of tubercle and fat-pad formation were, along with vitellogenin production, among the most sensitive indicators of reproductive effects from exposure to nonylphenol, a known endocrine disruptor (Harries et al. 2000). In male goldfish exposed to estradiol in the laboratory, tubercles declined coincident with decreases in the gonadosomatic index (GSI) and other indicators of male reproductive condition (Bjerselius et al. 2001). Reduced development of male secondary sex characteristics has also been reported for Great Lakes salmonid populations under stress (Leatherland 1993). Reduced expression of secondary sexual characteristics was also among several signs of reproductive impairment observed in fish exposed to paper and pulp mill effluents (McMaster et al. 1991; Munkittrick et al. 1991, 1992a, 1992b, 1992c).

The growth and development of the modified anal fin, or gonopodium, is a prominent male sexual characteristic in eastern mosquitofish *Gambusia holbrooki*. The gonopodium of mosquitofish is formed under the influence of testosterone and is critical for sperm transfer. Reductions in gonopodium length have been reported in male mosquitofish at a field site located downstream from a sewage treatment plant discharge point, suggesting the presence of androgen antagonists of endocrine-disrupting compounds in the discharges (Batty and Lim 1999). Male secondary sex characteristics are also reported to have been induced in female mosquitofish exposed to paper and pulp mill effluent (Bortone and Davis 1994), suggesting the presence of androgen agonists in these discharges. An example of a staging system using gonopodium development in female mosquitofish as an indicator of exposure to endocrine disruptors in paper and pulp mill effluent is shown in Figure 5 (modified from Ellis 2001).

Sexual Behavior

The effects of pollutants on the reproductive behavior of fishes is reviewed in Jones and Reynolds (1997). Various pesticides, including atrazin, carbofuran, diazinon (Moore and Waring 1996a, 1998), and cypermethrin (Moore and Waring 2001) appear to block the actions of the priming pheromone released by Atlantic salmon females to synchronize spawning. Brook trout *Salvelinus fontinalis* have been shown to avoid low pH areas in selection of spawning sites (Johnson and Webster 1977). In laboratory studies, nest-digging behavior was severely inhibited in salmon exposed to acid conditions (Kitamura and Ikuta 2000), suggesting that a significant cause of salmonid population declines during the early stages of acidification may be avoidance of slightly acidic water in selection of spawning sites or actual cessation of spawning behavior. Acid conditions also inhibit spawning behavior in the flagfish *Jordanella floridae* (Craig and Baski 1977).

Sexual behavior has been proposed as a reproductive effect biomarker in various fish species, including the guppy *Poecilia reticulata* (Bayley et al. 1999), goldfish, (Bjerselius et al. 2001), Atlantic salmon (Moore and Waring 2001), and medaka (Metcalf et al. 1999). In male goldfish, a suite of behavioral variables monitored in laboratory exposures to estradiol included pushing, courting, spawning, biting, and picking (digging) behaviors, with a sexual index proposed to sum up the results of critical behaviors (Bjerselius et al. 2001). In both the goldfish and guppy, exposure to estrogen or estrogenic compounds decreased male reproductive behaviors.

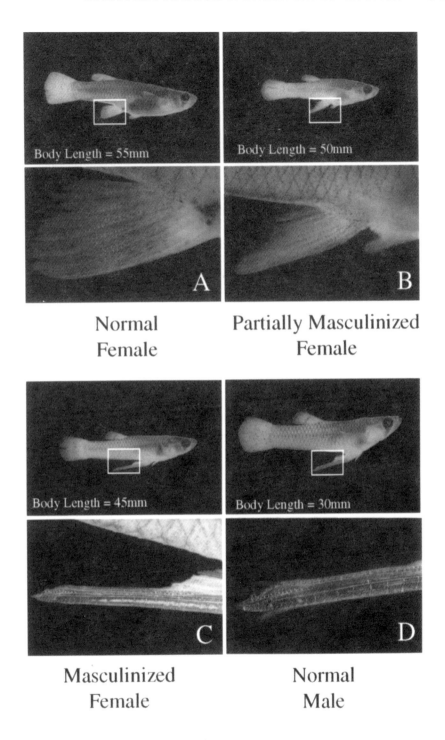

Figure 5. Mosquitofish anal fin morphology as a reproductive indicator of exposure to endocrine disruptors: (A) normal female, (B) partially masculinized female, (C) masculined female, and (D) normal male. Modified from Ellis 2001.

Gonadal-Related Indicators

Gonadosomatic Index

Determination of the GSI, or relative size of the gonad to body size, is a common method of rapidly estimating the reproductive status of fish (Nikolsky 1963). The GSI is simple and easily determined but provides little definitive information about actual gonadal condition. The GSI is most suitable for screening purposes, in part because of statistical concerns with the methodology (de Vlaming et al. 1982).

Alterations in fish GSI (generally downward) have been demonstrated with contaminant exposure in a number of field studies, including exposures to pulp and paper mill effluent (Munkittrick et al. 1994; Janz et al. 1997; van den Heuvel et al. 2002), aromatic hydrocarbons, PCBs, and other chlorinated compounds (Johnson et al. 1999). In laboratory exposures, reductions in GSI along with lowered plasma testosterone proved reliable indicators of exposure to estrogenic compounds, including DDT, octylphenol (Mills et al. 2001), and nonylphenol ethoxylates (Le Gac et al. 2001). Significant alterations in testicular morphology generally accompany decreases in the GSI, particularly inhibition of spermatogenesis or oogenesis. Decreased GSI is also often associated with decreased levels of circulating steroid hormones (Munkittrick et al. 1992, 1994). In male goldfish exposed to estrogen in the laboratory, the GSI was significantly reduced along with milt production and the appearance of secondary sexual characteristics (tubercles; Bjerlius et al. 2001).

Because of the rapidity and ease with which this parameter can be measured, the GSI is sure to remain one of the most widely used reproductive indicators of environmental stress. However, the GSI is best suited for use as part of a larger suite of indicators since it provides only limited data on reproductive condition. Also, natural variability in the GSI during and beyond the breeding season needs to be adequately controlled in the study design.

Fecundity

The production of viable eggs is among the most sensitive indicators of pollutant exposure and effects in the typical life-cycle tests of potential toxicants (Suter et al. 1987). Fecundity has been found to be very sensitive to xenobiotics in a number of laboratory studies (Tam and Payson 1986; Kovacs et al. 1995; Carlsson et al. 2000). In bluegill *Lepomis macrochirus* exposed to chlordane, fecundity was significantly reduced at the no-observed-effect concentration or NOEC (Cardwell et al. 1977) for other endpoints. In zebrafish exposed to 4-chloroaniline for three generations in the laboratory, fecundity was reduced at a 10-fold lower concentration as compared with growth and at a 1,000-fold lower concentration than the LC_{50} (concentration that is lethal to 50 percent of the organisms in a given time period; Bresch et al. 1990).

Numerous studies have also documented decreases in the fecundity of fish populations exposed to environmental contamination, including mixed industrial effluents (Adams et al. 1992a), mixed industrial and municipal wastes (Hose et al. 1989; Johnson et al. 1994), and pulp and paper mill effluent (McMaster et al. 1991; Adams et al. 1992b; Janz et al. 1997). An example of decreased fecundity in fish collected immediately downstream of a paper and pulp mill can be seen in Figure 2. Occasionally, the opposite trend has been observed, with increased egg production seen in bream exposed to contaminants in the Rhine River (Slooff and DeZwart 1983).

Fecundity is significantly affected by a variety of environmental conditions, such as temperature, salinity, and nutrition (Penzak 1985; Chappaz et al. 1987; Tanasichuk and Ware 1987). A study of English sole *Pleuronectes vetulus* sampled from several polluted sites in Puget Sound illustrates some of the difficulties of using fecundity as an indicator of environmental contamination at field locations (Johnson et al. 1997). In this study, elevated tissue PCBs concentrations were significantly correlated with increased egg numbers but also with decreased egg size, while elevated biliary fluorescent aromatic compounds, indicators of exposure to polycyclic aromatic hydrocarbons (PAH), were correlated with decreased egg number, increased egg weight, and increased ovarian atresia. Nutritional and other environmental factors appeared to play a significant modulating role in the apparent effects of these chemical contaminants on various reproductive parameters, including fecundity, potentially confounding the interpretation of the study results.

Fecundity can be measured by a variety of species-specific methods. Measurement of fecundity is very simple in fish with few eggs, since all eggs can be manually enumerated with little effort. In species with large single clutches of eggs, fecundity is often determined by taking a volumetric approach through enumerating the number of eggs in a small subset of the entire egg mass. Determining the fecundity of multiple spawners can be problematic. An approach adopted in our laboratory is to take a weighed percentage of an immediately prespawning ovary (determined either on the basis of biopsy results or appearance) and enumerate and stage all the oocytes to a certain small previtellogenic stage with the assistance of a dissecting microscope. When the size-frequency distribution of the developing oocytes is plotted, a "batch" fecundity can be determined by enumerating the leading clutch of synchronously developing oocytes. In addition, atretic oocytes can be enumerated as well, and the ovary can be classified on the basis of oocyte developmental stages. For recently spawned fish or for fish collected during the act of spawning, batch fecundity can be determined alternatively by counting the shed follicles remaining in a representative sample of ovary.

Fecundity is a reproductive metric with high ecological value and a proven relationship to environmental stress. However, because it can be affected by a wide variety of natural and anthropomorphic stressors, care must be taken in assigning causality to any specific environmental stressor strictly on the basis of alterations in this reproductive parameter.

Egg Size

Egg size often has a covariant relationship with fecundity. As with fecundity, egg size can be affected by a variety of environmental conditions including handling stress (Contreras-Sanchez et al. 1995), temperature, salinity, and nutrition (Tanasichuk and Ware 1987). Temperature and handling stress generally result in a shift to higher fecundity but lower egg size. Egg size and abundance can vary seasonally and with parental size (Buckley et al. 1991).

Egg size has been used previously as an indicator of reproductive condition in field assessments (i.e., Johnson et al. 1997). Perch *Perca fluviatilis* embryos environmentally exposed to paper and pulp mill effluent were observed to be smaller than reference embryos (Karås et al. 1991) and were associated with increased developmental abnormalities and decreased recruitment. Because egg size is dependent on so many environmental factors, it, like fecundity and many other reproductive measures, is best employed as part of a larger suite of reproductive indicators.

Fertility

Fertility is an important reproductive concept that is distinctly different from but complimentary to that of fecundity. Unlike fecundity, which is exclusively a measure of the reproductive potential of female fish, fertility relates to both male and female reproductive competence. Fertility can be measured either directly as hatchability (Nimrod and Benson 1998) or inferred as a consequence from adverse effects on other reproductive parameters such as gonadal condition.

Environmental stressors have long been known to adversely affect the fertility of fish eggs. Cypermethrin, a pesticide, was reported to reduce the number of fertilized eggs in Atlantic salmon (Moore and Waring 2001). Stress caused by confinement has also been observed to affect gamete quality in fish (Campbell et al. 1994b). Fish collected from field locations and spawned in the laboratory have been used in a number of studies to examine fertilization success, as well as developmental success (i.e., Johnson et al. 1997).

Fertility is not easily measured in a field situation. Fish are usually brought into the laboratory from field sites, often for days or weeks, to be spawned artificially for fertilization experiments. Females may require hormonal injections to complete development and maturation of the eggs, thereby introducing potentially confounding variables to the investigation (i.e., confinement stress). Therefore, caution is urged in the interpretations of results from such studies, and extra care should be taken in study design to control variables such as the length of confinement and the reproductive condition of the fish at the time of capture.

Gonadal and Gamete Abnormalities

Xenobiotics and other environmental stressors can effect gonadal development and function both directly and indirectly through primary effects on
the hypothalamic–pituitary–gonadal axis of endocrine regulation (Kime and
Nash 1999). In addition to oocyte atresia (which is considered separately
below), exposure to xenobiotics has been shown to have effects on various
aspects of both male and female gonadal structure and function, including
(1) distribution of PGCs prior to development of the gonad (Willey and
Krone 2001), (2) germ cell abundance (Gimeno et al. 1998), (3) retarded
testicular development or regression, (4) morphological changes such as
flattened seminiferous tubules, (5) degenerated sperm and spermatids (Wester
1991; Srivastava and Srivastava 1994; Zaroogian et al. 2001), (6) reduced
testicular growth and degeneration of testes, (7) abnormal lobular arrange-
ments, (8) degeneration of germ cells (Jobling et al. 1996; Christiansen et al.
1998), (9) altered oocyte development (Wannemacher et al. 1992; Matta et al.
1998; Olsson et al. 1999), and (10) reductions in sperm density and milt
volume (Jobling et al. 2002). Multinucleus oocytes and other gonadal abnormalities also have been reported in populations of fish exposed to thermal
stress (Lukšieng et al. 2000).

Sperm motility assays have been used to test the in vitro effects of chemicals on fish spermatozoa (reviewed by Kime and Nash 1999; also see
Cieresszko and Dabrowski 2000). Two criteria have found particular utility
as indicators of sperm dysfunction: the duration of sperm motility and the
velocity of sperm movement. In the laboratory, the fertilization rate of sperm
exposed to low concentrations of mercury was significantly reduced when
tested with low but normally effective concentrations of sperm (Rurangwa et
al. 1998).

Histopathological analysis of fish gonadal condition can be used to standardize and help explain the observed variation of other reproductive measures and, therefore, should be considered an essential component in any
tool chest of reproductive indicators. Methods for establishing gonadal maturation stage in fish by measuring a suite of plasma parameters could be
useful alternate bioindicators of reproductive condition when it is desired to
avoid the sacrifice of test or field-monitored animals (Johnson and Casillas
1991; Webb et al. 2002). However, a proposed scheme for establishing ovarian maturation stage based on biochemical measures in the English sole
(Johnson and Casillas 1991) proved only 68% effective in a field trial, which
seems, at best, marginally sufficient for the technique to be useful in a
biomonitoring situation.

Atresia

The presence of atretic oocytes in fish ovaries is representative of gamete
quality rather than quantity and may be one of the better reproductive indi-

cators of an actual pathological condition. Atresia is characterized by the degeneration and necrosis of developing oocytes and the subsequent infiltration by macrophages (Hinton et al. 1992). Examples of atresia in a teleost ovary are shown in Figure 6.

In mammals and birds, oocyte atresia is considered a natural degenerative process that occurs under the control of the endocrine system (Hsueh et al. 1994). However, atresia also is known to be induced by various environmental stressors including poor nutrition, elevated temperature, and exposure to xenobiotics (Saidapur 1978). Atresia is often cited as being relatively rare in healthy fish (Wallace and Selman 1981), but atresia has been observed to occur at specific times during or immediately following the breeding season of unstressed fish populations as well (Hunter and Macewicz 1985; Hsiao et al. 1994).

Many potential environmental contaminants, including PAHs (Krarup 1969; Mattison et al. 1983; Takizawa et al. 1984) and metals such as cadmium (Mattison et al. 1983) are known to specifically target and destroy developing oocytes in vertebrates other than fish. Elevated incidences of oocyte atresia have been reported in fish populations affected by effluents from paper and pulp mills (Sandström et al. 1988; Adams et al. 1992b) and, in some cases, have been associated with decreased concentrations of serum estradiol (Adams et al. 1992b). An example of elevated atresia downstream of a paper and pulp mill can be found in Figure 2. Atresia has also been observed in fish impacted by cooling water discharges (Lukšieng and Sandström 1994; Lukšieng et al. 2000); acid-induced stress (McCormick et al. 1987, 1989; Leino et al. 1990); disease (Winstead et al. 1991); confinement

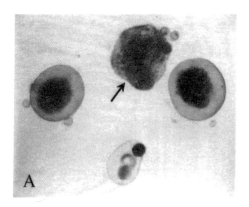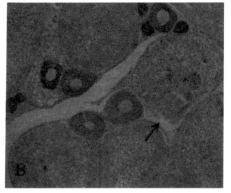

Figure 6. Large atretic oocytes in redbreast sunfish *Lepomis auratus* ovaries. (A) Whole fixed atretic oocyte (arrow) flanked by two normal vitellogenic oocytes; at the bottom of the picture is a parasitic cyst also present in the ovarian tissue. (B) Atretic oocyte (arrow) in histological section of ovary, surrounded by normal vitellogenic, cortical alveoli, and primary oocytes.

stress (Clearwater and Pankhurst 1997); pesticides (Kumar and Pant 1988; Rastogi and Kulshretha 1990; Sukumar and Karpagaganapathy 1992); PAHs (Johnson et al. 1997); mercury (Adams et al. 1999); a mixture of aromatic hydrocarbons, PCBs, and other chlorinated compounds (Johnson et al. 1999); and treated sewage effluent (Jobling et al. 2002).

Follicular atresia is associated in mammals and certain other vertebrates with the apotosis of follicular and other support cells (Tilly et al. 1991; Yang and Rajamahendran 2000) and with chemical-induced testicular injury (Richburg 2000). In fish, there is evidence both for (Weber and Janz 2001) and against (Wood and Van Der Kraak 2001) a role for apotosis in atresia. Until more is known of the potential involvement of apotosis in the mechanism of atresia in fish, apotosis should not be considered a proven indicator of atresia or toxicant-induced reproductive dysfunction in this vertebrate class.

Atresia has been cited elsewhere as the only histopathological biomarker to be adequately validated for use as a reproductive indicator of environmental stress on the basis of sufficient complimentary data derived from laboratory and field studies (Hinton et al. 1992). However, care must be taken with experimental design to account for the occasional natural occurrence of atresia in fish, particularly in multiple spawners. Furthermore, because atresia is a transitory phenomenon, frequent sampling may be needed to adequately characterize the occurrence or absence of this indicator at a specific study site or treatment group. The interpretation of atresia results also must be made with care, because a variety of conditions and stressors can cause this pathology. Finally, it should be recognized that atresia may not be a significant factor in reducing fecundity except in acute stress-related cases. For instance, nutritional deficiencies have been suggested to affect the quantity and quality of oocytes through a nonatresia mechanism in at least some fish species (Wootton 1979).

Size- and Age-at-Maturity

Size- and age-at-maturity in fish are determined by both genetic and environmental influences (Stearns and Crandall 1984). Along with growth and fecundity, age-at-maturity tends to integrate the effects of multiple environmental stressors on fish populations. Fish environmentally exposed to heavy metals were reported to demonstrate increased growth, increased fecundity, and earlier age of maturation but also reduced spawning success, reduced larval and egg survival, smaller egg size, and reduced longevity as compared with a reference population (McFarlane and Franzin 1978). Conversely, fish environmentally exposed to paper and pulp mill effluent were shown to grow more slowly and exhibited an increased age to maturity, decreased GSI, lower fecundity with age, an absence of secondary sexual characteristics in males, and a lack of an increase in egg size with age (Munkittrick et al. 1991). Flounder environmentally exposed to a mixture of aromatic hydrocarbons, PCBs, and other chlorinated compounds (Johnson et al. 1999) showed both a reduced age-at-maturity, as indicated by early vitellogenesis in a subset of the population, and a reduced percentage of sexual maturity in older females of reproducing age. Stress mimicked by feeding immature male common carp cortisol-containing feed pellets inhibited pubertal development, with both gonadal-somatic indices and the first wave of spermatogenesis significantly retarded (Consten et al. 2001).

Early Development

Mortality and the occurrence of developmental abnormalities are intimately associated with the effects of environmental stressors on the early development of fishes. In addition to lethality, numerous end points have been used as indicators of the effects of environmental stress on fish developmental processes (for a partial list, see review by Donaldson 1990) including (1) various gross developmental abnormalities or lesions in the embryo, (2) physiological measures such as heartbeat and respiration rate, (3) rates of development to specific developmental stages, (4) hatching success and timing, (5) size at hatch, (6) swim-up success and swimming speed, (7) histopathologies of the embryo and larvae, (8) altered sex ratio and gender differentiation (included here under the category of endocrine-related indicators), and (9) cytogenetic abnormalities.

A notable recent example of widespread mortality of fish during early developmental stages is the early-life-stage mortality problem of salmonids in the Great Lakes, the Finger Lakes of New York State, and the Baltic Sea (reviewed by Honeyfield et al. 1998). This syndrome, or series of related syndromes, is periodically manifested when feral fish are used for broodstock in hatcheries. Mortality generally occurs about the time of yolk sac absorption and the first feeding and is preceded by a number of characteristic symptoms. Although a thiamine deficiency of uncertain etiology is implicated as the immediate cause of the syndrome, interactions between thiamine deficiency and pollution are likely since the syndromes are most prevalent in contaminated areas. Exposure to persistent hydrophobic contaminants, particularly planar halogenated compounds (PHH), has been offered as another potential cause for this early-life-stage mortality (Wright and Tillitt 1999). Gross lesions associated with PHH exposure were seen prior to death, including yolk sac edema, craniofacial deformities, and hemorrhaging.

Acidification is another well-known environmental stressor known to adversely affect the early development of fish (Gunn 1986; Parker and McKeown 1986). Acid stress has been shown to affect both gametogenesis in adult fish and the incidences of developmental abnormalities in offspring (Peterson et al. 1982; Weiner et al. 1986; Ikuta and Kitamura 1995). Acid stress is often mediated by the effects of pH-dependent metals such as aluminum. Decreased swimming activity and reductions in the rates of ventilation, oxygen uptake, yolk absorption, and growth, as well as sodium influx were observed in northern pike *Esox lucius* and roach exposed to acid conditions and to lake water containing aluminum (Keinänen et al. 2000).

Numerous laboratory exposures of fish to environmental contaminants have shown the early life stages to be particularly sensitive to potential environmental stressors. Exposure of adult zebrafish to ethynylestradiol prior to spawning led to the arrest of the resulting embryos in the blastula stage of development (Kime and Nash 1999), although it is not yet clear whether this was due to a direct effect on the developing gametes or maternal loading of the embryos with this estrogenic compound. Methylmercury causes a variety of teratogenetic effects in fish including cyclopia, tail flexures, cardiac malformations, jaw deformities, twinning, and axial coiling (Weis and Weis 1977; Dial 1978). Thiobencarb, a herbicide previously associated with fish kills in the field, induced embryotoxicity in the medaka with a developmental stage specificity (Villalobos et al. 2000). Diazinon exposure resulted in decreased length of larvae (most sensitive indicator), hatching success, swim bladder inflation, and edema formation (Hamm and Hinton 2000).

The effects of environmental pollution on the early development of fish have been the subject of numerous field studies. Winter flounder *Pleuronectes americanus* from contaminated sites were shown to produce eggs with lower hatch rates and smaller larvae than fish from less contaminated sites (Nelson et al. 1991). Polychlorinated biphenyl concentrations in spawned eggs were inversely correlated with percent hatch in starry flounder *Platichthys stellatus* from sites in San Francisco Bay (Spies and Rice 1988). Following the *Exxon Valdez* oil spill in late 1989, developmental abnormalities, including genetic damage, physical deformities, and decreased larval size, were widely reported in Prince William Sound Pacific herring *Clupea pallasi* young (Hose et al. 1996). In a study of PCB-contaminated sites in New Bedford Harbor, Massachusetts, mummichog *Fundulus heteroclitus* embryo and larval survival was inversely related to the dioxin toxic equivalent concentrations of maternal liver tissue (Black et al. 1998).

Considerations in the Use of Reproductive Indicators

Some Potential Confounding Factors

Tolerance, Acclimation, and Resistance

The usefulness of reproductive and developmental measures as indicators of reproductive dysfunction in response to environmental stress may be tempered somewhat by the concepts of developed tolerance, acclimation, and resistance to chronic pollution. These terms are often used interchangeably, but in present usage, the first two terms refer to desensitization in individuals due to prior exposure, while the latter term refers to the genetic development of resistance to specific or generalized stressors in a population due to chronic prior exposure across generations. In field situations, all of these concepts may apply simultaneously and may not be easily distinguishable. Therefore, the terms tolerance and resistance will be used interchangeably here in discussing the responses of fish populations to environmental stressors that may have a basis in either or both concepts.

Examples of developed resistance in fish include populations of mummichog chronically exposed to mercury (Khan and Weis 1987; Weis and Weis 1989) and to polychlorinated biphenyls and furans (Prince and Cooper 1995). In both instances, populations were reported to have increased tolerance to these specific environmental contaminants as compared with reference populations. Partial resistance has been recognized in some fish, in which resistance is seen in certain developmental stages (i.e., embryonic and larval stages), but not in others (i.e., adults; Weis et al 2001). Multixenobiotic resistance has also been recognized in fish and other aquatic organisms, possibly mediated by the same P-glycoprotein mechanism involved in multidrug resistance (Bard 2000). Resistance can develop very rapidly in a fish population, possibly due to year-to-year changes in stressor exposure to the population (Weis and Weis 1984).

Resistant populations may have reduced reproductive effort as a tradeoff for diversion of resources in other directions (i.e., somatic growth), may have increased reproductive effort, or may exhibit no discernable changes at all. In any case, naïve or previously unexposed fish stressed in the laboratory or used as in situ sentinels in a field situation may not respond reproductively in an identical manner as native fish populations chronically stressed in the field. Therefore, it should be recognized that the absence in native fish of a reproductive effect anticipated from laboratory experiments (e.g., reductions in fecundity due to a chemical exposure) does not necessarily mean that the native population is not affected by the same stressor, but only that the two populations (naïve laboratory versus chronically stressed field fish) may be reacting differently to the same stressor, possibly as a result of developed resistance in the field population.

Natural Variability

Both inter- and intraspecies variation in reproductive measurements can be pronounced even in the absence of specific environmental stressors. Extreme examples of natural short-term variability in potential reproductive indicators of environmental stress can be found in some cyprinodontid fish that have semilunar reproductive cycles within their extended breeding seasons (Taylor et al. 1979; Greeley 1984; Greeley et al. 1986, 1988b). Examples of pronounced daily, and in some cases diel, fluctuations in various ovarian measures (GSI, egg abundance, oocyte stages, etc.) and plasma steroid concentrations (estradiol and testosterone) during the breeding seasons of these fish are demonstrated in Hsiao et al. (1994) and Greeley et al. (1988a), respectively. The magnitudes of these natural variations in reproductive indicators are at least as great as the potential effects anticipated from exposure to significant environmental stressors. Various reproductive hormones used as environmental stress indicators have also been shown to have diel fluctuations in other fish as well (Matsuyama et al. 1990; Jardine et al. 1996; Haddy and Pankhurst 1998; Nash et al. 2000). For most fish species and indicators, the normal ranges of natural variation in most reproductive indicators are not predetermined, so the collection of this essential information should be included in the study design if not already available.

Effects of Capture and Confinement Stress

Routine collecting procedures for fish, such as may result in fish spending approximately 12 hours in a hoop net, have been shown to affect the circulating levels of reproductive steroid hormones (McMaster et al. 1994). Plasma concentrations of androgens and estrogens in white sucker have been shown to be significantly influenced by both handling and confinement stress (Jardine et al. 1996). Furthermore, and of most concern, variation due to these short-term stresses is apparently not consistent from population to population or treatment group to treatment group (Jardine et al. 1996). Unfortunately, holding collected fish for a period of a few days before sampling to allow recovery from short-term stress has been shown to have no consistent effect on reproductive steroid levels in at least one study (Jardine et al. 1996.). These findings underscore the necessity to reduce capture and confinement stress as much as possible when long-term stress effects (such as from contaminant exposure) are of primary interest. Furthermore, when reductions in capture or confinement stress are not possible, the potential effects of such short-term stress on reproductive measures should be acknowledged. The potentially confounding effects of short-term procedural stress need to be further researched and strategies developed to better address the concerns.

Evaluation of Reproductive Indicators

Integrative Statistical Approaches

When measuring multiple reproductive indicators in fish populations or treatment groups, it is often useful to evaluate the integrated responses of the various indicators in a multivariate context. Appropriate multivariate approaches to the analysis of complex bioindicator data are discussed in detail elsewhere in this book (see Chapter 15). One recommended method for differentiating between different fish populations on the basis of their physiological or reproductive condition in field studies of contaminated areas is canonical variate analysis, also known as discriminant canonical analysis (Adams et al. 1994). An advantage of this statistical technique is that it provides summarizations of the data as new canonical variables that can be used to graphically illustrate variation in the reproductive responses of different fish populations or treatment groups, while also identifying which reproductive variables are most responsible for the discrimination between populations or groups. Figure 7 illustrates the results of a hypothetical canonical variate analysis on reproductive indicator data measured in fish from several reference sites and several polluted sites sampled along a spatial gradient downstream of a pollution point-source (modified from Adams et al. 2000). The reproductive indicators that account for the primary statistical differences among sites are listed in the figure. Less complex studies with fewer treatment groups or sites can be analyzed similarly and illustrated with 2-dimensional plots (for recent example, see Bernet et al. 2000) rather than the 3-dimensional figure shown here.

Modeling and Risk Assessment

Reproduction is an integral part of both stressor-response models (see Chapter 8) and the risk assessment process. Reproductive indicators are particularly useful for the risk assessment process because of their sensitivity and

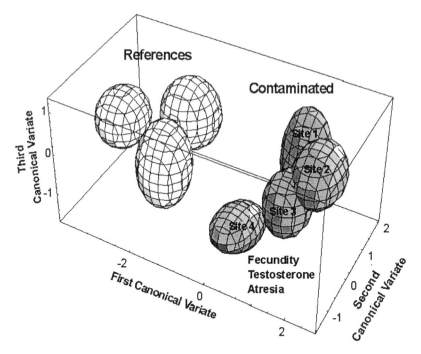

Figure 7. Hypothetical multivariate reproductive responses of fish populations at three reference sites and four locations along a contamination gradient, based on the incorporation of multiple reproductive indicator measurements into a canonical discriminant analysis procedure. Ellipsoids represent the mean integrated reproductive responses of the fish population at a site and the 95% confidence intervals of these responses. Overlap indicates similarity in integrative reproductive condition between sites. Listed are reproductive parameters that contribute most to discrimination between site responses. Modified from Adams et al. 2000.

intimate relationship to effects at higher levels of biological organizations (Adams et al. 2000). The continuing challenge is to integrate reproductive indicators at the molecular and biochemical levels with indicators such as fecundity and age-at-maturity that readily provide direct input into such bioassessment tools. Various models of fish reproduction have been proposed to link specific bioindicator measurements to reproductive outcomes at the population level (i.e., Arcand-Hoy and Benson 1998), but this topic remains an area in need of further research and development.

Standardized Tests

A number of standard laboratory methods are currently available to investigate the reproductive and developmental effects of chemicals in short-term embryo–larval or full or partial fish life cycle tests under controlled conditions (U.S. EPA 1982, 1989, 1994; OECD 1992). New tests are continually being proposed, especially in the rapidly evolving field of endocrine disruptor research (i.e., Patyna et al. 1999; Harries et al. 2000). These methods employ certain of the indicators discussed in this chapter and have the advantage of standardization and ease of measurement, which cannot be achieved easily in a natural environment. Such laboratory tests are well suited to the testing of the effects of chemicals or mixtures on fish reproduction and development under controlled laboratory conditions. They compliment, but do not supplant, field assessments of contaminated sites using selected suites of reproductive indicators measured in sentinel species.

Summary and Future Directions

Reproductive modes and strategies vary widely among fish species. The complexity of the reproductive process in fish ensures both an array of potential targets for stress-related reproductive dysfunction and also an abundance of potential reproductive indicators of environmental stress. Reproductive indicators currently in use in field and laboratory studies range from vitellogenin induction in male fish as a highly specific biomarker for endocrine-disrupting chemicals, to traditional life history parameters such as fecundity and age-at-maturity, which have high degrees of ecological relevance and utility for predicting reproductive and, ultimately, population-level success. New reproductive indicators continue to be developed: a recent example being zona radiata proteins measured in male fish as a specific indicator of exposure to endocrine disruptors (Arukwe et al. 1997b).

Because reproductive indicators with the greatest ecological relevance, such as fecundity, are generally integrative in nature and do not necessarily demonstrate great specificity for any particular type of environmental stressor, special care must be taken when attempting to relate cause with effect. In addition, since both inter- and intraspecies variation in reproductive measurements can be pronounced even in the absence of specific environmental stressors, care also must be taken in determining whether a response actually falls outside of a normal range of responses for each study species. Furthermore, in using reproductive indicators to evaluate the long-term effects of environmental stress, attempts must be made to minimize the potentially confounding short-term effects of collecting and confinement stress on the reproductive parameters being measured.

Molecular technology is opening up new avenues for exploration throughout the life sciences, including ecotoxicology.   Future reproductive indicators are already evolving from several of these recently developed biotechniques, including transgenic fish engineered to act as in situ sentinels of reproductive hazards (Legler et al. 2000), cDNA microarrays to assess and monitor global gene expression in response to environmental stressors, and global protein assays to examine changes in gene transcription products in fish exposed to reproductive stressors. Individual gene expression assays are becoming increasingly popular, in some cases substituting or supplementing traditional analyses. Beyond individual gene expression assays are methods to examine numerous genes simultaneously. Denslow et al. (2001) recently demonstrated a differential display reverse-transcriptase polymerase chain reaction assay to measure the induction of multiple genes in sheepshead minnows *Cyprinodon variegates* in response to exposure to natural estrogens and presumably other environmental pollutants with estrogenic characteristics. A good example of the probable future of reproductive indicator research in fish can be found in a recent study (Aquilar-Mahecha et al. 2001) where the expression of an array of 216 stress genes was used to examine selective susceptibility to stress in rat germ cells.   In Figure 8, a cDNA microarray, in this case, spotted with hundreds of zebrafish genes known to be responsive to estrogenic chemicals and other potential environmental stressors, is shown, along with dose–response curves for the occurrence of genes differentially expressed upon exposure to varying doses of 17β-estradiol. Similar arrays could be used in the near future to simultaneously monitor the responses of thousands of genes or even entire genomes to environmental pollutants.

Another way of looking at similar responses in a slightly different manner is to examine the global expression of the protein products of gene expression, as shown in Figure 9. In this example, the protein expression patterns resulting from the exposure of zebrafish embryos to estradiol and an estrogenic environmental contaminant, nonylphenol, are contrasted and compared through 2-dimensional gel electrophoresis. Although the specific proteins in this example are still largely undefined, we can anticipate, in the future, protein microarrays spotted with known protein substrates similar to the existing cDNA microarrays. These newly developed biomolecular tools can be expected to foster a vastly expanded understanding of the mecha-

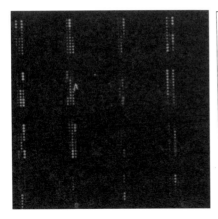
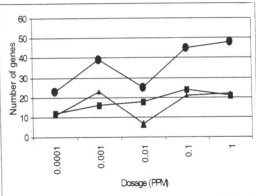

Figure 8. (A) Zebrafish cDNA microarray showing alterations in gene expression following exposure of zebrafish 7-d post-fertilization fry to 1 ppm 17β-estradiol (red = genes downregulated; green = genes upregulated; yellow = genes with no change); (B) numbers of genes differentially expressed in 7-d post-fertilization zebrafish fry in response to varying doses of 17β-estradiol (square = genes downregulated; triangle = genes upregulated; circle = genes with no change). Unpublished data of Hoyt, Doktycz, and Greeley; Oak Ridge National Laboratory.

nisms of stress-related effects on reproduction and development. Furthermore, when these new techniques are sufficiently standardized and validated against existing measures of reproductive competence, they could theoretically permit researchers to monitor and eventually predict the reproductive effects of particular environmental stressors on fish and other aquatic organisms using a single DNA or protein chip. However, it must be cautioned that our current understanding of the relationships between gene or even protein expression and consequences to the reproduction and survival of an organism is still rudimentary, and we know even less about the linkages between these molecular responses and consequences at the population level and beyond. Thus, although the promise of these new approaches may, at times, appear nearly unlimited, the more traditional reproductive indicators discussed in this chapter will continue to have important roles for the foreseeable future in predicting and evaluating the effects of environmental stress on aquatic organisms and the aquatic environment.

Acknowledgments

I thank Marshall Adams, Mike van den Heuvel, and Arthur Stewart for comments and suggestions that improved the manuscript. Evan Reddick assisted with the photomicrography. Personal research shown as examples in some figures was sponsored by the Laboratory Directed Research and Develop-

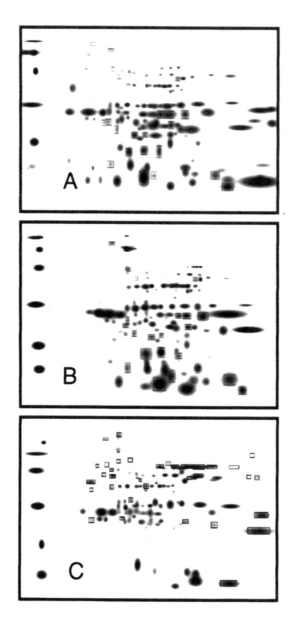

Figure 9. Common key proteins in the toxic responses of zebrafish embryos to 1 ppm 17-estradiol or nonylphenol: (A) proteins induced in fish embryos exposed to both treatments, shown on a nonylphenol treatment 2-D gel; (B) proteins induced in fish embryos exposed to both treatments, shown on estradiol treatment gel; and (C) proteins not expressed in either treatment, shown on the control gel. Unpublished data of Bradley, Schrader, and Greeley; University of Maryland and Oak Ridge National Laboratory.

ment Program of Oak Ridge National Laboratory, managed by UT-Battelle, LLC, for the U.S. Department of Energy under Contract No. DE-AC05-00OR22725.

References

Adams, S. M., M. S. Bevelhimer, M. S. Greeley, Jr., D. Levine, and S. J. Teh. 1999. Ecological risk assessment in a large river-reservoir: 6. Bioindicators of fish population health. Environmental Toxicology and Chemistry 18:628–640.

Adams, S. M., W. D. Crumby, M. S. Greeley, Jr., M. G. Ryon, and E. M. Schilling. 1992a. Relationship between physiological and fish population responses in a contaminated stream. Environmental Toxicology and Chemistry 11:1549–1557.

Adams, S. M., W. D. Crumby, M. S. Greeley, Jr., and L. R. Shugart. 1992b. Responses of fish populations and communities to pulp mill effluents: a holistic assessment. Ecotoxicology and Environmental Safety 24:347–360.

Adams, S. M., M. S. Greeley, Jr., and M. G. Ryon. 2000. Evaluating effects of contaminants on fish health at multiple levels of biological organization: extrapolating from lower to higher levels. Human and Ecological Risk Assessment 6:15–27.

Adams, S. M., K. D. Ham, and J. J. Beauchamp. 1994. Application of canonical variate analysis in the evaluation and presentation of multivariate biological response data. Environmental Toxicology and Chemistry 13:1673–1683.

Allen, Y. P. Matthiessen, A. P. Scott, S. Haworth, S. Fiest, and J. E. Thain. 1999. The extent of estrogenic contamination in the UK estuarine and marine environments: further surveys of flounder. The Science of the Total Environment 233:5–20.

Aquilar-Mahecha, A., B. F. Hales, and B. Robaire. 2001. Expression of stress response genes in germ cells during spermatogenesis. Biology of Reproduction 65:119–127.

Arcand-Hoy, L. D., and W. H. Benson. 1998. Fish reproduction: an ecologically relevant indicator of endocrine disruption. Environmental Toxicology and Chemistry 17:49–57.

Arukwe, A., T. Celius, B. T. Walther, and A. Goksøyr. 2000. Effects of xenoestrogen treatment on zona radiata protein and vitellogenin expression in Atlantic salmon (Salmo salar). Aquatic Toxicology 49:159–170.

Arukwe, A., L. Förlin, and A. Goksøyr. 1997a. Xenobiotic and steroid biotransformation enzymes in Atlantic salmon (Salmo salar) treated with an estrogenic compound, 4-nonyphenol. Environmental Toxicology and Chemistry 16:2576–2583.

Arukwe, A., F. R. Knudsen, and A. Goksøyr. 1997b. Fish zona radiata (eggshell) protein: a sensitive biomarker for environmental estrogens. Environmental Health Perspectives 105:418–422.

Bagenal, T. 1969. The relationship between food supply and fecundity in brown trout Salmo trutta. Journal of Fish Biology 1:167–182.

Bagenal, T. 1971. The interrelation of the size of fish eggs, the date of spawning, and the production cycle. Journal of Fish Biology 3:207–219.

Balon, E. K. 1975a. Reproductive guilds of fishes: a proposal and definition. Journal of the Fisheries Research Board of Canada 32:821–864.

Balon, E. K. 1975b. Terminology of intervals in fish development. Journal of the Fisheries Research Board of Canada 32:1663–1670.

Bard, S. M. 2000. Multixenobiotic resistance as a cellular defense mechanism in aquatic organisms. Aquatic Toxicology 48:357–389.

Barnthouse, L. W. 1993. Population-level effect. Pages 247–274 in G. W. Suter, editor. Ecological risk assessment. Lewis Publishers, Boca Raton, Florida.

Baroiller, J.-F., Y. Guiguen, and A. Fostier. 1999. Endocrine and environmental aspects of sex differentiation in fish. Cellular and Molecular Life Sciences 55:910–931.

Batty, J., and R. Lim. 1999. Morphological and reproductive characteristics of male mosquitofish (*Gambusia affinis holbrooki*) inhabiting sewage-contaminated waters in New South Wales, Australia. Archives of Environmental Contamination and Toxicology 36:0301–0307.

Bayley, M., J. R. Nielsen, and E. Baatrup. 1999. Guppy sexual behavior as an effect biomarker of estrogen mimics. Ecotoxicology and Environmental Safety 43:68–73.

Bernet, D., H. Schmidt-Posthaus, T. Wahli, and P. Burkhardt-Holm. 2000. Effects of wastewater on fish health: an integrated approach to biomarker responses in brown trout (*Salmo trutta* L.). Journal of Aquatic Ecosystem Stress and Recovery 8:143–152.

Bieniarz, K., P. Epler, M. Sokolowska-Mikolajczyk, and W. Popek. 1997. Reproduction of fish in conditions disadvantageously altered with the salts of zinc and copper. Archives of Polish Fisheries 5:21–30.

Bjerselius, R., K. Lundstedt-Enkel, H. Olsén, I. Mayer, and K. Dimberg. 2001. Male goldfish reproductive behavior and physiology are severely affected by exogenous exposure to 17b-estradiol. Aquatic Toxicology 53:139–152.

Black, D. E., R. E. Gutjahr-Gobell, R. J. Pruell, B. Bergen, L. Mills, and A. E. McElroy. 1998. Reproduction and polychlorinated biphenyls in *Fundulus heteroclitus* (Linnaeus) from New Bedford Harbor, Massachusetts, USA. Environmental Toxicology and Chemistry 17:1405–1414.

Blaxter, J. H. S. 1969. Development: eggs and larvae. Pages 177–252 in W. S. Hoar and D. J. Randall, editors. Fish physiology. Volume III. Academic Press, New York.

Blaxter, J. H. S. 1988. Pattern and variety in development. Pages 1–58 in W. S. Hoar and D. J. Randall, editors. Fish physiology. Volume XIA. Academic Press, San Diego, California.

Blazquez, M., A. Felip, S. Zanuy, M. Carrillo, and F. Piferrer. 2001. Critical period of androgen-inducible sex differentiation in a teleost fish, the European sea bass. Journal of Fish Biology 58:342–358.

Borg, B. 1994. Androgens in teleost fishes. Comparative Biochemistry and Physiology 109:219–245.

Bortone, S. A., and W. P. Davis. 1994. Fish intersexuality as an indicator of environmental stress. Bioscience 44:165–172.

Bresch, H., H. Beck, D. Ehlermann, H. Schlaszus, and M. Urbanek. 1990. A long-term toxicity test comprising reproduction and growth of zebrafish with 4-chloroaniline. Archives of Environmental Contamination and Toxicology 19:419–427.

Buckley, L. J., A. S. Smigielski, T. A. Halavik, E. M. Caldarone, B. R. Burns, and G. C. Laurence. 1991. Winter flounder *Pseudopleuronectes americanus* reproductive success. II: effect of spawning time and female size on size, composition, and viability of eggs and larvae. Marine Ecology Progress Series 74:125–135.

Campbell, P. M., T. G. Pottinger, and J. P. Sumpter. 1994a. Changes in the affinity of estrogen and androgen receptors accompanying changes in receptor abundance in brown and rainbow trout. General and Comparative Endocrinology 94:329–340.

Campbell, P. M., T. G. Pottinger, and J. P. Sumpter. 1994b. Preliminary evidence that chronic confinement stress reduces the quality of gametes produced by brown and rainbow trout. Aquaculture 120:151–169.

Cardwell, R. D., D. G. Foreman, T. R. Payne, and D. J. Wilbur. 1977. Acute and chronic toxicity of chlordane to fish and invertebrates. U.S. Environmental Protection Agency, EPA-600/3-77-019, Duluth, Minnesota.

Carlsson, G., S. Örn, P. L. Andersson, H. Söderstöm, and L. Norrgren. 2000. The impact of musk ketone on reproduction in zebrafish. Marine Environmental Research 50:237–241.

Cavaco, J. E. B., H. F. Vischer, J. G. D. Lambert, H. J. Th. Goos, and R. W. Schulz. 1997. Mismatch between patterns of circulating and testicular androgens in African catfish, Clarias gariepinus. Fish Physiology and Biochemistry 17:155–162.

Celius, T., T. B. Haugen, T. Grotmol, and B. T. Walther. 1999. A sensitive zonagenetic assay for rapid in vitro assessment of estrogenic potency of xenobiotics and mycotoxins. Environmental Health Perspectives 107:63–68.

Celius, T., and B. T. Walther. 1998. Differential sensitivity of zonagenesis and vitellogenesis in Atlantic salmon (Salmo salar L) to DDT pesticides. Journal of Experimental Zoology 281:346–353.

Chambers, R. C., W. C. Legget, and J. A. Brown. 1989. Egg size, female effects, and correlations between early life history traits of capelin (Mallotus villosus): an appraisal at the individual level. Fishery Bulletin 87:515–523.

Chang, J. P., R. E. Peter, C. S. Nahorniak, and M. Sokolowska. 1984. Effects of catecholaminergic agonists and antagonists on serum gonadotropin concentration and ovulation in goldfish: evidence for specificity of dopamine inhibition of gonadotropin secretion. General and Comparative Endocrinology 55:351–360.

Chappaz, R., G. Brun, and G. Olivari. 1987. Evidences for differences in feeding within a population of bleak Alburnus alburnus (L.) In the lake of St. Croix. The consequences for growth and fecundity. Annals of Limnology 23:224–252.

Christiansen, T., B. Korsgaard, and Å. Jespersen. 1998. Induction of vitellogenin synthesis by nonylphenol and 17b-estradiol and effects on the testicular structure in the eelpout Zoarces viviparus. Marine Environmental Research 46:141–144.

Cieresszko, A., and K. Dabrowski. 2000. In vitro effect of gossypol acetate on yellow perch (Perca flavescens) spermatozoa. Aquatic Toxicology 49:181–187.

Clearwater, S. J., and N. W. Pankhurst. 1997. The response to capture and confinement stress of plasma cortisol, plasma sex steroids and vittelogenic oocytes in the marine teleost, red gunard. Journal of Fish Biology 50:429–441.

Consten, D., J. Bogerd, J. Komen, J. G. D. Lambert, and H. J. Goos. 2001. Long-term cortisol treatment inhibits pubertal development in male common carp, Cyprinus carpio L. Biology of Reproduction 64:1063–1071.

Contreras-Sanchez, W. M., C. B. Schreck, and M .S. Fitzpatrick. 1995. Effect of stress on the reproductive physiology of rainbow trout, Oncorhynchus mykiss. Page 183 in Proceedings of the 5th International Symposium on the Reproductive Physiology of Fishes. University of Texas, Austin, Texas.

Copeland, P., J. P. Sumpter, T. K. Walker, and M. Croft. 1986. Vitellogenin levels in male and female rainbow trout Salmo gairdneri Richardson at various stages of the reproductive cycle. Comparative Biochemistry and Physiology 83B:487–493.

Coward, K., N. R. Bromage, and D. C. Little. 1997. Inhibition of spawning and associated suppression of sex steroid levels during confinement in the substrate-spawning Tilapia zillii. Journal of Fish Biology 52:152–165.

Craig, G. R., and W. F. Baski. 1977. The effects of depressed pH on flagfish reproduction, growth and survival. Water Resources 11:621–626.

Craik, J. C. A., and S. M. Harvey. 1984. Biochemical changes occurring during final
maturation of eggs of some marine and freshwater teleosts. Journal of Fish
Biology 24:599–610.

Crain, D. A., L. J. Guillette Jr., A. A. Rooney, and D. B. Pickford. 1997. Alterations in
steroidogenesis in alligators (*Alligator mississippiensis*) exposed naturally and
experimentally to environmental contaminants. Environmental Health Perspectives 105:528–533.

Das, S., and P. Thomas. 1999. Pesticides interfere with the nongenomic action of a
progestogen on meiotic maturation by binding to its plasma membrane receptor
on fish oocytes. Endocrinology 140:1953–1956.

Denslow, N. D., C. J. Bowman, R. J. Ferguson, H. S. Lee, M. J. Hemmer, and L. C.
Folmar. 2001. Induction of gene expression in sheepshead minnows (*Cyprinodon
variegatus*) treated with 17β-estradiol, diethylstilbestrol, or ethinylestradiol: the
use of mRNA fingerprints as an indicator of gene regulation. General and Comparative Endocrinology 121:250–260.

de Vlaming, V., G. Grossman, and F. Chapman. 1982. On the use of the gonosomatic
index. Comparative Biochemistry and Physiology 73A:31–39.

Dial, N. 1978. Methylmercury: some effects on embryogenesis in the Japanese medaka,
Oryzias latipes. Teratology 17:83–92.

Donaldson, E. M. 1990. Reproductive indices as measures of the effects of environmental stressors in fish. Pages 109–122 *in* S. M. Adams, editor. Biological
indicators of stress in fish. American Fisheries Society, Symposium 8, Bethesda,
Maryland.

Ellis, R. J. 2001. An assessment of endocrine disrupting potential of a New Zealand
pulp and paper mill effluent using rainbow trout (*Oncorhynchus mykiss*) and
mosquitofish (*Gambusia affinis*). Doctoral dissertation. University of Waikato,
Hamilton, New Zealand.

Emmersen, B. K., and I. M. Petersen. 1976. Natural occurrence and experimental
induction by estradiol-17-b of a phospholipoprotein (vitellogenin) in flounder,
Platichthys flesus (L). Comparative Biochemistry and Physiology 53B:443–446.

Fleming, I. A., and M. R. Gross. 1990. Latitudinal clines: a trade-off between egg
number and size in Pacific salmon. Ecology 71:1–11.

Folmar, L. C., N. D. Denslow, V. Rao, M. Chow, D. A. Crain, J. Enblom, J. Marcino,
and L. J. Guillette Jr. 1996. Vitellogenin induction and reduced serum testosterone concentrations in feral male carp (*Cyprinus carpio*) captured near a major
metropolitan sewage treatment plant. Environmental Health Perspectives
104:1096–1101.

Folmar, L. C., G. R. Gardner, M. P. Schreibman, L. Magliulo-Cepriano, L. J. Mills, G.
Zaroogian, R. Gutjahr-Gobell, R. Haebler, D. B. Horowitz, and N. D. Denslow.
2001. Vitellogenin-induced pathology in male summer flounder (*Paralichthys
dentatus*). Aquatic Toxicology 51:431–441.

Freemen, H. C., G. B. Sangalang, and J. F. Uthe. 1984. The effects of pollutants and
contaminants on steroidogenesis in fish and marine mammals. Pages 197–212 *in*
V. W. Cairns, P. V. Hodson, and J. O. Nriagu, editors. Contaminant effects on
fisheries. Advances in environmental science and technology. Volume 16. John
Wiley, New York.

Fukada, H., A. Haga, T. Fujita, N. Hiramatsu, C. V. Sullivan, and A. Hara. 2001. Development and validation of chemiluminescent immunoassay for vitellogenin in five salmonid species. Comparative Biochemistry and Physiology 130A:163–170.

Gagnon, M. M., J. J. Dodson, and P. V. Hodson. 1994. Ability of BKME (bleached kraft mill effluent) exposed white suckers (Catostomas commersoni) to synthesize steroid hormones. Comparative Biochemistry and Physiology 107C:265–273.

Gallagher, E. P., T. S. Gross, and K. M. Sheehy. 2001. Decreased glutathione S-transferase expression and activity and altered sex steroids in Lake Apopka brown bullheads (Ameriurus nebulosus). Aquatic Toxicology 55:223–237.

Garcia, E. F., R. J. McPherson, T. H. Martin, R. A. Poth, and M. S. Greeley, Jr. 1997. Liver cell estrogen receptor binding in prespawning female largemouth bass, Micropterus salmoides, environmentally exposed to polychlorinated biphenyls. Archives of Environmental Contamination and Toxicology 32:309–315.

Ge, W., R. E. Peter, and Y. Nagahama. 1997. Activin and its receptors in the goldfish. Fish Physiology and Biochemistry 17:143–153.

George, C. J. 1977. The implication of neuroendocrine mechanisms in the regulation of population character. Fisheries 2(3):14–19.

Giesy, J. P., S. L. Pierens, E. M. Snyder, S. Miles-Richardson, V. J. Kramer, S. A. Snyder, K. M. Nichols, and D. A. Villeneuve. 2000. Effects of 4-nonylphenol on fecundity, and biomarkers of estrogenicity in fathead minnows (Pimephales promelas). Environmental Toxicology and Chemistry 19:1368–1377.

Gilkey, J. C. 1981. Mechanisms of fertilization in fishes. American Zoologist 21:359–375.

Gimeno, S., H. Komen, A. G. M. Gerritsen, and T. Bowmer. 1998. Feminization of young males of the common carp, Cyprinus carpio, exposed to 4-tert-pentylphenol during sexual differentiation. Aquatic Toxicology 43:77–92.

Goodbred, S. L., R. J. Gilliom, T. S. Gross, N. D. Denslow, W. L. Bryant, and T. R. Schoeb. 1996. Reconnaissance of 17ß-estradiol,[2] 11-ketotestosterone, vitellogenin, and gonad histopathology in common carp of United States streams: potential for contaminant-induced endocrine disruption. U.S. Geological Survey Open-File Report 96-627.

Goetz, F. W. 1997. Follicle and extrafollicular tissue interaction in 17α-10β dehydroxy-4-pregnen-3-one-stimulated ovulation and prostaglandin synthesis in the yellow perch Perca flavescens ovary. General and Comparative Endocrinology 105:121–126.

Goetz, F. W., and M. Garczynski. 1997. The ovarian regulation of ovulation in teleost fish. Fish Physiology and Biochemistry 17:33–38.

Gray, M. A., A. J. Niimi, and C. D. Metcalf. 1999. Factors affecting the development of testis-ova in medaka, Oryzias latipes, exposed to octylphenol. Environmental Toxicology and Chemistry 18:1835–1842.

Greeley, M. S., Jr. 1984. Spawning by Fundulus pulvereus and Adinia xenica (Cyprinodontidae) along the Alabama Gulf coast is associated with the semilunar tidal cycles. Copeia 1984:797–800.

Greeley, M. S., Jr., D. R. Calder, and R. A. Wallace. 1986. Changes in teleost yolk proteins during oocyte maturation: correlation of yolk proteolysis with oocyte hydration. Comparative Biochemistry and Physiology 84B:1–9.

Greeley, M. S., Jr., H. Hols, and R. A. Wallace. 1991. Changes in size, hydration, and low molecular weight osmotic effectors during meiotic maturation of Fundulus oocytes in vivo. Comparative Biochemistry and Physiology 100A:639–647.

Greeley, M. S., Jr., C. G. Hull, and S. K. Sharp. 1998. Indicators of reproductive health. Pages 526–569 *in* R. L. Hinzman, editor. Third report on the Oak Ridge Y-12 Plant biological monitoring and abatement program for East Fork Poplar Creek. Oak Ridge Y-12 Plant, Y-TS-889, Oak Ridge, Tennessee.

Greeley, M. S., Jr., R. MacGregor III, and K. R. Marion. 1988a. Variation in plasma oestrogens and androgens during the seasonal and semilunar spawning cycles of female gulf killifish, *Fundulus grandis* (Baird and Girard). Journal of Fish Biology 33:419–429.

Greeley, M. S., Jr., R. MacGregor III, and K. R. Marion. 1988b. Changes in the ovary of the gulf killifish, *Fundulus grandis* (Baird and Girard), during seasonal and semilunar spawning cycles. Journal of Fish Biology 33:97–107.

Greeley, M. S., Jr., K. R. Marion, and R. MacGregor III. 1986. Semilunar spawning cycles of *Fundulus similis* (Cyprinodontidae). Environmental Biology of Fishes 17:125–131.

Grier, H. J. 1981. Cellular organization of the testis and spermatogenesis in fishes. American Zoologist 21:345–357.

Gunn, J. M. 1986. Behavior and ecology of salmonid fishes exposed to episodic pH depressions. Environmental Biology of Fish 7:385–388.

Guraya, S. S. 1986. The cellular and molecular biology of fish oogenesis. Karger, New York.

Haddy, J. A., and N. W. Pankhurst. 1998. Annual change in reproductive condition and plasma concentrations of sex steroids in black bream, *Acanthopagrus butcheri* (Munro) (Sparidae). Marine and Freshwater Research 49:389–397.

Hamm, J. T., and D. E. Hinton. 2000. The role of development and duration of exposure to the embryotoxicity of diazinon. Aquatic Toxicology 48:403–418.

Harries, J. E., A. Janbakhsh, S. Jobling, P. Matthiessen, J. P. Sumpter, and C. R. Tyler. 1999. Estrogenic potency of effluent from two sewage treatment works in the United Kingdom. Environmental Toxicology and Chemistry 18:932–937.

Harries, J. E., T. Runnalls, E. Hill, C. A. Harris, S. Maddix, J. P. Sumpter, and C. R. Tyler. 2000. Development of a reproductive performance test for endocrine disrupting chemicals using pair-breeding fathead minnows (*Pimephales promelas*). Environmental Science and Technology 34:3003–3011.

Harries, J. E., D. A. Sheahan, S. Jobling, P. Matthiessen, P. Neall, E. J. Routledge, R. Rycroft, J. P. Sumpter, and T. Tylor. 1996. A survey of estrogenic activity in United Kingdom inland waters. Environmental Toxicology and Chemistry 15:1993–2002.

Harries, J. E., D. A. Sheahan, S. Jobling, P. Matthiessen, P. Neall, P. Sumpter, T. Tylor, and N. Zaman. 1997. Estrogenic activity in five United Kingdom rivers detected by measurement of vitellogenesis in caged male trout. Environmental Toxicology and Chemistry 16:534–542.

Hinton, D. E., P. C. Baumann, G. R. Gardner, W. E. Hawkins, J. D. Hendricks, R. A. Murchelano, and M. S. Okihiro. 1992. Histopathological biomarkers. Pages 155–210 *in* R. J. Hugget, R. A. Kimerle, P. M. Mehrle, Jr., and H. L. Bergman, editors. Biomarkers: biochemical, physiological, and histological markers of anthropogenic stress. Lewis Publishers, Chelsea, Maryland.

Hobby, A. C., and N. W. Pankhurst. 1997. The relationship between plasma and ovarian levels of gonadal steroids in the repeat spawning marine fishes *Pagrus auratus* (Sparidae) and *Chromis dispilus* (Pomacentridae). Fish Physiology and Biochemistry 16:65–75.

Honeyfield, D. C., J. D. Fitzsimons, S. B. Brown, S. V. Marcquenski, and G. McDonald. 1998. Introduction and overview of early life stage mortality. Pages 1–7 *in* G. McDonald, J. D. Fitzsimons, and D. C. Honeyfield, editors. Early life stages mortality syndrome in fishes of the Great Lakes and Baltic Sea. American Fisheries Society, Symposium 21, Bethesda, Maryland.

Hose, J. E., J. N. Cross, S. G. Smith, and D. Diehl. 1989. Reproductive impairment in a fish inhabiting a contaminated coastal environment off Southern California. Environmental Pollution 57:139–148.

Hose, J. E., M. D. McGurk, G. D. Marty, D. E. Hinton, E. D. Brown, and T. T. Baker. 1996. Sublethal effects of the Exxon Valdez oil spill on herring embryos and larvae: morphologic, cytogenetic, and histopathological assessments, 1989–1991. Canadian Journal of Fisheries and Aquatic Sciences 53:2355–2365.

Hsiao, S.-M., M. S. Greeley, and R. A. Wallace. 1994. Reproductive cycling in female *Fundulus heteroclitus*. Biological Bulletin 186:271–284.

Hsueh, A. J. W., H. Billig, and A. Tsafriri. 1994. Ovarian follicle atresia: a hormonally controlled apoptic process. Endocrine Reviews 15:707–724.

Hunter, J. R., and B. J. Macewicz. 1985. Rates of atresia in the ovary of captive and wild northern anchovy, *Engraulis mordax*. Fishery Bulletin 83:119–136.

Ikuta, K., and S. Kitamura. 1995. Effects of low pH exposure of adult salmonids on gametogenesis and embryo development. Water, Air, and Soil Pollution 85:327–332.

Janssen, P. A. H., J. G. D. Lambert, A. D. Vethaak, and H. J. T. Goos. 1997. Environmental pollution caused by elevated concentrations of oestradiol and vitellogenin in the female flounder, *Platicichthys flesus* (L.). Aquatic Toxicology 39:195–214.

Janz, D. M., M. E. McMaster, L. P. Weber, K. R. Munkittrick, and G. Van Der Kraak. 1997. Elevated ovarian follicular apotosis and heat shock protein-70 expression in white sucker exposed to bleached kraft mill effluent. Toxicology and Applied Pharmacology 147:391–398.

Jardine, J. J., G. J. Van Der Kraak, and K. R. Munkittrick. 1996. Capture and confinement stress in white sucker exposed to bleached kraft mill effluent. Ecotoxicology and Environmental Safety 33:287–298.

Jobling, S., N. Beresford, M. Nolan, T. Rodgers-Gray, G. C. Brighty, J. P. Sumpter, and C. R. Tyler. 2002. Altered sexual maturation and gamete production in wild roach (*Rutilis rutilis*) living in rivers that receive treated sewage effluents. Biology of Reproduction 66:272–281.

Jobling, S., M. Nolan, C. R. Tyler, G. Brighty, J. P. Sumpter. 1998. Widespread sexual disruption in wild fish. Environmental Science and Technology 32:2498–2506.

Jobling, S., D. Sheanhan, J. A. Osborne, P. Marthiessen, and J. P. Sumpter. 1996. Inhibition of testicular growth in rainbow trout (*Oncorhynchus mykiss*) exposed to estrogenic alkylphenol chemicals. Environmental Toxicology and Chemistry 15:194–202.

Johnson, D. W., and D. A. Webster. 1977. Avoidance of low pH in selection of spawning sites by brook trout (*Salvelinus fontinalis*). Journal of the Fisheries Research Board of Canada 34:2215–2218.

Johnson, L. L., and E. Casillas. 1991. The use of plasma parameters to predict ovarian maturation stage in English sole *Parophrys vetulus* Girard. Journal of Experimental Marine Biology and Ecology 151:257–270.

Johnson, L. L., E. Casillas, T. K. Collier, B. B. McCain, and U. Varanasi. 1988. Contaminant effects on ovarian development in English Sole (*Parophrys vetulus*) from Puget Sound, Washington. Canadian Journal of Fisheries and Aquatic Sciences 45:2133–2146.

Johnson, L. L., S. Y. Sol, D. P. Lomax, G. M. Nelson, C. A. Sloan, and E. Casillas. 1997. Fecundity and egg weight in English sole *Pleuronectes vetulus* from Puget Sound, Washington: influence of nutritional status and chemical contaminants. Fisheries Bulletin 95:231–249.

Johnson, L. L., S. Y. Sol, G. M. Ylitalo, T. Hom, B. French, O. P. Olson, and T. K. Collier. 1999. Reproductive injury in English sole *(Pleuronectes vetulus)* from the Hylebos Waterway, Commencement Bay, Washington. Journal of Aquatic Ecosystem Stress and Recovery 6:289–310.

Johnson, L. L., J. E. Stein, T. K. Collier, E. Casillas, and U. Varanasi. 1994. Indicators of reproductive development in prespawning female winter flounder (*Pleuronectes americanus*) from urban and non-urban estuaries in the northeast United States. Science of the Total Environment 14:241–260.

Jones, J. C., and J. D. Reynolds. 1997. Effects of pollutants on reproductive behavior of fishes. Reviews in Fish Biology and Fisheries 7:463–491.

Karås, P., E. Neuman, and O. Sandström. 1991. Effects of pulp mill effluent on the population dynamics of perch, *Perca fluviatilis*. Canadian Journal of Fisheries and Aquatic Sciences 48:28–34.

Karels, A. E., M. Soimasuo, J. Lappivaara, H. Leppänen, T. Aaltonen, P. Mellanen, and A. O. J. Oikari. 1998. Effects of ECF-bleached kraft mill effluent on reproductive steroids and liver MFO activity in populations of perch and roach. Ecotoxicology 7:123–132.

Kavlock, R. J., G. P. Daston, C. DeRosa, P. Fenner-Crisp, L. E. Gray, S. Kaattari, G. Lucier, M. Luster, M. J. Mac, C. Maczka, R. Miller, J. Moore, R. Rolland, G. Scott, D. M. Sheehan, T. Sinks, and H. A. Tilson. 1996. Research needs for the risk assessment of health and environmental effects of endocrine disrupters: A report of the US-EPA-sponsored workshop. Environmental Health Perspectives 104:715–740.

Keinänen, M., S. Peuranen, M. Nikinmaa, C. Tigerstedt, and P. Vuorinen. 2000. Comparison of the responses of the yolk-sac fry of pike (*Esox lucius*) and roach (*Rutilis rutilis*) to low pH and aluminum: sodium influx, development, and activity. Aquatic Toxicology 47:161–179.

Khan, A. T., And J. S. Weis. 1987. Effect of methylmercury on egg and juvenile viability in two populations of killifish *Fundulus heteroclitus*. Environmental Research 44:272–278.

Khan, I. A., and P. Thomas. 2000. Lead and Aroclor 1254 disrupt reproductive neuroendocrine function in Atlantic croaker. Marine Environmental Research 50:119–123.

Kime, D. E., and J. P. Nash. 1999. Gamete viability as an indicator of reproductive endocrine disruption in fish. The Science of the Total Environment 233:123–129.

Kitamura, S., and K. Ikuta. 2000. Acidification severely suppresses spawning of hime salmon (land-locked sockeye salmon, *Oncorhynchus nerka*). Aquatic Toxicology 51:107–113.

Knudsen, F. R., A. Arukwe, and T. G. Pottinger. 1998. The *in vivo* effect of combinations of octylphenol, butylbenzylphthalate and estradiol on liver estradiol receptor modulation and induction of zona radiata proteins in rainbow trout: no evidence of synergy. Environmental Pollution 103:75–80.

Knudsen, F. R., and T. G. Pottinger. 1999. Interaction of endocrine disrupting chemicals, singly and in combination, with estrogen-, androgen-, and corticosteroid-binding sites in rainbow trout (*Oncorhynchus mykiss*). Aquatic Toxicology 44:159–170.

Knutsen, G. M., and S. Tilseth. 1985. Growth, development, and feeding success of Atlantic cod larvae *Gadus morhua* related to egg size. Transactions of the American Fisheries Society 114:507–511.

Koger, C. S., S. J. Teh, and D. E. Hinton. 2000. Determining the sensitive developmental stages of intersex induction in medaka (*Oryzias latipes*) exposed to 17β-estradiol or testosterone. Marine Environmental Research 50:201–206.

Korte, J. J., M. D. Kahl, K. M. Jensen, S. Pasha-Mumtaz, L. G. Parks, G. A. LeBlanc, and G. T. Ankley. 2000. Fathead minnow vitellogenin: complimentary DNA sequence, and messenger RNA, and protein expression after 17β-estradiol treatment. Environmental Toxicology and Chemistry 19:972–981.

Kovacs, T. G., S. J., Gibbons, L. A. Tremblay, B. I. Connor, P. H. Martel, and R. H. Voss. 1995. The effects of a secondary-treated bleached kraft mill effluent on aquatic organisms as assessed by short-term and long-term laboratory tests. Ecotoxicology and Environmental Safety 31:7–22.

Kramer, V. J., S. Miles-Richardson, S. Pierens, and J. P. Giesy. 1998. Reproductive impairment and induction of alkaline-labile phosphate, a biomarker of estrogen exposure, in fathead minnow (*Pimephales promelas*) exposed to waterborne 17β-estradiol. Aquatic Toxicology 40:335–360.

Krarup, T. 1969. Oocyte destruction and ovarian tumorigenesis after direct application of a chemical carcinogen (9:10-dimethyl-1:2-benzanthracene) to the mouse ovary. International Journal of Cancer 4:61–75.

Kumar, R. S., S. Ijiri, and J. M. Trant. 2000. Changes in the expression of genes encoding steroidogenic enzymes in the channel catfish (*Ictalurus punctatus*) ovary throughout a reproductive cycle. Biology of Reproduction 63:1676–1682.

Kumar, S., and S. C. Pant. 1988. Comparative sublethal pathology of some pesticides in the teleost *Puntius conchonius* Hamilton. Bulletin of Environmental Contamination and Toxicology 41:227–232.

Larsson, D. G. J., H. Hallman, and L. Förlin. 2000. More male fish embryos near a pulp mill. Environmental Toxicology and Chemistry 19:2911–2917.

Leatherland, J. F. 1993. Field observations on reproductive and developmental dysfunction in introduced and native salmonids from the Great Lakes. Journal of Great Lakes Research 19:737–751.

Lee, K. B. H., E. H. Lim, T. J. Lam, and J. L. Ding. 1992. Vitellogenin diversity in the perciformes. Journal of Experimental Zoology 264:100–106.

Le Gac, F., J. L. Thomas, B. Mourot, and M. Loir. 2001. In vivo and in vitro effects of prochloraz and nonylphenol ethoxylates on trout spermatogenesis. Aquatic Toxicology 53:187–200.

Legler, J., J. L. M. Broekhof, A. Brouwer, P. H. Lanser, A. J. Murk, P. T. Van der Saag, A. D. Vethaak, P. Wester, D. Zivkovic, and B. Van der Burg. 2000. A novel in vivo bioassay for (Xeno-) estrogens using transgenic zebrafish. Environmental Science and Technology 34:4439–4444.

Le Guével, F. G. Petit, P. Le Goff, R. Métivier, Y. Valotaire, and F. Pakdel. 2000. Inhibition of rainbow trout (*Oncorhyncus mykiss*) estrogen receptor activity by cadmium. Biology of Reproduction 63:259–266.

Leino, R. L., J. H. McCormick, and K. M. Jensen. 1990. Multiple effects of acid and aluminum on brood stock and progeny of fathead minnows, with emphasis on histopathology. Canadian Journal of Zoology 68:234–244.

Liley, N. R. 1969. Hormones and reproductive behavior in fishes. Pages 73–116 *in* W. S. Hoar and D. J. Randall, editors. Fish physiology. Volume 3. Academic Press, New York.

Loomis, A. K., and P. Thomas. 2000. Effects of estrogens and xenoestrogens on androgen production by Atlantic croaker testes in vitro: evidence for a nongenomic action mediated by an estrogen membrane receptor. Biology of Reproduction 62:995–1004.

Lukšieng, D., and O. Sandström. 1994. Reproductive disturbance in a roach (*Rutilis rutilis*) population affected by cooling water discharge. Journal of Fish Biology 45:613–625.

Lukšieng, D., O. Sandström, L. Lounasheimo, and J. Andersson. 2000. The effects of thermal effluent exposure on the gametogenesis of female fish. Journal of Fish Biology 65:37–50.

Ma, G., Lin, H., and W. Zhang. 1995. Effects of cadmium on serum gonadotropin and growth hormone in common carp (*Cyprinus carpio* L.). Journal of Fisheries of China 19:120–126.

MacLatchy, D. L., and G. J. Van Der Kraak. 1995. The phytoestrogen β-sitosterol alters the reproductive endocrine status of goldfish. Toxicology and Applied Pharmacology 134:305–312.

Makynen, E. A., M. D. Kahl, K. M. Jensen, J. E. Tietge, K. L. Wells, G. Van Der Kraak, and G. T. Ankley. 2000. Effects of the mammalian antiandrogen vinclozolin on development and reproduction of the fathead minnow (*Pimephales promelas*). Aquatic Toxicology 48:461–475.

Marteinsdottir, G., and K. W. Able. 1988. Geographic variation in egg size among populations of the mummichog, *Fundulus heteroclitus* (Pisces: Fundulidae). Copeia 1988:471–478.

Matsuyama, M., S. Adachi, Y. Nagahama, K. Maruyama, and S. Matsura. 1990. Diurnal rhythm of serum steroid hormone levels in the Japanese whiting, *Sillago japonica*, a daily-spawning teleost. Fish Physiology and Biochemistry 8:329–338.

Matta, M. B., C. Cairncross, and R. M. Kocan. 1998. Possible effects of polychlorinated biphenyls on sex determination in rainbow trout. Environmental Toxicology and Chemistry 17:26–29.

Mattison, D. R., K. Shiromizu, and M. S. Nightingale. 1983. Oocyte destruction by polycyclic aromatic hydrocarbons. American Journal of Industrial Medicine 4:191–202.

McCormick, J. H., K. M. Jensen, R. L. Leino, and G. N. Stokes. 1987. Fish blood osmolality, gill histology, and oocyte atresia as early warning acid stress indicators. Annales de la Societe Royale Zoologique de Belgique 117(Supplement 1):309–319.

McCormick, J. H., G. N. Stokes, and R. O. Hermanatz. 1989. Oocyte atresia and reproductive success in fathead minnows (*Pimephales promelas*) exposed to acidified hardwater environments. Archives of Environmental Contamination and Toxicology 18:207–214.

McFarlane, G. A., and W. G. Franzin. 1978. Elevated heavy metals: a stress on a population of white suckers, *Catostomus commersoni,* in Hamell Lake, Saskatchewan. Journal of the Fisheries Research Board of Canada 35:963–970.

McMaster, M. E., J. J. Jardine, G. Ankley, W. H. Benson, M. Greeley, T. Gross, L. Guilette, D. MacLatchy, G. Van Der Kraak and K. R. Munkittrick. 2001. An interlaboratory study on the use of steroid hormones in examining endocrine disruption. Environmental Toxicology and Chemistry 20:2081–2087.

McMaster, M. E., K. R. Munkittrick, and G. J. Van Der Kraak. 1994. Impact of low-level sampling stress on interpretation of physiological responses of white sucker exposed to effluent from a bleached kraft pulp mill. Ecotoxicology and Environmental Safety 27:251–264.

McMaster, M. E., G. J. Van Der Kraak, and K. R. Munkittrick. 1995. Exposure to bleached kraft pulp mill effluent reduces the steroid biosynthetic capacity of white sucker ovarian follicles. Comparative Biochemistry and Physiology 112C:169–178.

McMaster, M. E., G. J. Van Der Kraak, C. B. Portt, K. R. Munkittrick, P. K. Sibley, I. R. Smith, and D. G. Dixon. 1991. Changes in hepatic mixed-function oxygenase (MFO) activity, plasma steroid levels and age at maturity of a white sucker (*Catostomus commersoni*) population exposed to bleached kraft pulp mill effluent. Aquatic Toxicology 21:199–218.

McNabb, A., C. Schreck, C. Tyler, P. Thomas, V. Kramer, J. Specker, M. Mayes, and K. Selcer. 1999. Basic physiology. Pages 113–223 *in* R. T. Di Giulio and D. E. Tillit, editors. Reproductive and developmental effects of contaminants in oviparous vertebrates. Setac Press, Pensacola, Florida.

Metcalf, C. D., M. A. Gray, and Y. Kiparissis. 1999. The Japanese medaka (*Oryzias latipes*): an *in vivo* model for assessing the impacts of aquatic contaminants on the reproductive success of fish. Pages 29–52 *in* S. S. Rao, editor. Impact assessment of hazardous aquatic contaminants: concepts and approaches. CRC Press, Boca Raton, Florida.

Mills, L. J., R. E. Gutjarh-Gobell, R. A. Haebler, D. J. Borsay Horowitz, S. Jayaraman, R. J. Pruell, R. A. McKinney, G. R., Gardner, and G. E. Zaroogian. 2001. Effects of estrogenic (o,p'-DDT; octylphenol) and anti-androgenic (p,p'-DDE) chemicals on indicators of endocrine status in juvenile male summer flounder (*Paralichthys dentatus*). Aquatic Toxicology 52:157–176.

Mommsen, T. P., and P. J. Walsh. 1988. Vitellogenesis and oocyte assembly. Pages 347–406 *in* W. S. Hoar and D. J. Randall, editors. Fish physiology. Volume XIA. Academic Press, San Diego, California.

Monteiro, P. R. R., M. A. Reis-Henriques, and J. Coimbra. 2000. Polycyclic aromatic hydrocarbons inhibit in vitro ovarian steroidogenesis in the flounder (*Platichthys flesus* L.). Aquatic Toxicology 48:549–559.

Moore, A., and C. P. Waring. 1996a. Sublethal effects of the pesticide diazinon on olfactory function in mature male Atlantic salmon (*Salmo salar* L.) parr. Journal of Fish Biology 48:758–775.

Moore, A., and C. P. Waring. 1996b. Electrophysiological and endocrinological evidence that F-series prostaglandins function as priming pheromones in mature male Atlantic salmon (*Salmo salar* L.) parr. Journal of Experimental Biology 199:2307–2316.

Moore, A., and C. P. Waring. 1998. Mechanistic effects of a triazine pesticide on reproductive endocrine function in mature male Atlantic salmon (*Salmo salar* L.) parr. Pesticide Biochemistry and Physiology 62:41–50.

Moore, A., and C. P. Waring. 2001. The effects of a synthetic pyrethroid pesticide on some aspects of reproduction in Atlantic salmon (*Salmo salar* L.). Aquatic Toxicology 52:1–12.

Munkittrick, K. R., M. E. McMaster, C. B. Portt, G. J. Van Der Kraak, I. R. Smith, and D. G. Dixon. 1992a. Changes in maturity, plasma sex steroid levels, hepatic mixed-function oxygenase activity, and the presence of external lesions in lake whitefish (*Coregonus clupeaformis*) exposed to bleached kraft mill effluent. Canadian Journal of Fisheries and Aquatic Sciences 49:1560–1569.

Munkittrick, K. R., C. B. Portt, G. J. Van Der Kraak, I. R. Smith, and D. A. Rokosh. 1991. Impact of bleached kraft mill effluent on population characteristics, liver MFO activity, and serum steroid levels of a Lake Superior white sucker (*Catostomus commersoni*) population. Canadian Journal of Fisheries and Aquatic Sciences 48:1371–1380.

Munkittrick, K. R., G. J. Van Der Kraak, M. E. McMaster, and C. B. Portt. 1992b. Response of hepatic mixed function oxygenase (MFO) activity and plasma sex steroids to secondary treatment and mill shutdown. Environmental Toxicology and Chemistry 11:1427–1439.

Munkittrick, K. R., G. J. Van Der Kraak, M. E. McMaster, and C. B. Portt. 1992c. Reproductive dysfunction and MFO activity in three species of fish exposed to bleached kraft mill effluent at Jackfish Bay, Lake Superior. Water Pollution Research Journal of Canada 27:439–446.

Munkittrick, K. R., G. J. Van Der Kraak, M. E. McMaster, C. B. Portt, M. R. Van den Heuvel, and M. R. Servos. 1994. Survey of receiving-water environmental impacts associated with discharges from pulp mills. 2. Gonad size, liver size, hepatic EROD activity and plasma sex steroid levels in white sucker. Environmental Toxicology and Chemistry 13:1089–1101.

Munro, A. D., A. P. Scott, and T. J. Lam. 1990. Reproductive seasonality in teleosts: environmental influences. CRC Press, Boca Raton, Florida.

Nagahama, Y. 1987. Endocrine control of oocyte maturation. Pages 171–202 *in* D. O. Norris and R. E. Jones, editors. Hormones and reproduction in fishes, amphibians, and reptiles. Plenum, New York.

Nash, J. P., B. Davail-Cuisset, S. Bhattacharyya, H. C. Suter, F. Le Menn, and D. E. Kime. 2000. An enzyme linked immunosorbant assay (ELISA) for testosterone, estradiol, and 17α, 20β-dihydroxy-4-pregne-3-one using acetylcholinesterase as tracer: application to measurement of diel patterns in rainbow trout (*Oncorhynchus mykiss*). Fish Physiology and Biochemistry 22:355–363.

Nelson, D. A., J. E. Miller, D. Rusanowsky, R. A. Greig, G. R. Sennefelder, R. Mercaldo-Allen, C. Kuropat, E. Gould, F. P. Thurberg, and A. Calabrese. 1991. Comparative reproductive success of winter flounder in Long Island Sound: a three year study (biology, biochemistry and chemistry). Estuaries 14:318–331.

Nikolsky, G. V. 1963. The ecology of fishes. Academic Press, New York.

Nilsen, B. M., K. Berg, A. Arukwe, and A. Goksøyr. 1998. Monoclonal and polyclonal antibodies against fish vitellogenin for use in pollution monitoring. Marine Environmental Research 46:153–157.

Nimrod, A. C., and W. H. Benson. 1998. Reproduction and development of Japanese medaka following an early life stage exposure to xenoestrogens. Aquatic Toxicology 44:141–156.

Norris, D. O., and R. E. Jones. 1987. Hormones and reproduction in fishes, amphibians, and reptiles. Plenum, New York.

OECD (Organization for Economic Cooperation and Development). 1992. Section 2. Guideline 204. Fish early-life stage toxicity test. Guidelines for testing chemicals. Paris, France.

Oliveira, R. F., L. A. Carneiro, A. V. Canario, M. S. Grober. 2001a. Effects of androgens on social behavior and morphology of alternative reproductive males of the Azorean rock-pool blenny. Journal of Experimental Marine Biology and Ecology 261:137–157.

Oliveira, R. F., M. Lopes, L. A. Carneiro, A. V. M. Canario. 2001b. Watching fights raises fish hormone levels. Nature (London) 409:475.

Olsén, K. H., and N. R. Liley. 1993. The significance of olfaction and social cues in milt availability, sexual hormone status and spawning behaviour of male rainbow trout (*Oncorhynchus mykiss*). General and Comparative Endocrinology 89:108–118.

Olsson, P.-E., L. Westerlund, S. J. Teh, K. Billsson, A. H. Berg, M. Tysklind, J. Nilsson, L.-O. Eriksson, and D. E. Hinton. 1999. Effects of maternal exposure to estrogen and PCB on different life stages of zebrafish (*Danio rerio*). Ambio 28:100–106.

Oppen-Berntsen, D. O., E. Gram-Jensen, and B. T. Walther. 1992. *Zona radiata* proteins are synthesized by rainbow trout (*Oncorhynchus mykiss*) hepatocytes in response to oestradiol-17β. Journal of Endocrinology 135:293–302.

Oppen-Berntsen, D. O., S. O. Olsen, C. Rong, P. Swanson, G. L. Taranger, and B. T. Walther. 1994. Plasma levels of eggshell Zr-proteins, estradiol-17β, and gonadotropins during an annual reproductive cycle of Atlantic salmon (*Salmo salar*). Journal of Experimental Zoology 268:59–70.

Pakdel, F., S. Féon, F. Le Gac, F. Le Menn, and Y. Valotaire. 1991. *In vivo* estrogen induction of hepatic receptor mRNA and correlation with vitellogenin mRNA in rainbow trout. Molecular and Cellular Endocrinology 75:205–215.

Parker, D. B., and B. A. McKeown. 1986. The effects of low pH on egg and alevin survival of kokanee and sockeye salmon, *Oncorhynchus nerka*. Comparative Biochemistry and Physiology 87:259–268.

Parks, L. G., A. O. Cheek, N. D. Denslow, S. A. Heppell, J. A. McLachlan, G. A. LeBlanc, and C. V. Sullivan. 1999. Fathead minnow (*Pimephales promelas*) vitellogenin: Purification, characterization, and quantitative immunoassay for the detection of estrogenic compounds. Comparative Biochemistry and Physiology C 123:113–125.

Papoulias, D. M., D. B. Noltie, and D. E. Tillitt. 1999. An in vivo model fish system to test chemical effects of sexual differentiation and development: exposure to ethinyl estradiol. Aquatic Toxicology 48:37–50.

Patyna, P. J., R. A. Davi, T. F. Parkerton, R. P. Brown, and K. R. Cooper. 1999. A proposed multigeneration protocol for Japanese medaka (*Oryzias latipes*) to evaluate effects of endocrine disruptors. The Science of the Total Environment 233:211–220.

Penzak, T. 1985. Trophic ecology and fecundity of *Fundulus heteroclitus* in Chezzetcook Inlet, Nova Scotia. Marine Biology 89:235–243.

Peterson, R. H., P. G. Daye, G. L. Lacroix, and E. T. Garside. 1982. Reproduction of fish experiencing acid and metal stress. Pages 177–196 *in* R. E. Johnston, editor. Acid rain/fisheries. American Fisheries Society, Bethesda, Maryland.

Piferrer, F., and E. M. Donaldson. 1989. Gonadal ontogenesis in coho salmon *Oncorhyncus kisutch* after a single treatment with androgen or estrogen during ontogenesis. Aquaculture 77:251–262.

Potts, G. W. And R. J. Wootton. 1984. Fish reproduction: strategies and tactics. Academic Press, San Diego, California.

Prince, R., and K. R. Cooper. 1995. Comparisons of the effects of 2,3,7,8-tetrachlorodibenzo-p-dioxin on chemically impacted and nonimpacted subpopulations of *Fundulus heteroclitus*: I. TCDD toxicity. Environmental Toxicology and Chemistry 14:579–587.

Purdom, C. E., P. A. Hardiman, V. J. Bye, N. C. Eno, C. R. Tyler, and J. P. Sumpter. 1994. Estrogenic effects of effluents from sewage treatment works. Chemical Ecology 8:275–285.

Rastogi, A., and S. K. Kulshretha. 1990. Effect of sublethal doses of three pesticides on the ovary of a carp minnow, *Rasbora daniconius*. Bulletin of Environmental Contamination and Toxicology 45:742–747.

Redding, M. J., and R. Patino. 1993. Reproductive physiology. Pages 503–534 *in* D. H. Evans, editor. The physiology of fishes. CRC Press, Boca Raton, Florida.

Reis-Henriques, M. A., M. M. Cruz, and J. O. Pereira. 1997. The modulating effect of vitellogenin on the synthesis of 17-β-estradiol by rainbow trout (*Oncorhynchus mykiss*) ovary. Fish Physiology and Biochemistry 16:181–186.

Richburg, J. H. 2000. The relevance of spontaneous- and chemically-induced alterations in testicular germ cell apotosis to toxicology. Toxicology Letters 112–113:79–86.

Rodgers-Gray, T. P., S. Jobling, C. Kelly, S. Morris, G. Brighty, M. J. Waldock, J. P. Sumpter, and C. R. Tyler. 2001. Exposure of juvenile Roach (*Rutilis rutilis*) to treated sewage effluent induces dose-dependent and persistent disruption in gonadal duct development. Environmental Science and Technology 35:462–470.

Routledge, E. J., D. Sheahan, C. Desbrow, G. C. Brighty, M. Waldock, and J. P. Sumpter. 1998. Identification of estrogenic compounds in STW effluent. 2. *In vivo* responses in trout and roach. Environmental Science and Technology 32:1559–1565.

Ruby, S. M., D. R. Idler, and Y. P. So. 1993. Plasma vitellogenin, 17β-estradiol, T3 and T4 levels in sexually maturing rainbow trout *Oncorhynchus mykiss* following sublethal HCN exposure. Aquatic Toxicology 26:91–102.

Rurangwa, E., I. Roelants, G. Huyskens, M. Ebrahimi, D. E. Kime, and F. Ollevier. 1998. The minimum acceptable spermatozoa to egg ratio for artificial insemination and the effects of heavy metal pollutants on sperm motility and fertilization ability in the African catfish (*Clarias gariepinus* Burchell 1822). Journal of Fish Biology 53:402–413.

Saidapur, S. K. 1978. Follicular atresia in the ovaries of nonmammalian vertebrates. International Review of Cytology 54:225–244.

Sandström, O., E. Neuman, and P. Karås. 1988. Effects of a bleached pulp mill effluent on growth and gonad function in Baltic coastal fish. Water Science and Technology 20:107–118.

Sangalang, G. B., and H. C. Freemen. 1974. Effects of sublethal cadmium on maturation and testosterone and 11-ketotestosterone production in vivo in brook trout.
Biology of Reproduction 11:429–435.

Satoh, N., and N. Egami. 1972. Sex differentiation of germ cells in the teleost, *Oryzias
latipes*, during normal embryonic development. Journal of Embryology and
Experimental Morphology 28:385–395.

Scholz, S., and H. O. Gutzeit. 2000. 17-α-Ethinylestradiol affects reproduction, sexual
differentiation, and aromatase gene expression of the medaka (*Oryzias latipes*).
Aquatic Toxicology 50:363–373.

Selye, H. 1950. The physiology and pathology of exposure to stress. Acta, Montreal.

Sharp, S. K. 1994. Serum levels of 17β-estradiol and testosterone as indicators of
environmental stress in redbreast sunfish, *Lepomis auratus*. Master's thesis, University of Tennessee, Knoxville.

Singh, P. B., and D. E. Kime. 1995. Impact of γ-hexachlorocyclohexane on the in
vitro production of steroids from endogenous and exogenous precursors in the
spermiating roach, Rutilis rutilis. Aquatic Toxicology 31:231–240.

Singh, S., and T. P. Singh. 1980. Effect of two pesticides on ovarian ^{32}P uptake and
gonadotropin concentration during different phases of annual reproductive cycles
in the freshwater catfish, *Heteropneustes fossilis* (Bloch). Environmental Research
22:190–200.

Singh, S., and T. P. Singh. 1987. Evaluation of toxicity limit and sex hormone production in response to cythion and BHC in the vitellogenic catfish *Clarias batrachus*.
Environmental Research 42:482–488.

Sivarajah, K., C. S. Franklin, and P. Williams. 1978. The effects of polychlorinated
biphenyls on plasma steroid levels and hepatic microsomal enzymes in fish.
Journal of Fish Biology 13:401–409.

Slooff, W., and D. DeZwart. 1983. The growth, fecundity and mortality of bream
(*Abramis brama*) from polluted and less polluted surface waters in the Netherlands. Science of the Total Environment 27:149–162.

Smith, J. S., and P. Thomas. 1990. Binding characteristics of the hepatic estrogen
receptor of the spotted seatrout, *Cynoscion nebulosus*. General and Comparative Endocrinology 77:29–42.

Soimasuo, M. R., A. E. Karels, H. Leppänen, R. Santti, and A. O. J. Oikari. 1998.
Biomarker responses in whitefish (*Coregonus lavaretus* L. s.l.) experimentally
exposed in a large lake receiving effluent from pulp and paper industry. Archives of Environmental Contamination and Toxicology 34:69 80.

Solemdal, P. 1967. The effect of salinity on buoyancy, size and development of
flounder eggs. Sarsia 29:431–442.

Specker, J. L., and C. V. Sullivan. 1994. Vitellogenesis in fishes: status and perspective.
Pages 304–315 *in* K. G. Davey, R. E. Peter, and S. S. Tobe, editors. Perspectives in
comparative endocrinology. National Research Council of Canada, Ottawa.

Sperry, T. S., and P. Thomas. 1999. Identification of two nuclear androgen receptors
in kelp bass (*Paralabrax clathratus*) and their binding affinities for xenobiotics:
comparison with Atlantic croaker (*Micropogonias undulatus*) androgen receptors. Biology of Reproduction 61:1152–1161.

Spies, R. B., and D. W. Rice, Jr. 1988. Effects of organic contaminants on reproduction of the starry flounder (*Platichthys stellatus*) in San Francisco Bay. II. Repro-

ductive success of fish captured in San Francisco Bay and spawned in the laboratory. Marine Biology 98:191–200.

Spies, R. B., P. Thomas, and M. Matsui. 1996. Effects of DDT and PCB on reproductive endocrinology of *Paralabrax clathratus* in southern California. Marine Environmental Research 42:75–76.

Srivastava, A. K., and Srivastava, A. K. 1994. Effects of chlordecone on the gonads of freshwater catfish, *Heteropneustes fossilis*. Bulletin of Environmental Contamination and Toxicology 53:186–191.

Sukumar, A., and P. R. Karpagaganapathy. 1992. Pesticide-induced atresia in ovary of a fresh water fish, *Colisa lalia* (Hamilton-Buchanan). Bulletin of Environmental Contamination and Toxicology 48:457–462.

Stacey, N. E. 1987. The roles of hormones and pheromones in fish reproductive behaviour. Pages 28–69 *in* D. Crews, editor. Psychobiology of reproductive behaviour. Prentice-Hall, Englewood Cliffs, New York.

Stacey, N. E., and J. R. Caldwell. 1995. Hormones as sex pheromones in fish: widespread distribution among freshwater species. Pages 244–248 *in* F. W. Goetz and P. Thomas, editors. Proceedings of the Fifth International Symposium on the Reproductive Physiology of Fish. University of Texas Press, Austin, Texas.

Stearns, S. C., and R. E. Crandall. 1984. Plasticity for age and size at sexual maturity: a life-history response to unavoidable stress. Pages 13–33 *in* G. W. Potts and R. J. Wootton, editors. Fish reproduction: strategies and tactics. Academic Press, San Diego, California.

Suter, G. W., A. E. Rosen, E. Linder, and D. F. Parkhurst. 1987. Endpoints for responses of fish to chronic toxic exposures. Environmental Toxicology and Chemistry 6:793–809.

Takizawa, K., H. Yagi, D. M. Jerina, and D. R. Mattison. 1984. Murine starin differences in ovotoxicity following intraovarian injection with benzo(a)pyrene, (+)-(7R, 8S)-oxide, (-)-(7R,8R)-dihydodiol, or (+)-(7R,8S)-diol-(9S,10R)-epoxide-2. Cancer Research 44:2571–2576.

Tam, W. H., and P. D. Payson. 1986. Effects of chronic exposure to sublethal pH on growth, egg production, and ovulation in brook trout, *Salvelinus fontinalis*. Canadian Journal of Fisheries and Aquatic Sciences 43:257–280.

Tanasichuk, R. W., and D. M. Ware. 1987. Influence of interannual variations in water temperature on fecundity and egg size in Pacific herring (*Clupea harengus pallasi*). Canadian Journal of Fisheries and Aquatic Sciences 44:1485–1495.

Tata, J. R., W. C. Ng, A. J. Perlman, and A. P. Wolfe. 1987. Activation and regulation of the vitellogenin gene family. Pages 205–233 *in* A. K. Roy and J. H. Clark, editors. Gene regulation by steroid hormones III. Springer Verlag, New York.

Taylor, M. H., G. J. Leach, L. DiMichele, W. M. Levitan, and W. F. Jacob. 1979. Lunar spawning cycles in the mummichog, *Fundulus heteroclitus* (Pisces: Cyprinodontidae). Copeia 1979:291–297.

Thomas, P. 1989. Effects of Arochlor® 1254 and cadmium on reproductive endocrine function and ovarian growth in Atlantic croaker. Marine Environmental Research 28:499–503.

Thomas, P. 1993. Effects of cadmium on gonadotropin secretion from Atlantic croaker pituitaries incubated in vitro. Marine Environmental Research 35:141–145.

Thomas, P. 1994. Hormonal control of final oocyte maturation in sciaenid fishes. Pages 619–625 *in* K. G. Davey, R. E. Peter, and S. S. Tobe, editors. Perspectives in comparative endocrinology. National Research Council of Canada, Ottawa.

Thomas, P. 2000. Chemical interference with genomic and nongenomic actions of steroids in fishes: role of receptor binding. Marine Environmental Research 50:127–134.

Thomas, P., D. Breckenridge-Miller, and C. Detweiler. 1998. The teleost sperm membrane progesterone receptor: interactions with xenoestrogens. Marine Environmental Research 46:163–166.

Thomas, P., and I. A. Khan. 1997. Mechanisms of chemical interference with reproductive endocrine function in sciaenid fishes. Pages 29–52 *in* R. M. Rolland, M. Gilbertson, and R. E. Peterson, editors. Chemically induced alterations in functional development and reproduction of fishes. SETAC Press, Pensacola, Florida

Tilly, J. L., K. I. Kowalski, A. L. Johnson, and A. J. W. Hsueh. 1991. Involvement of apotosis in ovarian follicular atresia and postovulatory regression. Endocrinology 1991:2799–2801.

Todo, T., S. Adachi, F. Saeki, and K. Yamauchi. 1995. Hepatic estrogen receptors in Japanese eel, *Anguilla japonica*: characterization and changes in binding capacity during artificially-induced sexual maturation. Zoological Science 12:789–794.

Toomey, B. H., G. H. Monteverdi, and R. T. Di Giulio. 1999. Octylphenol induces vitellogenin production and cell death in fish hepatocytes. Environmental Toxicology and Chemistry 18:734–739.

Tremblay, L., and G. Van Der Kraak. 1999. Comparison between the effects of the phytosterol, β-sitosterol and pulp mill effluents on sexually immature rainbow trout. Environmental Toxicology and Chemistry 18:329–336.

Truscott, B., J. M. Walsh, M. P. Barton, J. F. Payne, and D. Idler. 1983. Effect of acute exposure to crude petroleum on some reproductive hormones in salmon and flounder. Comparative Biochemistry and Physiology 75C:121–130.

Tsai, M. J., and B. W. O'Malley. 1994. Molecular mechanisms of action of steroid/thyroid receptor superfamily members. Annual Review of Biochemistry 63:451–486.

Tyler, C. R., and J. P. Sumpter. 1996. Oocyte growth and development in teleosts. Reviews in Fish Biology and Fisheries 6:287–318.

Tyler, C. R., R. Van Aerle, T. H. Hutchinson, S. Madix, and H. Trip. 1999. An *in vivo* testing system for endocrine disruptors in fish early life stages using induction of vitellogenin. Environmental Toxicology and Chemistry 18:337–347.

U.S. EPA (U.S. Environmental Protection Agency). 1982. Users guide to conducting life-cycle tests with fathead minnows (*Pimephales promelas*). EPA 600/8-81-011. Technical Report. Duluth, Minnesota.

U.S. EPA (U.S. Environmental Protection Agency). 1989. Pesticide assessment guidelines. Subdivision E, hazard evaluation: wildlife and aquatic organisms. EPA 540/09–82–024. Technical Report. Washington, D.C.

U.S. EPA (U.S. Environmental Protection Agency). 1994. Short-term methods for estimating the chronic toxicity of effluents and receiving waters to freshwater organisms, 3rd edition. EPA 600/4-91-002. Technical Report. Cincinnati, Ohio.

U.S. EPA (U.S. Environmental Protection Agency). 1996. Strategic plan for the Office of Research and Development. EPA/600/R3-91-063. Technical Report. Washington, D.C.

U.S. EPA (U.S. Environmental Protection Agency). 1998. Endocrine Disruption Screening and Advisory Committee (EDSTAC) report. Office of Prevention, Pesticides and Toxic Substances, Washington, D.C.

USGS (U.S. Geological Survey). 1998. Investigations of endocrine disruption in aquatic systems associated with the National Water Quality Assessment (NAWQA) Program. Fact Sheet FS-081-98.

Van den Heuvel, M. R., R. J. Ellis, L. A. Tremblay, and T. R. Stuthridge. 2002. Exposure of reproductively maturing rainbow trout to a New Zealand pulp and paper mill effluent. Ecotoxicology and Environmental Safety 51:65–75.

Van Der Kraak, G. J., J. P. Chang, and D. M. Janz. 1998. Reproduction. Pages 465–88 *in* D. H. Evans, editor. The physiology of fishes. CRC Press, Boca Raton, Florida.

Van Der Kraak, G. J., K. R. Munkittrick, M. E. McMaster, C. B. Portt, and J. P. Chang. 1992. Exposure to bleached kraft mill effluent disrupts the pituitary-gonadal axis of white sucker at multiple sites. Toxicology and Applied Pharmacology 115:224–233.

Villalobos, S. A., J. T. Hamm, S. J. Teh, and D. E. Hinton. 2000. Thiobencarb-induced embryotoxicity in medaka (*Oryzias latipes*): stage-specific toxicity and the protective role of chorion. Aquatic Toxicology 48:309–326.

Wallace, R. A. 1985. Vitellogenesis and oocyte growth in nonmammalian vertebrates. Pages 127–177 *in* L. W. Browder, editor. Developmental biology, volume 1. Plenum, New York.

Wallace, R. A., and K. Selman. 1981. Cellular and dynamic aspects of oocyte growth in teleosts. American Zoologist 21:325–343.

Wallace, R. A., K. Selman, M. S. Greeley Jr., P. C. Begovac, Y.-W. Lin, and R. McPherson. 1987. The current status of oocyte growth. Pages 167–177 *in* D. R. Idler, L. W. Crim, and J. M. Walsh, editors. Proceedings of the Third International Symposium on the Reproductive Physiology of Fish. Memorial University of Newfoundland, St. John's, Canada.

Wannemacher, R., A. Rebstock, E. Kulzer, D. Schrenk, and K. W. Bock. 1992. Effects of 2,3,7,8-tetrachlorodibenzo-p-dioxin on reproduction and oogenesis in zebrafish (*Brachydanio rerio*). Chemosphere 24:1361–1368.

Ware, D. M. 1977. Spawning time and egg size of Atlantic mackerel, *Scomber scombrus*, in relation to the plankton. Journal of the Fisheries Research Board of Canada 34:2308–2315.

Webb, M. A. H., G. W. Feist, E. P. Foster, C. B. Schreck, and M. S. Fitzpatrick. 2002. Potential classification of sex and stage of gonadal maturity of wild white sturgeon using blood plasma indicators. Transactions of the American Fisheries Society 131:132–142.

Weber, L. P., and D. M. Janz. 2001. Effect of β-naphthofavone and dimethylbenz[*a*]anthracene on apotosis and HSP70 expression in juvenile channel catfish (*Ictalurus punctatus*) ovary. Aquatic Toxicology 54:39–50.

Weiner, G. S., C. B. Schreck, and H .W. Li. 1986. Effects of low pH on reproduction of rainbow trout. Transactions of the American Fisheries Society 115:75–82.

Weis, J. S., G. Smith, T. Zhou, C. Santiago-Bass, and P. Weis. 2001. Effects of contaminants on behavior: biochemical mechanisms and ecological consequences. Bioscience 51:209–217.

Weis, J. S., and P. Weis. 1984. A rapid change in methymercury tolerance in a population of killifish, *Fundulus heteroclitus*, from a golf course pond. Marine Environmental Research 13:231–245.

Weis, P., and J. S. Weis. 1977. Methylmercury teratogenesis in the killifish, *Fundulus heterociltus*. Teratology 16:317–326.

Weis, P., and J. S. Weis. 1989. Tolerance and stress in a polluted environment. Bioscience 39:89–95.

Wells, K., and G. Van Der Kraak. 2000. Differential binding of endogenous steroids and chemicals to androgen receptors in rainbow trout and goldfish. Environmental Toxicology and Chemistry 19:2059–2065.

Wester, P. W. 1991. Histopathological effects of environmental pollutants β-HCH and methylmercury on reproductive organs in freshwater fish. Comparative Biochemistry and Physiology C 100:237–239.

Willey, J. B., and P. H. Krone. 2001. Effects of endosulfan, and nonylphenol on the promordial germ cell population in pre-larval zebrafish embryos. Aquatic Toxicology 54: 113–123.

Winstead, J. T., D. P. Middaugh, and L. A. Courtney. 1991. Ovarian mycosis in the topsmelt *Atherinops affinis*. Diseases of Aquatic Organisms 10:221–223.

Wood, A. W., and G. J. Van Der Kraak. 2001. Apotosis and ovarian function: novel perspectives from the teleosts. Biology of Reproduction 64:264–271.

Wright, P. J. And D. E. Tillitt. 1999. Embryotoxicity of Great Lakes lake trout extracts to developing rainbow trout. Aquatic Toxicology 47:77–92.

Wu, T., H. Patel, S. Mukai, C. Melino, R. Garg, X. Ni, J. Chang, and C. Peng. 2000. Activin, inhibin, and follistatin in zebrafish ovary: expression and role in oocyte maturation. Biology of Reproduction 62:1585–1592.

Yadetie, F., A. Arukwe, A. Goksøyr, and R. Male. 1999. Induction of hepatic estrogen receptor in juvenile Atlantic salmon in vivo by the environmental estrogen, 4-nonylphenol. The Science of the Total Environment 233:201–210.

Yamamoto, T. 1969. Sex differentiation. Fish Physiology 3:117–175.

Yang, M. Y., and R. Rajamahendran. 2000. Involvement of apotosis in the atresia of nonovulatory dominant follicle during the bovine estrous cycle. Biology of Reproduction 63:1313–1321.

Zaroogian, G., G. Gardner, D. Borsay Horowitz, R. Gutjahr-Gobell, R. Haebler, and L. Mills. 2001. Effect of 17β-estradiol, *o, p'*-DDT, octylphenol, and *p, p'*-DDE on gonadal development, and liver, and kidney pathology in juvenile male summer flounder (*Paralichthys dentatus*). Aquatic Toxicology 54:101–112.

10
Assessing Fish Population Responses to Stress

Michael Power

Introduction

The advent of environmental risk assessment has shifted much of the emphasis in applied ecological research from measurement and discussion of individual-level effects to estimation and discussion of population- and community-level effects. Although population-level endpoints have been promoted (NRC 1981) and recognized (Adams 1990) as being more ecologically relevant than individual-based measurements, individual assessment endpoints still dominate in the scientific literature and regulatory requirements. The use of individual-level endpoints, however, assumes that observational conclusions based on a small group of individuals can be extrapolated reliably to populations and communities and that no-effect levels established using individual data accurately reflect any induced changes in the critical demographic parameters that ultimately determine population viability (Walthall and Stark 1997).

From both the ecological and social perspectives, populations and communities are of most concern, not individual organisms. There is not, necessarily, a simple or direct relationship between responses measured at the individual level and those detectable at population and community levels (Forbes and Calow 1999). Establishing such a relationship is difficult because individuals in their natural environment are subjected to numerous natural and anthropogenic stressors, including favorable and unfavorable physical and chemical factors, variations in community structure, predator–prey cycles, parasites, disease, food availability, and random catastrophic perturbations. Alone, or in combination, these stressors can trigger the reallocation of energy away from critical growth and reproductive functions and reduce the capacity of individuals to tolerate additional stress (Adams et al. 1993). When aggregated across the individuals in an identifiable group (e.g., populations), responses operating through individual-dependent energy or trophic pathways can induce effects having immediate relevance for individuals in the group (Barton 1997) because of the ways in which the distribution of individual growth, survival, and reproduction attributes may be altered. For example, the operation of homeostatic mechanisms can, but does not always, modify or obscure the aggregate responses of stressed individuals. Accordingly, there is no obvious connection between processes that govern individual-level responses and those that determine population-level consequences. Nevertheless, there is a need to understand how individual-level responses are translated to observed population- and community-level responses to stress.

Approaches to detecting, interpreting, and reporting the significance of stress for aquatic populations vary. In spite of general agreement that understanding the population-level effects of stressors is of paramount importance, relatively few investigators have attempted to systematically estimate those effects (Barnthouse 1993) or explicitly link changes at lower levels of biological organization to changes at the population level (Shuter and Regier 1989). Most investigations have followed the pioneering lead of Selye (1976), who defined stress as the response of an individual to a stressor, and are based on laboratory experimentation aimed at determining individual responses to acute or chronic stressors (Adams 1990). Although valuable for achieving the specific objectives of determining whether a measurable response exists, studies focused at the individual level lack the ecological realism (Power and McCarty 1997) necessary to appropriately predict or understand the consequent effects of stress at the population level.

The problem of quantifying the effect of stressor perturbations on populations is not new and has formed the core of wildlife and fisheries management research for many years (Barnthouse 1993). As with most population-based investigations, the objective of stress-related population research is to infer from individual sample data the characteristics of a well-defined grouping of like organisms (the population) and the dynamic responses of the group to stressors. In practice, ecologists have tended to equate stress with anthropogenic disturbances that induce responses outside the normal range of variation for the selected measurement endpoint (Evans et al. 1990). Although adoption of the normative range concept (Odum et al. 1979) dealt adequately with the duality of possible beneficial and harmful stressor effects at a theoretical level, inherent variability caused by the action of ecological stresses and the operation of natural compensation mechanisms has confounded the simple interpretation of stress–response relationships. In practice, the use of a normative range concept has not solved the problem of determining acceptable population-level response thresholds for toxic substances (Evans et al. 1990) or other stressors.

Understanding the responses of populations to stressors, therefore, requires knowledge of both the ways in which individuals respond to stress and the effect of collective individual responses on the processes that govern population dynamics. Individual responses to stress may be observed directly from experimental or field samples. The effect of collective individual responses on the processes that govern population dynamics, however, can only be inferred indirectly using individual response data to estimate sample means and variances. The processes governing population dynamics have been termed "emergent properties" (Kerr 1976). Although the emergent property concept follows naturally from the notion of biological hierarchy, general observations made from within the confines of a hierarchical level will have low predictive value in terms of the level above it (Kerr 1976, 1982). Low predictability stems, in part, from the effect of population regulating feedback mechanisms that operate through compensatory adjustments to birth and death rates (e.g., density-dependent recruitment success, compensatory growth and fecundity changes; McFadden 1977). Low predictability also results from the fact that it is the integration of all hierarchical subsystems (i.e., suborganismal, individual, population, and community) in the ecological system that determine the dynamic responses at any particular level of organization (Kerr 1982). The importance of the whole in determining the responses of ecosystem components implies that even detailed knowledge of the function of an organismal subsystem (e.g., endocrine) will not allow accurate prediction of individual success, much less the success of the population to which the individual belongs.

The likelihood that stressors will interact spatially and temporally with population compensation mechanisms further complicates attempts to evaluate the effect of stressors. The probability of stressor interactions necessitates additional understanding of the ways in which interactions between population-regulating mechanisms and stressors might influence observed population-level traits. Current ecological theory has tended to view each population as unique and so tightly integrated into its own particular ecosystem that the data collection programs necessary to produce credible assessments of stressor and stressor interaction effects would themselves significantly alter the populations under study (Rigler 1982). Furthermore, the sensitivity of manipulative experiments at the population level can be severely compromised because the number of replicate populations available for use in experimentation is limited, and local uniqueness among replicates is so high that variance estimates remain large (Walters et al. 1989). Thus, despite the many documented cases of population collapses (e.g., of fish stocks), it has typically been impossible to unambiguously select the specific causes of decline (Barnthouse et al. 1990). Nevertheless, attention will continue to focus on effects at the population level because of public concern for the fate of highly valued populations and legislative mandates requiring attempts to ensure the continued viability of threatened or endangered species and no net loss of habitat.

Against this background of ecological complexity, this chapter will examine field- and modeling-based approaches that have been developed to establish and understand the significance of population responses to stress. The population-regulating mechanisms and confounding biotic and abiotic factors critical to determining population-level responses to stress or limiting the general interpretation of measured population-level responses to stress will also be discussed. Discussion is largely, though not wholly, based on the detailed fisheries literature and any biases toward fish stem from the expertise of the author rather than any dearth of work completed with other aquatic fauna.

The Importance of Population Processes and Life History

For purposes of discussion, a population is defined as a group of individuals of the same species that occupy a definable geographic range and are reproductively self-sustaining. A population need not be isolated from the effects of immigration or emigration, but, for simplicity, such isolation is often assumed. Populations are governed by the fundamental processes of reproduction, mortality, and the somatic increases that render individuals capable of reproduction (Shuter 1990). To some extent these critical processes are moderated by density-independent and density-dependent adjustments that compensate for abnormal levels of numerical abundance. Ecologists have recognized, for some time, that populations could only persist if some form of compensatory response existed (e.g., Nicholson 1933). The same compensatory processes that allow populations to persist, however, are also partially capable of counteracting the adverse effects of stress at the population level (Nicholson 1954). The exact nature of density-dependence factors in the control of populations has been debated for many years (see reviews by Clark et al. 1967; Begon et al. 1990), but there is now widespread agreement that fluctuations in the abundance of persisting populations are the result of both density-dependent and density-independent processes (Hassell 1986; Elliott 1994).

Density-dependent processes operating on populations via feedbacks include competition for resources that lead to adjustments in growth (e.g., food and habitat), shifts in the sex ratio, changes in fecundity, predation, cannibalism, spawning habitat congestion, agonistic behavior, and variations in dispersal rates and disease events (Goodyear 1980; Hassell 1986; Calow and Sibly 1990; Barton 1997). The processes need not act in isolation, and the role of each mechanism can vary among different populations of the same species or within the same population between years (Goodyear 1980; Evans et al. 1990).

In fisheries ecology, density-dependent factors are viewed as the critical regulators of abundance. Knowledge of the underlying density-dependent model regulating survival (e.g., Ricker 1954; Beverton and Holt 1957), together with a predictive understanding of critical compensatory mechanisms, is believed to permit reliable predictions of abundance and characterization of the ability of a population to sustain harvesting (Frank and Leggett 1994) and other stressors. Determining the relative importance of compensatory mechanisms for a population is essential to understanding probable responses to stressors simply because populations with little or no compensatory capacity will be particularly vulnerable to stressor effects (Fogarty et al. 1991). The capacity of compensatory processes to offset increases in mortality is, however, limited. For example, under stress from fishing, fish populations often reduce age-at-maturity to compensate for reductions in population size (Trippel 1995). While such adjustments may, initially, be effective at maintaining population size, there is a physiological determined limit below which age-at-maturity cannot be reduced and, thus, a threshold below which the population is less able to respond positively to other stresses.

Density-independent factors are also widely believed to be important contributors to the highly variable abundances observed in many species. In the case of marine fishes, density-independent factors largely operate through effects on juvenile survival resulting from a match or mismatch between fixed spawning times and variability in the abiotic factors governing conditions necessary for growth and survival of larval fishes (Frank and Leggett 1994). In freshwater fishes, density independence operates, principally, through the random fluctuations in critical environmental variables controlling opportunities for growth, survival, and reproduction throughout the life cycle. While density-independent factors operate largely independent of population size, they may act indirectly through density-dependent processes, to attenuate or amplify the effective action of concurrently operating density-dependent processes (Evans et al. 1990; Power 1997). For example, changes in rainfall may affect usable habitat, which, in turn, exacts a density-dependent action on survival (Power and Power 1994). Or, if the probability of starvation is density dependent, the improper timing of larval hatching due to a density-independent event, such as late ice-out, may act to reduce available food sources and raise density-dependent induced mortality (Goodyear 1980).

One of the critical problems facing the detection and understanding of density-dependent effects has been sorting natural variability from the effect of the process itself. Variability may arise for reasons of demographic or environmental stochasticity. Sources of variability in demographic parameters can be broadly classified into those that depend on trophic interactions and those that depend on physical processes (Shuter 1990). These diverse considerations include biotic factors affecting variability in the energy available for growth and reproduction; the abundance and availability of prey items; the predator field that is encountered; and other factors such as disease and parasitism (Fogarty et al. 1991). Variability may also be mistaken for density-dependent processes resulting from within generation heterogeneity arising from mechanisms that render some individuals more susceptible to mortality than others (Hassell 1986). Such mechanisms include nonrandom mortality factors; the existence of refugia protecting individuals from a mortality-causing agent; and processes that give rise to temporal asynchrony (e.g., differential spawning or hatch dates).

Although much of the mainstream fisheries literature has concerned itself with addressing the implications of density-dependent factors for determining population success, traditional contaminant-based studies of possible population-level stressor effects have tended to ignore the issue altogether (Power and McCarty 1997). For example, density-dependent effects are not recognized among the traditional list of significant biotic-modifying factors (e.g., individual size and nutritional status), knowledge of which is necessary to appropriately interpret toxicity test results. The omission assumes density-related factors can be treated as experimental constants and denies the importance of population regulating mechanisms as determinants of the status of individuals in the population. Notable exceptions to the general trend include Vijayan and Leatherland (1988) who demonstrated a significant inverse relationship between stocking density and growth rate in brook trout *Salvelinus fontinalis* in a series of stress-response experiments and Arthur and Dixon (1993) who found rearing density to be a significant factor in the determination of chlorophenol toxicity for fathead minnows *Pimephales promelas*.

In addition to the density-dependent and density-independent processes that regulate population numbers directly, the choice of life history strategy holds obvious consequences for the realized effects of stress at the population level. Life history theory predicts a trade-off between energy allocated to reproduction and survival, the development of bet-hedging strategies (e.g., variable age-at-maturity and iteroparity) in stochastic environments, and a coupling of migration and life history traits (Frank and Leggett 1994). In North America, many fishes with relatively large body size tend to exhibit patterns of later maturity, high fecundity, smaller egg size, and reduced reproductive frequency when compared to fish with relatively small body size (Winemiller and Rose 1992). The broad conformity of fish life history traits with predictions of life history theory, together with direct evidence from those species that have been most intensely studied, indicate that life history traits are adapted in ways that lead to a natural dampening of the effects of environmentally induced variability in survival (Frank and Leggett 1994). For example, opportunistic life history strategy fishes tend to opt for early maturation, frequent reproduction over extended spawning seasons, rapid larval growth, and high population turnover rates (Winemiller and Rose 1992). This suite of characteristics allows opportunistic fishes to rebound quickly from local disturbances in the absence of predation or resource limitation, and they are, correspondingly, less sensitive to mortality imposed on the adult life stage than low fecundity, slow development K-selected species

(Barnthouse 1993). Acute stresses that impact critical life stages, however, have proportionally greater effect on opportunistic than K-selected species. This is largely because K-selected species tend to iteroparity (Wootton 1990) and adopt bet-hedging strategies that reduce the sensitivity of the population to the direct effects of a single stress event.

Field-Based Assessments of Population Stress

Realization of the importance of population processes and life history strategies for conclusions about fish population status has driven much of the development of traditional fisheries assessment and management practice. The resulting focus of work has been on the determination of population abundances, relative condition, and the ability of populations to withstand exploitation pressure within an analytical framework emphasizing the effects of population processes and life history strategies on field-obtained measures. A listing of common measures used for assessing fish population status is given in Table 1. Although developed largely to assess exploitation pressures, the techniques are useful for assessing the effects of other anthropogenic and natural stressors on population status (e.g., Healey 1978; Mills 1985; Mohr et al. 1990; Berlinsky et al. 1995). Reviews of the basic techniques required to estimate the measures listed in Table 1 are given in Murphy

Table 1. Common indicators of fish population status usable for assessing the effects of population-level stress. Indicators are based on or derived from field measurements. Indicators from each category should be included to appropriately infer any changes in the basic population processes of growth, survival, and reproduction likely to be indicative of stress-related effects. Parameters within each category are not necessarily independent of those in the other categories. For example, changes in mean length-at-age (growth) hold direct implications for fecundity (reproduction) and may affect age-specific survival rates (survival) in subsequent generations.

Survival related

- age-specific survival rates
- year-class strength
- age structure
- catch per unit effort
- density or abundance
- mean age
- maximum age
- recruitment indices

Growth related

- mean weight-at-age
- allometric relationships
- size structure
- condition factor
- proximate body composition
- mean length-at-age
- specific growth rates
- liver somatic index
- incidence of parasites

Reproduction related

- age-at-maturity
- reproductive life span
- gonad somatic index
- incidence of atresia
- fecundity
- sex ratio
- egg size
- spawning frequency

and Willis (1996) and Kohler and Hubert (1993) and will not be discussed here.

Many of the methods developed for fisheries assessment have been included in comprehensive programs attempting to combine information on both the stressor and its measurable biological effects (e.g., USEPA 1992). Several field-based procedures for integrating exposure and effects data exist and they may be grouped into three main approaches: compare and contrast, graded exposure–response, and sequential sampling methods. Compare and contrast methods may be further subdivided into before and after and reference site comparisons. Before and after comparisons monitor critical population endpoints before and after the application of a possible stressor to determine if change has occurred (e.g., Munkittrick and Dixon 1989). This approach assumes that the stressor is the cause of any measured change in the endpoints and that endpoint changes would not have occurred in the absence of the stressor. Reference site comparisons use data on population endpoints from affected and unaffected areas that are ecologically similar (e.g., Swanson et al. 1994). The comparisons assume differences between areas are attributable to the differential presence of the stressor and that no other systematic differences between areas exist.

Graded exposure–response approaches compare groups of individuals (possibly populations) in a series of graded stressor exposures and seek to establish a correlative link between the measured exposure (e.g., concentrations of a contaminant or degree of exploitation) and biological endpoints of each group (e.g., Adams et al. 1994). This assumes that any gradient observed in measured characteristics may be attributed to known differences in stressor intensity in the absence of other indicated causes of systematic differences between studied groups.

Sequential sampling methods use traditional field sampling approaches to study the long-term effects of exposure to a single stressor incident (e.g., Mills and Chalanchuk 1987; Mills et al. 2000). The approach assumes that critical variables relevant to describing or determining the effect of a known stressor action may be identified a priori and that sufficiently long-time series data will prove adequate to the empirical quantification of observed associations between population-level variables and environmental factors, including stressors. The approach also implicitly invokes the stability assumption that aspects of the environment not specifically measured do not change throughout the course of the study (Shuter and Regier 1989).

As a group, field-based attempts to assess population-level responses to stress and their possible ecological significance follow from the observation that fish in their natural environment are typically subjected to a number of stressors that alone, or in combination, are capable of triggering measurable physiological responses having population-level implications (Adams et al. 1993). Field-based measures are viewed as integrative and as one means of directly capturing the consequences of the complex interaction of environmental factors for studied populations, while, at the same time, avoiding the difficulties associated with the extrapolation of single stressor-response laboratory test data (Munkittrick and Dixon 1989). As a result, field-based studies have become a particularly popular means of assessing stressor impacts when questions about the larger scale or longer-term implications of stressors have been asked. A summary of the main field-based approaches for assessing possible population-level responses to stress is given in Table 2.

Compare and Contrast Approaches

Compare and contrast approaches to assessing population-level impacts have a long history (Cairns et al. 1984; Ryder and Edwards 1985; Munkittrick and Dixon 1989; Shuter 1990) and stemmed from the need to develop reliable methods capable of detecting the adverse effects of a wide variety of environmental stressors. Building on initial work by Colby (1984) and Ryder and Edwards (1985), Munkittrick and Dixon (1989) outlined a framework for comparatively assessing the effects of stressors at the population level. The framework proceeds by identifying population stress response patterns through an assessment of life history characteristics that provide indications of changes in age distribution, energy use, and energy storage. Generalized responses to stressors such as food limitation, overfishing, and hydrological impoundments (Munkittrick and Dixon 1989; Gibbons and Munkittrick 1994) are provided; responses are argued to be useful for identifying factors limiting the performance of fish, as well as in iterative programs designed to isolate the factors associated with observed population-level responses.

A basic assumption of the compare and contrast approach is that the ecosystem is the best indicator of its own status and that consequent changes in constituent populations will be associated with consistent characteristic responses measurable in terms of mean population age, fecundity, and condition factor. Accordingly, populations found to be growing, reproducing, and surviving within the limits of comparable reference populations are assumed to be free from the detrimental effects of possible stressors (Munkittrick and Dixon 1989). The Munkittrick and Dixon framework also assumes that populations are operating in isolation from the effects of interactions with other species; that density-dependent growth operates; and that the effects on population abundance, resulting from density-independent factors, do not mask density-dependent responses (Jaworska et al. 1997).

Measurements relating to the growth, reproduction, and survival of fish populations are collected and used for assessment. Because such measurements are not wholly independent, characteristic response patterns are defined on the basis of mean age, fecundity, and condition responses alone. Substitution for closely related parameters (e.g., maturation data for fecundity), however, is allowed where preferred data do not exist or cannot be obtained. Response patterns are identified in terms of the expected elicited effect on measured population parameters. Comparison of data from an impacted site with historical or reference site data are used to establish conformance to an expected pattern in the field. For routine assessment purposes, a small number of females may be examined for the age, condition, and fecundity measures required for completing comparative assessments (Munkittrick and Dixon 1989).

The framework has several admitted limitations (Munkittrick and Dixon 1989). It is static and provides only a snapshot of population status (Gibbons and Munkittrick 1994). The framework is also retrospective in the sense it can only detect changes once they have occurred. No allowance is made for movement between loosely connected subgroups within the population, and responses may be inadvertently detected or masked by movement of fish between impacted and nonimpacted sites. Care must be taken to ensure selection of a comparable reference site in terms of its carrying capacity,

Table 2. Major field-based approaches to assessing population-level effects of stress in fish. Key assumptions must be validated if the results are to be credible. Compare and contrast methods are best suited to rapid screening of stressor effects and may be used to rank stressors. Graded exposure-response methods have improved descriptive and statistical power, but are prone to inferential error if the complete set of environmental factors is not measured. Sequential sampling has good temporal resolution and can deal effectively with the detection of lagged effects, but requires expensive, long-term monitoring programs.

Approach	Key assumptions	Temporal resolution
Compare and contrast		
Before and after	Stressor addition causes differences in endpoint measures	Poor, period to period only
Reference site	Stressor alone causes differences in endpoint measures	Very poor, static point in time samples
Graded exposure-response	Response gradient attributable to differences in stressor intensity	Very poor, static point in time samples
Sequential data sampling	Critical response variables known a priori, stability in nonmeasured variables prevails	Good to excellent, depending on number and frequency of samples

habitat, and exploitation characteristics. Failure to do so may elicit responses that have more to do with inherent differences between sites, than stressor effects (Underwood 1989). The use of small sample sizes to estimate required average population response measures effectively reduces the approach to an individuals-based assessment where it is assumed obtained individual data will provide unbiased estimates of population-level responses to stress.

As a generic assessment method for populations, compare and contrast methods, like the framework approach suggested by Munkittrick and Dixon (1989), suffer from other obvious problems. Interpreting observed changes in summary measures for multiple year-class stocks may be difficult. Each year-class contributes to the production of new year-classes in ways that reflect relative population abundance (Trippel 1995) and age-specific fecundity. The resulting demographic complexity is typical of semelparous and iteroparous fish populations and must be accounted for when attempting to assess the response of summary measures to stress. For example, if populations are not in a demographic steady-state, year-class strengths will vary and consequent new year-class sizes will vary as a function of changes in the weighted average of reproductive year-class fecundities as moderated by any density-dependent mortality. The resulting variations in new year-class

Table 2. (continued).

Key advantage	Key disadvantage	Biological results relevance
Quick, removes problem of selecting comparable reference sites	Retrospective, does not consider response dynamics, discounts temporal change	Low and weakly descriptive
Quick, removes problem of considering temporal changes	Retrospective, does not consider response dynamics, discounts spatial and temporal differences	Low and weakly descriptive
Considers spatial aspects of effects explicitly	Weak diagnostic tool, cannot establish causation	Medium, helps describe population health, amenable to statistical testing
Considers temporal aspects of effects explicitly	Requires long-term data and consistency of sampling effort	Medium to high, very descriptive of rates of change and trends, amenable to statistical testing

size will require determination of the normal range of variability if the significance of any stressor induced change in size is to be established with any degree of statistical confidence.

Interdependencies between many of the measured variables also hold implications for the observed patterns of responses to stress detected by compare and contrast methods. For example, ration, growth, and temperature are not independent (Elliott 1994). Temperature and energy intake interact to affect gross conversion efficiency and growth rate through alterations in maintenance requirements and the scope for growth, with residual energy being allocated to reproduction. Thus, growth and reproductive effort are linked in complex ways that imply simple directional changes in growth or reproductive rate measures cannot be taken as reliable indicators of overall population status (Power 1997). Additionally, the small sample sizes suggested for compare and contrast methods (Munkittrick and Dixon 1989) cannot appropriately account for sampling biases that may preferentially exclude faster or slower growing fish (Ricker 1975).

Changes in recruitment are equally unlikely to be appropriately assessed using compare and contrast methods. Although stock–recruitment (SR) models may be used to characterize the resilience of populations to stress, application of the theory has been hindered by the variable rates of recruitment seen in most fish populations. Even when long-time series are available, uncertainty about the exact form of the stock–recruitment relationship has hampered efforts by fisheries workers to characterize the nature of the phenomenon (Fogarty et al. 1991). In the face of these problems, it is not reasonable to presume that the single season, small samples typically obtained for compare and contrast analysis will adequately characterize the complexity of stock dynamics. Nor is it reasonable to assert that assessed directional changes in recruitment will be truly representative of either the short-term or long-term effects of stressors on such a fundamental population process.

Oscillating populations in impacted and reference sites may converge or diverge by chance after the application of a stressor (Underwood 1989), and there is no means of excluding this possibility when simple static samples are used as a basis for comparison in compare and contrast studies. Responses to a stressor may be nonlinear, such that low stressor levels elicit a subsidence effect, followed by a return to the normative range and, ultimately, a large negative effect (Underwood 1989). The dynamic nature of such a response implies that static snapshots of the population would not necessarily detect an effect simply because of mistimed sampling. Attempts to validate compare and contrast frameworks using modeling (Jaworska et al. 1997) and multiple reference site (van den Heuvel et al. 1999) approaches have met with limited success. As van den Heuvel et al. (1999) have concluded, the physiological indices relied on by static frameworks may not be the most sensitive indicators of sublethal stressor exposure because impacts can be masked by compensatory responses. The conclusion suggests that compare and contrast frameworks may not be appropriate for predicting potential stressor risks to populations in the wide variety of chronic exposure cases that currently dominate discussions of environmental monitoring and regulation.

Graded Exposure–Response Approaches

The graded exposure–response approach to assessing population-level stresses is best developed in a series of papers by Adams (Adams et al. 1993; Adams and Ryon 1994; Adams et al. 1994) and others (Karas et al. 1991; Sandstrom 1994). The approach uses response information from multiple response endpoints on the assumption that environmental complexity is such that it is unlikely that single variable measures will accurately reflect responses to stress (Adams and Ryon 1994). In its initial form, qualitative methods for assessing the severity or degree of damage imposed by stressors in the field (Goede and Barton 1990) were adapted for quantitative analysis by assigning numerical weights to the qualitative biological response categories. The transformation of the data makes it usable in statistical analysis and reflective of the degree of damage being imposed by the suite of acting stressors (Adams et al. 1993).

Validation studies were conducted by comparing case study results to results obtained from assessing fish health using other biomonitoring approaches, including a suite of biomarkers of biochemical and physiological response, concentrations of contaminants in fish tissue, and direct measures of reproductive competence. In a more detailed comparison, Adams and Ryon (1994) assessed the relative detection power of individual response measures, integrative multivariate response indices, and integrative ecological responses at the population and community level. Each of the measures was found to lead to different conclusions concerning the interpretation and evaluation of the effects of contaminant stress on fish populations. It was concluded that a strategy employing the joint application of multiple methods was best. The approach was developed further with the application of canonical variate analysis to the evaluation of multivariate response data for fish populations exposed to contaminants (Adams et al. 1994). This approach takes interactions and associations between variables into account, and has demonstrated that integrated response patterns can be used to separate sites on the basis of the degree of contaminant exposure.

Although the procedure is useful in helping to identify factors that impair fish populations and the degree of difference between them, it is a weak diagnostic tool in the sense that it cannot establish the cause of the problem (Adams et al. 1993). Causation can only be established if no other environmental variable has a similar trend from place to place (Underwood 1989). Unfortunately, complete measurement of the possible set of environmental correlates is rarely, if ever, done owing to the practical limitations of cost.

This leads to the conclusion that causation based on a pattern of correlation between field-based measures and the intensity of a stressor (e.g., dose–response) may only be possible using a complex set of interlinked laboratory and field experiments.

Sequential Sampling Approaches

Few examples of sequential sampling studies on the effects of stress at the population level exist. Available studies invariably focus on a limited suite of field-related measures (e.g., growth, abundance, condition, survival, fecundity, and production). Nevertheless, such studies provide insights into the dynamics of population adjustments to stress. Among the more innovative sets of sequentially sampled data are those assembled from the Experimental Lakes area of northwestern Ontario. In a controlled acidification experiment, the pH of Lake 223 was systematically reduced from 6.5 to 5.0 over a five-year period from 1976 to 1981 (Mills et al. 1987). Species population responses to changing pH levels were monitored in lake trout *Salvelinus namaycush,* white sucker *Catostomus commersoni,* fathead minnow, slimy sculpin *Cottus cognatus,* and pearl dace *Semotilus margarita* (also known as *Margariscus margarita*) populations. Abundances of lake trout increased during the early years of acidification, reached a peak in 1981, and declined only after consecutive recruitment failures in 1980 and 1981. Recruitment resumed with the onset of pH recovery, but has remained consistently low despite the achievement of preexperiment pH levels by 1994 (Mills et al. 2000). Survival of age-1 and older individuals fluctuated until 1980 and declined consistently only after pH fell below a critical 5.59 threshold (Mills et al. 1987). During the initial years of pH recovery, survival increased and has since remained above 80% per year (Mills et al. 2000). Condition factor initially rose and, though variable, remained within the preacidification 95% confidence intervals until 1983 (Mills et al. 1987). With the onset of recovery, there was a significant rise in condition factor and a return to preacidification values (Mills et al. 2000). In spite of persistent declines in condition post-1978, growth did not decline and changes in condition during the acidification phase of the experiment were attributed to increases in population abundance rather than the acidification stressor (Mills et al. 1987). Low abundance of lake trout was expected to result in accelerated growth during the recovery phase of the experiment as a result of density-dependent effects. Although mean lengths of caught lake trout did increase gradually throughout the recovery period, increases did not correspond with expectations (Mills et al. 2000).

White sucker population numbers similarly increased to a peak in 1981, before rapidly collapsing in the face of recruitment failure. Parallel increases in chironomid and dipteran emergence occurred up to 1981, increasing prey availability. Survival showed moderate increases in the early years of the experiment and remained stable thereafter. Recruitment, though initially low, rose rapidly in the middle years of the experiment, before collapsing. Condition remained within the 95% bounds of the preacidification measures (Mills et al. 1987). During pH recovery, white sucker abundance was expected to decline as lake trout abundance rose. White sucker abundance, however, remained high throughout the recovery period, collapsing only in 1996 due to low annual survival and recruitment events that could not be attributed to lake trout abundance or predation (Mills et al. 2000). Populations of fathead minnow collapsed quickly following increased acidification and, after 1981, no fish could be sampled. Pearl dace, which had been rare prior to the experiment, increased dramatically in numbers through 1981. By 1983, only a few remained, as population numbers collapsed in the face of recruitment failure.

Apparent contradictory changes in abundance and growth among species appear to have been triggered by complex changes in the forage base of both lake trout and white sucker. Declines in some species were compensated for by increases in others (e.g., pearl dace). Initial increases in the forage base raised available ration and adult carrying capacity. Increased carrying capacity allowed abundance to rise, with population collapse occurring only when declines in pH initiated recruitment failure in essential prey species. The observed sequence of changes in abundance suggests the dynamics of population-level responses to environmental change must be tracked over considerable periods of time if a true picture of population-level responses to stress is to emerge.

The results of whole lake fertilization experiments, assessed using sequential sampling, have demonstrated similar temporal variability in population responses (Mills 1985; Mills and Chalanchuk 1987; Mills et al. 1998). Lake 226 was divided in half with the northeast basin receiving additions of phosphorus, nitrogen, and carbon and the southwest basin receiving additions of nitrogen and carbon only. Changes in lake whitefish *Coregonus clupeaformis* growth, condition, survival, fecundity, abundance, and production were measured for eight years (1973–1980) during and for three years (1981–1983) postfertilization. Although juvenile growth is generally considered to be a sensitive indicator of nutrient enrichment, no significant differences in growth were found for age-0 fish between basins or years (Mills 1985). Growth of older fish (Figure 1) was not significantly different between treatments in the initial years of experimentation, but increased significantly in the phosphorus-enriched basin from 1974 until 1982 (Mills and Chalanchuk 1987). Condition factor differences were generally insignificant, with the exception of two years (1979, 1982) when the condition of fish in the phosphorus-enriched basin was significantly greater. Survival of fish age-1 and older did not vary between basins. Survival of age-0 fish was greater in the phosphorus-enriched basin only for years with abundant year-classes, and the adjustment was suggested as a general response pattern of fish to lake fertilization (Mills and Chalanchuk 1987).

An analysis of the initial three years postfertilization predicted that population size structure in the phosphorous-fertilized basin would persist for several more years. The prediction was not confirmed. Lake whitefish population size structure changed rapidly with the sudden die-off of larger individuals (Mills et al. 1998). Results from these fertilization experiments point to the importance of sampling frequency in the detection of critical population-level responses to perturbation and the validation of response predictions. Continued tracking of population-level responses to changing ecosystem conditions allowed experimental predictions to be falsified and the development of new hypotheses to explain observed events. Failure to anticipate sudden size-structure changes related, in part, to the failure to appreciate the nature of lags relating reductions in phosphorous reductions to declines in prey abundance and subsequent survival. Where population characteristics respond too quickly to perturbations, responses will be detected only if sampling is conducted at regular and frequent intervals. Where population characteristics respond more slowly than the length of the sample interval, responses will be detected only if sampling is carried out repeatedly over significant periods of time (Underwood

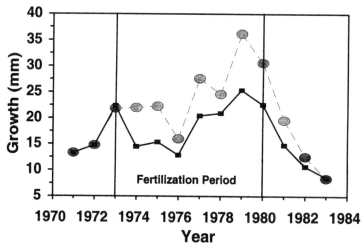

Figure 1. Yearly growth (mm/yr) of lake whitefish in the separate basins of Lake 226, Experimental Lakes area, northwestern Ontario, based on data from Mills and Chalanchuk (1987). Basins were fertilized with phosphorous, nitrogen, and carbon or nitrogen and carbon between 1973 and 1980. Values are the adjusted means from covariance analysis using back-calculated data. Growth in the basin receiving phosphorus, in addition to nitrogen and carbon, are plotted with gray circles. Growth in the basin receiving only nitrogen and carbon are plotted with solid squares. Differences in growth were significant ($P < 0.05$) from 1974 to 1982. In both basins, data demonstrate the dynamic response of the populations and the length of time required to adjust to a new equilibrium.

1989). In either case, significant benefit and increased understanding is derived from the pattern of persistent and consistent sampling required by sequential sampling approaches to assessing the population-level effects of stress.

Dynamic Lags

The many field-based attempts to quantify population-level stress responses have taught us much about the difficulties associated with measuring the extent and magnitude of stressor effects and understanding the ways in which an observed response may be functionally related to a known stressor. Among the more important lessons learned are those pertaining to the effect of dynamic time lags on observable responses and their implications for drawing conclusions about the significance of a stressor effect. Lags in the response of a population to stress pose obvious problems for field-based conclusions about the presence or absence of a stressor effect. It is also likely that a population responding slowly to a stressor will be similarly responding slowly to other simultaneously acting stressors. Under such conditions, the effect of any single stressor or episodic event will be difficult or impossible to identify on the basis of limited data (Underwood 1989). As Mills and Chalanchuk (1987) noted, after eight years of experimentation with a single fertilization event, there was still no evidence that a new equilibrium biomass of lake whitefish had been established.

Evans et al. (1990) and Shuter (1990) attempted to generalize the dynamics of staged responses to stress at the population level observed in field-based studies (e.g., Mills and Chalanchuk 1987). Evans et al. (1990) suggested a generalized ecosystem response to stress occurring in three stages: alarm, resistance, and extinction. Alarm is dominated by signs of changes in species dispersal rates and distribution and production or biomass changes. Presence or absence, relative frequency of occurrence, and individual growth rate studies are appropriate for determining the existence of alarm-related stressor effects. In the resistance phase, natural population-level regulating processes are triggered leading to expected changes in abundance, mortality, and survivorship. Decreases in abundance and survivorship, and increases in mortality, are expected to yield systematic changes in sampled population characteristics that are diagnostic of stress in general. In the extinction phase, population abundance declines rapidly to zero.

The pattern of expected responses parallels structured changes predicted by Shuter (1990) that begin with alterations in patterns of habitat occupation and use and end with shifts in the balance of population natality and mortality processes critical to determining overall abundance. Expected initial changes in habitat occupation and use patterns suggest field-based scientific and monitoring studies aimed at classifying stressors in terms of their spatial pervasiveness and defining population dispersal characteristics (Frank and Leggett 1994) are required to develop early warnings of stressor-related effects. Changes measured in individual-level characteristics as summarized in sample data (e.g., growth rate or condition factor) are expected to follow spatial shifts and be indicative of increasing stressor intensity. Ultimately, changes in abundance driven by changes in either age-at-maturity, fecundity, or the balance of other population vital rates will indicate stress has risen to the point where compensation processes can no longer offset stressor-related effects and further increases in stress can be expected to precipitate population-level collapses (Power 1997). A synthesis view of the structured responses likely to be observed in field-based measures is given in Figure 2. In addition to extinction-related population collapses, Figure 2 allows for the development of a residual population that may avoid extinction through alteration of population vital rates if the intensity of the stress is reduced prior to a minimum viable population threshold being reached.

Variability

The field detection and measurement of stresses acting on natural populations would appear to be beset with considerable difficulties. If it were already established that a population of interest was at equilibrium and the equilibrium level was known, then the detection of stress would simply involve determining that a known perturbation had caused the endpoint of interest (e.g., abundance) to deviate from the equilibrium value (Underwood 1989). The simple comparison, however, cannot be made because populations rarely, if ever, exist at equilibrium. Even if there are no temporal fluctuations in abundance, equilibrium cannot be established unless observations are available for a period of time exceeding the natural life expectancy of the species (Frank 1968). In cases where the sampling of natural populations has been carried out over a sufficiently long period, there are few indications that populations exist at equilibrium (Connell and Sousa 1983). Establishing the effects of stress on a population, therefore, requires more than demonstrating changes in population abundance endpoints. For stress to have had a significant effect, measured population-level changes must exceed those that would normally be expected based on knowledge of inherent population variability and the routine operation of population regulating mechanisms (Underwood 1989). Unfortunately, sampling imprecision complicates attempts to gather the required data. Census approaches to estimating population parameters are typically not feasible except in cases where population abundances have declined to the point where it is clear a population is threatened with local extinction. Accordingly, random sampling is used to establish a probabilistic parameter estimate that carries with it a variance and the associated problem of sampling error. If measurement variability is an integral part of any attempt to assess the significance of a possible impact, it should be directly incorporated into analyses attempting to determine the effects of stress at the population level. Therefore, the graded

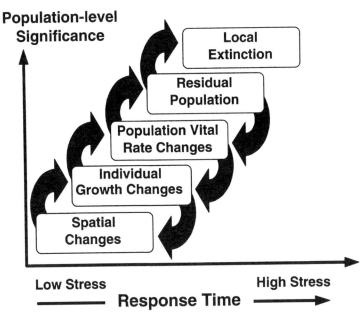

Figure 2. Sequential responses to stress plotted as a function of stressor intensity, response time, and population-level significance. Spatial changes reflective of distributional adjustments or presence or absence occur at low levels of stress as a result of avoidance behavior. Adjustments to individual somatic growth occur if spatial changes do not eliminate stressor effects. Accumulated individual changes in growth trigger changes in population vital rates (e.g., survival rates) via density dependence. Changes in population vital rates lead, initially, to a residual population, and, ultimately, extinction, when abundance falls below the minimum viable population threshold. Changes at any one stage below extinction can induce lagged adjustments both above and below the stage, making interpretation of observed field-based measures difficult. Responses at low stress levels occur quickly and have negligible impact on population-level parameters (e.g., size-structure, abundance). Responses at high stress levels occur slowly and have significant impacts on measured population-level parameters.

exposure–response (Adams et al. 1994) or sequential sampling (Mills 1985; Mills et al. 1987) approaches that specifically allow statistical tests of significance based on estimated variability are preferable to static compare and contrast approaches that eliminate consideration of variability on the pretense of diagnostic power.

Populations are known to vary over a wide range of densities in response to environmental fluctuations and disturbances and there are theoretical reasons for expecting the variance of life history characteristics to increase as a function of stress (Service and Rose 1985). Accordingly, variability itself might provide a convenient measure of the relative degree of population-level stress. Unfortunately, there seems to have been little effort focused in this area (Callaghan and Holloway 1999). Ryder (1990) reasoned that whatever the inherent complexity of an ecological system, all systems are characterized by a normative range of variability because of intrinsic homeostatic mechanisms. Variability outside the normative range, therefore, ought to provide an easily obtainable indicator of stress. DeAngelis et al. (1990) likewise suggested it was possible to compare, within species, variability at different times and under different conditions and use increases in variability to obtain evidence of population-level stress.

Marshall (1978) was among the first to demonstrate the responsiveness of variability to stress with a series of laboratory experiments that examined the effects of chronic cadmium exposure on *Daphnia* populations. Results showed an exponential rise in variability, measured as the coefficient of variation for population abundance (Figure 3). Interindividual physiological variability was subsequently suggested as a specific variability-based means of investigating stress effects (Depledge 1990). Forbes et al. (1995) examined the growth rate of gastropods to cadmium exposure and found that exposure to cadmium increased the variability in population growth rates. Results lent credibility to the use of variability as a potentially useful indicator of exposure to environmental stress. Power (1997) examined brook trout population responses to cumulative stresses within a modeling context using an index of stressor intensity and found broad support for the notion that variability increased as a function of stress. Measurement of the variance on an age-specific basis, however, indicated practical problems for field-based implementation. Rapid rises in age-0 variability and the difficulties associated with accurately assessing juvenile abundance in field situations implied that variability was unlikely to be of practical use for assessing stress on the most sensitive population age-classes. Correlations between stressor intensity and increases in the variability of adult abundance suggest that careful monitoring of variability might provide a convenient means of detecting when populations are at risk. However, the ability to detect such changes would depend on the availability of appropriate baseline data against which the significance of observed changes in variability could be assessed—a fact that emphasizes the need for increased monitoring and baseline data collection efforts.

The necessity of having adequate baseline data for drawing conclusions about population-level stress points to a chronic problem with field-based studies of population-level stress. For example, a Leslie matrix based analysis of the significance of reducing brook trout population biomass by increasing age-0 mortality by 50% indicated the first appearance of a significant change in yield statistics did not occur for at least 2–3 years (Jensen 1971a). Attempts to determine the number of years of data required to detect entrainment reductions in Hudson River white perch *Morone americana* year-class strength in light of sample data variability concluded that, at least, 20 years of data would be required to detect a greater than 50% reduction in mean year-class strength (Van Winkle et al. 1981; Vaughan and Van Winkle 1982). Power and Power (1995) demonstrated that, despite

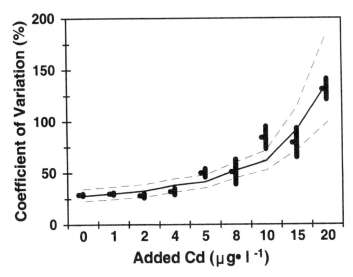

Figure 3. Coefficient of variation ±1 standard deviation of the average number of individuals from weekly samples of cadmium exposed populations of *Daphnia galeata mendotae* based on data from Marshall (1978). Changes in the coefficient of variation were modeled as an exponentially increasing function of stressor intensity ($r^2 =$ 0.915, *P*-value < 0.01) and show a significant rise in the variability of the population-level abundance parameter as stress levels increase. Regression 95% confidence intervals are plotted as dashed lines.

stressor induced monotonic declines in age-0 abundances, adult brook trout abundances displayed a subsidence response as predicted by Odum et al. (1979). Only when stressor intensities increased beyond the compensatory capacity of the population did population abundance display the significant negative changes typically associated with stressor effects. The subsidence induced lag between the initiation of stressor action and the ability to detect significant stressor-induced changes in an endpoint response poses a critical challenge to field-based assessment approaches, particularly if they do not include considerable detail on prior natural history and an understanding of the population-regulating processes (Underwood 1989).

Factors Modifying the Interpretation of Stressor Effects

In addition to including the effects of variations in natural and anthropogenic stress factors, measured population-level responses to stress will automatically integrate any induced compensatory responses that mitigate stress. Hutchinson (1957, 1978) proposed that all biotic and abiotic factors influencing individual survival in the environment could be described in terms of a multidimensional niche defined by bionomic axes reflecting variables directly involved in the somatic and reproductive success of individuals (e.g., food), and tolerance limit axes reflecting the abiotic physical and chemical (e.g., temperature and pH) variables that define the conditions under which an individual can survive. Shuter (1990) has argued along similar lines, noting that the biological effects of stressors fall naturally into categories defined by those that reflect changes in an environment's carrying capacity and those that reflect changes in population members' abilities to survive and reproduce in the local environment. Stressors that exceed the tolerance limits of an individual are normally lethal, and mortality may be brought about by exceeding a single physiological limit. Chronic stressors, which typically do not exceed the physiological limits of individuals, but which do elicit measurable physiological responses, are more common. Furthermore, because chronic stressors are applied over longer periods of time, they allow significant opportunity for interaction with bionomic axes variables, thereby complicating the interpretation of observed effect significance.

Although it is tempting to state that an individual's ultimate survival will be controlled by the niche variable for which the organism has the narrowest range of adaptability, variable interactions and compensatory feedbacks imply such a simplistic interpretation is unlikely to be tenable. Heterogeneous environments mean varying opportunity ranges for individuals in a given population. Reduced opportunities for some, leading to morbidity and mortality, change the opportunities for others on a variety of niche axes. For example, concentrations of a toxicant exceeding the tolerances of some individuals may increase the food resources available to others, thus, modifying the ability of the less sensitive individuals to cope with the stress (Sibly 1999). Understanding the responses of populations to stress, therefore, requires a knowledge of the important niche axes variables, the nature of stresses currently affecting the population, population-regulating mechanisms, and an understanding of the ways in which these may interact to determine measured population-level responses. The list of possible confounding factors to consider when interpreting field-based measures of responses to stress or when making population comparisons is long. Just about every environmental factor examined has been shown to influence the magnitude of measured population responses to stress (Barton 1997) including competition, density, dispersal, exploitation, indirect effects, life history traits, nutritional status, predator–prey relationships, salinity, and spatial heterogeneity.

Species Competition

Laboratory experiments with invertebrate populations have demonstrated the importance of competition between species for determining population-level responses to stress. Mixed-species experiments with the rotifer *Brachionus calyciflorus* and the cladoceran *Daphnia magna* indicated sublethal exposure to copper affected the ability of the rotifer to coexist and compete. Under exposure stresses, *Daphnia* were able to rapidly exclude and suppress rotifer populations in mixed-species cultures through exploitative competition for shared and limited food resources and through mechanical interference (Ferrando et al. 1993). Hanazato (1998) noted that relatively low insecticide exposure concentrations differentially affected zooplankton taxa on the basis of size and affected the population dynamics of coexisting zooplankton species through altered competitive relationships. Similar competitive effects on critical population-level parameters have been observed in field experiments with fish (Persson 1990). Intraspecific competition from roach *Rutilus rutilus* decreased the proportion of mature European perch *Perca fluviatilis* and the gonad weights of age-0 and age-1 males and females as the density of roach in ponds increased.

Density

Density has also been shown to directly affect the interpretation of stressor effects. Vijayan and Leatherland (1988) demonstrated a significant inverse linear relationship between stocking density and growth rate in brook char (also known as brook trout) *Salvelinus fontinalis.* Behavioral factors including social interactions, territoriality, and dominance hierarchies were implicated as being responsible for the observed relationship. Arthur and Dixon (1993) examined the effects of rearing density on growth responses of fathead minnow exposed to chlorophenols and found the variability in wet weight among fish held at the highest densities was three times that of fish held at the lowest densities. Such evidence suggests the problem with interpreting the impact of a stressor on growth and other density-dependent endpoints is that measured population-level responses may be related as much, or more, to density-dependent competition for food and space than to the effect of the stressor. Power (1997) demonstrated the general implications of density-dependent population regulating mechanisms for increasing the complications associated with determining lethal and sublethal stressor population effects. For example, high levels of stress, which exceed the ability of density-dependent mechanisms to maintain measured population parameters within normative ranges, can lead to sudden population collapses and make accurate prediction of incremental additions to stress difficult.

Exploitation

Exploitation undoubtedly represents the single most common stress for fish populations of commercial or sport interest. Accordingly, knowledge of interactions between nonexploitation and exploitation stresses is of particular interest (Barnthouse 1993). In theory, the rate of exploitation should increase the sensitivity of the population to additional stresses. Modeling studies of the combined effect of exploitation and other stressors has demonstrated that this is the case (Goodyear 1985; Barnthouse 1990; Power 1997). The accuracy with which combined stressor action may be predicted, or the effect of exploitation removed, however, is compromised by inherent nonlinearities in population-level responses to incremental stress (Shuter 1990; Power 1997).

Indirect Effects

Linkages between the processes (e.g., growth) determining either individual- or population-level status suggest that stressor-caused measurable effects on any single process can trigger measurable responses in other processes as a result of the connectivity between them. For example, Hanazato (1998) has shown that in the presence of chemical stressors, the average size of test organisms may be reduced, the efficiency of energy transfer between primary producer and top predator may be affected, and the length of the food chain elongated. Study results provide good evidence of the likelihood of indirect effects at both the population and community levels. Treatment of brook trout streams with insecticides at levels resulting in significant reductions in stream benthos, however, were found to have no causative effect on population age structure, movement patterns, and condition factor (Kreutzweiser 1990). Observed reductions in growth rates were attributed to coincidental increases in temperature that produced similar effects in control populations. These results suggest that the evidence for indirect effects on populations can be confounded by the operation of significant natural stressors that mask the effect of possible indirect stressor linkages through strong direct or interaction effects.

Marshall (1978) noted in a series of cadmium exposure experiments with *Daphnia* that compensation processes can trigger subsequent indirect effects that feed back through density-dependent adjustments to affect measured population-level responses. Power (1997) noted that time further complicates the stressor detection and measurement problem. Stressors that act directly hold immediate consequences for individuals or populations and will be detectable in changes in measured individual performance or shifts in characteristic population-level distributions. Stressors that act indirectly on either individuals or populations, through an intermediate process (e.g., reduction in the forage base) will have delayed effects that are not immediately detectable. In addition, compensation processes triggered by changes in abundance or local carrying capacity can only act with lags to influence measured population-level attributes. Accordingly, population responses to stressor action measured in the short-term may give poor estimates of long-term responses because long-term changes in population-level parameters will depend on feedbacks operating through lagged effect and lagged compensation mechanism responses (Ives 1995).

Life History Traits

Life history traits and the developmental stage at which the stress is applied also influence the extent to which a stress response is displayed at the population level (Barton 1997; Hanazato 1998). Mechanisms that increase stress tolerance may divert energy and other resources away from growth and reproduction under optimal conditions. Energy diversions lead to the expectation that tolerance of environmental stresses may show a negative correlation with many life history traits. Experiments with *Drosophila* have demonstrated that the correlation between stress tolerance and life history traits (fecundity) was significant and negative (Hoffmann and Parsons 1989). An examination of factors contributing to first-year recruitment failure in fish populations in acidified environments similarly concluded vulnerability may be based as much on life history (e.g., synchrony of sensitive or resistant life stages with stress challenges) as the physiological status of the organism (McCormick and Leino 1999).

Dispersal

Theoretical and field-based analyses of population-level stressor effects typically assume that individuals exist in a homogenous environment and are unconnected in any way to other, unexposed, populations. Frank and Leggett (1994) have noted population dispersal patterns are frequently ignored in population dynamics studies because populations are typically viewed to be either geographically isolated or self-reproducing. Recent analysis of dispersion in marine fishes has shown profound demographic consequences can result. Dispersal of late larval and juvenile stages beyond the normal distributional range results in population mixing that can continue until spawning (Frank 1992). Maurer and Holt (1996) argue the nondispersal assumption may be unrealistic for many population-level assessments given the significance of patch dynamics and habitat fragmentation for population viability. Using discrete and continuous time models, Maurer and Holt (1996) demonstrated that interacting populations groups (metapopulations) may be affected in predictable ways as a result of a single subgroup being exposed to stress. For example, if a single habitat patch maintains abundance levels in adjacent patches through emigration, any stressor that significantly increases the chances of extinction in the critical patch will also increase the likelihood of extinction in the other patches. The naturally lower rates of population increase in the patches not directly affected by the stressor will prevent overall population levels from recovering quickly because of their inability to maintain sufficient abundances in the absence of immigrant inflows from the critical patch. As a result, there will be a slow recolonization of the affected patch and long lags between the initial effects of the stressor and full recovery. In addition, the high rates of population decline in the affected patch relative to the rates of population increase in the initially unaffected patches will decrease the likelihood of metapopulation persistence by increasing the vulnerability of the metapopulation to the effects of additional stressors owing to the lengthened time over which patch abundances remain nearer their minimum viable thresholds.

The number of biotic and abiotic factors and mechanisms capable of modifying the effect of a single stressor action on measured population-level responses suggests that the interpretation or prediction of population-level responses will never be an easy task. Experience with actual case histories and theoretical analyses indicates that attempts should be made to document and consider the potential effects of co-occurring stressors and possible interactions when assessing population-level stressor effects (e.g., Figure 4). In addition, understanding the responses of populations to stress requires appropriate knowledge of species ecology, the mechanisms through which species population dynamics are affected, and the ways in which population regulating mechanism might interact to influence observed population-level variables. Particularly important will be knowledge of the nature of the compensation processes that aggregate and translate individual experience to the population-level measures. Taken together, these requirements imply that attempts to understand population-level stressor effects cannot divorce themselves from the basic ecological work necessary to describe affected populations in their natural environment.

Assessing the Effects of Stress with Models

A fundamental problem in elucidating the concept of stress in ecology is that the simplest measurements to make and understand are often those taken on individuals, while the primary interest is in effects at the population level (Nisbet et al. 1989). As a result, one of the major challenges facing those studying stress is evaluating the ecological consequences of observed individual-level responses to stress exposure (Maltby 1999). Addressing the challenge will be difficult without a detailed understanding of organism life history, the processes regulating population size, and a complete description of organism-specific effects. Reliance on field survey data alone has demonstrated the difficulties associated with attempts to establish the causes of observed changes in population parameters (Evans et al. 1990; Barnthouse 1993). For example, abundance changes with time and location, and factors like density dependence may also be affected by mechanisms acting through, or in conjunction with, changing levels of food supply, predation, parasitism, interspecific competition, migration, and other environmental variables (Sibly 1999). Awareness of the difficulties associated with integrating field-based data and controlled laboratory studies to interpret the consequences of stress at the population level in the face of factors modifying the interpretation of stressor effects has increased interest in the use of predictive mod-

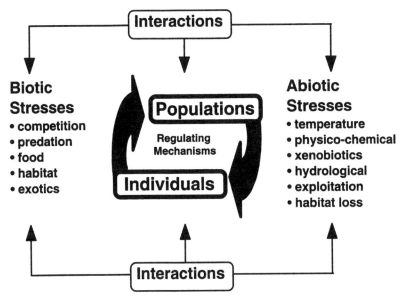

Figure 4. Population-level responses to stress integrate the numerous biotic and abiotic factors that act directly and indirectly on individuals or populations, or as a result of complicated interactions between stressors. Stressors affecting populations will trigger adjustments holding indirect consequences for individual probabilities of survival or success and visa versa. Connections between the two levels are reactive, implying observed changes will necessarily occur with lags and that understanding the temporal dynamics of responses to stressors is critical.

els that relate measured effects on individuals to changes in population dynamics (Emlen 1989a).

   Although modeling studies of populations can provide descriptions of the effects of stress, they do not, in themselves, provide information on causal mechanisms (Maltby 1999). Nevertheless, Minns (1992) argues that the complexity of ecosystems is such that modeling provides one of the few systematic means within which the dynamics of population-level responses to a suite of interacting anthropogenic and natural stimuli can be appropriately analyzed. All modeling methods aimed at assessing populations, however, resort to simplifying assumptions. Models that are too simple run the risk of omitting variables that are important for predicting the possible consequences of stressors. Models that are too complicated lack the perspicuity required of abstract analysis (DeAngelis et al. 1990). In that sense, modeling is a less direct approach than field studies (Sibly1999). Modeling does have the advantage of being less expensive and permitting investigation of potential population fluctuations over much longer time scales (Landahl et al. 1997). Furthermore, modeling has progressed to the point where identifying vulnerable life stages, ranking sources of stress on an effect basis, and comparing alternative mitigating strategies can now be easily accomplished (Vaughan et al. 1984; Evans et al. 1990; DeAngelis et al. 1990).

For decades, fisheries scientists have been occupied with attempts to understand the ramification of exploitation stresses on the continued viability of valued fish populations. To that end, numerous analytical frameworks suitable for determining the possible population-level effects of a wide variety of stressors have been developed. The modeling frameworks examine the responses of the basic population processes (mortality, growth, and reproduction) to stress at differing scales of complexity, disaggregation, and degree of individual organism detail and can be grouped into six families: stock–recruitment, Leslie matrix (including life table approaches), life cycle, physiological-based (including bioenergetic approaches), individual-based, and habitat supply. These basic model types have been arranged in Figure 5 by scale of complexity, degree of disaggregation, and degree of detail and are summarized in Table 3 with respect to their treatment of population regulating processes of growth, reproduction, and survival.

To some extent, the groupings are arbitrary as models in different categories often share similar features or directly incorporate portions of other model types. Within each model family, there are also a variety of closely related model forms, each representing a minor variant on the basic model. The suitability of a model type for predicting possible population-level effects will depend on the complexity of the question being asked, the suitability and availability of data for model building, the availability of stressor-effect data, the skill of the modeler in conceptualizing and representing complex processes within the constraints of simplifying model assumptions, and the perceived usefulness of model output to decision makers (Vaughan et al. 1984; Chambers 1993; Power and McKinley 1997).

Emlen (1989b) has argued that to appropriately address the effects of stressors at the population level, models must meet additional criteria. They must produce outputs describing endpoints of regulatory relevance. Models must also be able to incorporate available scientific information on stressor effects. Because much of this data will come from laboratory-based studies, models must be equally adept at incorporating individual sample effect data and field-based estimates of population parameters. Finally, models must be easily linked to available information on stressor exposure if attempts to understand the possible consequences of causal linkages are to be realistically assessed.

Stock–Recruitment Models

Stock–recruitment models describe the average relationship between the abundance of mature individuals in a population and the number of progeny recruited to the population in the next generation. In aquatic environments, recruitment is typically a multidimensional process governed by the combined action of many physical, chemical, and biological factors that determine growth and survival rates (Fogarty et al. 1991). As a result, SR models

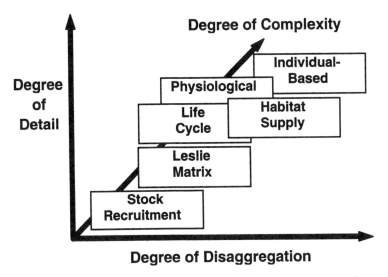

Figure 5. A hierarchal arrangement of models that may be used to evaluate the possible effects of stressors on fish populations. Models are classified, approximately, by degree of disaggregation, complexity, and detail. Less complex models often enter as submodels into more complex models. For example, stock–recruitment relationships are imbedded in many Leslie matrix models, and physiological models are incorporated into many individual-based models.

aggregate many different life history stages and influences that are typically studied separately. The need for aggregation is often driven by lack of appropriate data for more disaggregated modeling. Aggregation also depends on the fact that the spawning stock may be controlled by management action and it is important to have direct measures of the consequences of that control (Hilborn and Walters 1992).

Two SR models, one proposed by Ricker (1954) and the other by Beverton and Holt (1957), are well known and widely applied in fisheries studies.

$$\text{Ricker:} \quad R = \alpha S e^{-\beta S}$$

$$\text{Beverton-Holt:} \quad R = 1/(\rho + k/S)$$

R defines the number of recruits. S is the number of spawning adults or spawning stock biomass, and α, β, ρ and k are parameters estimated by the appropriate model. In the Ricker model, α defines the slope of the model at the origin and degree of density-independent mortality. The parameter β relates proportional survival to density and defines the strength of density-dependent survival within the population. In the Beverton-Holt model, ρ is measured in terms of reciprocal units such that $1/\rho$ defines maximal recruitment and k defines the steepness of the ascent to maximal recruitment. Example plots of both models are given in Figure 6.

The stock–recruitment curves associated with these models may be used to characterize the resilience of a population to harvesting or other stressors by calculating the slope of the recruitment curve at the origin. The slope defines recruits per unit stock or the rate of population increase at low populations levels (Ricker 1954). Low slopes indicate low rates of increase indicative of unfavorable conditions for population persistence (Fogarty et al. 1991). Stock–recruitment models were originally intended to represent changes in population numbers brought about by changes in natural and harvesting mortality factors. Recently they have been used to represent the effect of changes in other stressors (e.g., river discharge, species interactions, entrainment, or temperature) on population numbers (Stocker et al. 1985; Walters et al. 1986; Power and McKinley 1997) by adding specific terms for the additional stressor to the model (Hilborn and Walters 1992). For example, in the Ricker model environmental effects may be added by respecifying the model as

$$R = \alpha S e^{-\beta S + \Sigma b_i E_i}.$$

Table 3. Temporal and biological resolution in population processes included in commonly applied population modeling frameworks. Models are applied to the assessment of population-level effects of stress in fish populations depending on the availability of data and the type of question being asked.

Modeling approach	Population processes		
	Growth	Reproduction	Survival
Stock–recruitment (SR)	Not considered	Measured indirectly only after pre-recruitment mortality included	Average summed over all age-classes on an inter-generational basis
Leslie matrix	Not considered	Constant or stochastic age- or stage-specific reproductive rates	Constant or random age- or stage-specific values, may use SR model
Life cycle	May be included, but typically is not	Constant or stochastic age- or stage-specific reproductive rates	Constant or random age- or stage-specific values
Physiological or bioenergetic	Function of the energy budget	Function of the energy budget	Not generally considered
Individual-based	Function of bio-energetic based relationships	Function of growth and reproductive rate based relationships	Function of niche tolerances and rules for ecological interactions
Habitat supply	Density-driven empirical relationships	Habitat supply determines eventual hatching success	Density-dependent driven empirical relationships

where E_i defines the ith environmental factor included in the model (e.g., temperature) and b_i is the ith parameter estimate.

Despite their apparent flexibility, there are many factors that stock–recruitment models cannot take into account. These include the potential effects of changes in population size structure, the differential effects of stressors on varying age- or size-classes, and temporal variation in recruitment rates. The stock–recruitment relationship cannot, therefore, adequately describe the internal changes within a population that may be important early indicators of the adverse effects of stress. Application of SR models to studying the effects of stress on fish populations have also been hindered by the high levels of recruitment variability, short recruitment time series, and uncertainties about the form of the stock–recruitment relationship descriptive of recruitment dynamics in many populations (Fogarty et al. 1991). In addition, the potentially large number of interacting environmental factors affecting recruitment, the intrinsic nonlinear nature of the recruitment process,

Table 3. (continued).

Data requirements	Resolution	
	Temporal	Biological
Parental numbers or spawning biomass, recruit numbers	Low, generation based	Low, uses a single descriptive function
Age- or stage-specific fecundity and survival values	Medium, annual or seasonal	Medium, uses averages
Age- or stage-specific fecundity, survival, and life-history values	Medium, annual or seasonal	Medium, uses averages
Energy budget variables and physiological rates	High, daily to annual	High, uses individual detail
Detailed individual physiological rates and interaction probabilities	High, daily to annual	High, uses individual detail
Life-stage specific average rates and habitat inventories	Medium, seasonal or annual, but emphasis is spatial	High, uses both biotic and physical data

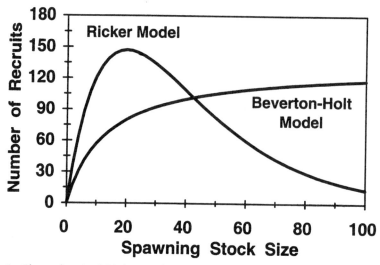

Figure 6. Plots of typical Ricker and Beverton–Holt stock–recruitment models. The Ricker model is characterized by its dome-shaped appearance and the Beverton–Holt model by its approach to an asymptotic value defining maximal recruitment. The exact shape of both models will vary as a function of changes in parameter values.

and biases in the collection of necessary data have made the development of reliable predictive recruitment models difficult (Walters et al. 1989; Koslow 1992). As such, SR models have only enjoyed limited success in the study of stress-related phenomenon at the population level.

Leslie Matrix Models

Where populations have been disturbed, age-structured models may be the only way to appropriately predict the consequences of stressor action. Stock–recruitment models implicitly assume that every individual within the population is identical (the average) and individual differences are not allowed to influence population dynamics. Age-structured models overcome this obvious problem by dividing the population into age-classes. Within an age-class, individuals are still treated as identical, but the characteristics of each individual in the population differ by age-class.

A principal analytical tool within the class of age-structured models is the Leslie matrix approach (Leslie 1945). The approach incorporates age-specific information on population birth, aging, reproduction, and death processes to describe population fluctuations through time. It begins with an initial age distribution for a population represented as a vector $N(t)$ divided into age-classes $n_k(t)$ representing the number of individuals of age k at time t as follows:

$$N(t) = \left[n_0(t), n_1(t), \dots n_k(t) \right]$$

The initial age distribution is then multiplied by a matrix containing age-specific survival probabilities and fecundity factors to determine the age distribution of the survivors and descendants in the next time period (t + 1) as follows:

$$N(t+1) = MN(t)$$

where M, whose elements are the survival probabilities and fecundity rates, is the Leslie matrix. It follows that the numbers alive in different age-groups at an arbitrary time (t) depend on the number initially alive, $N(0)$, as follows:

$$N(t) = M^t N(0)$$

The changing size and age distribution of the population is, thus, calculated by a process of successively multiplying the age distribution by the Leslie matrix, on the assumption that the age-specific survival and fecundity rates remain constant from period to period. As the population size increases, the relative numbers of individuals in each age-class will vary until a distribution, termed the stable age distribution, is reached (Manly 1990).

Once the stable age-distribution is reached, the proportion of individuals in each age-class will remain constant and the ratio of population values at successive intervals, [$N(t$ + 1)$/N(t)$], becomes constant. The ratio λ is known as the finite rate of population increase and is used to define the proportion by which the population will change in each successive time period. Values of $\lambda > 1$ are indicative of an exponentially expanding population. Values of $\lambda = 1$ define an equilibrium population and values of $\lambda < 1$ are characteristic of populations declining to extinction. Further details on the underpinning mathematical theory, construction, and estimation of matrix models can be found in Caswell (1989) or Manly (1990). Discussions of the use of matrix models in conservation biology are given in Burgman et al. (1993).

The age-structure detail in Leslie matrix models allows a population to be described in terms of its long-term abundance, intrinsic rate of natural increase, reproductive potential, resilience or extinction risks (Landahl et al. 1997). Application of Leslie models to the study of stressor effects requires that accurate age-specific survival and reproductive rates under various stressor scenarios are available. Generally, these rates have been derived based on water column toxicity test results and extrapolated from laboratory test species to the species of interest (Barnthouse et al. 1990). Questions about the validity of species-to-species extrapolation, however, exist (Power and McCarty 1997) and require that models be based on parameter estimates obtained from data collected for the species and stressor of interest. In constructing models, it is also important to remember that stressor-related impacts on survival and reproduction will be influenced by the extent to which stressor effects are mitigated by population regulating mechanisms such as emigration, immigration, or density dependence (Power 1997). Where migrations are likely to increment or decrement population numbers, age-specific mortality factors must be appropriately adjusted to reflect these effects. In addition, when recruitment processes are known to conform to standard stock–recruitment relationships (e.g., Beverton-Holt or Ricker) due consideration of these relationships in model construction should also be given.

A variety of fisheries-related techniques exist for deriving the required Leslie model parameters. Catch-curve analysis may be used to derive survival or mortality estimates for age-classes fully recruited to the fishery (Robson and Chapman 1961), but due attention must be paid to possible selective biases in the catch data (Ricker 1975). Estimates of age-0 and juvenile survival are, however, problematic. These are typically estimated by using data for a closely related species (Landahl et al. 1997) or by determining the values for age-0 and juvenile survival that yield an intrinsic rate of population increase equal to zero (Schaaf et al. 1987). Fecundity is best estimated directly for the species of concern from field data using a gravimetric (Bagenal and Braum 1978) or other suitable technique. When comparing populations or stressor scenarios for relative population effects, it is important to appropriately adjust for differences in average length between considered cases before using fecundity estimates from different samples in model construction (Bagenal 1978).

While the Leslie matrix approach provides useful summary information on populations, their age structures, and the risks of extinction, it is not without its drawbacks. The Leslie approach requires assuming stability in the population distribution over time. This may be unrealistic for populations subjected to random anthropogenic or environmental disturbances. Furthermore, a Leslie matrix model will be inappropriate when critical survival and reproductive values do not depend on age, but on physiological development that is independent of age. Variants of the Leslie model approach, based on stage-structured data, have been developed to deal with this particular exigency (Usher 1966, 1969; Lefkovitch 1965; Manly 1990).

Leslie matrix models have also been criticized for failing to include considerations of density dependence or temporal changes in critical survival parameters. As a result, matrix survival coefficient constants are now often viewed as random variables or made explicitly density dependent such that the value the coefficient takes on in any one time period is functionally dependent on abundance in the period (Goodyear 1985; Manly 1990). Drawing survival coefficients randomly from a distribution with a defined mean and variance allows the model to mimic the impact of varying environments on populations. And explicit consideration of density dependence accounts for the fact that natural populations typically fluctuate within defined bounds.

Temporal changes in fecundity can, likewise, be included through the substitution of equations for the appropriate matrix constants. This is typically done using a multistep process that relates length to age via a von Bertalanffy model, weight to length with a traditional allometric regression, and fecundity to length with an exponential model (Bagenal 1978). Estimation of each of the intermediate relationships and completion of the required substitutions allows the effects of nonlethal stressors on abundance, acting through changes in growth or reproduction, to be directly manipulated within the context of the Leslie model.

Although density dependence and temporal parameter changes may be accounted for in the Leslie framework, the adaptations come at the cost of increasing model complexity. Introduction of varying parameter values also limits the interpretative value of the intrinsic rate of population increase because repeated changes in parameter values imply that the stable age distribution on which the parameter depends mathematically will never be reached. Accordingly, a detailed description of the population growth rate parameter distribution in terms of its mean, variance, and empirical density function are necessary to interpret probable population-level effects when models are largely stochastic. Regardless of improvements, Leslie models still focus on abundance and do not produce information on other population parameters routinely used by biologists to assess the effects of environmental perturbations on populations, including rates of growth, length-frequency distributions, condition factors, and age-at-maturity.

Life Table Models

Life tables have been widely used as another means of summarizing the population-level effects of stress (Walthall and Stark 1997). The approach has been particularly favored by those studying the effects of chronic stressors on invertebrate cultures (e.g., *Daphnia* and *Ceriodaphnia*) in laboratory-based experiments (Levin et al. 1996). Life table models are closely related to the Leslie matrix models, but do not consider detailed fluctuations in age structure as a function of time. Instead survival probabilities and age-related abundances are used to calculate a single parameter, the intrinsic rate of population increase, using Lotka's equation as follows:

$$\Sigma l_x m_x e^{-rx} = 1$$

where l_x is the probability of surviving to age x, m_x is the number of female offspring per female born during the interval x to $x + 1$, and r is the intrinsic rate of population increase. The parameter r may be computed by solving the equation using iteration, through the use of Leslie population matrices or by sampling population sizes at different points in time. The intrinsic rate of population increase is argued to integrate the age at first reproduction, age-specific fecundity and survivorship, brood frequency, and longevity effects of stress in an examined population (Walthall and Stark 1997) and to account for the apparently contradictory effects elicited by compensatory effects (Daniels and Allan 1981). Forbes and Calow (1999) concluded that r was a better summary measure of responses to stressors than the measures of individual-level effect endpoints typically obtained from laboratory experimentation (e.g., acute survival and chronic survival) because it integrated potentially complex interactions among life history variables and provided a more ecologically relevant measure of population-level impacts than the weakly predictive individual endpoints. As a result, life table analysis can provide important insights into the population-level consequences of stressors and can be used to generate testable hypotheses to explain why certain species dominate in stressed habitats while others disappear (Forbes and Calow 1999). For example, Jensen (1971b) used life table models to compare the response of seven brook trout populations to exploitation. The analysis suggested fishing mortality was compensated by higher age-specific fecundities and that growth might be, indirectly, the most important variable in the adjustment of a trout population to stress.

With life table models, however, care must be taken to ensure that the intrinsic rate of population increases is calculated for a series of population densities. In most cases, studies are conducted at low densities (Sibly 1999). At higher densities, competition for resources (e.g., food) increases and, at least, some population members will do less well than others. Declines in resource acquisition for affected individuals will reduce physiological defenses against fluctuations in environmental factors or additional stressors and, through density-dependent feedback mechanisms, adjust the vital rates of the more competitively successful individuals. Such adjustments are particularly likely if mortality among competitively inferior individuals occurs. Overall, the consequences of changes in density for the ability of the population to cope with stress must be understood if the intrinsic rate of population increase is to be used as a realistic measure of characteristic population-level responses to stress.

Life Cycle Models

Life cycle models increase the detail, complexity, and disaggregation of the biological information required to obtain parameter estimates. Accordingly, life cycle models are more realistic representations of populations than SR or matrix-based models. The models divide the population into age-classes and are built around the framework of matrix models. Unlike traditional Leslie models, which use constant coefficients to represent the transfer of individuals from one age-class to the next, life cycle models employ detailed representations of survival from one age-class to the next.

Considerable biological detail on each age-class is included when building life cycle models. For example, reproduction and survival functions are often modeled separately and on a sex-specific basis. The degree of detail included will vary as a function of the information available to estimate model parameters and the modeling choices made by the modeler.

A variant of life cycle models, termed a network flow model, divides fish populations into a set of compartments, one for each stage of the life cycle. The models are constructed by specifying a list of life history stages, initial abundances in each life stage, and the fractions moving from one stage to another. Fractional coefficients generally represent survival probabilities but may also be used to split individuals at a given life stage into distinct subgroups representative of varying life history strategies within the same population. For example, as is the case with some populations of Arctic char *Salvelinus alpinus*, a proportion might chose to remain in freshwater, while the remaining proportion migrate to sea. Density dependence also can be introduced with the use of SR models to control the transition of individuals from one stage to the next. Network flow models use a set of difference equations, one for each life history strategy stage, which are solved analytically to obtain equilibrium solutions describing the number or proportion of individuals expected at each life history strategy stage. Changes to either the survival or proportional strategy parameters will affect the equilibrium solutions. The relative impact of parameter changes on equilibrium solutions is characterized using sensitivity analysis to compute parameter elasticities, which express the proportional change in the equilibrium solution resulting from a known proportional change in the parameter (Evans and Dempson 1986).

The network flow modeling approach has been applied to studying a number of stress-related phenomena, including the selection of management options and determination of the effects of exploitation and disease events. Evans and Dempson (1986) used a life cycle model to determine the relative efficacy of habitat protection measures versus changes in fishing regulations in terms of optimizing Atlantic salmon *Salmo salar* abundance. Hankin and Healy (1986) used the life cycle approach to determine the resilience of chinook salmon *Oncorhynchus tshawytscha* to high levels of exploitation. Similarly, des Clers (1993) used a network flow model to demonstrate the vulnerability of Atlantic salmon populations to high disease induced mortality at the parr life stage. Although they are highly flexible analytical frameworks, network flow models are closed and do not allow for immigration or emigration effects. This limits the approach to the study of stress in populations within a well-defined space (e.g., lakes) or populations known to have a limited home range. Computed results also refer only to mean population size, and predictions are accurate only when the effects of small parameter changes are studied (e.g., $\leq 20\%$).

Physiological-Based Models

The effects of stress on fish are manifested by changes in the primary biological processes of mortality, growth, and reproduction. Most modeling approaches attempt to incorporate aspects of all three processes. Physiologically based and bioenergetic models look only at growth under the assumption that growth integrates all the abiotic and biotic factors influencing the success of fish in the environment. Physiologically based models are designed specifically to bridge the gap between individual and population and community approaches to studying the effects of stress (Kooijman and Metz 1984). Connections between responses at the various levels of organization are established either by linking growth and reproduction directly to environmental factors that affect food supply using a fixed ratio rule for the energy allocated to reproduction and respiration (Kooijman and Metz 1984) or by tying survivorship to the scope for metabolic activity such that stresses lowering the metabolic scope also reduce the probability of survival (Calow and Sibly 1990).

Although theoretical connections between physiological processes and survivorship, physiological processes and fecundity, and physiological processes and growth can be readily established, experimental data detailing the precise nature of the relationships is currently lacking (Maltby 1999). The short-term measures of differences between energy absorbed from food and that lost via excretion and metabolism (scope for growth) have been shown to correlate well with longer term measures of growth and reproduction. The relationship between energy availability and physiological processes suggests that information on how stressors affect energy budgets can be incorporated into population models to predict the population-level responses to induced physiological changes. One of two approaches may be taken to predicting population-level responses from available physiological change data. Averaged experimental results may be used by treating all individuals within the population as replicates under the assumption that intrapopulation variability will be small. Alternatively, experimental results may be used to compute the mean and variance of a distribution describing individual responses and all individuals in the population may be subsequently treated as unique (e.g., DeAngelis and Gross 1992).

A second approach to linking individual responses to stress to population-level effects involves incorporating physiological change information into bioenergetic models, which examine the factors affecting individual growth (Elliott 1976; Kitchell et al. 1977; Elliott 1994). Changes in individual biomass are related to energy intake and use as defined by food consumption and the physiological processes of respiration, egestion, excretion, and reproductive effort. Energy intake and use, however, are also affected by the abiotic and biotic variables (e.g., temperature and ration) that define the environment in which individuals live. Accordingly, bioenergetic analyses must consider both the physiological processes that directly affect individual biomass changes and the environmental factors that indirectly affect individual biomass changes through the modifying influences they have on key physiological processes.

One of the major advantages of bioenergetic models is that they typically have a finer temporal resolution than traditional population models. Accordingly, they can model effects on a daily basis. This ability makes bioenergetic models particularly useful for modeling within-year dynamics (Vaughan et al. 1984) and has favored their inclusion as submodels in other modeling frameworks (e.g., individual based). Bioenergetic models may be scaled up to the population level either by aggregating the individual changes of all population members, as in the case of individual-based models (IBM), or by determining average cohort values and multiplying by the number of fish in each cohort (Rice 1990). In the latter case, where mortality factors are not explicitly included, adjustments for mortality losses to the population must be made.

Another advantage of bioenergetic models is that they allow the stressor responses to reflect a particular mode of action (Rice 1990). For example, stressors may act to affect individuals directly through changes in assimilation efficiency, increases in respiration, or lower reproductive output, or indirectly through changes in food availability and quality or alteration of thermal habitat. The ability to detail the mode of action allows stressor effects to be examined either individually or in combination. In addition, because individual biomass depends directly on body size, response rates can be made size dependent and the differential effects of stress at varying life stages, and its implications for the population, as a whole, can be modeled directly.

Generally, bioenergetic models use parameter values estimated from data on healthy, unstressed fish and have been used to examine questions concerning unstressed populations. However, the detail of physiological function included in bioenergetic models suits them for use in studying stress at the population level. Known effects of stressors on fish physiology can be incorporated directly into the models to determine effects on individual growth, which, in turn, can be used to estimate population-level abundance effects. Bioenergetic models are also versatile. They can incorporate chronic or transient stress, multiple effects, interactions, and indirect effects. They can also be used to define the range or distribution of individual responses to stress. The distribution of individual responses to stress, in turn, may be used to estimate key parameters for population-level model scenarios for examining the effect of physiological stressors on populations. The ability to connect bioenergetic models to population-level models provides a readily apparent means of integrating the study of stressor effects at more than a single level of biological organization.

As with all modeling approaches, there are practical constraints to the use of bioenergetic models. Substantial numbers of physiological measurements are required to estimate model parameters and opportunities to field validate the models are rare (Rice and Cochran 1984). Although parameter estimates obtained for one species may be used for other species within the same family, subtle variations in physiological responses to prevailing environmental conditions (e.g., temperature) suggest such substitutions must be treated with caution.

Individual-Based Models

Individual-based models have their roots in ecological theory and the idea that the properties of an ecosystem can be derived from the properties and relationships of the elements that compose the ecosystem (Lomnicki 1988). Individual-based models view interactions within and among individuals as the important processes from which population-level attributes are determined. Each individual in the population is represented as directly affecting and being affected by other individuals it encounters as a result of the rules governing individual location, movement, foraging, and reproductive and predator avoidance activities. Although complex and detailed in content, many argue that IBMs will free ecologists from the necessity of making unreasonable assumptions when modeling the behavior of complex systems (Huston et al. 1988). For example, the dependence of population-level descriptors on the summation of individual attributes in IBM models allows descriptors to adjust passively to the dynamic and heterogeneous nature of influences on constituent members in a manner reflective of actual ecological experience.

Individual-based models differ from other modeling approaches in the degree of detail they represent. In simple abundance models (e.g., SR models), individuals are aggregated to the population level and all individuals are represented as the "average." In age-structured models (e.g., Leslie matrix models), individuals are aggregated into categories such as age and size and are treated as identical within the defined category. There is no such aggregation in IBMs. The descriptive attributes of age, size, condition, fecundity, and age-at-maturity are defined for each individual in the population and, when taken together, define each member of the population as a distinct individual. Individuals may be aggregated for reporting purposes by shared attributes that include, but are not limited to, the categories typically reported for simple abundance or age-structured models (e.g., total abundance, age-specific abundance, and size-class). One of the benefits of the IBM approach is that it provides a framework within which to conceptualize the natural processes operating on populations through their actions on individuals. The design, analysis, and synthesis of field-based empirical studies within IBM models can be carried out in a synergistic manner that facilitates determining and testing the generality of many stress-related hypotheses (Minns 1992; Van Winkle et al. 1993).

The trend toward the inclusion of greater individual detail in predictive population models parallels the growth of knowledge on ecological processes and the vast amounts of data being collected by scientific investigations. The expectation is that more detailed models will aid in the identification of the factors responsible for population fluctuations, thereby, improving the accuracy of model predictions and the development of management strategies based on those predictions (Wright 1992). Though the requirement for increased biological detail comes at a cost, the rapid changes in computing technology have helped to substantially reduce the traditionally high costs of model design, construction, and testing.

Although IBMs have a promising research future, they face a number of important challenges. Most important among these challenges are the need to develop the ecological theory necessary to describe the importance of individual differences and the availability of appropriate and statistically valid data upon which to formulate model rules (Van Winkle et al. 1993). Collaboration among those studying the sources of individual variability and those developing IBMs will be critical to overcoming these challenges. In addition, the requirement for detailed data to estimate model parameters implies that model development cannot be divorced from the field and laboratory studies that generate the required data. Therefore, an iterative connection between field studies, empirical model building, and the construction and use of IBM models will be integral to the widespread acceptance and continued use of the IBM modeling approach.

Habitat Supply Models

One of the largest stressors on fish populations is habitat alteration and destruction (Thomas 1994). As a result, there has been increased emphasis on the role habitat plays in determining population success. Frameworks for assessing the possible impacts of habitat on fish populations have been developed. These include the HEP (habitat evaluation procedures; USFWS 1981) and UET (ultimate environmental threshold) ecosystem planning method (Kozlowski 1985). Both procedures determine overall habitat suitability by aggregating the life stage specific requirements of a studied population and embody a static approach to the derivation of fish habitat relationships. The aggregation of life stage specific requirements, however, takes no account of the importance or effect of potential interactions among life stages subject to differing habitat supply limits (Minns et al. 1996). An alternative approach that takes explicit account of how density-dependent effects operating at various life stages will be related to the differing supply of habitat types at each life stage has been proposed by Minns et al. (1996) as a means of connecting habitat changes to dynamic population responses. The modeling approach combines information on the supply of habitat types, parameters descriptive of key population attributes (e.g., minimum and maximum lengths at maturity, survival rates) and the age and sex structure of the population to model the effects of life stage habitat supply limits on population abundance. The resulting models are dynamic and based on the assumption that weighted areas of usable habitat derived from habitat suitability index models may be reliably used to estimate the habitat supply specific to each life stage in the population.

The development and application of dynamic habitat supply models is in its infancy. The possible effects of variability on the determination of abundance need to be addressed (Minns et al. 1996). Further development on incorporating community interactions and expansion of the indicators used to assess population impacts are also required. Although abundance dominates many of the public discussions of population-level impacts, knowledge of change in other population parameters is also required to effectively assess or mitigate possible stressor effects. Nevertheless, habitat supply model simulations have demonstrated how physical habitat variables like lake depth can affect population size and structure, and the models show promise in terms of their ability to guide restoration efforts in stressed ecosystems.

Summary

One of the major challenges facing applied ecological studies is evaluating the ecological consequences of observed individual-level responses to stressor exposure. Reductionist measures of effect (e.g., biomarkers of exposure) typically respond rapidly to stress and have high relevance for the development of causal mechanisms. Unfortunately, these same measures lack the ecological realism necessary for understanding or predicting population-level effects. The operation of homeostatic processes responsible for maintaining individual equilibrium and the compensatory mechanisms that offset fluctuations in biotic and abiotic environmental factors complicate the interpretation of reductionist measures at the population level by obscuring the simplicity of the links between cause and effect. The question of stressor-effect significance is difficult to answer without detailed knowledge of regulating mechanisms and the nature of any connections between observed individual- and population-level responses to stress. The effects of any given stressor, therefore, may only be interpretable if considerable prior natural history and a mechanistic description of the processes affecting the ecology of a species are well understood. In addition to more detailed laboratory studies aimed at elucidating the causal links between exposure and response at the organism level, well defined ecological monitoring programs recognizing the dynamic complexity, connectedness, and spatial and temporal heterogeneity of ecological systems will be required. Fisheries biologists and ecologists have much to offer in this regard. They must be more actively included in the development and design of monitoring programs and sampling and statistical approaches aimed at increasing sample precision and reducing sample bias, than has been the case for the development of ecotoxicological-driven stressor study programs to date.

Models are likely to help in identifying and quantifying the effects of changes in key population-level variables. There is a long history of model development and use, but the testing of models has not always been easy owing to the lack of appropriate data sets. It is likely that well-designed monitoring programs will provide a wellspring of information from which hypotheses of stressor cause and effect relationships may be generated for future field validation. In that respect, field and modeling studies cannot be viewed as separate approaches. Instead they must be viewed as mutually interdependent approaches requiring closely connected and parallel studies aimed at elucidating the effects of stress in complex environments.

Connecting field, modeling, and monitoring efforts requires the use of field studies to describe year-to-year variability in critical population attributes, the use of field data in the construction of empirically based models to predict population-level responses to change over the short-term, the construction of analytically oriented models capable of incorporating field and laboratory experimental data to estimate possible population-level impacts of a given stressor over the long-term and suggest testable hypotheses concerning possible change, and the use of monitoring data to test and validate all model predictions. The integration of field, model, and monitoring efforts, therefore, will require substantial interdisciplinary cooperation and, if done well, should improve our abilities to assess the ultimate effects of stressors on populations.

Acknowledgments

I thank Marshall Adams for having challenged me to address the difficult topic of assessing stressor effects on populations. Discussion of the topic owes much to the many population-stress conversations I have had with S. M. Adams, M. J. Attrill, L. S. McCarty, J. D. Reist, J. B. Dempson, D. G. Dixon, K. R. Munkittrick, and others. Particular thanks goes to G. Power and K. H. Mills who provided detailed and exacting comment on an initial draft of the chapter. Funding support for the work was provided by a Natural Sciences and Engineering Research Council of Canada research grant to M. Power.

References

Adams, S. M. 1990. Status and use of biological indicators for evaluating the effects of stress on fish. Pages 1–8 *in* S. M. Adams, editor. Biological indicators of stress in fish. American Fisheries Society, Symposium 8, Bethesda, Maryland.

Adams, S. M., A. M. Brown, and R. W. Goede. 1993. A quantitative health assessment index for rapid evaluation of fish condition in the field. Transactions of the American Fisheries Society 122:63–73.

Adams, S. M., K. D. Ham, and J. J. Beauchamp. 1994. Application of canonical variate analysis in the evaluation and presentation of multivariate biological response data. Environmental Toxicology and Chemistry 13:1673–1683.

Adams, S. M., and M. G. Ryon. 1994. A comparison of health assessment approaches for evaluating the effects of contaminant-related stress on fish populations. Journal of Aquatic Ecosystem Health 3:15–25.

Arthur, A. D., and D. G. Dixon. 1993. Effects of rearing density on the growth response of juvenile fathead minnow (*Pimephales promelas*) under toxicant-induced stress. Canadian Journal of Fisheries and Aquatic Sciences 51:365–371.

Bagenal, T. B. 1978. Aspects of fish fecundity. Pages 75–101 *in* S. D. Gerking, editor. Ecology of freshwater fish production. Blackwell Scientific Publications, Oxford, England.

Bagenal, T. B., and E. Braum. 1978. Eggs and early life history. Pages 166–198 *in* T. B. Bagenal, editor. IBP Handbook 3, Methods for assessment of fish production in fresh waters, 3rd edition. Blackwell Scientific Publications, Oxford, England.

Barnthouse, L. W. 1993. Population-level effects. Pages 247–274 *in* G. W. Suter, editor. Ecological risk assessment. Lewis Publishers, Boca Raton, Florida.

Barnthouse, L. W., G. W. Suter, and A. E. Rosen. 1990. Risks of toxic contaminants to exploited fish populations: influence of life history, data uncertainty and exploitation intensity. Environmental Toxicology and Chemistry 9:297–311.

Barton, B. A. 1997. Stress in finfish: past, present and future: a historical perspective. Pages 1–33 *in* G. K. Iwama, A. D. Pickering, J. P. Sumpter, and C. B. Schreck, editors. Fish stress and health in aquaculture. Cambridge University Press, Cambridge, England.

Begon, M., J. L. Harper, and C. R. Townsend. 1990. Ecology: individuals, populations and communities. Blackwell Scientific Publications, Oxford, England.

Berlinsky, D. L., M. C. Fabrizio, J. F. O'Brien, and J. L. Specker. 1995. Age-at-maturity estimates for Atlantic coast female striped bass. Transactions of the American Fisheries Society 124:207–215.

Beverton, R. J. H., and S. J. Holt. 1957. On the dynamics of exploited fish populations. U. K. Ministry of Agriculture and Fisheries, Fishery Investigations (Series 2) 19, London.

Burgman, M. A., S. Ferson, and H. R. Akçakaya. 1993. Risk assessment in conservation biology. Chapman and Hall, London.

Cairns, V. W., P. V. Hodson, and J. O. Nriagu. 1984. Contaminant effects on fisheries. Volume 16, Advances in environmental science and technology. John Wiley and Sons, New York.

Callaghan, A., and G. J. Holloway. 1999. The relationship between environmental stress and variance. Ecological Applications 9:456–462.

Calow, P., and R. M. Sibly. 1990. A physiological basis of population processes: ecotoxicological implications. Functional Ecology 4:283–288.

Caswell, H. 1989. Matrix population models: construction, analysis and interpretation. Sinauer, Sunderland, Massachusetts.

Chambers, R. C. 1993. Phenotypic variability in fish populations and its representation in individual-based models. Transactions of the American Fisheries Society 122:404–414.

Clark, L. R., P. W. Geier, R. D. Hughes, and R. F. Morris. 1967. The ecology of insect populations in theory and practice. Methuen, London.

Colby, P. J. 1984. Appraising the status of fisheries: rehabilitation techniques. Pages 233–257 *in* V. W. Cairns, P. V. Hodson, and J. O. Nriagu, editors. Contaminant effects on fisheries. Volume 16, Advances in environmental science and technology. John Wiley and Sons, New York.

Connell, J. H., and W. P. Sousa. 1983. On the evidence needed to judge ecological stability. American Naturalist 121:789–824.

Daniels, R. E., and J. D. Allan. 1981. Life table evaluation of chronic exposure to a pesticide. Canadian Journal of Fisheries and Aquatic Sciences 38:485–494.

DeAngelis, D. L., L. W. Barnthouse, and W. Van Winkle. 1990. A critical appraisal of population approaches in assessing fish community health. Journal of Great Lakes Research 16:576–590.

DeAngelis, D. L., and J. G. Gross. 1992. Individual-based models and approaches in ecology. Populations, communities and ecosystems. Chapman Hall, London.

Depledge, M. H. 1990. New approaches in ecotoxicology: can inter-individual physiological variability be used as a tool to investigate pollution effects? Ambio 19:251–252.

Des Clers, S. 1993. Modelling the impact of disease-induced mortality on the population of wild salmonids. Fisheries Research 17:237–248.

Elliott, J. M. 1976. The energetics of feeding, metabolism and growth of brown trout (*Salmo trutta* L.) in relation to body weight, water temperature and ration size. Journal of Animal Ecology 45:923–948.

Elliott, J. M. 1994. Quantitative ecology and the brown trout. Oxford University Press, Oxford, England.

Emlen, J. M. 1989a. Animal population dynamics: identification of critical components. Ecological Modelling 44:253–273.

Emlen, J. M. 1989b. Terrestrial population models for ecological risk assessment: a state-of-the-art review. Environmental Toxicology and Chemistry 8:831–842.

Evans, D. O., G. J. Warren, and V. W. Cairns. 1990. Assessment and management of fish community health in the Great Lakes: synthesis and recommendations. Journal of Great Lakes Research 16:639–669.

Evans, G. T., and J. B. Dempson. 1986. Calculating the sensitivity of a salmonid population model. Canadian Journal of Fisheries and Aquatic Sciences 43:863–868.

Ferrando, M. D., C. Janssen, E. Andreu, and G. Persoone. 1993. Ecotoxicological studies with the freshwater rotifer *Brancionus calyciflorus*. Resource competition between rotifers and daphnids under toxic stress. Science of the Total Environment (Supplement 1993, Proceedings of the Second European Conference on Ecotoxicology, Part 2):1059–1069.

Fogarty, M. J., M. P. Sissenwine, and E. B. Cohen. 1991. Recruitment variability and the dynamics of exploited marine populations. Trends in Ecology and Evolution 8:241–246.

Forbes, V. E., and P. Calow. 1999. Is the per capita rate of increase a good measure of population level effects in ecotoxicology? Environmental Toxicology and Chemistry 18:1544–1556.

Forbes, V. E., V. Møller, and M. H. Depledge. 1995. Intrapopulation variability in sublethal response to heavy metal stress in sexual and asexual gastropod populations. Functional Ecology 9:477–484.

Frank, K. T. 1992. Demographic consequences of age-specific dispersal in marine fish. Canadian Journal of Fisheries and Aquatic Sciences 49:2222–2231.

Frank, K. T., and W. C. Leggett. 1994. Fisheries ecology in the context of ecological and evolutionary theory. Annual Review of Ecology and Systematics 25:401–422.

Frank, P. W. 1968. Life histories and community stability. Ecology 49:355–357.

Gibbons, W. N., and K. R. Munkittrick. 1994. A sentinel monitoring framework for identifying fish population responses to industrial discharges. Journal of Aquatic Ecosystem Health 3:227–237.

Goede, R. W., and B. A. Barton. 1990. Organismic indices and an autopsy-based assessment as indicators of health and condition of fish. Pages 93–108 *in* S. M. Adams, editor. Biological indicators of stress in fish. American Fisheries Society, Symposium 8, Bethesda, Maryland.

Goodyear, C. P. 1980. Compensation in fish populations. Pages 253–280 *in* C. H. Hocutt, and J. R. Stauffer, editors. Biological monitoring of fish. Lexington Books, Lexington, Massachusetts.

Goodyear, C. P. 1985. Toxic materials, fishing and environmental variation: simulated effects on striped bass population trends. Transactions of the American Fisheries Society 114:107–113.

Hanazato, T. 1998. Response of a zooplankton community to insecticide application in experimental ponds: a review and the implications of the effects of chemical on the structure and functioning of freshwater communities. Environmental Pollution 1091:361–373.

Hankin, D. G., and M. C. Healey. 1986. Dependence of exploitation rates for maximum yield and stock collapse on age and sex structure of chinook salmon (*Oncorhynchus tshawytscha*) stocks. Canadian Journal of Fisheries and Aquatic Sciences 43:1746–1759.

Hassell, M. P. 1986. Detecting density dependence. Trends in Ecology and Evolution 4:90–93.

Healey, M. C. 1978. Fecundity changes in exploited populations of lake whitefish (*Coregonus clupeaformis*) and lake trout (*Salvelinus namaycush*). Journal of the Fisheries Research Board of Canada 35:945–950.

Hilborn, R., and C. J. Walters. 1992. Quantitative fisheries stock assessment: choice, dynamics and uncertainty. Chapman and Hall, London.

Hoffmann, A. A., and P. A. Parsons. 1989. An integrated approach to environmental stress tolerance and life-history variation: desiccation tolerance in *Drosophila*. Biological Journal of the Linnean Society 36:117–136.

Huston, M., D. L. DeAngelis, and W. Post. 1988. New computer models unify ecological theory. Bioscience 38:682–691.

Hutchinson, G. E. 1957. Concluding remarks Cold Spring Harbor Symposia. Quantitative Biology 22:415–427.

Hutchinson, G. E. 1978. An introduction to population ecology. Yale University Press, New Haven, Connecticut.

Ives, A. R. 1995. Predicting the response of populations to environmental change. Ecology 76:926–941.

Jaworska, J. S., K. A. Rose, and L. W. Barnthouse. 1997. General response patterns of fish populations to stress: an evaluation using an individual-based model. Journal of Aquatic Ecosystem Stress and Recovery 6:15–31.

Jensen, A. L. 1971a. The effect of increased mortality on the young in a population of brook trout, a theoretical analysis. Transactions of the American Fisheries Society 100:456–459.

Jensen, A. L. 1971b. Response of brook trout (*Salvelinus fontinalis*) populations to a fishery. Journal Fisheries Research Board of Canada 28:458–460.

Karas, P., E. Neuman, and O. Sandstrom. 1991. Effects of pulp mill effluent on population dynamics of perch, *Perca fluviatilis*. Canadian Journal of Fisheries and Aquatic Sciences 48:28–34.

Kerr, S. R. 1976. Ecological analysis and the Fry paradigm. Journal of the Fisheries Research Board of Canada 33:329–332.

Kerr, S. R. 1982. The role of external analysis in fisheries science. Transactions of the American Fisheries Society 111:165–170.

Kitchell, J. F., D. J. Stewart, and D. Weininger. 1977. Applications of a bioenergetics model to yellow perch (*Perca flavescens*) and walleye (*Stizostedion vitreum vitreum*). Journal of the Fisheries Research Board of Canada 34:1922–1935.

Kohler, C. C., and W. A. Hubert. 1993. Inland fisheries management in North America. American Fisheries Society, Bethesda, Maryland.

Kooijman, S. A. L. M., and J. A. J. Metz. 1984. On the dynamics of chemically stressed populations: the deduction of population consequences from effects on individuals. Ecotoxicology and Environmental Safety 8:254–274.

Koslow, J. A. 1992. Recruitment patterns in northwest Atlantic fish stocks. Canadian Journal of Fisheries and Aquatic Sciences 41:1722–1729.

Kozlowski, J. 1985. Threshold approach to environmental planning. Ekistis 311:146–153.

Kreutzweiser, D. P. 1990. Response of brook trout (*Salvelinus fontinalis*) population to a reduction in stream benthos following an insecticide treatment. Canadian Journal of Fisheries and Aquatic Sciences 47:1387–1401.

Landahl, J. T., L. L. Johnson, J. E. Stein, T. K. Collier, and U. Varanasi. 1997. Approaches for determining effects of pollution on fish populations for Puget Sound. Transactions of the American Fisheries Society 126:519–535.

Lefkovitch. L. P. 1965. The study of population growth in organisms grouped by stages. Biometrics 21:1–18.

Leslie. P. H. 1945. On the use of matrices in certain population mathematics. Biometrika 33:183–212.

Levin, L. A., H. Caswell, T. Bridges, C. DiBacco, D. Cabrera, and G. Plaia. 1996. Demographic responses of estuarine polychaetes to pollutants: life table response experiments. Ecological Applications 6:1295–1313.

Lomnicki, A. 1988. Population ecology of individuals. Princeton University Press, Princeton, New Jersey.

Maltby, L. 1999. Studying stress: the importance of organism-level responses. Ecological Applications 9:431–440.

Manly, B. F. J. 1990. Stage-structured populations: sampling analysis and simulation. Chapman and Hall, London.

Marshall, J. S. 1978. Population dynamics of *Daphnia galeata mendontae* as modified by chronic cadmium stress. Journal of the Fisheries Research Board of Canada 35:461–469.

Maurer, B. A., and R. D. Holt. 1996. Effects of chronic pesticide stress on wildlife populations in complex landscapes: processes at multiple scales. Environmental Toxicology and Chemistry 15:420–426.

McCormick, J. H., and R. L. Leino. 1999. Factors contributing to first-year recruitment failure of fishes in acidified waters with some implications for environmental research. Transactions of the American Fisheries Society 128:265–277.

McFadden, J. T., 1977. An argument supporting the reality of compensation of fish populations and a plea to let them exercise it. Pages 153–183 *in* W. Van Winkle, editor. Proceedings of a conference on assessing the effects of power-plant-induced mortality on fish populations. Pergamon, New York.

Mills, K. H. 1985. Responses of Lake Whitefish (*Coregonus clupeaformis*) to fertilization of Lake 226, the Experimental Lakes Area. Canadian Journal of Fisheries and Aquatic Sciences 42:129–138.

Mills, K. H., and S. M. Chalanchuk. 1987. Population dynamics of Lake Whitefish (*Coregonus clupeaformis*) during and after the fertilization of Lake 226, the Experimental Lakes Area. Canadian Journal of Fisheries and Aquatic Sciences 44 (Supplement 1):55–63.

Mills, K. H., S. M. Chalanchuk, and D. J. Allan. 2000. Recovery of fish populations in Lake 223 from experimental acidification. Canadian Journal of Fisheries and Aquatic Sciences 57:192–204.

Mills, K. H., S. M. Chalanchuk, L. C. Mohr, and I. J. Davies. 1987. Responses of fish populations in Lake 223 to 8 years of experimental acidification. Canadian Journal of Fisheries and Aquatic Sciences 44 (Supplement 1):114–125.

Mills, K. H., B. R. McCulloch, S. M. Chalanchuk, D. J. Allan, and M. P. Stainton. 1998. 1987. Growth, size, structure, and annual survival of lake whitefish (*Coregonus clupeaformis*) during eutrophication and oligotrophication of Lake 226, the Experimental Lakes Area, Canada. Archiv fuer Hydrobiologie Special Issues and Advances in Limnology 50:151–160.

Minns, C. K. 1992. Use of models for integrated assessment of ecosystem health. Journal of Aquatic Ecosystem Health 1:109–118.

Minns, C. K., R. G. Randall, J. E. Moore, and V. W. Cairns. 1996. A model simulating the impact of habitat supply limits on northern pike, *Exos lucius*, in Hamilton harbour. Canadian Journal of Fisheries and Aquatic Sciences 53 (Supplement 1):20–34.

Mohr, L. C., K. H. Mills, and J. F. Klaverkamp. 1990. Survival and development of lake trout (*Salvelinus namaycush*) embryos in an acidified lake in northwestern Ontario. Canadian Journal of Fisheries and Aquatic Sciences 47:236–243.

Munkittrick, K. R., and D. G. Dixon. 1989. Use of white sucker (*Catostomus commersoni*) populations to assess the health of aquatic ecosystems exposed to low-level contaminant stress. Canadian Journal of Fisheries and Aquatic Sciences 46:1455–1462.

Murphy, B. R., and D. W. Willis. 1996. Fisheries techniques, 2nd edition. American Fisheries Society, Bethesda, Maryland.

Nicholson, A. J. 1933. The balance of animal populations. Journal of Animal Ecology. 2 (Supplement 1):132–178.

Nicholson, A. J. 1954. Compensatory reactions of populations to stresses and their evolutionary significance. Australian Journal of Zoology 2:1–8.

Nisbet, R. M., W. S. C. Gurney, and W. W. Murdoch. 1989. Structured population models: a tool for linking effects at individual and population level. Biological Journal of the Linnean Society 36:79–99.

NRC (National Research Council). 1981. Testing for effects of chemicals on ecosystems. National Academy Press, Washington, D.C.

Odum, E. P., J. T. Finn, and E. H. Franz. 1979. Perturbation theory and the subsidy-stress gradient. Bioscience 29:349–352.

Persson, L. 1990. A field experiment on the effects of interspecific competition from roach, *rutilus rutilus* (L.), on age at maturity and gonad size in perch, *Perca fluviatilis* L. Journal of Fish Biology 37:899–906.

Power, M. 1997. Assessing the effects of environmental stress on fish populations. Aquatic Toxicology 39:151–169.

Power, M., and L. S. McCarty. 1997. Fallacies in ecological risk assessment practices. Environmental Science and Technology 31:370A-375A.

Power, M., and S. M. McKinley. 1997. Using fish population models in hydro project evaluation. Hydro Review 16:36–40, 69–71.

Power, M., and G. Power. 1994. Modeling the dynamics of smolt production in Atlantic salmon. Transactions of the American Fisheries Society 123:535–548.

Power, M., and G. Power. 1995. A modelling framework for analyzing anthropogenic stresses on brook trout (*Salvelinus fontinalis*) populations. Ecological Modelling 80:171–185.

Rice, J. A. 1990. Bioenergetics modeling approaches to evaluation of stress in fishes. Pages 80–92 *in* S. M. Adams, editor. Biological indicators of stress in fish. American Fisheries Society, Symposium 8, Bethesda, Maryland.

Rice, J. A., and P. A. Cochran. 1984. Independent evaluation of a bioenergetics model for large mouth bass. Ecology 65:732–739.

Ricker, W. E. 1954. Stock and recruitment. Journal of the Fisheries Research Board of Canada 11:559–623.

Ricker, W. E. 1975. Computation and interpretation of biological statistics of fish populations. Fisheries Research Board of Canada Bulletin 191. Ottawa, Canada.

Rigler, F. H. 1982. Recognition of the possible: an advantage of empiricism in ecology. Canadian Journal of Fisheries and Aquatic Sciences 39:1323–1331.

Robson, D. S., and D. G. Chapman. 1961. Catch curves and mortality rates. Transactions of the American Fisheries Society 90:181–189.

Ryder, R. A. 1990. Commentary. Ecosystem health, a human perception: definition, detection, and the dichotomous key. Journal of Great Lakes Research 16:619–624.

Ryder, R. A., and C. J. Edwards. 1985. A conceptual approach for the application of biological indicators of ecosystem quality in the Great Lakes basin. Great Lakes Fishery Commission, Windsor, Ontario, Canada.

Sandstrom, O. 1994. Incomplete recovery in a coastal fish community exposed to effluent from a modernized Swedish bleached kraft mill. Canadian Journal of Fisheries and Aquatic Sciences 51:2195–2202.

Schaaf, W. E., D. S. Peters, D. S. Vaughan, L. Coston-Clements, and C. W. Krouse. 1987. Fish population responses to chronic and acute pollution: the influence of life history strategies. Estuaries 10:267–275.

Service, P. M., and M. R. Rose. 1985. Genetic covariation among life-history components: the effect of novel environments. Evolution 39:943–945.

Selye, H. 1976. The stress of life. McGraw-Hill, New York.

Shuter, B. J. 1990. Population-level indicators of stress. Pages 145–166 *in* S. M. Adams, editor. Biological indicators of stress in fish. American Fisheries Society, Symposium 8, Bethesda, Maryland.

Shuter, B. J., and H. A. Regier. 1989. The ecology of fish and populations: dealing with interactions between levels. Pages 13–49 *in* C. D. Levings, L. B. Holtby, and M. A. Henderson, editors. Proceedings of the national workshop on effects of habitat alteration on salmonid stocks. Canadian Special Publication of Fisheries and Aquatic Sciences 105, Department of Fisheries and Oceans, Ottawa, Canada.

Sibly, R. M. 1999. Efficient experimental designs for studying stress and population density in animal populations. Ecological Applications 9:496–503.

Stocker, M., V. Haist, and D. Fournier. 1985. Environmental variation and recruitment of Pacific herring (*Clupea harengus pallasi*) in the Strait of Georgia. Canadian Journal of Fisheries and Aquatic Sciences 42 (Supplement 1):174–180.

Swanson, S. M., R. Schryer, R., Shelfast, P. J. Kloepper-Sams, and J. W. Owens. 1994. Exposure of fish to biologically treated bleached-kraft mill effluent. 3. Fish habitat and population assessment. Environmental Toxicology and Chemistry 13:1497–1507.

Thomas, C. D. 1994. Extinction, colonization and metapopulations: environmental tracking by rare species. Conservation Biology 8:373–378.

Trippel, E. A. 1995. Age at maturity as a stress indicator in fish. Bioscience 45:759–771.

Underwood, A. J. 1989. The analysis of stress in natural populations. Biological Journal of the Linnean Society 36:51–78.

USEPA (United States Environmental Protection Agency). 1992. Framework for ecological risk assessment. EPA/630/R-92/001. Environmental Protection Agency, Washington, D.C.

USFWS (United States Fish and Wildlife Service).1981. Standards for the development of habitat suitability index models. 103 ESM. Division of Ecological Services, U.S. Fish and Wildlife Service, Washington, D.C.

Usher, M. B. 1966. A matrix approach to the management of renewable resources, with special reference to selection forests. Journal of Applied Ecology 3:355–367.

Usher, M. B. 1969. A matrix model for forest management. Biometrics 25:309–315.

Van den Heuvel, M. R., M. Power, M. D. MacKinnon, and D. G. Dixon. 1999. Effects of oilsands related aquatic reclamation on yellow perch (*Perca falvescens*) II. Validation studies of chemical, and biochemical indicators of exposure to oilsands related waters. Canadian Journal of Fisheries and Aquatic Sciences 56:1226–1233.

Van Winkle, W., K. A. Rose, and R. C. Chambers. 1993. Individual-based approach to fish population dynamics: an overview. Transactions of the American Fisheries Society 122:397–403.

Van Winkle, W., D. S. Vaughan, L. W. Barnthouse, and B. L. Kirk. 1981. An analysis of the ability to detect reductions in year-class strength of the Hudson River white perch (*Morone americana*) population. Canadian Journal of Fisheries and Aquatic Sciences 38:627–632.

Vaughan, D. S., and W. Van Winkle. 1982. Corrected analysis of the ability to detect reductions in year-class strength of the Hudson River white perch (*Morone americana*) population. Canadian Journal of Fisheries and Aquatic Sciences 39:782–785.

Vaughan, D. S., R. M. Yoshiyama, J. E. Breck, and D. L. DeAngelis. 1984. Modelling approaches for assessing the effects of stress on fish populations. Pages 259–278 *in* V. W. Cairns, P. V. Hodson, and J. O. Nriagu, editors. Contaminant effects on fisheries. Volume 16, Advances in Environmental Science and Technology. John Wiley and Sons, New York.

Vijayan, M. M., and J. F. Leatherland. 1988. Effect of stocking density on the growth and stress-response in brook charr, *Salvelinus fontinalis* 75:159–170.

Walters, C. J., J. S. Collie, and T. Webb. 1989. Experimental designs for estimating transient responses to habitat alteration: is it practical to control for environmental interactions? Pages 13–20 *in* C. D. Levings, L. B. Holtby, and M. A. Henderson, editors. Proceedings of the national workshop on effects of habitat alteration on salmonid stocks. Canadian Special Publication of Fisheries and Aquatic Sciences 105, Department of Fisheries and Oceans, Ottawa, Canada.

Walters, C. J., M. Stocker, A. V. Taylor, and S. J. Westrheim. 1986. Interaction between Pacific cod (*Gadus macrocephalus*) and herring (*Clupea harengus pallasi*) in the Hecate Strait, British Columbia. Canadian Journal of Fisheries and Aquatic Sciences 43:830–837.

Walthall, W. K., and J. D. Stark. 1997. Comparison of two population-level ecotoxicological endpoints: the intrinsic (r_m) and instantaneous (r_i) rates of increase. Environmental Toxicology and Chemistry 16:1068–1073.

Winemiller, K. O., and K. A. Rose. 1992. Patterns of life-history diversification in North American fishes: implications for population regulation. Canadian Journal of Fisheries and Aquatic Sciences 49:2196–2218.

Wootton, R. J. 1990. Ecology of teleost fishes. Chapman and Hall, London.

Wright, S. 1992. Guidelines for selecting regulations to manage open-access fisheries for natural populations of anadromous and resident trout in stream habitats. North American Journal of Fisheries Management 12:517–527.

11

Behavioral Measures of Environmental Stressors in Fish

EDWARD E. LITTLE

Introduction

Behavior and Survival

Behavior of aquatic organisms is a sequence of quantifiable actions that operate through the central and peripheral nervous system (Keenleyside 1979). These patterns are the culmination of genetic, biochemical, and physiological processes and, as such, are sensitive to alterations in the steady state of the organism (Warner et al. 1966; Beitinger 1990). Although, seemingly, a random sequence of activities, behavior is a structured and predictable sequence of activities of significant adaptive value and is essential to the organism's existence (Little et al. 1993a). Behavioral responses are important for survival because they are necessary to perform essential life functions such as habitat selection, competition, predator avoidance, prey selection, and reproduction (Little et al. 1985). For example, selection of a habitat suitable for survival and reproductive success requires that the fish be capable of responding to appropriate environmental stimuli to locate beneficial habitats and avoid less favorable environmental conditions. Alteration of the competitive interaction among species and individuals by impairing a competitor's ability to exploit resources may have significant consequences for fish populations and communities (Schoener 1974). Changes in the ability of fish to detect, pursue, capture, and consume prey will have considerable influence on the fish's growth and survival (Brown et al. 1987). Conversely, impaired ability of a fish to detect and avoid predators can increase predation-induced mortality.

Behavior and Environmental Stressors

Because adaptive behavioral function is crucial, behavioral investigations are particularly relevant when evaluating the effects of hazardous substances and other environmental stressors such as temperature or low dissolved oxygen (DO) in the determination of environmental injury. Behavioral toxicity occurs when a contaminant or other stressful condition induces a behavioral change that exceeds the normal range of variability (Marcucella and Abramson 1978). The effects of contaminants on fish behavior have received increased attention over the past decade.

A diversity of quantifiable behavioral responses can be affected by hazardous substances. These responses vary in complexity, ranging from reflexive responses to complex social interactions. The primary criteria for behavioral measures are typically (1) well-defined endpoints that are practical to measure, (2) well understood relative to variables that affect them, (3) sensitive to a range of contaminants or other stressors, (4) adaptable to different species, and (5) ecologically relevant (Rand 1985). Several reviews (Little et al. 1985; Atchison et al. 1987; Beitinger 1990; Blaxter and Hallers-Tjabbes 1992; Birge et al. 1993) indicate that behavioral responses are usually quite sensitive to toxicants (Table 1) and are often the first indication of overt toxicosis, often more sensitive than observed changes in biochemical or physiological responses. In addition to being directly affected by environmental stressors, behavioral responses, such as avoidance–attractance or feeding, may also significantly influence the level or duration of exposure actually experienced by an organism through contact with its environment or diet (Beitinger 1990).

Behavioral responses may be affected simultaneously or consecutively during exposure to toxic substances, depending on the type of chemical, concentration, and duration of exposure. For example (Figure 1), during a 28-d exposure of juvenile rainbow trout Oncorhynchus mykiss to dioxin, 2,3,7,8 TCDD, five different behavioral responses were altered (Mehrle et al. 1988). Reduced feeding was evident after 7 d of exposure to 789 pg/L; hypoactivity and diminished responsiveness to external stimuli were evident by day 10 of exposure; abnormal swimming postures were noted by day 12; and severe lethargy occurred by day 19. The appearance of a particular abnormality varied with concentration. For example, reduced feeding appeared after 7 d of exposure at 789 pg/L; by day 14 at 176 pg/L; and by day 17 at 38 pg/L. Thus, as with other toxicological endpoints, the response and the magnitude or severity of behavioral response varies with the duration and concentration of exposure.

Physiology and Ecology: The Behavioral Link

Behavior provides a unique perspective between the organism and its environment, or between its physiology and ecology. Behavior is initiated when

Table 1. Comparison of behavioral responses relative to sensitivity, field use, ease of measure, ecological relevance, toxicological application, and sensitivity to other stressors.

Method	Sensitivity to contaminant	Observable in field	Ease of measurement	Ecological relevance	Toxicological application	Sensitive to other stressors
Avoidance	chemical-specific	yes	high with required apparatus	high	high	high
Feeding	high	limited	high	moderate	high	high
Predator avoidance	high	limited	moderate	moderate	high	low
Schooling	high	limited	moderate	moderate	species-specific	low
Aggression	high	limited	high	moderate	species-specific	low
Swimming activity	high	emerging	high	low	high	high
Swimming performance	variable	emerging	high with required apparatus	high	high	low
Reproduction	limited	low	high	limited	unknown	unknown

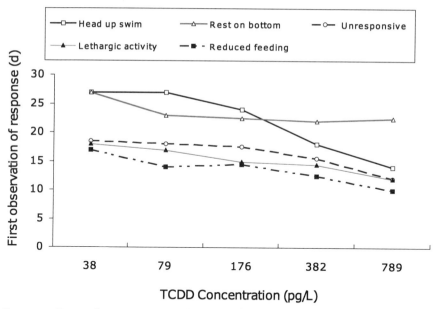

Figure 1. Days of exposure to induce a variety of behavioral changes in rainbow trout exposed as free swimming larvae to TCDD during a 28-d exposure (Mehrle et al. 1988).

a sensory cell encodes a stimulus from the environment as neural impulses and ends with the induction of response by muscle or another effector. Behavioral changes can be induced through direct effects to the nervous system, as well as indirect physiological alterations to respiration and metabolism, endocrine disruption, and immune function. Many pesticides target the nervous system and cause mortality by inhibiting neuromotor responses such as respiration and cardiac output through disruption of neuron to neuron and neuron to muscle events in the nervous system. Sublethal indicators of neurotoxicity are evident as a loss of coordination, control, or reaction time. There are numerous sites of action for toxicants to affect the nervous system, and interference at any of these sites can block or alter the sequence of neural responses and inhibit or alter behavioral function (Anthony et al. 1996). Although the relationship between the nervous system and behavior would appear to be intuitively obvious, there is sufficient redundancy in the nervous system to allow organisms to compensate (Olla et al. 1980), thus, complicating the determination of a direct linkage between a physiological lesion and a behavioral impairment. Available information reveals that changes in behavioral responses induced by toxicosis include impacts to sensory-mediated responses, as well as in neuromotor performance. Thus, it is adaptive for the organism to respond directly to the stimulus of environmental stressors such as water flow, temperature, turbidity, salinity; but, its response to such stimuli may become altered as a result of contaminant exposure.

Applications of Behavioral Testing in Toxicology

Behavioral responses can be rapidly altered by hazardous substances, thus, behavior can be used to signal the presence of hazardous conditions when monitoring effluents. Data generated in behavioral toxicity tests are often used to determine dose–response relationships of contaminant substances and to define no-effect concentrations for hazardous substances. Such information is useful in developing water quality criteria, as well as in appraising the mitigation of impacted habitats. Distinct behavioral toxicity syndromes or patterns of responses, although not indicative of specific toxins, are diagnostic of general classes of contaminants such as neurotoxins, metabolic inhibitors, and narcotic substances (Drummond and Russom 1990).

Behavioral data are also promising as predictive indices of population- and community-level effects. Impaired behavioral performance may be predictive of contaminant effects in the field when ecological consequences can be linked to impaired behavioral performance. Disruption of essential functions such as habitat selection, competition, predator–prey relationships, or reproduction can become ecologically apparent through loss of populations or changes in year-class strength when enough individuals are affected. Verification of behavioral effects in the field is an important step in understanding the causal relation between observed behavioral changes and the impact of contaminants on natural populations and communities (Sandheinrich and Atchison 1990). However, relating behavioral change to higher ecological organization in the field poses significant technical challenges for behavioral toxicology. Attempts to verify behavioral responses in the field have been few and usually rely on indirect methods, such as mark and recapture or telemetry, to infer habitat selection or avoidance responses, and stomach content analysis to support hypotheses about foraging behavior. In addition, there can be a considerable time lag between the behavioral response to the stressor and the manifestation of that response in the population or community. Experimental designs are difficult because of the mobility of the organisms, physical constraints of the environment, verification of exposure, and manpower required to accomplish such determinations (Little 1990).

General Considerations in the Design of Behavioral Tests

No behavioral procedures have been rigorously standardized to the extent, for example, that toxicity bioassays have been. There are, however, general and appropriate guidelines for exposing fish to chemical substances and for conducting certain behavioral measurements provided by the American Society for Testing and Materials (ASTM 1998a, 1998b, 1998c, 1998d) and original research publications. Precise, objective, operational definitions are required of behavioral endpoints measured during tests of chemical and nonchemical stressors. The primary data analyzed from behavioral observations made during laboratory tests usually include frequency, proportion, magnitude, or presence (or absence) of the behavioral response, as well as latency, duration, or sequence of activities. Measurements of variables of growth, mortality, reproduction, development, morphology, histology, and physiology or biochemistry, and the concentration of test materials in water, sediment, or the biota are often conducted in addition to behavioral observations.

Variables Influencing Behavioral Responses

A number of experimental variables can suppress, elicit, or alter behavioral responses and, thus, influence test results and their interpretation. The following factors should be considered when measuring behavioral responses during toxicity tests (ASTM 1998c, 1998d):

- Pretest handling of test organisms resulting from collection, transfer, and maintenance of culture environment
- Health, nutritional state, and physical condition of the organism
- Prior exposure to hazardous materials, environmental stresses, and pathogens
- Social status, such as dominance or sex of the individuals tested, and experiential factors, such as prior experience with predator or prey species
- Apparatus design and procedural sequence of the method of measurement
- Acclimation to the physical variables of the testing environment including water quality, temperature, water flow, light, cover, and substrate, as well as recovery from handling, acceptance of diet, and adjustment to novel testing chambers
- Furthermore, behavioral responsiveness usually varies by species, genetic strain, population, gender, and developmental stage of the organism. Cyclical changes (circadian, seasonal, annual, hormonal, reproductive) in behavioral responses can occur. Individuals tested in isolation may respond differently than when tested in groups. Consequently, the behavioral expression of toxic responses can be delayed or can subside over time.

Species Selection

When possible, species and life stages selected for study should include species known to occur at the affected site. If this is not possible, then a representative species should be selected, based on phylogenetic relationship, habitat (e.g., pelagic, benthic), or functional relationship within the aquatic community (e.g., predator, grazer). The species and life stage selected should be appropriate for the experimental setting and apparatus, tolerant of handling and confinement within a reasonable acclimation time, tolerant to site-relevant ecological conditions such as temperature or sediment grain size, and able to accept food under culture conditions. The organisms must be acclimated to testing conditions, including water quality (ASTM 1998c).

Organisms may be obtained from (a) laboratory cultures, (b) commercial, state, or federal institutions, or (c) natural populations from clean areas. Test organisms must not be diseased or injured. All organisms in a test must be from the same source and should be as uniform as possible in age and size-class. Laboratory cultures of test species are often preferable because their source, age, and history of contaminant exposure and disease are known. However, wild species can provide information about adaptation and tolerance to site-specific conditions and potential additivity among stressors. The relative health and quality of test organisms can be verified through an assessment of their behavioral repertoire and bioassays in response to reference toxicants. The taxonomic identity of test organisms obtained from field sites must be confirmed and may require local and state collecting permits. To maintain organisms in good condition and avoid unnecessary stress, the test organisms should not be crowded and should not be subjected to rapid changes in temperature or water quality characteristics (ASTM 1998c).

Study Design for Exposure to Chemical Substances

In accordance with ASTM Guidelines (ASTM 1998c), control treatments include dilution water or solvent water controls (or both) to which no test material has been added or which contain no sediments or effluents collected from reference sites. It is important that the reference sites have been sufficiently characterized to ensure that minimum contamination exists. In exposure tests, a geometric series of at least five concentrations are typically used, with each concentration being at least 50% of the next higher concentration. Tests employing numerous treatments over a broad range of concentrations are preferred since they also provide information on dose–response relationships.

In tests of single chemical compounds, the range of concentrations selected should also include sublethal concentrations that are expected to occur in the environment. When testing effluents, a 50% dilution series is used where water from an upstream or reference site is the diluent. Toxicity tests of field-collected sediments should include sediments collected in reference areas along with areas adjacent to contaminated sites. The diluent and sediment grain size should be of similar quality as that collected in the field. When limited information is available on the toxicity of the compound, sediment, or effluent, preliminary exposures should be conducted to establish relative lethality of the toxicant (ASTM 1998c).

Organisms should be randomly assigned to treatment groups, and individuals should be randomly sampled for behavioral responses during the exposure tests. The duration of exposure will depend on the chemical, species, and behavioral endpoint selected for study. Behavioral responses may be measured continuously or after selected intervals of exposure, especially during biomonitoring or during avoidance–attractance tests. Measurements of responses can be made during exposure, as well as during recovery to determine the stability of the response over time, as well as the extent to which the behavioral response recovers or that delayed effects occur (ASTM 1998c). The measurement of multiple endpoints will enhance the characterization of injury induced by the hazardous substance providing weight-of-evidence needed for environmental regulatory and management decisions.

Test Acceptance Criteria

Generally, excessive mortality among control organisms, high variability in the behavioral response of control individuals, disease, or variation in water quality or experimental parameters beyond acceptable limits are the basis for rejecting a behavioral test. The criteria for such limits will vary depending on the substance, species, and response being tested, as well as the objectives of the study. A behavioral toxicity test should be considered unacceptable if one or more of the following has occurred (ASTM 1998c, 1998d):

- All test chambers (and compartments) were not identical or were not treated as separate entities
- The exposure water was not acceptable to the test organisms
- Appropriate controls or effluents were not included in the test
- The solvent concentration in the control treatment affected survival, growth, or reproduction of the test organisms
- All animals in the test population were not obtained from the same source, were not all of the same species, or were not of acceptable quality
- Treatments were not randomly assigned to individual test chamber locations, and the individual test organisms were not impartially or randomly assigned to test chambers or compartments
- Each test chamber replicate did not contain the same amount of water determined either by volume or weight
- Temperature, DO, and concentration of test material were not measured or were not within the acceptable range
- Organisms exposed to negative controls and effluents did not survive, grow, or reproduce as required for test organisms
- Behavioral responses measured during the toxicity test were ambiguously defined
- More than 20% of the control organisms failed to respond or were abnormal in their behavior

- Variability of the behavioral measurement for controls exceeded 50% of the mean value

Behavioral Measures of Injury

The demonstration of environmental injury includes behavioral impairments that would clearly lead to reduced survival and diminished long-term viability of the population. Behavioral data can also demonstrate injury through adverse effects to reproduction, development, and viability by altering essential life functions that include habitat selection, competition, feeding, predator avoidance, and reproduction. Disruption of these essential functions can become apparent through loss of populations or reductions in reproductive success and viability of the progeny year-class. The following discussion describes the more common and established behavioral functions that have been measured on a variety of species or environmental stressors.

Avoidance Behavior

Appropriate habitat is critical to the survival of individuals and their offspring. Many aquatic species have specific habitat needs for survival and reproductive success. Habitat selection can be inhibited through physiological impairments in the ability of the organism to perceive and respond to environmental stimuli, as well as from altered environmental cues that result when a hazardous substance or condition masks, mimics, or inhibits responses to the habitat (Sutterlin 1974). Adaptive behavior allows fish to avoid areas containing unfavorable conditions and to occupy acceptable areas. Fish may respond to noxious, aversive habitat conditions in a manner that minimizes their exposure to them. In contrast, a chemical inducing attractance would increase exposure and probability of injury or death. Avoidance–attractance can significantly influence the level or duration of exposure experienced by the organism through contact with water, sediment, or diet (Beitinger 1990).

Many contaminants induce avoidance responses, and several reviews have been made of this extensive literature (Cherry and Cairns 1982; Giattina and Garton 1983; Hara et al. 1983; Beitinger 1990). The sensitivity of avoidance responses ranged from less than 3% of the LC50 (concentration lethal to 50% of the test population) for the herbicide, 2,4-D, to more than a 1,000 times the LC50 for the chlorinated biphenol, arochlor 1254 (Little et al. 1985). Generalizations regarding the avoidance of aquatic contaminants by fish are difficult to make because of the variety of species and experimental designs used to test behavioral responses, as well as variations in the modes and sites of action of the chemicals studied (Giattina and Garton 1983). In his review of published avoidance data for over 75 different chemicals, Beitinger (1990) found that roughly a third of the chemicals were avoided, whereas the others either failed to elicit a response or induced inconsistent responses. Many contaminants cause avoidance reactions, but some, including detergents (Hara and Thompson 1978) and certain metals (Timms et al. 1972; Kleerekoper et al. 1973; Black and Birge 1980), may attract aquatic organisms.

Behavioral avoidance studies are advised when the suspect hazardous substance or adverse environmental condition is known to cause avoidance responses. Avoidance studies are also recommended when fish are absent from sites adjacent to a suspected or known source of a hazardous substance or perturbation that would otherwise support such species. Avoidance tests should also be conducted to monitor mitigation and habitat recovery efforts. Representative discharge or receiving water samples must be collected from the site in order to characterize the chemical quality of the water, as well as to identify contaminants present.

Methods for Measuring Avoidance

Laboratory studies of avoidance behavior should be conducted, if possible, on species known to occur at the site, otherwise a representative species should be selected, based on phylogenetic relationships or functional relationships within the aquatic community.

A range of apparatuses can be used to evaluate the avoidance response, including countercurrent chambers, linear gradient, and circular gradient, and Y-maze apparatus (Rand 1985; Steele et al. 1996). The chemical concentration gradient throughout the apparatus must be fully characterized, because it is critical that unequivocal contaminant gradients are provided.

Although the testing of contaminants is emphasized here, the testing approaches described below can be used, with appropriate modifications, to test avoidance reactions to other environmental stressors associated with habitat alteration including temperature, DO, turbidity, light, salinity, substrate, rate of flow, electromagnetic fields, or noise. Application of the test substance or conditions should be sufficiently randomized so as to control for other variables that may influence the spatial selection of the test organism. This procedure may include randomized switching of the apparatus position into which the chemical or other perturbation is introduced and tests to determine position bias. The monitoring of fish position must be unintrusive, but must enable accurate location of the test organism within the test apparatus over time. Critical temporal parameters that are a basic feature of the experimental design include formation of a stable gradient, acclimation of the test organism, and stable behavioral response. The substance should be applied over a range of concentrations in order to understand the mode of response elicited. These tests determine the minimum concentrations that elicit an avoidance or attractance response. Such concentrations can then be compared with measured or estimated environmental concentrations.

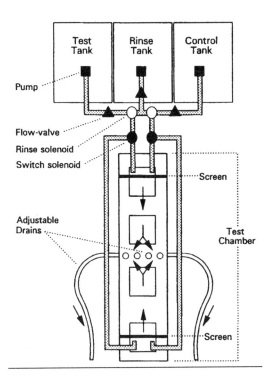

Figure 2. Counter-current avoidance apparatus used to determine the avoidance response of brown trout to metals contamination in the Clark Fork River (Montana). Reference and contaminated water are introduced a opposition ends of the chamber that drains from the center to form a steep gradient. (DeLonay et al. 1995).

DeLonay et al. (1996) used a cylindrical Plexiglas counter-current chamber (approximately 11 cm diameter × 92 cm long; Figure 2) similar to one described by Sprague (1964). Control and test solutions entered the chamber from opposite ends and drained from six adjustable drains in the center. Stainless steel screens placed at each end of the chamber reduced water surface disturbance and established a uniform laminar pattern of water flow. Spectrofluorometric studies with dye revealed that when properly adjusted and calibrated, this apparatus produced a uniform, steep gradient, closely confined to the central drain area. The bottom and sides of the chamber were covered with self-adhesive contact paper to provide a uniform, opaque background. Overhead and side fluorescent lighting eliminated shadows and provided uniform illumination of the chamber.

In the standardized procedure used by DeLonay et al. (1996), the avoidance testing sequence began with a 20-min rinse period to remove all residual traces of the previous test solution. After this period, a single rainbow trout was placed in the test chamber where it was allowed to acclimate for 40 min. The fish was monitored to ensure acclimation to the chamber. At the conclusion of the acclimation period, control and experimental solutions were introduced at randomly selected ends of the chamber during the 40-min test phase. The last 10 min of the acclimation period and the first 30 min of the test period were recorded on videotape. The response of fish was quantified from a 20-min sample of the video recording beginning 10 min after the initiation of the test period. The tests are replicated using a total of 8 to 10 fish. Measurements to quantify behavior included the number of times each fish crossed into the treatment end of the chamber, the average time per trip, and the cumulative percent time spent in the treatment and control ends of the chamber. Responses observed during the acclimation period ensured that the fish was appropriately responsive and provided data on spatial selection in the absence of the test substance. Avoidance was judged to have occurred when fish spent a significantly (statistically) smaller percentage of total time in the treated area than in the control area.

Variables Influencing the Avoidance Response

In addition to the factors listed in the section Variables Influencing Behavioral Responses, avoidance reactions are influenced by environmental variables such as contaminant concentration gradient, temperature, light, water flow, substrate, or cover (Kleerekoper 1976). Kleerekoper and his colleagues found that when gradients were shallow, goldfish *Carassius auratus* were attracted to copper concentrations between 11 and 17 mg/L (Kleerekoper et al. 1973), but when the gradient was steep, fish avoided concentrations as low as 5 mg/L (Westlake et al. 1974). Kleerekoper et al. (1973) examined the interaction of temperature and copper and found that goldfish no longer avoided 10 mg/L copper when the water temperature of this treatment was increased to 21.5°C. Giattina et al. (1981) studied the interactive effects of heated and intermittently chlorinated effluents on spotfin shiners *Cyrinella spilopterus* downstream from a power plant. They found the threshold chlorine concentration for avoidance increased from 0.24 mg/L to a maximum of 0.38 mg/L when fish acclimated to 12°C had the opportunity to simultaneously select a preferred temperature (21°C). Apparently, the preference of cold-acclimated fish for warmer water overrode the avoidance response and resulted in chlorine exposures that approached the LC50 values (0.41–0.65 mg/L) for residual chlorine. When a competing gradient between light and temperature was present during an avoidance test, fish remained in preferred darkened areas even at temperatures that had been avoided under uniform light conditions (Sullivan and Fisher 1942, 1954). After ingestion of mercury, walleye *Stizostedion vitreum* abandoned the shaded habitat they normally prefer (Scherer et al. 1975). Similarly, lake whitefish *Coregonus clupeaformis* avoided copper (1 mg/L), as well as lead and zinc (10 mg/L); however, when shade was provided in the test apparatus, higher test concentrations (72 mg/L copper; 3.2 mg/L lead; 1000 mg/L zinc) were required to induce avoidance (Scherer and McNicol 1998). The addition of shelter to breeding male fathead minnows *Pimephales promelas* increased by more the six-fold (from 0.28 mg/L to 1.83 mg/L) the threshold concentration at which they avoided zinc (Korver and Sprague 1989).

Organisms are often tested individually, but such isolation can be stressful for schooling fishes (e.g., early juvenile rainbow trout or fathead minnow) and, in such cases, the fish are likely to be more responsive if tested in small groups. When testing small groups, the group response is evaluated rather than the individual response. Care must also be taken to ensure that there are no aggressive individuals within the group that might influence spatial distribution.

Ecological Significance and Application

Avoidance reactions to contaminants are ecologically significant responses because the initiation of such responses can result in locally reduced biomass, loss of species diversity, diminished year-class strength, and loss of productivity and diversity for the impacted site (Lipton et al. 1997). Avoidance can be detrimental when organisms are displaced from preferred habitats to suboptimal areas where they may face greater competition and predation pressure or inadequate resource availability (Atchison et al. 1987).

The avoidance response has been confirmed in a number of field studies. Early studies by Sprague (1964) confirmed that avoidance reactions by Atlantic salmon *Salmo salar* observed in the laboratory tests of copper and zinc solutions also occurred in the field (Saunders and Sprague 1967). Geckler and coworkers (1976) also confirmed avoidance reactions in copper-treated streams. Gray (1990), using telemetry, documented the avoidance of oil-contaminated water and gas-supersaturated water by free-ranging fish. Hartwell et al. (1987) conducted integrated laboratory and field studies of avoidance and showed that fathead minnows avoided a blend of heavy metals (copper, chromium, arsenic, and selenium) that are representative of effluent from fly ash settling basins of coal-burning electrical plants. Fish avoided a 73.5 μg/L mixture of these metals in a natural stream and similarly avoided 34.3 μg/L in an artificial stream. Avoidance behavior by cutthroat trout *Oncorhynchus clarki* was also observed in laboratory and field tests of metals characteristic of the Coeur d'Alene River (Idaho) downstream from a large mining extraction operation. Telemetry studies conducted at the confluence with an uncontaminated tributary of the river, revealed a similar avoidance of the contaminated water within the concentration range that induced avoidance responses in laboratory studies (Woodward et al. 1997). DeLonay et al. (1996) conducted avoidance tests with brown trout *Salmo trutta* using a metals mixture representative of the Clark Fork River (Montana) to assess the impact of metal contamination from mining. Concentrations as low as 10% of the average metals concentration measured in the river induced avoidance (Figure 3).

The avoidance response is a recognized index of environmental injury under Natural Resource Damage Assessment Regulations (1986, NRDAR)

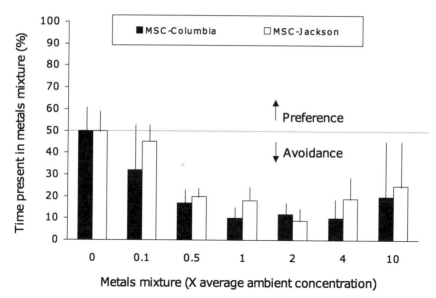

Figure 3. Brown trout avoidance reaction to various concentrations of metals mixture characteristic of the Clark Fork River (Montana). Note that the concentration designated as 1 is the ambient environmental concentration. Vertical lines above the bars represent 1 standard deviation. Avoidance studies were conducted a two different laboratories.

specified in the Comprehensive Environmental Response, Compensation and Liabilities Act (1980, CERCLA). Since the acceptance of avoidance behavior as a measure of injury, data on the avoidance response has been applied in NRDAR investigations, including Clark Fork River, Montana (Lipton et al. 1997), Coeur d'Alene River, Idaho (Woodward et al. 1997), and Panther Creek, Idaho (Marr et al. 1995). The avoidance response also provides effective measures of response to potentially noxious variables arising from habitat alteration. Such studies would be applicable to investigations required by the Endangered Species Act of 1973, as well as regulations addressing point source impacts.

Feeding Behavior

Survival of early life stage fish will decline if the development of foraging behavior is delayed or inhibited. Impaired feeding behavior also correlates with reduced growth, which may lengthen a fish's period of vulnerability to predation (Werner and Hall 1974) and impair overwinter survival (Oliver et al. 1979). Because changes in the ability of a fish to detect, pursue, capture, and consume prey will affect growth and survival, these behavioral variables may provide an ecologically relevant measure of contaminant-induced injury, as well as changes in other environmental variables.

Contaminants have been shown to affect numerous aspects of the feeding sequence, including detection of prey (Lemly and Smith 1987), prey capture (Little et al. 1990), handling time, and ingestion of prey (Sandheinrich and Atchison 1990), as well as general motivation to feed (Little et al. 1990). Thus, several variables of feeding behavior can be measured in toxicity studies (Sandheinrich and Atchison 1990). Such studies can be readily adapted to standard toxicity testing procedures (Mathers et al. 1985).

Feeding behavior is impaired by sublethal exposure to a diversity of contaminants. Reductions in feeding have been observed after sublethal exposure to metals (Atchison et al. 1987), organophosphates (Bull and McInerney 1974), dioxins (Mehrle et al. 1988), and petroleum hydrocarbons (Woodward et al. 1987). Feeding behavior can also be affected by other environmental variables, especially DO concentrations, turbidity, and temperature. Feeding, measured as frequency of prey capture, was a sensitive endpoint in defining no-effect concentrations of aluminum in acid-exposed brook trout *Salvelinus fontinalis* and was among the most sensitive indices of sublethal pH and aluminum exposure (Cleveland et al. 1989). Feeding behavior was altered among rainbow trout exposed to various agricultural chemicals at concentrations that ranged from less than 0.3% to 50% of the LC50 (Little et al. 1990). Inhibited motivation to feed seemed to be the predominant effect at higher concentrations, whereas reduced feeding efficiency and reduced strike frequencies were the predominant impairments after exposure to lower concentrations.

Methods for Measuring Feeding Behavior

Feeding behavior includes variables such as orientation to the food; movement toward, striking, or sucking activities used to capture the material; oral contact with; and acceptance of the material as indicated by consumption or rejection (spitting) of the material. The latency of response to prey or food material and the maximum distance from which the organism responds to prey are critical variables to measure for this test. In addition, prey selectivity, feeding efficiency, prey-handling time, and strike and capture frequencies are typically measured. Procedures described by Little and DeLonay (1996) provide for the measurement of feeding behavior at intervals during standard toxicity test protocols. These tests are usually conducted via overhead video recordings so as to minimize handling. Standard amounts of live prey are added to the exposure chamber and feeding behavior is recorded for intervals ranging from 5 to 30 min. Prey strike and capture frequencies are measured during analyses of video playback.

Testing procedures for reaction distances described by Sandheinrich and Atchison (1989) are conducted in observation chambers (100 cm long by 10 cm wide by 10 cm deep) with a ruled background where fish are observed for four days prior to and during contaminant exposure. Fish are acclimated for 5 min after being placed in the observation chamber each day, and then prey of different sizes are dropped in the opposite end of the chamber. After sighting the prey, bluegill *Lepomis macrochirus* normally stop movement, orient toward the prey, then move directly toward the prey and consume it. Response to 10 prey is measured each day and analysis is conducted from videotape. The time required to manipulate and swallow prey (handling time) is measured in 60 L aquaria with known prey composition. A littoral zone is modeled by the inclusion of a sand substrate and polypropylene ropes that serve as artificial vegetation. Fish are provided with live prey species, which range in abundance, size, and nutritive value. Measurements conducted during these tests include species captured, reaction distance, and prey-handling time. Data recorded from video recordings include number of prey attacked, number of prey captured, number of spits and misses, handling time per prey (time from prey capture until search is reinitiated), and capture efficiency (number of strikes per prey captured or time to capture prey per prey captured).

Variables Influencing Feeding Behavior

Culture conditions, health, and acclimation of the organism can significantly influence the responsiveness of the organism during feeding tests. Acclimation both to the water quality and temperature of the testing condition is essential as is the acclimation of the organism to the testing apparatus and sufficient recovery from handling. Generally, the resumption of species-typical locomotory activity and movement about the observation chamber is a good indication of acclimation. Satiated fish are not as likely to be responsive to food as food-deprived fish, therefore, a period of food deprivation ranging from 24 to 72 h, depending on the size and species of the test organism, is necessary. A prey item is often a living organism, such as another fish species or an invertebrate that is common to the diet of the fish. In some reports, prey items have also included fish food granules, flakes, or pellets, or pieces of organisms (worms, beef, etc). Fish generally require some experience with the prey item before it is readily accepted as food, so feeding them the prey item over time should precondition fish. The condition of the prey species may also influence the test. Prey organisms may be exposed prior to the feeding test as a part of the test design. Prey may also become affected by exposure when they are introduced to the test chamber during the test.

Ecological Significance and Application

Feeding behavior is of obvious significance relative to the development, fitness, and long-term viability of the organism. The response is clearly relevant to the assessment of environmental stressors such as sublethal contaminant exposure. Procedures for quantifying contaminant impacts on feeding behavior have been developed for numerous species. The response is clearly affected by a range of contaminants and other environmental factors, and the impact of inhibited feeding behavior on growth, development, viability, and subsequent reproductive success is clearly understood. Changes in feeding behavior can be translated to effects on important individual organism functions through application of bioenergetic models (See Chapter 8) or can be used to predict effects at the population level using individual-based models (See Chapter 10). Evaluation of the response is effectively assessed in laboratory investigations. Feeding responses in free-ranging fish are largely limited to an assessment of gut content. Future advances in dietary DNA analysis will greatly increase understanding of prey selection and increase the utility of feeding assessment as measures of resource injury.

Measures of the feeding response of fish can be used in determining no-effect concentrations of chemical substances useful in setting water quality criteria. Measures of feeding are effective in assessing the toxicity of complex mixtures in effluents during in situ or laboratory tests. Such responses can also be used to gauge the success of mitigation efforts. The recent biennial review of the newly revised National Resource Damage Assessments (NRDA; L. Cleveland, U.S. Geological Survey, personal communication) has recommended that feeding behavior be included as a weight of evidence of resource injury in NRDA investigations. Feeding tests are also applicable to the Endangered Species Act and other laws regulating point-source discharge.

Predator Avoidance

Practically all organisms are vulnerable to predation during some portion of their life cycle. Predation is a frequent cause of mortality in the field. Predation tests measure the ability of prey to escape predation. Increased vulnerability to predation may occur when a toxicant or other environmental condition alters the ability of fish to detect or respond to predators (Brown et al. 1985).

Assessment of predator–prey behaviors during toxicity tests provides not only a sensitive measure of toxicant effect, but also a measure of injury, since predation-induced mortality is affected by exposure. A range of contaminants has been shown to increase predation-induced mortality. Little et al. (1990) reported that different types of toxicants at concentrations as low as 2% of the lethal concentration significantly increased predation on exposed prey. The response did not consistently follow a dose–response relationship; however, because the behavioral aberrations resulting from the exposure sometimes made prey less conspicuous to predators due to inactivity or reduced mobility. In tests with bluegill exposed to methyl parathion (Little, personal observation), a 50% decrease in brain cholinesterase that occurred among exposed prey (bluegill) was associated with a 40% decrease in swimming activity and heightened rates of predation (Figure 4). Contaminant exposure may also cause aberrations or inhibitions of other defensive responses such as schooling or shelter-seeking that protect organisms from predation.

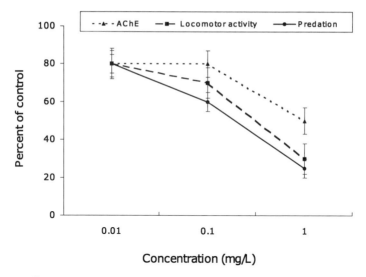

Figure 4. Effects (mean ± SE) of methyl parathion on acetylcholinesterase (AChE) and behavior of bluegill. Predation survival is reported as a percent of control.

Methods for Measuring Predator Avoidance

A common laboratory approach for measuring predation combines predators with contaminant-exposed and unexposed prey in testing chambers. The test chamber should not have corners, especially in glass chambers where fish are visually confused and easily trapped. In nature, such conditions would be rare. Substrate, vegetation, or physical structure should also be included to provide refugia and increase structural complexity and environmental realism of the experimental setting. Equal numbers of prey from each treatment group are added to the observation tanks. Dye marks, fin clips, or freeze branding differentiates prey from different treatment groups. Census of the surviving control and exposed prey are made when approximately 50% of the prey population has been captured (Little et al. 1990).

Variables Influencing Predator Avoidance

Success of predator avoidance tests is greatly affected by the appetite and experience of the predator. Therefore, it is necessary to have sufficient background data on the feeding response to ensure that the organism is sufficiently responsive to consume 50% of the prey organisms within the allotted time period, but not so efficient that the test cannot be terminated before more than 50% are eaten. Most tests are conducted with naïve prey fishes because experience with the predator significantly improves their ability to avoid predation. Observation of the predator–prey interaction is also important to the interpretation of results. For example, in tests with an organophosphate insecticide, we found that the high treatment group had better survival than other treatment groups because fish of the high treatment group remained motionless at the surface and apparently failed to attract the predator's attention (Little et al. 1990).

Ecological Significance and Application

Predator avoidance is most effectively evaluated in controlled investigations. Investigations of free-ranging organisms are largely limited to gut content analysis of predators and the appearance of grossly atypical responses among affected organisms that would preclude an increased vulnerability to predation. The biennial review of the NRDA Guidance for Injury to Fish and Wildlife (Cleveland, personal communication) recommends that laboratory observations of predation avoidance be included as weight of evidence measures of resource injury. Predation tests are also appropriate for assessing the impacts of nonindigenous or exotic species.

Social Behavior

Nonreproductive social interactions among fish range from highly synchronous schooling responses to isolated territoriality. These social responses are highly adaptive responses serving a variety of purposes ranging from protection to resource acquisition. Schooling and a less organized social aggregation, shoaling, are highly evolved social responses among fishes that provide protection from predators, as the number of fishes in an aggregate increases surveillance of the three-dimensional space of their habitat and provides shelter as individuals hide among one another. Schools are quick to respond to threatening situations by decreasing space between individuals. The increased movements of the schooling individuals may confuse the predator and hinder the predator's ability to focus and attack on a specific individual (Mueller 1972). The schooling response is also important for feeding in some species and for reproduction in others. Shaw (1978) estimates that over 50% of fish species exhibit a schooling response during some portion of their life cycle.

In contrast, social isolating mechanisms, including territorial or hierarchical responses, may be seen in other fish species. Aquatic organisms often share the same habitat, and competition among species or age-classes may result when resource needs overlap. One species may gain a competitive advantage over another by more efficient exploitation of those resources or through aggressive interactions (Schoener 1974). Changes in prey selection (Werner and Hall 1974), spatial distribution (Finger 1982), or diurnal patterns of activity (Reynolds and Casterlin 1976) may segregate competitors in time and space. Contaminant exposure may indirectly influence competition between and within species by reducing a resource such as food or vegetative cover. Loss of food, such as a contaminant-intolerant prey, may cause a shift in food preference and intensify competition. For example, decreases in the natural food supply have been shown to cause an increase in intraspecific aggression and the size of territories (Kalleberg 1958).

In many species, a range of social responses can be seen over the life cycle of the organism. Rainbow trout, for example, display shoaling response shortly after the onset of exogenous feeding, but the response subsides as they approach smoltification. As with other behavioral functions, social responses can be altered by environmental contamination. To the extent that such responses have significant survival value, such alterations may limit the long-term viability of the organism.

The measure of social behavior has not been broadly applied in toxicity tests nor have social responses been considered in many species. Schooling behavior declined following exposure of coho salmon *Oncorhynchus kisutch* to DDT (Besch et al. 1977) and also for fathead minnow to the herbicide 2,4-D (Holcombe et al. 1980). During tests of 400 different chemical substances, the schooling response of fathead minnow was consistently affected by exposure (Drummond et al. 1986).

Contaminant exposure can also increase or decrease aggression. Aggression among bluegill increased during sublethal exposure to a metals mixture (Henry and Atchison 1979) and to pesticides (Henry and Atchison 1984; Fairchild and Little 1999). Coho salmon became more aggressive during sublethal exposure to an organophosphorous pesticide (Bull and McInerney 1974). Aggression was reduced among bluegill exposed to a pyrethroid insecticide (Little et al. 1993b), and reduced aggression was associated with the loss of spawning territories among Atlantic salmon exposed to the pesticide fenitrothion (Symons 1973).

Methods for Measuring Schooling Behavior

Critical variables in the schooling response include distances between individuals, their orientation within the school, and the latency with which the school forms. The schooling response is readily measured in laboratory studies through photographic or video recording methods for measuring the spatial extent of the school and for measuring the rapidity with which the school forms following presentation of a threatening stimulus. Computer-assisted methods for measuring the schooling response of Atlantic silversides *Menidia menidia* are also described (Koltes 1985). Standardized procedures can be readily devised for specific species with measures of schooling responses being conducted in the laboratory setting.

Methods for Measuring Aggressive Behavior

Variables related to aggressive responses include changes in posture, coloration, body movements toward, or contact between conspecifics that results in the displacement of one individual. However, these variables are most commonly used to measure the frequency and magnitude of aggressive interactions. Aggressive interactions are interpreted as all interactions between two fish where the movement of one fish results in the movement or reorientation of another away from the first (Henry and Atchison 1979). Displacement can include rapid retreat from an area, change in position within the water column, or reduced individual distance (i.e., the characteristic three-dimensional volume of space surrounding an individual). Bodily contacts include bites, as well as nudging or pushing of one individual against another. An evaluation of aggressive behavior includes measuring the number of aggressive interactions during a limited observation period (5–10 min).

Aggression is a graded response, the intensity of which reflects varying levels of social stress (Henry and Atchison 1979). Approach is the lowest level of aggression and is the movement of one fish toward another that results in the retreat of the second fish. Chase is the rapid advance of one fish with concomitant rapid retreat of the second fish. Nudge is an intentional physical snout contact of an advancing fish against the body of another fish. Bite, the highest level of aggression, is open-jawed contact of an advancing fish with the body of another fish. Levels of aggression can be qualitatively scored to estimate the average intensity of interaction observed for a group of interacting fish, with 1 being a low intensity of aggression, with a predominance of approach–avoidance interactions, and 4 being a high intensity of aggression with a predominance of bites (Little et al. 1996).

Variables Influencing Social Responses

Social responses can be influenced by factors listed in the section Variables Influencing Behavioral Responses. The schooling response can be affected by changes in water quality including DO concentrations, ammonia concentrations, pH, and temperature. In some species, schooling follows diurnal periodicity and may be influenced by light intensity. The schooling response is frequently observed among early life stages of many species, but the response disappears as the organisms develop, thus, the developmental stage of the fish is a major variable to be considered in the environmental design. Aggressive behavior can be strongly influenced by the density of test organisms, the size distribution of test organisms, and the sex and social experience of individuals. Aggressive behavior will vary with species.

Ecological Significance and Application

The schooling aggregation affords early life stages of many fish species a degree of protection from predators. When the schooling response degrades, predators are more readily able to focus foraging attacks against individuals within the school. For example, during exposure to a petroleum product, early life stage inland silversides *Menidia beryllina* became less responsive to external stimuli (Little et al. 2000). Distances between individuals increased and individuals spent longer periods of time at the periphery of the school. Thus, they would be more conspicuous to the predator, less responsive to its approach, and more vulnerable to predation. Aggression is often a natural consequence of competition, serving to disperse individuals and to ensure access to habitat resources such as food or shelter. The loss of aggression can result in the abandonment of nests or failure to reproduce. Heightened aggression, on the other hand, can increase stress, reduce bioenergetic efficiency, cause physical injury and mortality, and affect species diversity. Measures of the impacts of contaminants on aggressive behavior have been limited to laboratory tests. However, territorial responses could be monitored in the field under favorable circumstances. For example, a major degradation of schooling may be evident at sites of discharge from direct visual or sonar observations, however, the means of measuring the response are not well developed. The consequences of altered aggression include loss of access to resources, increased stress, injury, and displacement to marginal habitats. However, a linkage of competition with resource injury is not sufficiently understood relative to direct impacts on growth, viability, or population status.

Measures of social behaviors are appropriate for evaluating no-effect concentrations of chemical substances or toxic effluents useful in setting water quality criteria. Measures of schooling or shoaling are effective in assessing the toxicity of complex mixtures in effluents during in situ or laboratory tests. Such responses can also be used to gauge the success of mitigation efforts. The evaluation of competition is recommended when competitive species are present at an impacted site or when there is an indication of aggressivity such as abandoned nesting areas or physical trauma. The evaluation of aggression is particularly important when conducting toxicity tests with socially aggressive species since the social stress associated with such interactions can interact with the stress of the toxicant exposure (Henry and Atchison 1979, 1984). The recent revision of the NRDA guidelines recommends the measure of social behaviors, including schooling responses and aggression, as weight-of-evidence for injury to fish and wildlife (Cleveland, personal communication).

Swimming Behavior

Locomotory activity is fundamental to every aspect of behavioral performance in fish. Swimming behavior includes such variables as frequency and duration of movements (Cleveland et al. 1989), speed and distance traveled (Miller et al. 1982), frequency and angle of turns (Rand 1977), position in the water column, form and pattern of swimming, and the orientation to water flow (Dodson and Mayfield 1979), as well as the capacity to swim against the current (Howard 1975; Kolok et al. 1998). The culmination of these variables constitutes a species-typical locomotory response that reflects a highly selective behavioral pattern fundamental to feeding, competing, avoiding predators, and reproduction. Pattern analysis of swimming behavior involves quantification of the component variables such as the radius of turns and frequency of turning behavior, the distance of linear travel between turns and the speed of travel, as well as the frequency and duration of movements (Kleerekoper

1974). Swimming behavior varies among species, as well as among life stages within a species; therefore, test methods must be tailored to a particular life stage of a single species.

Swimming behavior is a well-studied measure of toxicant effect and has become one of the most common behavioral measures in toxicity investigations. Changes in swimming behavior often result from sublethal exposure to contaminants and may impair the ability of fish to feed, avoid predation, or reproduce (Little et al. 1985). Such behavioral modifications provide an index of sublethal toxicity and also indicate the potential for subsequent loss of physiological performance and mortality. A review by Little and Finger (1990) revealed that the lowest behaviorally effective toxicant concentration that induced changes in swimming behavior of fish ranged from 0.1% to 5.0% of the LC50. Swimming behavior is affected by many chemical stressors, including metals, organochlorines, and industrial contaminants. When observations were made over time, changes in swimming behavior were commonly evident 75% earlier than the onset of mortality. Swimming behavior is also often affected before reductions in growth are detected. For example, swimming activity was significantly reduced in rainbow trout after a 96-h exposure to an organophosphorus defoliant at concentrations that only affect growth after 30 d (Little et al. 1990). Swimming performance or the physical capacity of fish to swim against water flow is also often significantly impaired by contaminant exposure (Little et al. 1990). Development of locomotory responses, frequency of movements, and duration of activity were significantly inhibited in brook trout alevins at lower aluminum concentrations under acidic conditions than those that affected survival or growth (Cleveland et al. 1991).

Because swimming variables are interrelated, several aspects of swimming behavior may be simultaneously affected by a toxicant, therefore, analysis of several variables of swimming behavior can be conducted concurrently to increase the sensitivity of the measurement. For example, a 28-d exposure of rainbow trout to dioxin (29 pg/L) reduced the frequency of activity, changed the posture of swimming, and affected positioning in the water column (Mehrle et al. 1988). Similarly, simultaneous declines in the distance traveled, frequency and duration of activity, and swimming capacity of brook trout occurred after exposure to aluminum (300 μg/L) in water with an acidity of pH 5.5 (Cleveland et al. 1989).

Methods for Measuring Swimming Behavior and Capacity

Grossly altered swimming behavior can be qualitatively assessed during toxicity tests. Such responses may be apparent as changes in water column position (e.g., surfacing, resting on bottom), swimming posture (e.g., head-up swimming), body movements (e.g., increased or decreased waveform of body movement), or swimming patterns (e.g., repetitive turns, spiraling) and, in extreme cases, a loss of coordination, convulsive movements or loss of equilibrium may occur. The all-or-none occurrence of these responses has been successfully used to qualitatively assess behavioral toxicity syndromes (Drummond et al. 1986) and to describe behavioral effects of dioxin exposure (Mehrle et al. 1988).

Hypoactivity and hyperactivity have been measured as the frequency of entries and exits through a dividing partition (Ellgaard et al. 1978) or number of grid lines crossed in a testing chamber. Frequency of activity of larval striped bass *Morone saxatilis* was measured as the number of visually detectable body movements that resulted in forward movement of the larva (Finger and Bulak 1989) and, also, as rate of change of direction (Smith and Bailey 1988). Duration of time in which fish are active (Little et al. 1990) or inactive (Finger et al. 1985) has also been measured to determine frequency of activity. In a study with rainbow trout (Little et al. 1990), duration of swimming activity was consistently reduced by each of 6 chemicals after 96-h sublethal exposures. This response was sensitive at 0.5% to 50% of the LC50 values for chlordane, pentachlorophenol, and the organophosphorus defoliant DEF.

The frequency of tremorous or convulsive movements can also be quantified (Figure 5). When bluegill received pulsed doses of the pyrethroid insecticide ES-fenvalerate, the first indication of toxicity was a tremorous tail motion as bluegill initiated movement (Little et al. 1993b). Exposure to sublethal concentrations of malathion resulted in convulsive movements in rainbow trout (Brewer et al. 1999). Stereotypic body movements that do not result in locomotion also reflect toxicant exposure. Fin flickering, a groom-

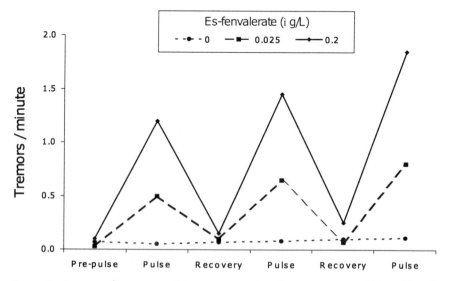

Figure 5. Average frequency of tremorous swimming responses of bluegill following pulsed insecticide (ES-fenvalerate) exposures. (Little et al. 1993).

ing motion of bluegill, showed strong dose–response relations with increased concentrations of methyl parathion (Henry and Atchison 1979). In addition, the frequency of stereotypic movements of the fathead minnow decreased during exposure to acidic water (Lemly and Smith 1987).

Pattern analysis of swimming behavior involves quantification of the component variables such as the radius of turns and frequency of turning behavior, the distance of linear travel between turns and the speed of travel, as well as the frequency and duration of movements (Kleerekoper 1974; Smith and Bailey 1988; Baatrup 1998). The assessment of these variables usually requires specialized equipment because measurements can be extensive. For example, Rand (1977) used a high resolution photodetection apparatus to determine that frequency of turning behavior and the angle of turns increased for bluegill and largemouth bass *Micropterus salmoides* during sublethal exposure to parathion. The catfish *Arius felis* also exhibited broader turning angles following 72-h sublethal exposures to copper (0.1 mg/L) (Steele 1983). Analysis of computer-processed video images by Smith and Bailey (1990) detected decreased speed of movement and angular velocity of steelhead trout (anadromous rainbow trout) within 7 min of exposure to phenol (8 mg/L).

Swimming capacity is commonly measured by subjecting fish in a current of water to incrementally increased water velocity until the fish become fatigued (Beamish 1978). Cutthroat trout exposed to Wyoming crude oil were unable to maintain position against water current and were swept downstream (Woodward et al. 1987). Sublethal exposure to fenitrothion (1.5 mg/L) reduced the stamina of brook trout (Peterson 1974). Altered swimming performance of gulf killifish *Fundulus grandis* was correlated with contaminant-induced changes in neurotransmitter concentrations in the brain after exposure to PCBs (Fingerman and Russell 1980). A number of apparatus have been developed for measuring this response. Modifications of a recirculating chamber designed by Brett (1964) are commonly used to measure swimming capacity. A system used for testing small fish consists of a 11.5 cm diameter, 1.5 m long clear Plexiglas swimming tunnel with two pressure-sealed access hatches for the addition and removal of test organisms (Figure 6). The upstream end of the chamber is fitted with an expansion cone followed by an 8-cm honeycomb of small diameter tubes to create a rectilinear velocity profile as water is recirculated through the chamber. A reduction cone is placed at the downstream end. Fine mesh screens at the upstream and down stream ends of the chamber confine individuals within the test portion of the swimming chamber. A centrifugal pump draws water from a temperature-controlled reservoir through the chamber. Pump speed is regulated to control water velocity. A programmable controller, interfaced with a digital display, allows the operator to monitor flow velocity and control water velocity and acceleration using a preprogrammed testing regime that subjects the fish to incrementally increasing water flow until fatigue occurs

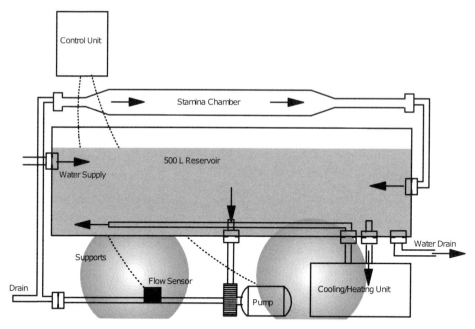

Figure 6. Diagram of swimming performance apparatus modified from Brett (1964).

and the fish are no longer able to swim. Swimming capacity is quantitatively scored by calculating the critical swimming speed (CSS) (Little et al. 1990 after Brett 1964):

$$CSS = u_i + (t_i / t_{iii} \times u_{ii})$$

where u_i is the highest velocity maintained for the last completed interval (cm/s), u_{ii} is the velocity increment (cm/s), t_i is the time fish swam at the "fatigue" or failure velocity in seconds, and t_{ii} is the prescribed length of the swimming interval in seconds. Critical swimming speed can also be reported in relative units or body lengths per second (LPS) by dividing CSS (cm/s) by total body length (cm) of the fish.

Variables Influencing Swimming Behavior

The influence of nonexperimental variables listed in the section Variables Influencing Behavioral Responses should be considered during the development of a behavioral method as a bioassay tool, with the goal of reducing variation and increasing precision of measurement. Maturity and physical condition of the test organism, the organism's social environment, and degree of exposure to other stressors, acclimation, and time of testing affect swimming behavior. Physical aspects of the testing situation such as temperature, water quality, illumination, or size and shape of the test chamber also affect results (Olla et al. 1980). Thus, tests involving behavioral responses require control treatments with rigorous uniformity of the test environment and handling regime of the test organisms. The response of each locomotory variable to a toxicant may change during exposure. Changes in orientation to water flow, spatial selection, and speed of movement may occur initially (Smith and Bailey 1988) because fish may be irritated by a toxicant or attempt to escape exposure. As these initial sensory reactions subside, more severe aberrations in locomotory behavior may occur as toxicants saturate detoxifying enzymes or alter neurochemical or metabolic processes (Drummond et al. 1986). Swimming capacity may also be affected by previous conditioning (Kolok et al. 1998).

Ecological Significance and Application

The ecological consequence of aberrant swimming behavior results primarily from impairment of associated adaptive behaviors such as migration, predation, or predator success. Hypoactivity and hyperactivity, as well as deviations in adaptive diurnal rhythmicity, can disrupt feeding and increase vulnerability to predation (Steele 1983). Heightened activity may increase an organism's vulnerability to predation, whereas reductions in swimming activity lessen the chance of encountering prey by reducing search areas (Laurence 1972). Deviations in swimming response can also interfere with the capture of prey or impair the efficiency of foraging behavior and decrease the amount of energy available for growth. Impairments in feeding behavior of fish associated with altered swimming behavior have also been observed after sublethal exposures to low pH and aluminum (Cleveland et al. 1989), metals (Drummond et al. 1973), organophosphates (Bull and McInerney 1974), dioxins (Mehrle et al. 1988), and petroleum hydrocarbons (Woodward et al. 1987), as well as DO concentrations and temperature. A decreased capacity in the physical ability of fish to swim has immediate bearing on their ability to pursue prey, avoid predators, avoid entrainment during high water flow events, and endure lengthy migrations. Swimming capacity is directly related to body size, thus, deficiencies in swimming capacity can be anticipated when growth is inhibited by contaminant exposure or degraded environmental conditions. Swimming capacity also reflects the bioenergetic efficiency of the organism to seek preferred flow regimes, depths, and temperatures that enable the most efficient use of energy.

For any aquatic species, the norm for swimming activity has not been defined in terms of a determined value or range of values. Thus, detection of abnormal activity is based on response comparisons of exposed fish either with activity measured during a baseline preexposure period or observations of fish under a control treatment. Recent advances in telemetry provide an opportunity to measure swimming performance in wild fish by monitoring muscular activity of the trunk muscles and differentiation of the response of red muscles used in burst swimming from the response of white muscles for continuous aerobic swimming (Jayne and Lauder 1994). The monitoring of muscular activity coupled with sensors for blood oxygen and heart rate should provide a comprehensive view of fish swimming capacity in the natural environment (Briggs and Post 1997).

Assessments of swimming activity and swimming capacity are useful in defining no-effect concentrations of chemical substances in the laboratory. Such responses can also be used to evaluate effluents and confirm mitigation of contaminated sites. Because locomotory responses such as frequency, posture, path, and sequence are most effectively evaluated in controlled laboratory studies, the NRDA guidelines recommend that data on swimming activity and swimming performance be included as weight-of-evidence for natural resource injury (Cleveland, personal communication). In the absent of evidence for disease, parasitism, or physical trauma, the appearance of grossly abnormal locomotory responses in the field should be regarded as environmental injury. Swimming behaviors should be highly effective in monitoring point source habitat alterations, as well as habitat restoration activities.

Reproductive Behavior

Reproduction includes a number of behavioral activities associated with the breeding sequence and, depending on the species, may include migration to reproductive habitats, establishment of territories or formation of breeding schools, preparation of nests, courtship, spawning, defense of nests, and parental care. As with other behavioral patterns, behaviors associated with reproduction are vulnerable to impairment by sublethal contaminant exposure and other environmental stressors. Behavioral impairments occurring at any point in the reproductive sequence will likely decrease reproductive success, which directly diminishes year-class abundance and viability.

Although reproductive failure is documented in natural habitats impacted by hazardous substances, the data are primarily from observations of year-class recruits or presence of nesting adults. Limited laboratory testing of reproductive behavior has been performed. Speranza et al. 1977 found that zinc exposure delayed spawning in zebrafish *Brachydanio rerio*. Aspects of courtship behavior were altered in the guppy *Poecilia reticulata* following sublethal exposure to contaminants (Schoder and Peters 1988). Weber (1993) found that a 30-d exposure of fathead minnow to lead resulted in a reduction of nest preparation and maintenance activities and, also, altered the sequence in which reproductive behaviors were performed. Also, lead exposure generally reduced the expression of secondary sexual characteristics and caused reductions in gametogenesis, deposited eggs, and developmental rate of resultant young. Atlantic salmon abandoned nesting territories following pesticide exposure (Symons 1973). Nests abandoned by spotted sunfish *Lepomis punctatus* during herbicide exposures resulted in extensive nest predation by other fish (Betolli and Clark 1992).

Measures of Reproductive Behavior

The measurement of behaviors associated with reproduction is, undoubtedly, among the most difficult of behavior functions to accomplish because of the complexities of reproductive physiology and synchronization of hormonal cycles and environmental cues. In addition, the study protocol must incorporate essential social and physical habitat conditions. Sufficient numbers of breeding adults must be available for study. This may be particularly challenging since the breeding pairs may not be synchronized in their breeding conditions. Weber (1993) described methods for conducting reproductive behavioral studies with the fathead minnow. Breeding pairs showing dimorphic breeding coloration were selected for each chemical exposure treatment and videotaped observations were conducted daily for 30 d. Time spent in territorial and nest preparation responses, and spawning and parental care behaviors were measured from the video recordings. The sequence of behaviors was also quantified in order to determine how frequently one behavior pattern followed another. Measures of gonadal development, gamete production, egg numbers, and viability and embryonic development were also made. Betolli and Clark (1992) measured defense of nesting areas in a field study of sunfishes exposed to a herbicide. Mortality of eggs and embryos resulting from nest predation readily occurred when the males abandoned their nests.

Variables Influencing Reproductive Behavior

In addition to variables discussed in the section Variables Influencing Behavioral Responses, the application of appropriate environmental variables including light and temperature regimes, water flow, substrate, shelter, and forage are likely to be critical for inducing the spawning response of many species. Reproduction can be inhibited by the ratio of males to females and spatial density of the test organisms. The age of the organism, the stage of gametogenesis at the time of the exposure, and the development of secondary sexual characteristics may also influence production.

Ecological Significance and Application

Among all of the behavioral functions, reproduction bears the most immediate relevance to predicting the ecological consequences of sublethal contamination to the natural populations compared with other behavioral functions because reproductive success is so closely linked with population success. Altered behaviors that result in the loss of eggs and young are as significant as infertility relative to reproductive success. The application of such data would be immediately applicable as a measure of environmental injury in NRDAR, as well as in the development of water quality criteria.

Discussion

Diversity Aspects of Behavioral Toxicology

A diversity of behavioral responses, species, and chemical substances has been considered in determining the consequences of sublethal contamination on the behavioral function of fishes. This rich diversity of studies provides an excellent basis for developing analytical tools for toxicological investigations and reflects the ingenuity and imagination of a large number of researchers. Behavioral tests have not been rigorously standardized for aquatic studies, yet generalized protocols are available for the measure of many behavioral functions. These protocols represent similar approaches that have been used independently in different laboratories or reflect extensive investigations of a particular behavioral response with a limited number of species. Certain behavioral approaches provide data that are predictive of contaminant disruptions to natural populations that have ecological significance relative to population success, reproduction progeny year-class strength and species diversity. The results of such efforts clearly indicate that behavioral measures provide highly sensitive indices of sublethal contamination that are often applicable to a range of chemical substances, as well as other environmental stressors and are adaptable to a range of species (Table 1). Other behavioral measures provide weight-of-evidence relative to potential impacts on essential life functions including habitat selection, competition, feeding, predator avoidance, and reproduction. In turn, these impacts could limit reproduction, development, and viability of the individual and its population.

Applications of behavioral data in environmental management include the determination of criterion concentrations for specific substances, early detection and monitoring of toxic water quality, confirmation of contaminant mitigation goals, and measures of environmental injury under CERCLA and the Oil Pollution Act of 1990. Behavioral data on habits and habitats are critical in understanding endangered species and defining critical habitats, as well as critical periods in their life cycle as mandated by the Endangered Species Act. Thus, behavioral data relating preferences or aversions for variables such as changes in temperature or flow regime, increases in turbidity, siltation, substrate, cover, or shelter are especially important in developing effective recovery plans. Behavioral data should prove especially useful in protecting species that are of heightened risk of contaminant injury such as those that form mass assemblages; species that change habitats periodically through narrow front migrations that concentrate migrants along particular routes; species exhibiting inflexible behaviors that must be performed in correct order at a correct time under specific environmental conditions; and species that are particularly dependent on specific resources such as critical prey or that perform keystone functions within their community (Ellis 1984). Challenges for future use of behavioral studies in environmental management include further understanding of sources of variation in behavioral responses, linkages of behavioral responses with physiological and ecological endpoints, and field verification of behavioral injury.

> Control Treatment (0.0 mg/L)

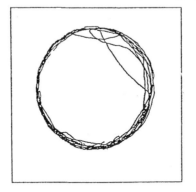

> Low Treatment (0.9 mg/L)

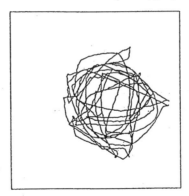

> High Treatment (3.6 mg/L)

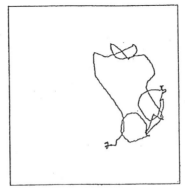

Figure 7. Swimming paths of juvenile rainbow trout following exposure to carbaryl were assessed with computer-assisted methods. (Brewer et al. 2001).

Advances in Computer-Assisted Analysis of Behavior

Behavioral data have largely been limited to manual data collection from direct visual or videotaped observations developed from ethological or experimental psychology studies. Such approaches are adequate and possibly the only way to quantify certain responses. However, data collection with sufficient replication and numerous treatments often results in a significant investment of time. Such methods depend upon the experience and judgment of the researcher and how well the response is defined. Subtle, more sensitive responses may be more likely to be biased by experimental design, including the inexperience of personnel, than more qualitatively obvious responses. Such discrepancies can lead to increased variability and decreased accuracy of the measurement. An increasing number of computer-assisted video analysis systems are being used in behavioral studies. The use of such instrumentation to quantify responses, such as swimming activity, has been shown to provide greater efficiency than direct observational methods. Computer-assisted methods employed to assess swimming paths of rainbow trout exposed to carbaryl (Figure 7) demonstrate the disruption of swimming behavior by exposure (Little and Brewer, in press). Instrumentation can also provide more types of measures not possible from direct observation and can increase objectivity and accuracy of the data collection process. Smith and Bailey (1990) have used computer-assisted technology to assess a range of swimming behavior variables continuously in real time among multiple individuals during chronic contaminant exposures, a task that would require monumental effort through manual means. Baatrup and Bayley (1998) have developed a video analysis system that measures not only locomotory variables, but also interactive responses between predator and prey and reproductive pairs. One can expect the applications of such systems for the collection of behavioral data will continue to expand.

Correlations with Physiological and Biochemical Processes

Correlations between behavioral and physiological toxicosis will continue to be an important research focus because such data may enhance the prediction of population-level responses from biomarker data of chemical exposure. The development of behavioral modes of action of chemicals is one approach that may be taken to link behavioral and physiological data. Drummond et al. (1986) divided behavioral and morphological responses into 10 categories that included a total of 40 responses monitored during acute exposures of fathead minnows to 139 single compounds. Statistical pattern recognition of the behavioral changes was induced during exposure, and three general responses were identified that correlated with three classes of contaminants. These correlations included narcosis-producing chemicals that depress central and peripheral nervous system activities, substances that disrupt metabolic activity, and neurotoxic chemicals. In a similar approach, Rice et al. (1997) observed unique behavioral and morphological abnormali-

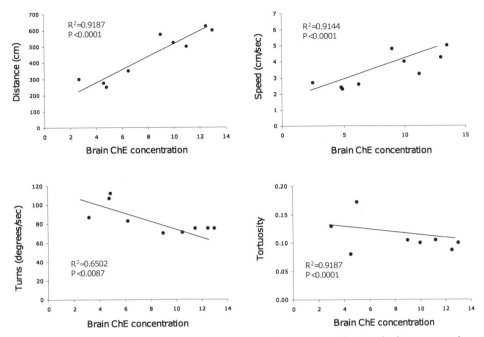

Figure 8. Correlations between locomotor behaviors and brain cholinesterase levels for rainbow trout exposed to malathion (Brewer et al. 2001).

ties for 4 compounds with different toxicological mechanisms among Japanese medaka *Oryzias latipes*. Exposure to a fifth substance, 2,4-dinitrophenol, an uncoupler of oxidative phosphorylation, displayed the fewest behavioral symptoms with a loss of equilibrium being the most common.

Recent studies have attempted to understand the extent to which behavior is altered by AChE inhibitions (Brewer et al. 2001). Rainbow trout were exposed to a concentration series of various insecticides, and locomotory responses were observed at 24 and 96 h of exposure, as well as after 48 h of recovery in uncontaminated water. Immediately after the behavioral observation, fish were removed from each exposure and brain tissues were taken for AChE measurements. After a 24-h exposure to malathion, speed of movement and distance traveled declined about 50%. Tortuosity of path (frequency of turns or ratio of turns to linear movement) also decreased as the organisms became more stereotypic in their movements. Brain AChE concentrations significantly declined as a result of inhibition and correlated strongly with the behavioral responses (Figure 8). Similar studies have been conducted with other neurochemicals such as serotonin (Fingerman and Russell 1980), and hormones such as thyroxin (Nichols et al. 1984).

The simultaneous observation of neurophysiological and behavioral responses is another promising means of correlating physiological responses with behavior. For example, the rapid escape response elicited by threatening stimuli is directed by Mauthner cells, which are paired interneurons located in the hindbrain of fish with axons that project the length of the spinal column. Stimulation of the Mauthner cells give rise to an externally measurable depolarization followed by excitation of motor neurons that in turn give rise to severe flexure of the body, which reorients and propels the organism away from the predator (Fetcho and Faber 1988). Carlson et al. (1998) found that a 24–96 h sublethal exposure of medaka to carbaryl and phenol delayed the stimulation of the motor neurons, whereas chlorpyrifos and 2,4-dinitrophenol significantly increased the latency of the neuromuscular response. Chlorpyrifos, carbaryl, phenol, 2,4-dinitrophenol, endosulfan, and 1-octanol significantly increased the magnitude to external stimulation required to induce the Mauthner cells response. The significance of these neural components to the escape response was confirmed in tests that showed that fish were more susceptible to predation following exposure to these chemicals.

Application of Behavioral Data in Environmental Monitoring

The application of behavioral measures in water quality monitoring was initiated early in the development of behavioral toxicology. In the 1970s, orientation against water flow was used to monitor municipal water treatment (Besch et al. 1977). Fish swimming activity assays were developed to provide early warning systems to alert for failures in water treatment (Cairns et al. 1973; Morgan 1979). Currently, the most developed and widely applied monitoring systems using fish as an environmental sensor are the fish respiration systems that monitor physiological changes in respiratory responses to hazardous substances (Shedd et al. 1986; Gruber and Diamond 1988; Diamond et al. 1990). Such systems provide an excellent model for the further development of behaviorally based monitoring approaches. Advances in telemetry applications have greatly increased the feasibility of behavioral tests in the field. The small size of available transmitters allows even small fish to be studied. In addition, the use of monitoring buoys, improved battery life, and the digital coding of individuals enable expanded study designs, as well as prolonged observations in the field. In the past, telemetry was used to mark the position of an organism; now various sensors allow for the determination of latitude and longitude coordinates of the fish position (Cote et al. 1998), cardiac and respiratory rates (Briggs and Post 1997), swimming activity (Demers et al. 1996), and reproduction (Kaseloo et al. 1996) in free-ranging fish and, as a result, greatly expands opportunities for studying the response of fish to environmental stressors.

In conclusion, the application and relevance of behavioral measures to investigations of environmental stress is expected to increase with increased automation, as linkages are established between physiological and population-level responses and with the continued ingenuity and creativity of behavioral scientists.

References

Anthony, D. C., T. J. Montine, and D. G. Graham. 1996. Toxic responses of the nervous system. Pages 463–486 *in* C. Klaassen, M. Amdur, and J. Doull, editors. Casarett and Doull's toxicology: the basic science of poisons. McGraw-Hill, New York.

ASTM, Annual Book of ASTM Standards. 1998a. Standard guide for conducting acute toxicity tests with fishes, macroinvertebrates, and amphibians. American Society for Testing and Materials, ASTM Guide E 729-96, West Conshohocken, Pennsylvania.

ASTM, Annual Book of ASTM Standards. 1998b. Standard guide for conducting early life-stage toxicity tests with fishes. American Society for Testing and Materials, ASTM Guide E 1241-92, West Conshohocken, Pennsylvania.

ASTM, Annual Book of ASTM Standards. 1998c. Standard guide for behavioral testing in aquatic toxicology. American Society for Testing and Materials, ASTM Guide E 1604-94, West Conshohocken, Pennsylvania.

ASTM, Annual Book of ASTM Standards. 1998d. Standard guide for measurement of behavior during fish toxicity test. American Society for Testing and Materials, ASTM Guide E 1711-95, West Conshohocken, Pennsylvania.

Atchison, G. J., M. G. Henry, and M. B. Sandheinrich. 1987. Effects of metals on fish behavior: a review. Environmental Biology of Fishes 18:11–25.

Baatrup, E., and M. Bayley. 1998. Animal locomotor behavior as a health biomarker of chemical stress. Archives of Toxicology 20:163.

Beamish, F. W. H. 1978. Swimming capacity. Pages 101–187 *in* W. S. Hoar and D. J. Randall, editors. Fish physiology, volume 7. Academic Press, New York.

Beitinger, T. L. 1990. Behavioral reactions for the assessment of stress in fishes. Great Lakes Research 16:495–528.

Besch, W. K., A. Kemball, K. Meyer-Waarden, and B. Scharf. 1977. A biological monitoring system employing rheotaxis of fish. Pages 56–74 *in* J. Cairns, Jr., K. L. Dickson, and G. F. Westlake, editors. Biological monitoring of water and effluent quality. American Society for Testing and Materials, ASTM STP 607, Philadelphia.

Betolli, P. W., and P. W. Clark. 1992. Behavior of sunfish exposed to herbicides: a field study. Environmental Toxicology and Chemistry 11:1461–1467.

Birge, W., R. Short, J. Black, M. Kercher, and W. Robinson. 1993. Effects of chemical stresses on behavior of larval and juvenile fishes and amphibians. Pages 55–65 *in* L. Fuiman, editor. Water quality and the early life stages of fishes. American Fisheries Society, Symposium 14, Bethesda, Maryland.

Black, J. A., and W J. Birge. 1980. An avoidance response bioassay for aquatic pollutants. University of Kentucky, Water Resources Research Institute Research Report 123. Lexington, Kentucky.

Blaxter, J., and C. T. Hallers-Tjabbes. 1992. The effect of pollutants on sensory systems and behaviour of aquatic animals, Netherlands. Journal of Aquatic Ecology 26:43–52.

Brett, J. R. 1964. The respiratory metabolism and swimming performance of young sockeye salmon. Journal of the Fisheries Research Board of Canada 21:1183–1226.

Brewer, S. K., E. E. Little, A. J. DeLonay, S. L. Beauvais, and S. B. Jones. 1999. The use of automated monitoring to assess behavioral toxicology in fish: linking behavior and physiology. Pages 370–386 *in* D. S. Henshel, M. C. Black, and M. C. Harrass, editors. Environmental toxicology and risk assessment, Volume 8, 1999. ASTM STP 1364. American Society for Testing and Materials, West Conshohocken, Pennsylvania.

Brewer, S. K., and five coauthors. 2001. Behavioral dysfunctions correlate to altered physiology in rainbow trout (*Oncorhynchus mykiss*) exposed to cholinesterase-inhibiting chemicals. Archives of Environmental Contamination and Toxicology 40:70–76.

Briggs, C. T., and J. R. Post. 1997. Field metabolic rates of rainbow trout estimated using electromyogram telemetry. Journal of Fish Biology 51:807–823.

Brown, J. A., P. H. Johansen, P. W. Colgan, and R. A. Mathers. 1985. Changes in the predator-avoidance behavior of juvenile guppies (*Poecilia reticulata*) exposed to pentachlorophenol. Canadian Journal of Zoology 63:2001–2005.

Brown, J. A., P. H. Johansen, P. W. Colgan, and R. A. Mathers. 1987. Impairment of early feeding behavior of largemouth bass by pentachlorophenol exposure: a preliminary assessment. Transactions of the American Fisheries Society 116:71–78.

Bull, C. J., and J. E. McInerney. 1974. Behavior of juvenile coho salmon (*Oncorhynchus kisutch*) exposed to Sumithion (fenitrothion), an organophosphate insecticide. Journal of the Fisheries Research Board of Canada 31:1867–1872.

Cairns, Jr., J., R. E. Sparks, and W. T. Waller. 1973. The use of fish as sensors in industrial waste lines to prevent fish kills. Hydrobiologia 41:151–167.

Carlson, R. W., S. P. Bradbury, R. A. Drummond, and D. E. Hammermeister. 1998. Neurological effects on startle response and escape from predation by medaka exposed to organic chemicals. Aquatic Toxicology 43:51–64.

Cherry, D. S., and J. Cairns Jr. 1982. Biological monitoring. V: preference and avoidance studies. Water Research 16:263–301.

Cleveland, L., E. E. Little, C. G. Ingersoll, R. H. Wiedmeyer, and J. B. Hunn. 1991. Sensitivity of brook trout to low pH, low calcium and elevated aluminum concentrations during laboratory pulse exposures. Aquatic Toxicology 19:303–317.

Cleveland, L., E. E. Little, R. H. Wiedmeyer, and D. R. Buckler. 1989. Chronic no-observed-effect concentrations of aluminum for brook trout exposed in low calcium dilute acidic water. Pages 229–246 *in* T. E. Lewis, editor. Environmental chemistry and toxicology of aluminum. Lewis Publishers, Chelsea, Michigan.

Comprehensive Environmental Response Compensation, and Liabilities Act. U. S. Code. Vol. 26, secs. 4611–4682.

Cote, D., and seven coauthors. 1998. A coded acoustic telemetry system for high precision monitoring of fish location and movement: application to the study of near shore nursery habitat of juvenile Atlantic cod (*Gadus morhua*). MTS Journal 32:54–62.

DeLonay, A. J., E. E. Little, J. Lipton, D. F. Woodward, and J. A. Hansen. 1996. Behavioral avoidance as evidence of injury to fishery resources: applications to natural resource damage assessment. Pages 268–280 *in* T. W. LaPoint, F. T. Price, and E. E. Little, editors. Environmental toxicology and risk assessment: fourth volume, ASTM STP 1262. American Society for Testing and Materials, West Conshohochen, Pennsylvania.

Demers, E., R. S. McKinley, and D. J. McQueen. 1996. Activity patterns of largemouth bass determined with electromyogram radiotelemetry. Transactions of the American Fisheries Society 125:434–439.

Diamond, J. M., M. J. Parson, and D. Gruber. 1990. Rapid detection of sublethal toxicity using fish ventilatory behavior. Environmental Toxicology and Chemistry 9:3–12.

Dodson, J. J., and C. I. Mayfield. 1979. Modification of the rheotropic response of rainbow trout (*Salmo gairdneri*) by sublethal doses of the aquatic herbicides diquat and simazine. Environmental Pollution 18:147–157.

Drummond, R. A., and C. L. Russom. 1990. Behavioral toxicity syndromes: a promising tool for assessing toxicity mechanisms in juvenile fathead minnows. Environmental Toxicology and Chemistry 9:37–46.

Drummond, R. A., C. L. Russom, D. L. Geiger, and D. L. DeFoe. 1986. Behavioral and morphological changes in fathead minnows, *Pimephales promelas*, as diagnostic endpoints for screening chemicals according to modes of action. Pages 415–435 *in* T. M. Poston and R. Purdy, editors. Aquatic toxicology. 9th Aquatic Toxicity Symposium. American Society for Testing and Materials, ASTM STP 921, Philadelphia.

Drummond, R. A., W. A. Spoor, and G. F. Olson. 1973. Some short-term indicators of sublethal effects of copper on brook trout, *Salvelinus fontinalis*. Journal of the Fisheries Research Board of Canada 30:698–701.

Ellgaard, E. G., J. E. Tusa, and A. A. Malizia Jr. 1978. Locomotor activity of the bluegill, *Lepomis macrochirus*: hyperactivity induced by sublethal concentrations of cadmium chromium, and zinc. Journal of Fish Biology 1:19–23.

Ellis, D. V. 1984. Animal behavior and pollution. Marine Pollution Bulletin 15:163–164.

Fairchild, J. F., and E. E. Little. 1999. Use of behavioral endpoints to determine protective concentrations of the insecticide fonofos for bluegill (*Lepomins macrochirus*). Pages 387–400 *in* D. S. Henshel, M. C. Black, and M. C. Harrass, editors. Environmental toxicology and risk assessment: standardization of biomarkers for endocrine disruption and environmental assessment: eighth volume, ASTM STP 1364. American Society for Testing and Materials, West Conshohocken, Pennsylvania.

Fetchco, J. R., and D. S. Faber. 1988. Identification of motorneurons and interneurons in the spinal network for escapes initiated by the Mauthner cell in goldfish. Journal of Neuroscience 8:4192–4213.

Finger, S. E., and J. S. Bulak. 1989. Toxicity of water from three South Carolina rivers to larval striped bass. Transaction of the American Fisheries Society 117:521–528.

Finger, S. E., E. E. Little, M. G. Henry, J. F. Fairchild, and T. P. Boyle. 1985. Comparison of laboratory and field assessment of fluorene, part I: effects of fluorene on the survival, growth, reproduction, and behavior of aquatic organisms in laboratory tests. Pages 120–133 *in* T. Boyle, editor. Validation and predictability of laboratory methods for assessing the fate and effects of contaminants in aquatic ecosystems. American Society for Testing and Materials, ASTM STP 865, Philadelphia.

Finger, T. R. 1982. Interactive segregation among three species of sculpins (Cottus) Copeia 1982:680–694.

Fingerman, S. W., and L. C. Russell. 1980. Effects of polychlorinated biphenyl Arochlor 1254 on locomotor activity and on the neurotransmitters dopamine and norepinepherine in the brain of gulf killifish, *Fundulus grandis*. Bulletin of Environmental Contamination and Toxicology 25:682–687.

Geckler, J. R., and five coauthors. 1976. Validity of laboratory tests for predicting copper toxicity in streams. U.S. Environmental Protection Agency, EPA-600/3-76-116, Environmental Research Laboratory, Diluth, Minnesota.

Giattina, J. D., D. S. Cherry, J. Cairns, and S. R. Larrick. 1981. Comparison of laboratory and field avoidance behavior of fish in heated chlorinated water. Transactions of the American Fisheries Society 110:526–535.

Giattina, J. D., and R. R. Garton. 1983. A review of the preference-avoidance responses of fishes to aquatic contaminants. Residue Reviews 87:43–90.

Gray, R. H. 1990. Fish behavior and environmental assessment. Environmental Toxicology and Chemistry 9:53–68.

Gruber, D., and J. Diamond. 1988. Automated Biomonitoring. Ellis Horwood, West Sussex, England.

Hara, T. J., S. B. Brown, and R. E. Evans. 1983. Pollutants and chemoreception in aquatic organisms. Pages 248–306 in J. O. Nriagu, editor. Aquatic toxicology. John Wiley and Sons, New York.

Hara, T. J., and B. E. Thompson. 1978. The reaction of whitefish, *Coregonus clupeaformis*, to the anionic detergent sodium lauryl sulphate and its effects on their olfactory responses. Water Research 12:893–897.

Hartwell, S. I., D. S. Cherry, and J. Cairns, Jr. 1987. Field validation of avoidance of elevated metals by fathead minnows (*Pimephales promelas*) following in situ acclimation. Environmental Contamination and Toxicology 6:189–200.

Henry, M. G., and G. J. Atchison. 1979. Influence of social rank on the behavior of bluegill, *Lepomis macrochirus*, exposed to sublethal concentrations of cadmium and zinc. Journal of Fish Biology 15:309–315.

Henry, M. G., and G. J. Atchison. 1984. Behavioral effects of methyl parathion on social groups of bluegill (*Lepomis macrochirus*). Environmental Toxicology and Chemistry 3:399–408.

Holcombe, G. W., J. T. Fiandt, and G. L. Phipps. 1980. Effects of pH increases and sodium chloride additions on the acute toxicity of 2,4-dichlorophenol to the fathead minnow. Water Research 14:1073–1077.

Howard, T. E. 1975. Swimming performance of juvenile coho salmon (*Oncorhynchus kisutch*) exposed to bleached kraft pulpmill effluent. Journal of the Fisheries Research Board of Canada 32:789–793.

Jayne, B. C., and G. V. Lauder. 1994. How swimming fish use slow and fast muscle fibers: implications for models of vertebrate muscle recruitment. Journal of Comparative Physiology A 175:123–131.

Kalleberg, G. H. 1958. Observations in a stream tank of territoriality and competition in juvenile salmon and trout. Institute of Freshwater Research Drottningholm Report 39:55–98.

Kaseloo, P. A., A. H. Weatherley, P. E. Ihssen, D. A. Anstey, and M. D. Gare. 1996. Electromyograms from radiotelemetry as indicators of reproductive activity in lake trout. Journal of Fish Biology 48:664–674.

Keenleyside, M. H. A. 1979. Diversity and adaptation in fish behavior. Zoophysiology, volume 11. Springer-Verlag, Berlin.

Kleerekoper, H. 1974. Effects of exposure to a subacute concentration of parathion on the interaction between chemoreception and water flow in fish. Pages 237–245 in F. J. Vernburg and W. B. Vernberg, editors. Pollution and physiology in marine organisms. Academic Press, New York.

Kleerekoper, H. 1976. Effects of sublethal concentrations of pollutants on the behavior of fish. Journal of the Fisheries Research Board of Canada 33:2036–2039.

Kleerekoper, H., J. B. Waxman, and J. Matis. 1973. Interaction of temperature and copper ions as orienting stimuli in the locomotor behavior of the gold fish (*Carassius auratus*). Journal Fisheries Research Board of Canada 30:725–728.

Kolok, A. S., E. P. Plaisance, and A. Abdelghani. 1998. Variation in the swimming performance of fishers: an overlooked source of variation in toxicity studies. Environmental Toxicology and Chemistry 17:282–285.

Koltes, K. H. 1985. Effects of sublethal copper concentrations on the structure and activity of Atlantic silverside schools. Transactions of the American Fisheries Society 114:413–422.

Korver, R. M., and J. B. Sprague. 1989. Zinc avoidance by fathead minnows (*Pimephales promelas*): computerized tracking and greater ecological relevance. Canadian Journal of Fisheries and Aquatic Sciences 46:494–502.

Laurence, G. C. 1972. Comparative swimming abilities of fed and starved larval largemouth bass (*Micropterus salmoides*). Journal of Fish Biology 4:73–78.

Lemly, A. D., and R. J. F. Smith. 1987. Effects of chronic exposure to acidified water on chemoreception of feeding stimuli by fathead minnows, *Pimephales promelas*: Mechanisms and ecological implications. Environmental Toxicology and Chemistry 6:225–238.

Lipton, J., E. E. Little, J. C. A. Marr, and A. J. DeLonay. 1997. Use of behavioral avoidance testing in natural resource damage assessment. Pages 212–223 *in* D. Bengtson and D.S. Henshel, editors. Environmental toxicology and risk assessment: biomarkers and risk assessment (5th Volume) ASTM STP 1306. American Society of Testing and Materials, Philadelphia.

Little, E. E. 1990. Behavioral toxicology: stimulating challenges for a growing discipline. Environmental Toxicology and Chemistry 9:1–2.

Little, E. E., R. D. Archeski, B. A. Flerov, and V. I. Kozlovskaya. 1990. Behavioral indicators of sublethal toxicity in rainbow trout. Archives of Environmental Contamination and Toxicology 19:380–385.

Little, E. E., and S. K. Brewer. 2001. Neural behavioral toxicity in fish. Pages 139–174 *in* W. Benson, and D. Schlenk, editors. Target organ toxicity in fishes. Taylor and Francis Publishers, London.

Little, E. E., L. Cleveland, R. Calfee, and M. G. Barron. 2000. Assessment of the photoenhanced toxicity of a weathered oil to the tidewater silverside. Environmental Toxicology, and Chemistry 19:929–932.

Little, E. E., and A. J. DeLonay. 1996. Measures of fish behavior as indicators of sublethal toxicosis during standard toxicity tests. Pages 216–233 *in* T. LaPoint, F. T. Price and E. E. Little, editors. Environmental toxicology and risk assessment: biomarkers and risk assessment: Fourth Volume, ASTM STP 1262. American Society of Testing and Materials, Philadelphia.

Little, E. E., F. J. Dwyer, and J. F. Fairchild, A. J. DeLonay, and J. L. Zajicek. 1993b. Survival and behavioral response of bluegill during continuous and pulsed exposures to the pyrethroid insecticide, ES-fenvalerate. Environmental Toxicology and Chemistry 12:871–878.

Little, E. E., J. F. Fairchild, and A. J. DeLonay. 1993a. Behavioral methods for assessing the impacts of contaminants on early life stage fishes. Pages 67–76 *in* L. A. Fuiman, editor. Water quality and the early life stages of fishes. American Fisheries Society, Symposium 14, Bethesda, Maryland.

Little, E. E., and S. E. Finger. 1990. Swimming behavior as an indicator of sublethal toxicity in fish. Environmental Toxicology and Chemistry 9:13–19.

Little, E. E., B. A. Flerov, and N. N. Ruzhinskaya. 1985. Behavioral approaches in aquatic toxicity: a review. Pages 72–98 *in* P. M. Mehrle Jr., R. H. Gray, and R. L. Kendall, editors. Toxic substances in the aquatic environment: an international aspect. American Fisheries Society, Water Quality Section, Bethesda, Maryland.

Marcucella, H., and C. I. Abramson. 1978. Behavioral toxicology and teleost fish, Pages 33–77 *in* D. I. Mostofsky, editor. The behavior of fish and other aquatic animals. Academic Press, New York.

Marr, J., and six coauthors. 1995. Fisheries toxicity injury studies. Blackbird Mine Site, Idaho. Report prepared for the National Oceanic and Atmospheric Administration, Washington, D.C. February.

Mathers, R. A., J. A. Brown, and P. H, Johansen. 1985. The growth and feeding behavior responses of largemouth bass (*Micropterus salmoides*) exposed to PCP. Aquatic Toxicology 6:157–164.

Mehrle, P. M., and nine coauthors. 1988. Toxicity and bioconcentration of 2,3,7,8-tetrachlorodibenzodioxin and tetrachlorodibenzofuran in rainbow trout. Environmental Toxicology and Chemistry 7:47–62.

Miller, D. C., W. H. Lang, J. O. B. Graeves, and R. S. Wilson. 1982. Investigations in aquatic behavioral toxicology using a computerized video quantification system. Pages 206–220 *in* J. G. Pearson, R. B. Foster, and W. E. Bishop, editors. Aquatic toxicology and hazard assessment: fifth conference. STP 766. American Society for Testing and Materials, Philadelphia.

Morgan, W. S. G. 1979. Fish locomotor behavior patterns as a monitoring tool. Journal of the Water Pollution Control Federation 51:580–589.

Mueller, H. 1972. Oddity and specific searching image more important than conspicuousness in prey selection. Nature (London) 233:345–346.

Natural Resource Damage Assessments: Final Rule. 1986. Federal Register 51:27674–27753.

Nichols, J. W., and six coauthors. 1984. Effects of freshwater exposure to arsenic trioxide on the parr-smolt transformation of coho salmon. Water Research 1:143–149.

Oil Pollution Act. U.S. Code. Vol. 33, secs. 2701–104 STAT 484 (1990).

Oliver, J. D., G. D. Holeton, and K. E. Chua. 1979. Overwintering mortality of fingerling smallmouth bass in relation to size relative energy stores and environmental temperature. Transactions of the American Fisheries Society 108:130–136.

Olla, B. L., W. H. Pearson, and A. L. Studholme. 1980. Applicability of behavioral measures in environmental stress assessment. Rapports et Proces-Verbaux des Reunions Commission Internationale pour l'Exploration Scientifique de la Mer Mediterranee Monaco 179:162–173.

Peterson, R. H. 1974. Influence of fenitrothion on swimming velocity of brook trout (*Salvelinus fontinalis*). Journal of the Fisheries Research Board of Canada 33:495–511.

Rand, G. M. 1977. The effect of subacute parathion exposure on the locomotor behavior of the bluegill sunfish and largemouth bass. Pages 253–268 *in* F. L. Mayer and J. L. Hamelink, editors. Aquatic toxicology and hazard assessment, ASTM STP 634. American Society for Testing and Materials, Philadelphia.

Rand, G. M. 1985. Behavior. Pages 221–256 *in* G. M. Rand and S. R. Petrocelli, editors. Fundamentals of aquatic toxicology: methods and applications. Hemisphere Publishing, New York.

Reynolds, W. W., and M. E. Casterlin. 1976. Activity rhythms and light intensity preferences of *Micropterus salmoides* and *M. dolomieui*. Transactions of the American Fisheries Society 105:400–403.

Rice, P. J., C. D. Drews, T. M. Klubertany, S. P. Bradbury, and J. R. Coats. 1997. Acute toxicity and behavioral effects of chlorophyrophos, permethrine, phenol, strychine and 2,4,-dinitro phenol to 30 day old Japanese medaka (*Oryzias atipes*). Environmental Toxicology and Chemistry 16:696–704.

Sandheinrich, M. B., and G. J. Atchison. 1989. Sublethal copper effects on bluegill, *Lepomis macrochirus,* foraging behavior. Canadian Journal of Aquatic Science 46:1977–1985.

Sandheinrich, M. B., and G. J. Atchison. 1990. Sublethal toxicant effects on fish foraging behavior: Empirical vs mechanistic approaches. Environmental Toxicology and Chemistry 9:107–120.

Saunders, R. L., and J. B. Sprague. 1967. Effects of copper-zinc mining pollution on a spawning migration of Atlantic salmon. Water Research 1:419–432.

Scherer, E., F. A. J. Armstrong, and S. H. Nowak. 1975. Effects of mercury-contaminated diet upon walleyes, *Stizostedion vitreum vitreum* (Mitchill). Environment Canada. Fisheries and Marine Services Developmental Report No. 597. Environment Canada, Freshwater Institute, Winnepeg, Manitoba.

Scherer, E., and R. E. McNicol. 1998. Preference-avoidance responses of lake whitefish (*Coregonus clupeaformis*) to competing gradients of light and copper, lead and zinc. Water Research 32:924–929.

Schoder, J. H., and K. Peters. 1988. Differential courtship activity of competing guppy males (*Poecilia reticulata* Peters) as an indicator for low concentrations of aquatic pollutants. Bulletin Environmental Contamination and Toxicology 40:396–404.

Schoener, T. W. 1974. Resource partitioning in ecological communities. Science 185:27–39.

Shaw, E. 1978. Schooling fishes. American Scientist. 66:166–175.

Shedd, T. R., W. H. van der Schalie, and M. G. Zeeman. 1986. Evaluation of an automated fish ventilatory monitoring system in a short-term screening test for chronic toxicity. U.S. Army Biomedical Research and Development Laboratory. Technical Report AD A172116, Fort Detrick, Maryland.

Smith, E. H., and H. C. Bailey. 1988. Development of a system for continuous biomonitoring of a domestic water source for early warning of contaminants. Pages 182–205 *in* E. Horwood, D. S. Gruber, and J. M. Diamond, editors. Automated biomonitoring: living sensors as environmental monitors, Chichester, England.

Smith, E. H., and H. C. Bailey. 1990. Preference/avoidance testing of waste discharges on anadromous fish. Environmental Toxicology and Chemistry 9:77–86.

Speranza, A. W., R. J. Seely, V. A. Seely, and A. Perlmutter. 1977. The effect of sublethal concentrations of zinc on reproduction in the zebrafish, *Brachydonia rerio*. Environmental Pollution 12:217–222.

Sprague, J. B. 1964. Avoidance of copper-zinc solutions by young salmon in the laboratory. Journal of the Water Pollution Control Federation 36:990–1004.

Steele, C. W. 1983. Effects of exposure to sublethal copper on the locomotor behavior of the sea catfish, *Arius felis*. Aquatic Toxicology 4:83–93.

Steele, C. W., D. H. Taylor, S. Strickler-Shaw. 1996. Perspectives in avoidance-preference bioassays. Pages 254–267 *in* T. LaPoint, F. T. Price and E. E. Little, editors. Environmental toxicology and risk assessment: biomarkers and risk assessment: fourth volume, ASTM STP 1262. American Society of Testing and Materials, Philadelphia.

Sullivan, C. M., and K. C. Fisher. 1942. Temperature selection and effects of light and temperature on movements in fish. American Physiology Society Federation Proceedings 6:213.

Sullivan, C. M., and K. C. Fisher. 1954. The effects of light on temperature selection in speckled trout, *Salvelinus fontinalis* (Mitchill). Biology Bulletin Marine Biology Laboratory Woods Hole 107:278–288.

Sutterlin, A. M. 1974. Pollutants and the chemical senses of aquatic animals: perspectives and review. Chemical Senses and Flavor 1:167–178.

Symons, P. E. K. 1973. Behavior of young Atlantic salmon (*Salmo salar*) exposed to or force-fed fenitrothion, an organophosphorous insecticide. Journal of the Fisheries Research Board of Canada 30:651–655.

Timms, A. M., H. Kleerekoper, and J. Matis. 1972. Locomotor response of goldfish, channel catfish, and largemouth bass to a "copper-polluted" mass of water in an open field. Water Resources Research 8:1574–1580.

Warner, R. E., K. K. Peterson, and L. Borgman. 1966. Behavioral pathology in fish: a quantitative study of sublethal pesticide toxication. Journal of Applied Ecology 3:223–247.

Weber, D. N. 1993. Exposure to sublethal levels of waterborne lead alters reproductive behavior patterns in fathead minnows (*Pimephales promelas*). Neurotoxicology 14:347.

Werner, E. E., and D. J. Hall. 1974. Optimal foraging and the size selection of prey of the bluegill sunfish (*Lepomis macrochirus*). Ecology 55:1042–1052.

Westlake, G. F., G. H. Kleerekoper, and J. Matis. 1974. The locomotor response of goldfish to a steep gradient of copper ions. Water Resources Research 10:103.

Woodward, D. F., J. N. Goldstein, and A. M. Farag. 1997. Cutthroat trout avoidance of metals and conditions characteristic of a mining waste site: Coeur d'Alene River, Idaho. Transactions of the American Fisheries Society 126:699–706.

Woodward, D. F., E. E. Little, and L. M. Smith. 1987. Toxicity of five shale oils to fish and aquatic invertebrates. Archives of Environmental Contamination and Toxicology 16:239–246.

12

Community-Level Indicators of Stress in Aquatic Ecosystems

Martin J. Attrill

Introduction

Community-level analysis is one of the most popular methods used to assess the effect of stressors on aquatic ecosystems. The community is viewed as not only an integrative multispecies indicator of stressors acting at a range of lower levels of organization but also the most ecologically relevant indicator because effects can be directly extrapolated to the ecosystem (Attrill and Depledge 1997). However, before the use of communities as bioindicators can be critically evaluated, it is useful to first define the term "community," which has been used rather carelessly in the past (Fauth et al. 1996), resulting in an array of aquatic studies that does not strictly address stress in communities.

In an ecological sense, a community can be defined as "the species that occur together in space and time" (Begon et al. 1996). Therefore, a true community-level investigation on, for example, an estuarine mudflat would assess stress effects on all interacting species, such as microbes, worms, crabs, birds, and fish. However, this is impractical (albeit desirable), and the vast majority of community-level studies only sample a proportion of this possible species set, with restrictions placed on which sets of species are selected. Examples of such restrictions are

1. Taxonomic restrictions, such as fish communities (e.g., Henderson 1989). Fish obviously interact with invertebrates, plants, and microbes, yet some fish (e.g., flatfish, herring) may rarely occur together, particularly as adults, due to their position in the water column or trophic level.
2. Size restrictions, such as meiofauna, defined as organisms that pass through a 500-μm sieve (e.g., Warwick et al. 1990). In comprising several taxo-

473

nomic groups (nematodes, copepods, etc.), the meiofauna clearly will be co-occurring in space and time with larger invertebrates.

3. Habitat restrictions, such as stream riffles and pools (e.g., Taylor 2000). Fish communities within these subhabitats are unlikely to remain separated in space and time.

4. Feeding guild restrictions, such as a grazer community (e.g., Cottingham and Schindler 2000). Perhaps the most narrow definition of a community. A single trophic level will be naturally interacting with other organisms.

Most often, communities are characterized in various studies by more than one of these restrictions (e.g., stream meiobenthic crustaceans, Rundle and Attrill 1995), so what is clear is that communities are operator defined rather than as they occur de facto in nature. Ecological communities, as they are perceived through a range of scientific studies, are, therefore, artificial abstractions from a continuum of population distributions, and this artificial abstraction should be taken into account when undertaking a community-level investigation and extrapolating results to a wider context. Strictly speaking, all community investigations will be targeting an assemblage of populations and, thus, should be termed as such (e.g., Underwood 2000). Within this context, "community level" still has a degree of meaning for bioindicator studies and, in such cases, the definition of Mills (1969) remains the most valuable and will be used to characterize communities for the purposes of this chapter:

> "...community means a group of organisms occurring in a particular environment, presumably interacting with each other and with the environment, and separable by means of ecological survey from other groups."

The aim of this chapter is to assess and evaluate the use of community-level bioindicators in aquatic systems and to provide practical recommendations and guidance in their use and application in environmental stress studies. The chapter is composed of three main sections:

1. Methodology—which biotic components of the aquatic system to target and their advantages and disadvantages; also, the importance of experimental design

2. Using community-level data to assess stress in aquatic systems; assessment of categories of statistical analysis

3. Additional community considerations; taxonomic resolution and meta-analysis; relationships with other levels of biological organization

Methodology

The focus of this chapter is primarily on field surveys and experiments because these are, by far, the most common methods of community-level investigation. Laboratory-based community investigations (i.e., mesocosms, microcosms), while useful in providing replicated, controlled, and repeatable conditions (Odum 1984), are inherently artificial when compared with field situations, and they have been the basis of heated scientific criticism and argument (e.g., Carpenter 1996; Huston 1997). Additionally, extrapolation from laboratory to field situations is fraught with dangers (see Munkittrick and McCarty 1995; Power and McCarty 1997; Parker et al. 1999). Mesocosm experiments can be valuable for addressing specific questions on the responses of organisms to stress, but any extrapolation of results to highly variable field situations should be treated with extreme caution. Therefore, such laboratory approaches will not be considered further; investigators interested in pursuing mesocosm studies should, as an excellent starting point, consult Petersen et al. (1999).

As is true of any methods or techniques employed to assess the impact of stress on aquatic systems, the community-level approach has both advantages and disadvantages. These aspects have been discussed extensively in previous papers (e.g., Warwick 1993; Clements and Kiffney 1994; Attrill and Depledge 1997; Adams et al. 2000), and the main points are summarized in Table 1 to prevent repetition of the arguments for and against community analyses. It is recommended, however, that before designing and implementing community-level studies, the investigator should be familiar with their merits compared to other approaches for evaluating stress.

An important consideration for any environmental assessment study, following the working definition for communities, is to identify which components of the biota are most suitable for determining the impact of the stressor under investigation. Generally, little thought is given to this aspect of experimental design for community analysis as the group of organisms chosen either reflects the expertise of the operator (e.g., meiofauna), or there is a predetermined requirement to sample a certain part of the community (e.g., fish). In aquatic systems, there is a relatively wide choice of biotic components, each with its own set of advantages and limitations, which are sum-

Table 1. Perceived advantages and disadvantages of community-level bioindicators.

Advantage	Disadvantage
1. Most ecologically relevant	1. Slow response time to provide "early warning"
2. Provide a multispecies response	2. Difficult to achieve proper randomized controls in field situation
3. Integrates conditions over long time period	3. Labor intensive or need for high level of taxonomic expertise
4. Integrates impact of complex mix of stressors	4. Endpoints can be equivocal
5. Nonexperimental—no need for advanced technology	

Table 2. Advantages and disadvantages of using different components of the biota for community-level impact investigations (from Warwick 1993; Attrill and Depledge 1997).

Component of the biota	Advantages	Disadvantages
Plankton (zoo- and phyto-)	1. Easy sampling 2. Taxonomy moderately easy 3. Useful for monitoring global changes or lake systems	1. Highly mobile—not useful for local effects in most marine and lotic situations
Soft-sediment marine macroinvertebrates (>500μm) (e.g., polychaetes, crustaceans, bivalves, etc.) Benthic freshwater macro-invertebrates (e.g., insect larvae, crustaceans, molluscs)	1. Comparatively nonmobile— useful for local effects 2. Taxonomy relatively easy 3. Quantitative sampling easy 4. Extensive literature on community responses	1. Subtidal sampling in sea requires moderately-sized vessel 2. Sample processing labor intensive 3. Can have long response time
Soft-sediment marine and benthic freshwater meiofauna (< 500 μm) (e.g., nematodes, copepods, mites, rotifers, etc.)	1. Nonmotile—useful for very local effects 2. Quantitative sampling easy 3. Fast response time (short generations)	1. Taxonomy difficult (experts or training needed) 2. Fewer documented responses in literature
Hard-bottom marine epifauna/algae (e.g., sponges, corals, kelps, tunicates, etc.)	1. Immobile—useful for local effects 2. 2-d structure, allows visual assessment of changes 3. Biomass measurement difficult	1. Quantitative remote sampling impossible 2. Problems with enumerating colonial organisms
Hard-bottom marine motile fauna (e.g., crabs, starfish, sea urchins, etc.)	1. High public and commercial interest 2. Quantification difficult 3. Very mobile—difficult to relate changes to local perturbations 4. Responses generally unknown 5. Variable habitat for controls	1. Remote sampling difficult
Fish (freshwater & marine)	1. Mobility allows use for large-scale regional effects (ecosystem health, e.g., Great Lakes) 2. Taxonomy easy—high level of local knowledge 3. High public profile and commercial importance 4. Multitrophic response—indicates health of underlying system	1. Mobility of marine fish limits use for local effects 2. Quantitative sampling difficult, particularly for marine fish 3. Slow response time to disturbance

marized in Table 2. Benthic macroinvertebrates tend to be the most widely employed component, particularly in marine situations and in freshwaters of Europe (e.g., Bargos et al. 1990; Smith et al. 1995), while fish communities have been targeted more often in studies of freshwater ecosystems in North America (e.g., Fausch et al. 1990; Detenbeck et al. 1992). This does not necessarily mean that these two groups of organisms are the most suitable for all scenarios (Attrill 2000), but, as a general rule, benthic macroinvertebrates probably have the greatest range of advantages for local community-level impact studies in both marine and most freshwater situations. For investigations involving larger-scale issues and impacts (e.g., global warming, Holbrook et al. 1997; eutrophication, Bonsdorff et al. 1997), and, in certain freshwater situations, the analysis of fish communities also has considerable benefits.

Perhaps the most important methodological consideration in any community-level investigation is that of experimental design, and many of the criticisms of community bioindicators (equivocal results, lack of controls) are related to an array of studies based on inadequate sampling designs, termed "mindless monitoring" by Underwood (2000). Primarily because of the work and guidance of Underwood (e.g., Underwood 1992, 1997, 1999, 2000), much more rigor is now apparent in the design of community impact studies in marine systems (e.g., Fagan et al. 1992; Lardicci et al. 1999), and lessons learned have gradually been applied to freshwater systems (e.g., Humphrey et al. 1995; Wright et al. 1995). It is essential that all community-level investigations include appropriate spatial and temporal replication (and arrangement of spatial replicates) and, in particular, replicated control locations. An inappropriate design (e.g., a disturbed site compared with a single undisturbed site, samples in a river above and below an outfall) will result in invalid conclusions, no matter how sophisticated the data analysis, as there is no consideration of environmental variation additional to the tested impact (i.e., multiple control sites). In order to construct suitably designed surveys or experiments and for further information, investigators should consult Underwood (1997, 2000) and Clarke and Green (1988).

Using Community Data to Assess Stress in Aquatic Systems

The general raw result from any community-level survey is a species–site data matrix containing a measure of the abundance or biomass of each species or taxon at each sampled location. This matrix should include an extensive suite of relevant and supporting environmental measurements taken at each sampling location. In addition, all naturally fluctuating environmental variables that are practicable to assess need to be measured so that their influence on community patterns can be accounted for in assessing the effects of target stressors. A relevant example is that of Somerfield et al. (1994), who investigated the effect of severe metal pollution on benthic communi-

ties in the Fal estuary system, United Kingdom. Through the use of replicated creek systems, environmental variables such as sediment grain size,
organic carbon, and salinity did not follow the gradient of metal concentrations, therefore allowing the effect of pollution to be segregated from these
other potentially influential environmental factors.

The community data contained within the species–site data matrix can
be analyzed in a vast array of different ways, but it is important to consider
what type of variability within the communities is to be measured in order to
determine the impact of stressors. A useful framework for impact analysis
within this context has been provided by Micheli et al. (1999) who defined
community variability into two categories: compositional variability (changes
in the relative abundance of component species) and aggregate variability
(changes in summary properties of the community, such as total abundance,
biomass, richness, etc.). Within this framework of assessing variability, four
types of responses are possible in impacted systems: a) no effect (no change
to either category); b) change in aggregate variability only (i.e., all species
exist in similar proportions, but in lower abundances); c) change in compositional variability only (i.e., community remains the same size, but there are
changes in the proportions of species within the community); and d) changes
to both categories. Techniques are therefore required that will identify any
significant change in either or both types of variability.

Data analyses for community-level metrics can be broadly divided into
three categories following Warwick & Clarke (1991) (individual techniques
will be developed later in this section).

1. Univariate analyses. The community is described by a single figure, such
    as total abundance, diversity indices, or biotic indices. Within this type of
    analysis, individual species identity is lost, potentially restricting the as-
    sessment of compositional variability. Although analysis of abundance–
    change of key species can also be used, this is really a population level
    analysis. This type of analysis is amenable to significance testing by, for
    example, analysis of variance (ANOVA) and regression with environ-
    mental variables, which is particularly useful for changes in aggregate
    variability.

2. Distributional or graphical approaches. These are pictorial representa-
    tions of aspects of the community that allow changes to be visualized.
    This includes techniques that keep species contributions separate (al-
    though identity is not important), such as k-dominance and abundance
    biomass comparison (ABC) curves, plus graphical techniques that por-
    tray other aspects of community structure, such as biomass size spectra.
    Testing for significance between curves is also possible.

3. Multivariate analyses. With this category of analysis, species identity is
    preserved, allowing an input into the analysis from each individual spe-
    cies. It generally involves classification (e.g., cluster analysis, TWINSPAN)

and ordination (e.g., principal components analysis [PCA], multidimensional scaling [MDS]) methods. Significance testing (e.g., ANOSIM) and correlation with environmental variables (BIOENV, CANOCO) are also possible and useful for detecting changes in compositional variability.

Before statistical analysis can be undertaken on raw data for community metrics, transformation of the data has to be considered. Parametric statistical techniques, such as ANOVA, assume normality in the data. However, community data tend to have skewed distributions and heterogeneous variances (Underwood 1997; Micallef and Schembri 1998), therefore, appropriate transformations can help to approximate normality or homogeneity in the data, as long as the technique is monotonic (i.e., sample means are left in the same rank order; Underwood 1997). If data do not conform to the assumptions of ANOVA (test using, for example, Cochran's test [heterogeneity of variances] or Shapiro-Wilk statistic [normality]), then a transformation is necessary. For community data, power transformations (e.g., square-root) or log (x + 1) tend to be the most suitable and are often desirable regardless of statistical necessity (see Downing 1979 and Underwood 1997 for further details).

When using nonparametric multivariate statistics, however, transformation of data are useful for a very different reason, namely to weight the contributions of common and rare species. Community data tend to have a couple of very abundant species, which will dominate most analyses. Transformations will reduce their individual influence and increase the importance of rarer species—generally a desirable feature for assessing overall changes in the community (Clarke and Warwick 1994a). However, choice of transformation is exceptionally important as the results of multivariate analyses can be vastly different depending on which form is chosen (Cao et al. 1999). Olsgard et al. (1998), for example, demonstrated that transformation type can have a greater effect on the results of analyses than the taxonomic level of identification (e.g., species, family, phylum). For most community studies, Thorne et al. (1999) recommend a 4th-root (i.e., $\sqrt{\sqrt{}}$, $x^{0.25}$) transformation as it allowed for effective discrimination of freshwater sites over a wide range of water qualities.

Univariate Analyses

Apart from simple measures of the size of the community (e.g., total abundance), univariate techniques applicable as community-level bioindicators tend to be grouped into either diversity indices (indications of the richness of the community, generally including measures of the number of species present and their distribution within the community), similarity indices (which give a value of how similar test communities are), or biotic indices (which tend to include a subjective score relating to environmental quality).

Diversity Indices

The simplest form of diversity index is the number of species present in the sample (*s*). Even though its inherent simplicity has great value because it is directly interpretable, the index is highly dependent on sample size (species–area relationship, *sensu* Preston 1962) and does not take into account the relative distribution of species (i.e., evenness or dominance) within a community. For example, a sample composed of two species, one with an abundance of 1,000 individuals and the other with only 1 individual, would be considered very different to another sample where the same two species both had an abundance of 5. However, the two samples would have identical values of *s,* so to counter this insensitivity, a rather bewildering number of diversity indices has been developed (some main indices are defined in Appendix 1, see also Washington 1984). Although much criticized (e.g., Green 1979; Kohl and Zingg 1996), diversity indices are regularly employed in biomonitoring studies, despite the existence of studies questioning their usefulness in discriminating impacted sites (e.g., Gray et al. 1990; Warwick and Clarke 1991; Cao et al. 1996; Drake et al. 1999). Some attempts, however, have been made to critically assess the performance of diversity indices (e.g., Rosenzweig 1997; Garcia-Criado et al. 1999; Mouillot and Lepretre 1999), the most suitable appearing to be the Shannon-Wiener index (H′) paired with Pielou's evenness index (J′), and Simpson's index of concentration (λ).

The Shannon-Wiener diversity index (H′, Shannon and Weaver 1949) integrates species richness and dominance, apparently being sensitive to changes in species abundance patterns (Mouillot and Lepretre 1999). In aquatic systems, this index is unlikely to be greater than 4.00 and can decrease in value with either a reduction in species richness or an increase in dominance. Wilhm and Dorris (1968) concluded that for stream invertebrate communities, a value of H′ greater than 3 indicates clean water, 1–3 moderate pollution, and less than 1 indicates heavily polluted conditions. While this may allow some comparison between very different stream sites, it is dangerous to extrapolate this categorization beyond this particular component (stream benthic invertebrates). Subtidal marine systems, for example, may have a much higher diversity and, thus, a H′ value of 2.0 could denote a significant impact (e.g., Gray et al. 1990). Conversely, estuaries are inherently low diversity systems, so a value of 1.5 may indicate "natural" conditions. Additionally, as the value of H′ can vary depending on the logarithm used (e.g., base 2, 10, e), care needs to be taken when comparing values between studies. It is, therefore, advised to limit the comparison of H′ to samples taken within the same study, using ANOVA for example, to identify a significant difference in the index at putative impacted sites. There are, however, many cases where H′ has been successfully (and suitably) employed as a bioindicator to assess stress in aquatic communities (e.g., Cornell et al. 1976; Thrush et al. 1998; Gyedu-Ababio et al. 1999).

When H' is used, it should be in conjunction with the calculation of Pielou's evenness index (J'; Pielou 1975). This index gives a direct measurement of evenness, which is unaffected by species richness and, therefore, provides additional information on the mechanisms behind any change in H'. Pielou's index varies from 0 to 1, the higher the index, the greater the evenness in the community. A dominance index can also be calculated (1-J'), which may be more intuitive when assessing the impact of stress. The use of H' and J' is particularly relevant for the responses of benthic invertebrates to organic enrichment, where a clear pattern of dominance is known (Hynes 1960; Pearson and Rosenberg 1978). However, the direct relationship between H' or J' and responses of other components (e.g., fish) are, perhaps, less clear. The impact of schooling fish or recruitment of juveniles, for example, would decrease both H' and J', but not necessarily equate to a stress response (Thomas 1998). A further problem with both H' and J' is that they are sensitive to sample size (Condit et al. 1996; Gimaret-Carpentier et al. 1998) and the number of individuals in a sample (Warwick and Clarke 1995). Ideally, H', for example, should only be used on random samples drawn from a large community where the total number of species is known (Krebs 1985), an unlikely situation in most aquatic impact studies! However, the nonindependent nature of both indices with sample size means that they cannot be compared between surveys using sampling devices of different areas.

A few indices appear to be sample-size independent, the most popular being Simpson's index of concentration (see Rosenzweig 1997). This measure of diversity was derived from probability theory (Krebs 1985) and gives the probability that two organisms picked at random from a community are the same species, the index decreasing with increasing diversity. Simpson (1949) published an unbiased form of his index (Appendix 1); its sample-size independence being subsequently verified (e.g., Nei and Roychoudhury 1974). This index is, therefore, potentially useful for comparing the impact of stress on diversity where semiquantitative methods are used for data collection (e.g., fish communities) or samples of different sizes are compared. For example, Simpson's index has been used to investigate the impact of lake acidification on both waterfowl (Elmberg et al. 1994) and phytoplankton (Korneva 1996). However, its sensitivity at smaller, local scales has been questioned (e.g., Vandervalk et al. 1994) as has its statistical resolution (Magnussen and Boyle 1995). Warwick and Clarke (1995) proposed two alternative sample-size independent indices for assessing stress in marine benthic systems (taxonomic diversity, Δ, and taxonomic distinctness, Δ^*), further developed by Clarke and Warwick (1998). These use a taxonomic approach by integrating the path–length linking species in a hierarchical classification, species that are taxonomically dissimilar therefore contributing more to the index. The indices were tested against a known pollution gradient (Warwick and Clarke 1995, 1998) and found to perform more consistently than standard diversity indices. However, the technique was not so successful when tested against other pollution gradients (Somerfield et al. 1997) or with fish communities (Hall and Greenstreet 1998). Additionally, as it is reliant on a full taxonomic description of the fauna, taxonomic distinctness is time consuming (and thus expensive) to calculate and rather limited to areas where such expertise exists. Nevertheless, by considering relatedness of species in a diversity estimate, the indices do have theoretical advantages over traditional measures of diversity.

Overall, it is recommended that comparative measurements of diversity are carefully considered at the experimental design stage. For most investigations (where comparisons are only made *within* the study), the simple species richness or Shannon-Wiener index is probably most suitable, with Simpson's index being used where semiquantitative or variable sample sizes are to be compared.

Similarity Indices

Several indices exist that give a measure of the similarity (or dissimilarity) between two sampled communities (Appendix 1). These are usually expressed as a proportion or percentage of similarity (e.g., sample A is 54% similar to sample B). This similarity refers to the taxonomic composition of the community and can be calculated using either the presence or absence of species within each community (e.g., Sorenson coefficient, Jaccard coefficient) or incorporates the abundance of each species (e.g., Bray-Curtis, or Czekanowski, coefficient, Canberra coefficient). As a bioindicator, similarity indices are useful as they can demonstrate that an impacted sample has become dissimilar from controls. For example, Ivorra et al. (1999) used the Bray-Curtis index in order to determine the impact of metal pollution on stream algal communities, while Rader and Richardson (1994) assessed the impact of nutrient enrichment on Florida everglade fish and invertebrate communities using the Sorenson coefficient. However, perhaps the main value of these indices is that they form the basis of many multivariate analytical techniques (see below). Clarke and Warwick (1994a) recommend the use of the Bray-Curtis index, with suitable transformation, for numerical abundance or biomass data as other indices (e.g., Canberra coefficient) can have a tendency to overcompensate for rare species. For presence or absence data, the Sorenson coefficient is recommended as this is a binary version of the Bray-Curtis index.

Biotic Indices

In order to aid in the detection and interpretation of stress effects on communities, a series of biotic indices has been developed, mostly in freshwater systems. These indices attempt to integrate measurements taken from the community (similar to other univariate measures) and incorporate a scoring system for perceived response to the perturbation under study. The scoring system, for example, can weight the importance of known sensitive indicator taxa, assess changes in trophic structure or include noncommunity-level responses, such as disease prevalence or percent of deformed organisms. Due to the specific nature of the scoring species included in such indices, they often tend to be site- or system-specific and, thus, a wide range of such biotic indices has been designed and reported (e.g., Madoni 1994; Stark 1998a; Simic and Simic 1999). However, two biotic indices have been more extensively employed and appear applicable to a comprehensive range of aquatic situations. These are the biological monitoring working party (BMWP) score (and derivatives) and the index of biotic integrity (IBI).

The BMWP score was developed to assess the quality of running water in the United Kingdom using benthic macroinvertebrates (see Hawkes 1998, for full details of the development of the system). The index relies on presence or absence data from semiquantitative sampling (e.g., a 3-min kick-sample, Furse et al. 1981) of all river habitats at any one site. Captured invertebrates are then identified to family level. Each family of freshwater invertebrate is allocated a score (from 1 to 10) depending on the family's tolerance to pollution, with very sensitive taxa (e.g., certain mayflies and stoneflies) scoring 10, while very tolerant oligochaetes score 1. The sum of scores for all families in the sample equals the BMWP score, with high scores indicating good water quality (Armitage et al. 1983). Changes in BMWP score over time can, therefore, be monitored in order to identify temporal changes in water quality. As the index is the sum of all scores in the sample, it can be influenced by the number of taxa in the system, so a companion index, the average score per taxon (ASPT) is also used in order to give a mean BMWP score for the sampled assemblage. Using these two indices, an indication of the water quality of a freshwater system can be obtained very quickly. An experienced investigator could sort and score the sample on site, providing a biotic index in a matter of minutes.

The value of the BMWP score is primarily to assess temporal changes in water quality or to compare sites within the same river. However, as freshwater sites naturally vary in their physicochemical parameters and in their diversity (and BMWP score), it is not possible to directly compare indices between contrasting river systems; a BMWP score of 100 might be low in a rich chalk stream, but high in an oligotrophic mountain brook. To counter this problem, a set of 370 unpolluted sites covering the full range of lotic environmental conditions has been sampled in order to construct a prediction system for running waters (e.g., Wright et al. 1984; Armitage et al. 1987). Using this technique (river invertebrate prediction and classification system [RIVPACS]; Wright et al. 1993), the maximum predicted BMWP score can be computed for any river from a set of environmental data (e.g., width, discharge, altitude, pH, substratum type, etc.). The measured BMWP, number of macroinvertebrate families, and ASPT indices can then be compared with the predicted value to obtain an indication of river degradation. This system is used by the U.K. Environment Agency to classify rivers in terms of their water quality (see Wright et al. 1993).

The BMWP scoring system has been successfully applied in a range of systems beyond U.K. streams because of its inherent simplicity, the relatively low level of expertise and time required, and the general consistency of freshwater fauna taxonomy at the family level (e.g., Rundle et al. 1993). Examples include Spanish rivers (Zamora-Munoz and Alba-Tercedor 1996), high-altitude streams in Ecuador (Jacobsen 1998), and Scandinavian lakes (Johnson 1998). In addition, it has been modified for use in assessing river water quality in Australia (Smith et al. 1999) and has been successfully employed to monitor the impact of drought on an upper-estuarine community (Attrill et al. 1996). Although recent improvements to score allocation have been suggested (Walley and Hawkes 1996), the BMWP system appears to have many features allowing it to be used as a simple bioindicator of stress in freshwater systems worldwide.

In parallel with the BMWP scoring system, the index of biotic integrity (IBI) was developed to assess environmental degradation in U.S. streams (e.g., Karr 1981; Fausch et al. 1984). However, the IBI is based on attributes of fish communities and integrates characteristics of community, population, and individual levels (Karr 1987). The original index was based on a series of 12 "metrics" (see Fausch et al. 1990) relating to aspects of species richness or composition (e.g., number of fish species, number and identity of sunfish species), trophic composition (e.g., proportion that are omnivores, proportion that are insectivorous cyprinids), and individual fish abundance or condition (e.g., proportion that are hybrids, proportion with disease). No absolute data are incorporated into the index, each metric being scored 1, 3, or 5 depending on preset criteria. Some criteria are generic (e.g., for proportion that are omnivores: <20% scores 5, 20–45% scores 3, >45% scores 1), but others (e.g., number and identity of sucker species, number of individuals) have to be scored considering stream size and region. Therefore, a high a priori level of knowledge of local river ecology is required to accurately define the metrics. A metric scoring of 5 approximates to an undisturbed site, with values of 3 and 1 deviating from this condition and the final IBI being the sum of the metric scores. The index has been successfully employed to detect aspects of environmental degradation in North American streams (e.g., Karr et al. 1987; Steedman 1988) and has more recently been modified for use in a wider range of systems, such as rivers in Mexico (Lyons et al. 2000), west Africa (Kamdem Toham and Teugels 1999), and France (Oberdorff and Hughes 1992). Additionally, the technique has been further modified for estuarine fish communities (Deegan et al. 1997) and to produce an IBI using both invertebrates (e.g., Burton et al. 1999) and periphyton (Hill et al. 2000). While the IBI has been shown to perform well and integrates bioindicators at a range of biological levels, it has a series of disadvantages (discussed by Fausch et al. 1990), not least is the problem of obtaining quantitative samples of the fish community (Attrill and Depledge 1997) and the high level of expertise and ecological knowledge required to score the index.

The two biotic indices discussed have proved successful in terms of their applicability outside the systems for which they were designed. The BMWP scoring or prediction system is inherently simple and needs minimal modification before application in other systems, allows direct comparison between systems, but is perhaps limited to certain freshwater types and categories of impact (mainly organic pollution). The IBI does not allow direct comparison between studies because it is logistically much more complicated and requires site-specific criteria to be included in the index. However, as an applied concept it is applicable and adaptable to a wide range of environmental situations and impact types.

Distributional and Graphical Techniques

A range of techniques has been developed, mainly from the marine environment, which attempt to present and describe communities in graphical form. Such techniques enable any stress-related changes in community structure (either compositional or aggregate variability) to be directly visualized. Most techniques use the distribution of abundance or biomass between species as the basis for visual construction, following the model response of marine benthic communities to pollution (Pearson and Rosenberg 1978), while others adopt a nontaxonomic approach. Early attempts at graphical distributional techniques for assessing stress included rarefaction (Sanders 1968) and lognormal plots (Gray 1979). Even though both of these techniques have been successfully employed as bioindicators of environmental stress, their relative merits have also been the subject of much discussion and criticism (e.g., Hughes 1984; Nelson, 1987; Warwick 1993) and have been superseded by more unequivocal techniques. Two of these are abundance biomass comparison (ABC) curves, a development of k-dominance curves, and benthic biomass size spectra (BBSS).

The original k-dominance curves were constructed by ranking all benthic macroinvertebrate species in order of their abundance and then plotting their cumulative proportional representation in the community (Lambshead et al. 1983; Figure 1). Following the Pearson and Rosenberg (1978) model, polluted marine benthic systems tend to be dominated by large numbers of r-selected species, so the resulting k-dominance curve would be very shallow (i.e., the first ranked species contributes a high proportion of the total abundance). Unpolluted situations would have a steeper curve (Figure 1), as there is a greater evenness in the community. While this technique is still used in impact studies (e.g., Kaiser et al. 2000), it is difficult to obtain unequivocal results and it also tends to lead to rather subjective conclusions. Warwick (1986) added an extra dimension to the k-dominance approach, by supplementing the distribution of abundance between species with an equiva-

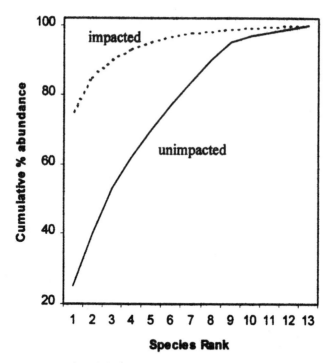

Figure 1. An example of the distributional or graphical technique using *k*-domi-nance curves (Lambshead et al. 1983) for impacted and unimpacted sites. Impacted sites tend to be dominated by one or two tolerant species, resulting in a much shal-lower slope than controls. For large species sets, it is standard to plot the *x*-axis on a logarithmic scale to prevent the plotting of a long tail of rare species.

lent curve for biomass, producing ABC curves. Following the above argu-ment, disturbed or polluted sites will have few large-bodied species, in con-trast to unimpacted, stable sites. By plotting both abundance and biomass lines on the same axis, a conclusion on the pollution status of the site can be drawn (Figure 2). In unpolluted systems, with a high evenness and several large-bodied species, the biomass curve will fall well above that for abun-dance, the reverse being true for heavily polluted situations dominated by high numbers of small species.

The ABC method has been extensively tested and validated for subtidal marine situations (e.g., Warwick 1993; Simboura et al. 1995) and has also been extended, with some reservations, to intertidal areas (Beukema 1988; Meire and Dereu 1990), brackish waters (Dauer et al. 1993; Reizopoulou et al. 1996), and demersal fish communities (Ungaro et al. 1998). The use of this technique has been advanced by the development of several ABC indi-ces (e.g., McManus and Pauly 1990; Meire and Dereu 1990), which give a quantitative measure of the difference between curves (Clarke 1990). This added robustness has encouraged the use of the ABC method in freshwaters, where it has been used to assess impacts on fish communities (Coeck et al. 1993; Penczak and Kruk 1999). Therefore, ABC curves and indices provide a

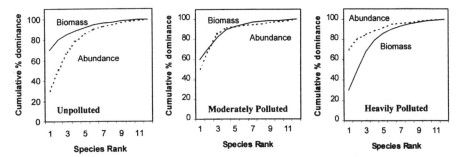

Figure 2. Hypothetical abundance biomass comparison (ABC) curves for marine benthic sites that are unpolluted, moderately polluted, and heavily polluted by organic waste (after Warwick 1986). The plots give interpretable results of pollution status without the need for historical data.

clear bioindicator of stress, particularly for organic pollution where increasing abundance of small organisms with increasing pollution is evident, although their use for determining the impact of other stressors where such a response is not apparent (e.g., metal pollution) may be more limited (Del Valls et al. 1998). Additionally, the technique will not necessarily be effective in systems where the theoretical decrease in body size with increasing stress does not apply (Lardicci and Rossi 1998), such as in freshwater macroinvertebrate communities.

An alternative graphical approach, which also, potentially, uses changes in biomass distribution, is the construction of biomass size spectra (BSS). Unlike previous bioindicators, size spectra do not require the identification of taxa, but instead categorize the community in terms of organism size, regardless of taxonomic composition, and can integrate organism response across a range of biotic components (microbes to large macrofauna). The resulting graph presents the proportion of individuals contributing to each defined size-class, generally a geometric progression based on $\log_2$ size intervals (for full details on theory see, for example, Sheldon and Parsons 1967; Schwinghamer 1981; Hanson 1990). While size spectra have been successfully used to compare plankton (e.g., Garcia et al. 1995; Cyr and Peters 1996) and fish communities (e.g., Duplisea and Kerr 1995; Macpherson and Gordoa 1996), the most extensive use of the technique has been in benthic assemblages where metazoan size can span 10 orders of magnitude (Poff et al. 1993). These spectra have been constructed for both marine (e.g., Saiz-Salinas and Ramos 1999), brackish (e.g., Drgas et al. 1998), and freshwater (e.g., Cattaneo 1993) benthic systems with the general conclusion that in comparative unperturbed situations, the spectra are remarkably consistent across systems (e.g., Schwinghamer 1985; Cattaneo 1993).

Benthic biomass size spectra (BBSS), therefore, have the potential to be a useful tool for assessing the influence of stress on the structure of a large part of the community (an additional facet of compositional variability) rather than simply changes in species composition or dominance of macroinvertebrates. Recent improvements in the methodology (e.g., Ramsay et al. 1997) have made construction of BBSS labor-efficient, even amenable to automation (Saiz-Salinas and Gonzalez-Oreja 2000). However, to date, few investigations have used BBSS as a bioindicator of environmental stress and existing studies have produced variable results. Schwinghamer (1988), Duplisea and Hargrave (1996), and Raffaelli et al. (2000) found no shifts in spectra associated with organic enrichment. Warwick, et al. (1986), using data from the literature, hypothesized that enriched communities should demonstrate a size convergence (and elevated biomass) at the meiofauna–macrofauna boundary, but found no empirical evidence to support this. Gonzalez-Oreja and Saiz-Salinas (1999), however, used BBSS to successfully document impact in a polluted estuary. While the ability of BBSS to detect the impact of organic pollution remains unclear, the technique appears more suitable as a bioindicator of metal pollution, a stressor where ABC curves appear less appropriate. Stark (1998b) reported a differential response of marine taxa to copper pollution with consequent impacts on certain size fractions of the community, while Abada (2000) reported a significant modification of size spectra in a metal polluted estuary when compared with controls. BBSS have many appealing advantages as a bioindicator, in both marine and freshwater systems, but Warwick's (1993) statement that the technique requires further testing remains pertinent.

Multivariate Analyses

Due to the development and accessibility of computing technology, multivariate statistical methods (which tend to be very computationally demanding) have become popular methods of analyzing community data. Multivariate methods compare the extent to which two or more samples share species, at comparable levels of abundance (Clarke and Warwick 1994a), so they are, therefore, the best techniques for determining the impact of stress on compositional variability within a community. Their popularity and perceived power to differentiate patterns in complex community data sets has led to their rather indiscriminate use in both marine and freshwater systems. In particular, there can be a tendency to ignore the rigors of good experimental design where a clear, well designed survey is essential in order to provide unequivocal, interpretable results (see Clarke and Green 1988; Underwood 1997, 2000). Taking a suite of samples from the environment and running the data through a computer program "to see what might come out" is not a scientifically defensible way to conduct environmental assessment studies. The early statement of Cormack (1971) that the "availability of …classification techniques has led to the waste of more valuable scientific time than any other 'statistical' innovation" still holds very true.

Nevertheless, used correctly, multivariate statistical techniques are most useful tools to employ in order to determine stress effects on aquatic systems. Many techniques that are available can be broadly separable into two categories: (a) classification (to define clusters of samples from a group that are more similar than samples in a different cluster), and (b) ordination (a pictorial representation of all samples, so that distance between samples equates to their similarity in community structure). It is beyond the scope of this chapter to review the range of techniques available, particularly for classification. For further information, see one of the many publications on the subject, for example, Everitt (1978), Clarke and Warwick (1994a), Manly (1994), and Micallef and Schembri (1998).

Ordination techniques are either nonmetric, based on a matrix of similarity indices (e.g., multidimensional scaling [MDS, Kruskal and Wish 1978]), or metric, using a "line of best-fit" approach from sample points (e.g., principal components analysis [PCA, Chatfield and Collins 1980]; detrended correspondence analysis [DECORANA, Hill and Gauch 1980]). Even though useful for environmental data, the suitability and stability of metric methods for the analysis of community data has been questioned (Van Groenewoud 1992; Clarke and Warwick 1994a; Oksanen and Minchin 1997; Micallef and Schembri 1998) and comparative tests have shown nonmetric MDS to be the most robust ordination technique for analyzing these data (Minchin 1987; Marchant 1990; Rydgren 1996). Therefore, nonmetric multivariate methods (and derivative statistical tests) based on similarity indices are considered further as the most suitable for bioindicator studies.

The starting point for nonmetric techniques is a similarity matrix, consisting of a calculated similarity index (e.g., Bray-Curtis) for each pair of samples in the species–site data array (following appropriate transformation, see above). Adapting the strategy of Clarke (1993), community-level responses to stress can be investigated in three main ways, following the construction of the similarity matrix:

1. Displaying the Community Pattern

The community pattern is a visual indication of how stressed and control sites are related in terms of their community composition. Two methods are used to portray the relationship between communities sampled: classification by cluster analysis resulting in a dendrogram (Figure 3a) and ordination by MDS producing a 2-dimensional "map" (Figure 3b). While dendrograms are useful for separating sites into distinct clusters, ordinations provide the best visual representation of community patterns, where distances between samples directly relate to their community similarity (Clarke 1993). As a method of demonstrating the effect of stress on communities, MDS ordinations have been widely employed (particularly in the marine environment) to demonstrate separation of impacted sites from controls (e.g., Dawson-Shepherd et al. 1992; Smith and Simpson 1995), portray the distribution of sites over a pollution gradient (e.g., Somerfield et al. 1994; Rundle and Attrill 1995), indicate temporal changes to a community following a perturbation (e.g.,

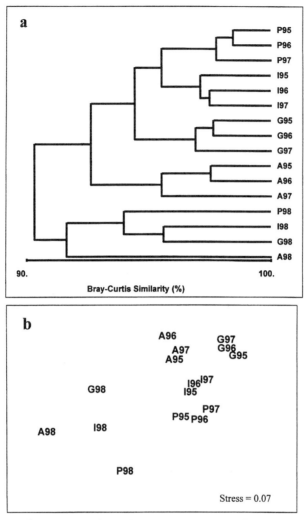

Figure 3. (a) Dendrogram resulting from classification of coral reef communities before and after effects of El Niño Southern Oscillation (ENSO). Cluster analysis demonstrates that post-ENSO sites are clustered separately in terms of the community composition, regardless of replicate reef. Prior to ENSO, reefs clustered together having similar communities regardless of year. (b) Ordination by MDS (see text), providing a 2-D map where distances between sites relate to their similarity in community composition: the further the sites are away, the more dissimilar the communities. Post-ENSO sites (1998) are clearly separated from preimpact sites.

Attrill et al. 1996; Kelmo and Attrill 2001, Figure 3), and analyze the results from experimental manipulations (e.g., Dahl and Blanck 1996; Morrisey et al. 1996).

2. Testing the Significance of Differences

If undertaking an impact study, it is important to determine whether the community at the putatively stressed site is significantly different from controls. Ordinations and dendrograms are very useful pictorial representations of community change, but a criticism of multivariate techniques has been that they lack endpoints open to significance testing. Traditional multivariate methods (e.g., principal components, MANOVA) are inappropriate for many community data sets (Clarke 1999), so Clarke and Green (1988) proposed a permutation test (analysis of similarities, [ANOSIM]; *sensu* Mantel 1967), which allows one- and two-way testing of significance between a priori sets of samples identified during the sampling design phase. Table 3 displays the ANOSIM results for the data from Figure 3, indicating that the composition of coral communities sampled in 1998 (post impact) was significantly different from previous years. The technique has been developed and extended (Clarke and Warwick 1994b; Chapman and Underwood 1999) and remains an important tool for biomonitoring studies using multivariate statistics.

3. Linking the Community Patterns to Environmental Variables

While significant differences may be detected in community data sets and given that natural environmental variables are controlled for as much as possible in the design, it is still desirable to determine which variable correlates most closely with the observed pattern in community composition ("best explains" the observed distribution; Clarke 1993). Identifying influential variables is most useful when investigating the impact of an environmental gradient, rather than discrete control and test sites. Clarke and Ainsworth (1993) describe a method of correlating the similarity matrices generated from both biotic and abiotic data sets (known as BIOENV). This produces a weighted Spearman coefficient for each single environmental variable and combinations of variables, defining which has the highest correlation with the community matrix (Table 3). Of course, the adage "correlation is not causation" applies here, but the results of this analysis can form a very plausible explanation for community change (Clarke 1993). An alternative technique for investigating the relationship between community and environmental patterns is (detrended) canonical correspondence analysis (CCA, or CANOCO). This method involves the fusion of species–sample and environmental data to produce a separate ordination method (e.g., ter Braak 1986; ter Braak and Verdonschot 1995), which, like BIOENV, is very useful for gradient analysis. The ordinations can be somewhat confusing if not carefully presented and the relationships with environmental variables somewhat subjective, but the method has been successfully used in biomonitoring studies, particularly in freshwater systems (e.g., Kinross et al. 1993; Ruse 1996).

Multivariate techniques are a powerful and useful tool in biomonitoring studies and are recommended, particularly in order to assess compositional variability of communities. The methods are also highly sensitive and can detect the impact of stress when other methods (e.g., univariate analyses) do not (Warwick and Clarke 1991; Cao et al. 1996; Lardicci and Rossi 1998). The use of multivariate versus univariate techniques is exemplified by the "classic" case study of the impact of Norwegian North

Table 3. Examples of further nonmetric multivariate statistical methods. Data correspond to the case study in Figure 3. Analysis of similarities (ANOSIM) results provide a global *r* value and *P*-value indicating that the reef top coral communities post–El Niño (1998) are significantly different from earlier years. Comparison of biotic and abiotic measurements (BIOENV analysis) indicates the environmental variables most closely correlated with the changes in community composition, both individually and in combinations, with mean temperature alone providing the best explanation for community patterns.

ANOSIM results	*Difference between years*: Global $r = 0.258$, $P = 0.019$
	Pairwise test: 1995–1997 communities not different (P all > 0.91)
	1998 community different from all other years (P all < 0.03)
BIOENV results (maximum correlation in bold)	*Single variables* (ρ_w): **mean temperature (0.58)**, temperature range (0.53), cloud cover (0.51)
	Combination of 2 variables (ρ_w): cloud cover and temperature range (0.53)
	Combination of 3 variables (ρ_w): irradiance, temperature range and cloud cover (0.53)

Sea oil drilling platforms on benthic communities (Gray et al. 1990, Figure 4). A series of sample sites was arranged in a star design (Figure 4a), allowing spatial separation of controls, and sites were categorized into four distance bands away from the oil platform. Univariate analysis of diversity (Figure 4b) indicated a significant impact less than 250 m from the platform but no detectable effect outside this distance. However, MDS ordination (Figure 4c) revealed that samples up to 3,000 m from the impact source had a different community composition from controls in which the significance was confirmed by the ANOSIM test. Use of multivariate techniques, therefore, detected a much more extensive impact on the community than if univariate methods were employed alone, which can have dramatic consequences for any management policy or environmental legislation. This example also highlights the importance of assessing both aggregate and compositional variability when using communities as biomonitors of stress.

A summary of all recommended techniques and approaches is provided in Appendix 2.

Additional Community Considerations

Taxonomic Resolution

Perhaps the most persistent criticism of community studies is that they are comparatively labor intensive and require a high level of taxonomic expertise to identify organisms to species level. While this is not necessarily true for fish communities, it certainly holds for the identification of benthic inver-

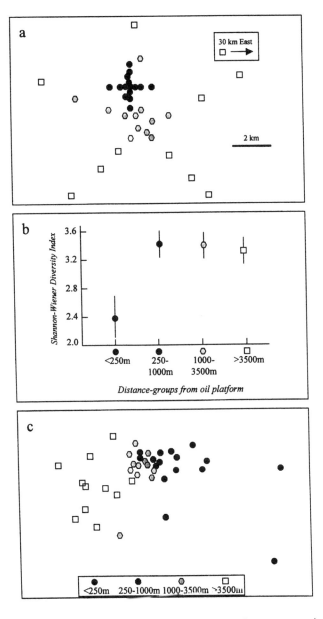

Figure 4. Results from the "Ekofisk" study (Gray et al. 1990) on the impact of an oil drilling platform, indicating the sensitivity of multivariate methods. (a) Sampling design, each symbol representing a sample point at successive distances from the platform (key on 4(c). Note spatial separation of "control" sites >3.5 km from the platform (•), including one site 30 km to the east. Platform is beneath black circles (sites <250 m from platform). (b) Univariate results indicate a significant impact (in terms of reduced diversity) up to 250 m from the platform. (c) MDS ordination of sites, using the symbols from 4a to denote distance of sites from the platform. Samples from up to 3,500 m from the platform have a significantly different community composition than controls (white squares), indicating that platform activity is having an impact on compositional variability not revealed by univariate analysis.

tebrates to species level, providing an employment lifeline for benthic ecologists world-wide! However, a solution to this problem is to reduce the taxonomic resolution of surveys by identifying organisms to, for example, family level or above. This approach is inappropriate for conservation or biodiversity assessment purposes, but has great potential for facilitating the use of invertebrate communities as bioindicators.

The development of the BMWP scoring system discussed earlier is a prime example of how decreased taxonomic resolution can be successfully employed to detect impact, with the training of operators to identify the relevant families being a comparatively quick process. The potential benefits of reduced levels of identification have been assessed in a range of aquatic systems, particularly in the comparatively diverse marine environment (see review by Olsgard and Somerfield 2000). Warwick (1988) reanalyzed the data from several well-described pollution studies and found, using multivariate techniques, that impact was detectable in many cases (including the Amoco Cadiz oil spill [Laubier 1980]) at the phylum level of identification. In such a case, a very low level of expertise is required in order to detect impact using communities (Attrill and Depledge 1997). Other similar studies have also reported the detection of impacts at higher taxonomic levels (e.g., Ferraro and Cole 1990; Olsgard et al. 1998; Drake et al. 1999), at times when impacts at the species level was not evidenced in univariate measures (Warwick 1993). Even though this technique has potentially great benefits for bioindicator studies, it is important to consider that all the cited cases in this paper were based on species-level baselines (as recommended by Olsgard and Somerfield 2000), and changes to taxonomic level identification and aggregation can considerably alter the multivariate description of the community (Bowman and Bailey 1997; Olsgard et al. 1998). However, as long as a consistent response to perturbation is detectable, this is not necessarily a problem for impact studies, but further testing in complex environments is advisable (Vanderklift et al. 1996).

An additional advantage of aggregation of data to higher taxonomic levels is that it allows direct comparison between geographically separated environmental assessment studies, which may have a vastly different species complement, but the same array of families or phyla. The main benefit of this version of the meta-analysis approach (cf. Gurevitch and Hedges 1993; Osenberg et al. 1999) is that it allows the relative severity of perturbations to be assessed, using comparative positions on ordinations. Warwick and Clarke (1993) used the data from four well-described impact studies, plus a bioturbation case study and three unpolluted situations, aggregating the data to phylum level. Using a calculated "production" approximate (based on a combination of biomass and abundance data), they plotted all sites on one MDS ordination. A clear gradient in pollution severity was apparent, the effects of organic enrichment being more severe than oil. This meta-analysis has provided a template allowing the comparative severity of other impacts to be assessed (e.g., Agard et al. 1993; Drake et al. 1999). Rundle and Attrill (1995) combined the data for two geographically separated stream systems in the United Kingdom, using family level identification of microcrustaceans, allowing the comparative severity of acidification to be explored. The meta-analysis approach allows direct comparison of community responses and is, therefore, a useful bioindicator of relative impact severity, which warrants further consideration, particularly on a global scale.

Relationships with Other Levels of Biological Organization

Biological systems are inherently hierarchical in nature (Rosen 1969), providing a framework for ecotoxicological investigation. Simplified diagrams linking the levels of biological organization may provide the raison d'être for ecotoxicological investigation, but oversimplification of the system to a linear model ignores one of the key characters of hierarchical organization, that of emergent properties. Emergent characteristics are a well-known phenomenon in other fields; for example, the properties of water (e.g., Kier and Cheng 1997) are unpredictable from the study of hydrogen and oxygen; the synchrony of an ant colony (e.g., Sole et al. 1993) could not be predicted from observation of individual ants, no matter how detailed the investigation. Similarly, communities demonstrate many emergent properties that only become apparent when the community itself is examined (e.g., Hatcher 1997). These emergent properties are often the cause of unpredictable variability (Drake 1991). Bown et al. (1999) studied a two-species microcosm, but concluded that the dynamics of this very simple system were an emergent property that "cannot be deduced from a study of the components in isolation." This has important consequences for bioindicator studies, as many studies aim to provide a method targeted at lower levels of biological organization that will predict stress responses at higher levels (e.g., communities). Due to the emergent properties of hierarchical systems, this is not feasible (F. E. J. Fry in Kerr 1976), particularly if the biomarker is aimed at causative links between levels of biological organization. Munkittrick and McCarty (1995) suggest that this view of causal linkages between lower levels and upper levels of biological organization may be limiting further theoretical developments in aquatic toxicology.

Biomarkers at lower levels of biological organization are potentially very useful for assessing stress effects, but they *must* be correlated and calibrated using accompanying investigations at an ecologically relevant level (McCarty and Munkittrick 1996), such as the community. Only when the emergent properties of a system are fully described can selection of variables and analysis at lower levels be undertaken (Kerr 1976). A biomarker that is calibrated and correlates with community response is, indeed, a valid bioindicator (McCarty and Munkittrick 1996), although not necessarily causative! However, studies solely employing low level biomarkers that correlate with contamination cannot be used to predict higher level response. By incorporating community-level studies into lower-level biomarker investigations, and accepting the restrictions imposed by biological hierarchies, investigators can improve the development and predictive capability of bioindicators.

References

Abada, A. E. A. 2000. From rivers to oceans: a comparison of contrasting aquatic ecosystems using benthic size spectra. Ph.D. thesis. University of Plymouth, Plymouth, England.

Adams, S. M., M. S. Greeley, and M. G. Ryon. 2000. Evaluating effects of contaminants on fish health at multiple levels of biological organization: extrapolating from lower to higher levels. Human and Ecological Risk Assessment 6:15–27.

Agard, J. B. R., J. Gobin, and R. M. Warwick. 1993. Analysis of marine macrobenthic community structure in relation to pollution, natural oil seepage and seasonal disturbance in a tropical environment (Trinidad, West Indies). Marine Ecology Progress Series 92:233–243.

Armitage, P. D., R. J. M. Gunn, M. T. Furse, J. F. Wright, and D. Moss. 1987. The use of prediction to assess macroinvertebrate response to river regulation. Hydrobiologia 144:25–32.

Armitage, P. D., D. Moss, J. F. Wright, and M. T. Furse. 1983. The performance of a new biological water-quality score system based on macroinvertebrates over a wide range of unpolluted running-water sites. Water Research 17:333–347.

Attrill, M. J. 2000. Stress in marine communities: an introduction. Journal of Aquatic Ecosystem Stress and Recovery 7:1–2.

Attrill, M. J., and M. H. Depledge. 1997. Community and population indicators of ecosystem health: targeting links between levels of biological organisation. Aquatic Toxicology 38:183–197.

Attrill, M. J., S. D. Rundle, and R. M. Thomas. 1996. The influence of drought-induced low freshwater flow on an upper-estuarine macroinvertebrate community. Water Research 30:261–268.

Bargos, T., J. M. Mesanza, A. Basaguren, and E. Orive. 1990. Assessing river water quality by means of multifactorial methods using macroinvertebrates: a comparative study of main water courses of Biscay. Water Research 24:1–10.

Begon, M., J. L. Harper, and C. R. Townsend. 1996. Ecology: individuals, populations, and communities. Blackwell Scientific, Oxford, England.

Beukema, J. J. 1988. An evaluation of the ABC-method as applied to macrozoobenthic communities living on tidal flats in the Dutch Wadden Sea. Marine Biology 99:425–433.

Bonsdorff, E., E. M. Blomqvist, J. Mattila, and A. Norkko. 1997. Long-term changes and coastal eutrophication. Examples from the Åland Islands and the Archipelago Sea, northern Baltic Sea. Oceanologica Acta 20:319–329.

Bowman, M. F., and R. C. Bailey. 1997. Does taxonomic resolution affect the multivariate description of the structure of freshwater benthic macroinvertebrate communities? Canadian Journal of Fisheries and Aquatic Sciences 54:1802–1807.

Bown, J. L., and seven coauthors. 1999. Evidence for emergent behaviour in the community-scale dynamics of a fungal microcosm. Proceedings of the Royal Society of London Series B–Biological Sciences 266:1947–1952.

Burton, T. M., and five coauthors. 1999. Development of a preliminary invertebrate index of biotic integrity for Lake Huron coastal wetlands. Wetlands 19:869–882.

Cao, Y., A. W. Bark, and P. W. Williams. 1996. Measuring the responses of macroinvertebrate communities to water pollution: a comparison of multivariate approaches, biotic and diversity indices. Hydrobiologia 341:1–19.

Cao, Y., D. D. Williams, and N. E. Williams. 1999. Data transformation and standardization in the multivariate analysis of river water quality. Ecological Applications 9:669–677.

Carpenter, S. R. 1996. Microcosm experiments have limited relevance for community and ecosystem ecology. Ecology 77:677–680.

Cattaneo, A. 1993. Size spectra of benthic communities in Laurentian streams. Canadian Journal of Fisheries and Aquatic Sciences 50:2659–2666.

Chapman, M. G., and A. J. Underwood. 1999. Ecological patterns in multivariate assemblages: information and interpretation of negative values in ANOSIM tests. Marine Ecology Progress Series 180:257–265.

Chatfield, C., and A. J. Collins. 1980. Introduction to multivariate analysis. Chapman and Hall, London.

Clarke, K. R. 1990. Comparisons of dominance curves. Journal of Experimental Marine Biology and Ecology 138:143–157.

Clarke, K. R. 1993. Non-parametric multivariate analyses of changes in community structure. Australian Journal of Ecology 18:117–143.

Clarke, K. R. 1999. Non-metric multivariate analysis in community-level ecotoxicology. Environmental Toxicology and Chemistry 18:118–127.

Clarke, K. R., and M. Ainsworth. 1993. A method of linking multivariate community structure to environmental variables. Marine Ecology Progress Series 92:205–219.

Clarke, K. R., and R. H. Green. 1988. Statistical design and analysis for a "biological effects" study. Marine Ecology Progress Series 46:213–226.

Clarke, K. R., and R. M. Warwick. 1994a. Changes in marine communities: an approach to statistical analysis and interpretation. Natural Environmental Research Council, U.K.:

Clarke, K. R., and R. M. Warwick. 1994b. Similarity-based testing for community pattern: the 2-way layout with no replication. Marine Biology 118:167–176.

Clarke, K. R., and R. M. Warwick. 1998. A taxonomic distinctness index and its statistical properties. Journal of Applied Ecology 35:523–531.

Clements, W. H., and P. M. Kiffney. 1994. Assessing contaminant effects at higher levels of biological organization. Environmental Toxicology and Chemistry 13:357–359.

Coeck, J. A., R. Vandelannoote, R. Yseboodt, and R. F. Verheyen. 1993. Use of the abundance/biomass method for comparison of fish communities in regulated and unregulated lowland rivers in Belgium. Regulated Rivers Research and Management 8:73–82.

Condit, R., and six coauthors. 1996. Species-area and species-individual relationships for tropical trees: a comparison of three 50-ha plots. Journal of Ecology 84:549–562.

Cormack, R. M. 1971. A review of classification. Journal of the Royal Statistics Society Series A 134:321–367.

Cornell, H., L. E. Hurd, and V. A. Lotrich. 1976. A measure of response to perturbation used to assess structural change in some polluted and unpolluted stream fish communities. Oecologia 23:335–342.

Cottingham, K. L., and D. E. Schindler. 2000. Effects of grazer community structure on phytoplankton response to nutrient pulses. Ecology 81:183–200.

Cyr, H., and R. H. Peters. 1996. Biomass size spectra and the prediction of fish biomass in lakes. Canadian Journal of Fisheries and Aquatic Sciences 53:994–1006.

Dahl, B., and D. Blanck. 1996. Pollution-induced community tolerance (PICT) in periphyton communities established under tri-n-butyltin (TBT) stress in marine microcosms. Aquatic Toxicology 34:305–325.

Dauer, D. M., M. W. Luckenbach, and A. J. Rodi. 1993. Abundance biomass comparison (ABC method): effects of an estuarine gradient, anoxic/hypoxic events and contaminated sediments. Marine Biology 116:507–518.

Dawson-Shepherd, A. R., R. M. Warwick, K. R. Clarke, and B. E. Brown. 1992. An analysis of fish community response to coral mining in the Maldives. Environmental Biology of Fishes 33:367–380.

Deegan, L. A., J. T. Finn, S. G. Ayvazian, C. A. Ryder-Kieffer, and J. Buonaccorsi. 1997. Development and validation of an estuarine biotic integrity index. Estuaries 20:601–617.

Del Valls, T. A., M. Conradi, E. Garcia-Adiego, J. M. Forja, and A. Gomez-Parra. 1998. Analysis of macrobenthic community structure in relation to different environmental sources of contamination in two littoral ecosystems from the Gulf of Cadiz (SW Spain). Hydrobiologia 385:59–70.

Detenbeck, N. E., P. W. Devore, G. J. Niemi, and A. Lima. 1992. Recovery of temperate-stream fish communities from disturbance: a review of case-studies and synthesis of theory. Environmental Management 16:33–53.

Downing, J. A. 1979. Aggregation, transformation and the design of benthos sampling programs. Journal of the Fisheries Research Board of Canada 26:1454–1463.

Drake, J. A. 1991. Community-assembly mechanics and the structure of an experimental species ensemble. American Naturalist 137:1–26.

Drake, P., F. Baldo, V. Saenz, and A. M. Arias. 1999. Macrobenthic community structure in estuarine pollution assessment on the Gulf of Cadiz (SW Spain): is the phylum-level meta-analysis approach applicable? Marine Pollution Bulletin 38:1038–1047.

Drgas, A., T. Radziejewska, and J. Warzocha. 1998. Biomass size spectra of nearshore shallow-water benthic communities in the Gulf of Gdansk (southern Baltic sea). Marine Ecology–PSZNI 19:209–228.

Duplisea, D. E., and B. T. Hargrave. 1996. Response of meiobenthic size-structure, biomass and respiration to sediment organic enrichment. Hydrobiologia 339:161–170.

Duplisea, D. E., and S. R. Kerr. 1995. Application of a biomass size spectrum model to demersal fish data from the Scotian shelf. Journal of Theoretical Biology 177:263–269.

Elmberg, J., K. Sjoberg, P. Nummi, and H. Poysa. 1994. Patterns of lake acidity and waterfowl communities. Hydrobiologia 280:201–206.

Everitt, B. 1978. Graphical techniques for multivariate data. Heinemann, London.

Fagan, P., A. G. Miskiewicz, and P. M. Tate. 1992. An approach to monitoring sewage outfalls: a case study on the Sydney deepwater sewage outfalls. Marine Pollution Bulletin 25:172–180.

Fausch, K. D., J. R. Karr, and P. R. Yant. 1984. Regional application of an index of biotic integrity based on stream fish communities. Transactions of the American Fisheries Society 113:39–55.

Fausch, K. D., J. Lyons, J. R. Karr, and P. L. Angermeir. 1990. Fish communities as indicators of environmental degradation. Pages 123–144 *in* S. M. Adams, editor. Biological indicators of stress in fish. American Fisheries Society, Symposium 8, American Fisheries Society, Bethesda, Maryland.

Fauth, J. E., and five coauthors. 1996. Simplifying the jargon of community ecology: a conceptual approach. American Naturalist 147:282–286.

Ferraro, S. P., and F. A. Cole. 1990. Taxonomic level and sample size sufficient for assessing pollution impacts on the southern California Bight macrobenthos. Marine Ecology Progress Series 67:251–262.

Furse, M. T., J. F. Wright, P. D. Armitage, and D. Moss. 1981. An appraisal of pond-net samples for biological monitoring of lotic macro-invertebrates. Water Research 15:679–689.

Garcia, C. M., F. Echevarria, and F. X. Niell. 1995. Size structure of plankton in a temporary, saline inland lake. Journal of Plankton Research 17:1803–1817.

Garcia-Criado, F., A. Tome, F. J. Vega, and C. Antolin. 1999. Performance of some diversity and biotic indices in rivers affected by coal mining in northwestern Spain. Hydrobiologia 394:209–217.

Gimaret-Carpentier, C., R. Pelissier, J. P. Pascal, and F. Houllier. 1998. Sampling strategies for the assessment of tree species diversity. Journal of Vegetation Science 9:161–172.

Gonzalez-Oreja, J. A., and J. I. Saiz-Salinas. 1999. Loss of heterotrophic biomass structure in an extreme estuarine environment. Estuarine Coastal and Shelf Science 48:391–399.

Gray, J. S. 1979. Pollution induced changes in populations. Philosophical Transactions of the Royal Society of London B Biological Sciences 286:545–561.

Gray, J. S. 1981. The ecology of marine sediments. Cambridge studies in modern biology: 2. Cambridge University Press, England.

Gray, J. S., K. R. Clarke, R. M. Warwick, and G. Hobbs. 1990. Detection of initial effects of pollution on marine benthos: an example from the Ekofisk and Eldfisk oilfields, North Sea. Marine Ecology Progress Series 66:285–299.

Green, R. H. 1979. Sampling design and statistical methods for environmental biologists. Wiley, Chichester, England.

Gurevitch, J., and L. V. Hedges. 1993. Meta-analysis: combining the results of independent studies in experimental ecology. Pages 378–398 *in* S. Scheiner and J. Gurevitch, editors. The design and analysis of ecological experiments. Chapman and Hall, New York.

Gyedu-Ababio, T. K., J. P. Furstenberg, D. Baird, and A. Vanreusel. 1999. Nematodes as indicators of pollution: a case study from the Swartkops River system, South Africa. Hydrobiologia 397:155–169.

Hall, S. J., and S. P. Greenstreet. 1998. Taxonomic distinctness and diversity measures: responses in marine fish communities. Marine Ecology Progress Series 166:227–229.

Hanson, J. M. 1990. Macroinvertebrate size distributions of two contrasting freshwater macrophyte communities. Freshwater Biology 24:481–491.

Hatcher, B. G. 1997. Coral reef ecosystems: how much greater is the whole than the sum of the parts? Coral Reefs 16(Supplement S):S77–S91.

Hawkes, H. A. 1998. Origin and development of the biological monitoring working party score systems. Water Research 32:964–968.

Henderson, P. A. 1989. On the structure of the inshore fish community of England and Wales. Journal of the Marine Biological Association of the United Kingdom 69:145–163.

Hill, B. H., and five coauthors. 2000. Use of periphyton assemblage data as an index of biotic integrity. Journal of the North American Benthological Society 19:50–67.

Hill, M. O., and H. G. Gauch. 1980. Detrended correspondence analysis, an improved ordination technique. Vegetatio 42:47–48.

Holbrook, S. J., R. J. Schmitt, and J. S. Stephens. 1997. Changes in an assemblage of temperate reef fishes associated with a climate shift. Ecological Applications 7:1299–1310.

Hughes, R. G. 1984. A hypothesis concerning the influence of competition and stress on the structure of marine benthic communities. Pages 391–400 *in* P. E. Gibbs, editor. Proceedings of the 19th European Marine Biology Symposium. Cambridge University Press, Cambridge, England.

Humphrey, C. L., D. P. Faith, and P. L. Dostine. 1995. Base-line requirements for assessment of mining impact using biological monitoring. Australian Journal of Ecology 20:150–166.

Huston, M. A. 1997. Microcosm experiments have limited relevance for community and ecosystem ecology: synthesis of comments. Ecology 80:1088–1089.

Hynes, H. B. N. 1960. The biology of polluted waters. Liverpool University Press, Liverpool, England.

Ivorra, N., and five coauthors. 1999. Translocation of microbenthic algal assemblages used for in situ analysis of metal pollution in rivers. Archives of Environmental Contamination and Toxicology 37:19–28.

Jacobsen, D. 1998. The effect of organic pollution on the macroinvertebrate fauna of Ecuadorian highland streams. Archiv Fur Hydrobiologie 143:179–195.

Johnson, R. K. 1998. Spatiotemporal variability of temperate lake macroinvertebrate communities: Detection of impact. Ecological Applications 8:61–70.

Kaiser, M. J., K. Ramsay, C. A. Richardson, F. E. Spence, and A. R. Brand. 2000. Chronic fishing disturbance has changed shelf sea benthic community structure. Journal of Animal Ecology 69:494–503.

Kamdem Toham, A., and G. G. Teugels. 1999. First data on an index of biotic integrity (IBI) based on fish assemblages for the assessment of the impact of deforestation in a tropical West African river system. Hydrobiologia 392:29–38.

Karr, J. R. 1981. Assessment of biotic integrity using fish communities. Fisheries 6:21–27.

Karr, J. R. 1987. Biological monitoring and environmental assessment: a conceptual framework. Environmental Management 11:249–256.

Karr, J. R., P. R. Yant, K. D. Fausch, and I. J. Schlosser. 1987. Spatial and temporal variability of the index of biotic integrity in three midwestern streams. Transactions of the American Fisheries Society 116:1–11.

Kelmo, F., and M. J. Attrill. 2001. Cnidarian community structure of coastal reefs from northern Bahia, Brazil. Bulletin of Marine Science 69.

Kerr, S. R. 1976. Ecological analysis and the Fry paradigm. Journal of the Fisheries Research Board of Canada 33:329–335.

Kier, L. B., and C. K. Cheng. 1997. A cellular automata model of the soluble state. Journal of Mathematical Chemistry 21:71–81.

Kinross, J. H., N. Christofi, P. A. Read, and R. Harriman. 1993. Filamentous algal communities related to pH in streams in the Trossachs, Scotland. Freshwater Biology 30:301–317.

Kohl, M., and A. Zingg. 1996. Applicability of diversity indices in long term studies on biodiversity in forest stands. Allgemeine Forst und Jagdzeitung 167:76–85.

Korneva, L. G. 1996. Impact of acidification on structural organization of phytoplankton community in the forest lakes of north-western Russia. Water Science and Technology 33:291–296.

Krebs, C. J. 1985. The experimental analysis of distribution and abundance, 3rd edition. Harper Collins, New York.

Kruskal, J. B., and M. Wish. 1978. Multidimensional scaling. Sage Publications, Beverly Hills, California.

Lambshead, P. J. D., H. M. Platt, and K. M. Shaw. 1983. The detection of differences among assemblages of marine benthic species based on an assessment of dominance and diversity. Journal of Natural History 17:859–874.

Lardicci, C., and F. Rossi. 1998. Detection of stress on macrozoobenthos: evaluation of some methods in a coastal Mediterranean lagoon. Marine Environmental Research 45:367–386.

Lardicci, C., F. Rossi, and F. Maltagliati. 1999. Detection of thermal pollution: Variability of benthic communities at two different spatial scales in an area influenced by a coastal power station. Marine Pollution Bulletin 38:296–303.

Laubier, L. 1980. The Amoco Cadiz oil spill: an ecological impact study. Ambio 9:268–276.

Lyons, J., and five coauthors. 2000. Development of a preliminary index of biotic integrity (IBI) based on fish assemblages to assess ecosystem condition in the lakes of central Mexico. Hydrobiologia 418:57–72.

Macpherson, E., and A. Gordoa. 1996. Biomass spectra in benthic fish assemblages in the Benguela system. Marine Ecology Progress Series 138:27–32.

Madoni, P. 1994. A sludge biotic index (SBI) for the evaluation of the biological performance of activated-sludge plants based on the microfauna analysis. Water Research 28:67–75.

Magnussen, S., and T. J. B. Boyle. 1995. Estimating sample-size for inference about the Shannon-Weaver, and the Simpson indexes of species diversity. Forest Ecology and Management 78:71–84.

Manly, B. F. 1994. Multivariate statistical methods: a primer. 2nd edition. Chapman and Hall, London.

Mantel, N. 1967. The detection of disease clustering and a generalized regression approach. Cancer Research 27:247–259.

Marchant, R. 1990. Robustness of classification and ordination techniques applied to macroinvertebrate communities from the La-Trobe River, Victoria. Australian Journal of Marine and Freshwater Research 41:493–504.

McCarty, L. S., and K. R. Munkittrick. 1996. Environmental biomarkers in aquatic ecology: fiction, fantasy or functional? Human and Ecological Risk Assessment 2:268–274.

McManus, J. W., and D. Pauly. 1990. Measuring ecological stress: variations on a theme by R. M. Warwick. Marine Biology 106:305–308.

Meire, P. M., and J. Dereu. 1990. Use of the abundance biomass comparison method for detecting environmental stress: some considerations based on intertidal macrozoobenthos and bird communities. Journal of Applied Ecology 27:210–223.

Micallef, R. M., and P. J. Schembri. 1998. The application of multivariate analytical techniques to the study of marine benthic assemblage: a review with special reference to the Maltese Islands. Xjenza 3:9–28.

Micheli, F., and nine coauthors. 1999. The dual nature of community variability. Oikos 85:161–169.

Mills, E. L. 1969. The community concept in marine zoology, with comments on continua and instability in some marine communities: a review. Journal of the Fisheries Research Board of Canada 26:1415–1428.

Minchin, P. R. 1987. An evaluation of the relative robustness of techniques for ecological ordination. Vegetatio 69:89–107.

Morrisey, D. J., A. J. Underwood, and L. Howitt. 1996. Effects of copper on the faunas of marine soft-sediments: an experimental field study. Marine Biology 125:199–213.

Mouillot, D., and A. Lepretre. 1999. A comparison of species diversity estimators. Researches on Population Ecology 41:203–215.

Munkittrick, K. R., and L. S. McCarty. 1995. An integrated approach to ecosystem health management: top-down, bottom-up or middle-out? Journal of Aquatic Ecosystem Health 4:77–90.

Nei, M., and A. K. Roychoudhury. 1974. Sampling variances of heterozygosity and genetic distance. Genetics 76:379–390.

Nelson, W. G. 1987. An evaluation of deviation from the lognormal distribution among species as a pollution indicator in marine benthic communities. Journal of Experimental Marine Biology and Ecology 113:181–206.

Oberdorff, T., and R. M. Hughes. 1992. Modification of an index of biotic integrity based on fish assemblages to characterize rivers of the Seine basin, France. Hydrobiologia 228:117–130.

Odum, E. P. 1984. The mesocosm. Bioscience 34:558–562.

Oksanen, J., and P. R. Minchin. 1997. Instability of ordination results under changes in input data order: explanations and remedies. Journal of Vegetation Science 8:447–454.

Olsgard, F., and P. J. Somerfield. 2000. Surrogates in marine benthic investigations–which taxonomic unit to target? Journal of Aquatic Ecosystem Stress and Recovery 7:25–42.

Olsgard, F., P. J. Somerfield, and M. R. Carr. 1998. Relationships between taxonomic resolution, macrobenthic community patterns and disturbance. Marine Ecology Progress Series 172:25–36.

Osenberg, C. W., O. Sarnelle, S. D. Cooper, and R. D. Holt. 1999. Resolving ecological questions through meta-analysis: goals, metrics and models. Ecology 80:1105–1117.

Parker, E. D., and 12 coauthors. 1999. Stress in ecological systems. Oikos 86:179–184.

Pearson, T. H., and R. Rosenberg. 1978. Macrobenthic succession in relation to organic enrichment and pollution of the marine environment. Oceanography and Marine Biology Annual Review 16:229–311.

Penczak, T., and A. Kruk. 1999. Applicability of the abundance/biomass comparison method for detecting human impacts on fish populations in the Pilica River, Poland. Fisheries Research 39:229–240.

Petersen, J. E., J. C. Cornwell, and W. M. Kemp. 1999. Implicit scaling in the design of experimental aquatic ecosystems. Oikos 85:3–18.

Pielou, E. C. 1975. Ecological diversity. Wiley, New York.

Poff, N. L., and seven coauthors. 1993. Size structure of the metazoan community in a Piedmont stream. Oecologia 95:202–209.

Power, M., and L. S. McCarty. 1997. Fallacies in ecological risk assessment practices. Environmental Science and Technology 31:A370–A375.

Preston, F. W. 1962. The canonical distribution of commonness and rarity. Ecology 43:185–215, 410–432.

Rader, R. B., and C. J. Richardson. 1994. Response of macroinvertebrates and small fish to nutrient enrichment in the northern Everglades. Wetlands 14:134–146.

Raffaelli, D., S. Hall, C. Emes, and B. Manly. 2000. Constraints on body size distributions: an experimental approach using a small-scale system. Oecologia 122:389–398.

Ramsay, P. R., and six coauthors. 1997. A rapid method for estimating biomass size spectra of benthic metazoan communities. Canadian Journal of Fisheries and Aquatic Sciences 54:1716–1724.

Reizopoulou. S., M. Thessalou-Legaki, and A. Nicolaidou. 1996. Assessment of disturbance in Mediterranean lagoons: an evaluation of methods. Marine Biology 125:189–197.

Rosen, R. 1969. Hierarchical organization in automata theoretic models of biological systems. Pages 179–199 in White, L. L., A. G. Wilson, and D. Wilson, editors. Hierarchical Structures. American Elsevier, New York.

Rosenzweig, M. L. 1997. Species diversity in space and time. Cambridge University Press, Cambridge, England.

Rundle, S.D., and M. J. Attrill. 1995. Comparison of meiobenthic crustacean community structure across freshwater acidification gradients. Archiv fur Hydrobiologie 133:441–456.

Rundle, S. D., A. Jenkins, and S. J. Ormerod. 1993. Macroinvertebrate communities in streams in the Himalaya, Nepal. Freshwater Biology 30:169–180.

Ruse, L. P. 1996. Multivariate techniques relating macroinvertebrate and environmental data from a river catchment. Water Research 30:3017–3024.

Rydgren, K. 1996. Vegetation–environment relationships of old-growth spruce forest vegetation in Ostmarka Nature Reserve, SE Norway, and comparison of three ordination methods. Nordic Journal of Botany 16:421–439.

Saiz-Salinas, J. I., and J. A. Gonzalez-Oreja. 2000. Stress in estuarine communities: lessons from the highly impacted Bilbao estuary (Spain). Journal of Aquatic Ecosystem Stress and Recovery 7:43–55.

Saiz-Salinas, J. I., and A. Ramos. 1999. Biomass size-spectra of macrobenthic assemblages along water depth in Antarctica. Marine Ecology Progress Series 178:221–227.

Sanders, H. L. 1968. Marine benthic diversity: a comparative study. American Naturalist 102:243–282.

Schwinghamer, P. 1981. Characteristic size distributions of integral benthic communities. Canadian Journal of Fisheries and Aquatic Sciences 38:1255–1263.

Schwinghamer, P. 1985. Observations on size-structure and benthic coupling of some shelf and abyssal benthic communities. Pages 347–359 *in* P. E. Gibbs, editor. Proceedings of the 19th European marine biology symposium. Cambridge University Press, Cambridge, England.

Schwinghamer, P. 1988. Influence of pollution along a natural gradient and in a mesocosm experiment on biomass size spectra of benthic communities. Marine Ecology Progress Series 46:199–206.

Shannon, C. E., and W. Weaver. 1949. The mathematical theory of communication. University of Illinois Press, Urbana, Illinois.

Sheldon, R. W., and T. R. Parsons. 1967. A continuous size spectrum from particulate matter in the sea. Journal of the Fisheries Research Board of Canada 24:909–915.

Simboura, N., A. Zenetos, P. Panayotidis, and A. Makra. 1995. Changes in benthic community structure along an environmental pollution gradient. Marine Pollution Bulletin 30:470–474.

Simic, V., and S. Simic. 1999. Use of the river macrozoobenthos of Serbia to formulate a biotic index. Hydrobiologia 416:51–64.

Simpson, E. H. 1949. Measurement of diversity. Nature (London)163:688.

Smith, J. A., G. E. Millward, N. H. Babbedge, M. J. Attrill, and M. B. Jones. 1995. Changes in benthic community structure caused by construction of a harbour impoundment scheme. Netherlands Journal of Aquatic Ecology 29:449–457.

Smith, M. J., and 12 coauthors. 1999. AusRivAS: using macroinvertebrates to assess ecological condition of rivers in western Australia. Freshwater Biology 41:269–282.

Smith, S. D. A., and R. D. Simpson. 1995. Effects of the Nella-dan oil-spill on the fauna of *Durvillaea antarctica* holdfasts. Marine Ecology Progress Series 121:73–89.

Sole, R. V., O. Miramontes, and B. C. Goodwin. 1993. Oscillations and chaos in ant societies. Journal of Theoretical Biology 161:343–357.

Somerfield, P. J., J. M. Gee, and R. M. Warwick. 1994. Soft-sediment meiofaunal community structure in relation to a long-term heavy-metal gradient in the Fal estuary system. Marine Ecology Progress Series 105:79–88.

Somerfield, P. J., F. Olsgard, and M. R. Carr. 1997. A further examination of two new taxonomic distinctness measures. Marine Ecology Progress Series 154:303–306.

Stark, J. D. 1998a. SQMCI: a biotic index for freshwater macroinvertebrate coded abundance data. New Zealand Journal of Marine and Freshwater Research 32:55–66.

Stark, J. S. 1998b. Effects of copper on macrobenthic assemblages in soft sediments: a laboratory experimental study. Ecotoxicology 7:161–173.

Steedman, R. J. 1988. Modification and assessment of an index of biotic integrity to quantify stream quality in southern Ontario. Canadian Journal of Fisheries and Aquatic Sciences 45:492–501.

Taylor, C. M. 2000. A large-scale comparative analysis of riffle and pool fish communities in an upland stream system. Environmental Biology of Fishes 58:89–95.

Ter Braak, C. J. F. 1986. Canonical correspondence analysis: a new eigenvector technique for multivariate direct gradient analysis. Ecology 67:1167–1179.

Ter Braak, C. J. F., and P. F. M. Verdonschot. 1995. Canonical correspondence analysis and related multivariate methods in aquatic ecology. Aquatic Sciences 57:225–289.

Thomas, R. M. 1998. Temporal changes in the movements and abundance of Thames estuary fish populations. Pages 115–140 *in* M. J. Attrill, editor. A rehabilitated estuarine ecosystem: the environment and ecology of the Thames estuary. Kluwer Academic Press, Dordrecht, Netherlands.

Thorne, R. S. J., W. P. Williams, and Y. Cao. 1999. The influence of data transformations on biological monitoring studies using macroinvertebrates. Water Research 33:343–350.

Thrush, S. F., and nine coauthors. 1998. Disturbance of the marine benthic habitat by commercial fishing: impacts at the scale of the fishery. Ecological Applications 8:866–879.

Underwood, A. J. 1992. Beyond BACI: the detection of environmental impact on populations in the real, but variable, world. Journal of Experimental Marine Biology and Ecology 161:145–178.

Underwood, A. J. 1997. Experiments in ecology: their logical design and interpretation using analysis of variance. Cambridge University Press, Cambridge, England.

Underwood, A. J. 1999. Trying to detect impacts in marine habitats: comparisons with suitable reference areas. Pages 279–308 *in* T. Sparks, editor. Statistics in Ecotoxicology. Wiley and Sons, Chichester, England.

Underwood, A. J. 2000. Importance of experimental design in detecting, and measuring stresses in marine populations. Journal of Aquatic Ecosystem Stress and Recovery 7:3–24.

Ungaro, N., G. Marano, R. Marsan, and K. Osmani. 1998. Demersal fish assemblage biodiversity as an index of fishery resources exploitation. Italian Journal of Zoology 65 (Supplement S):511–516,.

Vanderklift, M. A., T. J. Ward, and C. A. Jacoby. 1996. Effect of reducing taxonomic resolution on ordinations to detect pollution-induced gradients in macrobenthic infaunal assemblages. Marine Ecology Progress Series 136:137–145.

Vandervalk, A. G., L. Squires, and C. H. Welling. 1994. Assessing the impact of an increase in water level on wetland vegetation. Ecological Applications 4:525–534.

Van Groenewoud, H. 1992. The robustness of correspondence, detrended correspondence, and TWINSPAN analysis. Journal of Vegetation Science 3:239–246.

Walley, W. J., and H. A. Hawkes. 1996. A computer-based reappraisal of the biological monitoring working party scores using data from the 1990 river quality survey of England and Wales. Water Research 30:2086–2094.

Warwick, R. M. 1986. A new method for detecting pollution effects on marine macrobenthic communities. Marine Biology 92:557–562.

Warwick, R. M. 1988. The level of taxonomic discrimination required to detect pollution effects on marine benthic communities. Marine Pollution Bulletin 19:259–268.

Warwick, R. M. 1993. Environmental impact studies on marine communities: pragmatical considerations. Australian Journal of Ecology 18:63–80.

Warwick, R. M., and K. R. Clarke. 1991. A comparison of some methods for analyzing changes in benthic community structure. Journal of the Marine Biological Association of the United Kingdom 71:225–244.

Warwick, R. M., and K. R. Clarke. 1993. Comparing the severity of disturbance: a meta-analysis of marine macrobenthic community data. Marine Ecology Progress Series 92:221–231.

Warwick, R. M., and K. R. Clarke. 1995. New "biodiversity" measures reveal a decrease in taxonomic distinctness with increasing stress. Marine Ecology Progress Series 129:301–305.

Warwick, R. M., and K. R. Clarke. 1998. Taxonomic distinctness and environmental assessment. Journal of Applied Ecology 35:532–543.

Warwick, R. M., K. R. Clarke, and J. M. Gee. 1990. The effect of disturbance by soldier crabs *Mictyris platycheles* H. Milne Edwards on meiobenthic community structure. Journal of Experimental Marine Biology and Ecology 135:19–33.

Warwick, R. M., N. R. Collins, J. M. Gee, and C. L. George. 1986. Species size distributions in benthic an pelagic metazoa: evidence for interaction? Marine Ecology Progress Series 34:63–68.

Washington, H. G. 1984. Diversity, biotic and similarity indices: a review with special relevance to aquatic ecosystems. Water Research 18:653–694.

Wilhm, J. L., and T. C. Dorris. 1968. Biological parameters for water quality criteria. Bioscience 18:477–481.

Wright, I. A., B. C. Chessman, P. G. Fairweather, and L. J. Benson. 1995. Measuring the impact of sewage effluent on the macroinvertebrate community of an upland stream: the effect of different levels of taxonomic resolution and quantification. Australian Journal of Ecology 20:142–149.

Wright, J. F., M. T. Furse, and P. D. Armitage. 1993. RIVPACS–a technique for evaluating the biological quality of rivers in the U. K. European Water Pollution Control 3:15–25.

Wright, J. F., D. Moss, P. D. Armitage, and M. T. Furse. 1984. A preliminary classification of running-water sites in Great Britain based on macroinvertebrate species and the prediction of community type using environmental data. Freshwater Biology 14:221–256.

Zamora-Munoz, C., and J. Alba-Tercedor. 1996. Bioassessment of organically polluted Spanish rivers, using a biotic index and multivariate methods. Journal of the North American Benthological Society 15:332–352.

Appendix 1. Some commonly used indices of species diversity, evenness, and similarity.

Index	Formula	Useful reference
Margalef's species richness	$d = (S - 1)/\log_e N$ where S = no. of species, N = no. of individuals	Clarke and Warwick (1994a)
Shannon-Wiener diversity index	$H'e = -\Sigma pi(\log_e pi)$ where pi = proportion of total abundance arising from the ith species	Krebs (1985)
Pielou's eveness index	$J' = H'e$ (observed)/H'max where H'max = $\log_e S$	Krebs (1985)
Dominance index	$DI = (1 - J')$	Gray (1981)
Simpson's index (unbiased form)	$SI = \Sigma((n^2 - n)/(N^2 - N))$ where n = no. individuals of ith species $N = \Sigma n$	Rosenzweig (1997)
Bray-Curtis (Czekanowski) similarity coefficient	$$S_{jk} = 100\left\{1 - \frac{\sum_{i=1}^{p}\lvert y_{ij} - y_{ik}\rvert}{\sum_{i=1}^{p}(y_{ij} + y_{ik})}\right\}$$ where y_{ij} = abundance of the ith species in the jth sample, y_{ik} = abundance of the ith species in the kth sample	Clarke and Warwick (1994a)
Canberra coefficient	$$S_{jk} = 100\left\{1 - p^{-1}\sum_{i=1}^{p}\frac{\lvert y_{ij} - y_{ik}\rvert}{(y_{ij} + y_{ik})}\right\}$$	Clarke and Warwick (1994a)
Jaccard coefficient	$S_{jk} = 100.a/(a + b + c)$ where a = no. species present in both samples, b = number species present in jth sample, but absent in kth sample, c = number of species present in the kth sample, but absent from the jth	Clarke and Warwick (1994a)
Sorenson coefficient	$S_{jk} = 100.2a/(2a + b + c)$	Clarke and Warwick (1994a)

Appendix 2. Summary of recommended analytical techniques, together with major perceived advantages and limitations.

Community assessment technique	Primary use	Major advantage	Major limitation
Univariate analyses			
Diversity indices			
Shannon-Wiener & Pielou's evenness	Comparing α-diversity between control & test samples.	Integrates species richness and dominance.	Sample-size dependent. Only usable for comparisons within a study.
Simpson's index	Comparing α-diversity between control & test samples.	Independent of sample size. Can be used for semiquantitative data or comparisons between studies.	Questionable sensitivity at small spatial scales.
Similarity indices e.g., Bray-Curtis	Comparing the similarity in community composition between samples.	Take account of variations in species composition between samples.	Multisite matrix hard to interpret without further multivariate analyses.
Biotic indices			
BMWP	Assessing impact of organic pollution on river community.	Adaptable to a range of geographic locations. Easy to sample and calculate.	Valid only for organic pollution effects on river macroinvertebrates.
IBI	Assessing health of river systems using fish communities.	Adaptable to a wide range of stressors and habitats.	Construction requires extensive local knowledge and effort. Results only applicable to local system.
Distributional or graphical techniques			
ABC curves	Assessing impact of organic pollution on marine subtidal invertebrate communities.	No spatial or temporal controls required for interpretation.	Only applicable to situations where pollution results in dominance of small species.
Benthic biomass size spectra	Assessing impact on community structure using an ataxonomic approach.	No identification skills required, open to automation, integrates all sizes of organisms.	Not extensively tested and robustness equivocal from current results.
Multivariate analyses			
Ordination e.g., nonmetric MDS	Provides a 2-d map of samples relating to the similarity of their communities.	Enables difference in community composition to be easily visualised.	Interpretation can be subjective without further complementary techniques.
ANOSIM	Tests significance of differences in community composition between a priori sample groups.	Allows robust interpretations on impact to be drawn from multivariate results.	Use can be incorrect if design and a priori criteria are not correctly defined.
BIOENV/CANOCO	Relates changes in community composition to environmental variables.	Provides explanation for observed differences in community, so aiding interpretation.	BIOENV: no significance value possible. CANOCO: interpretation difficult without experience.

13

Integration of Population, Community, and Landscape Indicators for Assessing Effects of Stressors

Donald L. DeAngelis and John L. Curnutt

Introduction

This chapter evaluates some of the main methodologies for physical habitat assessment that have been used to indirectly assess effects of environmental stressors on fish populations, focusing on methodologies that apply at the landscape level. These methods include the long-established instream-flow method and habitat quality index methods. This chapter also describes a recent methodology that is being applied to the Everglades hydroscape, which represents the wetland hydroscape as a spatially explicit grid of pixels and calculates a suitability index for each spatial pixel. These suitability indices are then used to help evaluate the effects of stressors on fish populations.

Environmental monitoring is the periodic measurement of selected attributes of a system over a period of time in a given location or set of locations. Monitoring may be employed to detect effects of anthropogenic changes on various components and characteristics of ecological systems, such as the viability or size of key populations or species richness of a community. These aspects of populations and communities are difficult to measure, particularly in large aquatic ecosystems. For example, it is very difficult to directly measure population viability, and it is likewise difficult to count all species present in a community in order to calculate species richness. Therefore, if more easily measured empirical "indicators" were available, it would be more straightforward to relate environmental stressors to changes in populations or communities.

Important indicators to indirectly relate stressors to effects on organisms and populations include the quantity and quality of available habitat. Relating habitat quality to organism condition is possible because traits of organisms evolve in response to the properties of the habitats; that is, the habitat is the template on which life history characteristics are formed (Southwood 1977). The theoretical contours of this viewpoint have been fleshed out further, for example, by Townsend and Hildrew (1994) for the particular case of fish species in a river. These authors point out that predictions about the traits of individual species present, as well as such community-level characteristics as species richness, can be made on the basis of habitat.

Because the particular habitat or range of habitats used by species over evolutionary time shape their characteristics, species do best in habitats similar to those in which they evolved. Changes in habitat extent and characteristics can impose various levels of stress on populations. As Hayes et al. state (1996): "The habitat requirements for a fish population are directly related to the requirements for an individual's growth and survival; i.e., its niche. The production by a fish population is a function of how closely the niche of an individual is realized by the environmental conditions in which it resides." Therefore, changes in a habitat may be indicators of the well being or condition of specific species or communities.

One of the basic challenges associated with the approach of using habitat to evaluate the condition of populations and communities is to determine which variables best represent the habitat of the fish. Several physical and chemical indicators, such as water quality, hydrology, geomorphology, and availability of physical habitat (Maddock 1999), are obvious choices. An advantage of defining habitat in terms of physical and chemical features is that these are usually amenable to measurement in a systematic manner (Hayes et al. 1996). Physicochemical indicators may be static or may incorporate dynamic features. Static indicators include average flows and depths of water and habitat characteristics, whereas dynamic features include short-term events, such as storms and extreme water flows. For example, important dynamic hydrologic parameters that can be used to identify the degree to which alterations in water flow may be biologically relevant include the magnitude, the duration, and the timing of annual extreme conditions; the frequency and duration of high and low pulses; and the rate and frequency of a change in flow conditions (Maddock 1999).

Biotic features, such as vegetation, are also important in defining the habitat. For example, aquatic macrophytes provide cover and food organisms for many fish species that would not be able to survive in some areas without such cover when predators are present (Hayes et al. 1996). The presence and abundance of predators and symbionts may also indicate the presence and density of particular fish species.

Habitat and Landscape Indicators

Not only are there many possible types of habitat indicators that could point indirectly to stress on organisms, but their spatial and temporal scales can vary as well. An individual organism is affected most strongly by the local habitat characteristics in its home range or "ecological neighborhood" (Addicott et al. 1987). In Maddock's (1999) terminology, the local habitat of an organism can be divided into the microhabitat and the macrohabitat. The former includes depth, water velocity, substrate, presence of rocks, and overhanging branches, at a local point, which an individual might make use of at certain times. The latter term refers to the broader area of habitat, such as riffles or pools, in which an individual may spend much of its lifetime or which it may, at least, depend on for critical functions and periods of its life cycle, such as feeding and reproduction. Larger spatial units, such as a stream reach or a small lake, would be appropriate for a consideration of a subpopulation of a species, though they may also be appropriate scales for individuals of highly mobile species. Such units can be said to constitute landscape segments. At a still larger scale, entire watersheds or drainages are involved, which can be referred to as the landscape or regional scale. Because this spatial scale can encompass subpopulations, populations, and unique communities, the landscape (or "hydroscape," a more appropriate term for aquatic systems) scale is natural for considering stressors that affect whole populations and communities.

The use of measures of habitat at the landscape scale as an indicator of stress has been attractive to ecologists, including aquatic ecologists. It is not always possible or practical to measure important components of populations and communities, such as density and viability, particularly over wide areas; therefore, indicators are needed that indirectly reflect the effects of stressors at the population and community level. How does one develop indicators for a landscape? One approach is to develop a small set of habitat-based indicators for a whole region. These indicators would relate to broad geographic qualities such as geology, water quality, sediment load, topography, temperature, and rainfall. This set of indices could, at least, provide enough information to state whether this unique combination of habitat indicators would be conducive to persistence of a species in a particular region. This set of indicators might not in itself, however, provide sufficient data to conclude that the population definitely could persist or to say whether its condition would be optimal.

More specific habitat suitability information could be provided by indicators on a finer scale of resolution, such as a stream reach. Information on water flow velocity, channel size, and stream substrate should help to determine if conditions are conducive for the species to be present. These types of data may still not provide all the information necessary to specify the

condition of the population through this habitat quality assessment. Indicators at a still finer scale of resolution may be needed. Habitat indicators on
this finer scale might include the local microhabitat occurrence of water in
certain depth and velocity ranges, overhanging branches, or aquatic vegetation to provide favorable microhabitat conditions. Indices at the microhabitat
or macrohabitat scales are usually determined by sampling a hydroscape at a
number of different spatial locations. If enough such local sites are sampled,
the hydroscape can be characterized by some overall average rating. In some
cases, as discussed later, it may be possible to use geographic information
system (GIS) data, such as that from remote sensing, to go beyond the sampling of merely a few sites to computing indices for a high resolution grid of
points or pixels across a region.

Broad Scale Indices

Broad scale assessments are aimed at developing indices that apply to large
landscape units, such as whole lakes or river reaches; these can often be
determined rapidly from map information. These indices may be useful for
comparing lakes or river reaches within a region of relatively homogeneous
climate.

A "morphoedaphic" index was developed to predict fish yield from lakes
on the basis of a few abiotic factors (Ryder 1965, 1978; Jenkins 1967). The
index is N/z, where N is an indicator, based on edaphic and land use characteristics of a watershed, of nutrient loading that is relevant to the particular
lake, and z is the mean depth of the lake. These two measures could be
modified for particular situations, but they reflect the availability of nutrients
and warm temperature in the photic zone of a lake. Ryder (1982) shows that
empirical evidence supports this relationship when lakes of similar area are
compared, except when they differ greatly in perimeter or area ratios.

Another broad scale approach, designed to predict the presence and
well being of fish and other functional groups in a lake, is described by
Hakanson (1996). Hakanson derived predictive expressions for lake color,
pH, and total phosphorus for small glacial lakes based on catchment level
parameters such as percentage of mires, lakes, and forests, and lake level
parameters such as lake area, volume, mean depth. This approach allows a
relatively good prediction of pH, color, and total phosphorus in these types
of systems. Predictions of total phosphorus, in particular, can then be used
to provide first-order predictions of production of fish and other functional
groups (Peters 1986).

Analogous broad-scale methods have also been developed for assessing
river systems. These may be particularly important in planning restoration of
rivers and streams, as anthropogenic changes in the physical characteristics
of these systems may impose stresses on fish populations and communities.
Maddock (1999) discussed such classification schemes as the river habitat

survey used by the Environmental Agency in the United Kingdom (Fox et al. 1996). In this survey, the physical structure of the river, its banks, and the adjacent land, encompassing 500 m stretches along the river, are evaluated. This approach allows independent river reaches to be compared with those in the database and identified along the spectrum between natural and heavily impacted areas. Another reach-scale approach developed for appraisal of the success of river projects uses information on channel characteristics, floodplain land uses, and channel dynamics (Downs and Brookes 1994). This method requires information on geology, soil, and topography of the whole catchment. These assessment methods have the advantage of simplicity and application to large areas. They may provide indicators of possible stress effects that operate on broad spatial areas, such as low pH levels, widespread toxicants, or detrimental changes in river flow.

Macrohabitat and Microhabitat Scales

Maddock (1999) distinguished two major types of assessment methods that operate at the microhabitat and macrohabitat scales: the instream flow methodology, which attempts to determine physical habitat availability, and the habitat quality evaluation-regression approach, which attempts to relate biological characteristics to a variety of physical features.

Instream Flow Incremental Method

The instream flow incremental method developed from rule of thumb, such as the "Montana method" (Tennant 1976), for maintaining good stream habitat. These rule of thumb recommendations were derived from studies on hundreds of streams in the northern United States from the Atlantic Coast to the Rocky Mountains. In particular, the Montana method requires knowing the average annual flow at a point of interest on a stream and prescribing a minimum flow as a fraction of that average. Instantaneous flows lower than 10% of the average flow cause severe stress and degradation to the aquatic community. A base flow of 30% of the average was recommended to sustain good habitat.

An elaboration of this type of rule of thumb approach is called the instream flow incremental methodology (IFIM). This methodology, developed by the U.S. Fish and Wildlife Service, assesses incremental changes in the amount of usable habitat by relating it to stream flow. To implement this approach, it is necessary to have a description of the habitat, as well as to know the habitat preferences of the organisms of interest. A system of computer programs called physical habitat simulation, or PHABSIM (Milhous 1979), simulates the physical habitat of a stream as a function of stream flow. One of the simulation models combines water depth, water velocity, and substrate to compute the amount of usable habitat (Bovee et al. 1978). Computing usable habitat requires dividing the stream into spatial cells and conducting transect studies at suitable intervals to measure habitat across the stream. Streambed elevations are measured along transects at fixed spatial intervals to get cross-sectional profiles; hence, data are collected on a microhabitat scale. Also measured are depth, velocity, and substrate at the same intervals under different flows. From these measurements, linear regression equations are established for stage versus discharge and velocity versus discharge. The simulation model is, thus, calibrated for that stream reach, which can be used to simulate depth and water velocity for any flow. In general, only one sample reach is used for a stream.

Another computer program associated with this method calculates the amount of physical habitat weighted by its suitability for target species. For each spatial cell, a composite weighting factor, C_i, is obtained as follows: $C_i = f_v(V_i) \times f_d(D_i) \times f_s(S_i)$, where $f_v(V_i)$ is the suitability weighting factor for the velocity in cell i; $f_d(D_i)$ is the suitability weighting factor for the depth in cell i; and $f_s(S_i)$ is the suitability weighting factor for the substrate type in cell i. These suitability weighting factors are derived from curves of habitat suitability of particular target fish species. To obtain a weighted usable area (WUA) over a number of cells, C_i is multiplied by the amount of area in cell i and a summation is performed over all cells, $WUA = \Sigma_i \, C_i A_i$, where A_i is the area of the ith cell. The WUA can be evaluated at different flows, thus calculating the amount of usable area. Losses of area may be associated with stress to the population.

Some limitations of the instream flow approach have been discussed by Orth and Maughan (1982), Mathur et al. (1985), and Bowlby and Roff (1986). The assumption that depth, flow, and substrate are the most important variables, in general, for fish, has not been empirically demonstrated; hence, the emphasis on these variables may not always be appropriate. In addition, information on the habitat requirements of fish is not always available. Technical problems include the fact that the three habitat suitability functions are not, in general, independent, as assumed in the IFIM methodology. This lack of independence makes it impossible to properly write C_i as a product of these functions. Some studies using the instream flow approach have not obtained useful results. For example, Irvine et al. (1987) found no correlation of late summer and early winter biomass with the amount of usable habitat in the Waitaki River in New Zealand and noted that the method cannot predict fish biomass when the fish are not limited by space.

Habitat Evaluation Methods

The habitat evaluation approach uses empirical regressions to relate biotic variables of interest to measurable physical variables at specific locations, leading to a habitat suitability index (HSI). The HSI approach combines a number of habitat variables into a single index, reflecting conditions in a particular location throughout the year. The HSI, which can vary from 0.0 to 1.0 according to how suitable the habitat is rated for a species, can be multiplied by the area being evaluated to provide the total habitat units (HU) for the species being considered. Then the HU can be used to evaluate the change in stress to a population due to anthropogenic changes at a location.

Among the early applications of the HSI concept was that of Binns and Eiserman (1979), who developed a model to relate habitat to standing crop for coldwater trout streams. The study produced what they referred to as the habitat quality index (HQI), consisting of two regression models that related 11 habitat variables that represented food, shelter, streamflow variation, and maximum summer stream temperature to trout biomass density in Wyoming streams. Milner et al. (1985) applied a similar habitat evaluation approach to assessing the salmonid fishery in Wales. Their empirical model, HABSCORE, explains 93.6% of the variance between abundances in terms of habitat variables.

An example of an HSI computation is that of Pajak and Neves (1987) who tested the HSI approach for rock bass *Ambloplites rupestris* in fourth-order Virginia streams. The main categories of habitat variables they used were turbidity, temperature, current velocity, substratum, and gradient; within each of these five main categories were several variables, for a total of 17. The authors calculated HSIs for both 50-m and 250-m subsections in two streams, using habitat suitability curves for rock bass that had been previously established. When HSI predictions were compared in the streams against their own measurements of standing bass stocks, correct agreement of rankings were found only between the 250-m units. The lack of correlation between the 50-m HSIs and rock bass standing stock was apparently due to a combination of the homogeneity of the segments, the omission from the HSI of key variables such as water depth and cover and the fact that the ranges of individual fish movement patterns usually exceeded the 50-m units, so that the fish were integrative over larger areas larger than 50 m.

In many cases, it is important to include biotic components in addition to the physical variables in the HSI. Randall et al. (1996) show that indices of fish production and species richness are significantly higher in littoral habitats with abundant submerged macrophytes, as measured by percent bottom cover, than in adjacent areas with low macrophyte abundance. Bowlby and Roff (1986) examined relationships between the biomass of trout and physical and biological characteristics of streams in southern Ontario. At each site, habitat variables representing instream cover, substrate, stream morphology, velocity, temperature, and food availability were included. Regression and discriminant function analysis revealed that trout biomass in southern Ontario was related to microcommunity biomass, percent pool area, mean maximum summer temperature, biomass of small benthic invertebrates, the presence of piscivorous fish, and overhead cover.

Although widely used, HSI models have been criticized for their assumptions of linear responses of species to habitat parameters (Meents et al. 1983), their inability to incorporate spatial juxtaposition of habitat components (Cooperrider 1986), their lack of stochastic and temporal effects on habitat suitability (Pearsall et al. 1986), and insufficient statistical methodology (Dettmers and Bart 1999).

Hierarchy of Spatial Scales

It is clear that the well being of a fish population or community depends on habitat characteristics at several spatial scales. A number of habitat evaluation methods recognize this explicitly. For example, Imhof et al. (1996) outlined a hierarchical evaluation of fish habitat in a watershed. Three basic scales were considered: the watershed (linear scale of 10^5 m), the reach (scale of $10–10^4$ m), and the site (scale of 1–10 m). In addition, Imhof et al. noted that it might be useful to include two other scales in the habitat evaluation: a subwatershed scale of the order of 10^4 m and the habitat element scale at 0.1–1 m. Each of these five scales has its own set of causal-related factors. At the watershed level, climate, landform (slope and sediment type), and landcover are the key variables. At the reach level, the variables of interest are runoff flows, slope of pools and riffles, sediment load and size range, and vegetation community. At the site level, the operative variables include flow velocity and depth, slope of water surface, bank cohesion, types of debris, and successional stage of riparian vegetation. The habitat elements incorporate small scale variability in substrate and water flow and depth that have important functions for individual fish within their home ranges. The habitat element scale reflects variability of water velocity and depth, and of substrate that may be exploited by individual fish.

Because a variety of approaches must be used to effectively predict effects of stressors on aquatic ecosystems, the type of hierarchical arrangement described above provides a logical framework for asking the question of how well a particular population or community of organisms will do in a watershed. Merely taking local steps to decrease erosion in a stream channel, for example, may not address the problem of increased high flows at the watershed level, which cause the erosion. Marshall (1996), following the proposal of Ryder and Kerr (1989), also used a hierarchical approach to assess the ability of a lake to support a lake trout *Salvelinus namaycush* population. First, for the population to exist at all, certain threshold values of physical parameters had to be met in some fraction of the lake; such as an upper temperature limit and a lower limit on dissolved oxygen. The status of the lake trout population improved as the amount of lake area with optimal temperature and dissolved oxygen levels increased. Even under optimal conditions in these variables, however, other attributes at smaller spatial scales, such as substrate for spawning sites or cover, are generally required to reliably evaluate the effects of habitat modification on the health of wildlife species.

Spatially Explicit Species Index Model

The previously mentioned approaches either describe broad-scale indicators that characterize large spatial units, look for indices at the level of the micro- or macrohabitat scale, or use a hierarchical approach consisting of both. There is an alternative approach to all of these indices, which is to represent a large regional unit or landscape as a continuous grid of spatial cells or pixels at a spatial resolution relevant to the local habitat of the fish. With this method, entire aquatic systems, such as rivers or lakes, are covered by pixels, each of which has an index value representing its suitability for fish populations. This would appear to be an extraordinarily difficult task, but remote sensing makes this feasible in some cases, and it has the advantage of being able to provide information on any desired scale from the pixel on up to the size of the whole area covered. This approach is demonstrated here for a large wetland: the Florida Everglades.

Fish are recognized as being a key element in the ecology of the Everglades. Anthropogenic changes in hydrology are hypothesized to have caused major reductions in fish populations and, thus, to have influenced other species, such as wading birds, which rely on fish as prey (Loftus and Eklund 1994). It has been hypothesized that fish, crayfish, grass shrimp, and other aquatic organisms in Florida freshwater wetlands have declined in abundance and availability due to reduced spatial extent of wetlands, shortened hydroperiods, loss of aquatic refugia during dry seasons, and altered water recession patterns (Ogden 1994). A landscape index model called a spatially explicit species index (SESI) model has been developed as an attempt to provide a basis for quantifying this assessment. Here we will describe this approach and note some of its similarities to and differences from the other habitat approaches previously discussed.

Spatially explicit species index models are similar to HSI models in that population response is predicted by one or more a priori habitat relationships and in that habitat indices are quantified by the ratio of the value in a pixel to the maximum possible value. Thus, SESI values range from 0.0 to 1.0, with 1.0 indicating the maximum value. The SESI models, differ from traditional HSI models in that they (1) have a temporal component and, thus, incorporate both static and dynamic landscape features, (2) are based on a landscape structure which, once established, can be used to model the responses of any species in the system, (3) provide a landscape index map rather than just an index or set of indices, and (4) can provide a relatively easy method of comparing species responses with those responses predicted by more complex models including process models, size-structured population models, and individual-based models (DeAngelis et al. 1998). The addition of temporal factors differentiates the SESI approach from other recent spatially explicit habitat models (Dettmers and Bart 1999).

For SESI models, the landscape is divided into equal-sized spatial cells or pixels (500 × 500 m in the Everglades models), each pixel having a suite of values that correspond to the parameters included in the model. This scale of resolution was chosen because changes in elevation in the Everglades are sufficiently gradual that a 500 × 500 m area is relatively homogeneous in characteristics such as elevation, vegetation type, and hydroperiod. Habitat suitability for each pixel is determined by a set of rules. Generally, these rules are of two types: (1) binary (0/1 or yes/no) rules, which invoke known or estimated limits on the suitability of habitat or environmental conditions concerning a species; and (2) quantitative rules, whereby, having met the basic requirements of a species, the habitat parameters are given a value that reflects their relative potential for breeding, foraging, or both. As noted above, SESI models are able to incorporate spatial and temporal sequences. Spatially, this is based on the ability of the model to use the values of surrounding pixels when determining the value assigned to each pixel. This involves a method of spatial interpolation based upon the ecology of the species of concern, a form of knowledge-based interpolation.

The primary output of a SESI model is a visual representation of the landscape with color-coded values assigned to each pixel. One way to use the model is to compare different management scenarios. For example, the application of these models for assessing Everglades restoration plans involves comparing a baseline scenario that assumes no restoration with an alternative management plan. The scenarios differ only in hydrologic variables, primarily water depths across the landscape. Hydrological data for the base scenario assumes no changes to the current water management practices. The alternative scenarios incorporate proposed changes to management activities. The predicted effects of one scenario could be compared to those of the second by simply subtracting the index value for each pixel calculated under base conditions from the value for the same pixel calculated under the alternative scenario. In practice, three maps are generated representing the index values under the base scenario, the values under the alternative scenario, and the difference between the two (Figure 1). For the difference map (middle panel, Figure 1) the values range from −1 to 1. This type of methodology represents that of relative assessment. Relative assessment of the suitability of habitat under alternative management plans produces meaningful results, even when knowledge of ecological details are insufficient to assess habitat suitability of a pixel in an absolute sense (DeAngelis et al. 1998; Curnutt et al. 2000). The emphasis of this approach, therefore, is on comparing the spatial pattern of differences between two alternative management plans for forage availability for a predator.

In addition to these habitat maps, the SESI model can generate a time series of the mean suitability index attained each year under each management scenario. The time series values are spatial averages obtained from the sum of the suitability indices for all pixels divided by the number of pixels.

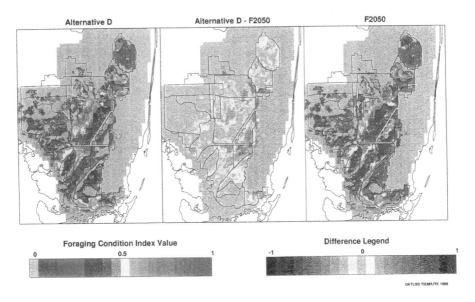

Figure 1. Output from the SESI model for forage fish availability to short-legged wading birds. Two hydrology scenarios, F2050 (right panel) and AltD (left panel), are compared. The index values for each 500 3 500 m pixel runs from 0 to 1 and are color coded. The central panel represents the differences between the values for the individual pixels. The particular comparison is done using rainfall for the year 1977, which was a typical rainfall year.

A SESI Model of Forage Fish Availability

One of the SESI models developed for Everglades restoration planning assesses the relative availability of small forage fish to wading birds in the Everglades under alternative hydrologic management scenarios (Curnutt et al. 2000). The model area includes the entire Everglades freshwater drainage, except agricultural and urban areas, from Lake Okeechobee south to Florida Bay. A landscape structure was developed for the model area using a GIS. The landscape structure includes the static elements of surface elevation and vegetation type from satellite imagery. It also includes the dynamic element of daily surface water levels, derived from a landscape hydrology model called the South Florida water management model (SFWMM). The basic landscape for the SESI model was made up of pixels, each with an elevation, vegetation type, presence or absence of small permanent ponds, and temporally varying marsh water level (Duke-Sylvester and Gross, personal communication).

The hydrological model outputs (daily water depths in each pixel) used in the SESI model were based on actual rainfall from 1965 through 1995. The use of these long-term model outputs enables the modeling of the response of species under historical hydrological conditions as a basis for calibrating the models with the limited historical species data available. These same rainfall data were used in both the base and the alternative management scenarios. The various hydrologic scenarios that were compared were generated by predicting the surface water conditions that would occur with changes in the amount and timing of water releases and the addition or removal of water control structures. The spatial resolution of the original SFWMM hydrologic data was 2 × 2 mi (3.2 × 3.2 km)—a resolution too coarse for many ecological investigations. A more reasonable spatial scale of resolution is the 500 × 500 m pixel, as this spatial resolution better captures the heterogeneity of the landscape, both physically and biologically. The finer resolution of scale required development of a high-resolution topography. This was done using techniques discussed elsewhere (see DeAngelis et al. 1998; Curnutt et al. 2000; and the ATLSS home page at http://atlss.org).

Development of a SESI model for fish availability to wading birds was important for predicting wading bird breeding success, as the availability of fish for foraging wading birds is critical to their successfully reproducing and raising nestlings. Generally, wading birds require increasing densities of prey throughout the breeding cycle to successfully fledge nestlings (Fleming et al. 1994). In the Everglades, water levels peak toward the end of the rainy season in the autumn and then slowly decline as the dry season progresses throughout the winter and spring. Populations of small fish expand during the wet season, but during the dry season, the gradual decline in water levels causes densities of small fish to increase as the fish concentrate in ever smaller pools. Some spatial pixels dry out, with the loss of most of the fish. However, if there are permanent ponds or solution holes in the limestone rock, many fish survive a drying down. Five basic conditions are used in the SESI model to quantify the quality of a pixel as foraging habitat for wading birds. First, the vegetation type is important, because it permits easy foraging for fish. Second, there should be high water during the wet season to permit the fish populations to expand. Third, during the dry season (wading bird breeding season) water levels should drop continuously to concentrate the fish and to be shallow enough for wading birds to forage. Fourth, there should be few reversals of the decline in water levels, which could disrupt the availability of fish. Finally, the pixel should maintain enough water in permanent ponds and other refuges to maintain a fish population if water levels fall below the surface elevation of the spatial cell. The degree to which each of these conditions is fulfilled in the model during a year can be quantified based on the daily water levels from the hydrologic model (SFWMM). From each pixel, an index of forage fish availability can be calculated.

For each individual 500 × 500 m spatial pixel to be suitable for wading bird foraging, surface water depth must be 5–35 cm and there must be a falling hydrograph (Bancroft et al. 1994; Ogden 1994). In order to compute an index of success for a particular breeding colony, it was necessary to integrate the fish availability indices for all spatial pixels within foraging range of the individual birds in the colony because colonial wading birds can forage over a broad area. The value of the colony foraging index for a colony in a particular pixel is, thus, a moving spatial average of the pixels surrounding it. If the mean suitable area surrounding a pixel decreased below 20% of the total area, the current wading bird breeding cycle for that pixel was terminated and calculations for a new cycle were not initiated until the area mean increased above 20%.

Reversals are increases in surface water depth that occur during the dry down. Reversals were included as a negative factor in the SESI model. However, since reversals do not necessarily occur simultaneously across all of the foraging area surrounding a colony, the impact of reversals is calculated based on the proportion of the currently available foraging area that is affected. While the colony foraging index does not include a mechanistic model of fish population dynamics, it does include a simple function for fish density based on the previous year's water depth. Thus, fish densities were assumed to be lower in a pixel during a wet year following a year in which that pixel completely dried down than during a wet year following a year in which the pixel did not dry down.

The SESI model for forage fish availability was applied to evaluate proposed water management scenarios. The motivation for the scenarios chosen was the Central and South Florida Project Review Study (or Restudy) of the U.S. Army Corps of Engineers (1999). A series of alternative water management strategies was compared with two baseline cases in order to determine the costs and benefits of each scenario. The two base scenarios are referred to as F2050 and C1995. Case C1995 was based on a 31-year time series of historical rainfall data, from 1965 to 1995, evaluated using current sea level, all existing control structures, human population level, and socioeconomic conditions existing in 1995. The F2050 scenario was based on the same rainfall data, but used projected sea level, population level, and socioeconomic conditions of 2050. These base cases were compared with a number of specific plans in which the water control structures were modified and additional structures added to create better water conditions in the wetlands. The index map for the base scenario F2050 and the alternative regulation plan AltD are shown in the right and left panels of Figure 1, and the difference map is shown in the central panel. Each pixel in the figure represents a measure of potential success of a breeding colony of birds, so the foraging conditions through the breeding season in surrounding pixels are taken into account. Whether or not there are good nesting sites (i.e., trees) within the pixels is omitted from these index values.

In itself, the wading bird SESI model does not provide an absolute indicator of stress to fish populations under different water management scenarios. It does, however, provide a relative comparison of habitat quality and, therefore, a comparison between the available fish across space and through time for the two scenarios. In interpreting SESI model output, the researcher or manager can identify spatial and temporal patterns that could adversely affect fish populations and, thus, other components of the Everglades food chain. In the Everglades, for example, a hydrologic regime that decreased hydroperiod to less than two years at a site would negatively affect fish density and fish species richness at that site (Trexler et al. 1996).

Landscape Simulation Approach

Simulation modeling has the same aim as the habitat index approaches described above, in that it attempts to relate certain measurable quantities of the habitat to stress on individuals, populations, or communities. However, the simulation model predicts the effects of the habitat in real time on a simulated wildlife population rather than producing an index that summarizes the influence of the habitat on the population. For example, an index model may relate average annual fish density to some measures of flow at critical periods of a year. A simulation model will attempt to represent the causal chain between flows and the fish population in a temporal manner, through quantitative descriptions of the processes of fish reproduction, growth, and mortality as functions of flow. Despite the obvious distinctions between these techniques, the simulation and index modeling approaches are not fundamentally different in their purposes. For both approaches, there is a reliance on empirical knowledge at some level, though simulation models typically are based on information at the process level. The simulation model may require a great deal of process type of input information, but, it also results in more detailed projections concerning the status of the population. Use of simulation techniques to predict effects of environmental changes on fish populations is demonstrated below using a simulation model, ALFISH (DeAngelis et al. 1997; Gaff et al. 2000). The model is designed to describe the dynamics of functional groups of fishes, particularly the smaller fishes, of the Everglades freshwater marshes. It is designed to describe how changes in the system, such as the hydrologic conditions, may affect the biomass dynamics of fish groups.

Description of the Simulation Model

As described earlier, during the rainy season in the Everglades, water levels rise and the populations of small fishes in the marsh (e.g., killifishes *Fundulus* spp. and mosquitofish *Gambusia* spp.) expand, supported by the seasonal pulse in primary production and the detrital food chain. During the dry season, the water levels drop. Fish are concentrated in ponds and shallow depressions and during this period are vulnerable to predation from wading birds. If the water level falls below the ground surface, some fish may still survive in the permanent ponds and solution holes. Some areas may dry out to the level that fish may have to invade from other areas to reestablish populations when those areas are reflooded.

Spatially explicit species index (SESI) models were described in the preceding section. To compute a landscape index map of fish availability, the SESI model uses several statistics derived from the dynamic simulation model of landscape hydrology (i.e., SFWMM), such as high and low water levels during the wading bird breeding season, water levels during preceding wet and dry seasons, and short term events such as reversals. This information serves as a basis from which an overall index of the fish population is computed. The simulation model differs primarily from the SESI model in projecting the daily fish population levels. The simulation model requires input from an adequate hydrologic model that can provide daily water depths as inputs to the simulation model. Fortunately, as mentioned above, the entire freshwater marshes of the Greater Everglades are included in a landscape hydrology model, SFWMM. This model simulates water flows and depths in the natural areas of South Florida on a 2×2 mi grid, based on rainfall and the operation of water regulation structures that alter the flow of water through the system. Additional analysis (Curnutt et al., in press) has enabled water depths to be resolved to a finer scale, 500×500 m, based on SFWMM output and other information.

Because the simulation model ALFISH actually projects population densities of fish rather than just using the hydrologic data to construct an index map, as does the SESI model, additional information is needed on the relationship of life cycle processes of fish, such as reproduction, growth, and survival, which are all related to changing water depths. In addition, estimates of the daily production of available prey (e.g., periphyton, meiofauna, and mesofauna) for the fish are needed. Prey availability is modeled as a seasonally varying function. In the current version of ALFISH, a broad functional group of fish is modeled, so the life cycle parameters for the fish represent those of an idealized "small fish."

With the simulation model, consider first the dynamics that occur within any given spatial pixel. The small fishes' functional group is modeled by tracking the number of fish within 1-month age-classes. The time step of the model is 5 d, so growth in size and mortality of these fish occur on 5-d time steps. Length, s, of fish in an age-class, Age, is described by the von Bertalanffy function

$$s = a_1[1.0 - exp(-a_2 Age + a_3)]$$

where a_1, a_2, and a_3 are constants and Age is age of the cohort in days.

With this model, there are three causes of mortality. The first is an age-dependent background mortality, the rate of which is described by the function, m_1;

$$m_1 = b_1/(1.0 + b_2 Age)^{b_3}$$

where b_1, b_2, and b_3 are constants. The second cause of mortality is starvation. This rate of mortality, m_1, is estimated by first calculating the energy needs for maintenance and growth of fish in each age-group, followed by calculating prey availability. Mortality rate is proportional to the ratio of energy needed to that which is available. Therefore, starvation mortality is density dependent and regulates the fish population.

The third type of mortality is due to the seasonal changes in water depths, which may lead to lowering the water level below ground surface in many areas. From the point of view of human impacts, this third type of mortality is particularly important, because it can be altered by employment of different water regulation strategies that affect water flow in the Everglades. In the model, if and when water depth in the pixel falls to a specific low level, certain fractions of the fish are assumed to follow the declining water levels and move into depressions in the pixel or into permanent ponds. Survivorship during the dry period then depends on a number of factors. First, in the model, it is assumed that for each pixel there is a hypsograph representing the microtopography of the pixel; that is, the depressions and solution holes in which some water may remain after the nominal surface of the pixel has dried out. The lower the water table during the dry season, the fewer fish that survive. Permanent ponds are also located randomly in some fraction of pixels, reflecting empirical data on the distributions of alligator holes. Some of the fish that manage to reach ponds can survive, although they are consumed, at estimated rates, by larger fish. When water levels rise again within pixels during the rainy season, fish are assumed to spread across the marsh areas and increase in numbers again. These ponds and small depressions represent the microhabitat variables associated with the small fish functional group.

Because detailed data on the microtopography across the whole Everglades are lacking and because it would be difficult in any case to compute accurate survivorship of fish throughout the dry period even if microtopography were well known, the modeling of the population in particular pixels has to be understood as an attempt to obtain approximate patterns of fish densities across the landscape and not as accurate projections at any particular place.

ALFISH generalizes this single-pixel model to operate on the entire Greater Everglades landscape. There are a total of 111,000 pixels, each 500×500 m, in the model, and the functional group dynamics described above occurs in all of them. Due to the present limitations on microtopography data, an identical generic hypsograph is used for every pixel, based on empirical data from several pixels, and 25% of the pixels are assumed to have permanent ponds. In addition to the seasonal within-pixel movements that occur for fish, the fish can diffuse across pixel borders. In this way, pixels that are completely dried, with loss of all fish, may recover, although this process takes time.

Simulations

It is impossible to forecast the rainfall pattern in the Everglades for any particular year and, therefore, even if ALFISH were a perfect model, it would be difficult to predict fish densities in advance for a particular year. Therefore, we cannot consider ALFISH, or models similar to it, as predictive. These models, for example, cannot predict the effects of a particular water regulation plan, because water levels through time depend as much on the particular pattern of rainfall as they do on the water regulation structures and operations. What such models can do, however, is to enable us to compare the effects of output of landscape hydrology models applied to a particular long-term set of rainfall data but with different alternative water regulation plans. This modeling strategy has two main advantages. First, by simulation over many years, the response of fish to a variety of annual rainfall patterns will be covered. Second, by emphasizing the comparison of two or more different alternative water management scenarios at a time, any systematic errors that occur in the simulation of one alternative will likely subtract out in the comparison of two or more of these scenarios.

The fish simulation model was used to determine the pattern of fish densities on the landscape for the same hydrology scenarios to which the SESI model was applied. Results are presented here for comparison between an alternative regulation plan called AltD and the base case F2050, as discussed earlier. Among the many changes in the water regulation plan AltD were the modifications of internal structures, such as levees, canals, and spillways, and construction of large storage areas for aquifer storage areas that allow water to be stored during periods of high rainfall and released during dry periods.

The ALFISH model computes the fish biomasses in each pixel in the Everglades landscape on 5-d time steps over a 31-year period. These computations are run separately for the AltD and F2050 scenarios. The spatial patterns of fish biomasses created by these different regulations can be compared based on particular days or on averages over any appropriate time period. Figure 2 shows the model fish biomass in each pixel in the Everglades ecosystem on July 1, averaged over all of the years in the 31-year period to produce an "average July 1." The left and right panels represent, respectively, outputs for the AltD plan and the F2050 based case scenarios, while the middle frame illustrates the differences between these two cases. In particular, for the AltD plan, the results show that higher mean fish biomasses will occur in much of the Everglades, particularly the eastern area, where longer hydroperiods will be maintained. The temporal patterns of these biomasses are also of interest. The fish densities can be spatially averaged over the entire region on each 5-d interval and plotted over either a particular wading bird breeding season (Figure 3) or the whole 31-year pe-

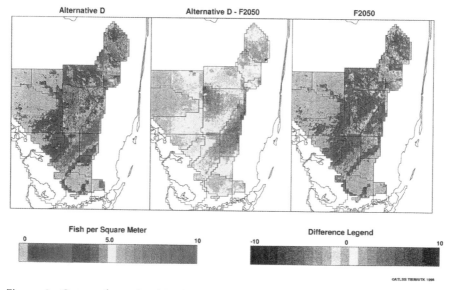

Figure 2. Output from the ALFISH computer simulation model for the functional group of small fishes. Two hydrology scenarios, F2050 (right panel) and AltD (left panel), are compared. The fish per square meter are simulated for each 500 ´ 500 m pixel with the density being color coded. The central panel represents the differences between the values for the individual pixels. This particular comparison averages over fish biomass in each pixel on July 1 of every year over the 31-year simulation by the model.

riod (Figure 4). One of the interesting temporal differences between the two scenarios is the high frequency of extremely low fish biomasses under the F2050 scenario (Figure 4).

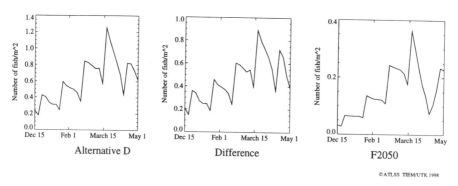

Figure 3. Mean number density of fish (fish/m²) computed from simulation model ALFISH, during the wading bird breeding season of a particular year, 1969. This mean density is obtained by averaging over all 111,000 pixels in ALFISH. The three panels show the density under the F2050 base case (right panel), the density under the alternative plan AltD (left panel), and the difference (central panel).

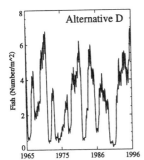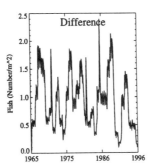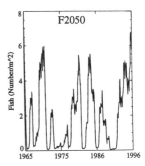

Figure 4. Mean number density of fish (fish/m²) computed from simulation model ALFISH, over a 31-year period. This mean density is obtained by averaging over all 111,000 pixels in ALFISH. The three panels show the density under the F2050 base case (right panel), the density under the alternative plan AltD (left panel), and the difference (central panel).

Discussion

Evaluating fish condition indirectly by habitat assessment can never be an ideal replacement for direct measurement of fish condition in the field. However, the long and continued interest in the use of habitat quality indicators to estimate the real or potential effects on fish populations and communities from environmental stressors testifies to a real need for this type of approach. Measuring changes of populations and communities in the field as they respond to environmental insults is difficult, and surrogate approaches, even if they are imperfect, are needed wherever possible to assess these changes. The bioindicators of effects addressed in this chapter (some of which are listed in Table 1) range from simple broad scale indices based on map data to approaches such as PHABSIM (physical habitat simulation models), which, because they require intensive data collection and modeling, are limited to a few locations. There is a need to combine broad areal coverage of ecosystems with high degrees of spatial resolution into predictive models. Rapid improvement of remote sensing techniques may serve this purpose of several groups of organisms, such as fish populations and communities, as has been demonstrated with the Everglades SESI and simulation models. In addition to remote sensing data, certain types of habitat information and relationships between habitat variables and fish populations must still be obtained through field studies. One of our main challenges, however, continues to be how to use, most productively, the tools provided by remote sensing and computational power (such as simulation models) in combination with more traditional studies, such as habitat surveys, to develop improved indicators of environmental stress on fish populations and communities.

Table 1. Summary of the major types of habitat index and simulation model for indirectly assessing the effects of environmental stressors on aquatic organisms.

Major index or model	Principal uses	Principal data needs	Main limitations	Selected references
Morphoedaphic index	Broad scale prediction of fish production	Lake watershed size and edaphic characteristics, lake depth	Explains only the broad scale variations in fish production	Ryder (1965, 1978); Jenkins (1967)
River habitat survey	Broad scale classification and prediction of stream health	Physical structure of river and banks (channel slope and pattern, geology, surrounding land use)	Detail and accuracy for local areas may be low	Fox et al. (1996)
Instream flow incremental methodology (IFIM)	Determination of physical habitat availability for fish and other aquatic organisms	Measurements of water depth, velocity, and substrate at fine scale resolution within a reach	Limits on data availability; restriction to effects of depth, velocity, and substrate; difficulty in extrapolating from one reach to a whole stream	Bovee et al. (1978); Milhous (1979)
Habitat evaluation methods (HSI, HQI)	Development of relationship between stream habitat characteristics and standing crop of fish	Relevant habitat variables (e.g., minimum and maximum temperatures by season, width and depth of stream, % pool area, substrate diversity	Difficulty in extrapolating from one stream system to others	Binns and Eiserman (1979); Milner et al. (1985)
Spatially explicit species index (SESI)	Development of relationships between spatial pattern of fish availability and spatially explicit information on habitat and hydrology	GIS data on habitat characteristics on a consistent resolution across hydroscape; fine temporal resolution hydrologic data; relationships between yearly fish abundance and habitat and hydrology	Restricted to environments and scales of resolution for which data are available	Curnutt et al. (2000)
Hydroscape simulation modeling (ALFISH)	Simulation of fish availability through time on a spatially explicit hydroscape	Data on spatiotemporal dynamics of hydroscape (water depths); data on fish life cycle, growth rates, and responses to water depth	Limited to situations for which data are available to develop simulation model of dynamic hydroscape and population or functional group model	Gaff et al. (2000)

Acknowledgments

This research was supported in significant part by the Department of Interior's Critical Ecosystem Studies Initiative, a special funding initiative for Everglades restoration administered by the National Park Service, and, in part, by the USGS's Florida Caribbean Science Center.

References

Addicott, J. F., J. M. Aho, M. F. Antolin, D. K. Padilla, J. S. Richardson, and D. A. Soluk. 1987. Ecological neighborhoods: scaling environmental problems. Oikos 49:340–346.

Bancroft, G. T., A. M. Strong, R. J. Sawicki, W. Hoffman, and S. D. Jewell. 1994. Relationships among wading bird foraging patterns, colony locations, and hydrology in the Everglades. Pages 615–657 *in* S. M. Davis and J. C. Ogden, editors. Everglades: the ecosystem and its restoration. St. Lucie Press, Delray Beach, Florida.

Binns, N. A., and F. M. Eiserman. 1979. Quantification of fluvial trout habitat. Transactions of the American Fisheries Society 108:215–228.

Bovee, K., J. Gore, and A. Silverman. 1978. Field testing and adaptation of a methodology to measure "in-stream" values in the Tongue River, Northern Great Plains Region. U.S. Environmental Protection Agency EPA-908/4-78-004A.

Bowlby, J. N., and J. C. Roff. 1986. Trout biomass and habitat relationships in southern Ontario streams. Transactions of the American Fisheries Society 115:503–514.

Cooperrider, A. Y. 1986. Habitat evaluation systems. Pages 757–776 *in* A.Y. Cooperrider, R. J. Boyd, and H. R. Stuart, editors. Inventorying and monitoring of wildlife habitats. U.S. Department of Interior, Bureau of Land Management Service Center, Denver.

Curnutt, J. L., J. Comiskey, M. P. Nott, and L. J. Gross. 2000. Landscape-level spatially-explicit species index models for Everglades restoration. Ecological Applications 10(6):1849–1860.

DeAngelis, D. L., L. J. Gross, M. A. Huston, W. F. Wolff, D. M. Fleming, E. J. Comiskey, and S. M. Sylvester. 1998. Landscape modeling for Everglades restoration. Ecosystems 1:64–75.

DeAngelis, D. L., W. F. Loftus, J. C. Trexler, and R. E. Ulanowicz. 1997. Modeling fish dynamics and effects of stress in a hydrologically pulsed ecosystem. Journal of Aquatic Stress and Recovery 6:1–13.

Dettmers, R., and J. Bart. 1999. A GIS modeling method applied to predicting forest songbird habitat. Ecological Applications 9:152–163.

Downs, P. W., and A. Brookes. 1994. Developing a standard geomorphological approach for the appraisal of river projects. Pages 299–310 *in* C. Kirby and W. R. White, editors. Integrated river basin development. John Wiley and Sons, Chichester, England.

Fleming, D. M., W. F. Wolff, and D. L. DeAngelis. 1994. Importance of landscape heterogeneity to wood storks in Florida Everglades. Environmental Management 18:743–757.

Fox, P. J. A., M. Naura, and P. Raven. 1996. Predicting habitat components for seminatural rivers in the United Kingdom. Pages 227–237 *in* M. Leclerc, H. Capra, S. Valentin, A. Boudreault, and I. Cote, editors. Proceedings of the 2nd International Symposium on Habitats and Hydraulics, volume B. Quebec.

Gaff, H., D. L. DeAngelis, L. J. Gross, R. Salinas, and M. Shorrosh. 2000. A dynamic landscape model for fish in the Everglades, and its application to restoration. Ecological Modeling 127:33–52.

Hakanson, L. 1996. Predicting important lake habitat variables from maps using modeling tools. Canadian Journal of Fisheries and Aquatic Sciences 53(Supplement 1):364–382.

Hayes, D. B., C. Ferreri, and W. W. Taylor. 1996. Linking fish habitat to their population dynamics. Canadian Journal of Fisheries and Aquatic Sciences 53(Supplement 1):383–390.

Imhof, J. G., J. Fitzgibbon, and W. K. Annable. 1996. A hierarchical evaluation system for characterizing watershed ecosystems for fish habitat. Canadian Journal of Fisheries and Aquatic Sciences 53(Supplement 1):312–326.

Irvine, J. R., I. G. Jowett, and D. Scott. 1987. A test of the instream flow incremental methodology for under-yearling rainbow trout, *Salmo gairdnerii,* in experimental New Zealand streams. New Zealand Journal of Marine and Freshwater Research 21:35–40.

Jenkins, R. M. 1967. The influence of some environmental factors on standing crop and harvest of fishes in U.S. reservoirs. Pages 298–321 *in* Reservoir fisheries resource symposium. American Fisheries Society, Southern Division, Reservoir Committee, Bethesda, Maryland.

Loftus, W. F., and A.-M. Eklund. 1994. Long-term dynamics of an Everglades small-fish assemblage. Pages 461–483 *in* S. M. Davis and J. C. Ogden, editors. Everglades: the ecosystem and its restoration. St. Lucie Press, Delray Beach, Florida.

Maddock, I. 1999. The importance of physical habitat assessment for evaluating river health. Freshwater Biology 41:373–391.

Marshall, T. R. 1996. A hierarchical approach to assessing habitat suitability and yield potential of lake trout. Canadian Journal of Fisheries and Aquatic Sciences (Supplement 1):332–341.

Mathur, D., W. H. Bason, E. J. Purdy Jr., and C. A. Silver. 1985. A critique of the instream flow incremental methodology. Canadian Journal of Fisheries and Aquatic Sciences 42:825–831.

Meents, J. K., J. Rice, B. W. Anderson, and R. O. Ohmart. 1983. Nonlinear relationships between birds and vegetation. Ecology 64:1022–1027.

Milhous, R. T. 1979. The PHABSIM system for instream flow studies. Pages 440–446 *in* Proceedings of the 1979 Summer Computer Simulation Conference, Toronto, Canada. Society for Computer Simulation International, San Diego, California.

Milner, N. J., R. J. Hemsworth, and B. E. Jones. 1985. Habitat evaluation as a fisheries management tool. Journal of Fish Biology 27(Supplement A):85–108.

Ogden, J. C. 1994. A comparison of wading bird nesting colony dynamics (1931–1946 and 1974–1989) as an indication of ecosystem conditions in the southern Everglades. Pages 533–570 *in* S. M. Davis and J. C. Ogden, editors. Everglades: the ecosystem and its restoration. St. Lucie Press, Delray Beach, Florida.

Orth, D. J., and O. E. Maughan. 1982. Evaluation of the incremental methodology for recommending instream flows for fishes. Transactions of the American Fisheries Society 111:413–445.

Pajak, P., and R. J. Neves. 1987. Habitat suitability and fish production: a model evaluation for rock bass in two Virginia streams. Transactions of the American Fisheries Society 116:839–850.

Pearsall, S. H., D. L. Durham, and D. C. Eager. 1986. Evaluation methods in the United States. Pages 111–133 *in* M. B. Usher, editor. Wildlife conservation evaluation. Chapman and Hall, London.

Peters, R. H. 1986. The role of prediction in limnology. Limnology and Oceanography 31:1143–1159.

Randall, R. G., C. K. Minns, V. W. Cairns, and J. E. Moore. 1996. The relationship between an index of fish production and submerged macrophytes and other habitat features at three littoral areas in the Great Lakes. Canadian Journal of Fisheries and Aquatic Sciences 53(Supplement 1):35–44.

Ryder, R. A. 1965. A method for estimating the potential fish production of north-temperate lakes. Transactions of the American Fisheries Society 94:214–218.

Ryder, R. A. 1978. Fish yield assessment of large lakes and reservoirs: a prelude to management. Pages 403–423 *in* S. D. Gerking, editor. Ecology of freshwater fish production. Blackwell Scientific Publications, Oxford, England.

Ryder, R. A. 1982. The morphoedaphic index: use, abuse, and fundamental concepts. Transactions of the American Fisheries Society 111:154–164.

Ryder, R. A., and S. R. Kerr. 1989. Environmental priorities: placing habitat in hierarchic perspective. Pages 2–12 *in* C. D. Levings, L. B. Holtby, and M. A. Anderson, editors. Proceedings of the National Workshop on Effects of Habitat Alteration on Salmonid Stocks, 6–8 May, 1987, Nanaimo, British Columbia, Canadian Special Publication of Fisheries and Aquatic Sciences 105.

Southwood, T. R. E. 1977. Habitat, the templet for ecological strategies? Journal of Animal Ecology 46:337–365.

Tennant, D. L. 1976. Instream flow regimens for fish, wildlife, recreation, and related environmental resources. Fisheries 1(4):6–10.

Townsend, C. R., and A. G. Hildrew. 1994. Species traits in relation to a habitat templet for river systems. Freshwater Biology 31:265–275.

Trexler, J., W. Loftus, O. Bass, and F. Jordan. 1996. High water assessment: the consequences of hydroperiod on marsh fish communities. Pages 103–124 *in* T. V. Armentano, editor. Proceedings of the Conference: Ecological Assessment of the 1994–1995 High Water Conditions in the Southern Everglades. Florida International University, Miami, Florida, August 22–23, 1996.

U.S. Army Corps of Engineers. 1999. Central and South Florida Comprehensive Review Study Final Integrated Feasibility Report and Programmatic Environmental Impact Statement. U.S. Army Corps of Engineers, Jacksonville District, and South Florida Water Management District.

14

Assessing Contaminant-Induced Stress Across Levels of Biological Organization

Lyndal L. Johnson and Tracy K. Collier

Introduction

Anthropogenic stressors are currently a major threat to marine fish populations. Marine fisheries are declining worldwide (Botsford et al. 1997; Musik et al. 2000) and, although overfishing is a major cause (Garcia and Newton 1997), habitat alteration and contaminants have also been implicated as contributing factors. As studies of the effects of stress move to higher levels of biological organization and toward higher environmental relevance, linking cause (stressor) to effect becomes more difficult (Figure 1). Consequently, while it is important to develop more accurate techniques for establishing these linkages between cause and effect, it is also critical to have a clear understanding of the limitations of our understanding and methods of risk prediction if we are to manage our aquatic resources in a prudent manner.

In this chapter, we present what we have learned about how the measurement of stress effects, in this case contaminant-induced stresses at different levels of biological organization, can improve our understanding of the risks posed to aquatic ecosystems and their causality. We describe how population modeling and multivariate analysis techniques, applied to data collected in field studies, can help track the effects of stress up through different levels of biological organization. Examples will be provided from case studies involving carcinogenesis and reproductive impairment in English sole *Pleuronectes vetulus* from Puget Sound, as well as other relevant studies with different species in other aquatic ecosystems.

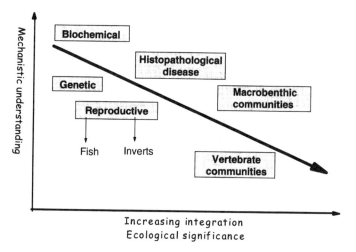

Figure 1. Relationship between mechanistic understanding and ease of establishing causality vs. biological integration and ecological significance. Adapted from SETAC presentation by S. M. Adams 2000.

Approaches for Linking Effects Across Levels of Biological Organization

Effects at the Organism and Population Levels

Analyses of the effects of stress across biological levels of organization can begin at various stages in the biological hierarchy. It is probably most common for investigations to begin with the observation of disease or other abnormalities in individuals in natural populations. In the case of Puget Sound flatfish, for example, our investigations were motivated by the discovery, in the late 1970s, of fish with liver tumors in industrial waterways near Seattle (Pierce et al. 1978). Since these initial observations, the occurrence of liver cancer and a spectrum of other related toxicopathic lesions have been well documented in English sole in Puget Sound (see Myers et al. 1998a, 1998b; and Chapter 7). Typically, between 25 and 40% of English sole sampled from urban embayments such as Elliott Bay and Commencement Bay exhibit neoplastic, preneoplastic, or unique degenerative liver lesions, as compared with 3–8% of fish from nonurban and moderately urbanized sites. Subsequent investigations revealed that English sole from Puget Sound sites also exhibited various types of reproductive abnormalities, including impaired ovarian development, alterations in egg size and number, inhibition of spawning, and reduced egg and larval viability (reviewed in Johnson et al. 1998). Similarly, comprehensive studies of the impacts of bleached kraft mill effluent on aquatic ecosystems in Canada (reviewed in Servos et al. 1996) were motivated by the discovery of altered reproductive traits in white sucker *Catostomus commersoni* from Jackfish Bay in Lake Superior (Munkittrick et al. 1992).

In some cases, studies to investigate cause and effect are initiated in response to a decline in the abundance of a commercially important species. A good example of this scenario is investigations on the effects of polychlorinated biphenyls (PCB) and dioxin on lake trout *Salvelinus namaycush* in Lake Ontario (reviewed in Ankley and Giesy 1998). The impetus for these studies was the decline of lake trout, which began in the 1940s, with virtual extinction of the population by the 1960s. Major contributing factors to the population decline initially were thought to be overfishing, habitat loss, and predation by sea lamprey *Petromyzon marinus*. However, stocking programs to rebuild lake trout populations had only limited success, and chronically poor reproduction suggested that chemical contaminants could be involved. Detailed studies over multiple levels of organization eventually showed that historical exposure to dioxin played an important role in the decline of lake trout populations. In still other cases, laboratory studies have indicated that certain contaminants may cause injury to organisms, and field studies were conducted to determine if problems occurred in areas in the environment where these chemicals were found.

Whatever the motivating factor for the initiation of such investigations, studies designed to establish linkages between anthropogenic stressors and biological effects across levels of organization typically involve one or more of the following efforts: (1) establishing a relationship between the stressor and the observed effect, (2) linking effects with earlier, more subtle, and more easily measured physiological changes that can be predictive of more severe impacts on the whole organism (e.g., disease condition, reduced survival, growth, reproduction on the individual level) and which also provide insight into mechanisms, (3) predicting how impacts on individuals (e.g., reduced growth, survival, reproduction) might affect abundance or population characteristics, or reconstructing a scenario to see if observed changes in abundance or population characteristics are consistent with impacts of contaminant stress on individuals, and (4) assessing community-level impacts of contaminant stress by examining indirect effects of pollutants on target species prey, predators, and competitors, or by measuring alternative indicators such as species diversity. Various approaches and techniques for conducting these types of studies (summarized in Table 1) will be discussed below.

Establishing Causality Between Stressors and Effects

When potential effects of contaminant stress on natural populations are observed, it is important to identify the risk factors that are associated with the change in health status and to determine whether or not this condition is actually related to contaminant stress. Criteria used to establish causal relationships from observational field studies include the strength, consistency, and specificity of the association; the biological credibility of the postulated

Table 1a. Summary of common applications, advantages, and disadvantages of various techniques used to establish linkages between stressors and effects across different levels of biological organization.

Technique	Application	Advantages	Disadvantages
Subcellular to organismal			
Logistic regression	Evaluate association between stressors and probability of injury or disease condition in individual or subpopulation; identify major risk factors	Can compare contribution of different stressors to injury; provides predictive measure of injury likelihood (odds ratio); can adjust for confounding factors; available in standard statistical packages	Results may be difficult to interpret if several stressors co-occur and co-vary; by itself, establish association, but not causation
Multiple linear regression	Evaluate association between stressors and change in level of some biological factor (e.g., hormone, enzyme) in individual	Can compare contribution of different stressors to injury; provides predictive model via regression equation; can adjust for confounding factors; available in standard statistical packages	Results may be difficult to interpret if several stressors co-occur and covary; association between stressor and biological factor may not fit linear model; by itself, establish association, but not causation
Nonlinear regression techniques	Evaluate association between stressors and change in level of some biological factor (e.g., hormone, enzyme) in individual	Can provide more realistic model of stressor–effect relationship than linear regression techniques	Models with multiple factors may be difficult to construct; interpretation of significance of association may not be straightforward with standard statistical packages; by itself, establish association, but not causation
Multivariate techniques, e.g., principal components or discriminant analysis	Identify complexes of co-occurring effects or stressors; evaluate associations between multiple stressors and impacts	Possible to work with covarying sets of stressors and effects, identify syndromes	Interpretation can be difficult or subjective; by itself, establish association, but not causation
Controlled laboratory trials	Direct test of cause and effect linkage between stressor and injury	More definitive than statistical techniques that establish association only; conditions may be controlled to avoid confounding factors	May not be possible to conduct tests with test organisms of interest; conditions may not reflect those in natural populations

Table 1b. Summary of common applications, advantages, and disadvantages of various techniques used to establish linkages between stressors and effects from the organism to the population level.

Technique	Application	Advantages	Disadvantages
Organism to population			
Leslie matrix models	Model population level responses based on age or stage specific reproductive and survival rates	Flexible and adaptable to a variety of population-specific applications	Results may not be accurate if realistic relationships among stressors and vital rates cannot be derived by other means; does not directly incorporate individual effects but uses averages
Individual-based models	Model population level responses based on characteristics of individual organisms	Greatest potential to directly link individual biochemical or physiological measures population-level responses; flexible enough to incorporate a wide range of scenarios	May be computation intensive, high data requirements; accuracy of model limited by data quality
Spatially explicit models	Model population responses based on geographically related reproductive and survival rates, metapopulation dynamics; both matrix and individually based forms can be used	Can incorporate metapopulation dynamics, habitat-related effects on survival and reproductive rates; increased biological realism	Similar to those of matrix and individual-based models, depending on form used; also requires accurate information on immigration and emigration rates.
Life-history based monitoring	Identify patterns of change in individuals in field populations that are likely associated with specific population responses	Focuses on natural populations; facilitates collection of data that is useful for predictive modeling	Predicted patterns may not always be found in natural systems due to trophic interactions and other complexities

Table 1c. Summary of common applications, advantages, and disadvantages of various techniques used to establish linkages between stressors and effects from the organism or population to the community level.

Organism or population to community

Technique	Application	Advantages	Disadvantages
Individual-based models	Model population-level responses based on characteristics of individual organisms	Can incorporate effects of trophic interactions (predation, food supply) on individual growth and reproductive rates into population model	May be computation intensive, high data requirements; accuracy of model limited by data quality; results can be equivocal or difficult to interpret
Mesocosms	Test cause and effect relationships in controlled experimental system where at least simple trophic interactions are incorporated	Possible to work with covarying sets of stressors and effects, identify syndromes	May not accurately simulate natural system of interest; results may be hard to interpret or difficult to extrapolate to field situation
Biotic indices	Combine a variety of biological changes associated with exposure to stressors into single semiquantitative index	Can evaluate a number of response alterations simultaneously to produce a more comprehensive indicator of community and ecosystem health	May not be effective at providing case-effect linkages
Multivariate techniques (e.g., discriminant analysis)	Identify groups of associated changes in response to stressors occurring at different levels of biological organization	Can integrate multiple types of responses; patterns of co-occurrence may be indicative of cause–effect linkages	Interpretation can be difficult or subjective; by itself, establishes association, but not causation

effect; the time sequence of exposure and the onset of the effect; and the demonstration of a dose–response gradient in the relationship between a health effect and its suspected cause (Breslow and Day 1980; Landahl et al. 1990).

Statistical Association

To better identify specific chemical and biological risk factors associated with alterations in health status or disease conditions in field organisms, various statistical methods can be employed. One of the most useful is logistic regression (Anderson et al. 1980; Schlesselman 1982). This technique is commonly used on binomial or proportional data in epidemiological and epizootiological studies (Breslow and Day 1980) and calculates the risk of an adverse health effect (for example, the risk of cancer or impaired gonadal growth) associated with a given contaminant exposure in an individual fish or fish subpopulation (e.g., fish collected from a particular sampling site). For effects that are not measured as a binomial or proportional outcome variable (e.g., changes in gonad size, larval survival, or fecundity), multivariate regression techniques can be used instead of logistic regression to identify factors that are most highly correlated with the observed abnormality. For additional discussions on these statistical techniques, see Chapter 15.

For many disease conditions and other changes in health status, including alterations in growth and reproductive function, a variety of biological factors unrelated to contaminant exposure can affect the likelihood of injury in a particular animal. In the case of liver cancer, for example, which requires some time to develop, fish age is a major factor affecting disease risk. Confounding biological factors are even more important in reproductive toxicology analyses because reproductive development is so strongly influenced by factors such as age, size, body weight, and season. A major advantage of logistic and multiple regression techniques is that they allow the significance of relationships between contaminant exposure and suspected toxicopathic injury to be determined while adjusting for factors such as fish age, sampling season, or fish condition.

We have used logistic regression in several studies with English sole and other Pacific Coast bottomfish to statistically relate prevalences of toxicopathic liver lesions and reproductive abnormalities to biological risk factors and measures of contaminant exposure (see reviews by Myers et al. 1998a and Chapter 7[1]; Johnson et al. 1998). Measures of exposure used in these analyses have typically included sediment polycyclic aromatic hydrocarbon (PAH), chlorinated hydrocarbon (CH), and metal concentrations; PAH and CH concentrations in stomach contents; CH concentrations in tissues; and concentrations of fluorescent aromatic compounds in bile, an indicator of exposure to rapidly metabolized PAHs (Krahn et al. 1993). Exposure to PAHs was a major risk factor for lesion development. Selected CHs, including PCBs and pesticides, which may play a role in lesion progression through their actions as tumor promoters, were also identified as risk factors for lesion development. More recent field studies (Myers et al. 1998a, 1998b; Johnson et al. 1999) have examined the relationship between biochemical alterations associated with contaminant exposure (cytochrome P4501A [CYP1A] induction and DNA adduct formation; e.g., Collier et al. 1995 and Reichert et al. 1998) and toxicopathic hepatic lesions and reproductive abnormalities in wild fish.

An example of a logistic regression analysis is shown in Table 2. In this case, the relationship between lesion prevalences at particular sampling sites and concentrations of PCBs and PAHs in sediments was examined, while adjusting for mean fish age. The data show that sediment PAH concentrations at those sampling sites can explain 45% of the variation in site-specific prevalences for toxicopathic liver lesions (i.e., neoplasms, preneoplasms, and specific degenerative lesions) in English sole, while sediment PCB concentrations account for an additional 8%. The data also indicate that fish age contributes significantly to the risk of toxicopathic lesions, explaining 24% of the observed variation. Similar analyses of factors influencing the likelihood of gonadal development, ovarian atresia, and inhibited spawning in English sole indicate significant effects of fish age, fish condition, and exposure to PAHs and selected CHs, including PCBs (Johnson et al. 1988, 1999).

Logistic regression analysis can also be applied to individual fish to generate dose–response curves that estimate the relative risk (as expressed by the odds ratio) of occurrence for a particular lesion or injury attributable to a chemical contaminant body burden or other exposure parameter. For example, the odds of an English sole from Eagle Harbor, Washington, exhibiting specific degeneration and necrosis (also called hepatocellular nuclear pleomorphism/megalocytic hepatosis in pathological terminology) or

Table 2. Associations between mean fish age, sediment PAH concentrations, and sediment PCB concentrations with site-specific prevalences of toxicopathic liver lesions (i.e., neoplasms, foci of cellular alteration, and specific degenerative and necrotic lesions) in English sole as determined by logistic regression ($P < 0.05$). The table indicates P-values and percent total variances in lesion prevalence explained by the risk factor (reduction in scaled deviance). Unpublished data from M. S. Myers and S. M. O'Neill, collected in conjunction with the Puget Sound Ambient Monitoring of the State of Washington.

Risk factor	df	Scaled deviance	%Total variance explained	P-value
Grand mean	58	546		0.001
ΣPAHs in sediment	1	299	45.2%	0.001
Mean fish age (yrs)	1	165	24.5%	0.001
ΣPAHs in sediment	1	119	8.4%	0.001
Residual			21.9%	

preneoplastic focal lesions increase by 1.05 and 1.03 times, respectively, for each unit increase in hepatic DNA adduct level (Myers et al. 1998a).

In our reproductive toxicology field studies with English sole, we have made extensive use of multiple regression techniques (see also Chapter 15) to examine relationships between indicators of contaminant exposure and variables such as plasma hormone concentrations, fecundity, egg weight, and gonadosomatic indices, which are continuously distributed variables (e.g., Johnson et al. 1988, 1997, 1999). These analyses suggest that exposure to PAHs and, to a lesser extent, CHs, is correlated with suppression of plasma E2 concentrations and reductions in gonadosomatic index (Johnson et al. 1988, 1999; Casillas et al. 1991). Indicators of PAH exposure also exhibit some correlation with reductions in fecundity (Johnson et al. 1997). Exposure to CHs, including PCBs, appears to be more highly correlated with reduced larval viability (Casillas et al. 1991), reduction in egg weight (Johnson et al. 1997), and precocious sexual development in subadult female sole (Johnson et al. 1999).

Similarly, Adams et al. (1989, 1994) used multivariate statistics to examine effects of contaminant exposure on redbreast sunfish *Lepomis auritus* in freshwater systems receiving inputs of complex contaminant mixtures containing PCBs, PAHs, heavy metals, and chlorine. Indicators such as mixed-function oxidase enzymes and DNA damage provided direct evidence of toxicant exposure, while condition indices and indicators related to lipid biochemistry and histopathology reflected impaired lipid metabolism, immune and reproductive system dysfunction, and reduced growth potential in fish subjected to contaminant stress.

Consistency of Association

An important element in establishing cause-and-effect relationships using statistical analyses of field data is consistency of association. This points out the importance of conducting multiple studies with various species and at various sites to establish a common pattern of cause-and-effect relationships in similar situations. In the case of liver cancer in fish, for example, logistic regression analyses of a number of different data sets representing different field situations have consistently identified PAH exposure as a significant risk factor for the development of hepatic lesions in English sole (Myers et al. 1987, 1994, 1998b; Rhodes et al. 1987), as well as other bottomfish species (Johnson et al. 1993; Stehr et al. 1997, 1998). While reproductive assessments have not been as extensive in English sole as studies for hepatic lesions, separate investigations at different sampling sites have identified alterations in gonadal development in female sole (Johnson et al. 1988, 1998, 1999). Logistic and multiple regression analyses applied to these different data sets have consistently identified exposure to PAHs as a major risk factor for impaired ovarian development and ovarian atresia in English sole, with CH exposure also showing some association with the development of this condition (Johnson et al. 1988, 1999).

Similarly, McMaster et al. (1996) reviewed a series of studies on the re-productive effects of bleached kraft pulp mill effluent on Great Lakes fish to evaluate whether white sucker and brown bullhead *Ameiurus nebulosus* exposed to different classes of organic compounds (nonchlorinated pulp mill effluent and polycyclic aromatic hydrocarbons) exhibited comparable alterations in reproductive function based on epidemiological criteria. They found that the criteria of probability, strength of association, and consistency with respect to time were satisfied with respect to these reproductive alter-ations, indicating that changes in reproductive performance represented a common response in fish exposed to pulp mill effluent and other organic contaminants, with associated biochemical or hormonal changes in affected fish being more variable.

Laboratory Studies

While correlative field studies are an important part of any ecotoxicological research program, the importance of laboratory exposure studies in linking indicators of exposure to biological effects cannot be overemphasized. Such controlled studies have played a key role in our research program with English sole. For example, a cause-and-effect relationship between PAHs and toxicopathic liver lesions in English sole has been confirmed in the laboratory by induction of degenerative, proliferative, and preneoplastic le-sions identical to those observed in field-collected fish (Schiewe et al. 1991). Similarly, results of the reproductive toxicology field studies are supported by laboratory experiments showing that pretreatment of gravid female En-glish sole with extracts of contaminated sediment or crude oil containing high levels of PAHs decreased levels of endogenous estradiol (Stein et al. 1991; Johnson et al. 1995). More recent experiments suggest that exposure to benzo[a]pyrene or sediment contaminated with PAHs or CHs may sup-press estradiol-induced vitellogenin production in English sole (Anulacion et al. 1997).

Laboratory exposure studies were also used extensively to establish cause-and-effect relationships between dioxin-like compounds and developmental effects found in lake trout from Lake Ontario (Ankley and Giesy 1998). For example, experiments were conducted to establish that the pathology ob-served in lake trout larvae in the field was similar to that produced by labo-ratory exposure to TCDD-like (2,3,7,8-tetrachlorodibenzo-p-dioxin) com-pounds (dioxins, furans, PCBs). Dose–response relationships were also determined for maternal exposure, egg contaminant levels, and fry mortality so that these contaminant levels could be compared to those found in fish and sediments in the environment where the effects were observed.

Prediction of Organism Effects From Lower Level Responses

With the observation of whole organism effects, efforts are typically made to identify earlier chemical and biochemical changes that could be predictive of injury development. These early warning parameters can include measurements such as changes in activities of xenobiotic metabolizing enzymes (e.g., 7-ethoxyresorufin-O-deethylase [EROD], aryl hydrocarbon hydroxylase [AHH], and glutathione S-transferase [GST] activities) that are primarily indicators of contaminant exposure, as well as subtle biochemical or physiological indicators of biological effects (see Chapter 2). Effect indicators include changes in plasma chemistry (see Chapter 4) or hormone levels (see Chapter 9) and various measures of DNA damage (see Chapter 5), such as DNA adducts. These types of studies that focus on the identification and validation of "bioindicators" of exposure and effect, which are easily measured and may have some potential to be predictive of higher level effects, predominated ecotoxicological research in the late 1980s and 1990s.

Recent field studies of English sole (Myers et al. 1998a, 1998b) have identified induction of hepatic AHH activity and elevated DNA adduct levels as significant risk factors for degenerative and preneoplastic lesions. These findings, particular the strong association between PAH-DNA adducts in liver and toxicopathic lesions, give further credence to the role of PAH exposure in the etiology of liver disease in English sole and suggest that xenobiotic-DNA adduct formation is integral to lesion induction and initiation of hepatocarcinogenesis.

Identifying Mechanisms of Toxic Action

In addition to their use in epizootiological analyses, biochemical parameters such as CYP1A activity and DNA adducts have provided insight into mechanisms of toxicant action and pathogenesis of toxicopathic hepatic lesions in other ways. For example, immunohistochemical localization of CYP1A activity, in combination with quantitation of PAH-DNA adducts, has recently been applied to investigate the role of resistance to cytotoxicity in liver neoplasia in English sole (Myers et al. 1998c). Immunohistochemical studies of English sole from PAH-contaminated sites show a consistent reduction in expression of CYP1A in hepatic neoplasms and most preneoplastic foci of cellular alteration. The reduction in CYP1A expression is accompanied by a significant and nearly parallel reduction in DNA adduct levels as compared to nonneoplastic liver tissue. These findings are consistent with the hypothesis, developed from studies with mice and rats, that neoplastic hepatocytes possess a "resistant" phenotype in which there is a reduced capacity to activate PAHs and related compounds to toxic and carcinogenic intermediates (Roomi et al. 1985).

English sole reproductive toxicology studies illustrate the linkages between biochemical indicators of effects and abnormalities that are observable at the whole organism level. In these studies, we have examined not only relationships between exposure indicators and effects, but also relationships between certain biochemical and physiological changes associated with the reproductive process (e.g., plasma hormone levels) and higher-level conditions such as inhibited spawning or reduced gonadal growth.

One example is a study we conducted on the impact of contaminant exposure on the spawning success of English sole (Casillas et al. 1991). In this study, gravid English sole from contaminated and reference sites were brought into the laboratory and induced to spawn with gonadotropin-releasing hormones, and eggs were collected and fertilized with pooled sperm from reference males. Ability to spawn, time required to spawn, and egg and larval viability were all assessed. Plasma estradiol 17-beta (E2) concentrations were also measured in these fish. Although all fish in the study had undergone gonadal development and had plasma E2 concentrations consistent with induction of vitellogenesis, E2 concentrations were significantly lower in fish from the contaminated study areas. Moreover, in logistic regression analysis, plasma E2 concentration was significantly associated with the probability that the fish would spawn in response to hormone treatment (Figure 2). While it is the progesterone-like maturation-inducing steroids, not estradiol, that trigger spawning in teleost fish, the association between plasma E2 levels and spawning ability is consistent with the role of E2 in upregulating pituitary production of the gonadotropic hormone (GTH II, now recognized as fish LH) that stimulates ovarian production of maturation-inducing steroids. This study did not establish a mechanistic link between depressed plasma E2 levels and inhibition of spawning, however, and there is evidence that contaminants may affect the hypothalamic-pituitary-gonadal axis through a variety of other pathways as well. Nevertheless, it does show how biochemical markers of effect can be linked to, and may serve as predictors of, more obvious types of injury on the whole organism level.

In another study, multiple regression techniques were used to examine complex relationships between contaminant exposure, nutritional factors such as plasma glucose and triglyceride levels, and fecundity and egg weight in English sole (Johnson et al. 1997). This analysis revealed that elevated PCB concentrations in tissues were associated with a pattern of higher fecundity, smaller egg size, elevated plasma triglyceride levels, and depressed plasma vitellogenin levels. Exposure to high concentrations of PAHs, on the other hand, was associated with increased ovarian atresia, increased egg weight, and reduced egg number.

A number of studies have emphasized the importance of measuring a suite of biochemical indicators in environmental pollution studies to provide a more comprehensive assessment of the overall health of individual fish, fish populations, and fish communities. By monitoring a suite of selected stress responses at several levels of biological organization, it may be possible to (1) assess the effects of sublethal stress on fish, (2) predict future trends (early warning indicators), and (3) obtain insights into causal relationships between stress and effects at the community and ecosystem level (Adams et al. 1989)

In studies attempting to establish causal relationships between stressors and multiple effects, multivariate techniques such as principal components

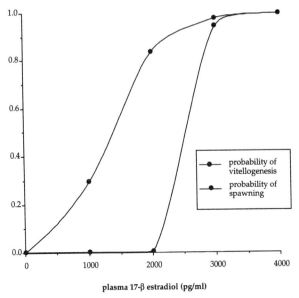

plasma 17-β estradiol (pg/ml)

Figure 2. Plasma estradiol 17-b concentration in female English sole vs. probability of vitellogenesis and probability of spawning. Relationships were determined though logistic regression. Data from Casillas et al. 1991; Johnson et al. 1998; and Johnson, unpublished data.

analysis and discriminant analysis have often been used to identify suites of indicators that are diagnostic of contaminant-associated alterations in fish health. An example of such a study with English sole is Casillas et al. (1985). In this investigation, a number of plasma parameters associated with alteration in liver function were measured in English sole from reference and contaminated sites and in fish with and without liver lesions. The results showed altered serum chemistry values in sole with liver disease, along with consistent changes in hematocrit and levels of serum albumin, calcium, and bilirubin. Similarly, Adams et al. (1994) used discriminant analysis and canonical variate analysis to characterize the integrated biological responses of freshwater fish species such as sunfish *Lepomis* spp. and largemouth bass *Micropterus salmoides* to chronic pollutant stress (Figure 3). Fish responses involved simultaneous changes in a variety of physiological functions including electrolyte homeostasis (indicated by alterations in serum sodium and potassium), carbohydrate metabolism (serum glucose), lipid metabolism (serum triglyceride and cholesterol) and protein metabolism (total serum protein and serum glutamate oxaloacetate transaminase), as well as alterations in condition factor, liver somatic index, and visceral somatic index. Analyses also showed that populations along a pollution gradient could most easily be discriminated by differences in the activity of the detoxification enzyme, EROD, indicators of organ dysfunction (such as serum electrolyte levels), and indicators of lipid metabolism. The use of discriminant analysis allows for the identification of distinctive response patterns in fish from pol-

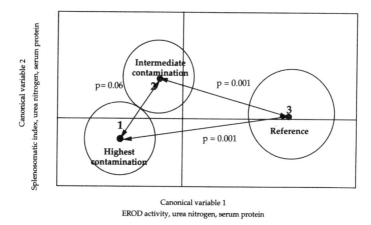

Figure 3. Multivariate two-dimensional responses, illustrated by circles, of large-mouth bass sampled from three sites with different levels of PCB contamination in Hartwell Reservoir on the South Carolina–Georgia border, USA. The numbers below the arrows represent significance values for differences between means of each response group. Adapted from Adams et al. 1994.

luted and control sites, which may be more informative than changes in single parameters alone. For additional discussion of statistical methods used to analyze multivariate field data see Chapter 15.

As an alternative to discriminant analysis, some investigators have used various individual measures of contaminant exposure and physiological and biochemical function to construct composite health or effect indices. Index variables are typically assigned numerical values based on the magnitude of the change for contaminant body burdens or physiological parameters or the severity of damage for disease conditions or lesions. The overall score is then used to evaluate the general health status of fish. Adams et al. (1993a) use a composite indicator that included gross pathology observations and fish condition information to rank fish populations in a wide range of reservoir types in the southern United States, including sites contaminated by PCBs and bleached kraft mill effluent. Stein et al. (1992) employed a bioeffects index including tissue contaminant concentrations, induction of a variety of xenobiotic metabolizing enzymes, and DNA adduct formation to rank toxicant responses in several species of flatfish from Puget Sound sites. The composite index proved to be more effective than any one individual measurement in differentiating fish populations from sites with various types and concentrations of contaminants in sediment.

Linking Individual Effects and Population-Level Effects

The expression of chronic pollutant stress in aquatic systems can result from both direct effects on organisms and indirect mechanisms operating within the aquatic food web (Adams et al. 1993b). Direct pollutant effects are usually expressed first at the molecular or cellular level and can be propagated upward through increasing levels of biological complexity to manifest themselves as changes in survivorship, growth, and reproduction. Indirect effects of pollutant stress are more typically expressed through the food chain via changes in the quality and quantity of energy available to organisms or through alterations in predator–prey dynamics.

Direct effects of pollutants, which impact growth, reproduction, and survivorship are the types of changes that can be most easily translated into population-level responses through standard population models. Such assessment techniques help bridge the gap between exposure and toxicological response at the suborganismal level and risk assessment at the population level (Rose et al. 1999a).

Population-level responses to stressors can be examined in various ways. In some cases, the existence of long-term data sets has made it possible to examine associations between contaminant stress or other anthropogenic activities and actual fish abundance. For example, Polgar et al. (1985) and Summers and Rose (1987) used long-term estuarine and coastal data sets to investigate the relationship between pollutant loadings and fish stock levels of striped bass *Morone saxatilis* and American shad *Alosa sapidissima* in northeastern United States estuaries. In both cases, they were able to demonstrate strong dependence of American shad stock levels over the period from the late 1920s to the mid-1970s on factors related to increased pollution and other anthropogenic activities (i.e., human population levels, sewage loading, dissolved oxygen, and biological oxygen demand). Striped bass stock level, on the other hand, was more highly correlated with climactic and hydrographic variables.

In many cases, however, long-term abundance data on fish stocks is unavailable, difficult to collect, or would not yield sufficient information about possible contaminant impacts in time to avoid placing affected populations at risk. One alternative, sometimes known as life-history-based monitoring (Munkittrick and Dixon 1989; Rose et al. 1999a), is to concentrate on measuring individual-level variables (e.g., growth or fecundity) rather than actual fish abundance and to use the distribution of individual responses to infer an ecological or population-level effect. For example, a study might show that a certain percentage of individuals fail to reproduce or a certain percentage of offspring had deformities and a population-level effect would be presumed from the fact that survival and reproductive rates had been altered. The changes observed in these individual-level variables may allow the investigator to make various inferences about likely population responses, as well as the nature of the stressors influencing the species. Munkittrick and Dixon (1989) describe a number of generalized response patterns in fish life history traits or population-level variables and their relationship to the type of stress (Table 3). For example, heavy exploitation of the fishery is likely to

Table 3. Generalized response patterns of fish populations to stressors. Adapted from Munkittrick and Dixon 1989, as presented in Rose et al. 1999.

Generalized pattern	Cause of changes	Follow-up study	Age distribution	Energy utilization	Energy storage
Exploitation	Decreased competition between adults associated with mortality or eutrophication	Examine food resource availability and population density	Shift to younger	Increased	Increased
Recruitment failure	Shift to older age-classes associated with decreased reproductive success	Detailed examination of spawning habitat, utilization, and reproductive development	Shift to older	No change	No change
Multiple stressors	Simultaneous impacts on food availability and reproductive success	Detailed studies of reproductive development and food resources	Shift to older	Decreased	Decreased
Food limitation	Increased competition associated with increased reproductive success or decreased food availability	Examine food resource availability and population density	No change	Decreased	Decreased
Niche shift	Modest increase in competition for forage base	Examine food base and competition aspects	No change	Decreased	No change
Metabolic redistribution	Inability to maximally utilize available food resources	Detailed physiological studies of energetics	Shift to younger	Mixed	Mixed
Chronic recruitment failure	Shift to population of older individuals	Detailed study of reproductive performance	Shift to older	Increased	Increased or decreased
Null response	No obvious changes		No change	No change	No change

be associated with a shift to a younger age distribution, increased energy utilization (e.g., increased growth rate, reproductive rate, or reduced age at maturity) and increased energy storage (e.g., high lipid levels or condition factor), whereas exposure to multiple stressors may result in a shift to an older age distribution, decreased energy utilization, and decreased energy storage.

Population-level responses may also be examined by using individual information on survival and reproductive rates to extrapolate to the dynamics of the target population. Examples of features that can be modeled include reductions in total abundance or recruitment, alterations in age or size structure, distortions in sex ratio, changes in population stability, and reductions in population resilience. The basic types of population models used in such analyses are aggregate, structured, and individual-based. Aggregate models predict the total number or biomass of individuals in the population. Structured models subdivide the total abundance into numbers in various classes based on age, life stage, or size. Matrix projection models (Caswell 1989) comprise one category of structured models that have been used to study population-level effects of contaminants (Grant et al. 1983; Emlen 1989; Slade 1994; Landahl et al. 1997). For more detailed descriptions of these types of population models and other techniques to assess the effects of environmental stressors on fish populations see Chapter 10.

We have employed an age-structured Leslie matrix model to explore the potential effects of pollution on English sole populations in Puget Sound (Landahl et al. 1997). An important part of this process is the estimation of vital rates (e.g., mortality rates and reproductive rates) in both control populations and also in populations affected by chemical contaminants. In this case, we used age-frequency data on English sole populations from sites throughout Puget Sound to estimate mortality rates in fish populations with different degrees of contaminant exposure and in fish with and without hepatic lesions (Johnson and Landahl 1994; Landahl et al. 1997). The results indicate that overall mortality rates were fairly similar throughout the sound, but, because of the lack of fishing pressure at industrialized sites, nonfishing mortality was likely somewhat higher at sites with high contaminant levels. Estimation of reproductive rates in fish subpopulations from contaminated and uncontaminated sites was more straightforward as we had collected data on the probability of sexual maturation, age-specific egg production, and larval production in the course of our reproductive toxicology studies. (i.e., Johnson et al. 1988, 1997; Casillas et al. 1991; Johnson and Landahl 1994). These survival and reproductive rates, along with supplementary values from the literature, were used to calibrate the model. The projections of this simple model indicated that contaminant effects, particularly on reproductive capacity, could substantially reduce the intrinsic rate of increase of English sole subpopulations in polluted areas if the loss of recruits were not offset by density-dependent changes in recruitment, immigration, or other compensating mechanisms.

As noted above, structured population models subdivide total abundance into classes based on age, life stage, or size. Individually based models take this process still further and represent every individual in the population. Probably the most commonly used individually based models in contaminant effect studies are configuration models that involve the simulation of the activities of hundreds or thousands of individual organisms. These models are typically computer simulations involving Monte Carlo methods, which allow practically any level of biological detail to be imposed on individuals (DeAngelis et al. 1993; Rose et al. 1999a).

Individually based models have proven to be useful for modeling toxicant impacts on population dynamics of fish populations because connections are easier to make between the effects of stressors at the individual level and effects at the population level. For example, Rose et al. (1993) used such a model to examine impacts of water quality on early life stages of striped bass. A set of model simulations of age-0 striped bass was used to assess effects of changes in temperature, toxics, and livable habitat at the population level. Effects of toxicants were simulated as increased mortality to egg yolk sac larvae, feeding larvae, and juvenile stages, and as the sublethal effects of reduced growth rate and reduced prey capture success. Of the variables tested, chronic toxic exposure had the greatest effect on population-level parameters.

Jaworska et al. (1997b) use a similar model to examine PCB effects on young of the year largemouth bass. The model simulated the daily development, growth, and survival of largemouth bass from the egg stage to the end of the first growing season, based on data collected from populations in two Tennessee River impoundments. In the model, individual fish mortality increased and growth decreased with increasing PCB exposure and PCB tissue concentrations (expressed as TCDD equivalents). Although the effects were not large relative to the normal variation in the model, number density and biomass density both decreased with increasing PCB levels.

In spite of the importance of evaluating population-level effects of contaminant stress on aquatic systems, detection and quantification of effects is often quite difficult, and modeling results may be equivocal or unconvincing to those who question the assumptions built into the analysis. This controversy is not completely surprising, as several major issues may confound such analyses (Rose et al. 1999). These include:

1. Density Dependence

In populations exhibiting compensatory density dependence, reproduction and survival rates increase with decreasing abundance. However, detecting and quantifying density dependence in the field and assessing its potential to modify stressor impacts is very difficult. Some of these issues are being examined for estuarine fish populations using individually based models (e.g., Rose et al. 1999b; Kimmerer et al. 2000), and the results suggest that density-dependent responses can easily mask population responses to contaminant-related stress. Consequently, it is important to remember that the lack of a clearly detectable decline in a contaminant-exposed population does not necessarily mean that the population is unaffected. Although its abundance may not change, its compensatory reserve may be lower, making it less resilient to additional stresses.

2. Spatial Scales and Dispersal

Many aquatic species, and marine fish in particular, have complex life history strategies and may use a variety of habitats and prey species that change with ontogeny. This can lead to wide variation in contaminant exposure at different life stages. Moreover, their life cycles often involve the dispersal of individuals over a broad geographical area, so an interbreeding population can consist of many subpopulations in different habitats and with different reproductive and survival rates. Large-scale migrations are also problematic. Immigration and emigration of young among nursery areas and intermixing of subpopulations make determination of localized contaminant effects difficult.

Our modeling efforts with English sole are a case in point. The simple Leslie matrix model that we initially employed for our analyses assumes a simple closed system within which local recruitment depends primarily on the reproductive output of adults residing at that site. In fact, the life history of English sole (Lassuy 1989) is more complex and may allow substantial migration of animals between urban and nonurban sites. Although adult English sole show high site fidelity for most of the year (Day 1976), they migrate to spawn at centralized sites that may draw animals from throughout the sound. Moreover, their eggs and larvae are planktonic and are dispersed by current flow. Larvae then settle and undergo metamorphosis in nearshore nursery areas, subsequently migrating to residential sites that will be their home territories throughout adulthood. Consequently, a realistic evaluation of the potential effect of urban pollution on English sole abundance must consider both the contribution of fish from urban sites to the greater central Puget Sound English sole population and the possible flux of recruits from nonurban to urban sites that could compensate for the reduced reproductive capacity of sole subpopulations from contaminated areas. The impacts of pollution and other anthropogenic activities on the distinctive habitats that are used by the English sole at different life stages must also be taken into account.

These spatial-scale problems can be addressed, in part, by the application of spatially explicit models (Dunning et al. 1995). Such models are being used more and more frequently in fisheries management for stock assessment (e.g., Punt et al. 2000) and to address issues such as the dynamics of fertilization success and larval distribution and production in free-spawning benthic species (Claereboudt 1999). These types of models have also been employed to analyze the importance of marine protected areas in the restoration of declining marine fish stocks (Tuck and Possingham 2000), to examine effects of hypoxia on estuarine larval fish growth and survival (Breitburg et al. 1999), and in studies of impacts of habitat quality on fish populations (Mason et al. 1995; Gaff et al. 2000). Such models would also be valuable for addressing the impact of pollution on the population dynamics of marine fish with open breeding systems, but, as yet, their application to this question has been limited. For English sole and other marine fish with similar breeding systems, an especially appropriate alternative might be spatially explicit, stage-based models (Hanski and Gilpin 1991; Gaines and Lafferty 1995) that would incorporate the recruitment and migration patterns of animals with multiple life stages into population analyses.

3. Stochasticity

The high interannual variability in population abundance that is especially characteristic of marine fish populations can make the detection of population-level contaminant effects difficult, even when effects on individuals are well established. This is illustrated by the study described earlier, conducted by Jaworska et al. (1997b), in which an individual-based model was used to examine the effect of PCBs on the growth of larval and juvenile largemouth bass. In the laboratory, growth, feeding, and mortality processes of PCB-exposed individuals were followed from spawning to 1 year of age; survival and growth clearly decreased with increasing PCB tissue concentration. The model was calibrated using these data and a Monte Carlo simulation was applied to simulate the effects of interannual environmental conditions that could also modulate growth. Under these conditions, predicted abundance of largemouth bass decreased with increasing PCB levels, but density-dependent processes largely offset effects at low exposure concentrations. Moreover, PCB-related mortality imposed on eggs and larvae led to increased growth and improved survival of juveniles because their densities were lower. The large overlap in the distributions of predicted abundance among exposed and unexposed populations implied that natural variability was large enough to effectively mask the adverse effects of PCBs, and the authors concluded that it would be difficult to identify PCB effects in field-based comparisons of young of the year bass densities.

4. Cumulative Effects

It is almost inevitable that natural populations are subject to cumulative anthropogenic and natural stressors, such as pollution, habitat alteration, fishing pressure, and climate change. The fact that such stressors can act simultaneously renders it more difficult to establish cause-and-effect relationships for individual effects. Moreover, the focus on individual stressors can lead to an underestimation of the magnitude of stress, as the cumulative effects of multiple stressors can be greater than the sum of independent effects. We are only beginning to deal with such issues today, and techniques such as individually based modeling may be useful tools for addressing these types of problems. An example of one attempt to look at cumulative effects is a study of striped bass in the Sacramento River–San Joaquin Delta (Rose et al. 1999a). The bass population in this region had been declining for some time; the suggested causes include reduced food supply, diversion of water for agricultural use, and reduced adult survival. When these factors were incorporated into the model individually, no one cause could explain the decline. However, when all three stressors were included simultaneously, there was a marked decline in bass abundance.

5. Community Interactions

The importance of indirect effects of various stressors other than contaminants on species abundance and community structure has been clearly established for aquatic ecosystems (Menge 1995; Kitchell and Carpenter 1996), and there is little doubt that the indirect effects of predators, prey, and interspecific competition can have a major influence on population responses. Consequently, population models are not realistic if these effects are not included. However, the realistic and accurate assessment of these types of interactions, using simulation modeling, remains a challenge. Some approaches that have been used to address the problem of community-level effects and interactions will be discussed in the following section.

Assessing Community-Level Impacts of Contaminant Stress

As noted above, modeling the population dynamics of a single species can be quite difficult, even without considering the added complexity of interspecific competition and the interactions of prey and predators. Indeed, in the opinion of many researchers, current methods for extrapolating individual-level effects on fish to levels of biological organization higher than the population level are not developed sufficiently to permit realistic assessments (Rose et al. 1999a). However, because of their importance, community-level assessments should be attempted. Those analyses that have been attempted in this regard highlight the degree to which indirect effects can modulate or alter the responses of populations exposed to environmental toxicants.

An analysis by Jaworska et al. (1997a) illustrates the potential modifying influences of community interactions on population dynamics. In this study, an individually based simulation model was used to evaluate generalized response patterns to stressors (Munkittrick and Dixon 1989). The model followed the daily growth, mortality, and spawning of individual yellow perch *Perca flavescens* and walleye *Stizostedion vitreum* through their lifetime and was calibrated using data from Oneida Lake perch and walleye populations.

Two versions of the model were used: a population model that simulated the dynamics of yellow perch populations only, and a community model that incorporated dynamic predation on yellow perch by walleye. Eight types of stressors were imposed on the population and community versions of the model with 100-year simulations being performed. The population model predicted population response patterns that were very similar to those proposed by Munkittrick and Dixon (1989); simulations using the community version of the model distorted the response patterns, either causing amplification, dampening, or reversal. These results suggest that even relatively simple food web interactions can substantially modify the responses of populations subjected to anthropogenic stressors.

Individually based models are being used more frequently to examine the influence of density-dependent compensatory processes and community interactions on fish population dynamics (e.g., Breitberg et al. 1999; McDermot and Rose 2000). These approaches could hold promise for analysis of indirect contaminant effects. The model developed by McDermot and Rose (2000), for example, tracks the daily feeding, growth, movement, reproduction, and mortality of individuals for up to six species for multiple generations. Breitburg et al. (1999) use a spatially explicit individual-based predation model to predict how effects of low dissolved oxygen on vertical distributions, predation rates, and larval growth might combine to influence the survival of estuarine fish larvae in the water column. While these models focus primarily on the impacts of factors such as temperature and dissolved oxygen, similar models could be developed to examine effects of environmental toxicant concentrations.

The models described in both Breitberg et al. (1999) and Rose and McDermot (2000) show strong effects of species interactions on survival and growth rates, emphasizing the influence of indirect effects on fish population responses to environmental stressors. Simulations conducted by Breitberg et al. (1999) suggest that eutrophication could have a large effect on larval fish survival and, possibly, recruitment through food-web-mediated effects, even in the absence of direct effects of low oxygen on larval mortality. Moreover, results of the Rose and McDermott model (2000) suggest that, in part, because of indirect effects, population responses to manipulations such as stock enhancement or fish die-off events were not straightforward and could have delayed or long-term impacts. The same would likely be true of stresses associated with chronic chemical contamination.

An example of a more comprehensive ecosystem model developed to assess ecological risks posed by toxic chemicals in Quebec rivers, lakes, and reservoirs is provided by Bartell et al. (1999). The model was designed to estimate the probability of changes in the biomass of multiple populations of primary producers and consumers as a function of the concentration of dissolved chemical contaminants and to permit the evaluation of both direct and indirect toxic effects. Hypothetical risk assessments were constructed for pentachlorophenol, copper, mercury, and diquat dibromide, which described the relative contributions of direct and indirect toxic effects on overall ecological risks estimated for functional guilds of producers and consumers in generalized Quebec river, lake, and reservoir ecosystems. This aquatic ecosystem model may become one component in a decision support system for assessing ecological risks.

As an alternative to simulation modeling, community impacts of contaminants have also been examined through measurement of endpoints such as species diversity or through various indices of environmental quality or ecosystem integrity, such as the index of biotic integrity (IBI) developed by Karr (1981). The IBI was originally developed to assess environmental quality in streams and typically incorporates a number of fish community variables (e.g., see Kestemont et al. 2000), including species diversity, richness, and trophic composition, fish abundance, presence or absence of sensitive or tolerant species and, in some cases, individual fish health metrics such as condition or gross pathological conditions. More recently, such indices have been constructed for estuarine environments (Hartwell 1997; Weisberg et al. 1997; Van Dolah et al. 1999). Such indices have been tested for their ability to identify contaminant-associated alterations in fish community health; some appear to be promising.

Hartwell et al. (1997), in a multivariate assessment of Chesapeake Bay watersheds that incorporated toxicity measurements and IBI metrics, found strong correlations between sediment toxicity measurements and fish community health metrics, particularly with bottomfish community diversity. A fish community IBI, incorporating measures such as species richness, composition, trophic position, and abundance, has been used to monitor recovery in fish communities following habitat remediation and restoration in Hamilton Harbor, Ontario, with some success (Smokorowski et al. 1998). Such techniques can be quite powerful for detecting contaminant-associated injury. For example, Van Dolah et al. (1999) applied a benthic IBI to rank a number of estuarine sites in the southeastern United States and found that the index was more effective than individual benthic measurements or bioassays in identifying sites where reduced health status of resident organisms or ecological injury would be expected based on sediment chemistry values. Schulz et al. 1999, on the other hand, found that a fish assemblage-based IBI was not very effective at estimating anthropogenic impacts in Florida lakes. Dyer et al. (2000), in a study of the application of an IBI to a number of Ohio streams, found that habitat factors were the best predictors and most positively related to the IBI and the number of fish species, whereas chemical factors, such as cumulative effluent, metals, ammonia, and biochemical oxygen demand, were more closely correlated with the proportion of fish observed with deformities, fin erosions, lesions, and tumors than with the IBI. Because of their composite nature, community health indices such as the IBI may not be the most appropriate measures for exploring cause-and-effect linkages across levels of biological organization. However, they provide a holistic assessment of conditions at sites impacts by contaminant stress and, by identifying sets of impacts that co-occur, may be useful for generating hypotheses about community interactions.

Mesocosms, which are replicated experimental ecosystems designed to simulate natural environments, provide a laboratory-oriented way to look at higher order effects in aquatic systems. Fairchild and Little (1993) review the use of mesocosm studies to examine direct and indirect impacts of water quality on early life stages of fishes. Such systems have obvious utility for examining the direct effects of toxicants on fish and have also been used successfully to examine the impacts of a wide range of indirect effects associated with reduced water quality, including food reduction, food-chain-mediated contaminant transfer, habitat alteration, changes in predation rates, and changes in competition. Studies in aquatic mesocosms offer some advantages because of increases in experimental control and statistical inference, and they are good vehicles for simulating potential community-level impacts and indirect effects of contaminants in aquatic systems. Also, because conditions in mesocosms are controlled, they may be especially helpful in identifying causal mechanisms. However, they are limited in their ability to simulate specific natural systems that may be of concern. More detailed discussion relevant to assessing stressor effects on communities and ecosystems can be found in Chapter 12 and Chapter 13.

Summary and Prospectus

Assessing the effects of stress across levels of biological organization is a broad task. Determining the relationship between biochemical or subcellular indicators and indications of effects at the cellular or tissue level is a relatively straightforward task. Examples of this include relationships between the presence of intracellular enzymes in blood and tissue damage, hormonal changes and alterations in growth and reproduction, and DNA damage and cancer development. However, most of the interest in linking the effects of stressors across levels of organization is focused on assessing effects at the individual level and beyond, up to the population and community levels. This interest stems from regulatory and management policies aimed at protecting the health of ecosystems, which generally do not focus on maintaining the health of individual organisms. Instead, many environmental protection policies focus on maintaining certain sizes of populations, such as is common in fishery management plans, or on maintaining functional ecosystems that sustain desirable species or species assemblages. While there is not a clear consensus about what to protect in ecological systems or at what organization level (Clark et al. 1999), the most widely accepted view is that adverse ecological effects at the population level and above will be the principle areas of concern when dealing with organisms in the environment (USEPA 1998). As a result, it is common for regulatory approaches to require evidence of impacts to populations or to communities before remedial actions or punitive measures are taken. Thus, there exists a strong desire to follow the effects of a particular stress up through the levels of organization to the population level or beyond, in order to support actions that could be taken to reduce or remediate that stress. For the experimental scientist, this presents a daunting challenge. In the real world, outside of laboratory microcosms and mesocosms, the intricate complexity of species, habitats, and multiple stresses makes it exceedingly difficult to track the effects of a single stressor to even the individual level, let alone the population level. Nonetheless, policies to protect ecosystems from human-induced stress can require that those determinations be made. As we have discussed in this chapter, some techniques and approaches do exist to help make the connections between the stressor of interest and effects at higher levels of biological organization (i.e., the population and the community). If properly applied, these methods can be useful tools for assessing the likelihood that anthropogenic activities will disrupt aquatic systems.

Even so, the view that population- and community-level effects must be demonstrated in order to warrant regulatory or management actions has distinct drawbacks, especially in the case of highly endangered species where population sizes are small and effects on the individual can have a disproportionate impact on future abundance. Indeed, the population level of protection is not considered adequate under the Endangered Species Act (ESA) for organisms that have been listed as threatened or endangered. Adverse impacts on individuals or their habitat or prey may be sufficient to trigger action (ESA section 3(19); ESA section 9[a][1]; 50 CFR 17.3; 50 CFR 222.102). We would assert that, in order to be proactive in the conservation of our aquatic resources, it may be appropriate to use regulatory approaches that protect the health of individual organisms from the effects of anthropogenic stress.

References

Adams, S. M., J. J. Beauchamp, and C. A. Burtis. 1989. A multivariate approach for evaluating responses of fish to chronic pollutant stress. Marine Environmental Research 24:459–464.

Adams, S. M., A. M. Brown, and R. W. Goede. 1993a. A quantitative health assessment index for rapid evaluation of fish condition in the field. Transactions of the American Fisheries Society 122:63–73.

Adams, S. M., M. S. Greeley Jr., and L. R. Shugart. 1993b. Responses of fish to pollutant stress: evaluating the relative importance of direct and indirect mechanisms. Marine Environmental Research 35:228–229.

Adams, S. M., K. D. Ham, and J. J. Beauchamp. 1994. Application of canonical variate analysis in the evaluation and presentation of multivariate biological response data. Environmental Toxicology and Chemistry 13:1673–1683.

Anderson, S., A. Auguier, W. W. Hauck, D. Oakes, W. Vandaele, and H. I. Weisberg. 1980. Statistical method for comparative studies. Wiley, New York.

Ankley, G. T., and J. P. Giesy. 1998. Endocrine disruptors in wildlife: a weight of evidence perspective. Pages 349–367 *in* R. Kendall, R. Dickerson, W. Suk, and J. Giesy, editors. Principles and processes for assessing endocrine disruption in wildlife. SETAC Press, Pensacola, Florida.

Anulacion B., D. Lomax, B. Bill, L. Johnson, and T. Collier. 1997. Assessment of antiestrogenic activity and CYP1A induction in English sole exposed to environmental contaminants. Page 137 *in* Proceedings of the SETAC 18th Annual Meeting, San Francisco, California, Society of Environmental Toxicology and Chemistry, Pensacola, Florida.

Bartell, S. M., G. Lefebvre, G. Kaminski, M. Carreau, and K. R. Campbell. 1999. An ecosystem model for assessing ecological risks in Quebec rivers, lakes, and reservoirs. Ecological Modeling 124:43–67.

Botsford, L. W., J. C. Castilla, and C. H. Peterson. 1997. The management of fisheries and marine ecosystems. Science 277:509–515.

Breitburg, D. L., K. A. Rose, and J. H. Cowan Jr. 1999. Linking water quality to larval survival: predation mortality of fish larvae in an oxygen-stratified water column. Marine Ecology Progress Series 178:39–54,

Breslow, D, and N. E. Day. 1980. Statistical methods in cancer research, volume I: the analysis of case control studies. International Agency for Research on Cancer, Lyon, France.

Casillas, E., D. A. Misitano, L. L. Johnson, L. D. Rhodes, T. K. Collier, J. E. Stein, B. B. McCain, and U. Varanasi. 1991. Inducibility of spawning and reproductive success of female English sole (*Parophrys vetulus*) from urban and nonurban areas of Puget Sound, Washington. Marine Environmental Research 31:99–122.

Casillas, E., M. S. Myers, L. D. Rhodes, and B. B. McCain. 1985. Serum chemistry of diseased English sole, *Parophrys vetulus* Girard, from polluted areas of Puget Sound, Washington. Journal of Fish Diseases 8:437–499.

Caswell, H. 1989. Matrix population models: construction, analysis, and interpretation. Sinauer Associates, Sunderland, Massachusetts.

Claereboudt, M. 1999. Fertilization success in spatially distributed populations of benthic free-spawners: A simulation model. Ecological Modeling 121:221–233.

Clark, J., K. Dickson, J. Giesy, R. Lackey, E. Mihaich, R. Stahl, and M. Zeeman. 1999. Using reproductive and developmental effects data in ecological risk assessments for oviparous vertebrates exposed to contaminants. Pages 363–401 *in* R. T. DiGiulio and D. E. Tillit, editors. Reproductive and developmental effect of contaminants in oviparous vertebrates. SETAC Press, Pensacola, Florida.

Collier, T. K., B. F. Anulacion, J. E. Stein, A. Goksoyr, and U. Varanasi. 1995. A field evaluation of cytochrome P4501A as a biomarker of contaminant exposure in three species of flatfish. Environmental Toxicology and Chemistry 14:143–152.

Day, D. S. 1976. Homing behavior and population stratification in central Puget Sound English sole *(Parophrys vetulus)*. Journal of the Fisheries Research Board of Canada 33:278–282.

DeAngelis, D. L., K. A. Rose, L. Crowder, E. Marshchall, D. Lika. 1993. Fish cohort dynamics: application of complementary modeling approaches. American Naturalist 142:604–622.

Dunning, J. B., D. J. Stewart, B. J. Danielson, B. R. Noon, T. L. Root, R. H. Lamberson, E. E. Stevens. 1995. Spatially explicit models: current forms and future uses. Ecological Applications 5:3–11.

Dyer, S. D., C. White-Hull, G. J. Carr, E. P. Smith, X. Wang. 2000. Bottom-up and top-down approaches to assess multiple stressors over large geographic areas. Environmental Toxicology and Chemistry 19:1066–1075.

Emlen, J. M. 1989. Terrestrial population models for ecological risk assessment: a state-of-the-art review. Environmental Toxicology and Chemistry 8:831–842.

Fairchild, J. F., and E. E. Little. 1993. Use of mesocosm studies to examine direct and indirect impacts of water quality on early life stages of fishes. Pages 95–103 in L. A. Fuiman, editor. Water quality and the early life stages of fishes. American Fisheries Society, Symposium 14, Bethesda, Maryland.

Gaff, H., D. L. DeAngelis, L. J. Gross, R. Salinas, M. Shorrosh. 2000. A dynamic landscape model for fish in the Everglades and its application to restoration. Ecological Modeling 127:33–52.

Gaines, S. D., and K. D. Lafferty. 1995. Modeling the dynamics of marine species: the importance of incorporating larval dispersal. Pages 389–412 in L. McEdward, editor. Ecology of marine invertebrate larvae. CRC Press, New York.

Garcia, S., and S. Newton. 1997. Current situation, trends, and prospects in world capture fisheries. Pages 3–27 in E. K. Pikitch, D. D. Hupper, and M. P. Sisenwine, editors. Global trends: fisheries management. American Fisheries Society, Bethesda, Maryland.

Grant, W. E., S. O. Fraser, and K. G. Isaacson. 1983. Effect of vertebrate pesticides on non-target wildlife populations: evaluation through modelling. Ecological Modeling 21:85–108.

Hanski, I., and M. Gilpin. 1991. Metapopulation dynamics: brief history and conceptual domain. Biological Journal of the Linnean Society 42:3–16.

Hartwell, S. I. 1997. Demonstration of a toxicological risk ranking method to correlate measures of ambient toxicity and fish community diversity. Environmental Toxicology and Chemistry 16:361–371.

Jaworska, J. S., K. A. Rose, and L. W. Barnthouse. 1997a. General response patterns of fish populations to stress: an evaluation using an individual-based simulation model. Journal of Aquatic Ecosystem Stress and Recovery 6:15–31.

Jaworska, J. S., K. A. Rose, and A. L. Brenkert. 1997b. Individual-based modeling of PCBs effects on young-of-the-year largemouth bass in southeastern USA reservoir. Ecological Modeling 99:113–135.

Johnson, L. L., E. Casillas, T. K. Collier, B. B. McCain, and U. Varanasi. 1988. Contaminant effects on ovarian development in English sole (*Parophrys vetulus*) from Puget Sound, Washington. Canadian Journal of Fisheries and Aquatic Science 45:2133–2146.

Johnson, L. L., and J. T. Landahl. 1994. Chemical contaminants, liver disease, and mortality rates in English sole (*Pleuronectes vetulus*). Ecological Applications 4:59–68.

Johnson, L. L., J. T. Landahl, L. A. Kubin, B. H. Horness, M. S. Myers, T. K. Collier, and J. E. Stein. 1998. Assessing the effects of anthropogenic stressors on Puget Sound flatfish populations. Journal of Sea Research 39:125–137.

Johnson, L. L., G. M. Nelson, S. Y. Sol, D. P. Lomax, and E. Casillas. 1997. Fecundity and egg weight in English sole *(Pleuronectes vetulus)* from Puget Sound, WA: influence of nutritional status and chemical contaminants. Fisheries Bulletin 92:232–250.

Johnson, L. L., C. M. Stehr, O. P. Olson, M. S. Myers, S. M. Pierce, C. A. Wigren, B. B. McCain, and U. Varanasi. 1993. Chemical contaminants and hepatic lesions in winter flounder (*Pleuronectes americanus*) from the northeast coast of the United States. Environmental Science and Technology 27:2579–2771.

Johnson, L. L., J. E. Stein, T. Hom, S. Y. Sol, T. K. Collier, and U. Varanasi. 1995. Effects of exposure to Prudhoe Bay crude oil on reproductive function in gravid female flatfish. Environmental Sciences 3:67–81.

Johnson, L. L., S. Y. Sol, G. M. Ylitalo, T. Hom, B. French, O. P. Olson, and T. K. Collier. 1999. Reproductive injury in English sole (*Pleuronectes vetulus*) from the Hylebos Waterway, Commencement Bay, Washington. Journal of Aquatic Ecosystem Stress and Recovery 6:289–310.

Karr, J. R. 1981. Assessment of biotic integrity using fish communities. Fisheries 6:21–27.

Kestemont, P., J. Didier, E. Depiereux, J. C. Micha. 2000. Selecting ichthyological metrics to assess river basin ecological quality. Archiv fuer Hydrobiologie (Supplement)121:321–348.

Kimmerer, W. J., J. H. Cowan Jr., L. W. Miller, K. A. Rose. 2000. Analysis of an estuarine striped bass *(Morone saxatalis)* population: influence of density-dependent mortality between metamorphosis and recruitment. Canadian Journal of Fisheries and Aquatic Science 57:478–486.

Kitchell, J. F., and S. R. Carpenter. 1996. Cascading trophic interactions. Pages 1–14 *in* S. R. Carpenter and J. F. Kitchell, editors. The trophic cascade in lakes. Cambridge University Press, Cambridge Studies in Ecology, New York.

Krahn, M. M., G. M. Ylitalo, J. Buzitis, S. L. Chan, and U. Varanasi. 1993. Rapid high-performance liquid chromatographic methods that screen for aromatic compounds in environmental samples. Journal of Chromatography 642:15–32.

Landahl, J. T., L. L. Johnson, T. K. Collier, J. E. Stein, and U. Varanasi. 1997. Marine pollution and fish population parameters: English sole *(Pleuronectes vetulus)* in Puget Sound, Washington. Transactions of the American Fisheries Society 126:519–535.

Landahl, J. T., B. B. McCain, M. S. Myers, L. D. Rhodes, and D. W. Brown. 1990. Consistent associations between hepatic lesions (including neoplasms) in English sole (*Parophrys vetulus*) and polycyclic aromatic hydrocarbons in bottom sediment. Environmental Health Perspectives 89:195–203.

Lassuy, D. R. 1989. Species profiles: life histories and environmental requirements of coastal fish and invertebrates (Pacific Northwest): English sole. U.S. Army Corps of Engineers, Report 82(11.101)/TR EL-82–4, Slidell, Louisiana.

Mason, D. M., A. Goyke, and S. B. Brandt. 1995. A spatially explicit bioenergetics measure of habitat quality for adult salmonines: comparison between Lakes Michigan and Ontario. Canadian Journal of Fisheries and Aquatic Sciences 52:1572–1583.

McDermot, D., and K. A. Rose. 2000. An individual-based model of lake fish communities: application to piscivore stocking in Lake Mendota. Ecological Modeling 125:67–102.

McMaster, M. E., G. J. van der Kraak, and K. R. Munkittrick. 1996. An epidemiological evaluation of the biochemical basis for steroid hormonal depressions in fish exposed to industrial wastes. Journal of Great Lakes Research 22:153–171.

Menge, B. A. 1995. Indirect effects in marine rocky intertidal interaction webs: patterns and importance. Ecological Monographs 65:21–74.

Munkittrick, K. R., and D. G. Dixon. 1989. A holistic approach to ecosystem health assessment using fish population characteristics. Hydrobiologia 188–189:123–135.

Munkittrick, K. R., M. E. McMaster, C. B. Portt, G. J. Van der Kraak, I. R. Smith, and D. G. Dixon. 1992. Changes in maturity, plasma sex steroid levels, hepatic mixed-function oxygenase activity, and the presence of external lesions in lake whitefish (Coregonus clupeaformis) exposed to bleached kraft mill effluent. Canadian Journal of Fisheries and Aquatic Science 49:1560–1569.

Musik, J. A., M. M. Harbin, S. A. Berkeley, G. H. Burgess, A. M. Eklund, L. Findley, R. G. Gilmore, J. T. Golden, D. S. Ha, G. R. Huntsman, J. C. McGovern, S. J. Parker, S. G. Poss, E. Sala, T. W. Schmidt, G. R. Sedberry, H. Weeks, and S. G. Wright. 2000. Marine, estuarine, and diadromous fish stock at risk of extinction in North American (exclusive of salmonids). Fisheries 25:6–30.

Myers, M. S., B. French, W. L. Reichert, M. J. Willis, B. F. Anulacion, T. K. Collier, and J. E. Stein. 1998c. Reductions in CYP1A expression and hydrophobic DNA adduct concentration in liver neoplasms in English sole: further evidence supporting the "resistant hepatocyte" model of hepatocarcinogenesis in this species. Marine Environmental Research 46:1–5.

Myers, M. S., L. L. Johnson, T. Hom, T. K. Collier, J. E. Stein, and U. Varanasi. 1998a. Toxicopathic hepatic lesions in subadult English sole (Pleuronectes vetulus) from Puget Sound, Washington, U.S.A.: relationships with other biomarkers of contaminant exposure. Marine Environmental Research 45:47–67.

Myers, M. S., L. L. Johnson, O. P. Olson, C. M. Stehr, B. H. Horness, T. K. Collier, and B. B. McCain. 1998b. Toxicopathic hepatic lesions as biomarkers of chemical contaminant exposure and effects in marine bottomfish species from the Northeast and Pacific coasts, U.S.A. Marine Pollution Bulletin 37:92–113.

Myers, M. S., L. D. Rhodes, and B. B. McCain. 1987. Pathologic anatomy and patterns of occurrence of hepatic neoplasms, putative preneoplastic lesions, and other idiopathic hepatic conditions in English sole (Parophrys vetulus) from Puget Sound. Journal of the National Cancer Institute 788:333–363.

Myers, M. S., C. M. Stehr, O. P. Olson, L. L. Johnson, B. B. McCain, S. L. Chan, and U. Varanasi. 1994. Relationships between toxicopathic hepatic lesions and exposure to chemical contaminants in English sole (Pleuronectes vetulus), starry flounder (Platichthys stellatus), and white croaker (Genyonemus lineatus) from selected marine sites on the Pacific Coast, U.S.A. Environmental Health Perspectives 102:200–215.

Pierce, K. V., B. B. McCain, and S. R. Wellings. 1978. Pathology of hepatomas and other liver abnormalities in English sole (Parophrys vetulus) from the Duwamish River Estuary, Seattle, Washington. Journal of the National Cancer Institute 60:1445–1453.

Polgar, T. T., J. K. Summers, R. A. Cummins, K. A. Rose, and D. G. Heimbuch. 1985. Investigation of relationships among pollutant loadings and fish stock levels in northeastern estuaries. Estuaries 8:125–135.

Punt, A. E., F. Pribac, T. I. Walker, B. L. Taylor, J. D. Prince. 2000. Stock assessment of school shark, Galeorhinus galeus, based on a spatially explicit population dynamics model. Marine and Freshwater Research 51:205–220.

Reichert, W. L., M. S. Myers, K. Peck-Miller, B. French, B. F. Anulacion, T. K. Collier, J. E. Stein, and U. Varanasi. 1998. Molecular epizootiology of genotoxic events in marine fish: linking contaminant exposure, DNA damage, and tissue-level alterations. Mutation Research 411:215–225.

Rhodes, L. D., M. S. Myers, W. D. Gronlund, and B. B. McCain. 1987. Epizootic characteristics of hepatic and renal lesions in English sole (Parophrys vetulus) from Puget Sound, Washington. Journal of Fish Biology, 31:395–408.

Roomi, M. W., R. K. Ho, D. S. F. Sarma, and E. Farber. 1985. A common biochemical pattern in preneoplastic hepatocyte nodules generated in four different models in the rat. Cancer Research 45:564–571.

Rose, K. A., L. W. Brewer, L. W. Barnthouse, G. A. Fox, N. W. Gard, M. Mendonca, K. R. Munkittrick, and L. J. Vitt. 1999a. Ecological responses of oviparous vertebrates to contaminant effects on reproduction and development. Pages 225–281 in R. T. DiGiulio and D. E. Tillit, editors. Reproductive and developmental effect of contaminants in oviparous vertebrates. SETAC Press, Pensacola, Florida.

Rose, K. A., J. H. Cowan Jr., M. E. Clark, E. D. Houde, and S.-B. Wang. 1999b. An individual-based model of bay anchovy population dynamics in the mesohaline region of Chesapeake Bay. Marine Ecology Progress Series 185:113–132.

Rose, K. A., J. H. Cowan Jr., E. D. Houde, and C. C. Coutant. 1993. Individual-based modeling of environmental quality effects on early life stages of fishes: a case study using striped bass. Pages 125–145 in L. A. Fuiman, editor. Water quality and the early life stages of fishes. American Fisheries Society, Symposium 14, Bethesda, Maryland.

Schiewe, M. H., D. D. Weber, M. S. Myers, F. J. Jacques, W. L. Reichert, C. A. Krone, D. C. Malins, B. B. McCain, S.-L. Chan, and U. Varanasi. 1991. Induction of foci of cellular alteration and other hepatic lesions in English sole (Parophrys vetulus) exposed to an extract of an urban marine sediment. Canadian Journal of Fisheries and Aquatic Science 48:1750–1760.

Schlesselman, J. J. 1982. Case-control studies: design, conduct, and analysis. Oxford University Press, New York.

Schulz, E. J., M. V. Hoyer, and D. E. Canfield Jr. 1999. An index of biotic integrity: a test with limnological and fish data from sixty Florida lakes. Transactions of the American Fisheries Society 128:564–577.

Servos, M. R., K. R. Munkittrick, J. Carey, and G. Van der Kraak, editors. 1996. Environmental effects of pulp mill effluents. St. Lucie Press, Boca Raton, Florida.

Slade, N. A. 1994. Models of structured populations: age and mass transition matrices. Pages 189–199 in R. J. Kendall and T. E. Lacher, editors. Wildlife toxicology and population modeling. Lewis, Boca Raton, Florida.

Smokorowski, K. E., M. G. Stoneman, V. W. Cairns, C. K. Minns, R. G. Randall, and B. Valere. 1998. Trends in the nearshore fish community of Hamilton Harbour, 1988 to 1997, as measured using an index of biotic integrity. Canadian Technical Report of Fisheries and Aquatic Sciences No. 2230.

Stehr, C. M., L. L. Johnson, and M. S. Myers. 1998. Hydropic vacuolation in the liver of three species of fish from the U.S. West Coast: lesion description and risk assessment associated with contaminant exposure. Diseases of Aquatic Organisms 32:119–135.

Stehr, C. M., M. S. Myers, D. G. Burrows, M. M. Krahn, J. P. Meador, B. B. McCain, and U. Varanasi. 1997. Chemical contamination and associated liver diseases in two species of fish from San Francisco Bay and Bodega Bay. Ecotoxicology 6:35–65.

Stein, J. E., T. K. Collier, W. L. Reichert, E. Casillas, T. Hom, and U. Varanasi. 1992. Bioindicators of contaminant exposure and effects: studies with benthic fish in Puget Sound, WA. Environmental Toxicology and Chemistry 11:701–714.

Stein, J. E., T. Hom, H. R. Sanborn, and U. Varanasi. 1991. Effects of exposure to a contaminated-sediment extract on the metabolism and disposition of 17b-estradiol in English sole (*Parophrys vetulus*). Comparative Biochemistry and Physiology 99C:231–240.

Summers, J. K., and K. A. Rose. 1987. The role of interactions among environmental conditions in controlling historical fisheries variability. Estuaries 10:255–266.

Tuck, G. N., and H. P. Possingham. 2000. Marine protected areas for spatially structured exploited stocks. Marine Ecology Progress Series 192:89–101.

USEPA (U.S. Environmental Protection Agency). 1998. Guidelines for ecological risk assessment. U.S. EPA Office of Research and Development, EPA/630/R-95/002F, Washington D.C.

Van Dolah, R. F., J. L. Hyland, A. F. Holland, J. S. Rosen, and T. R. Snoots. 1999. A benthic index of biological integrity for assessing habitat quality in estuaries of the southeastern USA. Marine Environmental Research 48:269–283.

Weisberg, S. B., J. A. Ranasinghe, D. M. Dauer, L. C. Schaffner, R. J. Diaz, and J. B. Frithsen. 1997. An estuarine benthic index of biotic integrity (B-IBI) for Chesapeake Bay. Estuaries 20:149–158.

15

Statistical Considerations in the Development, Evaluation, and Use of Biomarkers in Environmental Studies

Eric P. Smith

Introduction

Accurate assessment and evaluation of environmental stressors on biological systems requires careful consideration and selection of appropriate statistical procedures. Environmental stress information can be in the form of physical, chemical, biological, as well as social data. Because of the multivariate nature of this information, it is often necessary to try to summarize the data in terms of a single measure, and biomarkers represent one method for this summarization. There are many possible biomarkers that may be selected and measured in a given experimental study. To be useful in environmental studies, they must meet criteria of statistical acceptability. Biomarkers may be evaluated in terms of biological realism, cost of measurement, relationship with environmental stress, measurement properties, as well as other criteria. In this chapter, statistical criteria are explored, first from the view of development of markers to serve as early warning signals of stress and, second, in a comparative context. Properties of well-designed studies are discussed. A design that is common to many biomarker studies is a nested design. This design results when a primary unit, say fish, has multiple measurements made on each unit. The problem of sample size and allocation of effort between primary and secondary units is investigated, and the selection of optimal sample sizes for a fixed cost is described.

The quality of a biomarker study depends on the statistical design that is used and the choice of an appropriate method for analysis of the resulting data. Several commonly used methods are described for data from univariate and multivariate studies. Univariate methods are appropriate when emphasis is on a single biomarker or a single measurement. When interest is in a set of biomarkers or measurements, then multivariate methods are appropriate. Some commonly used multivariate methods for the analysis of biomarker data are presented.

Criteria for Selection of Biomarkers

Statistical criteria for evaluating biomarkers of environmental stress may be categorized in terms of descriptive and predictive properties. Descriptive properties are the summary statistics related to the biomarker and its relationship to environmental stress. Two common measures are variance and sensitivity (Anderson 1990). These are measures related to the signal and noise properties of the markers.

To develop a benthic indicator of environmental stress that includes these statistical properties, Smith and Voshell (1997) developed several indicators of stress based on ecological and statistical considerations. These measures combine numerical information on different taxa and are referred to as metrics. Benthic metrics such as the percentage of mayflies were compared for variation and ability to separate impacted and unimpacted sites. Metrics with small variance and high separation were considered useful for building a "multimetric" (MM) or combination of metrics. The multimetric is to be used in a biomonitoring program to measure the "health" of the ecosystem or study population.

Consider the information displayed in Figure 1 for two different multimetrics. These indices are based on different combinations of benthic metrics. Note that the scale on the graph is different for the different multimetrics. The difference between the reference site and impact site means is 1.14 for MM10 and 0.90 for MM9. However, this does not mean that MM10 is better than MM9. The variance (pooled over all sites) is 0.71 for MM10 and 0.34 for MM9. Thus, the variance is considerably smaller for MM9. A measure of the sensitivity of the metric is given by the standardized mean difference (SMD):

$$SMD = \frac{\bar{y}_R - \bar{y}_I}{s_p}$$

where $\bar{y}_R$, $\bar{y}_I$ are sample means for reference and impact and s_p is the pooled standard deviation.

The SMD for MM9 is 1.55, while for MM10 is 1.35 indicating that MM9 is slightly better for the given data set. Based on this information, MM9 would be recommended for inclusion in a field biomonitoring program. Among the metrics considered, it has the best signal (as measured by difference with the reference) relative to the noise (measured by the standard deviation of the data).

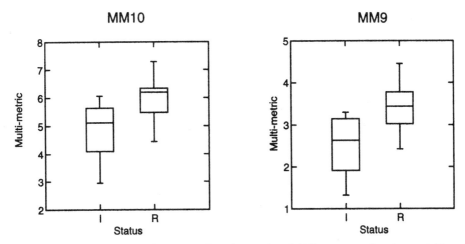

Figure 1. Graph of distributions of multimetrics (MM) measured at impact (I) and reference (R) sites.

Another approach to evaluate the usefulness of biomarkers in bioassessment is based on their use as predictive tools. There are three important properties of biomarkers that deserve consideration in this regard. These are predictive power, relevance, and consistency (McMillan et al. 1994). A biomarker's predictive power is the probability that an affected animal or sample displays the marker. The predictive power is related to the ability to make correct decisions in the presence of uncertainty. Thus, for example, a biomarker with high predictive power will correctly signal when pollution is present and also correctly signal when pollution is not present. Table 1 illustrates the four possible decisions that may be made.

In the context of biomarkers related to a presence of a biological signal in an animal (such as a tumor or particular enzyme), positive predictive power for a biomarker would be defined as the proportion of animals that have the marker present given they are from polluted sites or Prob(PP|MP). The negative predictive power is measured by Prob(PA|MA). Ideally, we would like to have a biomarker with both high positive and high negative predictive power. This need not always be the case as the presence of a marker may be a precursor to further hazard that occurs with additional stress.

Note that predictive power is different from an association between a marker and pollution. Predictive power implies that the presence of the

Table 1. Decision table for biomarker analysis. The presence of the marker is used as a signal for a pollution effect.

	Pollution present (PP)	Pollution absent (PA)
Marker present (MP)	Ok	Positive error
Marker absent (MA)	Negative error	Ok

marker may be used to predict if pollution is present. Even if a study establishes that the marker is correlated with another measure of pollution, what is important is how well the marker predicts pollution. Thus, an appropriate measure is not the strength of the correlation between stress and the biomarker for a set of samples but the percentage of correctly predicted samples or the stress predicted from the biomarker.

An example of the evaluation of the predictive power of the multimetrics is given in Table 2. The predictive power of two multimetrics was compared based on classification errors. Crossvalidation error rates were calculated for the different multimetrics based on the observed data. The crossvalidation method removes one (i.e., holds out one observation from the data) observation, builds a model, then predicts the held out observation. The observation is placed back into the data set and the process repeated on another observation.

The multimetrics have the same overall error rate of 26% (i.e., 12 sites are incorrectly classified). Multimetric 9 is slightly better at predicting reference sites, while MM10 is slightly better at predicting impact sites. Based on sensitivity, MM9 is better; however, classification rates indicate that the two measures are close with a moderately high error rate.

The second important aspect for consideration of biomarkers as predictive tools is relevance. Relevance refers to the strength of the relationship between the presence of a marker and the likelihood of stress effects being manifested at increasing levels of biological organization. If we observe that a particular enzyme occurs prior to the formation of a tumor in fish, then the enzyme would be relevant or function as a sensitive indicator of short-term stress. Indices are usually relevant at different levels of temporal and ecological resolution (Adams 1990). For example, an index that may be relevant at a species level may not be relevant for a community. An indicator may be quite useful for short-term stress but not useful in the long term. Evaluation of relevance with respect to time and ecology is possible by comparing the temporal and ecological pattern of marker prevalence to measures of hazard at different levels of biological organization.

The third property of interest relevant to biomarkers is consistency. If a marker is to be used as a general indicator of pollution, it must be consistent across the organisms in the environment and the degree of pollution. If the marker is intended to be a surrogate for a measure of stress such as mortal-

Table 2. Summaries of predictions for two multimetrics. The number of correctly predicted sites are in bold.

	MM10		MM9	
	Impact	Reference	Impact	Reference
Site predicted as impact	**7**	8	**6**	7
Site predicted as reference	4	**27**	5	**28**
Totals	11	35	11	35

ity, then the relationship needs to be with mortality and not with the dose or species used. For example, if the biomarker increases linearly with dose but mortality increases exponentially with dose, then using the biomarker as a linear predictor of mortality is not advisable. Reporting a strong relationship with dose is not adequate to assess the consistency of the biomarker.

Using Biomarkers in Environmental Stress Studies

Perhaps as important as the choice of the biomarker is the study design for the use of the biomarker in environmental stress studies. Field collected data may be highly variable over time, space, and organism (organisms respond differently to natural factors). Without careful planning, a study may reach an incorrect conclusion due to poor study design and inadequate sampling (Peterman 1990). It is especially important that appropriate attention be paid to the planning and implementation of the study. Some guidelines are presented below, with a focus on biomarkers. Additional descriptions of principles for good design are given in Green (1979), Eberhardt and Thomas (1991), and National Research Council (1990).

Develop Objectives

Study objectives are important for developing well-defined experimental designs and increasing the likelihood of obtaining defensible results. Studies that do not define objectives often produce vague results. In defining objectives, it is important to consider how information will be summarized and compared, how the statistical and biological populations will be defined, and how to identify the spatial and temporal constraints that are imposed.

Consider the problem of comparing two sites, one that is polluted with one that is considered a reference site. A poorly proposed hypothesis would be that values of the biomarker do not differ between the two sites. This hypothesis is vague, because information to be summarized is not included in the hypothesis. The lack of summarization information will lead to confusion in planning. A somewhat more refined hypothesis that might be considered is that the mean of a biomarker is the same at both sites versus a difference in the means between the sites. This, however, also is a vague hypothesis. Because the mean represents the average value in the "population," we are effectively implying the existence of a population. The population might represent all values of the biomarker in all fish over the entire site over some period of time. The population might represent measurements at a particular temperature, time of the year, and zone in the site (i.e., littoral zone). Or, a certain age-class of the fish might further characterize the population. Thus, the population and, hence, mean of the population are constrained by factors associated with space, time, and possibly other constraints (political, social, etc.). A hypothesis that is more refined would be that age-1 fish in the littoral zone of the reference and polluted sites have the same mean value of the biomarker during the first week of April. This hypothesis defines the population in terms of the age of fish, the space, and the time, and specifies how the data collected on the biomarker will be compared.

Define the Unit of Measure

In experimental design, statisticians view an experimental unit as the unit receiving the treatment of interest. A necessary property of experimental units is that two units must be able to receive two different treatments. In practice, the unit that receives the treatment might not be the same as the unit on which measurements are made. It is valuable to recognize that different types of units exist and these types need to be distinguished. Suppose, for example, that fish are treated with a chemical and the objective is to determine the effect of the chemical on a particular enzyme. If a pollutant is applied to fish in the tank, the treatment is the concentration of the pollutant. The experimental unit is the tank or the group of fish in the tank because this is what received the treatment. Individual fish are not considered the experimental unit as the two fish in a tank cannot receive different concentrations of the pollutant and would not be independent.

It is typical in the study of markers that there is often a hierarchy associated with the measurement of markers. For example, in the study of biomarkers in fish at two lakes, there will be selections made concerning sites to sample, fish within the sites, fish organs within a fish, tissue samples within fish organs, and readings within a tissue sample. Very often the larger unit (the site in this case) is the primary unit of analysis, while the smaller unit (reading) is the measurement unit. Very often the primary unit is related to the goal of the study and the statistical model that will be used to analyze the data. These types of studies lead to what are known as nested or subsampling designs. These designs are quite useful for understanding the sources and importance of variability associated with different components of the measurement process.

Understand Types and Sources of Variability

The general goal of a study is often to identify definitive signals in the presence of noise. The signal may be the presence of an enzyme or an index of a particular value. In field studies, there is always noise present that is generally manifested as various sources of natural variability. These sources of variation occur for a number of reasons and often the focus is on the wrong source of variation.

From a measurement perspective, it is often useful to think in terms of natural variation and variation in the measurement process. Natural variation might be due to spatial or temporal variation or to variation in the unit (e.g., fish, benthos) that is selected. Variation in the measurement process results from the method of selecting and processing the unit and those sources of variation due to laboratory and technician differences. The distinction between the two types of variation is that variation from the measurement process is often controllable through refinement of the process. Natural variation is not controllable through changes in the measurement process but is often viewed as controlled through optimal selection of sample size. By collecting information on the causes of variation, one can often sort the components of variation into those that are large and those that are relatively small or insignificant. With this approach, some of the components of variability can be ignored in order to reduce the greatest sources of variation and focus on the important sources.

In environmental laboratory studies, one major source of variation is due to the method that is used to select the units. Typically in a designed experiment, variation enters through the randomization process, the manner in which the units are assigned to the treatment. The role of statistical experimental design is to provide different methods of allocation to attempt to reduce variation due to the unit. In field studies, the variance often enters through the sampling design rather than through randomization. In field and laboratory studies, there is a need to recognize how the sampling or experimental design contributes to the variance and to select a design that would lead to appropriate variance estimates.

To illustrate this point, consider a study that focuses on comparing a polluted site with a reference site. We plan to sample fish and measure a biomarker on each of the fish. There are many ways to sample fish in each location and different sampling approaches will lead to different variance estimates depending on how the heterogeneity of the fish biomarkers relates to different factors. For example, one approach is to sample fish in an area at a given time. A second approach is to sample over different locations and times. The different sampling approaches lead to different estimates of variation. As the estimate of variance is used in statistical tests and confidence intervals, appropriate variance terms are required.

The use of an inappropriate error term or variance is sometimes referred to as pseudoreplication (Hurlbert 1984). The term implies that the replication of what is viewed as the treatment does not represent a true replicate (i.e., a new experimental unit receiving the treatment) but is distorted in some manner. The amount of distortion is important and affects the ability to obtain representative samples. If the samples represent the sites, then the characteristics of the population will be represented in the units. In the above example, choosing to sample at one time and one site results in a representative sample but not one that matches the objective of the study. The population is restricted to the particular sites selected and time of sampling. Distortion may be introduced due to the properties of the site and time other than pollution. The true hypothesis that is tested is not that the sites are different but that the sites at a given time and location are different. This may not be relevant to the purpose of the study.

For example, consider the problem of comparing two locations using a fish biomarker. A poor design would select one fish in each location and make ten (for example) readings on a fish organ. This would be an example of pseudoreplication as the repeated measurements represent replicates of the fish, not the location. Appropriate replicates would represent the location and not a single fish within each location. An approach that would minimize pseudoreplication would be to select fish, perhaps at randomly selected sites within the location and a time frame of interest. The site and time define a replication and the variation of interest is the variation in the biomarker over site and time. How the biomarker varies within the fish or between fish within the site and time is not relevant for comparing locations.

To summarize, it is important to choose a well-defined objective, then identify the appropriate experimental or sampling unit. One approach is to relate treatment factors associated with design, measurement, or sampling factors to the response through an appropriate statistical model. The model is then used to evaluate how information will be summarized and analyzed.

Understand How the Biomarker Relates to Study Goals

It is also important to consider how the biomarker will be used in a particular study. In any application, there will often be a model used to evaluate the data. The model may be quite simple or complex, depending on the application. For evaluating studies, statistical models are often a good starting point.

Statistical models are designed to include both deterministic and stochastic terms. Thus, many of the models may be represented as

$$\text{response} = \text{deterministic effect} + \text{stochastic component}$$

Choice of the deterministic components is often clear and related to the study goals and objectives. In simple models of change, the deterministic part is usually a mean or means. For example, in a study with a reference site and a potentially impacted site, one might believe that the differences between the sites may be described in terms of differences in the mean values of the biomarker measurement. In studies of change associated with some variable (as opposed to a factor such as site) that is believed to explain the change, the model is often a regression or dose–response model. The stochastic component of the model relates to the sources of variability and determinants of the variation in the measurements. By combining the mean and variation together, the response of measurement is characterized. For a given application, many models might be possible, and it is important to consider how the model is to be used and how it relates to the hypotheses or relevant parameters.

For example, if one is simply considering estimation of an effect or change in some simple parameter for a single location or population, a simple model is

$$Y_i = \mu + \varepsilon_i,$$

where μ is the mean level of the biomarker in the population and ε_i represents measurement error. This model suggests that all organisms have the same level of response for a biomarker and the major sources of variability are due to investigator effects such as field sampling or laboratory processing. Another model might represent the data in terms of a common mean, a difference associated with an individual fish and measurement error. One representation might be

$$Y_i = \mu + \varepsilon_i + \gamma_i,$$

where μ represents a common mean value in a population, ε_i represents a deviation of an individual fish from the mean of the population, and γ_i is measurement error.

Note that models typically differ based on the purpose of the study and the manner in which the study is designed and implemented. For example, the above model would suggest that the study be implemented by selecting one unit from the population, then taking one reading on the biomarker. The variability results from the unit that is measured and the associated measurement process. If multiple measurements are made, then the model would be a subsampling or nested model (Sokal and Rohlf 1995), such as

$$Y_{ij} = \mu + \varepsilon_i + \gamma_{j(i)}.$$

The indexing allows for multiple measurements on the unit. The notation ( ) indicates that the measurements are nested within the unit. This means simply that we distinguish measurements made on two different units within the experimental system. Recognizing that the measurements are nested within the units is very important for the modeling process as it describes the sources of variation. For example, if m measurements are made on n units for a total of $N=nm$ measurements, then the variance of the mean is expected to be

$$V(\bar{Y}) = \frac{\sigma_\varepsilon^2}{n} + \frac{\sigma_\gamma^2}{nm},$$

where σ_ε^2 is the variance associated with unit to unit differences and σ_γ^2 is the variance associated with the laboratory or field measurement process. The variance of the mean is a sum of variances associated with different components and the variance components are weighted differently, with the denominator for the first term (between unit) variance being larger than the second term (within unit). If the between unit variance is large relative to the within unit variance (as might be expected if the measurement process is well designed), then the variance of the mean is not decreased much by decreasing the number of measurements but by decreasing the number of units. Thus, we are often better off measuring more units than making many measurements on each unit.

In a field situation, such a model might arise if sampling units are selected within each location and measurements are made on subunits. For example, in a location, there may be multiple sites (n) and several fish sampled (m). If a measurement is made on each fish, then differences in the fish measurements represent within unit variation while differences between sites represents between unit variation.

A model that focuses on the measurement process should contain terms associated with the variability in all stages of the measurement of the biomarker. When interest is on change in the biomarker associated with pollution concentration, the components of the variability may no longer be as relevant. Hence, the model for the measurement Y_{ij} may be represented:

$$Y_{ij} = \mu + \alpha_i + \varepsilon_{ij}.$$

Here ε_{ij} is a global term that combines all the sources of error, μ is the overall mean, and α_i is the effect of concentration i.

This type of model would occur in the above example if one fish is selected at each site or if the fish are combined in some manner (composited) prior to analysis.

These types of models must account for the deterministic and stochastic components. Stochastic components include variation associated with sampling and selection of units. Some variation may be controlled through careful implementation of the measurement and sampling process. The protocol is a tool that helps control potential errors.

Develop a Protocol for the Study

Protocols are written guidelines for implementing a study. Protocols may not be important for simple studies that involve a single researcher. As the experimental complexity increases or if more than one individual is involved, experiment protocols become essential. The role of the protocol is to reduce variability by establishing a consistent set of rules for the experimental process. Variability results from differences in the manner that an individual selects, processes, and measures the unit of interest. By providing a set of rules, the experimental variance due to these differences should be diminished.

The protocol must be simple yet comprehensive. Simplicity is required to ensure that the number of rules that must be followed is at a minimum. Complex experiments may require training of researchers and hence an evaluation of the training. Protocols also need to be comprehensive. A comprehensive protocol results in fewer decisions made by researchers. By removing the collector of the data from making decisions, variation due to differences in decisions is removed. This may be especially important in field sampling studies where selection of units is important. Therefore, in experimental design, to the extent possible, all major sources of variability should be accounted for and minimized to increase the probability of detecting differences in measurements between treatments, if, in fact, these differences exist.

Run a Pilot Study

A pilot study is a valuable component of an experimental investigation. The pilot study allows for evaluation of the sampling protocol, determining optimal sample sizes, and estimating variation and information that is useful for improving the design. The protocol for a new study may be overly simplistic or complicated. A pilot study will indicate where problems will occur in the running of the study and will suggest changes to the program. By collecting data in the pilot study, variance associated with different components of the measurement process may be estimated. This estimated variation might be used to indicate where additional resources should be applied to reduce the variation. The data that results may also be useful for initial model evaluation. Given the model and initial data, sample sizes may be calculated for the actual study based on power in testing situations or width of confidence intervals in estimation situations.

Evaluate the Ability of the Biomarker to Meet the Needs of the Study

An evaluation also needs to be done relative to the model, the use of the biomarker, and the sampling or data collection process. Does the model account for important deterministic and stochastic components that affect the biomarker? Will the model parameters adequately relate to the relevant biomarker parameters? Can model terms be estimated and used in hypothesis tests? An evaluation of the biomarker as it will be used in actual analysis should be made. If multiple measurements are made on a fish organ, how will this information be incorporated into the model? If there are different approaches for the biomarker measurement, is there a best approach? It is important to focus on the important aspects of variation and not all sources. By important, we mean those that relate to the important level of units. It is not sensible to focus on fine-tuning the laboratory processes if field sampling is never done properly. If the biomarker is used to compare two sites, it is better to focus on how the variability associated with the site may be reduced.

For example, suppose we are interested in comparing two locations. As part of the design, we might select fish at random from the locations, remove an organ, and make measurements on a particular enzyme. We recognize that there might be several sources of error in the laboratory process as it involves a fairly complicated set of steps. Because of this, we decide to take 10 enzyme measurements on each of the selected organs (at each location). Is this wise? Well, it depends on the model and purpose of the study, the cost of sampling and analysis, and what contributes to the variability.

Consider several approaches that result in 10 measurements:

Case 1. We select 2 fish in each location, slice the organs into 5 sections and assay each section.

Case 2. We select 2 fish in each location, select one slice from the organs and take 5 readings.

Case 3. We select 10 fish in each location, select one slice from the organs and take 1 reading.

Although all the studies result in 10 observations, the observations differ in the sense that there are different components to the variability of the different measurements and, hence, different statistical models are required to analyze the results. Three important sources of variability are involved in these study cases. There is variation due to the organ tissue, variation due to the laboratory equipment used, and there is variability associated with the fish itself. The most efficient design is the one that will address the question of interest with the smallest cost and least amount of variation. In this case, the question of interest is whether or not there are differences in the locations. Thus, what is important is what contributes to the variance associated with the estimated mean of each location. The variance of the mean will have three sources according the above information: variation due to fish σ_f^2, organ slice within the fish σ_s^2, and reading (laboratory analysis) (σ^2). An observation is represented by y_{ijkm} for location i, fish j, organ i, and reading m. Thus, a model for this situation would be

$$y_{ijkm} = \mu + \alpha_i + fish_{j(i)} + organ_{k(ij)} + reading_{m(ijk)},$$

where α_i is the effect of location on the mean (i.e., by how much will the mean change if we sample in location i?), *fish* measures the effect on the mean of a randomly selected fish (the notation $j(i)$ means that the jth fish is nested within location i), *organ* represents the effect of the selected organ, and *reading* describes the effect of the selected reading. Readings are nested in organs, organs are nested in fish, and fish nested within locations. The factors *fish, organ, and readings* are treated as random, which means that they are chosen in a probabilistic manner and contribute variance to the measurements. We calculate the mean for a location by averaging, over all, the observations (assuming that we have a balanced set of measurements). This means we add over the n_r readings, the n_s slices, and n_f fish. The variability in the mean for location i is given by

$$Var(\bar{y}_i) = \frac{\sigma^2}{n_r n_s n_f} + \frac{\sigma_s^2}{n_s n_f} + \frac{\sigma_f^2}{n_f}.$$

Note that the mean is obtained by dividing by the total number of observations while the variance depends on the numbers associated with the different components. Thus, all components are treated equally in the evaluation of the variance. The important term (generally) here is the last one that is associated with the variance between fish σ_f^2. This is often the largest term in the analysis (variation between fish is often greater than variation between organs) and is the important term to reduce. Note that increasing the number of fish will not only decrease the variance term associated with fish but also that of the other components.

The models used to evaluate the biomarker will differ based on objectives and sampling. It is important to recognize the connection between the model and the data collection process and how changes in the data collection process will change the model. For example, if the purpose of the study is to compare different groups, then the focus should be on group differences. The measurement process becomes nuisance information that we try to control. On the other hand, if the interest is in developing a new biomarker, then the measurement process becomes the important aspect of the study, and we seek to estimate and evaluate the variance and its sources.

Nested Designs and Optimal Allocation of Effort

The above example suggests that the different components of variance and the sample sizes should be evaluated in a study. Perhaps the most common type of study with biomarkers is a nested design. The design we consider here includes a treatment factor, then measurements nested within the units selected to receive the treatments (or observed having the treatment). An example is given in Zhou et al. (1999) who studied thyroid cells in fish in a polluted site and in a reference site. Fish were collected in one polluted site (15 fish) and one reference site (8 fish). Thyroid tissue was collected from the fish. The tissue is composed of follicles (epithelial cells, vacuoles, and colloid). Because of variability in the size of follicles, 10 of the largest follicles per fish were selected. Then measurements were made on five epithelial cells in each of the follicles. This is a nested design with cells nested within follicles that are nested in fish. Location is viewed as the treatment factor. The model for measurement involves the average cell height considered over all groups and an effect due to the site, an effect due to fish differences within the site, an effect due to follicle differences within each fish, and an effect due to cell differences within follicles. The model would be

$$Y_{ijklm} = \mu + \alpha_i + Fi_{j(i)} + Fo_{k(ij)} + Cell_{l(ijk)} + \epsilon_{ijklm},$$

where Y_{ijklm} is the mth measurement (or transformed measurement) on cell l in follicle k from fish j from site i, μ is the mean of the measurements, α represents the site effect, Fi is the effect of the fish that is selected, Fo represents an effect due to the selection of the follicle, *Cell* is the effect due to the selection of the cell, and ϵ represents error due to the measurement.

The factors for fish, follicles, and cells are treated as random effects. This means they are not fixed factors with levels chosen by the experimenter, but rather they represent randomly selected levels. If the experiment were repeated, there would be different fish selected, hence, different follicles and cells. The selection process has an important effect on the way the data are analyzed. When the design is balanced (same number of observations per combination of effect levels), analysis of variance (ANOVA) tests require testing the importance of an effect using the variance associated with the term nested within the effect. Thus, to test for a site effect, the mean square for fish within a site is used as the error term. The error term that is used has tremendous implications for planning an analysis. In this example, the error term for fish has 21 degrees of freedom. This is in contrast to the error term for follicles that has 916 degrees of freedom. The implication is that, given the design, there is much more information for follicles than there is for sites. The ANOVA table associated with this example is given in Table 3 with expected sum of squares, assuming that there are an equal number of fish, follicles, and cells. For details on the complications that occur when the design is not balanced, see Sokal and Rohlf (1995).

A relevant question is whether this is an efficient design. To evaluate the design would require information on variances associated with the different factors, as well as costs. Given this information, quantities of interest, such as power or variance of means, may be calculated and compared for different designs at a fixed cost. To simplify the above example, suppose that effects are associated with sites, fish, follicles, and cells. Assume that there are (*a*) sites, that an equal number of fish (*b*) are selected at each site, the same number of follicles measured on each fish (*c*), and the same number of cells measured from each follicle (*n*). The total number of measurements is then $a \times b \times c \times n$.

The variance of the mean for a site is given by

$$s\frac{2}{y} = \frac{s^2}{ncb} + \frac{s^2_{follides(fish)}}{cb} + \frac{s^2_{fish(sites)}}{b}.$$

Different designs may be compared by calculating the variance from different sample allocations, given information on the different variances. If a majority of the variance is associated with factors in the field rather than in the laboratory, then a simple view is possible. In this case, a good strategy is to note that by increasing *b*, the other denominators also increase, and it is better to have a large number of fish with few follicles and cells than it is to have a small number of fish with a larger number of follicles and cells.

A further consideration in this nested example is the cost associated with the different procedures, for example, as the cost of field sampling

Table 3. Analysis of variance table for nested design example.

Source	Df	Expected mean square
Site	$a-1$	$\sigma^2 + n\sigma^2_{follicles(fish)} + nc\sigma^2_{fish(site)} + \dfrac{ncb}{a-1}\sum\alpha_i^2$
Fish within site	$a(b-1)$	$\sigma^2 + n\sigma^2_{follicles(fish)} + nc\sigma^2_{fish(site)}$
Follicles within fish	$ab(c-1)$	$\sigma^2 + n\sigma^2_{follicles(fish)}$
Cells within follicles	$abc(n-1)$	σ^2

may be expensive relative to that of laboratory analysis of additional follicles or cells. If costs and variance can be estimated, then estimates of appropriate sample sizes for different levels of nesting may be calculated. The cost of the experiment may be estimated by (assuming equal number of fish are used at each site)

$$\text{Cost} = \text{cost(initial)} + a \times b \times \text{cost(fish)} + a \times b \times c \times \text{cost(follicle)} + a \times b \times c \times n \times \text{cost(cell)},$$

where cost(initial) represents the costs not associated with measurement and sampling, cost(fish) represents costs associated with sampling and processing individual fish (i.e., how much would it cost to add an additional fish to the study), and cost(follicle) and cost(cell) are costs associated with processing and analyzing follicles and cells, respectively.

Sokal and Rohlf (1995) indicate that the optimal strategy (minimal cost and minimal variance) is to select the number of measurements per follicle, n, to be

$$n = \sqrt{\frac{cost(follicle)s^2}{cost(cell)s^2_{follicle(fish)}}}$$

and the number of follicles per fish, b, to be

$$c = \sqrt{\frac{cost(fish)s^2_{follicle(fish)}}{cost(follicle)s^2_{fish}}}.$$

The final quantity, b, or the number of fish, is found by solving the cost equation. In some cases, the calculated value of n is smaller than 1. In this case, measuring multiple cells is not informative relative to adding follicles.

It is not advisable to measure multiple cells within follicles if

1. The costs are equal and the follicle variance is higher than the variance between cells.

2. The variances are equal and the cost of cell sampling is great relative to the cost of obtaining a new follicle.

Costs associated with finding follicles and measuring cells may be relatively small, whereas, the cost of obtaining and processing fish may be large. If the fish at particular sites are heterogeneous with respect to biomarker responses, then it is advisable to add more fish to the study rather than locating more follicles unless the cost of adding a fish is considerably more than that of finding another follicle.

The example in Table 4 illustrates these ideas. Suppose that there is $10,000 available to study a lake. Interest is in estimating the mean value of a biomarker; it is desired to have the variance of the mean to be around 0.10 (this might be based on a desired confidence limit). We believe that the lake is heavily polluted so it is expected that biomarker measurements will exhibit low variance between fish and high variance associated with follicles and cells for illustration. Values selected are given in Table 4. Costs of selection of fish are high as sampling and processing of fish is required and cost of follicle preparation is assumed to be small relative to cost of measuring the cell. Adding fish to the study will increase cost but will decrease the contribution of variance from all sources. Measuring more cells will increase

Table 4. Example calculations for cost of study and variance of mean for different sampling schemes for a heavy pollution scenario. It is assumed that the cost of the study needs to be below $10,000 and the initial costs are $2,000.

Scenario	1	2	3
Fish			
Cost of a fish: cost(fish)	100	100	100
Number of fish: b	20	10	10
Variance associated with fish: $s^2_{(fish)}$	1	1	1
Follicle			
Cost of a follicle: cost(follicle)	10	10	10
Number of follicles: c	7	2	4
Variance associated with a follicle: $s^2_{follicle(fish)}$	4	4	4
Cell			
Cost of a cell: cost(cell)	20	20	20
Number of cells: n	1	7	6
Variance of cells: $s^2_{(cells)}$	4	4	4
Variance of the mean	0.107	0.326	0.217
Initial cost	2,000	2,000	2,000
Cost of sampling	6,200	4,000	6,200
Total cost	8,200	6,000	8,200
Total N	140	140	240
Optimal n	0.707	0.707	0.707
Optimal c	6.325	6.325	6.325

the cost by a small amount but will not decrease the variance due to fish or follicles, only the variation due to cells. Given the cost and variance structure, it is better to add more fish.

Given the parameters for Table 4, we observe the following. If we take a sample of 20 fish with 7 follicles selected per fish and 1 cell measured per follicle, we obtain the required variance and are below cost. The total number of measurements is 140. This is optimal in terms of cost and precision. If we take the same number of samples with a different allocation of sample size, the result will be, perhaps, a lower cost ($6,000), but the variance of the mean will increase (0.326). If we fix the total cost, we can take a larger sample, but, given the allocation, we do not achieve the low precision from the first case. The reason the precision cannot be obtained results from not being able to allocate enough sampling effort to fish. Note that the number of fish will reduce the variance by 20 and by 10 in the other cases. Thus, in the later cases, (given that the variance associated with fish is 1.0) it is not possible to reduce the variance below 0.10 regardless of how many measurements are made.

Note that a potential problem with this particular design is whether or not the fish "represent" the locations. If locations are considered the treatment, then the fish selected need to represent the fish in the location. Potential problems leading to lack of representativeness include schooling of fish, age structure, and movement such as immigration.

Reducing Variability

The success of a study program depends on the development and use of sensitive biomarkers. A simple but generally expensive method for increasing the sensitivity of experiments is to increase the sample size of the study. As seen in the above examples, one must recognize that there may be more than one type of sample size in the study and that changing different sample sizes does not necessarily have the same effect on measures of sensitivity. The researcher must recognize what the important unit is, relative to the lowest and highest sensitivity and control sample size associated with this unit.

Another approach to reducing variation is to refine the laboratory and data collection process. Data collected in a pilot study indicate the magnitude of the variation associated with different processes. If processes with large variation are studied, it may be possible to control the process to reduce variation. In fish studies, variation may result from the transportation of the fish from the field to the laboratory. By exerting more control over the handling and transportation of the fish, it is often possible to manage this type of variation.

Use a Cheap Method to Reduce Variability Associated with Other Sources

There are also a variety of statistical schemes available to reduce variation, and they are often inexpensive to implement. One technique is the process of blocking. We recognize that the units that we measure differ according to one or more factors. If a treatment were applied to units, we would expect differences in measurements due to the treatment, as well as within the units selected. The idea behind blocking is that if treatments are applied to sets of similar units, then the variation due to the selection process is made small and that source of variation may be removed from the error variance. By making the error variance smaller, the hypothesis tests are more sensitive.

Another inexpensive method for reducing variation is compositing. If there is variation in the measurement process from different sources, then, by physically combining measurement units, this variation might be reduced. For example, in studying benthic organisms, it is commonly found that the organisms are in clusters. Suppose we wish to estimate proportional abundance of a particular taxa in a stream section. The section is divided into units and each unit is subsampled by taking 10×0.1 m^2 sections of soil and counting organisms in each section. The protocol may call for counting 100 organisms. An approach based on compositing would take the same samples but mix the soil together prior to counting. Hence, the 10 samples of soil are mixed and then 100 organisms are counted. In the first approach, one must count 1,000 total organisms versus 100 in the second case. The second approach is almost always better as it is less costly and allows the researcher to focus on obtaining more of the important information (between unit variation rather than within unit variation). As noted in the analysis of nested designs, the information from lower levels of nesting is averaged when evaluating variation in the higher levels. In composite sampling, the variation associated with differences due to the lower levels (sections) is physically averaged rather than numerically averaged. Compositing is not expensive if the cost associated with taking samples at the lower levels is small compared with costs at higher levels. When the cost is high at the lower level, this type of sampling will not be cost effective.

Concluding Remarks about Designs for Studies Using Biomarkers

Designing studies using biomarkers is not a simple adventure. To design appropriate studies that yield definitive and reliable results, we should evaluate how data from the study will be collected and used to evaluate hypotheses of interest. It is also important to not only focus on the laboratory analysis of samples but also on the statistical analysis. The variability of a measurement is quite important and should be the focus of initial planning of the study. It is important to recognize different types of units and variances associated with the different components. Potential models for analyzing the data need to be considered and matched to the data and sampling or experimental processes. By paying attention to several statistical principles, it should be possible to use biomarkers as a sensitive metric for biological stress. It is important to avoid excessive measurement and to focus on connecting the sampling process, the objectives of the study, and the model for the analysis to optimize the relationship between study costs and definitive results.

Statistical Analysis of Biomarkers

The interpretation of information collected from studies using biomarkers depends on the objectives of the study, the design of the study, and the type of data that is collected. While a detailed description of the methods used for analysis of biological data is outside the scope of this paper, a summary of common methods used in laboratory and field studies is presented. The discussion addresses both univariate methods for single response measurements and multivariate methods when multiple endpoints are measured. For additional details on many of these methods see Sparks (2000).

Univariate Methods

Proper analysis of data from environmental studies requires that the appropriate statistical model be chosen and that the assumptions of the analysis are valid. Poor choice of a statistical model may result in incorrect conclusions regarding the data being analyzed, the parameters of the model, or poor predictive power.

The most common approach for determining whether the appropriate model has been used in the analysis involves evaluation of the residuals of the model. Statistical models are generally of the form

$$response = deterministic\ part + stochastic\ part$$

The deterministic part describes the relationship between a response variable and the explanatory variables. The stochastic part explains what remains, that source of variation unexplained by the model. If problems exist with the model, there will often be patterns in the residuals. Examples of problems include variables that are important but are not in the model, models that are missing quadratic or nonlinear terms, and odd observations. Descriptive methods focus on both the deterministic and stochastic parts of the model. Displays of the response are often used to suggest or verify a model. Common displays include boxplot displays for evaluating separation of groups, scatterplot displays to illustrate relationships between response and explanatory variables, and plots of means and standard deviations. Displays of the residuals illustrate patterns in the stochastic component. Common displays are plots of residuals versus variables that are in or not in the model to check for missed relationships, odd values, and homogeneity of variance. Normal probability plots are commonly used to check for normality and odd values.

The methods for analysis of the data differ in terms of the model that is used and the type of data. In ANOVA and regression analysis, the response variable is continuous. In ANOVA, the interest is in testing for the difference between groups or the importance of an effect. Estimation of the differences and determination of group differences is also important. In regression analysis, the focus is on relating a response to measurements using a model based on explanatory variables. Regression analysis and ANOVA are similar, and the same computer package may be used to perform both analyses. They differ in focus more than in computational method. In ANOVA, the response is explained in terms of group differences. In regression analysis, the response is explained by a relationship with a variable. The relationship is typically linear (in simple and multiple linear regression); but, in more complex models, the relationship may be nonlinear. A summary of the primary types of univariate models is provided in Table 5.

Multivariate Methods

Many studies involving biomarkers select several responses to evaluate the effects of stressors on biological systems. Ecological effects may occur on

Table 5. Summary of the principle types of univariate statistical methods.

Method	Model	Examples	Assumptions
Descriptive methods	No model, but methods are typically chosen to suggest a model Models should be checked using methods applied to residuals	French and Lindley 2000	Good graphical display requires proper selection of type of display and careful selection of scales for axes
Analysis of variance	Response variable is modeled in terms of means or effects associated with factors or groups	Tyler et al. 1999	Normality Independence of units Homogeneity of variance
Regression	Response variable is modeled as a linear or nonlinear function of explanatory variables	Lange et al. 1993	Normality Independence of units Homogeneity of variance Model is correct
Logistic regression	Binary response variable	Field et al. 1999	Independence of units, binomial probability model

different biological levels, hence, investigating several levels may be useful. Multivariate statistics provide tools for evaluation of sets of multiple biomarkers in a single analysis. In analyzing all the measures together, relationships among the responses are taken into account. Multivariate methods are capable of detecting the patterns that are apparent in a multiple univariate analysis, as well as in the less apparent relationships, when multiple responses are involved. Multivariate analysis is not appropriate when the effect of the stressor is associated with a single biomarker. The multivariate approach is more appropriate when there are several biomarkers involved that reflect similar stressor responses, particularly at one level of biological organization.

There are numerous multivariate methods that may be applied to the analysis of biomarker data. The methods differ in terms of purpose, data type, and assumptions. Some of the major types of multivariate methods are presented in Table 6.

Principal components analysis, factor analysis, multidimensional scaling, and correspondence analysis are commonly used to explore the data and provide graphical displays of the important patterns in the data. The goal with these methods is typically to produce a combination of the information that may be used to reduce the interpretation problem from a large dimensional problem to a smaller one. In principal components analysis and factor analysis, linear relationships between the measurements are used to find combinations of the measurements (called components or factors) that summarize or explain basic relationships among all the measurements. These components then may be interpreted and used instead of the original measurements. Principal components analysis and factor analysis are used when the relationships are linear, and a correlation or covariance matrix provides a reasonable summary of these relationships.

Multidimensional scaling and correspondence analysis are useful when the relationships are not linear or the data are not continuous. In multidimensional scaling, the focus of the analysis is the array or matrix of distances between the observations. The goal of the analysis is to find observations in fewer dimensions (usually 2) that have a similar distance matrix. Starting with the original distance between objects, new "measurements" are obtained that result in the same distance matrix (using Euclidean distance). The distances may be the actual data (as in the case of mapped distances) or may be derived from data (using for example, the Bray-Curtis distance [Romesburg 1984]). Correspondence analysis is used when data are in the form of counts or are collected and summarized as a contingency table.

When comparing several populations of organisms, some interesting questions are: which populations are different, what is the magnitude of difference between the populations, and what is the basis of differences between the populations? In this case, it is better to use an inferential method to test for and describe differences among populations. Multivariate ANOVA

Table 6. Summary of the principle types of multivariate statistical methods.

Method	Purposes	Examples	Assumptions/ comments
Principal components	Reduce dimension of data Plot data on reduced dimension to find patterns Outlier detection Find groups of variables	Beltman et al. 1999; Genter and Lehman 2000	Linear relationships between variables
Factor analysis	Find factors that describe the data	MacKenzie et al. 1982	Linear relationships between variables
Multidimensional scaling	Find points in lower dimensions that preserve distance properties	Gray et al. 1990	Depends on choice of distance, use when relationships are nonlinear
Correspondence analysis	Reduce dimension of count data	Wright et al. 1984	Chi-square distance, use with nonlinear relationships
MANOVA	Test for group separation	Genter and Lehman 2000	Multivariate normality Independence Homogeneity of covariance matrices
Canonical discriminant analysis	Describe group separation using linear combinations of variables	Adams et al. 1994	Multivariate normality Independence Homogeneity of covariance matrices
Discriminant analysis	Predict the group that an observation comes from	Hall et al. 1995; Wright et al. 1984	Varies with method
Canonical correlation analysis	Relate two sets of variables, typically biomarkers with environmental variables	Garcia and Fernandez 1994	Relationships are linear
Canonical correspondence analysis	Relate a set of counts to environmental variables	Stevenson et al. 1989; Snoeijs and Prentice 1989	Relationships are nonlinear, following a Gaussian shape
Cluster analysis	Find groups of observations or variables Find outliers	Wright et al. 1984	Results depend on distance measure chosen

(MANOVA) is the commonly used method for testing for differences between groups. Canonical discriminant analysis is similar to principal components analysis although the focus is not on explaining the data but rather identifying the separation of groups. Combinations of variables are derived that may be used to summarize the separation. If the separation is concentrated in a few dimensions, then canonical discriminant analysis is able to find new variates that are combinations of the variables that account for the separation. These may then be used in a graphic display of the separation.

Discriminant analysis is similar to canonical variate analysis, however, the focus is on predicting the group an object belongs to rather than the separation between groups. In linear discriminant analysis, regression type models are developed to predict group membership. Other forms of discriminant analysis include quadratic discriminant analysis, which is used when variances are heterogeneous, and nonparametric methods, which focus on the distance from observations to groups.

Cluster analysis is a distance-based method that forms groups of observations. Given a set of observations, a distance matrix is created and the distances are used to indicate objects that are similar or different. Objects with similarities may then be grouped together. Cluster algorithms group objects together that are similar and separate sets of objects that are different. There are many algorithms used for cluster analysis, as well as summarization of results (see Romesburg 1984).

In many field studies, information is collected, at a number of sites, on biological variables, as well as the influential environmental variables. In this situation, one objective is to relate the biological data with the environmental information. Useful multivariate methods for investigating these relationships are canonical correlation analysis and canonical correspondence analysis. The approaches are similar in that combinations of the biological variables are found that relate to combinations of the environmental variables. In canonical correlation analysis, the biological data are typically continuous measurements, while in canonical correspondence analysis, species counts are used as the biological endpoints. In both cases, combinations are formed to maximize relationships between biological responses and the influential environmental variables. These combinations are then used for interpretation or graphic displays.

As in univariate analysis, the multivariate methods are dependent on assumptions, and a careful analysis is required to produce defensible and accurate results. It is important to not only apply the methods but to check the quality of the analysis. For example, in principal components analysis, an important assumption is that the correlation or covariance matrix provides a reasonable summarization of the relationship between the variables. This assumption may be evaluated by graphic displays of the data in a scatterplot matrix (plots of all pairs of variables in one display). It is important to check for nonlinear relationships, as well as extreme observations, as linear relationships are often required for use with multivariate methods. By investing time in the evaluation of the quality of the analysis, we often produce better results.

Concluding Remarks

The investigator must pay attention to the quality of the study for biomarkers to be defensible. For many, quality is synonymous with laboratory quality. While it is very important that good laboratory practices be adhered to, there are other aspects of a study that require attention. We have focused on the statistical component of design. Several rules should be followed to improve the design of a study:

1. Develop objectives.
2. Define the unit of measure.
3. Understand types and sources of variability.
4. Understand how the biomarker relates to study goals.
5. Develop a protocol for the study.
6. Run a pilot study.
7. Evaluate the ability of the biomarker to meet the needs of the study

If these rules are adhered to, we believe that good designs may be obtained. If these rules are ignored, we believe that the resulting design will have problems and results will not be as defensible.

There are many methods that are available for the analysis of data resulting from biomarker studies. For any given set of data, it may be possible to apply several methods. Choice of method depends on the properties of the data (in relation to validity of assumptions) and the goals of the study (as defined through the statistical parameters and hypotheses). Univariate methods may be summarized as descriptive methods, methods for comparing groups, and regression methods. When multiple measurements are made in biomarker studies, multivariate methods are appropriate. While many multivariate methods are extensions of univariate analysis, the general goal is to combine the multiple information to simplify (make sense of the multiple endpoints) and possibly enhance the analysis (increase the power of the study). A number of multivariate methods are discussed with a focus on descriptive methods, as well as inferential methods. Again, choice of method depends on the goals of the study and the properties of the data. As in the design of the study, paying attention to quality is important. The quality of the analysis is improved by matching the goals of the study to the analysis that is chosen and to check that the assumptions of the analysis are reasonably met. An important part of the study that was not discussed but that is quite important is data entry and management. Errors in data values are often easy to detect if checks are made. Too often these checks are skipped.

References

Adams, S. M. 1990. Status and use of biological indicators for evaluating the effects of stress on fish. Pages 1–8 *in* S. M. Adams, editor. Biological indicators of stress in fish. American Fisheries Society, Symposium 8, Bethesda, Maryland.

Adams, S. M., K. D. Ham, J. J. Beauchamp. 1994. Application of canonical variate analysis in the evaluation and presentation of multivariate biological response data. Environmental Toxicology and Chemistry 13(10):1673–1683.

Anderson, D. P. 1990. Immunological indicators: effects of environmental stress on immune protection and disease outbreaks. Pages 38–50 *in* S. M. Adams, editor. Biological indicators of stress in fish. American Fisheries Society, Symposium 8, Bethesda, Maryland.

Beltman, D. J., W. H. Clements, J. Lipton, and D. Cacela. 1999. Benthic invertebrate metals exposure, accumulation, and community-level effects downstream from a hard-rock mine site. Environmental Toxicology and Chemistry 18(2):299–307.

Eberhardt, L. L., and J. M. Thomas. 1991. Designing environmental field studies. Ecological Monographs 61:53–74.

Field, L. J., D. D. MacDonald, S. B. Norton, C. G. Severn, and C. G. Ingersoll. 1999. Evaluating sediment chemistry and toxicity data using logistic regression modeling. Environmental Toxicology and Chemistry 18(6):1311–1322.

French, D., and D. Lindley. 2000. Exploring the data. Pages 33–68 *in* T. Sparks, editor. Statistics in ecotoxicology. John Wiley and Sons, Chichester, U.K.

Garcia, F. J. G., and J. V. L. Fernandez. 1994. Recreational use model in a wilderness area. Journal of Environmental Management 40:161–171.

Genter, R. B., and R. M. Lehman. 2000. Metal toxicity inferred from algal population density, heterotrophic substrate use, and fatty acid profile in a small stream. Environmental Toxicology and Chemistry 19(4):869–878.

Gray, J. S., K. R. Clarke, R. M. Warwick, and G. Hobbs. 1990. Detection of the initial effects of pollution on marine benthos: an example from the Ekofisk and Eldfisk oilfields, North Sea. Marine Ecology Progress Series 66:285–299.

Green, R. H. 1979. Sampling design and statistical methods for environmental biologists. John Wiley and Sons, Chichester, U.K.

Hall, J. R., S. M. Wright, T. H. Sparks, J. Ullyett, T. E. H. Allott, and M. Hornung. 1995. Predicting freshwater critical loads from national data on geology, soils, and land use. Water, Air and Soil Pollution 85:2443–2448.

Hurlbert, S. J. 1984. Pseudoreplication and the design of ecological field experiments. Ecological Monographs 54:187–211.

Lange, T. R., H. E. Royals, and L. L. Connor. 1993. Influence of water chemistry on mercury concentration in largemouth bass in Florida lakes. Transactions of the American Fisheries Society 122:74–84.

MacKenzie, D. I., S. G. Sealy, and G. D. Sutherland. 1982. Nest site characteristics of the avian community in the dune-ridge forest, Delta Marsh, Manitoba: a multivariate analysis. Canadian Journal of Zoology 60:2212–2223.

McMillan, A., A. S. Whittemore, A. Silvers, and Y. DiCiccio. 1994. Use of biological markers in risk assessment. Risk Analysis 14:807–813.

National Research Council. 1990. Managing troubled waters: the role of marine environmental monitoring. National Academy Press, Washington, D.C.

Peterman, R. 1990. Statistical power analysis can improve fisheries research and management. Canadian Journal of Fisheries and Aquatic Sciences 47:2–15.

Romesburg, C. H. 1984. Cluster analysis for researchers. Wadsworth, Belmont, California.

Smith, E. P., and R. Voshell. 1997. Studies of benthic macroinvertebrates and fish in streams within EPA Region 3 for development of biological indicators of ecological condition. Virginia Polytechnic Institute and State University, Blacksburg.

Snoeijs, P. J. M., and I. C. Prentice. 1989. Effects of cooling water discharge on the structure and dynamics of epilithic algal communities in the northern Baltic. Hydrobiologia 184:99–123.

Sokal, R. R., and F. J. Rohlf. 1995. Biometry: the principles and practices of statistics in biological research. Freeman, New York.

Sparks, T., editor. 2000. Statistics in ecotoxicology. John Wiley and Sons, Chichester, U.K.

Stevenson, A. C., H. J. B. Birks, R. J. Flower, and R. W. Battarbee. 1989. Diatom-based pH reconstruction of lake acidification using canonical correspondence analysis. Ambio 18(4):228–233.

Tyler, C. R., R. van Aerle, T. H. Hutchinson, S. Maddix, and H. Trip. 1999. An in vivo testing system for endocrine disruptors in fish early life stages using induction of vitellogenin. Environmental Toxicology and Chemistry 18(7):337–347.

Wright, J. F., D. Moss, P. O. Armitage, and M. T. Furse. 1984. A preliminary classification of running water sites in Great Britain based on macroinvertebrate species and the prediction of community type using environmental data. Freshwater Ecology 14:221–256.

Zhou, T., H. B. John-Alder, P. Weis, and J. S. Weis. 1999. Thyroidal status of mummichogs (*Fundulus Heteroclitus*) from a polluted versus a reference habitat. Environmental Toxicology and Chemistry 18(12):2817–2823.

16

Biomarkers and Bioindicators in Monitoring and Assessment: The State of the Art

Peter V. Hodson

Introduction

As described in Chapter 1, the purpose of this book is to provide information and guidance to improve our ability to monitor, evaluate, and predict the effects of environmental stressors on aquatic ecosystems, especially fish populations. The various chapters of this book have presented methods and approaches for measuring a variety of biological characteristics or responses at levels of organization ranging from the molecular (biomarkers) to the community (bioindicators) that might be useful in research, monitoring, or assessment. This final chapter will discuss why we are interested in biomarkers and bioindicators, what their relative value and state of development are, which measures are ready to be applied in a systematic way, which are still at the research stage, and what research needs emerge from the current state of the art. Other chapters provide an excellent overview of the major issues associated with design and implementation (e.g., Johnston and Collier, Chapter 14), and will not be considered in detail here. A large proportion of the biomarkers and bioindicators reviewed in this book have evolved from an interest in the impacts of chemicals on ecosystems, so most references to "stressors" in this chapter will be related to studies of chemical toxicity.

The importance of the tools and approaches described in this book is derived from the potential utility and application of the tools and approaches in programs of environmental management. As with any other type of management, the results must be examined closely—did the management action achieve its goal? As Dr. John Sprague explained in his Founder's Award address at the 1989 Annual Meeting of the Society of Environmental Toxicology and Chemistry, there are three main strategies for managing the quality or health of aquatic ecosystems: effluent regulations to limit the discharge of hazardous substances to amounts or concentrations that should not harm aquatic ecosystems; environmental quality criteria (water, sediment, tissue) to limit the occurrence of hazardous conditions in aquatic ecosystems; and environmental-effects monitoring, as a quality control check to verify that the first two strategies have been effective. Current science is also driven by a regulatory need for proof of damage and cause: "It is common for regulatory approaches to require evidence of impacts to populations or to communities before remedial actions or punitive measures are taken" (Johnston and Collier, Chapter 14). Finally, where remediation has been necessary to restore damaged ecosystems, monitoring will measure the success of restorative measures and indicate whether additional work is required.

Assessment is the collection of data on specific characteristics of a system to describe its current condition in relation to known or unknown stressors. Monitoring is the repeated assessment of a system to detect changes or trends in its state, so that corrective actions may be applied should there be a change that is considered undesirable. Bioassesment and biomonitoring measure the biological characteristics that we wish to protect. Therefore, bioassessment and biomonitoring of aquatic ecosystems provide an ongoing series of measurements on the state of those systems to answer some very practical and important questions:

- Is the system healthy or degraded? That is, does it exhibit characteristics indicating that all components are present at the expected frequency or density and interacting as they should? Do the components exhibit their expected productivity? Expectations (hypotheses) are based on history (what was observed in the past), theory (ecosystem structure and productivity), and environmental management goals (objectives for structure or productivity).

If the system is degraded at any level of organization,

- where is the degradation evident and what components are affected?
- what is the apparent cause, and what is the source and nature of that cause?
- what is the extent and intensity of degradation, in terms of the geographic area, and the proportion of a specific resource affected?
- is the system getting better or worse? What are the temporal trends in the extent, nature, and intensity of the degradation, and of the components being degraded?

Therefore, biomonitoring and bioassessment embody two broad objectives: to describe the state of the environment and to link cause and effect where there is degradation. To meet these objectives, programs designed to monitor effects of stressors can choose from an impressive array of tools that vary from highly specific organismal-level responses (biomarkers) that aid in diagnosing cause and effect, to relatively nonspecific bioindicators of environmental state that are highly relevant to the well-being of ecosystems (Johnston and Collier, Chapter 14). Arguments as to whether biomarkers are more useful than bioindicators for monitoring create a false dichotomy; clearly the choice of method depends on the questions being asked and on practical issues such as resource limitations (see Johnston and Collier, Chapter 14, for a detailed discussion). The utility of both biomarkers and bioindicators increases with the number of measures that can be applied in concert to create a "weight-of-evidence." However, a major impediment to integrated monitoring is a lack of understanding of the links among different levels of organization. The inability to reliably predict responses across different levels of biological organization limits the utility of results of biomonitoring and constitutes a consistent theme throughout this book. Fortunately, some systems have been studied sufficiently (e.g., Puget Sound) to provide guidance for monitoring and assessment in less well-known systems.

Given that the responsibility for managing large aquatic ecosystems (e.g., Great Lakes) as a public resource lies primarily with government agencies, monitoring is primarily a government function. The critical role of government is reinforced by the high cost of monitoring large ecosystems and the need for a long-term commitment of resources to collect sufficient data in a consistent and standardized way. Without such a long-term commitment, the value of monitoring may be lost if programs are canceled or radically changed before trends in environmental quality are evident. Given the high costs, some industries are now responsible for monitoring the effects of their own discharges under government-supervised programs (e.g., Canada's Environmental Effects Monitoring [EEM] program for the pulp and paper industry; EC and DFO 1992a, 1992b). Large-scale monitoring is not typically conducted by academic scientists because they have fewer resources than government agencies. While academic scientists might collect a very focused set of data to answer a specific question, they are more likely to do research that supports monitoring and assessment (e.g., studies of mechanisms) than monitoring itself.

Although governments are the prime agents of monitoring, there are relatively few large-scale biological-monitoring programs. Monitoring is often regarded negatively because it has been associated with expensive programs, ill-defined objectives, and few useful outcomes (see Chapter 12, Attrill, quote of Underwood 2000 re: mindless monitoring), particularly when programs are canceled or reduced in scope before trends in data are established. There is often an attitude that monitoring is not "real science." Effective monitoring programs have clearly defined questions, measure appropriate indicators derived from well-characterized relationships between the indicator and the state of the ecosystem, and generate data that can be translated into easily understood indices that trigger specific management actions. An excellent analogy is the measurement of blood alcohol in drivers. Readings above widely accepted thresholds accurately predict the capacity of an individual to drive a vehicle and lead to immediate regulatory action. When collected systematically, blood alcohol data also provide a useful overview of the success of programs to curb drinking and driving.

Therefore, biomonitoring programs must start with clear management objectives that define the questions to be answered, the geographic scope, frequency and intensity of sampling, and the nature of the measurements to be made. Of prime importance is the design of appropriate sampling strategies to describe spatial and temporal trends, to statistically discriminate differences in response parameters among sites and times, and to help associate cause with effect. Experimental design and approaches to biomonitoring are well addressed in Chapter 12 (Attrill—community), Chapter 14 (Johnston and Collier—integration), and Chapter 15 (Smith—statistics). Guidance is required on sampling designs, including randomization, definition of reference sites and controls, replication, and criteria for target species (e.g., residency, sensitivity, size, uniformity, density, and tolerance at affected sites [tough old survivor syndrome]). Overall, there are many options for experimental designs and appropriate statistical analyses, and the design of monitoring and assessment programs can be considered a relatively mature discipline.

Criteria for Useful Biomarkers and Bioindicators

How can useful tools be identified for assessing exposure and effects that are ecologically relevant at the population and community level (bioindicators)? Given that biomonitoring of natural environments shares many characteristics with epidemiology, Fox (1991) applied criteria commonly used in epidemiology to show how the relative strength of proposed cause and effect relationships could be tested in programs of monitoring and assessment. These criteria also define some of the characteristics desired in useful biomarkers and bioindicators. Because cost is always an issue, and because monitoring methods would be applied repetitively under standardized conditions, they could be evaluated against the following useful characteristics:

Ability to Define Cause–Effect Relationships

Where the intent of a biomonitoring program is to link cause and effects (or more properly exposure and response), biomarkers based on known mechanisms of effects (e.g., carbamate pesticide inhibition of ACh-ase) and with a high degree of specificity would be one important component. Establishing cause and effect with bioindicators can be more difficult than with biomarkers but, where possible, should be included to reinforce the importance of the cause–effect relationship and its relevance to regulatory and management decisions.

The choice of biomarkers and bioindicators to apply in biomonitoring will depend on

- Specificity of the association. Specificity refers to the relationship between a stressor and an effect or between a response at one level of organization and an effect at the next higher level. For stressors, is there specificity of cause for an effect (only a limited number of causes for a unique effect)? Is there specificity for effects of a cause (i.e., does a stressor cause a limited range or a wide range of effects)? To describe the overall state of the ecosystem, bioindicator responses should be related to a specific, predictable impairment at the next higher level of organization and should be based on known relationships to degradation (e.g., algal species composition as an indicator of eutrophication).
- Strength of association (statistical). This criterion is self-evident.
- Consistency. The association between cause and effect, or between responses and effects at higher levels of organization, should be repeatable among situations, and among monitoring programs, with a demonstrated utility in at least three published field studies.
- Mechanistic and biological plausibility. Methods should be supported by controlled laboratory experiments that define the mechanism of stressor effects and that demonstrate their consequences for the next higher level of organization.
- Temporal sequence. Exposure should precede effects, and effects should decline when exposure is removed.
- Clear exposure–response relationships. The intensity of biomarker or bioindicator responses should increase with the intensity of exposure to a stressor.
- Sensitivity. The indicator should be appropriately sensitive and provide early warning, that is, responding at exposure levels typical of stressors encountered in degraded ecosystems and in advance of ecologically significant effects.

Relevance

Where biomonitoring is applied to assess the quality, health, or state of an ecosystem or some component (e.g., an exploitable resource), bioindicators would be the predominant tool applied. The results of monitoring should be relevant to environmental and resource managers and expressed in an understandable form with specified thresholds that trigger management action. Monitoring should include methods relevant to the stressors known to be acting on the ecosystem to ensure that environmental management is directed at the most important source of stress. For example, if the objective is to verify whether the rate of cancer-induced early mortality of adult fish is declining following site remediation of a coal tar deposit, measures of population structure (bioindicator) and biochemical, genetic, and histopathology responses (biomarkers) would form a highly relevant and coherent biomonitoring program (e.g., effects on brown bullhead *Ameiurus nebulosus* populations, Black River, Ohio; Baumann et al. 1996). The biomarkers of specific stages of cancer development would provide a mechanistic link between polyaromatic hydrocarbon (PAH) exposure and cancer effects. For example, impacts at the population level would be indicated by changes in the age-frequency distribution.

Many highly relevant parameters discussed in this book are relatively nonspecific as to cause and effect. Nevertheless, they can still be applied systematically to deduce specific causes by describing spatial and temporal patterns of stressors and responses and the characteristics of affected species that make them particularly susceptible to a specific stressor. When resources limit the scope of monitoring, biomarkers of exposure or effects may be more cost-effective than bioindicators. Highly specific organism-level responses generally have lower variance and require smaller sample sizes to discriminate differences between sites compared with population or community-level responses that reflect the influence of a wide array of stressors. The trade-off, however, is reduced relevance to the state of the environment.

State of Method Development

For widespread application to monitoring, methods should be "off-the-shelf"; that is, little development or research should be needed to apply them to new situations or samples. Methods also should be technically feasible, cost effective, and within the capability of standard biological or chemical laboratories. As Rice states in Chapter 6, "In order to determine how stress affects immune function, there must be a balance between the complexity of assays, and, thus, the inherent training requirements of personnel, the expense, and how much detailed information is required."

- Quality control procedures should be well established and an important part of the overall monitoring design. Quality control protocols should include standardized methods for sample collection, storage and analysis, standards, blanks, positive and negative controls, and an expected variance based on interlaboratory comparisons and round-robins (Munkittrick et al. 1991, 1993) The frequency of false positives and negatives must be defined, as well as the sources of analytical error or artifacts. Recommended sample sizes should be designed to detect a specified difference between responses at exposed and reference sites and with a specified level of confidence (see Chapter 15).
- Covariates or other confounding factors that may bias the interpretation must be defined from laboratory and field studies and from an indepth knowledge of the biology of the species. Common covariates that might be considered include

- the organism: age, size, sex, growth rate, sexual maturity, migration, species, disease
- the environment: diet, season, depth, temperature, pH
- the collection methods: collection gear; time of day; holding time and conditions; changes among sites in depth, light, temperature; susceptibility to capture of healthy or diseased fish
- other stressors: fisheries, physical habitat changes; change in fish community (predator and prey abundance)

Expression of Results

Results must be easily understood by the public and by the user community (e.g., environmental managers and regulators) or be readily translated into simple indices. Before monitoring begins, there must be clear guidance on how information can be used and interpreted once available. This understanding should be expressed in interpretive documents describing what constitutes "normal," how deviations from normal might be interpreted in light of specific stressors, and what confounding variables might bias the interpretation. Expert opinion and additional research should not be needed for interpretation. For complex arrays of data, consideration should be given to "intelligent systems" as a means of rapidly generating an interpretation.

The above list might appear onerous, but the extensive research and development underpinning such simple metrics as blood glucose used in medical diagnosis demonstrates the challenge in generating data that leads to unambiguous conclusions. Clearly, not all the requirements would be met in every situation. For example, pilot studies might be needed on local species to establish ranges and variability of normal states, and labs would have to acquire equipment, adopt and validate standard methods and train personnel on how to collect and analyze samples. Planning also would be required to identify factors that might affect the outcome of the analyses so that appropriate covariates could be included in the monitoring design.

State of the Art

As is clear from the layout of the chapters and the discussions in each paper, the methods presented in this book fall on a continuum relating ecological relevance to mechanistic understanding (Figure 1, Chapter 12). Biomarkers are generally of little direct relevance to ecosystem health unless validation studies are conducted to demonstrate the links, causal or coincidental, among the different levels of biological organization (Adams et al. 2000, McCarty and Munkittrick 1996). Conversely, measures of ecosystem health (e.g., diversity indices) are general indicators of state and provide little information in themselves about the causes or determinants of that state.

It is obvious that measuring one bioindicator or biomarker will not answer all questions about environmental quality. The lack of specificity, even at the biochemical level, means that no one method can stand alone. A successful biomonitoring program will measure multiple biomarkers to describe syndromes that suggest cause and effect relationships for individual species. Similarly, several bioindicators applied together will generate "weight-of-evidence" conclusions about the state of ecosystems. At the design stage, multiple indicators must be carefully chosen from different levels of biological organization to ensure that the effects of anticipated stressors can be detected and applied in concert to create recognizable and reliable patterns of response (see Chapter 14, Johnson and Collier; Chapter 15, Smith). The following synopsis reviews the different biomarkers and bioindicators presented in this book, compares them to the criteria listed above, and assesses their relative stage of development for field application. Are they simply good ideas but not yet well studied? Do they show promising early results or good solid progress? Are they mature methods, ready to be applied in broad-scale programs of assessment and monitoring?

Biochemical Methods Including Molecular Genetics (Chapters 2 and 3)

Biochemical and genetic responses are generally well studied, well understood, and widely applied, particularly in the study of the effects of chemical toxicity. They are usually based on detailed models of chemical interactions at the cellular and molecular level, and are often highly specific due to protein receptor specificity for chemical structure (e.g., the interaction between organophosphate [OP] pesticides and acetylcholinesterase enzymes). As a consequence, specificity of response is characteristic of families of chemicals sharing specific properties. For example, only chlorinated or unchlorinated PAH with a relatively narrow range of properties (planarity, hydrophobicity, aromaticity, shape, MW 200–400) can bind with the arylhydrocarbon receptor protein to cause cytochrome P450 (CYP1A) induction. Therefore, biochemical methods provide the majority of biomarkers to help establish relationships between chemical causes and biological effects. Nevertheless, the methods described in these chapters represent a wide array of stages of development and suitability for monitoring. The biochemical methods reviewed in Chapters 2 and 3 are assessed in order of decreasing readiness for application in biomonitoring.

Cytochrome P450 induction and bile fluorescent aromatic compounds (FAC) have been widely applied in tandem in laboratory and field studies. They are considered reliable biomarkers of PAH exposure and of possible toxicity due to cancer (unchlorinated PAH such as benzo(a)pyrene) or to recruitment failure (chlorinated PAH such as dioxins, furans, polychlorinated biphenyls). Cytochrome P450 induction by PAH or increased concentrations of biliary FACs are biologically and toxicologically plausible components of well-established models for cancer. Relevance for ecological effects is fairly strong, as shown by studies of fish populations in which population-level impacts (early mortality of adults) were associated with all aspects of the cancer model. Many covariates of these responses have been identified and controlled for, and a variety of technically simple methods and bioassays have been standardized for enzymatic indices of induction (e.g., Hodson et al. 1991, 1996), including interlaboratory comparisons (Munkittrick et al. 1991, 1993). The results are readily understood, although there are as yet no commonly accepted thresholds of CYP1A activity or bile metabolites of PAH predicting biologically important effects such as cancer. These represent mature methods, ready to apply in programs of biomonitoring.

Cholinesterase inhibition is a highly specific and sensitive response to OP and carbamate exposure. Inhibition is associated with cardiovascular, respiratory, and gastrointestinal dysfunction; hyperactivity or lethargy; unconsciousness; muscle weakness; and respiratory collapse. There is a high degree of ecological relevance because inhibition is often associated with acute mortality, although there are exceptions that cause uncertainty in deducing cause and effect. While the toxic effects are neurological, blood cholinesterases are practical surrogates for cholinesterases in nerve tissue. The acute nature of this response can be problematic in that both tissue concentrations of pesticides and enzyme inhibition are relatively transient (hours to days) due to the rapid metabolism of OP and carbamate pesticides. Cholinesterase inhibition may be more long-lasting than tissue pesticide concentrations. This means that monitoring must coincide with pesticide spraying, unless there is a chronic source of contamination. Cholinesterase inhibition is a direct indicator of effect but is highly variable and may be reversible (carbamates). There are simple spectrophotometric methods available, including adaptations for plate readers, and the methods have been applied by many laboratories, both in the field and laboratory. Inhibition is easily interpreted, and this method is highly suited for monitoring specific situations such as pesticide applications.

Other biochemical and genetic responses have been well studied in the laboratory and frequently applied in the field, but are still in the realm of research. For example, *Phase II or conjugation enzyme activities*, have been used as biomarkers at a number of contaminated sites, but induction of activity in response to chemical exposure is low, being twofold at most. Hence, a "response" is difficult to separate from "noise." Phase II enzymes are also much less specific than CYP1A enzymes, metabolizing a wider array of endogenous and xenobiotic substrates. Hence, specificity, exposure–response, and sensitivity are all relatively low for Phase II enzymes. A variety of analytical methods are available, but they are not as well characterized or tested as measures of CYP1A induction.

Biomarkers of *oxidative stress* and of *antioxidant defense systems (enzymatic, nonenzymatic)* are also relatively nonspecific in that they reflect a broad array of potential effects (damage to DNA, protein, lipids, and cellular function). While there are many methods for assessing the occurrence and effects of oxidative stress, "no single measure by itself confirms the occurrence of oxidative stress or prestress conditions" (Schlenk and DiGiulio, Chapter 2), and many measures are not consistently useful. Hence, a battery of methods applied simultaneously will provide the most reliable picture of oxidative stress, although multiple measures increase costs and sample sizes. Methods for oxidative stress are within the capacity of most laboratories, but the results are difficult to interpret. There is still little known about the basic biology and toxicology of the oxidative stress response and the influence of covariates, and the link to ecological effects is unclear.

Metallothioneins (MT) are widespread in plants and animals, have well-conserved genetic sequences, and are induced in response to exposure to a few metals (e.g., Cd, Cu, Zn, Hg, Pb, Ag), although they also scavenge oxyradicals. They have been applied widely in studies of metal-contaminated ecosystems, both in vertebrates and invertebrates. However, results are not always repeatable or understandable, in part because MTs respond to starvation and oxidative stress, and also because induction can be downregulated or inhibited by factors such as handling and caging stress. Hence, MT should "never be used alone without another indicator of cellular damage, primarily oxidative damage" (Schlenk and DiGiulio, Chapter 2). Metallothioneins have been linked to models of metal metabolism and toxicity, but many aspects of their role in biology and toxicity remain poorly understood. Several analytical methods for MT are available, but they are often technically difficult, require radioisotopes, or are relatively expensive.

Vitellogenin (VG) concentrations in male fish plasma, as indicated by antibody or mRNA assays, is emerging as a very reliable biomarker of exposure to estrogenic compounds. It is relatively specific for a class of effects (estrogenic effects or feminization) but somewhat less specific for a class of stressors; that is, many chemicals cause the same effect. The method is highly plausible in that male fish have the gene for VG synthesis but do not normally express it unless exposed to estrogen. The VG response is a highly sensitive and reliable predictor of exposure to estrogenic compounds, as demonstrated in a variety of laboratory studies and an increasing number of field studies. However, the response is somewhat less relevant to effects, in that the toxicological and ecological consequences of elevated VG concentrations are not yet well understood. Nevertheless, a VG response precedes reproductive impairment in chronic toxicity studies with experimentally exposed fish, so it may have ecological relevance. Accurate measurement of VG requires relatively sophisticated techniques, special training, and special equipment. There have been few interlaboratory comparisons of methods and results, but there is a growing database on factors that affect the response. The results are clear and unequivocal and can be readily interpreted.

While a relatively new biomarker, VG could be added to monitoring programs if methods can be simplified.

Genotoxicity is detectable by a wide array of techniques that measure the integrity of DNA. However, because there are many compounds that can damage DNA, either as the parent material, as a metabolite, or by generating reactive oxygen species, cause–effect is often difficult to discern without some indicator of exposure. Possible exceptions are the adducts of PAH, which can be detected by their fluorescence characteristics. For PAH, there is a well-defined model relating PAH metabolism to the formation of adducts and the disruption of normal cell replication. Numerous laboratory and field studies of fish populations sampled near a source of PAH demonstrated a high prevalence of cancer and high concentrations of DNA adducts. However, the most commonly applied analytical techniques (e.g., ^{32}P-postlabeling) are often difficult and expensive and require highly specialized personnel and sophisticated equipment for measurement and interpretation. Some tools (e.g., cytogenetic analysis) may be relatively routine in human health monitoring, but there is little information for other species about "normal" and about factors affecting responses. Hence, methods for detecting adducts and relating them to chemical exposure and to the development of cancer tend to be restricted to research laboratories.

Porphyrins are the product of heme destruction, a normal part of the turnover of red blood cells and any heme-containing protein. Production rates can be enhanced by disease and by oxidative stress (e.g., following CYP1A induction). While a relatively nonspecific response, they have been proposed as indicators of exposure to chlorinated hydrocarbons, pulp mill effluents, and PAHs, based on laboratory and field studies of fish and birds from contaminated sites. The diagnostic strength of porphyrin analysis could be improved by measuring physiological covariates (e.g., bilirubin, heme), but, by itself, porphyrin analysis is more an indicator of organism health than of chemical exposure or ecological effects. "Little is known about baseline values and what effects species, gender, development, or season may have on production" (Schlenk and DiGiulio, Chapter 2). Analysis is usually by high-pressure liquid chromatography with a fluorescence detector (HPLC-fluorescence), which is within the capacity of most laboratories

A third group of biochemical responses is still at the idea stage, and these responses have not been widely tested or developed. *P-glycoproteins* are chemical pumps that move compounds across cell membranes and are rapidly induced by many xenobiotics. They are easily measured by fluorescent dye transport and can be inhibited by the drug Verapamil, although these measurements require intact cells and may not be suitable for widespread application to samples from the field. They respond to environmentally relevant concentrations of chemicals, are associated with the prevalence of neoplasms, and show homology among species. However, they show a lack of selectivity among metals and organics and do not provide a specific test of causal relationships by themselves, although they are sensitive to any compound affecting ATP synthesis.

Retinoic acid (vitamin A) and its various metabolites may be depleted or changed in relative concentrations in the tissues of fish and birds following exposure to chlorinated aromatics and PAH. Their concentrations are correlated to CYP1A induction and the prevalence of embryonic malformations, so there is some specificity and relevance to both chemical exposure and ecological effects. They can be measured by HPLC, but little is known about thresholds, baseline values, and the role that these changes play in the mechanisms of toxicity. While some laboratory and field studies of fish and wildlife have identified changes in retinoid profiles that are correlated to chemical contamination, the interpretation of these results is not well developed and requires considerable expertise.

Gene expression is an alternate method for assessing genotoxicity. As reviewed above, one of the best understood biomarkers is the induction of the *cyp1A* gene. The measurement of mRNA to demonstrate gene expression offers additional advantages of greater sensitivity (greater fold induction relative to CYP1A protein concentrations or enzyme activity). Many factors have been identified that can modulate expression, thereby increasing the specificity of this biomarker if appropriate covariates are monitored. The opposite is true for p-glycoprotein. Although mRNA provides a means of assessing gene expression, there are few data on its application to monitoring and some major problems with specificity and confounding factors. There may also be technical barriers to application of mRNA measurements to previously untested, indigenous species.

Overall, biochemical and molecular biomarkers hold great promise for relating cause and effect, but their weakness is often nonspecificity for effects of a cause and a low relevance to ecological effects. While most have been demonstrated in fish species, much less is known about invertebrates, which limits the choice of indicator species. The responses are generally easily understood. However, except for CYP1A induction and biliary FACs, there have been too few laboratory or field studies to establish relevance to chemical exposure and ecological effects and to identify covariates that might affect interpretation. Hence, many of these methods are still at the research stage.

Physiological and Condition Indices (Chapter 4)

Measurements of the physiological status of an organism indicate more about the health of the organism than about the determinants of health. Physiological and condition indices are very nonspecific indicators of fish health, particularly as it relates to stress. However, environmental relevance increases from primary to secondary to tertiary stress responses (i.e., changes in fish performance), with tertiary responses providing a functional link between physiological stress and population-level impacts. Therefore, physiological measurements represent a highly relevant and plausible transition from biomarkers (highly specific indicators of chemical exposure and toxicity) to bioindicators (measures of actual impacts on populations, communities, and ecosystems). However, because of their nonspecificity for cause, there are many uncertainties about interpretation, and the large numbers of environmental and biological factors that can influence a response have generated inconsistencies among studies. Hence, the diagnostic capacity of physiological measurements can be strengthened by measuring covariates (age, sex, size, season, habitat characteristics, etc.) and the temporal and spatial distribution of responses and stressors.

Methods for physiological measures (hematology, brain neurotransmitters, heat shock proteins, stress hormones, measures of energy storage and depletion, etc.) are generally well developed, inexpensive, and easily applied, and some have been subjected to interlaboratory comparisons. However, somatic indices are relatively insensitive, responding slowly and at relatively high levels of a stressor. Overall, physiological measures are suitable for rapidly assessing fish health and for providing answers that are understandable by nonscientists. They can be incorporated into biomonitoring programs but only if a well thought-out array of covariates is included in the design to aid in interpretation.

Genetic Responses (Chapter 5)

Genetic responses, like physiological stress responses, provide information about the health of the organism and are relatively nonspecific. Their value is their relevance to population-level impacts. An array of established and emerging technologies is available to assess chemical effects on heterozygosity and changes in genetic structure within and between populations. Genetic responses can provide earlier warning of effects and require less sampling effort than population or community analysis. However, they also respond to natural causes that create selective pressures, such as migration (e.g., dams affecting fish movement and gene flow). Stressors may reduce diversity (temporary reduction in population size, e.g., fish kills), increase mutation rates (persistent mutations are usually rare, lethal, and recessive), and alter selective pressures (select for tolerant organisms). To detect stressor effects with a nonspecific genetic signal, multiple exposed and reference populations must be sampled, and hypotheses of stressor-induced genetic damage must be supported with additional lines of evidence to establish biological and mechanistic plausibility (e.g., changes in fecundity or survival combined with indices of stressor exposure).

Of the genetic tools presented, most come from medical research and few have been applied to laboratory or field studies of fish or invertebrates. Hence, the results may appear inconsistent, and it is difficult to judge if proposed cause–effect associations are plausible. Plausibility is also weak where the molecular basis for banding in electrophoretic gels is unknown (e.g., randomly amplified polymorphic DNA/amplified fragment length polymorphisms [RAPD/AFLP]). Sensitivity can be relatively low (e.g., allozymes) when the genetic structure of a population changes following fish kills or reproductive failure or very high where there is a direct measure of genetic mutations (microsatellites). Due to polymerase chain reaction gene amplification, most methods require only very small sample sizes of blood or tissue and allow retrospective analysis of stored tissues. For some methods, automation is available to reduce costs and to increase reliability. However, many methods require specialized facilities such as radioisotope laboratories, are relatively complex, or require expensive equipment, and the analysis of data and interpretation of results can be speculative and difficult to understand. Results may be sensitive to the conditions of the assay (e.g., RAPD/AFLP), so that rigorous quality control is required, and there is a need for species-specific primers for some methods (e.g., microsatellites). Hence, there have been few environmental studies to date that incorporate genetic diversity indices and few interlaboratory comparisons of methods. Because of the rapid evolution of methods in molecular genetics, this class of response holds great promise for monitoring, and emerging technologies such as gene array chips may provide screening tests for prioritizing study sites. Overall, genetic responses show a great potential as useful bioindicators, but, at the moment, they are primarily research tools.

Immunology (Chapter 6)

Measurements of the immunocompetence of organisms assess their capacity to resist disease, assuming that an impairment of immunity results in an increased risk of disease. Hence, immune responses may be relevant indicators of organism and population health and may help define ecological impacts of different stressors. The mechanisms by which immunity is affected by sexual maturation, social interactions, aquaculture conditions, exposure to pathogens, and alterations in food supply and habitat quality are not yet clear. There is little specificity of cause for effect, although immunosuppression has been associated with increased susceptibility to predation and increased mortality rates of fish exposed to pathogens or to chemical pollutants.

Immune dysfunction may be a primary response to chemical toxicity or a secondary response to effects on some other physiological system, so that biological and toxicological plausibility is often weak. There have been many substances consistently associated with immune responses, among them PAH, chlorinated hydrocarbons, pulp mill effluents, metals, and pesticides, but toxicity varies among compounds and species, and mechanisms are rarely understood. A variety of field studies have shown immune responses in organisms exposed to chemicals, and some have demonstrated corresponding changes in susceptibility to disease. However, while field studies in urban areas have emphasized chemical pollutants, there has been little attention to the role of other stressors such as sewage and associated pathogens.

While many methods are available to assess immune responses, "no single assay, regardless of expense or complexity, is able to assess total immune function in the host" (Rice and Arkoosh, Chapter 6). It was recommended that methods be applied in tiers, with Tier I tests being simple, low cost, and requiring little training. While the methods are reasonably well developed and standardized, it is unclear whether Tier I responses predict risks of poor health or mortality. Tier II tests are more complex but examine specific targets in the immune system and provide information about possible mechanisms. This approach was designed originally for mammals, and there is a lack of reagents (e.g., antibodies) for fish species due to poor interspecies cross-reactivity. Tier II methods are more complex and difficult to interpret, and there is less understanding of the role of many covariates. "Immunology is a complicated discipline requiring significant education and training to understand the implications of immunomodulation by environmental stress" (Rice and Arkoosh, Chapter 6). Overall, Tier I tests could be applied in biomonitoring, as has been demonstrated in Puget Sound, but Tier II tests are still more of a subject for research than for monitoring.

Histopathology (Chapter 7)

Histology is a tool that provides a direct measure of the actual state of tissues and of the organism and is a direct measure of injury. Histology can target specific tissues and even proteins within tissues, so there is a potential for great specificity for effects, and histopathology can be quite predictive of disease and mortality. Observed effects are often quite consistent in laboratory and field studies, but many natural and methodological factors can affect the appearance of tissues and the diagnosis of pathology. In contrast, specificity for cause is generally low, because few lesions are uniquely distinctive of a disease, although patterns of change can form syndromes. In fish, the most studied tissue is the liver due to its central role in the metabolism and excretion of a large number of compounds causing cell proliferation and cancers. Kidney and gills show obvious changes in architecture with pathology, but causes are numerous and analyses of suspected stressors are needed to support cause–effect relationships. Chloride cell proliferation may indicate acidification, reduced alkalinity, and ion loss, but gills are particularly subject to histological artifacts. Skin pathology is confounded by diseases and parasites but is not seen in all species, although pigment cell neoplasms may be specific to pulp and paper effluents in marine systems. Skeletal deformities have many known causes with some etiologies well understood (e.g., cadmium, lead, pesticides, diet, parasites, electrical shock), but again, there are many confounding factors (diet, age). Fin erosion lacks specificity, and there are few studies of the brain, nervous system, and eye, although, potentially, they may be quite sensitive.

In field studies, tissue pathology is quite common but the causes and etiologies are unclear (e.g., macrophage aggregates, oocyte atresia). Other conditions, such as intersex, ovarian maturation, and testicular abnormalities, may be specific to a broad class of endocrine disruptors such as environmental estrogens. Many case studies have successfully related cancer in fish to exposure to PAH. In all cases, conclusions about cause and effect and the role of procarcinogens were strengthened considerably by measuring biomarkers that demonstrated the co-occurrence of all stages of a chemical-carcinogen model. Histological evidence of cancer also has been correlated to increased mortality rates of older fish and changes in population structure, and chemical carcinogenesis has also been replicated in laboratory studies of sediment extracts and contaminated sediments.

Methods for histology and for interpreting pathology are quite well standardized, although interpretation requires considerable judgment, training, and expertise. "Slides should be read by an experienced fish histopathologist with formal training in human and veterinary or comparative pathology"; "each slide reader must complete a training period of 3–9 months under the teaching and supervision of the chief histopathologist before the diagnostic data from that slide reader begins to be incorporated into the database" (Myers and Fournie, Chapter 7). Results are often subjective and nonnumerical so that standardized criteria for classifying "normal" and "pathological" are needed, as well as blind sampling of slides. Many texts and atlases are now available, and numerical indices for proportions, cell proliferation, cellular localization of proteins, intensity and area of staining, and relative cell dimensions have been proposed. Some automation is possible, but image analysis may still be too labor intensive for large scale monitoring. For statistically valid data, costs are affected by the numbers of fish, tissues per fish, and sections per tissue needed, and study designs should evaluate the advantages of assessing a few critical tissues compared with a complete necropsy.

Overall, histopathology is a mature discipline that can be applied effectively in biomonitoring and bioassessment, particularly when combined with other biomarkers or bioindicators. As the authors state: "When effectively used in integrated studies involving multiple disciplines such as analytical chemistry, biochemistry, immunology, reproductive biology, and fish behavior, one can often identify the general stressors (e.g., classes of chemical contaminants) that are the likely etiological agents for the lesions detected" (Myers and Fournie, Chapter 7).

Energetics (Chapter 8)

Energy is the common currency of life. The application of energetics to bioassessment and biomonitoring is derived from the general adaptation syndrome (see also Chapter 4) in which the metabolic or energetic costs of alarm, resistance, and exhaustion have consequences for survival, growth, and reproduction. In this perspective, stress is "the change in the energy budget resulting from the combined effects of specific and nonspecific responses to a stressor" (Beyers et al. 1999, cited by Beyers and Rice, Chapter 8). Energy budgets may be affected indirectly (e.g., increased energy cost of tissue repair) or directly (e.g., through impairments in feed consumption, appetite, conversion efficiency, growth, and lipid metabolism).

For species subjected to different environmental conditions and stressors, bioenergetics models predict changes in energy intake and allocation to digestion, maintenance, reproduction, growth, and excretion. The application of energetics models to biomonitoring requires model development from laboratory studies of energetics in experimentally stressed fish and from studies of rates of survival, growth, and reproduction of populations in the field. Data must be collected to match the life stage and the spatial and temporal distribution of the target species. Models must be thoroughly tested (e.g., sensitivity analysis) over the range of expected environmental variables to ensure a reasonable accuracy of predictions, a process that could be expensive and time consuming. Data availability is a critical issue, given the species and site-specific information needed to run such models. Energetics may also appear as a trivial response when the primary effect is cancer, which could occur well in advance of changes in ecosystem energetics. In contrast, recruitment success of a spring-spawning species at the northern edge of its range may depend on stored energy after overwintering of reproductive age fish.

A major advantage of energetics modeling is the integration of effects across different levels of organization and among different natural and anthropogenic stressors. However, the strength of association between cause and effect is difficult to judge, as is consistency on replication. While actual measurements of energetics are widely applied in research and fisheries management, each ecosystem is unique, and studies are not often replicated. Model simulations provide an estimate of the magnitude and pattern of responses of fish to different stressor scenarios, but there are few applications to real situations. Because models are simplifications of reality, they are best used to generate hypotheses to drive the design of biomonitoring programs.

In general, energetics can be used to establish time order, in that responses will follow stress, and recovery will follow removal of stress. However, energetics may integrate stressors slowly over long periods of time, so time order may not be readily apparent. Hence, energetics responses do not appear very sensitive and provide no early warning of stress. The extent of energetics responses increases with severity of exposure to stressors, but many mitigating factors unique to each ecosystem influence the magnitude and direction of responses and create the need for modeling to understand their relative importance. However, more data are needed for species and site-specific models. Generating this data requires fairly complex laboratory studies and the involvement of expert fish physiologists. While the importance of changes in growth of organisms is easy to understand, other endpoints will require interpretation by experts.

In summary, the application of energetics and energetics models to biomonitoring design has some good precedents in fisheries management. In well-studied ecosystems, an energetics approach could be applied directly or added to existing biomonitoring programs. For less well-studied ecosystems and for species without the requisite specific model data, considerable research will be needed before energetics can be applied routinely.

Population-Level Indicators (Chapter 10)

Changes in the structure and function of populations are highly relevant to ecological and socioeconomic impacts of stressors. The status of a population represents the integration of many natural and anthropogenic stressors acting simultaneously, modified by compensatory responses of the population and its associated aquatic community. Hence, the whole is greater than the sum of its parts, and the unique characteristics of populations in different ecosystems are shaped by the diversity of variables experienced by each population. As a consequence, there are no direct or simple relationships between responses of individuals and of populations. This justifies monitoring populations rather than simply predicting population effects from organism-level responses. However, Power (Chapter 10) states that "Understanding the responses of populations to stressors, therefore, requires knowledge of both the ways in which individuals respond to stress and the effect of collective individual responses on the processes that govern population dynamics."

The compensatory responses of populations to stress are controlled by density-dependent and density-independent factors that often interact. The capacity of a population to compensate for stress can be a valuable indicator of its resistance to stressors. However, it is often difficult to distinguish natural variability from stressor effects when life-history strategies are adapted in ways that dampen the effects of stressors. Hence, population monitoring indicates the state of the population and of the ecosystem, but population responses have little inherent specificity for a stressor, except that which is implied through sampling design. The stressor of interest may not be the only factor separating reference and exposed populations, and the nature and intensity of natural stressors may vary with time.

There are several possible designs for biomonitoring populations. *Compare and contrast* designs (exposed versus reference populations) measure life-history traits (e.g., mean population age, fecundity, condition factor) to indicate changes associated with a known stressor. It is the least expensive approach because a few sites are visited only once and sample sizes are small. The approach is static, retrospective, assumes no migration, and the selection of reference sites is critical. This approach cannot discriminate between multiyear classes and may be biased by changes in year-class strength. Measures of growth and reproduction are not independent variables, and impacts can be masked by compensatory responses. *Exposure–response* designs sample population indicators over gradients of a stressor, and large sample sizes and regression analysis provide statistical power to relate cause and effect. To avoid bias, other natural and anthropogenic stressor gradients must also be described. The cost of this approach is affected by multiple sample sites and endpoints. *Sequential sampling* designs assess populations before and after a stressor is applied or removed. Experience with whole-lake manipulation of pH and nutrients in the Experimental Lakes Area of Ontario, Canada, demonstrated that effects on fish communities "must be tracked over considerable periods of time if a true picture of population-level responses to stress is to emerge" (Power, Chapter 10). The frequency of sampling must also be tailored to the anticipated rate of change. These designs require a long-term commitment to sampling, which can be a serious weakness.

All three sampling designs are inherently insensitive, particularly with respect to early warning. The time lags between exposure and response may obscure cause–effect relationships but may also provide characteristic patterns of response that could be modeled. The elements of a response correspond to alarm, resistance, and extinction, and if models can build on characteristic sequences of response, they could provide logical and useful frameworks for monitoring design. Considerable research will be needed to demonstrate if this approach can be applied.

Variability in population characteristics can also obscure patterns of response. However, variance itself is a characteristic that can be monitored, so that patterns of variance outside normative bounds can indicate that a stress is acting. This approach would be very data intensive and best applied retrospectively to long-time series. More research using existing long-term data on fish and wildlife populations is needed to demonstrate the best indices of variance and how they might be applied to detect change.

Population modeling to describe patterns of response and potential cause–effect relationships is a less direct but less expensive and faster way to describe the status of populations than actually collecting data. Sensitivity analysis can be used to identify vulnerable life stages, to rank stressors, and to compare mitigative strategies, assuming that the data and the model are comprehensive. Population models encompass different scales of complexity and reality but all require significant amounts of data to generate predictions, and predictions, in turn, should be verified with additional data. Modeling is not a biomonitoring tool by itself but can identify hypotheses that will direct biomonitoring design and sampling.

For all population monitoring, strength of associations between cause and effect is often weak due to confounding factors acting on the population, particularly for the compare and contrast approach. There is greater strength of association for exposure gradient and sequential studies due to the nature of their study designs. Consistency of population responses on replication tends to be poor because each ecosystem and population is different. However, the compare and contrast approach of Canada's Environmental Effects Monitoring (EEM) program for the pulp and paper industry (EC and DFO 1992a, 1992b) is rapidly building a large database that will provide more information about consistency. The plausibility of proposed cause-and-effect mechanisms is generally weak in population studies, but there are some exceptions, such as reproductive impacts predicted from exposure to endocrine disruptors. In contrast, ecological plausibility is high because population-level changes can be linked to important ecological effects.

Sampling and analytical protocols for fish population monitoring are well established due to their widespread application in fisheries management, and Canada's EEM program (EC and DFO 1992a, 1992b) is highlighting quality control and descriptions of "normal." Covariates that affect population responses are also well-known, although that knowledge is not always applied. In the compare and contrast approach, study designs limit the extent to which covariates are assessed to keep costs and complexity down, and the trade-off is increased uncertainty about interpretation. Exposure–response, sequential, and variance studies are more data intensive and expensive than compare and contrast designs, while modeling is potentially less expensive but can be highly complex. Where modeling is integrated with monitoring and requires real data, it can become quite expensive to do well. The results of population monitoring are usually easily understood and converted to indices, although outputs of some models (predictions of stock abundance, recommended catch rates) seem to be controversial. Variance may be a difficult concept to incorporate in indices, and there are too few examples to illustrate the best means of data presentation.

Overall, population monitoring is well established in fisheries management. However, the uncertainties of fisheries management (e.g., of Atlantic cod *Gadus morhua* stocks) suggest that much research remains to provide stronger links between responses at lower and higher levels of organization.

Behavioral Indices (Chapter 11)

This chapter makes a good case for the relevance of behavior to the well-being of individuals, populations, and ecosystems. Intuitively, disturbance of such behavioral traits as migration, habitat selection, competition, predator–prey relationships, and reproduction should have important ecological consequences. The approaches reviewed are primarily laboratory-based bioassays of different behavioral endpoints, although there are a few notable field studies demonstrating chemical or pollution effects on fish behavior. Behavioral responses can be rapid and sensitive if fish are healthy but may be less predictive if fish are diseased or affected by compensatory reactions such as acclimation to the test chamber. Behavior can be correlated to physiological responses in some cases, with an established etiology relevant to a specific chemical stressor (e.g., cholinesterase inhibition or other neurological responses to toxicity). However, most responses are nonspecific and simply measure the "state" of the organism, and it is often difficult to connect laboratory and field responses. Hence, there are few reports linking behavior to actual ecological impacts.

Some standardized procedures for behavior testing are available, including two guidelines published by the American Society of Testing and Materials. Recent advances in computer technology have reduced the costs of laboratory studies, but, in field studies, physical constraints, labor requirements, and movements of organisms make observations more difficult. Behavioral studies should include related measurements of organism performance, environmental variables, and stressor exposure. A number of covariates may affect results, such as methodology (handling of organisms, health and nutritional status, prior exposure to other stressors, apparatus design, acclimation to environmental variables and test substances) and environmental and biological variables (strain, social status, and species, age, sex). The most advanced of the methods are tests of chemical avoidance, in both the laboratory and field. In the field, avoidance can be measured by movement of fish through traps, or by telemetry, to track the reaction of individual fish to the presence of a plume, contaminated site, or other stressor. Nevertheless, a weakness in behavior monitoring is quantifying changes; it is also labor intensive and often subjective.

Other behavioral indicators include feeding behavior, which is relatively sensitive to stressors and partially standardized. Predator avoidance, social behavior (shoaling, aggregations, schooling, territorial and hierarchical responses), swimming behavior (pattern analysis of hyperactivity, periodicity of activity, duration of swimming and convulsions, and swim tunnel performance), and reproductive behavior (spawning migrations, territoriality, schooling, nest building, courtship, spawning, nest defense, and parental care) all respond to chemical exposure. However, with the exception of nesting behavior, there have been few field studies to demonstrate how such behaviors might be monitored. Overall, avoidance is well established as a useful method in laboratory studies of potential stressors and in field studies of the distribution and movements relative to potential stressors. Other methods, however, are still at the research stage.

Community-Level Indicators (Chapter 12)

Community-level indicators measure the state of an ecosystem. They are metrics that describe the diversity and relative abundance of different species in an ecosystem or segment of an ecosystem in relation to natural and anthropogenic stressors. As with population monitoring, the properties of the whole cannot be readily predicted from the characteristics of its parts; interactions create new emergent and unpredictable properties. "The emergent properties resulting from biological hierarchies prevent prediction of community effects from lower levels of organization, so all biomarker studies should be validated against parallel community-level investigations" (Attrill, Chapter 12).

As Attrill points out (Chapter 12), communities are not just one taxa; they include all species, plant and animal, and any restriction is an artificial abstraction. Therefore, community analyses are, by definition, broad-scale and complex. Narrowing the scope of a study by selecting a segment of an ecosystem can make monitoring more practical, but the risk is that other community-level effects may be overlooked. Nevertheless, Attrill recommends monitoring benthic invertebrate communities in aquatic ecosystems because they provide the greatest amount of information about the state of aquatic communities for a given level of effort. Another simplifying approach to monitoring communities is to reduce the level of taxonomic resolution. This reduces complexity, cost, and degree of expertise needed, with a trade-off of a potential loss of sensitivity. Sampling designs are critical in community surveys and require adequate spatial and temporal replication, including multiple reference sites and relevant supporting environmental measurements. Knowledge of the effect of natural variables can assist in the statistical discrimination of stressor effects.

There are three categories of community indices: univariate statistics (total abundance, diversity index, biotic index), distribution or graphical approaches (k-dominance, ABC curves, biomass size spectra), and multivariate analyses (cluster analysis, principle components analysis). None of the community methods are specific to any stressor. As general indicators of the status of ecosystems, they provide evidence of cause only through appropriate sampling designs (location of sampling sites, measurements of environmental covariables) that generate patterns of response associated with a cause. If community monitoring is properly designed, the strength of association between cause and effect can be quite high, particularly for multivariate approaches where sufficient information was collected to segregate stressor effects from effects of natural variables. Consistency on replication tends to be poor because each ecosystem is different. There may also be considerable observer bias, so that results from one laboratory are not easily compared with results from another. However, some indices and approaches consistently produce good results, suggesting that the methods are robust. Community indices have little plausibility for cause–effect mechanisms, but considerable biological or ecological plausibility, particularly where the presence or absence of indicator species reinforce associations between cause and effect (e.g., mayfly sensitivity to hypoxia). Community indices are strengthened by time-order, in that degradation of ecosystems can be detected following a pollution stressor, and recovery follows removal of the stressor. The extent of response increases with the severity of exposure, but exposure–response relationships may be modified by many natural factors unique to each ecosystem. Nevertheless, community indicators have low sensitivity, are slow to respond, and provide no early warning of impending environmental effects of a stressor.

Methods for community indices are well established and are widely and routinely applied. Many options are available for sampling, taxonomy, and statistical analysis, and there is abundant knowledge about environmental covariates that affect community-level responses. Monitoring with community indexes can be quite complex and expensive. It requires expertise in taxonomy and statistics, large data sets, and, in some cases, extensive computer power. Cost and effort are least for univariate indices and greatest for multivariate approaches. Simple univariate indices are also the easiest to understand and express, but perhaps the least informative. Multivariate results can often be quite complex and difficult to understand (e.g., principle components analysis) and often require experts to interpret the data for the public and for environmental managers. Overall, univariate indices are well established and equivalent to "off-the-shelf" methods for biomonitoring. The other approaches are not far behind and catching up rapidly due to new computing power.

Integration-Landscape Level (Chapter 15)

Examination of aquatic ecosystems on large scales provides an alternative, indirect approach to assessing effects of stressors on populations and communities. This approach describes physical habitat and defines the available area of specific habitat type or quality as a basis for estimating production of different species having those specific habitat requirements. By comparing observed fish production to reference conditions derived from "theory," this approach can identify populations under stress from forces unrelated to habitat. This approach also allows the testing of alternate hypotheses about cause and effects by identifying physical habitat characteristics that are not responsible for observed effects. Several examples where this approach was used successfully include the habitat suitability index, and the river habitat survey (U.K.). This approach is less successful where there is a nonlinear relationship between habitat and fish production. In general, these approaches are unable to consider the juxtaposition of habitat components, they lack stochastic and temporal resolution, and they are not supported by sufficient statistical methodology. A newer approach is the spatially explicit species index (SESI), which incorporates temporal change. However, SESI models rely heavily on remote-sensing data and require considerable effort in ground-truthing. There are fewer examples where this approach has been applied.

Spatially explicit species index models cannot test cause–effect relationships directly, but instead rely on a process of elimination. They are excellent tools for understanding how physical habitat features affect fish populations and communities but are indirect in that they predict productivity without actually measuring it. Therefore, they are not a bioindicator of the state of the ecosystem so much as a management tool for comparing different management scenarios that can affect physical habitat.

Biomarkers and Indicators of Reproductive Effects (Chapter 9)

Chapter 9 in this book stands out in dealing with a single class of effects: impairment of reproduction. This issue arises from a concern for compounds classed as "endocrine disruptors." They have captured the public's attention because of their potential impacts on the reproduction and development of human beings and of other species that might act as sentinels for human beings. Chapter 9 is presented at the end of this review because it is an example of how one class of effects that has great ecological relevance can be investigated in an integrated way from the molecular to the population level in fish, and it provides a useful model for other classes of effects. Highly specific biomarkers can be used to investigate molecular, cellular, or physiological dysfunctions induced by known chemical agents (e.g., disrupted regulation of steroid hormones by estrogenic compounds), while well-established bioindicators can be used to describe the reproductive capacity of populations (e.g., fecundity).

There are a variety of biomarkers of chemical toxicity and effects for specific aspects of reproductive physiology, and many are relatively new to monitoring and assessment. Nevertheless, serum steroid sex hormone concentrations have proven to be reliable indicators of pulp mill effluent effects on fish reproductive function. While vitellogenin concentrations in the serum of male fish (reviewed above) is a particularly sensitive indicator of exposure to compounds that mimic estrogen and cause disruptions in sexual maturation. Hormone concentrations can be measured by well-established techniques derived from human medicine for the common hormones such as estradiol. However, for fish-specific hormones such as 11-ketotestosterone, analytical methods must often be developed, often on a species-by-species basis. Methods for measuring serum VG concentrations are relatively new, and improvements appear frequently, providing greater sensitivity and reliability, simply because there are so many laboratories with an interest in VG as a biomarker. Many of these methods have been subjected to recent interlaboratory comparisons and are considered quite reliable.

As the mechanisms of chemical toxicity to reproductive systems are explored in greater detail, new biomarkers become evident. These include enzymes involved in steroid hormone synthesis, hormones that control reproductive cycles (e.g., gonadotropins), and concentrations of hormone receptors. However, they are not yet widely applied in fish toxicology, and analytical methods are much less developed and tested, and may be expensive or require special techniques and expertise. Other new methods, such as the measurement of the concentrations of zona radiata proteins, protein expression, and gene expression, are very promising but still at an early stage of development in which the responses to environmental stressors such as chemical exposures are being determined.

At the physiological and population level, methods to assess histopathology, oocyte atresia, egg size, gonad somatic indices, age and size at maturity, sex ratios, and fecundity are all well established in fisheries science. They are relatively inexpensive, and widely applied in both research and monitoring of stressors such as pulp mill effluent (e.g., Munkittrick 1992; EC and DFO 1992a, 1992b). When applied in a systematic way, and in combination with biomarkers of chemical exposure, these parameters can be used to distinguish the effects of gradients of pulp mill effluent on the reproductive capacity of fish from the effects of natural gradients of ecosystem characteristics (e.g., Gagnon et al. 1995). Other bioindicators are less well developed, specifically the measurement of fertility, sexual behavior, and development. For histopathology, responses are widely recognized but causes are not well worked out. While these indicators may be highly relevant and sensitive, there has been too little experience with their measurement to recommend them for widespread application in monitoring and assessment.

In any study of stressors, Greeley (Chapter 9) has wisely recommended that multiple indicators be used to compare reproductive status among exposed and unexposed populations. There should be careful standardization of techniques and selection of reference conditions, and seasonality of reproductive cycles must be recognized. Finally data should be examined with statistical analyses that emphasize pattern recognition (e.g., discriminant function analyses) to create a weight-of-evidence approach to establishing cause and effect. There have been many environmental factors identified that affect reproductive performance (temperature, pH, metals, organic chemicals), and, for some, there may be specific patterns of response that would suggest a cause. For the wide array of reproductive indices reviewed, there is the additional advantage that the hierarchy of responses from the molecular to the population level can provide mechanistic explanations of proposed associations between cause and effect.

Research Needs and Recommendations

1. For all biomonitoring and assessment methods reviewed, there is a critical need for field, hatchery, and laboratory studies that link responses at one level of organization to another. Links to higher levels are needed to understand the consequences of stressors and to lower levels to understand the mechanisms by which effects are propagated: "Current meth-

ods for extrapolating individual-level effects on fish to levels of biological organization higher than the population are not developed sufficiently to permit realistic assessments" (Johnson and Collier, Chapter 14). This need represents a current challenge to monitoring and a very active area of research.

a. There is a need to develop laboratory toxicity tests at the organismal level within the context of population models so that appropriate methods and data can link individual responses to population responses.

b. There is a need to further develop population-level techniques and to integrate them with community-level measures. Monitoring must go beyond abundance to more complex interrelations of species distribution. The success of population monitoring will depend on a better understanding of both the ecosystem and populations under management.

2. There is a critical need to better characterize exposure to stressors so that the nature and intensity of exposure can be used to predict the nature and intensity of effects. For example, chemical exposure must be characterized according to environmental concentrations, exposure–response curves, thresholds, and internal doses. The role of metabolism and complexation must be understood to define the active agent and its dose at critical receptor sites.

3. Research is needed to standardize methods and develop quality assurance–quality control procedures. These include interlaboratory comparisons, "solid-gold" standards, standard reference materials, internal performance standards unique to each analytical laboratory, operational detection limits, anticipated variance, and minimum sample sizes (amount of tissue and numbers of organisms) needed to achieve detection limits and to establish statistical significance for comparisons among sites.

4. Interpretive documents are needed for each biomarker and bioindicator that provide

a. ranges of normal, partitioned according to major biological (species, age, sex, maturity) and environmental (temperature, season, salinity, etc.) variables.

b. a review of pathogenic factors that cause deviations from normal, the direction of deviation, and the extent of deviation.

c. an interpretation of response in terms of possible ecological consequences, expressed as a risk where possible, and of possible causes and associated mechanisms.

d. the types of covariates that should be measured to assist with this interpretation.

5. There should be increased interactions between modeling and monitoring to provide iterative feedback that will improve both. This will require substantial interdisciplinary interactions.

6. Many of the biomarkers and bioindicators and much of the research on monitoring and assessment have been initiated by researchers generally unconnected with monitoring programs. In effect, they have developed "tools looking for an application," when monitoring and assessment programs often have applications that are in great need of appropriate tools. Methods development without any connection to users, in the absence of any programs or objectives for biomonitoring, and without guidance on what is needed for good methods, may be wasted effort. The most productive research will arise from interactions between the scientists responsible for monitoring and assessment, and those with the opportunity to do research on biomarkers, bioindicators, and mechanisms.

   There is a need to build on existing population-monitoring programs. An increased number of interdisciplinary studies could have a more-than-additive enhancement of the value of each component. Advantage must be taken of mandated monitoring programs to analyze the data from methodological perspectives and to use very large databases to develop models that improve our predictive capacity.

7. Throughout this book, and inherent in this list of research needs, is a clear need for a more thorough knowledge of the structure and function of aquatic ecosystems. Without knowledge of the biology of aquatic species, their adaptations to the innumerable niches available in aquatic ecosystems, and the physical, chemical and biological processes that govern the distribution of species and the stressors that act upon them, the selection and interpretation of biomarkers and bioindicators is severely limited. The promise of new monitoring methods will only be realized when we can interpret the signals we measure in terms of predictable changes in the state of aquatic ecosystems and the benefits we derive from them. If this goal can be realized, government agencies will be more willing to pay the costs of monitoring and assessment and to take action to eliminate stressors or to reduce their impacts.

Summary

This book has provided a broad perspective and excellent conceptual framework for the biological monitoring and assessment of stress in fish. The development of biomarkers and bioindicators is a very active area of research, as scientists seek appropriate tools to describe and understand the impacts of anthropogenic stressors on aquatic ecosystems. This field of study brings together a wide array of disciplines, and the greatest challenge to scientists and research managers is to find ways of integrating observations, understanding, and theory among disciplines and across different levels of biological organization. The understanding of mechanisms of toxicity is advancing very rapidly, particularly at the molecular level and in areas highly relevant to human health (e.g., endocrine disruption). The development and application of bioindicators to understand the ecological consequences of stressors appear to be developing somewhat more slowly. Research on bioindicators is building on established fisheries techniques that describe the structure and productivity of fish populations, and on the principles of epidemiology developed from human health management. Progress is slow, however, due to the great complexity of natural ecosystems, the size and cost of population and ecological studies, and the difficulty in applying reductionist approaches to natural systems. Progress in linking observations at one level of organization to another is limited, so that biomonitoring and assessment may not attain the predictive capability needed to link bioindicators of ecological effects to specific causes, and to understand the relevance and early warning signals derived from biomarkers of stress in individuals. This is the challenge for the next 10 years, and, hopefully, the area of greatest advance for the next edition of this book.

References

Adams, S. M., M. S. Greeley, and M. G. Ryan. 2000. Evaluating effects of contaminants on fish health at multiple levels of biological organization: extrapolation from lower to higher levels. Human and Ecological Risk Assessment 6:15–27.

Baumann, P. C., I. R. Smith, and C. D. Metcalfe. 1996. Linkages between chemical contaminants and tumors in benthic Great Lakes fish. Journal of Great Lakes Research 22:131–152.

Beyers, D. W., J. A. Rice, W. H. Clements, and C. J. Henry. 1999. Estimating physiological cost of chemical exposure: integrating energetics and stress to quantify toxic effects in fish. Canadian Journal of Fisheries and Aquatic Sciences 56:814–822. (Not seen, cited in Beyers and Rice, this volume.)

EC and DFO. 1992a. Aquatic Environmental Effects Monitoring Requirements. Published by Environment Canada and the Department of Fisheries and Oceans, Ottawa, Ontario. Publication EPS 1/RM/18.

EC, and DFO. 1992b. Aquatic Environmental Effects Monitoring Requirements at Pulp and Paper Mills and Off-Site Treatment Facilities Regulated under the Pulp and Paper Effluent Regulations of the Fisheries Act, May 20, 1992. Published by Environment Canada, and the Department of Fisheries and Oceans, Ottawa, Ontario. Environment Canada. 1992. 27 p. Annex 1 to: Aquatic Environmental Effects Monitoring Requirements. Published by Environment Canada and the Department of Fisheries and Oceans, Ottawa, Ontario. Publication EPS 1/RM/18.

Fox, G. A., 1991. Practical causal inference for epidemiologists. Journal of Toxicology and Environmental Health 33:359–373.

Gagnon, M. M., D. Bussieres, J. J. Dodson, and P. V. Hodson. 1995. White sucker (*Catostomus commersoni*) growth, maturation and reproduction in pulp mill contaminated and reference rivers. Environmental Toxicology and Chemistry 14:317–327.

Hodson, P. V., S. Efler, J. Y. Wilson, A. El-Shaarawi, M. Maj, and T. G. Williams. 1996. Measuring the potency of pulp mill effluents for induction of hepatic mixed function oxygenase activity in fish. Journal of Toxicology and Environmental Health 49:101–128.

Hodson, P. V., P. J. Kloepper-Sams, K. R. Munkittrick, W. L. Lockhart, D. A. Metner, L. Luxon, I. R. Smith, M. M. Gagnon, M. Servos, and J. F. Payne. 1991. Protocols for measuring mixed function oxygenases of fish liver. Canadian Technical Report of Fisheries and Aquatic Sciences 1829.

McCarty, L. S., and K. R. Munkittrick. 1996. Environmental biomarkers in aquatic toxicology: fiction, fantasy or functional? Human and Ecological Risk Assessment. 2:268–274.

Munkittrick, K., M. R. van den Heuvel, J. J. Stegeman, D. A. Metner, W. L. Lockhart, J. F. Paine, P. V. Hodson, S. Kennedy, J. Bureau, I. R. Smith, M. Adams, J. A. Miller, and P. Martel. 1991. An interlaboratory comparison of the ethoxyresorufin-o-deethylase assay for MFO activity. Pages 123–127 *in* Proceedings of the Eighteenth Annual Aquatic Toxicity Workshop: September 30–October 3, 1991, Ottawa, Ontario. Department of Fisheries and Oceans.

Munkittrick, K. R., 1992. A review and evaluation of study design considerations for site-specifically assessing the health of fish populations. Journal of Aquatic Ecosystem Health 1(4):283–293.

Munkittrick, K. R., M. R. van den Heuvel, D. A. Metner, W. L. Lockhart, and J. J. Stegeman. 1993. Interlaboratory comparison and optimization of hepatic microsomal ethoxyresorufin-o-deethylase activity in white sucker (*Catostomus commersoni*) exposed to bleached kraft pulp mill effluent. Environmental Toxicology and Chemistry 12:1273–1282.

Index